Study Guide

to accompany

The Science of Biology

TENTH EDITION

KRISTINE F. NOWAK
Kennesaw State University

BETTY MCGUIRE
Cornell University

MARK A. SARVARY
Cornell University

LAUREL L. HESTER
Keuka College

JOSEPH A. BRUSEO
Holyoke Community College

THOMAS M. KOVAL
Johns Hopkins University

MEREDITH G. SAFFORD
Johns Hopkins University

EDWARD W. AWAD
Vanier College

SINAUER ASSOCIATES, INC.

Cover photograph © Alex Mustard/Naturepl.com.

Study Guide to accompany ***Life: The Science of Biology,*** **Tenth Edition**

Address editorial correspondence to:
Sinauer Associates, Inc.
23 Plumtree Road
Sunderland, MA 01375 U.S.A.
Fax: 413-549-1118
Internet: www.sinauer.com; publish@sinauer.com

Address orders to:
MPS/W.H. Freeman & Co. Order Department
16365 James Madison Highway, U.S. Route 15
Gordonsville, VA 22942 U.S.A.
Examination copy information: 1-800-446-8923

This SFI label applies to the text paper.

ISBN 978-1-4641-2365-8
Printed in U.S.A.

4 3 2 1

To the Student

Biology is an incredibly exciting field of study, but in order to appreciate new discoveries and discussions, it is necessary to have a firm grasp of the underlying concepts and ideas. Your textbook is designed to give you a comprehensive overview of important biological phenomena. It will also serve as a resource to you in future studies. Together with your instructor, your textbook will provide you with invaluable information for beginning your study of biology.

This Study Guide is designed to supplement, not replace, your textbook and your instructor. It was written for you, the student, in language that you can understand, but it does emphasize proper usage of biological terminology. Important concepts and ideas have been synthesized into easy-to-read text that provide an overview of the biological concepts discussed in your textbook. Each Study Guide chapter includes three review elements: (1) The Big Picture, an introductory overview of the topics covered in the chapter; (2) Study Strategies, tips for effective ways to study the material and master common problem areas; and (3) Key Concept Review, a summary of each numbered concept in the chapter interspersed with short answer and diagram questions. These will help you preview a chapter before reading it, check your understanding, and review the chapter later.

In addition to the Key Concept questions, each Study Guide chapter also includes Test Yourself Questions. The Key Concept Review questions are integrated into the Key Concept summary section, allowing you to test your factual and conceptual knowledge of each chapter while reading the pertinent review text. In some cases, diagrams are provided and you are asked to label them or answer questions based on them, while in other chapters, you are asked to create your own diagrams based on a question or series of questions. Key Concept Review questions ask you to apply the knowledge you have gleaned from a chapter to answer questions that are more open-ended. These questions require you to have assimilated several concepts and to think beyond what you have just read. Your instructor may ask questions similar to these on exams, or may use an entirely different approach to assess your knowledge, but these questions will be a good check of how well you understand the material. Test Yourself questions are multiple choice style questions, designed to determine if you have retained information from a chapter, and if you can put together various concepts in order to answer questions. Answers to both types of questions are provided at the end of each chapter of the Study Guide. These answers are not exhaustive, but are instead designed to point you to the correct concepts in the textbook. Because of the nature of many of the Key Concept Review questions, your answers should be more expansive than the short explanations given in this Study Guide.

Strategies for Studying Biology

Each individual has his or her own unique study pattern. However, there are some successful study strategies that are universal. We recommend that you first preview a chapter in your textbook. In this initial preview, it is important to note the organization of the chapter and the main points, and to go over the chapter summary at the end. By referring to the Study Guide at this point, you will further understand the organization of the material. If you follow these steps, you might have some specific questions in your mind that you will expect your reading to answer.

Reading a textbook is an active process. We recommend that you always have a pencil and paper available. Jotting notes in margins serves as an excellent mental trigger when it comes time to review. As you read each section of your textbook, see if you can summarize it in your own words. Compare your summaries to those provided at the end of each section and at the end of the chapter. You want to assure yourself that the main points you are noting match those that the author has selected. Refer back to those questions that came up as you previewed the chapter. You should be able to answer your own questions by the time you have finished your reading.

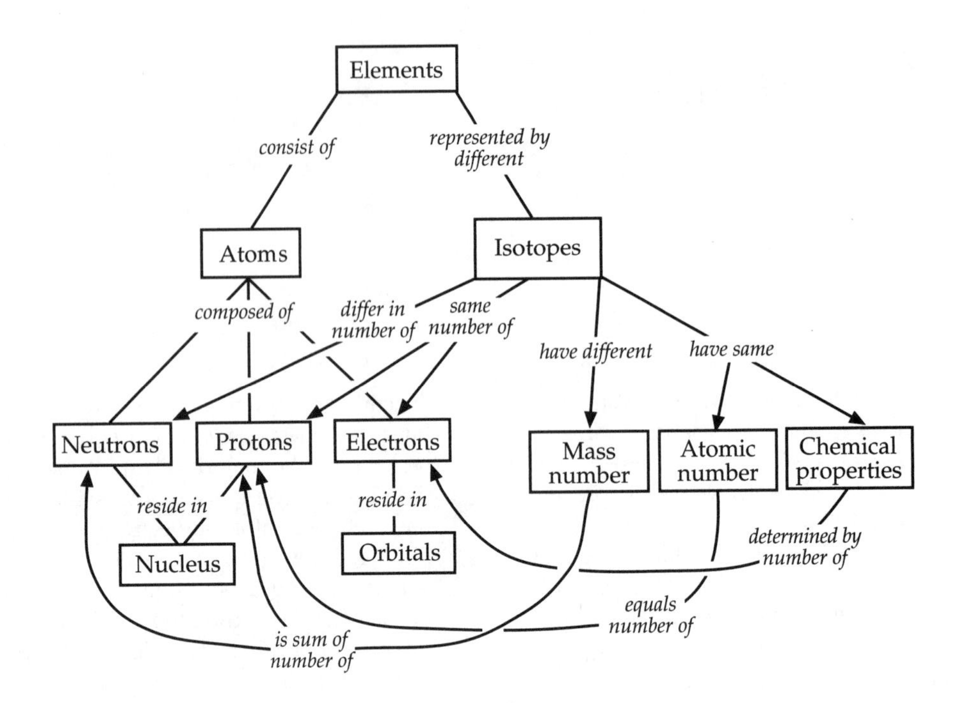

As you are reading, take time to review all the figures and tables. Your textbook makes use of figures to illustrate points, describe pathways, and give visual representation to complex topics. Often these figures are more helpful than the paragraphs of written explanation. In places where figures might be beneficial to you but are not provided in the textbook, try to draw them yourself. This is especially important when attempting to understand structures and pathways. You will find that the Study Guide also refers you to specific animated tutorials and activities, both of which are available within BioPortal, at yourBioPortal.com (the animated tutorials and activities are also available directly via the QR codes and Web addresses listed in the textbook). Each of these tools will aid in your understanding of the material. In addition, BioPortal includes many other study and review resources, some of which your instructor may assign.

Once you have read the material, summarized it in your own words, and reviewed all of the figures and tables, it is time to do another quick review. Scan the summary in your textbook again and read over the Key Concept Review section of this Study Guide. Additionally, the textbook includes many features to help you learn the material as you read and review, balloon captions in figures, section recaps, Working with Data Exercises, chapter summaries, end-of-chapter review questions, and a statistics primer appendix. (Refer to the inside front cover of the textbook for more about these features.) We recommend further that you construct a "concept map" of the main points of a chapter. This will assure you that you see how concepts are interrelated and that you understand the organization of the material. An example of a concept map is shown above. Main concepts are in boxes, and arrows are used to connect concepts and show relationships.

Finally, when you are comfortable with chapter content, work on the questions section of this Study Guide (both from the Key Concept Review and Test Yourself sections). We recommend working through all of the questions before checking your answers. Any questions that you do not understand or that you answer incorrectly should be flagged as material you need to review. Remember that your goal is to understand the material completely, not just to answer these particular questions.

Experiments have suggested that the average attention span of a reader is approximately 12 minutes. You cannot expect to sit down and master an entire chapter in one sitting. Break up your study into short segments of approximately 30 minutes each. This will give you time to get down to business, maximize your attention, and learn without becoming frustrated or drained. At the end of this time, move to some other activity or area of study. When you return to biology, you will find your mind is ready to absorb more.

This entire process should take place well in advance of your exams. You cannot learn even one chapter adequately the day before an exam. Mastering biology requires daily study and review. If you keep up with learning the concepts as your course proceeds, you will find that reviewing the textbook summaries and the Study Guide, along with the online review tools, is adequate preparation. The key to learning biology well is slow and diligent daily work.

Doing Well on Biology Exams

Your instructors are your best resource for doing well on exams. They are there to assist you in learning the material that is presented in your textbook. They are experts in their fields and understand how the concepts presented are vital to your biology education. Follow their lead in preparing for exams. Exam performance is directly linked to classroom attendance and daily study. Sit front and center in your lecture hall and pay close attention to what your instructor writes and provides in presentations. Your instructor will give you guidance about the most important points to study. You should take careful notes and ask for clarification when necessary.

You should always compare your classroom and lecture notes to your textbook and the Study Guide. Take note of the points your instructor emphasizes. Were there specific questions that the instructor brought up in class? Do these correspond to the questions in the Study Guide? If so, chances are you will see them again on an exam. This practice should be part of your daily study regimen.

Approach the exam itself with confidence. Read the directions carefully and read the questions completely. One of the biggest mistakes students make on exams has to do with not following instructions. We recommend that you read carefully through the entire exam even before you begin to answer the first question. You may find that early questions are answered partially in later ones, and that an overview of the whole exam helps to trigger your memory of the material.

If your exam contains multiple choice questions, treat each of the answer options as if it were a separate true/false question. In other words, mentally fill in the question with each answer and ask yourself if the resulting statement is true or false. Be sure to read each of the options, even if the first one strikes you as the correct one. You may be dealing with a question that has multiple correct answers or asks you to select "all of the above" or "none of the above." If you are unsure about how to answer a question, begin by ruling out answers that you know are incorrect. By eliminating some possibilities, your chance of selecting the right one is greatly improved! If you narrow the choices down to two possible answers and just can't decide between them, go with your gut response. You just may be right. And when you go over your answers, change your response only if you know you answered a question incorrectly. If you aren't sure, it may be best to leave the answer as-is. Your first instincts are often correct.

Essay questions require you not only to think through the concepts, but also to decide how you will present them. Organized, concise essay answers are always preferred to rambling answers with little direction. Take a look at the point value of each question. This is often an indication of how many "points" you need to make in your discussion. A 10-point question rarely can be answered with a single sentence. Look carefully at what your instructor is asking. If you are required to "discuss" a concept or a problem, do not present a list or some scattered phrases. Write in complete sentences and paragraphs unless you are asked specifically to list or itemize the material. Be sure that you address all points in the question. Essay questions frequently have multiple parts to test your understanding of the links between various concepts.

Always be aware of the time you have to complete your exam. Answer the questions you know swiftly, and allow yourself time to go back and really think about those you are struggling with. Be sure to concentrate on the questions with the highest point value. Those are the questions testing the most critical concepts and they are the ones that will have the most influence on your grade.

Once you think you have completed the exam, take a few minutes to go back over it. If there are any questions about which you are still unclear, be sure to ask your instructor for clarification.

Learn from Your Mistakes

After your instructor returns your exam or posts the answers, review the mistakes you have made. Take this opportunity to clarify misunderstandings and re-learn material when necessary. You will find this helpful not only for the final exam, but also as you approach upper-level courses. Material reviewed, corrected, and re-learned is more likely to be retained in the long run.

Get Help before It Is Too Late

Many students come to college without adequate study habits and/or are ill prepared for the rigor of biology at the college level. But this is a problem that can be solved fairly easily. Many colleges and universities have academic centers that assist with locating tutors, teaching study habits, and acquainting you with a variety of other resources. Make good use of these facilities. They are there for you. Also, for many additional study tips and strategies, see the *Survival Skills* document that is available within BioPortal.

Good luck in your study of introductory biology!

Table of Contents

Table of Contents (continued)

Studying Life

The Big Picture

- Biology is the study of living things. Living things are composed of cells, which contain genetic material, require nutrients for energy, maintain a constant internal environment, and interact with one another. All available evidence points to an origin of life some 4 billion years ago. Natural selection has led to the current host of organisms inhabiting the planet. There are specific characteristics of life, and the evolutionary roots of these characteristics can be traced through time.
- All life on Earth is organized in the tree of life based on molecular evidence. When studying organisms from an evolutionary perspective, scientists are concerned with evolutionary relatedness and common ancestors.
- Science follows a specific hypothesis–prediction method. The necessity of a testable and falsifiable hypothesis sets science apart from other methods of inquiry. Control of variables and repeated testing lend credibility to the method.

Study Strategies

- It is important to remain open-minded when beginning your study of evolutionary biology. Neither your instructors nor this book are asking you to set aside your beliefs. We are asking you to learn science. We encourage you to examine the evidence supporting evolution without prejudice, as you would any other scientific subject.
- Refrain from learning the hypothesis–prediction method merely as a series of steps. Think about how each step in the process follows from the preceding step. Two characteristics of science set it apart from other modes of inquiry. One is the setting of hypotheses. The other is the continual process by which it is carried out. With each conclusion leading to a new set of hypotheses, science has the capacity to alter our overall base of knowledge.
- You may find that understanding the hypothesis–prediction approach makes understanding some of the evidence behind evolution easier. It may make sense to turn to Section 1.2 ("How Do Biologists Investigate Life") before studying the rest of the chapter.
- Drawing a timeline or examining the calendar model of Figure 1.2 will help you understand the chronology of the major evolutionary events discussed in your textbook.
- Go to the Web addresses shown to review the following animated tutorial and activities. You can also find them in BioPortal (yourBioPortal.com), along with many additional learning resources.

  Animated Tutorial 1.1 Using Scientific Methodology (Life10e.com/at1.1)

  Activity 1.1 The Major Groups of Organisms (Life10e.com/ac1.1)

  Activity 1.2 The Hierarchy of Life (Life10e.com/ac1.2)

Key Concept Review

1.1 What Is Biology?

Life arose from non-life via chemical evolution

Cellular structure evolved in the common ancestor of life

Photosynthesis allows some organisms to capture energy from the sun

Biological information is contained in a genetic language common to all organisms

Populations of all living organisms evolve

Biologists can trace the evolutionary tree of life

Cellular specialization and differentiation underlie multicellular life

Living organisms interact with one another

Nutrients supply energy and are the basis of biosynthesis

Living organisms must regulate their internal environment

Biology is the study of living things, called organisms, all of which have descended from a common origin of life (see Figure 1.1). Living things share many characteristics that allow them to be distinguished from the nonliving world. Organisms are made up of a common set of chemical components, including particular carbohydrates, fatty acids, nucleic acids, and amino acids, among others. The building blocks of most organisms are cells—individual structures enclosed by plasma membranes. The cells of living organisms convert molecules obtained from their environment into new biological molecules. Cells extract energy from the environment and use it to do biological work. Organisms contain genetic information that uses a nearly universal code to specify the assembly of proteins. Organisms share similarities among a fundamental set of genes and replicate this genetic information when reproducing themselves. Organisms exist in populations that evolve through changes in the frequencies of genetic variants within the populations over time. Living organisms self-regulate their internal environments, thus maintaining the conditions that allow them to survive.

All life forms are believed to have evolved from a common ancestry. However, some life forms do not fit all of the characteristics used to distinguish living organisms from the nonliving world. For example, viruses lack a cellular structure and must parasitize cellular organisms to carry out their functions. However, they share similarities among a fundamental set of genes and replicate this genetic information when reproducing themselves.

Life arose approximately 4 billion years ago. If we imagine the history of Earth as a 30-day calendar, life first appeared around the end of the first week (see Figure 1.2). The random aggregation of complex chemicals led to the existence of the first biological molecules. These initial simple molecules led to molecules that could reproduce themselves and act as templates for larger, more complex molecules. The critical step for evolution of life was the appearance of nucleic acids, which can serve as templates for the synthesis of proteins.

The next step in the origin of life probably involved the enclosure of these molecules in a membrane enclosure, allowing all the necessary complex biological molecules to interact. Water-insoluble molecules called fatty acids can form films, which can form spherical structures. The creation of an internal environment within these spheres concentrated the reactants and products of chemical reactants. These early unicellular organisms, called prokaryotes, were simple, with no internal membrane-enclosed compartments, and existed for the first billion years of cellular life (see Figure 1.3A). Two main groups of prokaryotes that developed, bacteria and archaea, still exist today. Eukaryotic cells are thought to have arisen when certain types of prokaryotic cells merged. Eukaryotic cells have intracellular membrane-enclosed compartments, known as organelles, that carry out specific cellular functions (see Figure 1.3B). The permanent aggregations of eukaryotic cells, which evolved about 1 billion years ago, made it possible for cells to specialize and for organism size to increase. These larger multicellular organisms were able to gather resources more efficiently, which enabled adaptation to specific environments.

All cells require nutrients for survival. These nutrients are used to carry out cellular work, such as the work of building the cell, or mechanical work, such as the moving of molecules from one location to another. Metabolism or metabolic rate is a measure of all the chemical transformations and work done in a cell or organism. Metabolism requires a source of energy. The earliest prokaryotes took in small molecules from the environment to obtain chemical energy. The ability to carry out photosynthesis arose approximately 2.7 billion years ago.

Photosynthesis involves converting light energy from the sun into usable chemical energy, with oxygen as a by-product. Early photosynthetic cells were probably related to present-day cyanobacteria (see Figure 1.4). Large numbers of photosynthetic organisms increased atmospheric oxygen levels, allowing for the evolution of aerobic metabolism. This is a far more efficient biochemical process than anaerobic metabolism, which does not utilize oxygen. Photosynthetic organisms also contributed to the insulating ozone layer that shields Earth from radiation and modifies temperature. This shield eventually made life on land possible.

The genome of a cell contains all the genetic information in the form of DNA, which is composed of four different nucleotide subunits. Segments of DNA make up genes, which are used to produce proteins, which control chemical reactions within cells and are also important structural components (see Figure 1.5). All the cells in an organism contain the same genome, but they express different genes. Replication of the genome is not perfect, introducing errors, known as mutations. Most mutations occur spontaneously during replication, but some are induced by extracellular factors, including chemicals and radiation. Most mutations are harmful or have no effect, but occasionally mutations are beneficial.

A species can be defined as a group of organisms that are similar. Organisms from one species living together make up a population. Evolution is the change in the genetic makeup of biological populations through time. Based on observations, Charles Darwin proposed that all living things are related to one another and that species have evolved through the process of natural selection. Modern-day molecular genetic evidence supports and explains Darwin's theory. Any animal or plant population displays variation, and humans have used artificial selection to develop domesticated animals with specific traits, such as pigeons with unusual feather patterns, beak shapes, and body sizes (see Figure 21.5). Natural selection works because individuals vary in their traits, and the characteristics of traits possessed by some of the individuals of a population may provide them with a better chance of survival and reproduction.

In nature, only a small percentage of an individual's offspring will survive to reproduce. Natural selection can produce adaptations, which are structural, physiological, or behavioral traits that enhance the chance of survival of an organism in a given environment (see Figure 1.6). In addition to natural selection, other processes, including sexual selection and genetic drift, can contribute to the rise of diversity.

Populations of the same species that become isolated from one another can evolve differences that lead to different species which, accordingly, can no longer successfully breed and produce offspring. Species that have recently evolved share a common ancestor. Biologists can identify, analyze, and quantify similarities and differences among species to construct phylogenetic trees.

Biologists study the evolutionary relationships among all organisms. Organisms are classified by a binomial system that identifies their genus followed by their species names. For example, the human species is denoted by *Homo sapiens*. Our closest relative in the genus Homo is the extinct species *Homo neanderthalensis*. Scientists are determining the branching pattern of the tree of life based on genome sequencing and other molecular techniques, which they use to augment evolutionary knowledge based on the fossil record. Knowledge about a species' placement on the tree of life reveals a great deal about its biology. Information about a new species is provided by comparisons with closely related species and allows the compilation of phylogenetic trees, which support the development of an overarching tree of life (see Figure 1.7).

All life is placed into one of three major domains: Archaea, Bacteria, and Eukarya. The mitochondria and chloroplasts of eukaryotic cells are believed to have originated from the endosymbiosis of bacteria. Plantae, Fungi, and Animalia are three major groups of multicellular eukaryotes that evolved from different unicellular protists. Initially all life was unicellular, and unicellular organisms remain highly successful. Within multicellular organisms, cells differentiate to carry out specific functions and lose many of the functions carried out by single-cell organisms, thereby creating many levels of organization—a biological hierarchy (see Figure 1.8). A group of similar cells in a multicellular organism that work together to accomplish a specialized task are known as a tissue. A number of different tissues can be organized into an organ that accomplishes specific functions, and multiple organs work together to form an organ system.

Different organisms interact with one another. Organisms from one species living together make up a population. The many populations of different species in an area make up a community. Communities and their abiotic environment make up ecosystems. Interactions between individuals in animal populations (such as competition for food and other resources) and cooperative behavior (which includes the formation of social units) result in the evolution of social behaviors. Plants interact with their external environment and depend on relationships with other plants, fungi, bacteria, animals, and microorganisms for such purposes as obtaining nutrients and producing and dispersing fertile seeds. Plants also experience competition for resources and suffer predation. These interactions within ecosystems are major evolutionary forces that produce specialized adaptations. Ecology is the study of these interactions within and between ecosystems.

All cells require nutrients for survival. These nutrients are used to carry out cellular work. Mechanical work involves moving molecules from one location to another, moving cells or tissues, or even moving the organism as a whole (see Figure 1.9A). Nutrients can be used to build structures or stored for future utilization (see Figure 1.9B). The electrical work that occurs within an organism, such as the processing of information in the nervous system, is another form of work. Lack of integration and control of these complex network systems can lead to malfunction and disease.

Multicellular organisms must carefully regulate their internal environment, which is made up of extracellular fluids. Homeostasis, the process of maintaining a stable environment, maintains the extracellular fluid within a narrow range of physical and chemical conditions. Homeostasis enables the cells of a multicellular organism to function even when conditions outside of the organism are unfavorable for cellular processes.

Question 1. Cellular life is divided into three major lineages of life. What are these three lineages and what are their key similarities and differences?

Question 2. What is the hypothesis for the evolution of eukaryotic cells?

Question 3. The study of biology can be organized from the most basic unit, the molecule, up to the biosphere. Describe how each level is connected with the level below it.

Question 4. Multicellular organisms must regulate their internal environment. Explain the concept of homeostasis, why it is necessary, and how organisms maintain homeostasis.

1.2 How Do Biologists Investigate Life?

Observing and quantifying are important skills

Scientific methods combine observation, experimentation, and logic

Good experiments have the potential to falsify hypotheses

Statistical methods are essential scientific tools

Discoveries in biology can be generalized

Not all forms of inquiry are scientific

Biologists have always studied life by means of observation, but many new technologies are enhancing their ability to observe life. Many of these new tools allow scientists to quantify the data they collect as they make observations. Applying mathematical and statistical calculations to the data is essential. Previously, biologists classified organisms simply based on qualitative attributes. Today, the quantification of molecular and physical differences among species is combined with mathematical models, thereby enabling quantitative analyses of evolutionary history and comparative investigations of other aspects of an organism's biology.

The scientific method, or hypothesis–prediction method, is the basis of most scientific investigations. This method involves making observations and asking questions (see Figure 1.10). These questions lead to the formation of hypotheses, which are provisional answers to the questions. Predictions are made in regard to the hypotheses, and then they are tested.

After posing a question, a scientist may use inductive logic to propose a possible answer. This involves creating a new proposition, or hypothesis, that is compatible with observations or facts. Tyrone Hayes used inductive logic to hypothesize that exposure to the herbicide atrazine was causing the rapid decline of frog populations. Next, the scientist applies deductive logic, which begins with the hypothesis and then predicts what facts would also have to be true to be compatible with the hypothesis. Using deductive logic, Hayes predicted that frog tadpoles exposed to atrazine would show adverse effects of the chemical as adults.

Science depends on a hypothesis that is testable and that can be rejected by direct observation and experiments. Observations must also be reproducible and quantifiable for a conclusion to be considered scientific. A hypothesis can be tested by either comparative or controlled experiments. A controlled experiment involves isolating variables of interest while keeping other variables that may influence the outcome as steady as possible. The variable that is manipulated is the independent variable, and the response that is measured is the dependent variable. A controlled experiment conducted by Tyrone Hayes was the quantification of the effects of atrazine on male frogs in a laboratory setting (see Figure 1.11). Observations are then made to test the hypothesis. A comparative experiment predicts that there will be differences between samples or groups and involves gathering data from multiple groups or conditions to examine the patterns found in nature. In comparative experiments, the variables cannot be controlled. Hayes and his colleagues conducted a comparative study in which they collected samples from various different sites with varying levels of atrazine (see Figure 1.12)

Statistical methods are used to determine if the results of an experiment are significant. Typically, these statistical tests start with a null hypothesis stating that there are no differences. Results are considered significant if statistical tests show that the result is not merely the result of random variation in the sample tested.

Because all life is related, biologists can use model systems for research and apply, with thought and care, the knowledge gained from their investigations to other life forms. Research on bacteria and other model systems that include sea urchins, frogs, chickens, roundworms, mice, and fruit flies has led to basic understandings of biochemical reactions and development that are applicable to the cells of more complex organisms, including humans. Much of what we know about photosynthesis has come from the study of model systems such as *Chlorella*, a unicellular green alga, and much of our understanding of plant development is based on research done using *Arabidopsis thaliana*, a relative of the mustard plant.

Other areas of scholarship make observations and ask questions, but scientific research is distinguished by obtaining quantifiable data and subjecting it to appropriate statistical analysis, which is critical in evaluating hypotheses. Science depends on a hypothesis that is testable and that can be rejected by direct observation and experiments. Observations must also be reproducible for a conclusion to be considered scientific.

Question 5. The study of biology is advanced by using the scientific method. Create a flow chart showing the steps involved in the scientific method.

Question 6. Scientists interested in human biology typically perform experiments with other model systems. Why do scientists use model systems in this way?

Question 7. Discuss how the process of scientific inquiry is different from other forms of inquiry. Include in your discussion a description of the hypothesis–prediction approach.

1.3 Why Does Biology Matter?

Modern agriculture depends on biology

Biology is the basis of medical practice

Biology can inform public policy

Biology is crucial for understanding ecosystems

Biology helps us understand and appreciate biodiversity

Human beings exist in and depend on a world of living organisms. Understanding biological principles is essential to maintain a functioning Earth upon which humans depend.

Increased knowledge of plant biology has resulted in huge boosts in the amount and quality of food production (see Figure 1.13), as well as new strains of crops that are resistant to pests or can tolerate drought.

Biology informs us about how living organisms work. This knowledge aids in the development of treatments or cures for infectious and genetic diseases. An effective response to the resurgence of tuberculosis, due to the evolution of antibiotic-resistant strains, requires understanding aspects of molecular, physiological, microbial, and evolutionary biology. Many microbial organisms have short generation times and high mutation rates. The high rate of evolution of the influenza virus requires the generation of a new influenza vaccine on a yearly basis. An understanding of evolutionary biology, molecular biology, and ecology informs medical researchers in their development of vaccines and other strategies to control epidemics (see Figure 1.14).

Biological science has an impact on every human being. Newfound abilities to manipulate the genomes of organisms present vast new possibilities to control human diseases and increase agricultural productivity, but they also raise ethical questions and policy issues. Science alone cannot provide all of the answers to ethical and policy issues, but accurate scientific information is required for sound public policy decisions. Policy makers are increasingly calling on biologists to provide scientific knowledge to the assessment and formulation of public policy (see Figure 1.15). However, policy makers often have to take into consideration other factors in addition to scientific knowledge and recommendations.

The world is constantly changing, but human activity is resulting in an unprecedented rate of change in the world's ecosystems. Anthropogenic (human-generated) changes, such as increasing atmospheric carbon dioxide levels through mining and consumption of fossil fuels, put stress on ecosystems (see Figure 1.16) and are leading to the extinction of large numbers of species. Biological knowledge is crucial for understanding the causes of changes in ecosystems and for devising policies to deal with these changes.

Humans have a "need to know," a natural curiosity that may be an adaptive trait that selected for individuals who were motivated to learn about their surroundings. Human curiosity leads to new discoveries and greater knowledge motivates biologists to learn more and to constantly collect new information (see Figure 1.17). An understanding of the natural history of a group of organisms facilitates observation and the development of hypotheses based on these observations. Many popular human activities depend upon biodiversity and support the growing industry of ecotourism.

Question 8. Modern biological scientific research has greatly increased the cure rate for many diseases. How has biological research improved the development of disease treatments?

Question 9. Find an article in a newspaper or news site that describes an issue that is informed by scientific data. How are the policy decisions influenced by the scientific evidence? What nonscientific issues influence the policy decisions related to this issue?

Test Yourself

1. Life arose on Earth approximately _______ years ago.
   a. 4 billion
   b. 4 million
   c. 4,600
   d. 1.5 billion
   e. 4 trillion
   Textbook Reference: *1.1 What Is Biology?*

2. Which of the following is the feature or component of organisms that allows for life in such a wide variety of environments on Earth?
   a. Prokaryotic cells
   b. Eukaryotic cells
   c. Homeostasis
   d. Adaptation
   e. Model systems
   Textbook Reference: *1.1 What Is Biology?*

3. Which of the following is *not* a characteristic of most living organisms?
   a. Regulation of internal environment
   b. One or more cells
   c. Ability to produce biological molecules
   d. Ability to extract energy from the environment
   e. All of the above are characteristics of most living organisms.
   Textbook Reference: *1.1 What Is Biology?*

4. Photosynthesis was a major evolutionary milestone because
   a. photosynthetic organisms contributed oxygen to the environment, which led to the evolution of aerobic organisms.
   b. photosynthesis led to conditions that allowed life to arise on land.
   c. photosynthesis is the only metabolic process that can convert light energy to chemical energy.
   d. photosynthesis provides food for other organisms.
   e. All of the above
   Textbook Reference: *1.1 What Is Biology?*

5. Which of the following is *not* an attribute of homeostasis in a multicellular organism?
   a. Maintaining a stable internal environment
   b. Maintaining the extracellular fluid within a range of physical conditions
   c. Maintaining a stable external environment
   d. Maintaining the extracellular fluid within a range of physical conditions
   e. None of the above
   Textbook Reference: *1.1 What Is Biology?*

6. A group of cells that work together to carry out a similar function is known as a(n)
   a. tissue.
   b. organ system.
   c. unicellular organism.
   d. protein.
   e. gene.
   Textbook Reference: *1.1 What Is Biology?*

7. Which of the following does *not* contribute to adaptation in the wild?
   a. Artificial selection— not in wild
   b. Genetic drift
   c. Natural selection
   d. Sexual selection
   e. All of the above contribute to adaptation in the wild.
   Textbook Reference: *1.1 What Is Biology?*

8. Which of the following is *not* considered part of the natural history of a group of organisms?
   a. How the organisms behave
   b. How the organisms interact with other organisms
   c. How the organisms get their food
   d. How the organisms reproduce
   e. A natural history includes considerations of all of the above.
   Textbook Reference: *11.3 Why Does Biology Matter?*

9. The information needed to produce proteins is contained in
   a. nutrients.
   b. tissues.
   c. evolution.
   d. organs.
   e. genes.
   Textbook Reference: *1.1 What Is Biology?*

10. Evolution is
    a. not important to the study of biology.
    b. the change in the genetic makeup of a population through time.

c. the change in protein expression of a population through time.
d. not influenced by natural selection.
e. None of the above
Textbook Reference: *1.1 What Is Biology?*

11. In a model experiment, researchers subjected frogs to various levels of atrazine while keeping all other variables constant. This is an example of a _______ experiment.
a. controlled
b. repeated
c. laboratory
d. comparative
e. None of the above
Textbook Reference: *1.2 How Do Biologists Investigate Life?*

12. For a hypothesis to be scientifically valid, it must be _______ and it must be possible to _______ it.
a. testable; prove
b. testable; reject
c. controlled; prove
d. controlled; reject
e. testable; control
Textbook Reference: *1.2 How Do Biologists Investigate Life?*

13. Eukaryotic cells differ from prokaryotic cells in that eukaryotic cells have
a. genes.
b. proteins.
c. organelles. -has own lipid bilayer
d. membranes.
e. All of the above
Textbook Reference: *1.1 What Is Biology?*

14. In the names of organisms, the _______ is placed first and the _______ is placed second.
a. species; genus
b. genus; domain
c. domain; genus
d. genus; species
e. domain; species
Textbook Reference: *1.1 What Is Biology?*

15. Metabolism refers to
a. natural selection.
b. the chemical transformations and work of a cell.
c. communities.
d. mutations in DNA.
e. cellular structure.
Textbook Reference: *1.1 What Is Biology?*

16. Which of the following factors are taken into consideration in the biological classification of organisms?
a. Physical characteristics
b. Fossil records
c. Molecular characteristics
d. All of the above
e. None of the above
Textbook Reference: *1.2 How Do Biologists Investigate Life?*

17. Which of the following is *not* a step in the scientific method?
a. Observation
b. Quantifying data
c. Asking questions
d. Formulating a hypothesis
e. All of the above are steps in the scientific method.
Textbook Reference: *1.2 How Do Biologists Investigate Life?*

18. The term "anthropogenic" refers to
a. human-caused fires.
b. the study of insects.
c. human-generated effects upon the environment.
d. the study of human biology.
e. the study of agriculture.
Textbook Reference: *1.3 Why Does Biology Matter?*

19. Which of the following is *not* a domain on the tree of life?
a. Archaea
b. Plantae
c. Eukarya
d. Bacteria
e. All of the above are domains on the tree of life.
Textbook Reference: *1.1 What Is Biology?*

Answers

Key Concept Review

1. The three major lineages of life are Bacteria, Archaea, and Eukarya. Both Bacteria and Archaea are prokaryotes—unicellular organisms that have an outer membrane but lack membrane-bound organelles. Members of Eukarya are defined by their DNA being contained within a nuclear membrane and containing other membrane-bound organelles, including mitochondria and, within some members, chloroplasts. The Eukarya domain includes unicellular eukaryotic cells (e.g., yeast) and multicellular organisms that include plants, protists, animals, and fungi.
2. Eukaryotic cells contain a number of membrane-bound organelles. It is hypothesized that organelles such as mitochondria and chloroplasts evolved from engulfed prokaryotic organisms that were not digested, but began a mutual relationship with the new host cell.
3. The cell is composed of many different types of molecules. Cells with the same function and coordination are grouped together to form tissues. Several tissue types work together to form a functioning organ. A complex multicellular organism is composed of organs and organ systems. Many organisms of the same

species living together make up a population. A community encompasses all of the populations within a given area. An ecosystem is composed of many communities in the same geographical area. All of the ecosystems on Earth make up the biosphere.

4. Homeostasis is the state of maintaining an organism's internal environment within a narrow range of conditions. To survive in environments that are unfavorable to cellular processes, multicellular organisms require the ability to maintain homeostasis. To do so, organisms have developed regulatory systems that can sense the conditions of the internal and external environments on a continuous basis. The regulatory system integrates this information and modulates the internal physiological systems to maintain homeostasis.

5.

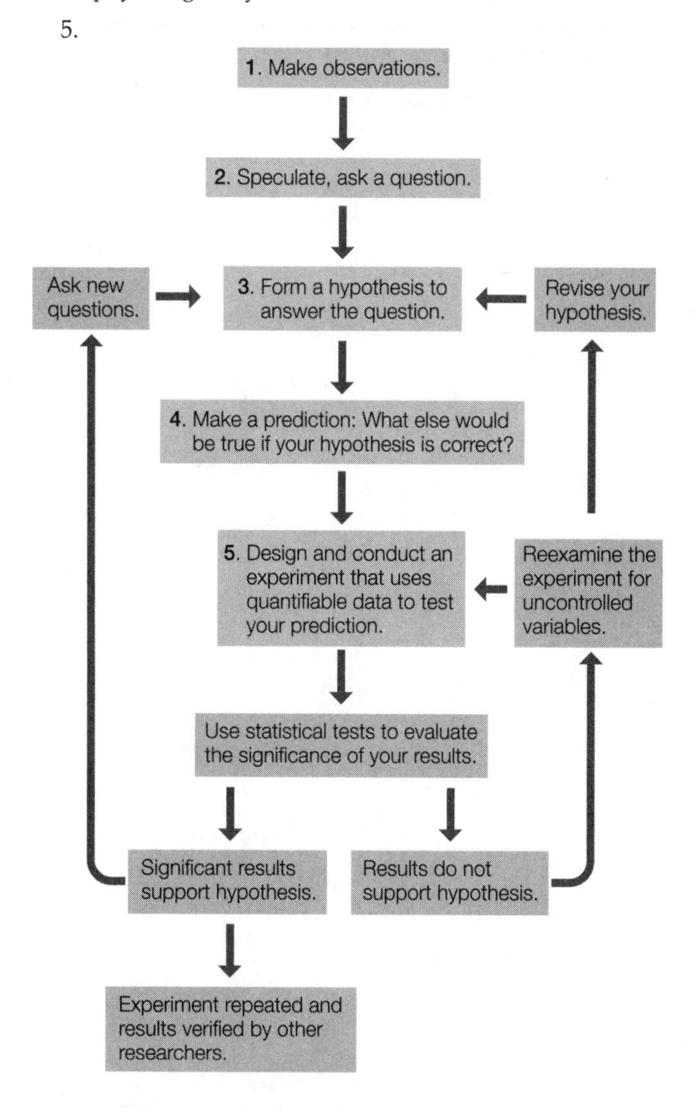

6. Model systems are useful in the study of biology because all organisms have evolved from a common ancestor. Therefore, cellular pathways in, for example, bacteria and fruit flies are very similar to those found in humans. Model systems are valuable because in many cases they can be manipulated experimentally.

7. The process of scientific inquiry is unique in that a hypothesis must be testable, and it must be possible to reject it. The hypothesis–prediction approach begins with observations that lead to questions. From the questions, hypotheses are formed that are probable explanations for the observed phenomena. Predictions are formed from the hypotheses and tested. Conclusions are drawn from the tests. These conclusions may, in turn, lead to additional hypotheses.

8. Biological research has assisted the development of medical treatments by providing basic understandings of the causes of diseases, such as the role of genetics and environmental factors in their development. This knowledge enables the development of treatments that are more efficacious and have fewer side-effects than treatments that are discovered through trial and error, and with no understanding of how the treatments work against the disease.

9. Many public policies rely on sound biological findings. Examples include medical issues, environmental quality issues, and issues related to the ecology of local wild animal populations. The policies developed in response to these issues, however, may be influenced by factors other than scientific findings, including economics, politics, and religion.

Test Yourself

1. **a.** Available evidence places the beginning of life at about 4 billion years ago.

2. **d.** Adaptations are the differences found in organisms that allow them to live in an environment.

3. **e.** Most living organisms are composed of cells, have the ability to make biological molecules, can regulate their internal environment, and get energy from the external environment.

4. **e.** Photosynthesis caused the accumulation of oxygen in the atmosphere and contributed to the formation of an ozone layer. The presence of oxygen made the evolution of aerobic organisms possible, and the ozone layer shielded Earth from harmful radiation. These two phenomena contributed to the evolution of terrestrial life.

5. **c.** Homeostasic mechanisms can maintain a stable internal environment but cannot regulate the external environment of the organism.

6. **a.** Multicellular organisms have tissues that are formed from many similar cells.

7. **a.** Adaptations arise from genetic drift, natural selection, and sexual selection. Artificial selection can produce adaptations, but they occur through human selection and not in the wild.

8. **e.** A natural history of a group of organisms includes all observations that can be used to develop a better understanding of how the organism lives in its environment and has evolved.

9. **e.** Genes are specific sequences of DNA that contain the information used to make proteins.
10. **b.** Evolution is a change in the frequency of genes within a population over time and can occur through natural selection. This is the major unifying principle of biology.
11. **a.** An experiment is said to be controlled when in one or more test groups one variable, the independent variable, is manipulated whereas all other variables are kept constant. The data from the test group(s) is/are compared with the data generated from a "control" group, in which the independent variable is not manipulated.
12. **b.** Scientific hypotheses are set apart from mere conjecture by being testable and falsifiable.
13. **c.** Both eukaryotic and prokaryotic cells contain genes, proteins, and some sort of membrane. Only the eukaryotes have organelles.
14. **d.** Examples are *Homo sapiens* and *Homo neanderthalensis*.
15. **b.** The metabolism, or metabolic rate, of an organism is the sum of all the chemical transformations and work that it performs.
16. **d.** Biologists classify organisms using a wide range of methods, which include physical characteristics, the fossil record, and molecular characteristics including the genome and gene expression. Today, mathematics and statistics are applied to the data, leading to more informative analyses.
17. **e.** The scientific method is a process that includes observations, asking questions, formulating hypotheses, and quantifying data.
18. **c.** Impacts upon the environment that are caused by human behavior are anthroprogenic effects.
19. **b.** Plantae is a kingdom found in the domain Eukarya.

domains have kingdoms

Small Molecules and the Chemistry of Life

The Big Picture

- Understanding the chemical building blocks of all matter is essential to understanding the biology of organisms. Atoms, which contain protons, neutrons, and electrons, form elements or combine to form molecules.
- Atoms in molecules may be held together with covalent or ionic bonds. Molecules with polar covalent bonds may form hydrogen bonds within or between molecules. Although weak, hydrogen bonds are involved in stabilizing the three-dimensional conformations of giant molecules such as DNA and proteins. Hydrophobic interactions between nonpolar molecules are enhanced by van der Waals forces.
- Chemical reactions change reactants to products without creating or destroying matter and involve energy transformations.
- Water has unique properties that make it biologically important, and hydrogen bonds are important in determining these properties, which include high heat capacity, high heat of vaporization, cohesion, and high surface tension.
- An essential property of water is its ability to act as a solvent. How a substance ionizes in water determines whether it is acidic or basic. Acids have a high concentration of H^+, whereas bases have a low concentration of H^+. The pH scale allows us to compare the concentrations of H^+ in a variety of solutions. Weak acids acting as buffers are necessary for an organism to maintain homeostasis.

Study Strategies

- You are likely to be familiar with the basic notions of chemistry from your high school years or first-year college chemistry courses. However, it is common to fail to realize that living organisms are part of the natural world and therefore are subject to the laws of physics and chemistry. This means that it is difficult to understand how living organisms function without a good understanding of chemistry. Therefore, it is useful to review some of the basic notions of chemistry and how these notions can be applied to understanding living organisms.
- Learning quantitative terminology and how solutions are made is often difficult. Completing practice problems is the best way to learn this material. Also, take advantage of any laboratory experience in which you make solutions. Every good biologist makes many solutions over time.
- You may be overwhelmed by the periodic table. Consult Figure 2.2 in the textbook to make sure you understand its components and how they are arranged. Do not try to memorize the table, but keep it handy as a reference.
- Use the figures in your book to help you visualize how atoms and molecules are represented and how they interact.
- It is easy to become overwhelmed by the different chemical bonds. Use Table 2.1 to understand the differences among the various types of bonds and interactions. Keep it handy as a reference.
- pH tends to be confusing. Remember that it is a concentration of H^+ ions. Those that have a higher concentration are acidic and have a low pH, and those that have a lower concentration are basic with a high pH. Also remember that pH is measured on a logarithmic scale. This means that if you decrease the pH from 8 to 7, you increase the concentration of H^+ ions by a factor of 10.
- Be careful not to get lost in terminology. Make a list of definitions of the chapter's key terms and focus on understanding the concepts associated with each term to help you memorize them.
- Go to the Web addresses shown to review the following animated tutorial and activity. You can also find them in BioPortal (yourBioPortal.com), along with many additional learning resources.

  Animated Tutorial 2.1 Chemical Bond Formation (Life10e.com/at1.1)

  Activity 2.1 Electron Orbitals (Life10e.com/ac2.1)

Key Concept Review

2.1 How Does Atomic Structure Explain the Properties of Matter?

An element consists of only one kind of atom

Each element has a unique number of protons

The number of neutrons differs among isotopes

The behavior of electrons determines chemical bonding and geometry

All matter is composed of the same chemical elements. Atoms combine to form all matter. The nucleus of atoms contains a defined number of positively charged protons and neutral neutrons. Negatively charged electrons move in their outer shells. An element is made up of only one type of atom. There are more than 100 elements, and they are grouped in the periodic table according to certain characteristics. Six elements compose most of the mass of every living organism: carbon (C), hydrogen (H), oxygen (O), nitrogen (N), phosphorus (P), and sulfur (S).

The number of protons in an element determines its type and is known as its atomic number. Neutrons are found in every element except hydrogen. The total number of protons and neutrons equals the mass number of the atom. Isotopes of an element have the same number of protons but differ in the number of neutrons present. Radioisotopes are unstable isotopes that give off energy. This decay eventually causes the atom to change to a new element. Radioisotopes are important in the medical field. Protons and neutrons have mass and contribute to atomic weight. Atomic weights on a periodic chart are averages of all isotopes of an element.

Biologists are interested in chemical reactions and the interactions of electrons. Interactions between atoms involve the associations of their electrons. Electrons continuously orbit the nucleus of an atom in a defined space. The orbital of an atom is the space in which an electron is found at least 90 percent of the time. These patterns of orbitals form a series of electron shells or energy levels, each with a specific number of electrons. The first shell (*s* orbital) can contain up to two electrons. The second shell can have as many as four orbitals (*s* orbital and three *p* orbitals), each containing two electrons. The number of electrons in the outermost shell determines how the atom interacts with other atoms. Many biologically important atoms, such as carbon and nitrogen, are stable when they have eight electrons in the outermost shell. This phenomenon is referred to as the octet rule. Two or more linked atoms compose a molecule, resulting in stabilization of the outer electron shell of the atoms.

Question 1. Draw the electron shells and place electrons in the appropriate shells based on the atomic numbers of the elements carbon ($_6$C), nitrogen ($_7$N), sodium ($_{11}$Na), chlorine ($_{17}$Cl), argon ($_{18}$Ar), and oxygen ($_8$O).

Question 2. Examine the periodic table in Figure 2.2. What are the names of the elements K and Si, what are their atomic numbers and mass numbers, and how many protons, neutrons, and electrons do they have?

2.2 How Do Atoms Bond to Form Molecules?

Covalent bonds consist of shared pairs of electrons

Ionic attractions form by electrical attraction

Hydrogen bonds may form within or between molecules with polar covalent bonds

Hydrophobic interactions bring together nonpolar molecules

van der Waals forces involve contacts between atoms

Chemical bonds hold elements together to make molecules. Atoms share pairs of electrons to stabilize their outer shells in covalent bonding. This type of bond is very stable and strong and can be broken only with a great deal of energy. All molecules have a three-dimensional shape, and the interactions between a given pair of atoms always have the same length, angle, and direction. A molecule's shape affects how it behaves.

Molecules containing two or more elements in a fixed ratio make a compound. The molecular weight of a compound is the sum of the atomic weights of all atoms in the compound.

Multiple covalent bonds may exist. Although a single covalent bond involves one pair of shared electrons, double bonds involve two pairs of shared electrons. Triple bonds share three pairs, but are rare. Covalent bonding between atoms of the same element results in equal sharing of electrons. The attractive force an atom exerts on electrons is known as electronegativity. Two atoms that have the same electronegativity form nonpolar covalent bonds.

Bonds between different elements generally result in polar covalent bonds with unequal sharing of electrons. Unequal sharing of electrons results in partial (δ) charges in molecules because one nucleus is more electronegative than the other, attracting the electrons more strongly. One atom of a molecule may be partially negative, whereas the other is partially positive. This balance of partial charges results in a polar molecule with a δ^- pole and δ^+ pole.

Ions are formed when atoms lose or gain electrons, resulting in a net positive or negative charge. Cations are positively charged and have fewer electrons than protons. Anions are negatively charged and have more electrons than protons.

More than one electron can be gained or lost from an atom or molecule. A group of covalently bonded atoms may also gain or lose electrons to form complex ions.

Ionic bonds form between ions of opposite charge. Ionic bonds are not as strong as covalent bonds and they also attract partial charges of polar molecules.

Partial charges allow hydrogen bonding of molecules. This occurs often between molecules of water where the positive hydrogen is attracted to the negative oxygen. This is a weak bond that is easily broken because no electrons are shared, but this type of bond is important in stabilizing the three-dimensional shape of large molecules such as proteins and DNA.

Polar molecules are hydrophilic, or "water loving," because of the partial charges. Nonpolar molecules are hydrophobic, or "water hating," and they have hydrophobic interactions. Attractions of nonpolar molecules are enhanced by van der Waals forces. Polar molecules tend to aggregate with other polar molecules, and nonpolar molecules aggregate with other nonpolar molecules.

Question 3. Calcium has an atomic number of 20. Draw structures for Ca and Ca^{2+}. What is the difference between these structures?

Question 4. Examine the molecule of the amino acid glycine shown below. Explain how you would determine the type of bonds between the atoms in this molecule (polar or nonpolar covalent bonds, ionic bonds, and hydrogen bonds). Refer to the table below shown below and shown on p. 28 (Table 2.3) of the textbook for a description of chemical bonds and electronegativity.

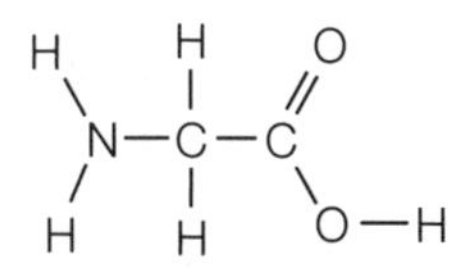

Some Electronegativities

Element	Electronegativity
Oxygen (O)	3.5
Chlorine (Cl)	3.1
Nitrogen (N)	3.0
Carbon (C)	2.5
Phosphorus (P)	2.1
Hydrogen (H)	2.1
Sodium (Na)	0.9
Potassium (K)	0.8

2.3 How Do Atoms Change Partners in Chemical Reactions?

Chemical reactions occur when atoms combine and form new bonding partners. Chemical reactions involve reactants that are altered to produce products. Matter is neither created nor destroyed, but changed. Energy is the capacity to do work. Some reactions produce energy and release it as heat, whereas others require an input of energy. Chemical bonds hold potential energy. When these bonds are broken, release of this energy is important to living systems.

Question 5. Consider the following reaction:

$$C_6H_{12}O_6 + O_2 \rightarrow CO_2 + H_2O + \text{energy}$$

Balance this chemical equation and determine the reactants and the products.

Question 6. Which of the two molecules shown below has more potential energy? Explain.

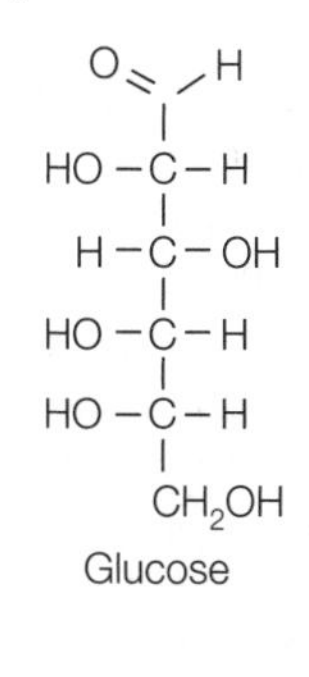

Glucose

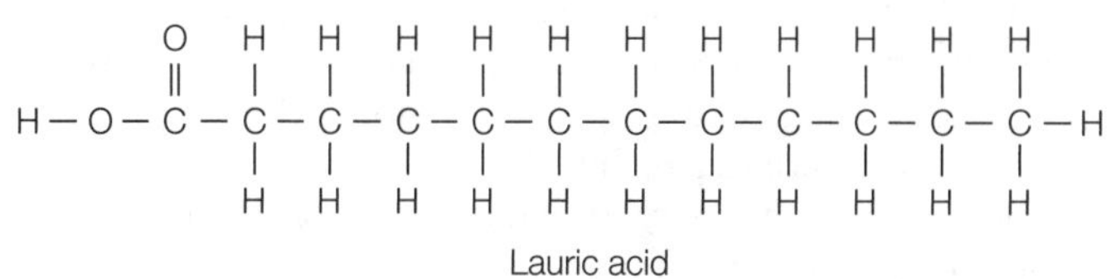

Lauric acid

2.4 What Makes Water So Important for Life?

- Water has a unique structure and special properties
- The reactions of life take place in aqueous solutions
- Aqueous solutions may be acidic or basic

Water is a biologically important molecule with many unique and crucial properties. Water is a polar molecule with the ability to form hydrogen bonds. Due to the four pairs of electrons in the outer shell of oxygen, water has a tetrahedral shape. Ice floats because of the less-dense pattern of hydrogen bonding that results when water molecules are frozen. Ice insulates ponds and lakes in the winter, allowing life to continue beneath the frozen layer. As ice melts, breaking the hydrogen bonds requires the input of a great deal of heat energy. In contrast, heat energy is lost during the freezing of water.

The amount of energy needed to raise the temperature of 1 gram of a substance by 1°C is known as the specific heat. Liquid water has a high specific heat, which provides it with a high heat capacity and the ability to moderate temperatures. Water has a high heat of vaporization which means a lot of heat is required to change water from its liquid to its gaseous state. This property makes evaporating water an effective coolant. The polar nature of the water molecule and the formation of hydrogen bonds contribute to the cohesive strength of water and its surface tension, both of which are biologically important. Solutions are formed when a substance is dissolved in water or another liquid. An aqueous solution has water as the solvent.

Quantitative terms are important for the understanding of both biology and chemistry. Concentration is the amount of a substance in a given amount of solution. A mole is the amount of a substance (in grams) that is numerically equal to its molecular weight. Avogadro's number relates the number of molecules of any substance to its weight and is 6.02×10^{23} molecules per mole. A 1 molar (1 *M*) solution is one mole of substance dissolved in water to make 1 liter.

Acids and bases are substances that dissolve in water and are biologically important. Acids release H^+ ions, while bases attach to H^+ ions. Acids and bases do not all ionize in the same way. Some ionize easily and completely and are called strong acids or bases. Others ionize only partially or reversibly and are called weak acids or bases. When strong acids and bases are ionized, the reaction is irreversible. Ionization of weak acids and bases can be reversible reactions. Water is a very weak acid that requires two water molecules for ionization into the base hydroxide ion and the acid hydronium ion. Though ionization of water is uncommon, it is biologically important.

Solutions are referred to as acidic or basic, whereas compounds and ions are referred to as bases or acids. The pH of a solution refers to the concentration of H^+ ions and is measured as the $\log_{10}$ of the H^+ concentration, $pH = -\log_{10}[H^+]$. Lower pH indicates a higher H^+ concentration or a more acidic solution. A higher pH indicates a lower H^+ concentration or a more basic solution.

Homeostasis is the maintenance of a constant internal environment. A buffer is a mixture of a weak acid and its corresponding base. Buffers prevent fluctuations in the pH of a solution and play an important role in maintaining pH homeostasis.

Question 7. Water is a polar molecule. This property contributes to cohesion and surface tension. Draw six water molecules. In your drawing, indicate how hydrogen bonding between molecules contributes to cohesion and surface tension. Be sure to include appropriate covalent bonds in each molecule.

Question 8. Rank the following solutions in order from the solution with the highest concentration of H^+ ions to the solution with the lowest concentration of H^+ ions: lemon juice, pH = 2; Mylanta, pH = 10; Sprite, pH = 3; drain cleaner, pH = 15; seawater, pH = 8. Of the preceding, which solution is the most acidic? Which solution is the most basic?

Question 9. If you have 12 moles of a substance, how many molecules do you have of that substance? Suppose the substance has a molecular weight of 342. How many grams of the substance would you have to dissolve in a liter of water to make a 12 *M* solution?

Test Yourself

1. Atomic number is determined by the number of _______ in an atom.
   a. protons and neutrons
   b. electrons
   c. neutrons
   d. protons
   e. protons, neutrons, and electrons
   Textbook Reference: *2.1 How Does Atomic Structure Explain the Properties of Matter?*

2. Which of the following statements concerning electrons is *false*?
   a. Electrons orbit the nucleus of an atom in defined orbitals.
   b. The outer shell of all atoms must contain eight electrons.
   c. An atom may have more than one valence shell.
   d. Electrons are negatively charged particles.
   e. All of the above are true, none is false.
   Textbook Reference: *2.1 How Does Atomic Structure Explain the Properties of Matter?*

3. The element with which of the following atomic numbers would be most stable?
   a. 1
   b. 3
   c. 12
   d. 15
   e. 18
   Textbook Reference: *2.1 How Does Atomic Structure Explain the Properties of Matter?*

4. What is the difference between an element and a molecule?
   a. Molecules may be composed of different types of atoms, whereas elements are always composed of only one type of atom.
   b. Molecules are composed of only one type of atom, whereas elements are composed of different types of atoms.
   c. Molecules are elements.
   d. Molecules always have larger atomic weights than elements have.
   e. Molecules do not have electrons, whereas elements do.
   Textbook Reference: *2.1 How Does Atomic Structure Explain the Properties of Matter?*

5. The strongest chemical bonds occur when
   a. two atoms share electrons in a covalent bond.
   b. two atoms share electrons in an ionic bond.
   c. hydrogen bonds are formed.
   d. van der Waals forces are in effect.
   e. there are hydrophobic interactions.
   Textbook Reference: *2.2 How Do Atoms Bond to Form Molecules?*

6. Any molecule that is hydrophilic
   a. cannot form hydrogen bonds.
   b. is a polar molecule.
   c. is a nonpolar molecule.
   d. has a partial positive region and a partial negative region.
   e. Both b and d
   Textbook Reference: *2.2 How Do Atoms Bond to Form Molecules?*

7. The stability of the three-dimensional shape of many large molecules is dependent on
   a. covalent bonds.

b. ionic bonds.
c. hydrogen bonds.
d. van der Waals attractions.
e. hydrophobic interactions.
Textbook Reference: *2.2 How Do Atoms Bond to Form Molecules?*

8. The molecular weight of glucose is 180. If you added 180 grams of glucose to a 0.5 liter of water, what would be the molarity of the resulting solution?
a. 18
b. 1
c. 9
d. 2
e. 0.5
Textbook Reference: *2.4 What Makes Water So Important for Life?*

9. Ice floats in water because
a. ice is less dense than water.
b. there are no hydrogen bonds in ice.
c. ice is denser than water.
d. water has a higher heat capacity than ice.
e. ice has more covalent bonds than water.
Textbook Reference: *2.4 What Makes Water So Important for Life?*

10. Cola has a pH of 3; blood plasma has a pH of 7. The hydrogen ion concentration of cola is therefore _______ than the hydrogen ion concentration of blood plasma.
a. 4 times greater
b. 4 times smaller
c. 400 times greater
d. 10,000 times greater
e. 30,000 times greater
Textbook Reference: *2.4 What Makes Water So Important for Life?*

11. Solution A has a pH of 2. Solution B has a pH of 8. Which of the following statements about these solutions is true?
a. A is basic and B is acidic.
b. A is acidic and B is basic.
c. A is a base and B is an acid.
d. A has a greater $[OH^-]$ than B.
e. None of the above
Textbook Reference: *2.4 What Makes Water So Important for Life?*

12. What is the weight of one mole of glucose ($C_6H_{12}O_6$)?
a. 180 grams
b. 42 atomic mass units
c. 96 grams
d. 342 grams
e. 6.02 grams
Textbook Reference: *2.4 What Makes Water So Important for Life?*

13. Which of the following statements about buffers is true?
a. They allow the pH of a solution to vary widely.
b. They make solutions basic.
c. They maintain pH homeostasis.
d. They disrupt pH homeostasis.
e. They make solutions more acidic.
Textbook Reference: *2.4 What Makes Water So Important for Life?*

14. Which of the following statements about water is true?
a. Water has a low heat of vaporization.
b. Water has a high specific heat.
c. When water freezes, it gains energy from the environment.
d. All of the above
e. None of the above
Textbook Reference: *2.4 What Makes Water So Important for Life?*

15. Which of the following statements about chemical reactions is true?
a. The bonding partners of atoms remain constant.
b. All reactions release energy as they proceed.
c. The bonding partners of atoms change.
d. All reactions consume energy as they proceed.
e. None of the above
Textbook Reference: *2.3 How Do Atoms Change Partners in Chemical Reactions?*

16. Covalent bonds form when
a. atoms of opposite charge are attracted to each other.
b. hydrogen and oxygen interact.
c. hydrophilic molecules bind hydrophobic molecules.
d. electrons of nonpolar substances interact.
e. atoms share electrons.
Textbook Reference: *2.2 How Do Atoms Bond to Form Molecules?*

Answers

Key Concept Review

1.

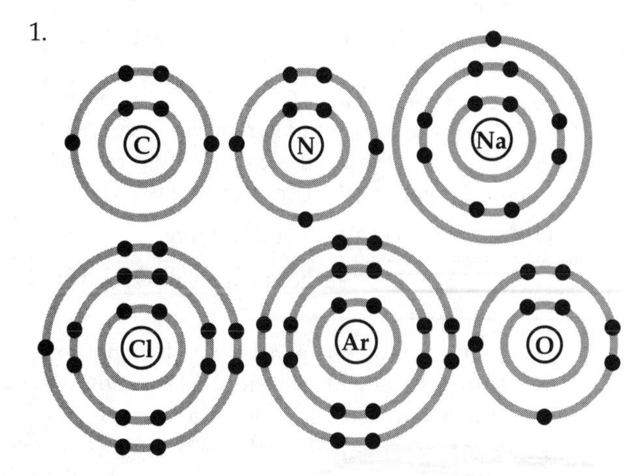

2.

Element				Number of		
Symbol	Name	Atomic number	Mass number	Protons	Neutrons	Electrons
K	Potassium	19	40	19	21	19
Si	Silicon	14	28	14	14	14

3.

The difference between these structures is that calcium (Ca) has two electrons in its outer shell. Calcium ion (Ca^{2+}) has lost its two outer electrons and therefore has a positive charge of 2 because it has two more protons than electrons.

4. The best way to determine the types of bonds present in molecules is to determine each of the atom's electronegativity. As a general rule, the closer the outer electron shell of an atom to saturation, the higher is its electronegativity. In addition, the fewer the electron shells in an atom, the greater its electronegativity. Once you know the electronegativity of the atoms that make up a certain molecule, you next determine the difference in electronegativity between the atoms of this molecule. If the difference in electronegativity is less than 0.5, the bond is likely to be nonpolar covalent. If the difference is between 0.5 and 1.6, the bond is likely to be polar covalent. Differences greater than 1.6 generally result in ionic bonds.

   Therefore, the electronegativity values in Table 2.3 allow us to determine the following: (1) the bond between C and H is nonpolar covalent (difference in electronegativity is 0.4); (2) the bond between C and N is polar covalent (difference in electronegativity is 0.5); (3) the bond between N and H is polar covalent (difference in electronegativity is 0.9); (4) the bond between C and O is polar covalent (difference in electronegativity is 1.0); and (5) the bond between O and H is polar covalent (difference in electronegativity is 1.4).

5. $C_6H_{12}O_6 + 6O_2 \rightarrow 6CO_2 + 6H_2O$ + energy

   $C_6H_{12}O_6$ and O_2 are the reactants, while $6O_2$ and H_2O are the products.

6. Lauric acid has more potential energy than glucose. Bonds between elements are a form of potential energy that is stored in molecules. Lauric acid has a much larger number of covalent bonds in its molecule than glucose has.

7.

Hydrogen bond

The partially positive hydrogens of one water molecule are attracted to the partially negative oxygens of another water molecule. This attraction tends to cause water molecules to "stick" together, creating surface tension.

8. The order is the following: (1) lemon juice, pH = 2 (highest concentration of H^+ ions and therefore the most acidic); (2) Sprite, pH = 3; (3) seawater, pH = 8; (4) Mylanta, pH = 10; (5) drain cleaner, pH = 15 (lowest concentration of H^+ ions and therefore the most basic).

9. You would multiply Avogadro's number (6.02×10^{23}) by 12 to determine the number of molecules. This gives you 7.224×10^{24} molecules of this substance. A 12 *M* solution would be made by multiplying the formula weight of the molecule by 12 and adding that many grams of the substance to one liter of water. The formula weight of this molecule is 342, thus you would need 4104 g.

Test Yourself

1. **d.** The atomic number refers to the number of protons in an atom. The atomic weight can vary with the isotope and is the total of the number of protons and neutrons in an atom.

2. **b.** The first energy shell requires only two electrons to be full; subsequent shells may hold eight or more electrons.

3. **e.** The element with atomic number 18 (Argon) is the most stable because its outermost electron shell is full.

4. **a.** An element is always composed of only one type of atom. Molecules contain many types of atoms that are bonded together.

5. **a.** Covalent bonds, resulting from the sharing of a pair (or more) of electrons, form the strongest types of bonds. Ionic bonds, in which electrons are donated or received, are the second strongest. Hydrogen bonds are the weakest bonds and result from the attraction of partial charges to one another. Van der Waals forces are not bonds but attractions between nonpolar molecules. Hydrophobic interactions involve nonpolar substances and water.

6. **e.** Hydrophilic molecules are polar and have a partial positive and a partial negative pole. These molecules are said to be hydrophilic because they easily form hydrogen bonds with water.

7. **c.** Hydrogen bonds, though weak individually, are quite effective in large numbers and are responsible for maintaining the structural integrity of many large molecules such as proteins and DNA.
8. **d.** The molarity of the solution is 2. A 1 *M* solution is made by adding 180 grams to 1 liter of water. Adding the same amount to half a liter would result in a solution with twice the molarity.
9. **a.** When water forms ice, it becomes more structured. However, the resulting form is less dense than water in the liquid state, allowing ice to float on water.
10. **d.** pH is measured on a logarithmic scale. Each increase or decrease in pH value is by a factor of 10. Therefore, a difference between pH 3 and pH 7 would be a difference of 10 × 10 × 10 × 10, or 10,000.
11. **b.** A solution with a pH of 2 is acidic, and one with a pH of 8 is basic. The lower the pH value, the higher the concentration of H^+ ions.
12. **a.** Each carbon atom weighs 12 atomic mass units, each hydrogen atom is 1 atomic mass unit, and each oxygen atom is 16 atomic mass units. Adding these yields a molecular weight of 180 atomic mass units. Because one mole is equal to the atomic weight in grams, one mole of glucose equals 180 grams.
13. **c.** Buffers are solutions that contain a weak acid (carbonic acid) and the corresponding base (bicarbonate ions). Buffers help maintain the pH of a solution when either an acid or base is added.
14. **b.** Water has both a high heat of vaporization and a high specific heat. In addition, when water freezes it releases a great deal of energy to the environment.
15. **c.** Chemical reactions involve the combining or changing of bonding partners of the reactants to produce the products. During this process matter is neither created nor destroyed. These reactions can either release energy or require energy.
16. **e.** Covalent bonds form when atoms share electron pairs.

12 6.02×10^{23}

Covalent bonds stronger than Ionic.

Proteins, Carbohydrates, and Lipids

The Big Picture

- Polymers are biologically important molecules constructed by the covalent bonding of their building blocks, called monomers. Polymers with molecular weight exceeding 1,000 are called macromolecules. There are four basic classes of biologically important macromolecules: proteins, carbohydrates, nucleic acids, and lipids.
- Macromolecules all have regular primary structures held by covalent bonds, but many, especially proteins, can be folded into complex shapes held by hydrogen bonds, van der Waals forces, or ionic attractions.
- Macromolecules provide for specific biological activities and contribute to the cell's energy and information storage, structure, catalytic activity, and other functions. All macromolecules are synthesized from smaller molecules through condensation and can be broken down through hydrolysis.

Study Strategies

- Since proteins, carbohydrates, and lipids are all complex molecules, first focus on each group separately.
- Make a table comparing their monomers, the bonds connecting them, and their functions. Make sure that you also include examples; they will help you remember the different categories. A table will also help you see the patterns of similarity and the differences.
- Avoid the temptation to memorize structures. If you understand condensation and hydrolysis and know the basic components of the macromolecules, memorizing the structures of the large macromolecules is unnecessary.
- Go to the Web addresses shown to review the following animated tutorial and activities. You can also find them in BioPortal (yourBioPortal.com), along with many additional learning resources.

  Animated Tutorial 3.1 Macromolecules (Life10e.com/at3.1)

  Activity 3.1 Functional Groups (Life10e.com/ac3.1)

  Activity 3.2 Features of Amino Acids (Life10e.com/ac3.2)

  Activity 3.3 Forms of Glucose (Life10e.com/ac3.3)

Key Concept Review

3.1 What Kinds of Molecules Characterize Living Things?

Functional groups give specific properties to biological molecules

Isomers have different arrangements of the same atoms

The structures of macromolecules reflect their functions

Most macromolecules are formed by condensation and broken down by hydrolysis

Molecules have specific properties that enhance their biological activity. Macromolecules are very large polymers with a wide range of biological functions. Functional groups are small groups of atoms that confer specific properties to the molecules they are attached to. In chemical structures, the functional group is often shown bonded to an R (for "remainder") to indicate that the functional group can be attached to a wide variety of carbon skeletons.

Isomers are molecules that have the same number and kinds of atoms but in which the atoms are arranged differently. Structural isomers differ in the way their atoms are linked together (*cis*- or *trans*-). In optical isomers, the carbon has four different groups or atoms attached to it in differing orders.

All organisms are composed of four major biological macromolecules: proteins, nucleic acids, carbohydrates, and lipids. In all life forms the specific types of macromolecules share similar roles that are dependent on the chemical properties of their component monomers. Macromolecules provide energy storage, structural support, protection, catalysis, transport, defense, regulation, movement, and information storage.

Polymers are made by the covalent bonding of monomers in a process called condensation (see Figure 3.4A). Polymers may be broken down into their constituent monomers through hydrolysis (see Figure 3.4B). The size and structure of each polymer is related to its biological function.

Question 1. Describe the difference between optical isomers and structural isomers.

Question 2. In the figure below, you can see how macromolecules are formed and broken down. Identify the two types of reactions (Reaction A and Reaction B) and describe which reaction requires energy and which reaction releases energy.

Reaction A

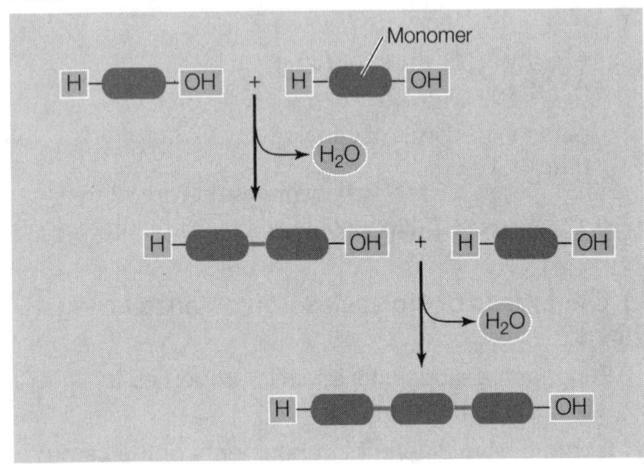

Reaction B

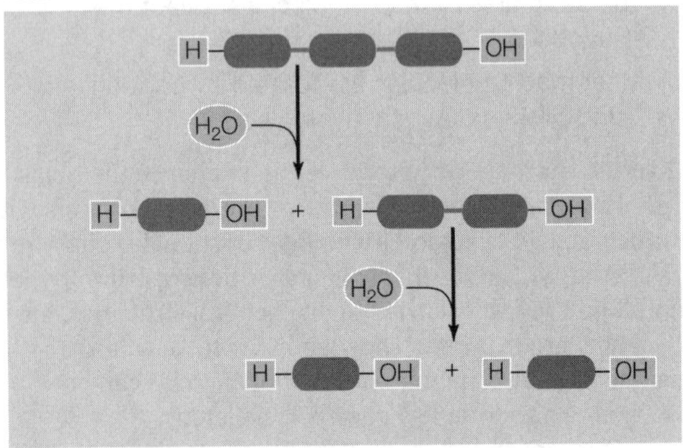

3.2 What Are the Chemical Structures and Functions of Proteins?

Amino acids are the building blocks of proteins

Peptide linkages form the backbone of a protein

The primary structure of a protein is its amino acid sequence

The secondary structure of a protein requires hydrogen bonding

The tertiary structure of a protein is formed by bending and folding

The quaternary structure of a protein consists of subunits

Shape and surface chemistry contribute to protein function

Environmental conditions affect protein structure

Protein shapes can change

Molecular chaperones help shape proteins

Proteins, which are polymers of amino acids, function in structural support, protection, transport, catalysis, defense, regulation, and movement. Proteins have many functions in an organism, including enzymatic activity, defense, hormonal regulation, storage, structural support, transport, and genetic regulation. Proteins range in size from small proteins (with 50–60 amino acids) to large proteins (with more than 26,000 amino acids). Proteins may be made of single chains of amino acids or multiple interacting chains.

Proteins are composed of 20 amino acids, each of which has a central carbon atom with a hydrogen atom, amino group, and carboxyl group attached to it.

The shape and structure of proteins is influenced by the properties of the side chains of the constituent amino acids. Five amino acids have charged side chains, which attract water; five have polar hydrophilic side chains, which form hydrogen bonds; seven have hydrophobic, nonpolar side chains; and three (cysteine, glycine, and proline) have special hydrophobic functions. The side chain of cysteine, —SH, can form disulfide bridges with other cysteines and influences protein chain folding. Glycine is small because of its single hydrogen atom side chain and fits into folds of proteins. Proline has limited bonding ability and often contributes to looping in proteins.

Amino acids are chained or polymerized together during condensation. The resulting bonds between the amino and carboxyl groups are called peptide bonds or linkages (see Figure 3.6). Peptide chains always begin with the N terminus and end with the C terminus. The C—N peptide linkages are relatively rigid, limiting the extent of protein folding. Hydrogen bonds are favored within the folded proteins, helping maintain the form and function of proteins.

Proteins have four levels of structural organization:

(1) A protein's primary structure refers to the basic sequence of amino acids. The primary structure is held together by covalent bonds. All higher levels of structure are due to the specific amino acid sequence in the primary structure of the protein.

(2) Secondary structure refers to regular, repeated spatial patterns of the amino acid chain stabilized by hydrogen bonds. The α helix consists of a right-handed coiling of a single polypeptide chain. The β pleated sheet is formed from two or more hydrogen-bonded chains. A single chain can also fold over and form hydrogen bonds in a similar manner.

(3) Tertiary structure refers to an amino acid chain that is bent and folded into a more complex pattern. This structure is stabilized by disulfide bridges, hydrophobic interactions, van der Waals forces, ionic interactions, and hydrogen bonds. The sequence of R groups in the primary structure is responsible for this level of folding.

(4) Quaternary structure refers to the three-dimensional interactions of multiple protein subunits.

Protein shape and surface chemistry influence function. The ability of proteins to bond noncovalently to other molecules allows them to carry out their specific function in the cell. Binding sites in proteins have characteristic shapes, allowing for selective and specific interactions with other molecules. The particular orientation and position of chemical groups within the binding site and their ability to interact with other molecules are a result of the protein's three-dimensional shape and the properties of its constituent amino acids. Because a protein's three-dimensional structure is stabilized by relatively "weak" bonds, environmental conditions such as temperature, pH changes, or altered salt concentrations can change the shape of the protein into an inactive form. This is called denaturation.

Chaperones are a class of proteins that prevent other proteins from denaturing by inhibiting inappropriate bonding. Heat shock proteins (HSPs) are a group of chaperones that are found in most eukaryotic cells.

Question 3. List at least four different categories of proteins and briefly describe their functions.

Question 4. Amino acids are the building blocks of proteins. The 20 amino acids can be distinguished by their side chains, but they all share a common general structure. Describe the chemical structure that can be found in every amino acid in living organisms.

Question 5. Protein structure can be affected by environmental conditions, which eventually can lead to denaturation. Differentiate between four environmental effects that can denature proteins and explain how they affect protein structure.

Question 6. Although all four structures of proteins are important in determining their functions, one of the structures provides the backbone of proteins. Discuss this structure. Why is it considered the most important of the four?

3.3 What Are the Chemical Structures and Functions of Carbohydrates?

Monosaccharides are simple sugars

Glycosidic linkages bond monosaccharides

Polysaccharides store energy and provide structural materials

Chemically modified carbohydrates contain additional functional groups

Sugar polymers, called carbohydrates, are a primary energy source and make up the carbon skeleton in biological systems. Carbohydrates have the general formula $C_n(H_2O)_n$. They are used to store energy, transport energy, or as carbon skeletons. Monosaccharides are simple sugar monomers that are used in the synthesis of complex carbohydrates. More complex arrangements include disaccharides (2 monosaccharides), oligosaccharides (3–20 monosaccharides), and polysaccharides (large, complex carbohydrates).

Simple sugars (monomers) are biologically important as an energy source (glucose) and as a structural backbone for RNA and DNA (ribose and deoxyribose, two pentoses). Simple sugars typically have three to six carbon atoms and may exist as different isomers. Dehydration synthesis between monomers results in glycosidic linkages. Oligosaccharides are composed of several monosaccharides that have been bound by glycosidic linkages.

Polysaccharides are very large chains of monomers that provide energy storage or structural support. The specific structure of polysaccharides contributes to their function. For instance, starch, glycogen, and cellulose are all chains of glucose, but the glycosidic bonds are in different orientations and yield very different chemical properties. Added functional groups, such as a carboxyl group, give carbohydrates altered functions. Sugar phosphates and amino sugars, such as chitin, have significant biological importance. Chitin and cellulose are among the most abundant substances in the living world.

Question 7. Fructose and glucose are very important monosaccharides, present in most living organisms. They can form a single glycosidic linkage and produce sucrose (common table sugar). Draw the two monosaccharides and the reaction that produces the disaccharide sucrose.

Question 8. Identify the carbohydrates that are shared by living organisms and differentiate among those that are specific for plants, fungi, or animals.

Question 9. Complex carbohydrates should be a mainstay of one's diet. What are some properties that make them excellent food sources?

3.4 What Are the Chemical Structures and Functions of Lipids?

Fats and oils are triglycerides

Phospholipids form biological membranes

Some lipids have roles in energy conversion, regulation, and protection

Lipids are chemically diverse but water insoluble, and serve many biological functions. They are insoluble in water due to many nonpolar covalent bonds. They aggregate because of their nonpolar nature and are held together by van der Waals forces. They play a wide role in the biology of organisms; they provide energy storage, play structural roles, they help plants capture light, play regulatory roles as hormones and vitamins, provide electrical and thermal insulation, and repel water.

Fats and oils (also known as triglycerides or simple lipids) are composed of fatty acids and glycerol and function primarily in the storage of energy. At room temperature, fats are solid and oils are liquid. Each triglyceride contains a glycerol bound to three fatty acids with carbon atoms that are all single-bonded (saturated fatty acids) or contain double bonds (unsaturated fatty acids). During condensation, three fatty acids are covalently linked to the glycerol by ester linkages. Whether the fatty acids are saturated or unsaturated determines the shape of the molecule and its melting point. The greater the saturation of the fatty acid chains, the higher the melting point.

Phospholipids form cellular membranes. In phospholipids, the hydrophobic fatty acid of a typical lipid is replaced with a hydrophilic phosphate group. This allows phospholipids to be amphipathic. That is, they have two opposing chemical properties: a hydrophilic, "water-loving" head and hydrophobic, "water-hating" tails. In an aqueous environment, the hydrophobic tails of the phospholipids tend to aggregate, with the phosphate heads facing out. This bilayer effect allows for the establishment of a hydrophobic inside surrounded by an aqueous environment.

Carotenoids and steroids are lipids that do not have the glycerol–fatty acid structure. Carotenoids trap light energy during photosynthesis. Steroids serve as signaling molecules and are synthesized from cholesterol. Vitamins A, D, E, and K are all lipids, and their deficiency can lead to serious health symptoms. The nonpolar nature of lipids makes them water repellent. Wax coatings on hair, feathers, and leaves are all lipid-derived.

Question 10. Explain how the fluidity of fatty acids is related to their structure.

Question 11. Draw a diagram of a phospholipid bilayer and identify the hydrophobic tails and the hydrophilic heads. Where in the body can you find such a structure?

Test Yourself

1. Which of the following statements about polymers is *false*?
   a. Polymers are synthesized from monomers during condensation.
   b. Polymers are synthesized from monomers during dehydration. X
   c. Polymers consist of at least two types of monomers.
   d. Both a and c
   e. Both b and c
   Textbook Reference: *3.1 What Kinds of Molecules Characterize Living Things?*

2. A macromolecule with many hydrogen and peptide bonds is most likely a
   a. carbohydrate.
   b. lipid.
   c. protein.
   d. nucleic acid.
   e. vitamin.
   Textbook Reference: *3.2 What Are the Chemical Structures and Functions of Proteins?*

3. An α helix is an example of the ______ level of protein structure.
   a. primary
   b. secondary
   c. tertiary
   d. quaternary
   e. hepternary
   Textbook Reference: *3.2 What Are the Chemical Structures and Functions of Proteins?*

4. Which of the following statements about isomers is true?
   a. They all have different chemical formulas but the same arrangement. X
   b. They are found only in proteins. X
   c. They can only be structural. X
   d. They all have the same chemical formula but different arrangements.
   e. None of the above
   Textbook Reference: *3.1 What Kinds of Molecules Characterize Living Things?*

5. Cellulose and starch are composed of the same monomers but have structural and functional differences. Which of the following is the characteristic that accounts for those differences?
   a. Different types of glycosidic linkages
   b. Different numbers of glucose monomers
   c. Different types of bonds holding them together
   d. A linear shape in one versus a ring shape in the other
   e. None of the above
   Textbook Reference: *3.3 What Are the Chemical Structures and Functions of Carbohydrates?*

6. Which of the following statements about proteins is *false*?
   a. Enzymes are proteins.
   b. Proteins are part of the phospholipid bilayer.
   c. Some hormones are proteins.
   d. Proteins are structural components of the cell.
   e. All of the above are true of proteins.
   Textbook Reference: *3.2 What Are the Chemical Structures and Functions of Proteins?*

7. A disulfide bridge is formed by
   a. two cysteine side chains.
   b. two glycerol linkages.
   c. two proline side chains.
   d. condensation.
   e. hydrolysis.
   Textbook Reference: *3.2 What Are the Chemical Structures and Functions of Proteins?*

8. Triglycerides are synthesized from ______ and ______.
   a. glycerol; amino acids
   b. amino acids; cellulose
   c. steroid precursors; starch
   d. cholesterol; glycerol
   e. fatty acids; glycerol
   Textbook Reference: *3.4 What Are the Chemical Structures and Functions of Lipids?*

9. Proteins consist of amino acids linked together by
   a. noncovalent bonds.
   b. peptide bonds.
   c. phosphodiester bonds.
   d. van der Waals forces.
   e. Both a and b
   Textbook Reference: *3.2 What Are the Chemical Structures and Functions of Proteins?*

10. Which of the following characteristics distinguishes carbohydrates from other macromolecule types?

a. Carbohydrates are constructed of monomers that always have a ring structure.
b. Carbohydrates never contain nitrogen.
c. Carbohydrates consist of a carbon bonded to hydrogen and a hydroxyl group.
d. Carbohydrates contain glycerol.
e. None of the above

Textbook Reference: *3.3 What Are the Chemical Structures and Functions of Carbohydrates?*

11. Which of the following statements about carbohydrates is *false*?
a. Monomers of carbohydrates have six carbon atoms.
b. Monomers of carbohydrates are linked together during dehydration.
c. Carbohydrates are energy-storage molecules.
d. Carbohydrates can be used as carbon skeletons.
e. All of the above are true.

Textbook Reference: *3.3 What Are the Chemical Structures and Functions of Carbohydrates?*

12. One could predict that the R groups of amino acids located on the surface of protein molecules embedded in the interior of biological membranes would be
a. hydrophobic.
b. hydrophilic.
c. polar.
d. able to form disulfide.
e. electrically charged.

Textbook Reference: *3.2 What Are the Chemical Structures and Functions of Proteins?; 3.4 What Are the Chemical Structures and Functions of Lipids?*

13. The characteristic of phospholipids that allows them to form a bilayer is their
a. hydrophilic fatty acid tail.
b. hydrophobic head.
c. hydrophobic fatty acid tail.
d. hydrophilic glycogen acid tail.
e. All of the above

Textbook Reference: *3.4 What Are the Chemical Structures and Functions of Lipids?*

14. A five-carbon sugar is known as a
a. glutamine.
b. glucose.
c. hexose.
d. pentose.
e. None of the above

Textbook Reference: *3.3 What Are the Chemical Structures and Functions of Carbohydrates?*

15. Which of the following statements about lipids is *false*?
a. Lipids are a major component of the phospholipid bilayer.
b. Lipids provide waterproofing for the surfaces of organisms.
c. Steroid hormones are lipids.
d. A number of vitamins are lipids.
e. All of the above statements are true, none is false.

Textbook Reference: *3.4 What Are the Chemical Structures and Functions of Lipids?*

Answers

Key Concept Review

1. Both types of isomers are based on the order of atoms linked together in a molecule. Structural isomers differ in how their atoms are joined together, so if the atoms can be linked together in different orders, then the molecules will have different isomers. Optical isomers occur when a carbon atom has four different groups attached to it. The four groups can link to the carbon in two different ways, creating mirror images of each other.
2. Reaction A is condensation. Polymers are formed from monomers using covalent bonds. A molecule of water is released when a covalent bond is formed, and energy is stored in these bonds. Reaction B is hydrolysis. This reaction results in the breakdown of polymers and the release of energy.
3. Enzymes can catalyze biochemical reactions. Structural proteins provide physical stability. Defensive proteins react to foreign substances. Signaling proteins control physical processes. Receptor proteins respond to chemical signals. Membrane transporters regulate substance movement across cell membranes. Storage proteins reserve amino acids for later use. Transport proteins carry substances. Gene-regulating proteins affect gene expression.
4. Each amino acid has an α carbon atom with a hydrogen atom, and a carboxyl and amino functional group. The α carbon also has a side chain, or R group. However, this R group differs from one amino acid to another.
5. Increased temperature can break hydrogen bonds and hydrophobic interactions. The change of pH can disrupt the ionic attractions and repulsions. Increase of polar substances disrupts the hydrogen bonding. Nonpolar substances disrupt hydrophobic interactions.
6. The backbone of the polypeptide chain is its primary structure. The seemingly infinite number of protein configurations made possible by the biochemical properties of the 20 amino acids has driven the evolution of life's diversity. The properties associated with each functional group in the side chains determine how the protein will twist and fold. The other structures of a protein would not be able to exist without the primary structure (the primary structure specifies the tertiary structure, as was shown by Anfiensen et al. [1961]). Also, the primary structure is the last one to denature because it is maintained by strong covalent bonds. The primary structure is the foundation of the other structures and therefore of the function of each protein.

7.

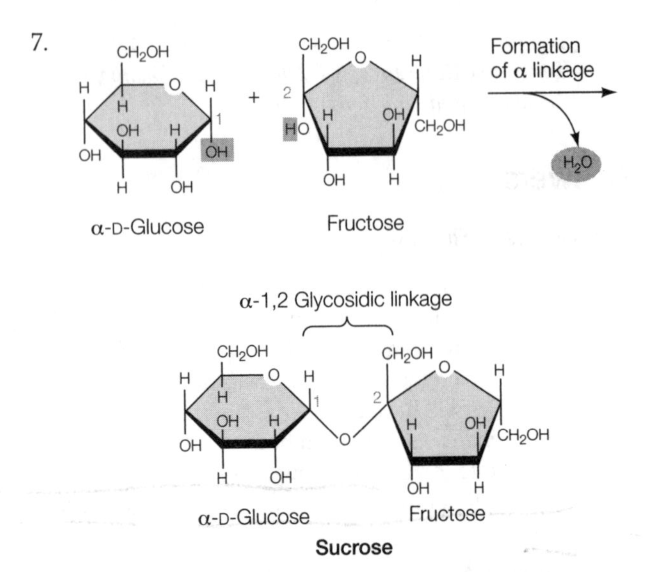

8. All living cells contain the monosaccharide glucose. Starch, such as amylose (a polysaccharide) is the principal energy storage compound of plants, while in animals another polysaccharide called glycogen has the same function. Amino sugars (carbohydrates with a phosphate group added) such as chitin are present in insects and in the cell walls of fungi. Plant cell walls often contain the polysaccharide cellulose instead of chitin.

9. Complex carbohydrates are easily broken down into glucose monomers, which provide cellular energy. Storage of glucose monomers in large carbohydrates allows the osmotic strain on any given cell to be reduced without sacrificing the availability of energy.

10. The kinks in fatty acid molecules are important in determining the fluidity and melting points of lipids. The kinks are caused by the double bonds in the hydrocarbon chain. Because saturated fatty acids have only single bonds between the carbon atoms in the chain, when they form triglycerides, they are packed tightly together. Therefore, they are solid at room temperature and have high melting points (e.g., animal fats). In contrast, triglycerides with unsaturated fatty acids that contain double bonds and kinks cannot pack together as tightly and therefore have a lower melting points (e.g., oils).

11. Phospholipid bilayers can be found in biological membranes. See Figure 3.22.

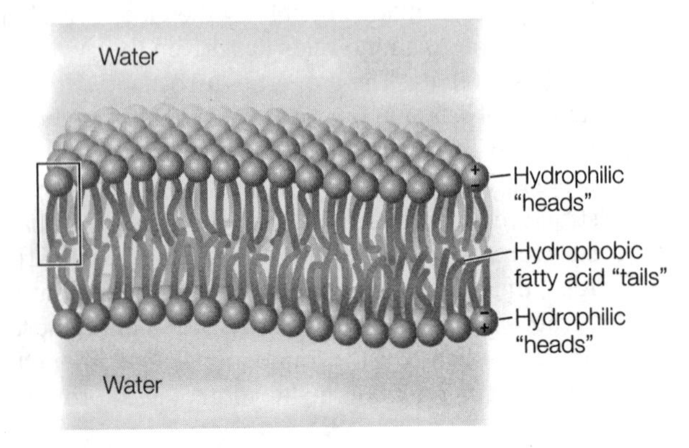

Test Yourself

1. **e.** Polymers are synthesized via condensation and broken down via hydrolysis. They may be made of a single type of monomer or from multiple types of monomers.
2. **c.** Proteins consist of amino acids joined by peptide bonds. Higher levels of structure are stabilized by hydrogen bonds.
3. **b.** α helices are examples of secondary structure and are maintained by hydrogen bonds between the amino acid residues.
4. **d.** Proteins, carbohydrates, or lipids can have the same chemical formula, but the isomer molecules are arranged differently.
5. **a.** Starch has β-glycosidic linkages, whereas cellulose has α-glycosidic linkages. This difference accounts for structural and functional differences between the two macromolecules.
6. **b.** Proteins do not make up the phospholipid bilayer, which is composed of phospholipids. Proteins do perform all of the other functions.
7. **a.** Disulfide bridges are formed when the —SH groups of two cysteines interact.
8. **e.** Triglycerides are formed from one glycerol and three fatty acid molecules.
9. **b.** The peptide bond that links amino acids to form proteins is a type of covalent bond.
10. **c.** Carbohydrates always have carbon atoms bonded to hydrogen atoms and hydroxyl groups. They may have a variety of other associate molecules in addition to these.
11. **a.** Carbohydrate monomers may have different numbers of carbon atoms. Five-carbon monomers (pentoses) and six-carbon monomers (hexoses) are both common.
12. **a.** The interior of the plasma membrane is hydrophobic; therefore, an embedded protein would have hydrophobic residues.
13. **c.** Phospholipids are composed of a hydrophilic head and a hydrophobic tail. When they are placed in water, the hydrophobic tails come together in the interior of the bilayer, surrounded by the hydrophilic heads facing outward.
14. **d.** Five-carbon sugars are known as pentoses and form the backbones for RNA and DNA.
15. **e.** Lipids play a role in all of the functions listed.

Nucleic Acids and the Origin of Life

The Big Picture

- Nucleic acids are built from nucleotide monomers, each of which consists of a phosphate group, a pentose sugar, and a nitrogenous base.
- Sequences of nucleic acids are either single-stranded (such as in RNA) or double-stranded (such as in DNA) and contain the genetic information needed to code for the proteins required for life. The information stored in DNA as sequences of nitrogenous bases allows for a copying mechanism of DNA and for the transcription of DNA into RNA, which is then translated into specific amino acid sequence in proteins.
- The small molecules required for life first originated in water. One hypothesis for the origin of life involves meteorites containing the molecules required for life. An alternative hypothesis is that of chemical evolution, which states that the atmospheric environment of early Earth may have been conducive to the formation of organic molecules such as amino acids, nucleic acid bases, and simple sugars.
- Polymers, such as proteins and nucleic acids, may have evolved in concentrated droplets with the help of some nucleic acid polymers with catalytic functions such as RNA.
- The cell may have evolved from a simple protocell in which a lipid bilayer enclosure permitted the maintenance of chemical conditions that were different from the external environment.

Study Strategies

- Avoid the temptation to memorize structures. It is unnecessary to memorize the structures of nitrogenous bases. Table 4.1 and Figures 4.2–4.4 in your textbook will help you understand the structural differences between DNA and RNA.
- The directionality (5′and 3′ ends) of DNA and RNA may present a challenge. An easy way to remember which end is which is to recall that the 5′ end of a DNA/RNA strand has the free phosphate group while the 3′ end has the free hydroxyl (OH) group of ribose (RNA) or deoxyribose (DNA).
- The different hypotheses regarding the origin of life can be confusing. Remember that the monomers required for life had to arise first, and this was followed by the ability to replicate and carry out metabolic processes.
- This chapter contains a large amount of terminology. Making a terminology/vocabulary list may be helpful.
- Go to the Web addresses shown to review the following animated tutorials and activities. You can also find them in BioPortal (yourBioPortal.com), along with many additional learning resources.

  Animated Tutorial 4.1 Pasteur's Experiment (Life10e.com/at4.1)

  Animated Tutorial 4.2 Synthesis of Prebiotic Molecules (Life10e.com/at4.2)

  Activity 4.1 Nucleic Acid Building Blocks (Life10e.com/ac4.1)

  Activity 4.2 DNA Structure (Life10e.com/ac4.2)

Key Concept Review

4.1 What Are the Chemical Structures and Functions of Nucleic Acids?

Nucleotides are the building blocks of nucleic acids

Base pairing occurs in both DNA and RNA

DNA carries information and is expressed through RNA

The DNA base sequence reveals evolutionary relationships

Nucleotides have other important roles

The nucleic acids (DNA and RNA) store, transmit, and use genetic information. The primary role of nucleic acids is to store and transmit hereditary information. DNA (deoxyribonucleic acid) and RNA (ribonucleic acid) are the two different types of nucleic acids. Nucleic acids are made up of nucleotide monomers. Each nucleotide monomer consists of a pentose, or five-carbon, sugar (either deoxyribose or ribose), a phosphate group, and a nitrogenous base. The bases of nucleotides contain either a single ring with six members (pyrimidine) or a fused double ring (purine). The backbone of

DNA and RNA consists of alternating sugars and phosphates with the bases projecting outward. Four types of nitrogenous bases are found in DNA: two purines (adenine and guanine) and two pyrimidines (thymine and cytosine). In RNA, thymine is replaced by uracil.

In both RNA and DNA, nucleotides are joined together by phosphodiester linkages formed between the OH group of carbon 3′ on one nucleotide and the phosphate group on carbon 5′ of another nucleotide. In a nucleic acid strand, nucleotides are added to the free 3′ end of the strand, thus growing in the 5′ to the 3′ direction. Nucleic acids are either short oligonucleotides (such as RNA molecules used for DNA replication or regulation of gene expression) or longer polynucleotides (such as DNA).

In DNA, bases are capable of complementary base pairing, with adenine pairing with thymine and cytosine pairing with guanine. Complementary base pairing is facilitated by the size and shape of the molecules, the geometry of the sugar–phosphate backbone, and hydrogen bonding. DNA is typically found as a double helix with two complementary strands paired and twisted together. The two strands in DNA are stabilized by hydrogen bonds and run in opposite directions. Most RNA is single stranded, with the capacity to fold back on itself or to interact with other RNA molecules through complementary base pairing to produce different three-dimensional structures.

The sequences of bases in DNA provide a means of storing information in the cell. To ensure the correct use of this information, DNA is replicated precisely and can be transcribed into RNA, which in turn is translated into specific polypeptides. Replication and transcription rely on properties of the five base pairs. The base pairs A-T and G-C in DNA are replaced with their appropriate RNA counterpart (A-U and G-C) during transcription. When a molecule of DNA is replicated, the entire DNA molecule is copied. The genome contains a complete set of an organism's DNA. The specific portions of the DNA that encode for proteins are genes; they are transcribed only when the cell requires the specific proteins.

Similarities in base sequences of DNA can help determine evolutionary relationships among organisms. Closely related species tend to have greater similarity in base sequences compared to distantly related species. The increased ability of scientists to sequence DNA is leading to insights into the evolutionary relationships of organisms that were not available based on anatomical and behavioral comparisons.

Question 1. If you were to transcribe both strands of the DNA sequence below, what two RNA sequences would result? Make sure to label to the 3′ and 5′ ends of the resulting RNA molecules.

5′-AAGCGTC-3′
3′-TTCGCAG-5′

Question 2. In the diagram below, use the base-pairing rules for DNA and RNA to label the strand of RNA (right) that is complementary to the single strand of DNA (left), where C = cytosine, G = guanine, A = adenine, T = thymine, and U = uracil. Then circle and label an example of a nucleotide and a nucleoside, showing the double-stranded DNA and RNA hybrid. Based on the orientation of the sugar molecules, label the four ends of the molecule as 3′ or 5′, showing the double-stranded DNA and RNA hybrid.

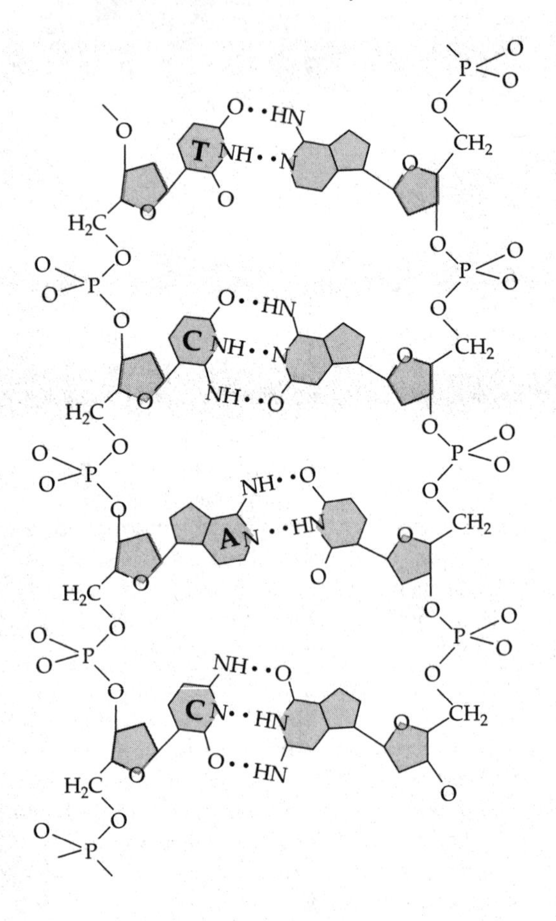

Question 3. What properties of DNA structure make it well suited for its function as an informational molecule?

Question 4. Suppose you want to study the chemical composition of a type of cell. You culture a group of these cells in the presence of radioactive tracers to label DNA, RNA, and proteins. After allowing the cells to multiply and incorporate the radioactive tracers, you prepare extracts of DNA, RNA, and proteins from the cells and store them in three different tubes. Which radioactive elements and/or compounds should you use in order to label one compound but not the other so that you can differentiate between, and therefore identify, the three different types of molecules?

4.2 How and Where Did the Small Molecules of Life Originate?

Experiments disproved the spontaneous generation of life

Life began in water

Life may have come from outside Earth

Prebiotic synthesis experiments model early Earth

Two alternative hypotheses exist for the beginning of life on Earth. In 1668, the Italian physician Francesco Redi conducted experiments with flies that showed that life does not arise by spontaneous generation. Louis Pasteur conducted similar experiments to show that microorganisms arise from other microorganisms.

Approximately 4.6 billion years ago, the solar system was formed from an exploding star, producing our sun and other bodies known as planetesimals. The first signs of life on Earth date back 4 billion years. Water high in the atmosphere condensed on the surface of Earth as it cooled, and simple chemical reactions might have led to life. But the large impact of meteorites and comets hitting Earth would have released large amounts of energy, and likely destroyed any initial molecules necessary for life.

One hypothesis of the origin of life on Earth is based on the idea that water and some molecules characteristic of life may have come from space. In 1969 a meteorite containing purines, pyrimidines, sugars, and ten amino acids was found in Murchison, Australia. In addition, more than 90 meteorites from Mars have been recovered on Earth, some of which showed signs that they contained water.

Another hypothesis of the origin of life on Earth, known as chemical evolution, states that the conditions on Earth billions of years ago led to the emergence of biological molecules, either physically or chemically.

Originally, the atmosphere of Earth had very little oxygen, which began to accumulate only as single-celled organisms began to carry out photosynthesis about 2.5 million years ago.

Stanley Miller and Harold Urey produced a primitive atmosphere containing hydrogen gas, ammonia, methane gas, and water, and by means of electrical charges passed through these gasses (the so-called hot experiment) were able to produce numerous organic molecules such as amino acids, which are the building blocks of proteins. This experiment provided support for the chemical origin of life on Earth.

Stanley Miller was also able to produce amino acids and nucleotide bases from ammonia gas, water vapor, and cyanide in a cold environment. These initial molecules were placed in a sealed tube, which was frozen at –78°C for 27 years and then reopened. In pockets of liquid water within the ice, high concentrations of the starting materials had accumulated. This experiment provided support for the notion that cold water within ice pockets on ancient Earth and other celestial bodies may have provided environments for the formation of molecules required for the initial evolution of living forms.

Other experiments modeling the conditions of the atmosphere of ancient Earth have provided additional evidence for the origin of monomers needed for the production of biological polymers (such as proteins and nucleic acids).

Question 5. In what ways did Miller and Urey's experiment simulate the conditions thought to have existed in the atmosphere of ancient Earth?

Question 6. Based on the evidence supporting the hypothesis of chemical evolution, is it reasonable to expect to find carbon-based life forms in very hot or very cold extraterrestrial environments? Explain.

4.3 How Did the Large Molecules of Life Originate?

Chemical evolution may have led to polymerization

RNA may have been the first biological catalyst

Formation of polymers may have been the result of chemical evolution. Several models have been proposed by scientists that simulate the conditions that may have resulted in the polymerization of simple monomers into polymers. All of these models are based on observations and assumptions that catalytic reactions occured, resulting in the polymerization of monomers. These reactions occurred on the surfaces of minerals in clay and hot metals in hydrothermal vents, and in hot pools at the edges of oceans where evaporation favored the polymerization of concentrated monomers.

RNA may have been the first biological catalyst. It is believed that a key event in the evolution of life on Earth was the appearance of catalysts. While enzyme proteins are the main biological catalysts today, there are several lines of evidence supporting the hypothesis of an "RNA world," which suggests that RNA or an RNA-like nucleic acid was the first catalyst to evolve on ancient Earth.

The presence of ribozymes, catalytic RNA that exists in cells, supports the "RNA world" hypothesis since this type of RNA catalyzes the formation of peptide linkages. On ancient Earth, RNA may have acted as a self-replicating catalyst for the polymerization of amino acids to form proteins. RNA contains genetic code, making it a potential precursor for the evolution of DNA.

Question 7. Which of the following statements supports the "RNA world" hypothesis? Justify your answer.

a. RNA contains the sugar ribose instead of thymine.
b. Unlike DNA, RNA can act as both an information-storing molecule and a catalyst of chemical reactions.
c. RNA contains uracil instead of thymine.
d. DNA is double-stranded, whereas RNA is single-stranded.

Question 8. According to the "RNA world" hypothesis, DNA could have evolved from self-replicating catalytic RNA. What are the steps postulated by the "RNA world" hypothesis that led to the evolution of DNA from RNA?

4.4 How Did the First Cells Originate?

Experiments explore the origin of cells

Some ancient cells left a fossil imprint

Evolution of the first cells involved the formation of a cell membrane providing a barrier between the internal and external environments. It is possible that the amphipathic nature of fatty acids may have formed a lipid bilayer protocell that trapped water and other molecules. Large molecules such as DNA and RNA cannot cross this protocell bilayer, but single nucleic acids and sugars can. Thus, if a self-replicating RNA strand were to be in the protocell, it could produce new polynucleotide chains using external nucleic acids.

The fossil record indicates that early cells may have evolved 3.5 billion years ago. Some of the fossil imprints of

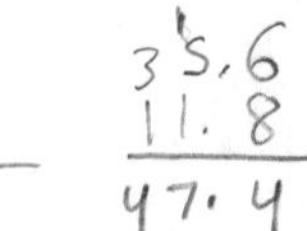

such cells bear a striking resemblance to modern filamentous cyanobacteria.

Question 9. One of the first steps leading to a cell that could reproduce may have been the formation of protocells. What characteristics may have allowed protocells to be the precursors to the living cell?

Question 10. What evidence allowed J. William Schopf to prove that the 3.5-billion-year-old fossil he discovered in Australia was that of ancient cells?

Test Yourself

1. DNA utilizes the bases guanine, cytosine, thymine, and adenine. In RNA, _______ is replaced by _______.
   a. adenine; arginine
   b. thymine; uracil
   c. cytosine; uracil
   d. cytosine; arginine
   e. cytosine; thymine
   Textbook Reference: *4.1 What Are the Chemical Structures and Functions of Nucleic Acids?*

2. The pairing of purines with pyrimidines to create a double-stranded DNA molecule is called
   a. complementary base pairing.
   b. phosphodiester linkages.
   c. antiparallel synthesis.
   d. dehydration.
   e. the genome.
   Textbook Reference: *4.1 What Are the Chemical Structures and Functions of Nucleic Acids?*

3. The end product of transcription is _______ and the end product of translation is _______.
   a. proteins; DNA
   b. proteins; RNA
   c. RNA; DNA
   d. DNA; RNA
   e. RNA; proteins
   Textbook Reference: *4.1 What Are the Chemical Structures and Functions of Nucleic Acids?*

4. Which of the following is *not* a nucleotide?
   a. Adenosine triphosphate
   b. Guanosine triphosphate
   c. Cyclic adenosine monophosphate
   d. Thymine
   e. All of the above are nucleotides.
   Textbook Reference: *4.1 What Are the Chemical Structures and Functions of Nucleic Acids?*

5.–6. Suppose you obtain four different samples of nucleic acids. You analyze the nitrogenous base composition of each one and you get the following results (the numbers indicate percentage of each nitrogenous base in the samples):

Nucleic acid	Adenine	Thymine	Cytosine	Guanine	Uracil
A	40.1	39.1	9.5	9.7	0.0
B	35.6	0.0	12.3	11.8	34.9
C	39.4	0.0	11.2	25.6	18.9
D	21.	0.0	18.9	11.2	29.2

5. Which of these nucleic acids is single-stranded RNA?
   a. Nucleic acid A
   b. Nucleic acid B
   c. Nucleic acid C
   d. Nucleic acid D
   e. Both nucleic acid C and D are single-stranded RNA.
   Textbook Reference: *4.1 What Are the Chemical Structures and Functions of Nucleic Acids?*

6. Which of these nucleic acids is double-stranded DNA?
   a. Nucleic acid A
   b. Nucleic acid B
   c. Nucleic acid C
   d. Nucleic acid D
   e. Both nucleic acid C and D are single-stranded RNA.
   Textbook Reference: *4.1 What Are the Chemical Structures and Functions of Nucleic Acids?*

7. Consider a fragment of DNA that has a total of ten base pairs. Four of these contain adenine and six contain guanine. How many hydrogen bonds would there be in this fragment between purine/pyrimidine pairs?
   a. 10
   b. 16
   c. 20
   d. 26
   e. 30
   Textbook Reference: *4.1 What Are the Chemical Structures and Functions of Nucleic Acids?*

8. You obtain a sample of dried DNA. You add water to the sample and find that it dissolves in water. Why is DNA water soluble?
   a. DNA has a hydrophobic interior.
   b. The presence of hydrogen bonds in the nitrogenous bases makes DNA water soluble.
   c. DNA has a hydrophilic exterior consisting of chains of polar deoxyribose sugar and negatively charged phosphate.
   d. Cells consist mostly of water.
   e. None of the above
   Textbook Reference: *4.1 What Are the Chemical Structures and Functions of Nucleic Acids?*

9. An individual human's genome
   a. consists only of the genes that code for specific proteins.
   b. is identical for every individual.
   c. is the complete set of the individual's DNA.
   d. is the complete set of the individual's RNA.

e. None of the above

Textbook Reference: *4.1 What Are the Chemical Structures and Functions of Nucleic Acids?*

10. Francesco Redi's experiment with flies disproved the concept of
    a. the genome.
    b. translation.
    c. an "RNA world."
    d. spontaneous generation.
    e. the protocell.

    Textbook Reference: *4.1 What Are the Chemical Structures and Functions of Nucleic Acids?*

11. Life is thought to have originated
    a. on land.
    b. in the air.
    c. underground.
    d. in water.
    e. All of the above

    Textbook Reference: *4.2 How and Where Did the Small Molecules of Life Originate?*

12. Which of the following was/were *not* synthesized in the "hot chemistry" experiments conducted by Miller and Urey?
    a. Uracil
    b. Amino acids
    c. 6-carbon sugars
    d. 3-carbon sugars
    e. Polypeptides

    Textbook Reference: *4.2 How and Where Did the Small Molecules of Life Originate?*

13. The first biological catalyst may have been similar to
    a. a ribozyme.
    b. DNA.
    c. a protein.
    d. glucose.
    e. a lipid.

    Textbook Reference: *4.3 How Did the Large Molecules of Life Originate?*

14. Prebiotic water-filled structures that are defined by a lipid bilayer are known as
    a. lipases.
    b. protocells.
    c. genomes.
    d. neocells.
    e. lipocells.

    Textbook Reference: *4.4 How Did the First Cells Originate?*

15. In the origin of life, _______ may have evolved first, followed by _______.
    a. proteins; amino acids
    b. DNA; RNA
    c. RNA; DNA
    d. RNA; nucleic acids
    e. All of the above appeared at the same time.

    Textbook Reference: *4.3 How Did the Large Molecules of Life Originate?*

16. In DNA and RNA, nucleotides are joined by
    a. phosphodiester linkages.
    b. hydrogen bonds.
    c. peptide linkages.
    d. glycosidic linkages.
    e. All of the above

    Textbook Reference: *4.1 What Are the Chemical Structures and Functions of Nucleic Acids?*

17. Which of the following subunits make(s) up a nucleotide?
    a. Pentose sugar
    b. Phosphate group
    c. Nitrogenous base
    d. Amino acid.
    e. a, b, and c

    Textbook Reference: *4.1 What Are the Chemical Structures and Functions of Nucleic Acids?*

Answers

Key Concept Review

1. RNA sequence for top: 3′-UUCGCAG-5′;
   RNA sequence for bottom: 5′-AAGCGUC-3′
2.

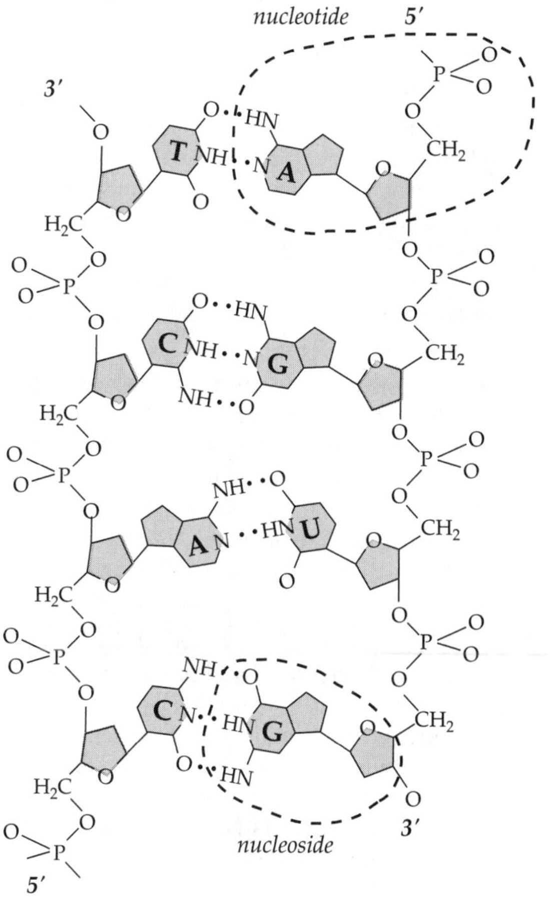

3. DNA structure is highly regular and conserved, which makes it usable by all cells, while the different bases allow it to encode specific genetic information.
4. DNA contains the nitrogenous base thymine. RNA contains uracil but not thymine. In addition, DNA contains the pentose sugar deoxyribose, whereas RNA contains ribose sugar. To differentiate between DNA and RNA, you could incubate the cells with radioactive thymine, uracil, ribose, and deoxyribose. As cells multiply, they incorporate the radioactive tracers. In this way, ribose and uracil would specifically label newly synthesized RNA, whereas deoxyribose and thymine would specifically label newly synthesized DNA. To differentiate between nucleic acids (RNA and DNA) and proteins, you could incubate the cells with radioactive sulfur. Culturing cells with the radioactive form of this element would allow you to selectively label sulfur-containing amino acids but not DNA or RNA, since nucleic acids do not contain sulfur in their structures.
5. Refer to Figure 4.8. Miller and Urey's experimental setup was designed to create a microcosm of ancient Earth. The heated round flask represented the "oceanic" compartment in which heat, presumably from volcanic activities, resulted in the release of methane, ammonia, hydrogen gas, and water vapor into the atmosphere. The second round flask represented the "atmospheric" compartment in which electrical sparks simulated lightning, which provided the energy for the synthesis of new compounds. The cold water condenser allowed for the condensation of atmospheric gases and their precipitation back into the "oceanic" compartment, thus simulating rain.
6. "Hot chemistry" experiments (such as those conducted by Miller and Urey) provided evidence for the possibility of organic compounds having formed from simple inorganic molecules under hot environmental conditions. Miller's "cold chemistry" experiments showed that organic molecules such as amino acids and nucleotide bases could have formed in ice pockets under very cold conditions (78°C).
7. Only statement **b** supports the "RNA world" hypothesis. The fact that ribozymes (RNA) can catalyze the formation of peptide linkages supports the idea that life was kick-started by a self-replicating catalyst. In the laboratory, a synthetic ribozyme was shown to catalyze the polymerization of shorter RNA strands into longer ones that are identical to the synthetic ribozyme, thus indicating that it may have once been possible.
8. Review Figure 4.9. The "RNA world" hypothesis states that in a world before DNA, RNA was formed from simple nucleotides. Some of this RNA gained the ability to replicate itself, and some of the self-replicating RNA molecules had a catalytic function that allowed them to make enzymes (catalytic proteins). The evolution of enzymes helped RNA to become more efficient in replicating itself and in making new proteins. With the help of enzymes, RNA became double-stranded and ultimately evolved into double-stranded DNA.
9. Protocells are lipid bilayers that allow for the organization of molecules in an environment distinct from the external environment. Experiments have shown that sugars and nucleotides can enter these protocells, while larger molecules such as DNA and RNA cannot exit. In the presence of a self-replicating nucleic acid strand, protocells may produce polynucleotide chains. Thus, structures similar to protocells could have been the precursors to living cells.
10. To prove that the 3.5-billion-year-old cyanobacteria-like fossil was once alive, Schopf examined both chemical and structural evidence. He showed that the Australian fossil had a carbon radioisotope signature (^{13}C:^{12}C) typical of photosynthetic organisms. He also showed that the chainlike structures of the fossil had internal complex structures that are characteristics of living cells.

Test Yourself

1. **b.** RNA and DNA differ by one oxygen molecule in their ribose sugar and in the substitution of uracil (RNA) for thymine (DNA).
2. **a.** Complementary base pairing results from the attraction of charges and the ability of purines to pair with pyrimidines through hydrogen bond formation.
3. **e.** Transcription involves making RNA from DNA, and translation involves making proteins from RNA.
4. **e.** These are all examples of nucleotides. ATP and GTP both serve as energy sources. cAMP is involved in signal transduction of hormones and the nervous system.
5. **e.** Both nucleic acids C and D are RNA samples since they both contain uracil, which is found only in RNA. Both nucleic acids C and D are single stranded because the number of purines (adenine and guanine) is not equal to the number of pyrimidines (cytosine and uracil).
6. **a.** Nucleic acid A is double-stranded DNA because it contains thymine, which is found only in DNA and the number of purines (adenine and guanine) is equal to the number of pyrimidines (cytosine and thymine).
7. **d.** Four adenine form complementary base pairs with four thymine via two hydrogen bonds between each pair, resulting in a subtotal of eight hydrogen bonds. Six guanine form complementary base pairs with six cytosine via three hydrogen bonds between each pair, resulting in a subtotal of 18 hydrogen bonds. The total number of hydrogen bonds in this DNA fragment is: $8 + 18 = 26$.
8. **c.** The exterior-facing backbones of double-stranded DNA consist of chains of alternating deoxyribose sugars and phosphate groups that are negatively charged and hydrophilic.

9. **c.** A person's genome is the complete set of that individual's DNA, including the portions that code for genes and those that do not code for genes.
10. **d.** Redi was an Italian scientist who did experiments with flies and meat to show that life comes from other life, rather than arising by spontaneous generation.
11. **d.** Life is thought to have originated in water.
12. **e.** The "hot chemistry" experiments conducted by Miller and Urey resulted in the five nucleotide bases, amino acids, and 3- and 6-carbon sugars, but not polypeptides.
13. **a.** Ribozymes are catalytic RNA molecules and may have been the first catalytic molecules.
14. **b.** These prebiotic cells are known as protocells, and similar structures may have been involved in trapping water and other molecules necessary for the evolution of life.
15. **c.** The "RNA world" hypothesis for the origin of life suggests that RNA may have evolved first, followed by DNA. This is plausible because RNA is catalytic and also stores information.
16. **a.** The sugar of one nucleotide and the phosphate of the next are joined by phosphodiester linkages in both DNA and RNA.
17. **e.** A nucleotide consists of the following three subunits: a phosphate group, a pentose sugar, and a nitrogenous base.

Cells: The Working Units of Life

The Big Picture

- Cell theory states that all living things are composed of cells and all cells come from preexisting cells. Cells are studied with microscopes and exist in two basic forms: the structurally simpler prokaryotic cell and the more complex eukaryotic cell containing membrane-bound organelles.
- The organelles of eukaryotic cells have a variety of functions, including information storage and transmission, modification and packaging of cellular products, and energy transformation. See Figure 5.7 to understand the organization, structure, and function of each component.
- Both prokaryotic and eukaryotic cells have a selectively permeable plasma membrane and a cytoskeleton. Eukaryotic cells may have originated when plasma membrane infoldings surrounded internal structures and/or smaller cells that eventually became organelles for the larger cell.

Study Strategies

- Terminology is the most difficult part of this material. The vocabulary is easier to learn if you understand the concepts before you memorize the terms.
- Refer to Figure 5.7 while you are studying the organelles and their components in order to visualize how their functions relate to their structure. Then quiz yourself using either the BioPortal Chapter 5 flashcards or notecards you make yourself. Sometimes physically writing out the terms and definitions helps you retain the information.
- See if you can come up with your own analogies for the functions cell structures perform. For example, you might compare the Golgi apparatus to the post office or the lysosome to a waste-treatment and recycling center.
- Go to the Web addresses shown to review the following animated tutorials and activities. You can also find them in BioPortal (yourBioPortal.com), along with many additional learning resources.

  Animated Tutorial 5.1 Eukaryotic Cell Tour (Life10e.com/at5.1)

  Animated Tutorial 5.2 The Golgi Apparatus (Life10e.com/at5.2)

  Activity 5.1 The Scale of Life (Life10e.com/ac5.1)

  Activity 5.2 Know Your Techniques (Life10e.com/ac5.2)

  Activity 5.3 Lysosomal Digestion (Life10e.com/ac5.3)

Key Concept Review

5.1 What Features Make Cells the Fundamental Units of Life?

Cell size is limited by the surface area-to-volume ratio

Microscopes reveal the features of cells

The plasma membrane forms the outer surface of every cell

Cells are classified as either prokaryotic or eukaryotic

The cell theory states that the cell is the basic unit of life, that all organisms are composed of cells, and that all cells come from preexisting cells.

This means that when we study cell biology we are studying life, and that life is a continuous. All cells in your body come from cells going back in time to the first living cell.

Cell size is limited by surface area-to-volume ratio. Very large cells are not feasible because as volume increases, surface area increases at a slower rate. A cell's capacity for chemical activity is related to the cell volume. However, cells with large volumes are unable to take up enough material across their surface to support their metabolic activity. Thus, all large organisms are made of many small cells. Because of the small size of cells, they can only be visualized with microscopes. Light microscopes, which use lenses and light, can resolve objects as small as 0.2 μm. Electron microscopes, which use electron beams, can resolve objects as small as 0.2 nm, but the visualization process kills the cells. Staining techniques assist in visualizing cellular components.

All cells are surrounded by a plasma membrane composed of a phospholipid bilayer with many embedded and protruding proteins. The plasma membrane and the associated proteins are selectively permeable, permitting some small molecules to pass through but preventing others from doing so. The membrane allows for maintenance of the internal cellular environment and is responsible for communication and interactions with other cells. Prokaryotes, which include the Archaea and Bacteria, lack a nucleus and other membrane-bound internal compartments. Eukaryotes, which are composed of the domain Eukarya, have organelles and a nucleus containing the cell's DNA. The organelles compartmentalize the different functions occurring in the cell.

Question 1. Why are cells small?

Question 2. What are the functions of a cell membrane?

5.2 What Features Characterize Prokaryotic Cells?

Prokaryotic cells share certain features

Specialized features are found in some prokaryotes

In terms of numbers and diversity, prokaryotes are the most successful cell type on Earth. Prokaryotic cells do not have internal membrane compartments and are generally smaller than eukaryotic cells. Prokaryotes are one-celled organisms in the domains Archaea and Bacteria. Each prokaryotic cell has a plasma membrane and a region containing the genetic material, which is not membrane-enclosed. The cytoplasm of the cell consists of the liquid cytosol and insoluble particles that act as subcellular machinery (i.e., ribosomes for protein synthesis). The cytoplasm is in constant motion to ensure that reactants come together to bring about biochemical reactions.

Most prokaryotes have a rigid cell wall outside the plasma membrane that supports the cell and determines its structure. Bacterial cell walls contain peptidoglycan. Some prokaryotes have a second outer membrane or produce a slimy layer, or capsule, that is rich in polysaccharides and is located outside the cell wall. Some prokaryotes have an internal membrane system that provides partial compartmentalization for photosynthesis or other energy-releasing reactions. Prokaryotes can also have structures that help with movement or support. Flagella are tiny protein machines containing the protein flagellin that spin to help move the cell. Pili are shorter structures that allow cells to stick to a surface or to one another (e.g., during genetic exchange). Prokaryotes also have protein filaments that form a cytoskeleton that helps shape the cell or contributes to cell division.

Question 3. In your biology course, your professor gives you a sample of cells and tells you that they can only be prokaryotic cells. What characteristics identify them definitively as prokaryotic cells?

Question 4. Compare flagella and pili with respect to both structure and function.

5.3 What Features Characterize Eukaryotic Cells?

Compartmentalization is the key to eukaryotic cell function

Organelles can be studied by microscopy or isolated for chemical analysis

Ribosomes are factories for protein synthesis

The nucleus contains most of the genetic information

The endomembrane system is a group of interrelated organelles

Some organelles transform energy

There are several other membrane-enclosed organelles

The cytoskeleton is important in cell structure and movement

Biologists can manipulate living systems to establish cause and effect

Both eukaryotes and prokaryotes have a plasma membrane, a cytoskeleton, a cytoplasm, and ribosomes, but eukaryotic cells are usually larger than prokaryotic cells and have membrane-enclosed internal organelles. Each organelle has its own internal environment uniquely suited to its function. Microscopes reveal organelles and other cell structures. Their chemical composition can be studied by using stains for specific structures or cell fractionation for chemical analysis. Ribosomes are made of ribosomal RNA (rRNA) and many different proteins. They act as information transcription centers and guide the synthesis of proteins from the messenger RNA nucleic acid blueprints. Prokaryotic ribosomes are smaller than eukaryotic ribosomes and are located in the cytoplasm. Eukaryotic ribosomes are often attached to the endoplasmic reticulum. The nucleus stores DNA, is the site of DNA duplication, and regulates DNA transcription into RNA. Its nucleolus region is involved in ribosome assembly and RNA synthesis. Inside the nucleus, DNA and proteins combine to make chromatin. The long thin threads of chromatin are called chromosomes and are attached to a protein network called the nuclear lamina. Just before a cell divides, chromosomes condense and become visible with the use of light microscopy. The nucleus is bounded by a double lipid bilayer called the nuclear envelope. Small openings in the nuclear envelope, called nuclear pores, control the passage of larger molecules into and out of the nucleus.

The endomembrane system consists of the nuclear envelope, the endoplasmic reticulum (ER), the Golgi apparatus, lysosomes, and the plasma membrane. Membrane-bound vesicles move various substances within the endomembrane system. The ER is a complex of membrane sacs throughout the cell that connects to the nuclear envelope. Rough endoplasmic reticulum (RER) is studded with active ribosomes and is involved in the synthesis, storage, transport, and modification of new proteins. Many of the cell's membrane-bound proteins are produced in the RER. Carbohydrate groups are added to proteins to make glycoproteins in the RER. These carbohydrate groups help, for example, in identifying proteins and ensuring that they reach the correct destinations within the cell. Ribosomes do not attach to the smooth endoplasmic reticulum (SER), which helps modify toxic chemicals taken in by the cell and plays a role in protein modification and transport. The SER stores calcium ions that serve as cell signals and is the site of glycogen hydrolysis and of lipid and steroid synthesis.

The Golgi apparatus is an organelle composed of flattened membrane stacks called cisternae and small vesicles. It contributes to further modification, packaging, and concentration of proteins received from the RER. In plants, it is the site of cell wall polysaccharide synthesis. The three different regions of the Golgi have different enzymes and functions. The *cis* region is closest to the nucleus or RER, the *trans* region is closest to the cell surface, and the medial region lies in between. Proteins are released from the RER in vesicles and are transported to the *cis* region, where the vesicles fuse with the Golgi membrane releasing their contents to the Golgi's interior. Vesicles containing proteins pinch off of the *trans* Golgi for transport to the plasma membrane or lysosomes. Lysosomes are "digestion centers" within a cell that break down proteins, polysaccharides, nucleic acids, and lipids into their monomer components. Primary lysosomes released from the Golgi contain a host of powerful enzymes in a slightly acidic environment; these break down cellular waste and molecules engulfed by phagocytosis. In phagocytosis, the cell takes up material in a phagosome, which fuses with a primary lysosome to make a secondary lysosome, which digests the enclosed material. Through the process of autophagy, lysosomes digest organelles such as mitochondria, breaking them down to monomers for reuse in new organelles. Lysosomal storage diseases occur when lysosomes fail to digest internal components. Plant cells do not contain lysosomes but have a vacuole that serves many of the same functions.

Two organelles, mitochondria and chloroplasts, are involved in harvesting energy. During cellular respiration, mitochondria convert potential chemical energy stored in glucose into adenosine triphosphate (ATP), a form of energy readily usable by the cell. Cells can contain anywhere from one to a few hundred thousand mitochondria. Mitochondria have two membranes—an outer smooth membrane and a highly folded internal membrane. The folds are called cristae, and the remaining internal space is called the matrix. The matrix contains ribosomes, DNA, and many enzymes. Protein complexes used during cellular respiration are embedded in the cristae. Plastids are found in plants and protists, but not in animals. Chloroplasts are the site of photosynthesis, where light energy is converted to chemical energy. Chloroplasts have two membranes. The innermost membrane contains circular compartments, or thylakoids, which are folded into stacks called grana. The thylakoid membranes have a distinctive lipid composition and contain the photosynthetic pigment chlorophyll and enzymes for photosynthesis. The fluid content of the chloroplast is called the stroma and contains ribosomes and DNA. Other plastid types include chromoplasts, which contain pigments involved in flower color, and starch-storing leucoplasts.

Other specialized organelles include peroxisomes, glyoxysomes, and vacuoles. Peroxisomes have a single membrane and function to break down harmful peroxide by-products. Glyoxysomes, which are found in young plants, also have a single membrane and function to convert lipids to carbohydrates. Vacuoles, found in plants, fungi, and protists, have various functions, depending on the organism, including storage of waste products, structural support, digestion of stored protein, and concentration of pigments that help plants reproduce (by attracting pollinators or seed dispersers). Contractile vacuoles of freshwater protists function in water regulation.

The cytoskeleton is involved in cell support, position and movement of organelles, cytoplasmic streaming, and anchoring of the cell. Microfilaments are made from actin monomers arranged in a double helix with a "plus" and a "minus" end in a structure that can easily disband and reform. Microfilaments stabilize cell shape and are involved in cell division, cytoplasmic streaming, and the formation of pseudopodia. In muscle cells, the motor protein myosin interacts with actin to produce muscle contraction. The keratin proteins that make up hair and fingernails are one type of intermediate filament. Intermediate filaments are fairly permanent and act to anchor the nucleus and other organelles. They are found in the nuclear lamina and in desmosomes, where they help resist tension placed on cells. Microtubules are long hollow tubes of the dimer protein tubulin that contribute to the rigidity of the cell and act as a framework for the movement of motor proteins. Like microfilaments, they have a "plus" and a "minus" end and can be quickly lengthened or shortened at the "plus" end. Other functions of microtubules include the arrangement of cellulose fibers in plant cell walls, the movement of chromosomes during cell division and functioning as tracks for motor proteins that transport materials across cells.

Cilia and flagella of eukaryotes are movable cell projections. Cilia are smaller than flagella and tend to move fluid over a cell, whereas the longer flagella tend to move the cell itself. Both have the same basic internal "9 + 2" structure: nine microtubule doublets linked by the nexin protein form a cylinder with spokes to two inner microtubules. Movement of cilia and flagella occurs when microtubule doublets slide past one another. The sliding is caused by an ATP-driven shape change in dynein motor proteins that span the space between adjacent microtubule doublets. Cytoplasmic dynein moves vesicles along microtubule tracks toward the microtubule's "minus" end, whereas the motor protein kinesin moves vesicles toward the "plus" end. Experiments demonstrating a drug's effect on microfilament formation and cell movement provide a good example of how biologists use inhibition of a process to demonstrate a link between cause and effect. Biologists also use mutant cells that lack a gene for some structure to investigate causal links between cell structures and their functions.

Question 5. Eukaryotic cells possess organelles where specific metabolic functions occur. What are the benefits of compartmentalization to a cell?

Question 6. If we can assume that form follows function, what would be the explanation for the structural similarities between mitochondria and chloroplasts?

Question 7. The role of a certain cell in an organism is to secrete a protein. Create a flow chart in which you trace the production of that protein from the nucleus through all necessary organelles to the point of release from the cell.

Question 8. Explain how microtubules and dynein function to make cilia and flagella move.

Question 9. As you examine a cell under the microscope, you notice what appear to be small membrane-bound units moving along a path within the cell. Based on your knowledge of filaments in a cell, describe what you are observing.

5.4 What Are the Roles of Extracellular Structures?

The plant cell wall is an extracellular structure

The extracellular matrix supports tissue functions in animals

Plants have semirigid cell walls composed of fibrous cellulose and a gel-like polysaccharide and protein matrix. The cell wall functions to support and protect the cell and can grow when cells expand. Small holes through cell walls called plasmodesmata allow connections between cells through which substances can diffuse. Animal cells instead have an extracellular matrix composed of fibrous proteins such as collagen linked by other proteins to a matrix of proteoglycans. This extracellular matrix functions to hold cells together in tissues, filter materials, orient cell movement during embryonic development and tissue repair, and assist with chemical signaling. Some cells, such as bone cells, secrete an elaborate and rigid matrix.

Question 10. Which is more variable in structure and function: the plant cell wall or the animal extracellular matrix? Explain your answer.

5.5 How Did Eukaryotic Cells Originate?

Internal membranes and the nuclear envelope probably came from the plasma membrane

Some organelles arose by endosymbiosis

After 2 billion years of solely prokaryotic life, the innovation of compartmentalization gave birth to the eukaryotic lineage. The evolution of compartmentalization allowed a grèater variety of biochemical functions to coexist within a single cell. Because certain prokaryotes have plasma membranes that fold inward, one theory is that an elaboration of this inward folding eventually created the endomembrane system and nuclear envelope. The separation of the genetic material from protein synthesis may have permitted increased control of gene expression. Concentration of certain chemicals within organelles may have increased the efficiency of chemical reactions. Mitochondria and plastids may have originated when one cell engulfed another and formed a symbiosis instead of digesting the phagocytosed cell. These organelles eventually transferred many of their original genes to the nucleus but still retain some DNA within each organelle. This endosymbiotic theory explains the similarity between chloroplasts and cyanobacteria. Additional support for this theory comes from modern day examples of algae that live within other eukaryotes.

Question 11. In symbiotic relationships, each organism provides the other with something. Use your knowledge of organelle function to suggest what the original chloroplast and mitochondrial symbionts might have provided for and received from the host eukaryotic cell.

Question 12. Both mitochondria and chloroplasts have an inner membrane and an outer membrane. In an endosymbiotic event, which membrane comes from the engulfing cell and which from the ingested cell?

Test Yourself

1. A mass of cells is found in the sediment surrounding a thermal vent in the ocean floor. The salinity in the area is quite high. Microscopic examination of one of the cells reveals no evidence of membrane-enclosed organelles. This cell would be classified as a
   a. eukaryotic cell.
   b. prokaryotic cell.
   c. member of domain Archaea or Bacteria.
   d. Both a and c
   e. Both b and c
   Textbook Reference: *5.2 What Features Characterize Prokaryotic Cells?*

2. Centrifugation of a cell results in the rupture of the cell membrane and the compacting of the contents into a pellet in the bottom of the centrifuge tube. Bathing this pellet with a glucose solution yields metabolic activity, including the production of ATP. One of the contents of this pellet is most likely which of the following?
   a. Cytosol
   b. Mitochondria
   c. Lysosomes
   d. Golgi bodies
   e. Thylakoids
   Textbook Reference: *5.3 What Features Characterize Eukaryotic Cells?*

3. Which of the following is *not* one of the tenets of cell theory?
   a. All living things are composed of cells.
   b. Cells are the fundamental units of life.
   c. All cells come from preexisting cells.
   d. All cells contain mitochondria.
   e. None of the above is an element of cell theory.
   Textbook Reference: *5.1 What Features Make Cells the Fundamental Units of Life?*

4. Though science fiction has produced stories like "The Blob," we do not see many large, single-celled organisms. Which of the following tends to limit cell size?
   a. The difficulty of maintaining a continuous large membrane
   b. The difficulty of reproduction in a large cell
   c. Surface area-to-volume ratios
   d. All of the above
   e. None of the above
   Textbook Reference: *5.1 What Features Make Cells the Fundamental Units of Life?*

5. Which technique would be best suited to a study of normal cell migration during embryonic development?
   a. Direct visual observation
   b. Light microscopy
   c. Electron microscopy
   d. Cell fractionation
   e. Experimentation on mutants

   Textbook Reference: *5.1 What Features Make Cells the Fundamental Units of Life?*

6. The cellular function of the RER is
   a. DNA synthesis.
   b. photosynthesis.
   c. cellular respiration.
   d. protein synthesis.
   e. mRNA degradation.

   Textbook Reference: *5.3 What Features Characterize Eukaryotic Cells?*

7. Photosynthesis occurs in the
   a. chloroplast.
   b. mitochondria.
   c. Golgi apparatus.
   d. nucleus.
   e. RER.

   Textbook Reference: *5.3 What Features Characterize Eukaryotic Cells?*

8. Lysosomes are involved in
   a. DNA synthesis.
   b. the breakdown of phagocytized material.
   c. protein folding.
   d. pigment production.
   e. cell membrane production.

   Textbook Reference: *5.3 What Features Characterize Eukaryotic Cells?*

9. The packaging of proteins to be used outside the cell occurs in the
   a. nucleus.
   b. SER.
   c. Golgi apparatus.
   d. chromoplast.
   e. nuclear pore.

   Textbook Reference: *5.3 What Features Characterize Eukaryotic Cells?*

10. A hospital nurse notices a slick spot on an IV needle and suspects bacterial contamination. What bacterial structure might have helped these cells attach to one another and stick to the needle?
    a. Capsule
    b. Cell wall
    c. Cytoplasm
    d. Flagella
    e. Ribosomes

    Textbook Reference: *5.2 What Features Characterize Prokaryotic Cells?*

11. Movement of cells in both prokaryotes and eukaryotes is accomplished by which of the following structures?
    a. Cilia
    b. Pili
    c. Dynein
    d. Cell membranes
    e. Flagella

    Textbook Reference: *5.3 What Features Characterize Eukaryotic Cells?*

12. Which of the following statements about mitochondria and chloroplasts is true?
    a. Animal cells produce chloroplasts.
    b. Both mitochondria and chloroplasts may be found in the same cell.
    c. Mitochondria and chloroplasts cannot be found in the same cell.
    d. In certain conditions, chloroplasts can revert to mitochondria.
    e. None of the above

    Textbook Reference: *5.3 What Features Characterize Eukaryotic Cells?*

13. Which of the following cell structures is paired with the molecule type most similar to its actual chemical components?
    a. Cilia – nucleic acid
    b. Thylakoids – peptidoglycan
    c. Ribosomes – phospholipids
    d. Microfilaments – cellulose
    e. Plant cell wall matrix – proteoglycans

    Textbook Reference: *5.4 What Are the Roles of Extracellular Structures?*

14. Nuclear DNA exists as a complex of proteins called _______, which condense(s) into _______ during cellular division.
    a. chromosomes; chromatin
    b. chromatids; chromosomes
    c. chromophors; chromatin
    d. chromatin; chromosomes
    e. None of the above

    Textbook Reference: *5.3 What Features Characterize Eukaryotic Cells?*

15. Rough endoplasmic reticulum and smooth endoplasmic reticulum differ
    a. only in terms of the presence (RER) or absence (SER) of ribosomes.
    b. in their function, and also in terms of the presence (RER) or absence (SER) of ribosomes.
    c. only in terms of their microscopic appearance.
    d. only in their function.
    e. None of the above

    Textbook Reference: *5.3 What Features Characterize Eukaryotic Cells?*

16. Which of the following depicts an endosymbiosis? helps
    a. Some nitrogen-fixing bacteria live inside legume root nodules, releasing fixed nitrogen and absorbing plant carbohydrates.
    b. Some birds follow army ant raids across the forest, eating insects that fly up in their attempt to escape the ants.
    c. Barnacles can be found living on whales, where they may impede the whale's movement but do not otherwise harm it.
    d. Certain mites live in hair follicles, where they occasionally cause skin irritation but are otherwise benign.
    e. The influenza or flu virus replicates itself inside its host's cells, using the host cell's transcription and translation machinery.

    Textbook Reference: *5.5 How Did Eukaryotic Cells Originate?*

Answers

Key Concept Review

1. Cells are small because as a cell's size increases, its surface area-to-volume ratio decreases. A cell will no longer be able to grow when it reaches the point where it can no longer move enough materials across the plasma membrane to support increased metabolic activity. must have surface are
2. The cell membrane allows for the enclosure of biochemical functions within a defined space and acts as a selectively permeable barrier. It allows a cell to maintain homeostasis and is important in communication with adjacent cells and in receiving signals from the environment. In addition, it plays an important structural role and affects cell shape.
3. A prokaryote has a nucleoid region instead of a membrane-bound nucleus and is likely to be smaller than a eukaryote. Certain prokaryotic cells also have a peptidoglycan cell wall, a second outer membrane and/or a capsule—all traits not found in eukaryotes
4. Both pili and flagella project from the surface of certain bacterial cells, but flagella are longer than pili. Whereas flagella spin and typically function to move the cell, pili do not spin and function primarily in cell adhesion (e.g., when cells stick to one another or to a surface).
5. Organelles allow different metabolic environments to exist in the same cell. This partitioning of jobs allows for greater specialization.
6. Both mitochondria and chloroplasts are involved in energy-transformation activities that require many enzymes. The stacking or folding of membranes provides enzymatic activity centers for these reactions.
7. 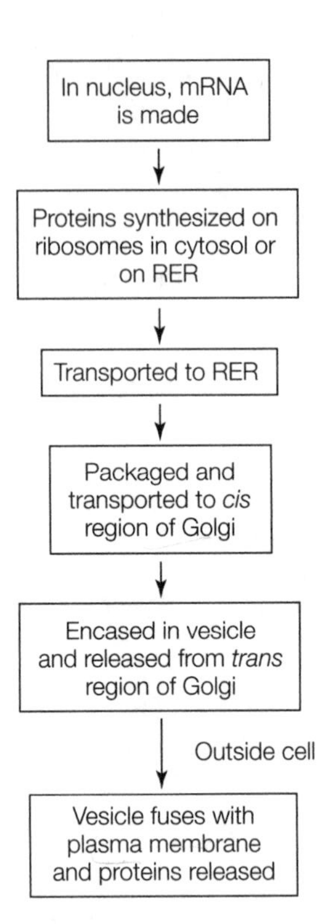

8. Dynein molecules bind to pairs of microtubules in the flagella or cilia. With the addition of cellular energy, the dynein molecules undergo a conformational change that results in the microtubules sliding past one another, resulting in a whiplike motion of the flagella.
9. The moving units are vesicles that are travelling along microtubules within the cell. The vesicles are attached to the microtubules by either kinesin or dynein, depending on the direction of movement (toward the "plus" end for kinesin and toward the "minus" end for dynein).
10. The extracellular matrix of an animal cell is more variable in structure and function. For example, all plant cell walls contain cellulose, but the fibrous proteins in animal extracellular matrix vary, as does the function—ranging from chemical signaling to filtration to rigid support. The plant cell wall primarily serves as a barrier and as rigid support.
11. Both mitochondria and chloroplasts are involved in energy transformations and could have provided ATP or other high-energy compounds to a host cell. They could have received protection, specialized proteins, and/or a unique environment from their host.
12. The outer membrane comes from the engulfing cell, and the inner membrane comes from the ingested cell (see Figure 5.23).

Test Yourself

1. **e.** Several characteristics suggest that this is a prokaryote. It survives in high salinity and high heat, although the true indication is that it contains no membrane-enclosed organelles. Prokaryotes are in the domain Archaea and Bacteria.
2. **b.** The pellet is undergoing cellular respiration, a function that occurs in the mitochondria. You can also assume that if the single membrane of the cell itself is ruptured, other organelles enclosed in single membranes would be ruptured as well.
3. **d.** The cell theory states that cells are the fundamental units of life, all living things are composed of cells, and all cells come from preexisting cells.
4. **c.** As volume increases, the surface area available for exchange does not increase proportionally. Eventually the surface is not large enough for maintenance of the metabolic activity of the cell.
5. **b.** Light microscopy would be best since it can be used on living, moving cells. Almost all cells are too small to be easily observed directly, but using electron microscopy or cell fractionation would kill the cell, preventing it from performing normal embryonic migration. Similarly, mutants might not show normal cell movements.
6. **d.** The RER is the site of protein synthesis.
7. **a.** The chloroplasts are the organelles involved in photosynthesis.
8. **b.** Lysosomes are organelles that contain digestive enzymes used to break down macromolecules taken in by phagocytosis.
9. **c.** The Golgi apparatus packages proteins for both internal and external use.
10. **d.** Flagella contribute to the movement and adhesion of prokaryotic cells. The slick on the IV needle is most likely due to bacteria that possess flagella.
11. **e.** Though the flagella have different structures, they serve the same role in prokaryotes and eukaryotes.
12. **b.** Mitochondria and chloroplasts may be found in the same cell. Almost all eukaryotic cells contain mitochondria.
13. **e.** The plant cell wall matrix is made of polysaccharides and proteins, and proteoglycans are polysaccharides connected to proteins (as in the animal extracellular matrix). Thylakoids are membranes, so they are made of lipids. Ribosomes are made of rRNA complexed with many proteins. Microfilaments are made of the protein actin, whereas cellulose is a carbohydrate.
14. **d.** The complex of proteins and DNA is called chromatin. Chromatin takes the form of chromosomes only during cell division.
15. **b.** Both the structure and the function of RER and SER differ.
16. **a.** Like mitochondria and chloroplasts, the nitrogen-fixing bacteria live inside another organism, giving the plant something it needs (fixed nitrogen), and in turn receiving something from its host (carbohydrates).

6 Cell Membranes

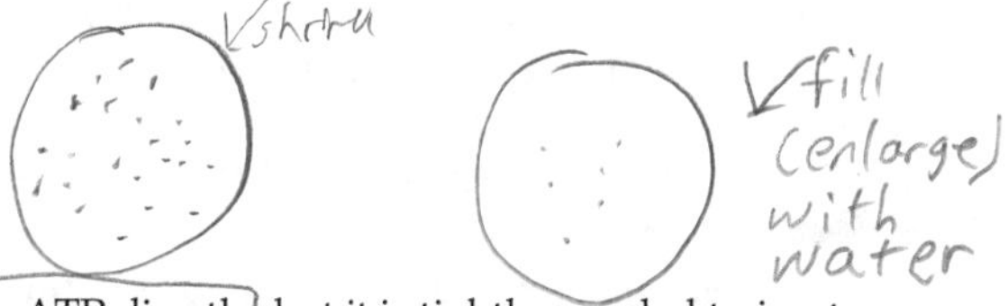

The Big Picture

- Cellular membranes are a dynamic composition of a phospholipid bilayer, integral and peripheral proteins, and carbohydrates. The nature of the constituent phospholipids allows for the formation of a barrier that is semipermeable. Small hydrophilic molecules can traverse the membrane via simple diffusion, water can cross the membrane through aquaporins by osmosis, small charged ions can pass through protein channels, and carrier proteins shepherd through other select molecules. In addition to these passive processes, active transport of molecules against their concentration gradients is possible but requires energy input, either directly from ATP or via processes coupled to ATP-driven transport. Larger substances depend on endocytosis and exocytosis for transport into and out of the cell.
- Plasma membranes also help cells recognize each other and stick together. For example, similar cells usually have a particular type of glycolipid, glycoprotein, or proteoglycan that projects from the membrane surface and allows cells to bind to each other. Other specialized cell junctions include tight junctions, desmosomes, and gap junctions. Tight junctions allow multiple cells to form a sheet that directs movement of substances in a single direction. Desmosomes enhance structural support of sheets of cells that are prone to abrasion. Gap junctions allow communication by connecting adjacent cells. Animal cells can also connect to the extracellular matrix using integrin proteins.

Study Strategies

- In the study of membranes, the processes of diffusion and osmosis are the most difficult to understand. It is very easy to get the terminology confused, especially the terms "hypertonic" and "hypotonic." A consideration of the Latin roots of the words is helpful. "Hyper-" generally means excess, and "hypo-" generally means "less than." "Tonic" refers to solute concentration. Therefore, "hypertonic" means excess solutes, and "hypotonic" means fewer solutes.
- Secondary active transport also tends to be confusing. Remember that secondary active transport does not use ATP directly, but it is tightly coupled to ion transport, which does require ATP.
- When studying the fluid mosaic model, think about the properties of the constituent molecules. This will make understanding the membrane's structure much easier. Draw a diagram of the plasma membrane and all the potential components.
- For the study of diffusion and osmosis, draw diagrams of the movement of water and solutes; this will help you visualize what is happening across a membrane.
- Go to the Web addresses shown to review the following animated tutorials and activities. You can also find them in BioPortal (yourBioPortal.com), along with many additional learning resources.

  Animated Tutorial 6.1 Lipid Bilayer Composition (Life10e.com/at6.1)

  Animated Tutorial 6.2 Passive Transport (Life10e.com/at6.2)

  Animated Tutorial 6.3 Active Transport (Life10e.com/at6.3)

  Animated Tutorial 6.4 Endocytosis and Exocytosis (Life10e.com/at6.4)

  Activity 6.1 The Fluid Mosaic Model (Life10e.com/ac6.1)

  Activity 6.2 Animal Cell Junctions (Life10e.com/ac6.2)

Key Concept Review

6.1 What Is the Structure of a Biological Membrane?

Lipids form the hydrophobic core of the membrane

Membrane proteins are asymmetrically distributed

Membranes are constantly changing

Plasma membrane carbohydrates are recognition sites

Biological membranes are composed of lipids, proteins, and carbohydrates. Lipids provide the structure and barrier functions for the membrane. Proteins are involved in creating channels and transporting materials across the lipid barrier. Carbohydrates found on the outside of the plasma membrane are involved in signaling and adhesion.

The fluid mosaic model describes how lipids interact to produce a fluid membrane with which proteins and carbohydrates interact. Phospholipids have both hydrophilic ("water-loving") phosphorus heads and hydrophobic ("water-hating") nonpolar fatty tails. They arrange themselves into a bilayer, with the hydrophobic tails touching and the heads extending into the aqueous environment inside and outside the cell. This arrangement allows for the fluid movement of the two layers on top of each other and the sealing of any disruptions to the membrane. Though the basic structure of a bilayer is always the same, the inner and outer halves differ in lipid composition and thus have slightly different properties.

Phospholipids differ in their length, degree of unsaturation, and degree of polarity. Phospholipids that are saturated can be packed together more closely in the membrane, while unsaturated phospholipids have kinks in their fatty acids that make them more fluid. Cholesterol can constitute up to 25 percent of the lipid content in the plasma membrane of animal cells. The amount of cholesterol present and the degree of fatty acid saturation (membrane kinks) influence the fluidity of the membrane. Increases in cholesterol and fatty acid saturation make a membrane less fluids as do temperature decreases. Many organisms change the lipid composition of their membranes in response to a change in temperature to maintain appropriately fluid membranes.

The typical ratio of proteins to phospholipid molecules in a plasma membrane is 1 to 25, although this varies by cell type. Integral membrane proteins have hydrophobic domains that allow them to embed in or extend across the internal hydrophobic part of membranes. Other regions of a membrane protein contain hydrophilic amino acid side chains and extend out from the membrane. Peripheral membrane proteins do not have exposed hydrophobic domains and, instead, attach to the cytosolic or extracellular side of a membrane through interactions between polar or charged regions.

The phospholipids and proteins are independent of each other, allowing for movement throughout the membrane (as demonstrated in cell fusion experiments). Most membrane proteins and lipids move freely along the membrane because the constituents interact noncovalently, although some membrane proteins are anchored to the cytoskeleton, which restricts their movement. Special types of integral proteins, called transmembrane proteins, span the entire bilayer. These embedded proteins can be seen sticking out of the middle of a membrane that has been freeze-fractured for electron microscopy. Proteins are distributed in membranes asymmetrically "as needed," and the numbers of proteins and their placement vary greatly based on cell type. The "inside" and "outside" of a membrane often have different properties due to the different characteristics of the proteins on the two sides.

Membranes vary over time as well as from cell to cell. Phospholipids are produced on the surface of the smooth endoplasmic reticulum and distributed to the Golgi as vesicles. From the Golgi they move to the plasma membrane. The membranes of the different organelles also differ. Cell membranes differ in their carbohydrates as well as in their proteins and lipids. In fact, many membrane carbohydrates serve as recognition sites for other cells and molecules. Carbohydrates are frequently covalently bound to lipids or proteins, forming glycolipids or glycoproteins, respectively. These carbohydrates form many different shapes, some of which bind only to similar carbohydrates on adjacent cells. The binding of oligosaccharides on membrane glycoproteins causes cell–cell adhesion.

Question 1. Diagram a cell membrane and label the phospholipid bilayer, integral proteins, peripheral membrane proteins, and carbohydrates. Describe the fluid mosaic model with reference to your diagram.

Question 2. You are examining the membrane of a cell and notice many carbohydrates in the plasma membrane. What role do these carbohydrates most likely play in the cell?

6.2 How Is the Plasma Membrane Involved in Cell Adhesion and Recognition?

Cell recognition and adhesion involve proteins and carbohydrates at the cell surface

Three types of cell junctions connect adjacent cells

Cell membranes adhere to the extracellular matrix

Carbohydrates and proteins on a cell's surface allow it to recognize and adhere to other cells. Cell adhesion is essential for tissue formation in all multicellular organisms. Often the proteoglycans, glycolipids, or glycoproteins on two cells that adhere to each other are the same; these molecules and the bonds they form are referred to as homotypic. Occasionally the two cells have different proteins with complementary binding sites. This type of binding, called heterotypic binding, characterizes the binding of different cell types, such as an egg and a sperm.

Cell junctions allow animal cells to seal intercellular spaces, reinforce attachments to one another, and communicate with one other. The three main types of animal cell junctions are tight junctions, desmosomes, and gap junctions. Tight junctions are found specifically in epithelial cells and function to prevent substances from moving through the spaces between cells. They also restrict migration of membrane proteins and phospholipids that would otherwise be free to float around in the "fluid" plasma membrane. Desmosomes are structural connections between cells that hold them tightly together while still allowing free movement of materials through the extracellular space. Desmosome connections are found in tissues like skin that must resist physical stress. Gap junctions facilitate communication between cells. They are made of protein connexins that span adjacent plasma membranes and act as channels.

Cell membranes form attachments to the extracellular matrix using transmembrane proteins called integrins. These proteins bind to actin filaments inside the cell as well, so they contribute both to cell structure and to adhesion. Because integrin binding is noncovalent, these bonds can easily be broken. The breaking and reforming of integrin bonds plays an important role in cell movement (see Figure 6.8). New attachments keep being formed at the "front" of the cell (toward the direction of movement), tethering the front even as the attachments at the "back" are removed. This frees the back end to move toward the front. The removed integrins are taken up in vesicles that can be moved to the front of the cell, where the integrins are reused.

Question 3. In some animals (such as sponges), the cells that make up tissue can be separated mechanically but then will re-form themselves over time. How are the separated cells able to reorganize themselves?

Question 4. Why would tight junctions be more important than desmosomes in creating and maintaining cell polarity?

6.3 What Are the Passive Processes of Membrane Transport?

Diffusion is the process of random movement toward a state of equilibrium

Simple diffusion takes place through the phospholipid bilayer

Osmosis is the diffusion of water across membranes

Diffusion may be aided by channel proteins

Carrier proteins aid diffusion by binding substances

Cell membranes are selectively permeable: some substances can cross and others cannot. A membrane is said to be permeable to those substances that can pass through it and impermeable to those that cannot.

Passive transport does not require any energy input; it requires only a concentration gradient. Diffusion is the net movement of a substance from an area of greater concentration to an area of lesser concentration. It is a random process that moves toward equilibrium. If a substance can pass through a membrane, it will diffuse until concentrations on either side of the membrane are equal (equilibrium). At equilibrium, the molecules of the substance continue to move across the membrane, but the net movement of molecules in both directions is equal.

The rate of diffusion depends on the size of the diffusing substance, the temperature of the solution, and the concentration gradient. Diffusion within small areas such as single cells may occur rapidly, but diffusion occurs more slowly with increasing distance. In simple diffusion, small molecules pass directly through a membrane. The more lipid-soluble the molecules are, the faster they diffuse across the membrane. Charged and polar molecules do not readily pass through a membrane because of the formation of many hydrogen bonds with water and the hydrophobic nature of the internal layer of the membrane.

Osmosis is the diffusion of water across a membrane, typically through channels in the membrane. Water will move across a membrane from areas of low solute (and high water) concentration to areas of high solute (and low water) concentration. This movement of water molecules equalizes solute and water concentrations on either side of the membrane. Solute concentrations separated by a membrane are classified as isotonic, hypertonic, or hypotonic. An isotonic solution has the same solute concentrations on both sides of a membrane, and thus there is no net water movement across the membrane. A hypertonic solution has a solute concentration that is higher than the concentration on the other side of the membrane, and a hypotonic solution has a concentration that is lower. Since water moves from areas of low solute to areas of high solute, a red blood cell placed in a hypotonic solution will burst as water moves into the cell. Cells with cell walls do not burst in hypotonic solutions but merely build up turgor pressure within the cell wall.

Facilitated diffusion is the passive movement of a substance across a membrane with the help of membrane-bound proteins that act as channels or carriers. A substance can cross the membrane by facilitated diffusion through protein channels running through the plasma membrane. The pores of these proteins have polar amino acids that allow polar molecules and ions to cross the membrane. The best-studied channels are ion channels. Most ion channels are either ligand-gated or voltage-gated, allowing the passage of ions to be controlled depending on the cellular environment. Ion channels are very specific about the ions they allow through. Other specific channels called aquaporins allow water to cross the membrane by osmosis. Water can also diffuse through ion channels. Carrier proteins also facilitate diffusion by binding to a substance (e.g., a sugar or an amino acid) and transporting it across the membrane. Diffusion in this case is limited not only by the concentration gradient, but also by the number of available membrane carrier proteins. When all of the carrier proteins are bound to the substance, they are saturated, and the rate of diffusion is limited.

electric gates

Question 5. A marathon runner has just arrived in the emergency room with severe dehydration, and the physician must decide which type of solution to pump into his veins: pure water, 0.9 percent saline, or 1.5 percent saline. Blood cells are approximately 0.9 percent saline. Describe what is likely to happen to the blood cells when they are exposed to each solution.

Question 6. The plasma membrane is a good barrier to the movement of water. How, therefore, does water typically cross the plasma membrane into a cell?

6.4 What Are the Active Processes of Membrane Transport?

Active transport is directional

Different energy sources distinguish different active transport systems

Cells can maintain different concentrations across their membranes, but moving substances against their concentration gradients costs energy. Three types of proteins perform active transport: (1) the uniporter, which moves one substance in only one direction; (2) the symporter, which moves two substances together in the same direction; and (3) the antiporter, which moves one substance in one direction and a second substance in the opposite direction. Because symporters and antiporters move two substances together, they are also called coupled transporters.

Coupled = uni and sym porters

Primary active transport uses the energy stored in ATP to move ions against their concentration gradient. For example, the sodium–potassium pump is an integral membrane glycoprotein found in animal cells that uses ATP to transport two K^+ ions into the cell and three Na^+ ions out of the cell. Secondary active transport uses ATP indirectly by coupling solute transport with an ion concentration gradient established by primary transport. Diffusion of an ion down its concentration gradient powers the movement of another substance against a concentration gradient. For example, primary transport establishes a sodium gradient across the membrane (high sodium outside the cell and low sodium inside it). A symporter can then bring glucose into the cell against its concentration gradient by allowing Na^+ to diffuse back into the cell alongside the glucose. See Table 6.1 for a summary of membrane transport mechanisms.

Question 7. You are examining a membrane that contains a number of proteins that appear to be involved in transport of an ion across that membrane. Design an experiment that will allow you to determine if these transporters are passive transporters or active transporters.

Question 8. The sodium–potassium pump is important for maintaining a gradient of Na^+ and K^+ between the inside and the outside of the cell. Describe how these gradients can function in secondary active transport to move another molecule, such as glucose.

6.5 How Do Large Molecules Enter and Leave a Cell?

Macromolecules and particles enter the cell by endocytosis

Receptor-mediated endocytosis is highly specific

Exocytosis moves materials out of the cell

Some macromolecules are unable to cross the plasma membrane because of their size or charge, or because they are polar. Endocytosis is the process by which a cell brings these substances into the cell. This is accomplished by the cell membrane, which folds around the substance to form an endocytotic vesicle. Large substances and even entire cells are engulfed in the process of phagocytosis (as described in Chapter 5). The cell takes up liquids in small vesicles from the outside in the process of pinocytosis. Animal cells use receptor-mediated endocytosis to capture specific macromolecules, such as cholesterol, from the environment. Receptors for specific macromolecules cluster together on the cell surface in coated pits containing the protein clathrin. Upon binding of the specific molecule to the receptors, the coated pit invaginates to form a vesicle. The resulting vesicle becomes clathrin-coated until it is well inside the cell, where it loses its coat and fuses with a lysosome.

Exocytosis moves materials in vesicles out of the cell. Binding proteins found on the surface of the cell-produced vesicles bind with receptor proteins on the cytoplasmic side of the cell membrane. The vesicle membrane fuses with the cell membrane, and the contents of the vesicle are released outside the cell membrane. Sometimes the vesicle merely forms a pore as it touches the cell membrane. Sweat is expelled from cells in sweat glands using this "kiss and run" process.

Question 9. Cells have the ability to take in large molecules by endocytosis and secrete them to the environment by exocytosis. Describe each process and explain why both are important for the cell.

Question 10. Is endocytosis more similar to active processes or passive processes of membrane transport? Explain your reasoning.

Test Yourself

1. Which of the following statements regarding cellular membranes is *false*?
   a. The hydrophobic nature of the phospholipid tails limits the migration of polar molecules across the membrane.
   b. Integral proteins and phospholipids move fluidly throughout the membrane.
   c. Membrane phospholipids flip back and forth from one side of the bilayer to the other.
   d. Glycolipids and glycoproteins serve as recognition sites on the cell membrane.
   e. All of the above are true; none is false.

   Textbook Reference: *6.1 What Is the Structure of a Biological Membrane?*

2. Which of the following contributes to differences in the two sides of the cell membrane?
   a. Differences in peripheral proteins
   b. Different domains expressed on the ends of integral proteins
   c. Differences in phospholipid types
   d. Differences in the carbohydrates attached to membrane proteins
   e. All of the above

   Textbook Reference: *6.1 What Is the Structure of a Biological Membrane?*

3. Which of the following cell membrane components serve as recognition signals for interactions between cells?
   a. Cholesterol
   b. Glycolipids
   c. Phospholipids
   d. Carrier proteins
   e. All of the above

   Textbook Reference: *6.1 What Is the Structure of a Biological Membrane?*

4. Which of the following would most likely disable a cell's ability to move without affecting other cellular processes?
   a. Removing the carbohydrate parts from cell glycoproteins

b. Breaking the desmosome linkages between cells
c. Preventing the formation of new integrin actin bonds
d. Disabling the gap junctions
e. Dissolving the extracellular matrix surrounding the cell

Textbook Reference: *6.2 How Is the Plasma Membrane Involved in Cell Adhesion and Recognition?*

5. You are monitoring the diffusion of a molecule across a membrane. Of the options listed below, the fastest rate of diffusion would result from an internal concentration of _______ and an external concentration of _______.
   a. 5; 60
   b. 35; 40
   c. 50; 40
   d. 50; 50
   e. 100; 120

   Textbook Reference: *6.3 What Are the Passive Processes of Membrane Transport?*

6. If a red blood cell with an internal salt concentration of about 0.85 percent is placed in a saline solution that is 4 percent, the
   a. cell will lose water and shrivel.
   b. cell will gain water and burst.
   c. turgor pressure in the cell will increase greatly.
   d. turgor pressure in the cell will decrease greatly.
   e. cell will remain unchanged.

   Textbook Reference: *6.3 What Are the Passive Processes of Membrane Transport?*

7. Solution X is hypotonic relative to solution Y if solution X has a solute concentration that is _______ solution Y.
   a. greater than that of
   b. lower than that of
   c. the same as that of
   d. Both a and c
   e. Both b and c

   Textbook Reference: *6.3 What Are the Passive Processes of Membrane Transport?*

8. Which of the following statements about osmosis is *false*?
   a. Osmosis refers to the movement of water along a concentration gradient.
   b. In osmosis, water moves to equalize solute concentrations on either side of the membrane.
   c. The movement of water across a membrane can affect the turgor pressure of some cells.
   d. If osmosis occurs across a membrane, then diffusion is not occurring.
   e. Water moves through membrane channels during the process of osmosis.

   Textbook Reference: *6.3 What Are the Passive Processes of Membrane Transport?*

9. Channel proteins allow ions that would not normally pass through the cell membrane to pass through via the channel. The property of the channel protein that makes this possible is its pore, which is composed of
   a. polar amino acid groups.
   b. hydrophobic amino acid groups.
   c. Ca^{2+}.
   d. carbohydrates.
   e. None of the above

   Textbook Reference: *6.3 What Are the Passive Processes of Membrane Transport?*

10. Which of the following affects the movement of molecules by means of carrier-mediated facilitated diffusion?
    a. The concentration gradient
    b. The number of carrier molecules
    c. The availability of carrier molecules
    d. Temperature
    e. All of the above

    Textbook Reference: *6.3 What Are the Passive Processes of Membrane Transport?*

11. Active transport differs from passive transport in that active transport
    a. uses up energy.
    b. always involves direct coupling of ATP hydrolysis as molecules cross the membrane.
    c. moves molecules from high concentrations to low concentrations.
    d. Both a and c
    e. Both b and c

    Textbook Reference: *6.4. What Are the Active Processes of Membrane Transport?*

12. Single-celled animals such as amoebas engulf entire cells for food. This manner of "eating" is called
    a. exocytosis.
    b. endocytosis.
    c. facilitative transport.
    d. active transport.
    e. osmosis.

    Textbook Reference: *6.4. What Are the Active Processes of Membrane Transport?*

13. Many cells have sodium–potassium pumps in their plasma membranes. In order to function, sodium–potassium pumps require
    a. ATP.
    b. a channel protein.
    c. the absence of a concentration gradient.
    d. ADP.
    e. All of the above

    Textbook Reference: *6.4. What Are the Active Processes of Membrane Transport?*

14. Bacterial cells are often found in very hypotonic environments. Which of the following characteristics prevents them from taking in too much water from their environment?
   a. The presence of a cell wall, which allows for a buildup of tonic pressure, preventing additional water from entering the cell
   b. The presence of a cell wall, which allows for a buildup of turgor pressure, preventing additional water from entering the cell
   c. The capacity of the cell to expel water as quickly as it takes it up
   d. The presence of an active water pump
   e. None of the above

   Textbook Reference: *6.3 What Are the Passive Processes of Membrane Transport?*

Turgor pressure limits osmosis

15. A researcher is investigating the movement of signaling molecule X into a cell. The concentration of X outside the cell is manipulated over a wide range of concentrations, and the rate at which X enters the cell is measured. The graph below represents the data collected. Based on this data, which of the following processes is most likely responsible for X's entrance into the cell?

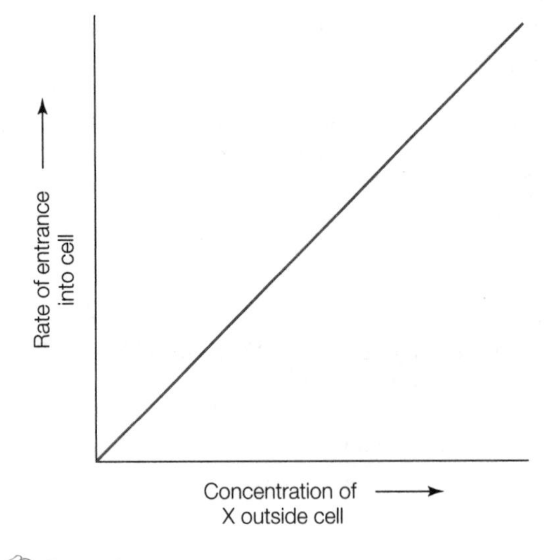

   a. Osmosis
   b. Active transport
   c. Exocytosis
   d. Facilitated diffusion
   e. Diffusion

   Textbook Reference: *6.3 What Are the Passive Processes of Membrane Transport?; 6.4 What Are the Active Processes of Membrane Transport?; 6.5 How Do Large Molecules Enter and Leave a Cell?*

Answers

Key Concept Review

1.

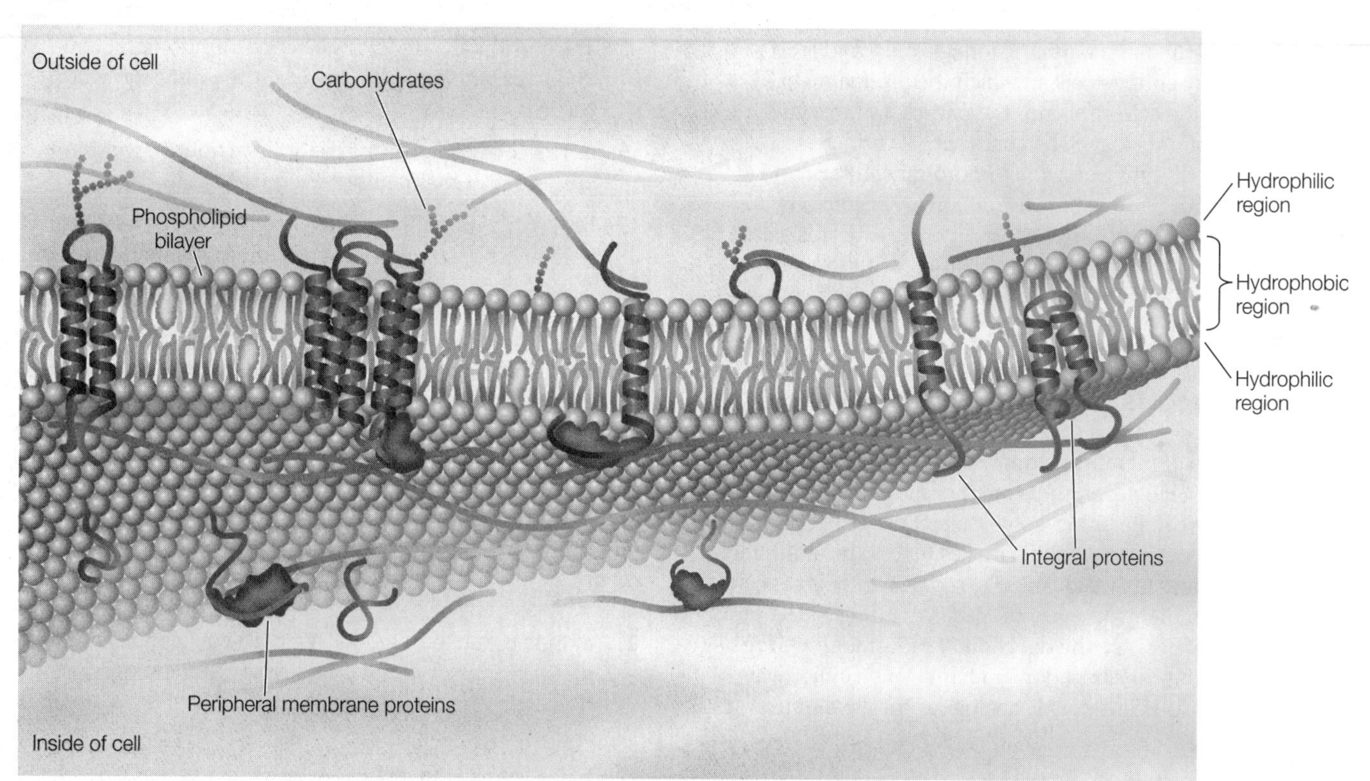

The fluid mosaic model states that the phospholipid bilayer allows the embedded proteins to float freely through the bilayer.

2. Carbohydrates located on the surface of the plasma membrane are important in recognizing specific molecules and also serve as signaling sites.

3. The cells in the sponge have the ability to recognize and adhere to other cells from their original body. This cell recognition is tissue-specific (allowing the tissues to re-form correctly) and species-specific (so that only cells of one species adhere to each other). This ability to recognize and adhere to other cells to re-form the sponge is due to specific proteoglycans on the plasma membrane that bind to one another. In other cells, glycolipids or glycoproteins cause cell adhesion.

4. Tight junctions limit movement of membrane proteins and material through the extracellular space, whereas desmosomes hold cells together tightly. Therefore, it is the tight junctions that help maintain differences in membrane composition and in extracellular fluid content between two sides of a cell (e.g., between the apical and basal sides of epithelial sheets).

5. In pure water, the blood cells will take on water through osmosis, swell, and eventually rupture. In 0.9 percent saline, the cells should neither gain nor lose a significant amount of water. In a 1.5 percent saline solution, the cells should lose water and shrivel.

   In order to rehydrate the runner, a solution isotonic to the patient's blood cells, 0.9 percent, should be infused into his bloodstream.

6. Many plasma membranes contain special protein channels known as aquaporins, which allow for the movement of water by osmosis into and out of the cell.

7. The main difference between active and passive transport is that active transport goes against a concentration gradient and requires energy, whereas passive transport diffuses passively and does not require energy. To test this, you would measure the movement of the ions across a membrane. One side of the membrane should have a high concentration of the ion and the other, a low concentration. If ions only move from an area of high concentration to low concentration, then the transporter is a passive transporter. If the addition of ATP is needed for movement to occur, it is an active transporter.

8. The sodium–potassium pump sets up a gradient of Na^+ and K^+, with more Na^+ outside the cell and more K^+ inside the cell. A secondary active transport mechanism has a transport protein for the specific molecule, which in this case is glucose. This transporter passively transports Na^+ into the cell by means of the

Na^+ gradient and brings along the glucose. If the K^+ gradient were to be used instead, there would have to be an antiporter that allowed K^+ to leave the cell down its concentration gradient as glucose was brought in against its concentration gradient.

9. Cells take up large particles, foreign cells, and food sources by endocytosis, in which the plasma membrane of the cell surrounds the particle to form an endocytotic vesicle. Substances such as undigested material, digestive enzymes, neurotransmitters, and material for plant wall construction are secreted by the cell by means of exocytosis. During exocytosis, the membrane of a secretory vesicle fuses with the plasma membrane and the contents are released to the outside of the cell.
10. Endocytosis is a complex process that requires energy use on the cell's part to make a vesicle and bring it into the cell. In this way it is more similar to active processes of membrane transport than it is to passive processes. However, movement of molecules by endocytosis can be either down their concentration gradients (for example, food particles tend to be at higher concentrations outside of cells) or against their concentration gradients (by way of specific binding to receptor proteins). Because the direction of movement with respect to the concentration gradient can be either way, endocytosis is not exactly analogous to active transport.

Test Yourself

1. **c.** Because of the hydrophobic tails and hydrophilic heads of the phospholipids, it is impossible for them to flip back and forth from one side of the membrane bilayer to the other.
2. **e.** The cell membrane is asymmetric and has different properties and functions on the cytoplasmic side versus the extracellular side. These properties arise from differences in the constituents of the membrane.
3. **b.** Glycolipids serve as recognition signals (as do glycoproteins).
4. **c.** Integrins are involved in cell movement, whereas glycoproteins, desmosomes, and gap junctions are not. Preventing formation of new integrin–actin bonds would not affect current integrin connection between actin and the extracellular matrix, but it would prevent new attachments from forming. Dissolving the extracellular matrix would affect general cell structure and disrupt many cell processes but would not prevent cell movement.
5. **a.** Diffusion always follows a concentration gradient. The larger the gradient, the faster the diffusion will occur.
6. **a.** The cell will lose water as solute concentrations on both sides of the membrane equalize.
7. **b.** A solution that has a lower solute concentration than another solution is hypotonic in comparison.
8. **d.** Diffusion and osmosis are not mutually exclusive and may take place at the same time.
9. **a.** The charged, or polar, lining of the channel proteins allows passage of polar and charged molecules.
10. **e.** Anything that affects the rate of diffusion will affect carrier-mediated facilitated diffusion. Thus, temperature can be a limiting factor. Carrier-mediated facilitated diffusion also relies on the number of carrier molecules and their availability (whether or not they are already saturated with the solute being transported).
11. **a.** Active transport works against a concentration gradient and requires energy to do so. That energy does not always have to be supplied directly in the form of ATP.
12. **b.** Cells carry out cellular eating by phagocytosis, which is a type of endocytosis.
13. **a.** Sodium–potassium pumps are forms of primary active transport and require energy in the form of ATP.
14. **b.** Turgor pressure limits osmosis, and once a cell is turgid, no more water may be taken on. Most bacteria living in hypotonic environments have cell walls.
15. **e.** A linear graph suggests that diffusion is responsible. Osmosis is the diffusion of water, but water is not a signaling molecule. The other processes would all show a slowed rate of entering the cell at high concentrations of X, as the carriers or cell vesicles all were in use such that no more could be used to bring increasing amounts of X into the cell.

7 Cell Communication and Multicellularity

The Big Picture

- Signal transduction is the means by which cells receive information from the environment or other cells and react to those signals. Transduction is a highly regulated series of events that depends on the binding of a signal ligand to a receptor protein. The signal binding must cause a change in the shape of the receptor protein, which causes a responder protein to initiate events in the cell that change its function. The effects of signals are often mediated by secondary messengers.
- Cells in animals, plants, and other multicellular organisms have specialized intercellular junctions and coordinate their responses to the surrounding environment so that they can work as a whole.

Study Strategies

- It is easy to become overwhelmed by the different examples of signal transduction. Try to focus on particular details of the systems. For instance, you could compare plasma membrane receptors with cytoplasmic receptors, or list three kinds of secondary messengers. When you think you understand the details, put them together by picking one of the signal transduction examples and creating a table that includes the signal, the receptor, the transduction (responders and amplification), and the effect in the cell. Expand your table to include other examples of signal transduction.
- A number of other strategies can be useful for studying signal transduction. Create a flow chart or a diagram of the signal transduction pathways. Review chapter figures and then explain the processes depicted in them aloud to a friend with your book closed. Create a list of the different secondary signals and provide an example of how each signal is activated and how it alters cell function.
- Both gap junctions and plasmodesmata physically connect adjacent cells, but the details of their structures differ in ways that affect their functions. Create concept maps for these structures and their functions and compare the maps.
- Go to the Web addresses shown to review the following animated tutorials and activities. You can also find them in BioPortal (yourBioPortal.com), along with many additional learning resources.

  Animated Tutorial 7.1 A Signal Transduction Pathway (Life10e.com/at7.1)

  Animated Tutorial 7.2 Signal Transduction and Cancer (Life10e.com/at7.2)

  Activity 7.1 Chemical Signaling Systems (Life10e.com/ac7.1)

  Activity 7.2 Concept Matching (Life10e.com/ac7.2)

Key Concept Review

7.1 What Are Signals, and How Do Cells Respond to Them?

Cells receive signals from the physical environment and from other cells

A signal transduction pathway involves a signal, a receptor, and responses

Cells receive signals from their environment (chemicals, light, temperature, touch, or sound) and from other signals (usually chemicals). This process of receiving a signal and communicating it to the cell is called signal transduction. Multicellular organisms receive signals from their environment, from other cells, or from extracellular fluid. Autocrine signals are local signals that affect the cells making the signal, whereas juxtacrine signals affect cells next to the signal-producing cell. Paracrine signals affect nearby cells, and hormones are longer-distance signals that travel through the circulatory system to affect more distant cells. Only those cells that have the correct receptor will respond to a given chemical signal. The response may occur through a short-acting enzyme or longer-term gene expression. All these signal transduction pathways include a signal, a receptor, and a response.

Cell response typically involves a change in the activity of specific proteins, usually enzymes or transcription factors. For example, phosphorylation can change an enzyme's shape, making it more or less active. One activated enzyme may, in turn, activate and/or deactivate many other proteins to cause many cell responses. Crosstalk occurs when multiple signal transduction pathways interact to cause the cell response.

Question 1. Which of the types of signals described in this section are not found in prokaryotes? Why?

Question 2. Based on the generalized depiction of a signal transduction pathway below, describe two different ways in which short-term changes stimulated by the signal (e.g., enzyme activation, cell movement) could be prevented from occurring within a cell, even in the presence of the signal molecule.

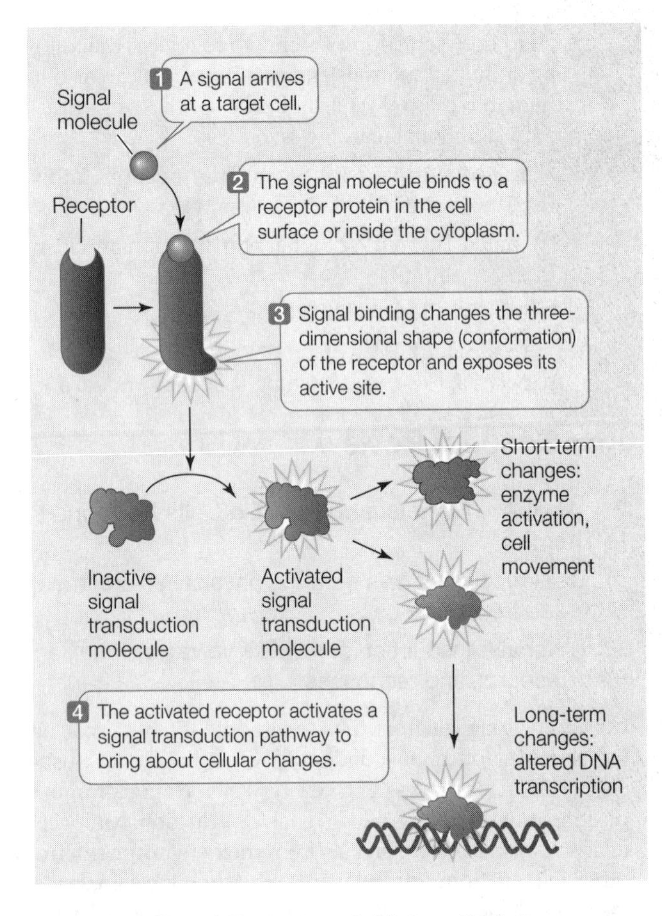

7.2 How Do Signal Receptors Initiate a Cellular Response?

Receptors that recognize chemical signals have specific binding sites

Receptors can be classified by location and function

Intracellular receptors are located in the cytoplasm or the nucleus

A cell's response to a signal is dependent upon the presence of proteins that function as receptors. Binding to the receptor requires the ligand (the signal) to fit into a receptor binding site. After the ligand (L) binds, the receptor (R) undergoes a conformational change, becoming RL, to cause an effect. The ligand is not changed in the binding process, and binding of the signal to the receptor is reversible. Thus, the rate of binding depends on the concentration of the receptor, the concentration of the ligand, and a rate constant: K_1[R][L]; and the rate of dissociation equals the concentration of the receptor-ligand complex multiplied by the dissociation rate constant: K_2[RL]. At equilibrium, the binding rate equals the dissociation rate; K_2/K_1 = [R][L]/[RL] = a dissociation constant K_D, an equation that summarizes how easily the receptor binds the ligand. Cells that are very sensitive to a particular ligand have receptors with low dissociation constants for that ligand. Information about receptor-ligand binding is particularly important in designing and determining dosage levels for drugs that bind to hormone receptors.

There are two major classes of receptors: membrane receptors, which bind large and/or polar ligands that cannot cross the plasma membrane, and intracellular receptors, which bind small nonpolar ligands that diffuse across the plasma membrane. Three important types of membrane receptors are: (1) ion channel receptors, (2) protein kinase receptors, and (3) G protein-linked receptors. Ion channel receptors are ion channels that open when a signal binds to them (e.g., the acetycholine receptor of muscle cells). When a signal binds to a protein kinase receptor (e.g., the insulin receptor; see Figure 7.6), the receptor adds phosphates to itself and/or other proteins to change the protein's shape and function. The G protein-linked receptors are the most complex group (see Figure 7.7). Receptors in this group respond to many different types of external and internal signals and initiate a wide range of possible cell responses. G protein receptors have seven transmembrane domains and they all respond to their specific signal by changing shape and binding to a G protein. First, the bound G protein exchanges a bound GDP for GTP. The activated G protein then travels through the membrane until it interacts with an effector protein, which causes some type of cell response. Eventually the G protein's GTP loses a phosphate and returns to its inactive state.

Finally, intracellular receptors are often transcription factors that affect which mRNAs (and therefore which proteins) a cell makes. Some of these intracellular receptors are found in the nucleus, and others only enter the nucleus after the signal binds to them or to an attached chaperone protein.

Question 3. Describe which variables need to be larger and which variables need to be smaller in order for a receptor to have a high dissociation constant (K_D). Would a receptor bind easily to its ligand if it had a high K_D?

Question 4. Glucagon is a polar ligand that initiates a series of events involving several sequential phosphorylations. Based on the properties of this ligand, what types of receptors might it be interacting with? Why?

Question 5. Describe the steps involved in a signaling cascade acting through a G protein-linked receptor.

7.3 How Is the Response to a Signal Transduced through the Cell?

A protein kinase cascade amplifies a response to ligand binding

Second messengers can amplify signals between receptors and target molecules

Signal transduction is highly regulated

Signal transduction can be very complex. Knowledge of entire signal transduction pathways can illuminate pathological conditions and likely, targets for cures. For example, many bladder cancers have an abnormal version of the G protein Ras. This abnormal Ras is continuously bound to GTP and thus continuously promotes cell division. This particular pathway combines a protein kinase receptor with a G protein. When activated, Ras binds to and activates another protein, Raf, which is also a protein kinase. Activated Raf phosphorylates/activates many molecules of MEK, another protein kinase which, in turn, phosphorylates and activates many MAP kinases, allowing them to enter the nucleus and further direct the cell response. A series like this, in which protein kinases sequentially activate one another, is called a protein kinase cascade. This type of signal transduction amplifies the signal at each step and carries the signal throughout the cell (e.g., into the nucleus). The enzymes activated are specific to the signaling pathway, but the cell's response can vary depending on the specific target proteins present in the cell.

Cell signals can also be magnified and transmitted throughout the cell using small nonprotein molecules called second messengers, which diffuse throughout the cytoplasm. For example, the hormone epinephrine cannot directly activate the enzyme glycogen phosphorylase to cause the release of glucose from stored glycogen. Instead, the G protein activated by the epinephrine receptor activates the enzyme adenylyl cyclase. This produces the second messenger cyclic AMP (cAMP), which transmits epinephrine's signal to cytoplasmic enzymes. Because one activated adenylyl cyclase makes many molecules of cAMP, second messengers distribute and amplify the initial signal. The same second messenger may be produced following activation of different receptors and can usually activate more than one signaling pathway. This convergence of signaling pathways is a type of crosstalk.

One important phospholipid-derived second messenger is produced when the phosphatidyl inositol-bisphosphate (PIP_2) is hydrolyzed to produce the hydrophobic diacylglycerol (DAG) and hydrophilic inositol trisphosphate (IP_3). As with cAMP, a G protein-activated enzyme catalyzes production of these second messengers, both of which cause activation of protein kinase C (PKC). PKC then phosphorylates yet more enzymes and proteins to cause the cell response. Calcium ions (Ca^{2+}) can also be used as second messengers because they are at extremely low concentrations inside most cells due to the action of ion pumps. Ion channels in the cell membrane or the ER open to allow Ca^{2+} into the cytosol, where it can activate PKC or other proteins. The gas nitric oxide (NO), which causes relaxation of smooth muscles, is another type of second messenger. Many medical drugs act on signal transduction pathways to affect second messengers. For example, lithium blocks PIP_2 hydrolysis and nitroglycerine releases NO.

Along with the activation of G proteins and kinases and the production of second messengers, the inactivation of G proteins and kinases and the sequestering or breakdown of second messengers are equally important in modulating cell function. Cells can fine-tune signal transduction pathway responsiveness by synthesizing or breaking down enzymes or modifying enzyme activity for specific pathway steps.

Question 6. Discuss the role of secondary messengers in a signaling pathway. How are they different from ligands and receptors? What roles do secondary messengers and signals have in common?

Question 7. Describe how G proteins and protein kinases can interact in a signal transduction cascade.

Question 8. Why are signal transduction pathways highly regulated? What would happen if they were not?

Question 9. Create your own signal transduction pathway and sketch it. Include a protein kinase and/or a G protein-linked receptor, at least one point at which the signal is amplified, and at least one second messenger.

7.4 How Do Cells Change in Response to Signals?

Ion channels open in response to signals

Enzyme activities change in response to signals

Signals can initiate DNA transcription

Signaling transduction pathways cause many different types of cell responses. Some signal transduction pathways open ion channels. This can occur as one of many steps or it can be the final endpoint of the signal transduction process, as in signaling by the nervous system (e.g. sense of smell, Figure 7.17). Others cell responses end in altered enzyme function, either through inhibition or activation (as with epinephrine stimulation). Sometimes many enzymes are affected, with some enzymes being activated and others inhibited. Signal transduction pathways can also influence the expression of genes by regulating transcription. In this case, the signal transduction pathway typically ends by activating or inactivating transcription factors—proteins that bind to specific DNA sequences and either promote or prevent transcription of a particular gene or set of genes (e.g., MAPK in the Ras signaling cascade).

Question 10. Describe three different ways in which the cascade of reactions leading to altered enzyme activity (as in Figure 7.18) could be inhibited.

Question 11. What is the likely reason that signals affecting DNA transcription are especially important during early development?

7.5 How Do Cells in a Multicellular Organism Communicate Directly?

Animal cells communicate through gap junctions

Plant cells communicate through plasmodesmata

Modern organisms provide clues about the evolution of cell–cell interactions and multicellularity

Multicellular organisms must have cells that effectively coordinate their activities to the advantage of the whole organism. In addition to the intercellular signals discussed earlier, some cells communicate more directly through specialized intercellular junctions. In animals, these specialized intercellular junctions are called gap junctions. Gap junctions are protein-lined (connexon protein) channels that allow the passage of small molecules, including signal molecules and ions, between adjacent cells. Plasmodesmata are membrane-lined channels that connect adjacent plant cells. They are filled with tubules derived from the endoplasmic reticulum called desmotubules, which allow small metabolites and ions to move between plant cells. Development of intercellular communication was a key event in the evolution of multicellular organisms. Cell signaling makes functional specialization possible because it allows cells to coordinate the functions of specialized tissues (e.g., evolution of *Volvox*).

Question 12. You are looking at a group of cells from an unknown organism and notice specialized channels between cells that are lined with membrane. What are these structures?

Question 13. Why does increased cell specialization require increased coordination among cells within the organism?

Test Yourself

1. Histamine triggers a local inflammatory response in the presence of certain pathogens. Because it is inactivated very quickly, histamine does not usually reach the circulatory system. What type of signaling best describes this mechanism of action?
   a. Autocrine
   b. Juxtacrine
   c. Paracrine
   d. Pathocrine
   e. Hormonal
   Textbook Reference: *7.1 What Are Signals, and How Do Cells Respond to Them?*

2. When you smell vanilla flavoring, the vanilla scent molecules bind to a G protein found in the membrane of certain olfactory neurons. This G protein then changes shape and activates the enzyme adenylyl cyclase, which in turn catalyzes the formation of a small molecule called cAMP, which causes cell ion channels to open, eventually causing an electrical signal to be sent to your brain. In this scenario, what role does the G protein play?
   a. It is the ligand.
   b. It is the receptor.
   c. It is the signal transduction molecule.
   d. It is the altered DNA transcript.
   e. It is the signal.
   Textbook Reference: *7.1 What Are Signals, and How Do Cells Respond to Them?*

3. Autocrine signals
   a. act on the cells that made them.
   b. move through the blood and act on cells far from their source.
   c. act on cells that are near to those that secrete them.
   d. do not act through receptors.
   e. require large concentrations of the signaling molecule to function.
   Textbook Reference: *7.1 What Are Signals, and How Do Cells Respond to Them?*

4. Intracellular receptors bind
   a. small signals that can diffuse through the plasma membrane.
   b. secondary messengers, such as cAMP.
   c. hydrophilic molecules.
   d. G proteins
   e. All of the above
   Textbook Reference: *7.2 How Do Signal Receptors Initiate a Cellular Response*

5. Which of the following statements about receptors is true?
   a. Receptors are found only on the surface of cells.
   b. Receptors are specific to the signal ligand.
   c. Most receptors can bind with many signal ligands.
   d. All receptors can act as ion channels.
   e. All of the above
   Textbook Reference: *7.2 How Do Signal Receptors Initiate a Cellular Response*

6. Which of the following represents the correct order of signal transduction?
   a. Binding of signal, release of secondary messenger, alteration of receptor conformation, alteration of cellular function
   b. Binding of signal, release of secondary messenger, alteration of receptor conformation, transcription of gene
   c. Binding of signal, activation of target protein by responder, alteration of receptor conformation, release of secondary messenger, transcription of gene
   d. Binding of signal, alteration of receptor conformation, alteration of cellular function, release of second messenger
   e. Binding of signal, alteration of receptor conformation, activation of target protein by responder, alteration of cellular function
   Textbook Reference: *7.2 How Do Signal Receptors Initiate a Cellular Response?*

7. Which of the following statements about secondary messengers is true?
   a. They amplify the signal.
   b. They bind to the active site of the receptor.
   c. They result in multiple effects from a single signal.
   d. Both a and c
   e. All of the above
   Textbook Reference: *7.3 How Is the Response to a Signal Transduced through the Cell?*

8. Which of the following statements about protein kinase cascades is true?
   a. Amplification can occur at each step in the path.
   b. Information at the plasma membrane can be communicated to the nucleus.
   c. The multiple steps allow for the specificity of the process.
   d. Different targets can produce variation in the cellular response.
   e. All of the above

   Textbook Reference: *7.3 How Is the Response to a Signal Transduced through the Cell?*

9.–10. Examine the figure below.

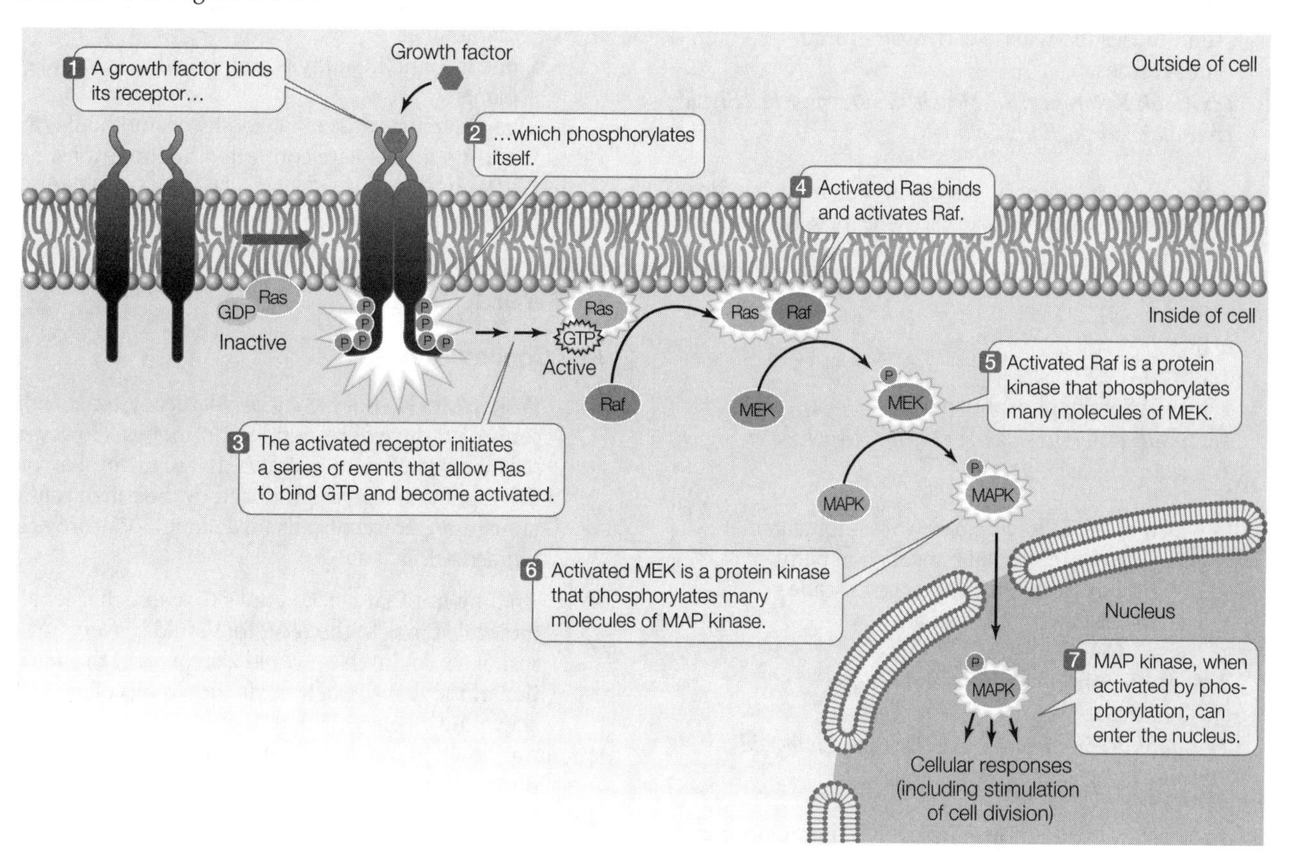

9. Consider a signal transduction pathway in which each time a growth factor binds to its receptor, two Ras G proteins are activated, and each Ras G protein activates one Raf protein. If each Raf then activates 10 MEK molecules and each activated MEK activates 10 molecules of MAP kinase, the original signal been amplified by a factor of
   a. 20
   b. 22
   c. 100
   d. 200
   e. 1,000

   Textbook Reference: *7.3 How Is the Response to a Signal Transduced through the Cell?*

10. Which of the following would have the greatest inhibitory effect on the signal transduction pathway shown in the figure?
   a. Increased synthesis of GTP from GDP within the cell
   b. Synthesis of a protein that binds to Raf and prevents it from binding to Ras
   c. Activation of a MAP phosphatase (an enzyme that removes phosphates from MAPK)
   d. Inhibition of the growth factor receptor phosphatase (an enzyme that removes phosphates from the receptor)
   e. Release of Ca^{2+} from intracellular calcium stores

   Textbook Reference: *7.3 How Is the Response to a Signal Transduced through the Cell?*

11. Which of the following scenarios is the best example of crosstalk between signal transduction pathways?
    a. One tyrosine kinase is activated by cAMP, but inactivated by increased intracellular Ca^{2+} caused by the release of IP_3 from PIP_2.
    b. Increased intracellular Ca^{2+} activates PKC, working with DAG to cause cellular responses.
    c. Sperm entry into an egg causes increased intracellular Ca^{2+}, which in turn helps trigger cell division.
    d. Lithium blocks the hydrolysis of PIP_2 into IP_3 and DAG. Because IP_3 binds to and opens ion channels that release Ca^{2+} from the ER, average intracellular calcium levels decrease in the presence of lithium.
    e. In some cells, intracellular Ca^{2+} both activates PKC and causes the release of neurotransmitter-containing vesicles.

    Textbook Reference: *7.3 How Is the Response to a Signal Transduced through the Cell?*

12. In the IP_3/DAG second messenger system, two hydrophilic components that can carry the message through the cytosol are _______ and _______.
    a. IP_3; DAG
    b. PIP_2; DAG
    c. IP_3; Ca^{2+}
    d. PKC; Li^+
    e. the G protein; phospholipase C

    Textbook Reference: *7.3 How Is the Response to a Signal Transduced through the Cell?*

13. In which of the following pathways is ion channel opening part of the signal transduction pathway?
    a. The growth factor protein kinase cascade
    b. The Ras protein kinase cascade
    c. The IP_3/DAG pathway
    d. The epinephrine cascade of reactions
    e. All of the above

    Textbook Reference: *7.4 How Do Cells Change in Response to Signals?*

14. According to your knowledge of cell communication, would it be more beneficial for heart cells to coordinate contraction via a G protein receptor signal transduction pathway or using ion flow through gap junctions?
    a. Heart cells should use the G protein receptor pathway to coordinate contraction in order to efficiently inhibit heart contractions when necessary.
    b. Heart cells should use the G protein receptor pathway to coordinate contraction because it can more easily regulate gene transcription.
    c. Heart cells should use the G protein receptor pathway for heart contraction because it causes simple and speedy cell effects.
    d. Heart signals should be targeted to gap junctions, because muscle fibers pass those locations.
    e. Heart cells should use gap junctions to coordinate contractions because ion flow through gap junctions will communicate the signal in a speedy and direct manner.

    Textbook Reference: *7.4 How Do Cells Change in Response to Signals?*

15. Gap junctions and plasmodesmata differ in that
    a. gap junctions are connected by protein tubules called connexons, whereas plasmodesmata are connected by extensions of the plant's plasma membrane.
    b. gap junctions allow much larger molecules to pass through them.
    c. gap junctions have no real physical connection but are rather the space between adjacent cell membranes.
    d. one is found in animals and the other is found in plants.
    e. gap junctions are connected by desmotubules and plasmodesmata are connected by connexons.

    Textbook Reference: *7.5 How Do Cells in a Multicellular Organism Communicate Directly?*

Answers

Key Concept Review

1. Prokaryotes do not have a circulatory system, so they cannot produce or respond to hormones. However, recent research demonstrates that many prokaryotes do respond to signals produced by their own cells (autocrine), adjacent cells (juxtacrine), or nearby cells (paracrine).
2. An inhibitor that blocked the site where the signal molecule binds to the receptor would prevent any cell response. An inhibitor could also prevent the activation of the first signal transduction molecule (e.g., by activating an enzyme that dephosphorylates a signal transduction molecule that is activated by phosphorylation), or it could specifically prevent the cell responses (e.g., by binding to a motor protein in a way that prevented cell movement). The earlier in the signal transduction pathway the inhibitor acts, the more likely it is to prevent all of the cell's response.
3. In order to have a high dissociation constant, the concentrations of the receptor and the ligand have to be large relative to the concentration of the ligand–receptor complex, and the rate constant for the dissociation of the ligand–receptor complex must be large relative to the rate constant for the binding of the receptor and ligand. A receptor with a high dissociation constant would not bind easily with its ligand. This is why there is so little of the ligand–receptor complex, especially since the few complexes made come apart so easily (high rate of dissociation).
4. Because glucagon is a large polar ligand, it will not be able to cross the cell membrane on its own and it must

bind to a membrane receptor. Because it acts by causing a series of phosphorylations, it is unlikely that its receptor is an ion channel. Protein kinase receptors always involve some protein phosyphorylation as part of their response, but G protein-linked receptors can also activate kinase enzymes. Thus, while ion channel receptors and intracellular receptors can be ruled out, either a G protein or a protein kinase receptor are possibilities. (In fact, glucagon binds to a G protein-linked receptor but one usually cannot tell this from information about the signal's chemistry.)

5. A G protein-linked receptor is stimulated by hormone binding to its external binding site. This in turn activates the G protein, which activates an effector protein. The effector protein converts reactants into signaling products, amplifying the signal from the single hormone.
6. Secondary messengers function to amplify and spread signals. They do not bind to receptors and do not act like signals in the cascade. Signals result in a change in cell function; secondary messengers are part of the pathway that results in the change in cell function.
7. Sometimes, as in the Ras pathway, a G protein is activated by a protein kinase receptor. Also, an activated G protein may act to activate (or inactivate) a protein kinase as part of the signal transduction cascade.
8. Signal transduction pathways are highly regulated because they are temporary events within a cell that help to regulate the function of the cell. If they were not highly regulated, the response of the cell would lag behind the changes in the environment. This would potentially lead to cell death.
9. There are many possible correct answers to this question, but Figure 7.17 fulfills the listed criteria. It includes a G protein-linked receptor, amplification at the adenylyl cyclase stage (once activated, adenylyl cyclase can make many cAMP molecules), and two second-messengers (both cAMP and Ca^{2+}).
10. Four possible ways in which the cascade of reactions leading to altered enzyme activity could be inhibited are: (1) the binding of epinephrine to its receptor could be inhibited or blocked; (2) a GTPase could be activated; (3) phosphodiesterase could be activated so that any cAMP made is quickly converted back to AMP; (4) a protein phosphatase could be activated that removes any phosphates added by the listed kinases.
11. During early development, cells are specializing for different functions. As part of this specialization, some genes have to be turned on and others have to be turned off so that the cell has the correct types of proteins to carry out its designated function.
12. The junctions between these cells are plasmodesmata because they have membrane-lined tubules (desmotubules) joining the cells.
13. If cells were specialized but not coordinated in their functions, an organism might, for example, put all of its energy into moving and none into reproduction, or all of its energy into eating and none into digestion. Without communication, some cells would rapidly run out of energy or perform functions that were not helpful for a particular situation and the organism as a whole would suffer.

Test Yourself

1. **c.** Because the histamine does not travel in the bloodstream, it is not a hormone. Since it affects "local" cells rather than just the originating cell (autocrine) or cells immediately adjacent (juxtacrine), it is a paracrine signal.
2. **b.** The G protein is the receptor, because when the signal molecule (the vanilla scent) binds, it changes its shape and causes activation of a signal transduction molecule (the enzyme adenylyl cyclase).
3. **a.** Autocrine signals affect the cells that make them, while paracrine signals affect nearby cells. Both types of signals bind to receptors as part of their signal transduction pathway.
4. **a.** To bind to an intracellular receptor, the ligand must be able to pass through the plasma membrane.
5. **b.** Receptors interact only with specific ligands to bring about a response in the cell.
6. **e.** Signals must bind their receptors, the conformation of the receptor is altered, a responder activates a target protein, and cell function changes.
7. **d.** Secondary messengers both amplify the signal and have multiple effects within a cell.
8. **e.** Amplification, regulation, specificity, and variation are all roles of protein kinase cascades.
9. **d.** Two Ras are activated, each of which activates only one Raf (giving a total of two Raf). Since each Raf activates 10 MEK, there will be 20 active MEK and 200 MAPK (20×10).
10. **b.** The pathway shown does not involve Ca^{2+}, and increased synthesis of GTP would tend to activate the pathway (if anything) by making it easier for Ras to bind GTP. Inhibition of the growth factor receptor phosphatase would also tend to activate the pathway by preventing the removal of phosphates from the activated receptor (keeping the receptor in its activated state). Synthesis of a protein that prevents Raf from binding to Ras and activation of a MAP phosphatase would both have inhibitory effects. However, because Raf occurs earlier in the pathway, inhibition of this step will have a greater impact. For every one Raf molecule that is unable to bind to Raf and become active, there are many MEK and MAP enzymes that will never be activated by phosphorylation. The action of the MAP phosphatase will only affect the MAPK and can be counteracted by the MAP kinase activity.

11. **a.** As shown in Figure 7.13, there is no role for cAMP in the PIP_2 signaling pathway that produces IP_3. However, cAMP is an important second messenger in other pathways. Thus, the activation/inactivation of this tyrosine kinase is an example of crosstalk between two different signal transduction pathways. The activation of PKC, on the other hand is just another part of the same PIP_2 signal transduction pathway. Sperm entering an egg is another example of Ca^{2+} use as a signal, but it does not indicate interaction between two pathways. Inhibition by lithium is an example of drug function, rather than normal cell crosstalk, and the release of neurotransmitter is an additional step in the cell's response to a single signal rather than crosstalk between pathways.
12. **c.** IP_3 and Ca^{2+} are both hydrophilic and can diffuse through the cytosol. Lithium is also hydrophilic, but it interferes with this pathway. The other answer options are all at least partly hydrophobic and are found associated with the cell membrane rather than loose in the cytosol.
13. **c.** In the IP_3/DAG pathway, IP_3 opens Ca^{2+} channels.
14. **e.** Heart cells do use gap junctions to coordinate heart contractions. The contraction of the heart is not a process that should be inhibited too much (heart rate does slow down sometimes, but the contractions still have to occur). Also, for the chambers of the heart to contract in coordination, speedy transmission of the signal is important. All the steps in the G protein pathway would slow down the signal transmission between cells, making rapid coordination difficult or impossible.
15. **d.** Gap junctions are found only in animals and plasmodesmata are found only in plants. The other statements all have at least one inaccuracy (e.g., the membrane lining of plasmodesmata is derived from the endoplasmic reticulum, not from the plasma membrane).

10/3/2014
Kwadwo Assensoh

8 Energy, Enzymes, and Metabolism

· Detergents are supplemented with enzymes

· Energy is transferred or transformed

The Big Picture

- Cells need energy to carry out their functions. Metabolism is a series of energy-transferring reactions that fuel the processes of the cell. All reactions either give off energy (exergonic) or require energy to proceed (endergonic). Energy may be stored in chemical bonds (potential energy) and released when the bonds are broken to do work (kinetic energy). ATP serves as the energy shuttle for many metabolic processes.
- Enzymes aid biological reactions by lowering the activation energy required to start the reaction. Each enzyme has a specific three-dimensional conformation that interacts specifically with the substrate. Interaction between the enzyme and substrate results in an optimal orientation for a reaction to take place. Enzymes are proteins and are affected by temperature, pH, reactants, activators, and inhibitors.
- Metabolism is regulated by enzymes. Most metabolic pathways are under allosteric control. Enzyme complexes allow for interactions and regulation of adjacent active sites. Often the final product of a specific pathway regulates the commitment step of the pathway itself.

Study Strategies

- The best strategy for this chapter is to take "small bites." You may find the concepts to be very unfamiliar; therefore, make sure you understand each section before proceeding. This chapter contains a large amount of terminology. Making a terminology/vocabulary list may be helpful. Think through examples when studying enzymes and their roles in reactions. This will help you understand and remember the processes.
- The chemistry and physics involved in this material are often overwhelming, and you may question why you are learning them in biology. Remember that biological processes are based on physical and chemical properties. To understand the function of organisms, it is necessary to understand the basics of chemistry and physics as they pertain to energy transfer.
- When thinking about the equation $\Delta G = \Delta H - T\Delta S$, you may find it helpful to convert the concepts into familiar terms. For example, your net pay is equal to your gross pay minus income taxes and other deductions. Similarly the energy in the raw materials (ΔH) is reduced by the entropy "tax" ($T\Delta S$) imposed by the universe on all chemical reactions, resulting in a final amount of energy that can be used to do work (ΔG).
- When studying the energy equations, be sure that you understand how any change in either free energy, unusable energy, or absolute temperature will affect the total energy of the system. Also, focus on what a larger or smaller ΔG value means in terms of how reactions proceed and whether they require or release energy.
- Allosteric regulation can be confusing because the terms "allosteric regulation" and "allosteric enzyme" are related but slightly different. Remember that an allosteric enzyme is an enzyme with more than one binding site: an active site (for binding and acting on substrates) and one or more allosteric sites. When allosteric regulators are bound to the allosteric site, they control whether or not the enzyme can bind substrate.
- Remember that in enzyme-mediated reactions, any alteration on the left side of the equation will lead to changes in the amount of product formed and that enzymes are conserved across the equation.
- Go to the Web addresses shown to review the following animated tutorials and activities. You can also find them in BioPortal (yourBioPortal.com), along with many additional learning resources.

  Animated Tutorial 8.1 Enzyme Catalysis (Life10e.com/at8.1)

  Animated Tutorial 8.2 Allosteric Regulation of Enzymes (Life10e.com/at8.2)

  Activity 8.1 ATP and Coupled Reactions (Life10e.com/ac8.1)

  Activity 8.2 Free Energy Changes (Life10e.com/ac8.2)

Key Concept Review

8.1 What Physical Principles Underlie Biological Energy Transformations?

There are two basic types of energy

There are two basic types of metabolism

The first law of thermodynamics: Energy is neither created nor destroyed

The second law of thermodynamics: Disorder tends to increase

Chemical reactions release or consume energy

Chemical equilibrium and free energy are related

Energy is the capacity to do work. No living cell manufactures energy. The energy needed for life must be obtained from an environmental energy source and transformed to a usable form. Energy transformations are linked to chemical transformations in cells. Kinetic energy is energy in motion and does work. Potential energy is stored energy. Energy can be stored biologically in chemical bonds of fatty acids and other molecules. Breaking the bonds converts the energy to kinetic energy.

Metabolism is all of the chemical reactions occurring in a living organism. Anabolic reactions, (collectively anabolism) link simple molecules to create complex molecules and store energy in the resulting bonds. Catabolic reactions, (collectively catabolism) break down complex molecules and release stored energy.

The first law of thermodynamics states that energy is neither created nor destroyed. Energy can be converted from one form to another. The second law of thermodynamics states that not all energy can be used. When energy is converted from one form to another, some becomes unusable. Total energy (enthalpy, H) = usable energy (free energy, G) + unusable energy (entropy, S) × absolute temperature (T).

We cannot measure absolute energy; we can only measure the change (Δ) in energy. We determine the change in usable energy as follows: $\Delta G = \Delta H - T\Delta S$. If ΔG is negative, free energy is released, and the reaction is *exergonic*. If ΔG is positive, free energy is required, and the reaction is *endergonic*. With each conversion, entropy (S) tends to increase; therefore, processes tend toward disorder and randomness. All biochemical reactions must either release ($-\Delta G$) or take up ($+\Delta G$) usable energy. Reactions that are exergonic in one direction are endergonic in the other. The direction of the reaction depends on concentrations of reactants versus products. The point at which the forward reaction occurs at the same rate as the reverse reaction is termed chemical equilibrium.

The further toward completion a reaction's equilibrium lies, the more free energy is given off (larger $-\Delta G$). A larger $+\Delta G$ means equilibrium favors the reverse reaction (equilibrium falls toward the reactant). A ΔG near zero indicates that both products and reactants have free energy.

Question 1. It is estimated that approximately 90 percent of energy that passes between levels in a food web is "lost" at each level. Explain the first law of thermodynamics and discuss why this apparent loss of energy does not contradict the law.

Question 2. You decide to purchase a new water heater and start looking at the energy efficiency ratings. You find one unit that is labeled as 100 percent energy efficient, and the salesperson says that the more efficient the appliance is, the more money you will save. However, you don't trust that the store is providing accurate information, and you do not buy the product. Was this decision correct?

8.2 What Is the Role of ATP in Biochemical Energetics?

ATP hydrolysis releases energy

ATP couples exergonic and endergonic reactions

ATP (adenosine triphosphate) is the primary energy currency of living cells. ATP is needed to capture, transfer, and store free energy (usable energy) in cells to do work. Additionally, ATP can donate a phosphate group to different molecules by phosphorylation. Hydrolysis of ATP to ADP and P_i releases a large amount of free energy (ΔG = –7.3 to –14 kcal/mol). The equilibrium falls toward ADP production. The formation of ATP is endergonic, requiring as much energy as is released by ATP hydrolysis. This results in an "energy-coupling cycle" of the endergonic and exergonic reactions. ATP is formed when coupled with the exergonic reactions of cellular respiration. ATP is hydrolyzed to fuel endergonic reactions like protein synthesis (see Figure 8.6).

Cells use the energy released by ATP hydrolysis to fuel endergonic reactions, for active transport across membranes, for movement, or to produce light (bioluminescence). ATP molecules are "consumed" immediately after their formation, and each one undergoes more than 10,000 cycles of hydrolysis and synthesis daily.

Question 3. What is the result of the hydrolysis of ATP? Draw the products using the reactants structures given in the figure below as a guide.

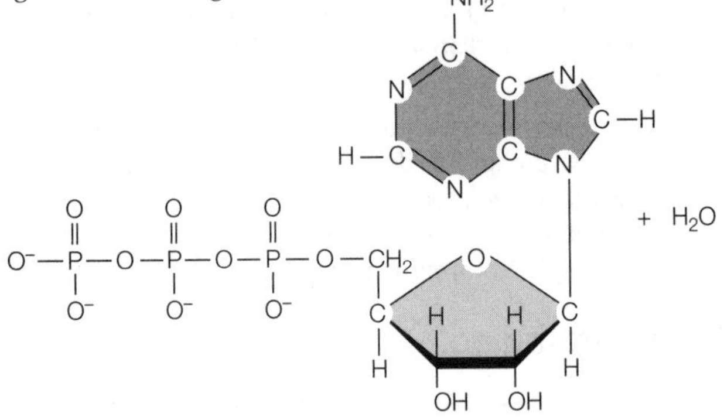

Question 4. The ultimate goal of metabolism is to drive ATP synthesis. ATP is considered the energy currency of the cell. Discuss how ATP couples endergonic and exergonic reactions and why it is so important in cellular functions.

8.3 What Are Enzymes?

To speed up a reaction, an energy barrier must be overcome

Enzymes bind specific reactants at their active sites

Enzymes lower the energy barrier but do not affect equilibrium

Enzymes catalyze biochemical reactions by lowering activation energy. Catalysts are substances that speed up reactions without themselves being permanently altered. Catalysts do not cause reactions to occur but they increase the rate of the reactions both forward and backward. Enzymes do not alter the equilibrium of a reaction.

Energy barriers exist between reactants and products that slow down reactions. Energy must be added to get past this barrier; it may be thought of it as a little "shove" to get the reaction going. This is called activation energy (E_a) (see Figure 8.8). For reactions to take place, the reactants must be slightly destabilized and turned into "transition-state intermediates" with higher free energy than that of either the reactants or the products. This is accomplished with the help of a catalyst.

Enzymes are highly specific biological catalysts that act on specific reactants or substrates. Most biological catalysts are protein enzymes. They have specific substrate-binding areas on their surfaces called active sites. The enzyme's specificity comes from the shape of its active site. The names for enzymes typically end in the suffix "–ase." Only specific substrates can fit in and bind with an enzyme's active site. Once bound, the site is referred to as an enzyme–substrate complex (ES). The enzyme alters the conformation of the substrate, thus lowering the activation energy for the reaction. From this, the product is formed, and the enzyme remains unchanged. Such reactions can be represented as follows:

$$E + S \rightarrow ES \rightarrow E + P$$

Question 5. Label the graph below with the following: Activation (E_a) energy for catalyzed reaction, activation energy for uncatalyzed reaction, ΔG for catalyzed reaction, ΔG for uncatalyzed reaction, free energy of reactants, free energy of products, least stable state on graph, most stable state on graph.

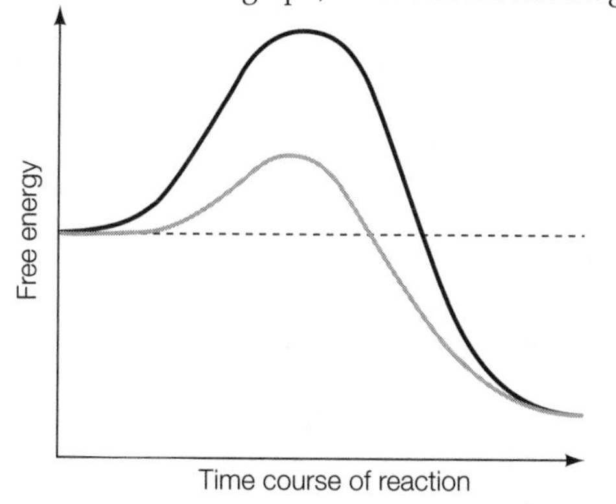

Question 6. The first enzymes that evolved were ribozymes, but now cells use proteins rather than RNA to catalyze reactions. Why are proteins better catalysts than enzymes?

8.4 How Do Enzymes Work?

Enzymes can orient substrates

Enzymes can induce strain in the substrate

Enzymes can temporarily add chemical groups to substrates

Molecular structure determines enzyme function

Some enzymes require other molecules in order to function

The substrate concentration affects the reaction rate

The structure of an enzyme determines its function. Enzymes speed up reactions by orienting substrates for maximum chemical interactions, by inducing strain on the substrate, or by adding charges to substrates.

Enzymes are very large macromolecules with small active sites for binding small substrates, while the rest of the macromolecule provides the framework, provides binding sites for regulatory molecules, or helps change the protein shape. Binding of small molecules to the active sites depends on H-bonds, charge interactions, and hydrophobic interactions.

Binding of substrate by an enzyme can change the enzyme's shape. The enzyme's shape can be altered by the substrate to produce an "induced fit." The large size of an enzyme helps it position the correct amino acids at the active site and helps regulate shape and allow for the induced fit. The rate of a reaction increases as the substrate concentration increases until all enzyme active sites are occupied. At this saturation point, no amount of additional substrate will increase the reaction rate because no more enzyme is available for catalysis. Enzyme efficiency is measured in turnover number, or how fast an enzyme can convert substrate to product and free up its active site. In the human body the turnover number ranges from 0.5 to 40 million substrates per second.

Cofactors, coenzymes, and prosthetic groups all contribute to an enzyme's activity. Cofactors are inorganic ions such as zinc, copper, or iron. Coenzymes are organic molecules that bind in the active site and can be thought of as a cosubstrate. Coenzymes are chemically changed during the reaction and regenerated in other pathways. Prosthetic groups are permanently bound to the enzyme.

Question 7. Sucrase is an enzyme that catalyzes the hydrolysis of sucrose. Suppose that in the laboratory you are trying to produce glucose and fructose from sucrose. You add sucrase to a sucrose solution and start to measure the fructose and glucose levels. Although you do not have access to more sucrase, you keep adding sucrose to the reaction because you want to produce fructose and glucose much faster. Draw a curve with the sucrose concentration on the x-axis and the reaction rate on the y-axis, and explain what will happen in your solution as you gradually increase the sucrose concentration.

Question 8. You run a lab experiment and notice that the following reaction catalyzed by hexokinase reaches its saturation level at the maximum reaction rate.

$$\text{Glucose} + \text{ATP} \rightarrow \text{glucose } 6 - \text{phosphate} + \text{ADP}$$

Can you still increase the reaction rate above the current level? How?

8.5 How Are Enzyme Activities Regulated?

Enzymes can be regulated by inhibitors

Allosteric enzymes are controlled via changes in shape

Allosteric effects regulate many metabolic pathways

Many enzymes are regulated through reversible phosphorylation

Enzymes are affected by their environment

All life must maintain stable internal conditions, or homeostasis, and this is accomplished by the regulation of enzymes. The chemical reactions occurring in an organism are organized into specific metabolic pathways, in which the product of one reaction is the reactant for the next. These pathways interact, and each reaction is catalyzed by a specific enzyme.

Inhibitors are substances that bind with enzymes to inhibit their function. Irreversible inhibition occurs when an inhibitor forms a covalent bond with an enzyme and permanently destroys the active site. Reversible inhibition occurs when an inhibitor binds to and alters the enzyme active site but the inhibition is reversible (depending on the concentration of the inhibitor and substrate). Some reversible inhibitors are called competitive inhibitors because they compete with the substrate for the active site. Others are called noncompetitive inhibitors because they bind elsewhere but alter the active site so that the substrate cannot bind, or the rate of binding is reduced.

Allostery is the regulation of an enzyme by a molecule that binds to the enzyme at a site other than the active site, resulting in a change in the enzyme shape. Noncompetitive inhibitors are an example of inhibitors that work allosterically. Most allosteric enzymes exist naturally in an active form and an inactive form, and they can switch back and forth between the two forms. When an allosteric regulator binds to the enzyme, the enzyme becomes locked in either its active form or its inactive form.

Allosteric interactions help control metabolism. The first step in an enzyme-mediated pathway is referred to as the commitment step. Once initiated, the pathway is followed to completion. Frequently, the final product allosterically inhibits the enzyme of the commitment step, preventing overproduction of the final product. This process is called end-product or feedback inhibition.

Enzyme function is influenced by the cellular environment. Changes in pH influence charges of amino and other groups on the enzyme or substrate. This may change the folding of the enzyme or its ability to interact with the substrate and drastically alter its catalytic ability. Enzymes typically have an ideal pH at which they function best. Increases in temperature may aid in reduction of activation energy by adding kinetic energy; however, this may also result in denaturing of enzymes. Each enzyme tends to have an optimal temperature. Organisms may produce different forms of the same enzyme, called isozymes, which have different optimal temperatures.

Question 9. Amylase is a digestive enzyme secreted in human saliva that breaks down starch. Amylase functions well in the mouth but ceases to function in the stomach. Explain why.

Question 10. Figure 8.17 shows the behavior of an allosteric enzyme that has binding sites for a negative regulator. Evaluate the behavior of an enzyme with binding sites for a positive regulator instead of a negative regulator.

Question 11. Enzyme X has an optimal pH of 9.0 and is completely inactive at pH 7.0 and pH 11.0. Using the graph below, draw the predicted enzyme activity and determine if this enzyme prefers an acidic, a neutral, or a basic environment.

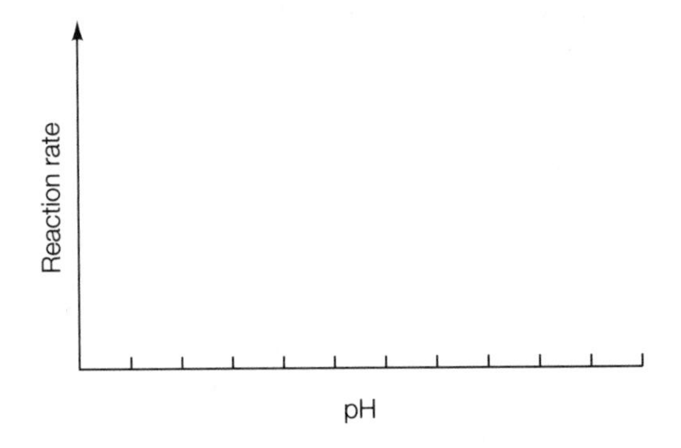

Test Yourself

1. ATP is necessary for the conversion of glucose to glucose 6-phosphate. Splitting ATP into ADP and P_i releases energy into what form that is used by the cell?
   a. Potential energy
   b. Kinetic energy
   c. Entropic energy
   d. Enthalpic energy
   e. Heat energy
   Textbook Reference: *8.2 What Is the Role of ATP in Biochemical Energetics?*

2. Before ATP is split into ADP and P_i, it holds what type of energy?
   a. Potential energy
   b. Kinetic energy
   c. Entropic energy
   d. Enthalpic energy
   e. Physical energy
   Textbook Reference: *8.1 What Physical Principles Underlie Biological Energy Transformations?*

3. Which of the following statements concerning energy transformations is true?
   a. Increases in entropy reduce usable energy.
   b. Energy may be created during transformation.
   c. Potential energy increases with each transformation.
   d. Increases in temperature decrease total amount of energy available.
   e. Decreases in entropy reduce usable energy.
   Textbook Reference: *8.1 What Physical Principles Underlie Biological Energy Transformations?*

4. A reaction has a ΔG of –20 kcal/mol. This reaction is
   a. endergonic, and equilibrium is far toward completion.
   b. exergonic, and equilibrium is far toward completion.
   c. endergonic, and the forward reaction occurs at the same rate as the reverse reaction.
   d. exergonic, and the forward reaction occurs at the same rate as the reverse reaction.

e. of an indeterminate nature, according to the information provided.
Textbook Reference: *8.1 What Physical Principles Underlie Biological Energy Transformations?*

5. ATP hydrolysis is
   a. endergonic.
   b. exergonic.
   c. chemoautotrophic.
   d. anabolic.
   e. None of the above
   Textbook Reference: *8.2 What Is the Role of ATP in Biochemical Energetics?*

6. Enzymes are biological catalysts and function by
   a. increasing the free energy in a system.
   b. lowering the activation energy of a reaction.
   c. lowering entropy in a system.
   d. increasing the temperature near a reaction.
   e. altering the equilibrium of a reaction.
   Textbook Reference: *8.3 What Are Enzymes?*

7. Which of the following contributes to the specificity of enzymes?
   a. Each enzyme has a wide range of temperature and pH optima.
   b. Each enzyme has an active site that interacts with many substrates.
   c. Substrates themselves may alter the active site slightly for optimum catalysis.
   d. Enzymes are more active at higher temperatures.
   e. All of the above
   Textbook Reference: *8.4 How Do Enzymes Work?*

8. Coenzymes and cofactors, as well as prosthetic groups, assist enzyme function by
   a. stabilizing three-dimensional shape.
   b. assisting with the binding of enzyme and substrate.
   c. maintaining active sites in an active configuration.
   d. reversibly binding to the enzyme to regulate the enzyme's activity.
   e. a, b, and c only
   Textbook Reference: *8.4 How Do Enzymes Work?*

9. Which of the following statements about enzymes is true?
   a. They are consumed by the enzyme-mediated reaction.
   b. They are not altered by the enzyme-mediated reaction.
   c. They raise activation energy.
   d. They can be composed of RNA or proteins.
   e. They are only rarely regulated.
   Textbook Reference: *8.3 What Are Enzymes?*

10. Ascorbic acid, which is found in citrus fruits, acts as an inhibitor to catecholase, the enzyme responsible for the browning reaction in fruits such as apples, peaches, and pears. One explanation for the inhibiting function of ascorbic acid could be its similarity, in terms of size and shape, to catechol, the substrate of the browning reaction. If this explanation is correct, then this inhibition is most likely an example of _______ inhibition.
    a. competitive
    b. indirect
    c. noncompetitive
    d. allosteric
    e. feedback
    Textbook Reference: *8.5 How Are Enzyme Activities Regulated?*

11. Metabolism is organized into pathways that are linked in which of the following ways?
    a. All cellular functions feed into a central pathway.
    b. All steps in the pathway are catalyzed by the same enzyme.
    c. The product of one step in the pathway functions as the substrate in the next step.
    d. Products of the pathway accumulate and are secreted from the cell.
    e. Different substrates are acted on by the same enzyme.
    Textbook Reference: *8.5 How Are Enzyme Activities Regulated?*

12. Which of the following represents an enzyme-catalyzed reaction? (E = enzyme, P = product, S = substrate)
    a. $E + P \rightarrow E + S$
    b. $E + S \rightarrow E + P$
    c. $E + S \rightarrow P$
    d. $E + S \rightarrow E$
    e. $P + S \rightarrow E$
    Textbook Reference: *8.3 What Are Enzymes?*

13. In the pathway $A + B \rightarrow C + D$, enzyme X facilitates the reaction. If compound D inhibits enzyme X, one would conclude that
    a. enzyme X is an allosteric inhibitor of the above reaction.
    b. compound D is an allosteric stimulator of the above reaction.
    c. compound D is a competitive inhibitor of the above reaction.
    d. enzyme X is subject to feedback stimulation.
    e. compound D is a coenzyme in the above reaction.
    Textbook Reference: *8.5 How Are Enzyme Activities Regulated?*

14. Suppose you are studying a new species that has never been studied before. It lives in acidic pools in volcanic craters where temperatures normally stay above 90°C and often reach 100°C. You determine that the species has a surface enzyme that catalyzes a reaction leading to its protective coating, and you decide to study this enzyme in the laboratory. At what temperature would optimal activity of this enzyme most likely be found?
    a. 0°C
    b. 37°C

c. 55°C
d. 95°C
e. 105°C
Textbook Reference: *8.5 How Are Enzyme Activities Regulated?*

15. Enzymes alter the _______ of a reaction.
a. ΔG value
b. activation energy
c. equilibrium
d. rate
e. Both a and b
Textbook Reference: *8.3 What Are Enzymes?*

16. You fill two containers with identical amounts of reactants A and B and enzymes 1–4. If product D inhibits enzyme 2 and product F is an allosteric stimulator of enzyme 1, what will be the final result if you add extra product D to the second container? (Assume that the experiment lasts long enough for the reactions in both containers to go to completion.)
a. The concentration of product C will increase compared to the first container and there will be no change in the concentration of product F compared to the first container.
b. The concentration of reactants A and B will increase relative to the first container.
c. The concentration of product F will increase in the second container because more of D is converted back to C.
d. The concentration of products E and F will both increase in the second container, since D inhibits enzyme 2.
e. The concentration of product F will increase relative to the first container, since enzyme 2 will have been inhibited from converting as much of C into D.
Textbook Reference: *8.5 How Are Enzyme Activities Regulated?*

17. Suppose you are given an unlabeled enzyme and told to add a compound to the container that will irreversibly bind to the enzyme and increase its function. You ask for information about the enzyme, but your instructor simply hands you a list of possible compounds. Based on what you have learned about the enzyme partners below, which one is the best choice?
a. Coenzyme A
b. Zinc (Zn^{2+})
c. Flavin
d. ATP
e. NAD
Textbook Reference: *8.4 How Do Enzymes Work?*

Answers

Key Concept Review

1. The first law of thermodynamics states that energy cannot be created or destroyed, but that it may be converted from one form to another. In the transfer between levels of a food web, approximately 90 percent of the energy is converted to unusable heat energy. There is a net loss of usable energy during each conversion, but the total amount of energy (usable and unusable) remains the same.

2. The decision was correct. An appliance with 100 percent energy efficiency is not possible. The second law of thermodynamics indicates that whenever energy is transformed, some is lost in the form of entropy. An appliance with no energy lost to entropy therefore does not exist.

3. The hydrolysis of ATP to ADP breaks the bond between the phosphate groups and releases energy.

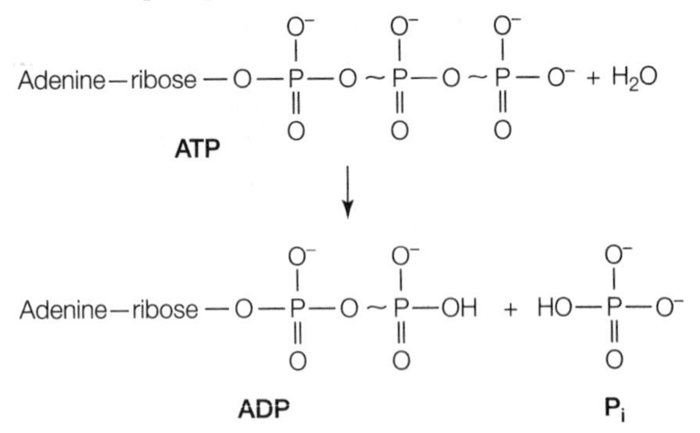

4. The conversion of ATP to ADP and P_i releases approximately 7.3 kcal/mol of energy. This energy release fuels (endergonic) reactions in the cell. Equilibrium of the reaction is far to the right and favors the formation of ADP. In the converse situation, the formation of ATP from ADP and P_i is energy intensive and can be coupled to highly exergonic reactions within the cell. Thus, ATP functions as an energy shuttle between endergonic and exergonic reactions. The small size of the molecule and its ubiquitousness allow it to be available and move freely within the cell.

5.

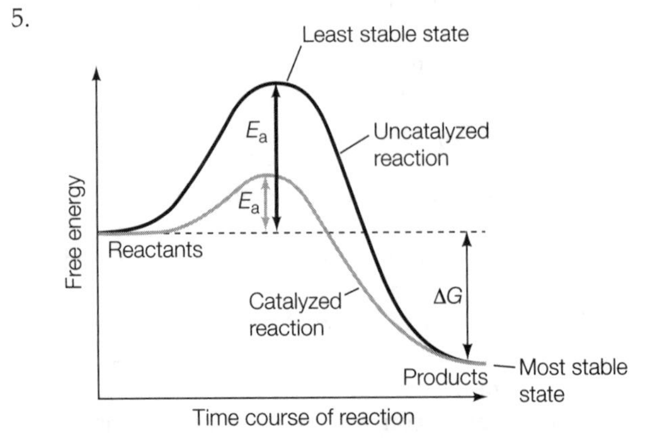

6. Proteins have a greater diversity in their three-dimensional structures and they have a great variation of functional groups that make them very versatile. Since catalysts are scaffolds within which chemical reactions take place, a greater variety of structure and reaction sites will increase effectiveness.

7. Refer to Figure 8.13. At low sucrose levels the hydrolysis will be fast, as there are accessible sucrase enzyme molecules in the solution. If the sucrose concentration is increased continuously, a saturation level eventually will be reached. At this maximum rate all enzyme molecules are occupied, so despite the further increase of sucrose, the rate at which fructose and glucose are produced will not increase.
8. Increasing the concentration of hexokinase would eventually result in the increase of the maximum reaction rate because of the highter number of available enzymes to act as catalysts.
9. The pH optimum of amylase is approximately 7.0. At that pH, the protein has the three-dimensional shape that allows starch to bind to its active site and catalyze its hydrolysis. When it is at stomach pH (approximately 2.0), the protein is denatured, and its three-dimensional shape and active site are lost; therefore, it can no longer catalyze the reaction.
10. A positive regulator would stabilize the enzyme in its active form. In the absence of the regulator, the enzyme would alternate between its inactive and active forms. When it encountered substrate, it would bind it only if it happened to be in its active form. When bound to the regulator, the enzyme would be fixed in its active form, and it would bind substrate at a greater rate.
11.

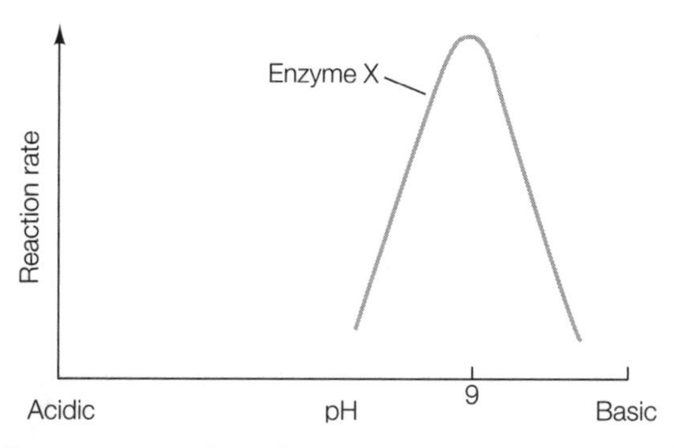

The enzyme prefers a basic environment. Your graph should be symmetrical around a peak at pH 9.0.

Test Yourself

1. **b.** The released energy is available to do work; therefore, it is kinetic energy. Some energy is lost in the form of entropy, but it will not be used by the cell to do work.
2. **a.** Potential energy is energy held within chemical bonds that may be converted to working kinetic energy.
3. **a.** Total energy = free energy + entropy × temperature. Any increase in entropy is necessarily going to reduce free energy.
4. **b.** A negative ΔG indicates an exergonic reaction with energy being liberated. A large ΔG indicates that equilibrium lies toward completion.
5. **b.** ATP hydrolysis is exergonic, resulting in a ΔG of –7.3 kcal/mol.
6. **b.** Enzymes reduce activation energy and speed up reactions.
7. **d.** Enzymes are specific to particular substrates that may actually "adjust" the fit of the active site. They also function in specific, narrow optimum ranges of pH and temperature.
8. **e.** Cofactors, coenzymes, and prosthetic groups assist with the maintenance of an enzyme's three-dimensional shape and the conformation of the active prosthetic groups are permanently bound to the enzyme.
9. **b.** Enzymes are not consumed or altered in any way during an enzyme-mediated reaction, and they function to lower the activation energy of a reaction. Ribozymes (composed of RNA) are catalysts, but they are not true enzymes.
10. **a.** Competitive inhibitors compete for the active site with the substrate.
11. **c.** Within a given pathway, the products of the preceding step act as substrates for subsequent steps.
12. **b.** Substrate is converted to product and the enzyme is unchanged.
13. **a.** Products of a reaction that inhibit the enzymes via feedback inhibition are allosteric inhibitors.
14. **d.** Enzymes typically work at a maximal rate at a particular temperature or within a range of temperatures. This optimal temperature tends to be correlated with the body temperature of the organism.
15. **b.** The free energy levels of the reactants and products are not changed by enzymes. Only the activation energy is altered. See Figure 8.10.
16. **e.** The addition of more D to the second container will reduce the activity of enzyme 2. The pathway from A + B going all the way to F will be the predominant reaction that takes place, leading to a greater final concentration of F in the second container. The first container will convert some of C into D, and thus the level of intermediate E and final product F will be lower than in the second container. (Hint: Draw a diagram of the experiment.)
17. **c.** Of the five compounds listed, only one is a prosthetic group, which is irreversibly bound to its target enzyme.

9 Pathways That Harvest Chemical Energy

The Big Picture

- Metabolic processes occur in pathways and are regulated by enzymes. These pathways may be further controlled by compartmentalization into organelles in eukaryotic cells. Because metabolic processes are energy-transferring reactions, ATP and NAD are necessary for shuttling energy and electrons between steps of the pathways. Redox reactions are the basis of metabolism. As one compound is oxidized, another is reduced.
- Glucose provides cellular energy. It may be metabolized in the absence of oxygen through glycolysis and fermentation. In the presence of oxygen, it is metabolized through glycolysis and cellular respiration in the form of the citric acid cycle, electron transport, and oxidative phosphorylation. Compared to aerobic respiration, anaerobic respiration produces significantly less cellular energy (in the form of ATP) from the same amount of glucose.
- Each step of the catabolic pathway is regulated by the products of subsequent pathways and depends on the presence of oxygen as the final electron acceptor. This level of control is achieved through allosteric regulation.
- Glycolysis is the process of converting glucose to pyruvate and usable energy. The citric acid cycle oxidizes acetate to CO_2 and forms $FADH_2$, NADH + H^+, and ATP. The metabolic pathways for the production and breakdown of lipids and amino acids are tied to those of glucose metabolism.

Study Strategies

- Do not memorize pathways. Focus on understanding the beginning and end products of each pathway, why the pathway is present, and how the pathway does what it does. Make sure you understand how the pathways are connected and how changes in one pathway can alter another. Once you conceptually understand this material, move on to learning the pathways themselves.
- This chapter is very visual in nature. You should spend a significant amount of time studying the figures.
- Students often fail to see how regulation of the pathways occurs. By understanding how the presence or absence of intermediates affects the entire pathway, you will better understand the pathways themselves.
- Remember that ATP and NAD are energy currencies: they move energy from pathway to pathway. In order to understand the pathways completely, think about the roles of ATP and NAD in redox reactions.
- Go to the Web addresses shown to review the following animated tutorials and activities. You can also find them in BioPortal (yourBioPortal.com), along with many additional learning resources.

  Animated Tutorial 9.1 Electron Transport and ATP Synthesis (Life10e.com/at9.1)

  Animated Tutorial 9.2 Two Experiments Demonstrate the Chemiosmotic Mechanism (Life10e.com/at9.2)

  Activity 9.1 Glycolysis and Fermentation (Life10e.com/ac9.2)

  Activity 9.2 Energy Pathways in Cells (Life10e.com/ac9.1)

  Activity 9.3 The Citric Acid Cycle (Life10e.com/ac9.3)

  Activity 9.4 Respiratory Chain (Life10e.com/ac9.4)

  Activity 9.5 Energy Levels (Life10e.com/ac9.5)

  Activity 9.6 Regulation of Energy Pathways (Life10e.com/ac9.6)

Key Concept Review

9.1 How Does Glucose Oxidation Release Chemical Energy

Cells trap free energy while metabolizing glucose

Redox reactions transfer electrons and energy

The coenzyme NAD^+ is a key electron carrier in redox reactions

An overview: Harvesting energy from glucose

All living organisms require a source of energy to survive. Most organisms use glucose ($C_6H_{12}O_6$) as a metabolic fuel source. Cells oxidize glucose for energy: $C_6H_{12}O_6 + 6\ O_2 \rightarrow 6\ CO_2 + 6\ H_2O$ + energy. The energy is typically captured in ATP and is used to carry out cellular work.

The metabolism of glucose involves up to three metabolic processes. In the first pathway, glycolysis, glucose is converted to pyruvate with a small net energy release. In the second pathway, cellular respiration, the pyruvate from glycolysis is converted to CO_2 and energy in the form of ATP. Cellular respiration occurs only in the presence of O_2 and is thus an aerobic process. Complete conversion of glucose to CO_2 in the presence of O_2 releases 686 kcal/mol of energy. In the absence of O_2, pyruvate is converted to lactic acid or ethanol with a small net energy release in the process of fermentation. Both fermentation and glycolysis occur in the absence of O_2, making them anaerobic processes. Incomplete breakdown through fermentation produces substantially less energy than the aerobic conversion of glucose.

In redox reactions energy is transferred from compound to compound through transfer of electrons. The gain of electrons or hydrogen atoms is called reduction, and the loss of electrons or hydrogen atoms is called oxidation. Oxidizing agents accept electrons and become reduced, while reducing agents donate electrons and become oxidized. In metabolism, O_2 is the oxidizing agent and glucose is the reducing agent. Oxidation and reduction are always coupled, because in order for a material to lose an electron or a hydrogen atom, another material must accept it. The net ΔG of a redox reaction is negative; therefore, energy is liberated as heat.

Just as ATP is an energy shuttle in metabolism, NAD (nicotinamide adenine dinucleotide) and FAD (flavin adenine dinucleotide) are electron carriers in the redox reactions of metabolism. NAD may exist as either NAD^+ or $NADH + H^+$ and thus acts as an electron carrier. Two electrons are transferred in the reduction of NAD^+ to $NADH + H^+$. The oxidation of NADH by O_2 is exergonic and liberates –52.4 kcal/mol.

Question 1. Examine the figure below, and answer the following questions:

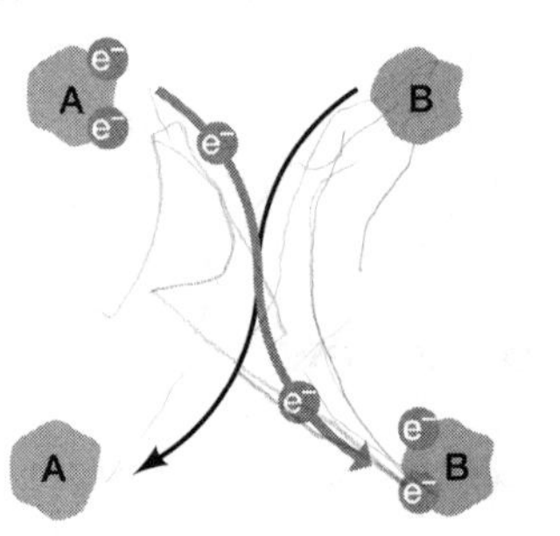

a. Which is the oxidizing agent and which is the reducing agent?

b. At the end of the reaction, which compound is oxidized and which is reduced?

c. If this figure represents the metabolism of glucose, which letter (A or B) is the glucose? Which letter represents O_2?

Question 2. Describe the major energy pathways that exist in the cytoplasm of eukaryotic cells and prokaryotic cells.

9.2 What Are the Aerobic Pathways of Glucose Catabolism?

In glycolysis, glucose is partially oxidized and some energy is released

Pyruvate oxidation links glycolysis and the citric acid cycle

The citric acid cycle completes the oxidation of glucose to CO_2

Pyruvate oxidation and the citric acid cycle are regulated by the concentrations of starting materials

Glycolysis, which occurs in the cytosol of the cell, is the process of converting glucose to pyruvate and usable energy. It is part of both aerobic and anaerobic cell respiration. During glycolysis, glucose is converted to two molecules of pyruvate, with a net production of two ATP and two NADH.

There are several key points to remember about glycolysis. (See Figure 9.5 for a representation of the 10 steps in the process.) It begins with a 6-C (6-carbon) sugar that is ultimately converted to two 3-C pyruvate molecules. During the initial three energy-intensive steps, two phosphate groups from two ATP molecules are added to the 6-C sugar to produce the 5-C fructose-1,6-biphosphate (FBP). During steps 4 and 5, the carbon ring is opened and split into two 3-C phosphorylated molecules: glyceraldehyde 3-phosphate (G3P).

The conversion of glyceraldehyde 3-phosphate (G3P) to 1,3-bisphosphoglycerate (BPG) during step 6 is significant because it is an oxidation process, and energy is stored in two $NADH + H^+$ coenzymes after this conversion. The cell has only small amounts of NAD, so it must be recycled for glycolysis to continue.

During steps 7 and 10, the phosphate groups are transferred to ADP to produce ATP in the process of substrate-level phosphorylation (addition of a phosphate group to a molecule). The end product of glycolysis is pyruvate. Though glycolysis results in a net gain of two ATP, the initial steps are endergonic and require energy input.

In eukaryotes, pyruvate is transported from the cytosol into the mitochondria. Aerobic conditions allow the oxidation of pyruvate to acetate, which, with the help of the pyruvate dehydrogenase complex, is combined with coenzyme A to form acetyl CoA on the inner mitochondrial membrane. During this process CO_2 is liberated, and $NADH + H^+$ and acetyl CoA are produced. Each glucose molecule yields two acetyl CoA.

Acetyl CoA is the starting molecule of the citric acid cycle, which consists of a total of eight reactions (see Figure 9.6). The concentrations of the intermediate molecules involved in the citric acid cycle are maintained at a constant or steady state. In step 1, citrate is formed by the combination of acetyl CoA and oxaloacetate. During this step, CoA is liberated and recycled for combination with another pyruvate. In steps 3 and 4, one CO_2 molecule and one $NADH + H^+$ molecule are formed at each step. Step 5 involves the conversion of GDP + P_i into GTP, which is then used to produce ATP. Step 6 yields one $FADH_2$. The last reaction in the cycle results in the formation of one $NADH + H^+$ molecule and oxaloacetate, which reenters the cycle at the beginning. So one turn of the cycle produces two CO_2 molecules, three $NADH + H^+$ molecules, one $FADH_2$, and one ATP.

The NADH produced by glycolysis and the citric acid cycle must be reoxidized during fermentation in the absence of oxygen or by oxidative phosphorylation in the presence of oxygen.

Question 3. One of the by-products of aerobic cellular respiration is CO_2. Assuming you begin with labeled glucose, trace the fate of that molecule until CO_2 is released.

Question 4. What is the energy yield of each glucose molecule by the end of the citric acid cycle?

Question 5. When you consume a packet of pure sugar and burn it for energy, where does the carbon in the sugar ultimately go? Is this true for all of the 6 sugar atoms?

9.3 How Does Oxidative Phosphorylation Form ATP?

The respiratory chain transfers electrons and protons, and releases energy

Proton diffusion is coupled to ATP synthesis

Some microorganisms use non-O_2 electron acceptors

The process of oxidative phosphorylation takes places in the mitochondria in eukaryotes and on the plasma membrane in prokaryotes. The two components of oxidative phosphorylation are: (1) electron transport, which involves the passing of electrons through a series of membrane-associated electron carriers, and (2) chemiosmosis, which involves the movement of protons with the synthesis of ATP.

In electron transport, the respiratory chain is composed of four enzyme complexes (I, II, III, and IV) plus cytochrome *c* and ubiquinone (Q). The first two complexes shuttle the electrons of NADH + H^+ and $FADH_2$ to Q. Energy is liberated from reduced NAD^+ and FAD through electron transport in the respiratory chain. The third complex moves electrons from Q to cytochrome *c*. The final complex passes the electrons on to O_2, the ultimate electron acceptor, resulting in H_2O as a by-product. (Figures 9.7 and 9.8 detail the respiratory chain.) The purpose of the electron shuttling is to pump protons (H^+) across the mitochondrial membrane against a concentration gradient. This movement of electrons by these proton pumps results in the establishment of a proton gradient across the inner mitochondrial membrane. This also polarizes the membrane by creating a charge differential across it. These coupled effects are called the proton-motive force.

In chemiosmosis (ATP synthesis), ATP is produced as protons diffuse across the mitochondrial membrane. An ATP synthase in the inner mitochondrial membrane couples the movement of protons with the synthesis. ATP synthase is a protein made of two parts. The F_0 unit spans the membrane and acts as the channel for H^+, and the F_1 unit is the site of ATP synthesis. (Chemiosmosis is not limited to the inner mitochondrial membrane and is also found in chloroplasts and some bacterial membranes; this concept will be explained again in later chapters.) Because the ATP synthase reaction may go in either direction, ATP is removed from the mitochondrial matrix immediately to keep the ATP concentration low and hence to favor ATP formation. Also, the proton gradient is maintained to favor ATP synthesis. If the proton flow is uncoupled from ATP synthase, the resulting energy is lost as heat. This phenomenon is seen in thermoregulation.

Question 6. Why is oxygen necessary for aerobic respiration?

Question 7. Explain how the proton-motive force drives chemiosmosis.

9.4 How Is Energy Harvested from Glucose in the Absence of Oxygen?

Cellular respiration yields much more energy than fermentation

The yield of ATP is reduced by the impermeability of mitochondria to NADH

In the absence of O_2, ATP is generated via fermentation. Fermentation occurs in the cell cytosol and reduces pyruvate (or a metabolite) before regenerating NAD^+ to keep glycolysis operating. This allows the cell to continue to produce small amounts of ATP by glycolysis.

There are two types of fermentation based on the final end products produced. In lactic acid fermentation, pyruvate serves as the electron acceptor and lactic acid is produced. In alcoholic fermentation (in yeast and in some plant cells), ethyl alcohol is produced as the end product.

Fermentation and cellular respiration have different energy yields. Fermentation (both lactic acid fermentation and alcoholic fermentation) yields a net total of two ATP for every one glucose molecule. Aerobic respiration yields a total of 32 ATP for every one glucose molecule. Two ATP are produced in glycolysis, two from the citric acid cycle, and 28 from the respiratory chain.

The general formula for aerobic cellular respiration is:

$$C_6H_{12}O_6 + 6\,O_2 \rightarrow 6\,O_2 + 6\,H_2O + 32\text{ ATP}$$

Question 8. Draw a diagram of glycolysis and cellular respiration that includes the following steps: glucose, pyruvate, pyruvate oxidation, citric acid cycle, electron transport, and ATP synthesis.

Question 9. Draw the general formula for cellular respiration and fermentation, in each case starting from a simple sugar molecule. Compare the energy yields of both processes.

Question 10. Cyanide kills by inhibiting cytochrome oxidase in the mitochondria so that oxygen can no longer be utilized and the electron transport chain is halted. However, many cells in the human body are capable of lactic acid fermentation. Since cyanide does not inhibit glycolysis and fermentation, what explains cyanide's lethal effect in humans?

Question 11. What accounts for the presence of bubbles in beer?

9.5 How Are Metabolic Pathways Interrelated and Regulated?

Catabolism and anabolism are linked

Catabolism and anabolism are integrated

Metabolic pathways are regulated systems

Glycolysis and cellular respiration occur while other metabolic processes are occurring. Components of the metabolic pathways are also components of other metabolic processes, including catabolism and anabolism.

Catabolism is the breaking down of molecules to release energy. Polysaccharides, lipids, and proteins all feed into the metabolic pathways at different points. Polysaccharides are broken down into glucose and enter at glycolysis. Lipids are broken down and can enter at glycolysis or the citric acid cycle as acetyl CoA. Proteins enter glycolysis and the citric acid cycle as amino acids. Just as intermediates can be broken down (catabolized) to release energy, they can also be synthesized in anabolism for storage or use by the cell.

Metabolism—both anabolism and catabolism—is regulated by enzymes to maintain stable concentrations of intermediates for metabolic homeostasis. Metabolic enzymes are allosterically controlled to regulate the production of ATP. Metabolism responds to both positive and negative feedback provided by intermediates and end products of the process. Each process has a specific control point for regulation. The control point for glycolysis is phosphofructokinase, which is inhibited by ATP. This allows glycolysis to speed up during fermentation and slow down during cellular respiration. The main control point for the citric acid cycle is isocitrate dehydrogenase. NADH + H^+ and ATP inhibit this enzyme to slow down the process, and NAD^+ and ADP act as activators. If the citric acid cycle slows due to abundant ATP, glycolysis slows as well. If excess acetyl CoA is produced and ATP is abundant, acetyl CoA can be shuttled to fatty acid synthesis for storage.

Question 12. The fate of acetyl CoA differs according to how much ATP is present in the cell. Explain what happens to acetyl CoA when ATP is limited, and compare this to what happens when acetyl CoA is abundant.

Question 13. Cellular respiration occurs simultaneously with many other cellular processes. Draw a figure that shows, in general, how cellular respiration interacts with other cellular metabolic events.

Question 14. During exercise, catabolic and anabolic processes interact to maintain the glucose level in the blood stream. Explain how.

Question 15. Explain why humans may become overweight if they eat too much.

Test Yourself

1. Which of the following cellular metabolic processes is active in all cells, regardless of the presence or the absence of oxygen?
   a. The citric acid cycle
   b. Electron transport
   c. Glycolysis
   d. Fermentation
   e. Pyruvate oxidation
   Textbook Reference: *9.1 How Does Glucose Oxidation Release Chemical Energy?*

2. Which of the following statements regarding glycolysis is *false*?
   a. A 6-C sugar is broken down to two 3-C molecules.
   b. Two ATP molecules are consumed.
   c. It requires oxygen.
   d. A net sum of two ATP molecules is generated.
   e. It occurs in the cytosol.
   Textbook Reference: *9.2 What Are the Aerobic Pathways of Glucose Catabolism?*

3. During which process is most ATP generated in the cell?
   a. Glycolysis
   b. The citric acid cycle
   c. Electron transport coupled with chemiosmosis
   d. Fermentation
   e. Pyruvate oxidation
   Textbook Reference: *9.4 How Is Energy Harvested from Glucose in the Absence of Oxygen?*

4. One purpose of the electron transport chain is to
   a. cycle NADH + H^+ back to NAD^+.
   b. use the intermediates from the citric acid cycle.
   c. break down pyruvate.
   d. increase the number of protons in the mitochondrial matrix.
   e. consume excess ATP.
   Textbook Reference: *9.3 How Does Oxidative Phosphorylation Form ATP?*

5. Cellular respiration is allosterically controlled. Which of the following acts as an inhibitor at the various control points?
   a. ATP
   b. NADH
   c. Both ATP and NADH
   d. ADP
   e. Both ADP and NADH
   Textbook Reference: *9.5 How Are Metabolic Pathways Interrelated and Regulated?*

6. Which of the following statements about the inner mitochondrial membrane is true?
   a. It acts as an anchor for the membrane-associated enzymes of cellular respiration.
   b. It allows for the establishment of a proton gradient.
   c. It separates the mitochondria's environment from that of the cytosol.
   d. It anchors enzymes, allows for the establishment of the proton gradient, and is involved in separating the contents of the mitochondria from the cytosol.
   e. It anchors enzymes and allows for the establishment of the proton gradient, but it is not involved in separating the contents of the mitochondria from the cytosol.
   Textbook Reference: *9.3 How Does the Oxidative Phosphorylation Form ATP?*

7. In the following redox reaction, _______ is oxidized and _______ is reduced.

   Glyceraldehyde 3-phosphate (G3P) + NAD^+ + H^+ + P_i → 1,3-Bisphosphoglycerate (BPG) + NADH

a. G3P; NAD^+
b. BPG; $NADH + H^+$
c. G3P; $NADH + H^+$
d. NAD^+; $NADH + H^+$
e. The equation does not show a redox reaction.
Textbook Reference: *9.2 What Are the Aerobic Pathways of Glucose Catabolism?*

8. Which of the following statements about redox reactions is true?
a. Oxidizing agents accept electrons.
b. Oxidizing agents donate electrons.
c. A molecule that accepts electrons is said to be oxidized.
d. A molecule that donates electrons is said to be reduced.
e. Oxidizing agents accept electrons and are reduced in the process.
Textbook Reference: *9.1 How Does Glucose Oxidation Release Chemical Energy?*

9. If cyanide poisoning inhibits aerobic respiration at cytochrome *c* oxidase, which of the following should *not* be a result of cyanide poisoning at the cellular level?
a. Reduction of oxygen to water
b. Cessation of ATP synthesis in the mitochondria, because electron transport is never completed
c. Switching of cells to anaerobic respiration and fermentation if possible
d. Continuation of glycolysis as long as NAD^+ is available
e. Less acidic pH of the intermembrane space
Textbook Reference: *9.3 How Does Oxidative Phosphorylation Form ATP?*

10. Which of the following is matched correctly with its catabolic product?
a. Polysaccharides → amino acids
b. Lipids → glycerol and fatty acids
c. Proteins → glucose
d. Polysaccharides → glycerol and fatty acids
e. Nucleic acids → monosaccharides
Textbook Reference: *9.5 How Are Metabolic Pathways Interrelated and Regulated?*

11. The main function of cellular respiration is the
a. conversion of energy stored in the chemical bonds of glucose to an energy form that the cell can use.
b. recovery of NAD^+ from NADPH.
c. conversion of kinetic to potential energy.
d. creation of energy in the cell.
e. elimination of excess glucose from the cell.
Textbook Reference: *9.1 How Does Glucose Oxidation Release Chemical Energy?*

12. Which of the following statements concerning the synthesis of ATP in the mitochondria is *false*?
a. ATP synthesis cannot occur without the presence of ATP synthase.
b. The proton-motive force is the establishment of a charge and concentration gradient across the mitochondrial membrane.
c. The proton-motive force drives protons back across the membrane through channels established by the ATP synthase channel protein.
d. The ATP synthase protein is composed of two units.
e. The intermembrane space is more basic than the mitochondrial matrix.
Textbook Reference: *9.3 How Does Oxidative Phosphorylation Form ATP?*

13. Which of the following does *not* occur in the mitochondria of eukaryotic cells?
a. Fermentation
b. Oxidative phosphorylation
c. The citric acid cycle
d. The electron transport chain
e. The creation of a proton gradient
Textbook Reference: *9.1 How Does Glucose Oxidation Release Chemical Energy?*

14. Which of the following is recycled and reused in cellular metabolism?
a. ADP
b. NAD^+
c. FAD
d. P_i
e. All of the above
Textbook Reference: *9.2 What Are the Aerobic Pathways of Glucose Catabolism?*

15. For each molecule of glucose, how many ATPs are synthesized in fermentation?
a. 0
b. 1
c. 2
d. 3
e. 4
Textbook Reference: *9.4 How Is Energy Harvested from Glucose in the Absence of Oxygen?*

16. If additional malate is added to a cell undergoing cellular respiration, there will be
a. an increase in CO_2 production but no increase in ATP synthesis.
b. an increase in CO_2 production and a decrease in ATP synthesis.
c. an increase in both CO_2 production and ATP synthesis.
d. a decrease in both CO_2 production and ATP synthesis.
e. no change in the rates of CO_2 production or ATP synthesis.
Textbook Reference: *9.2 What Are the Aerobic Pathways of Glucose Catabolism?*

Answers

Key Concept Review

1.
   a. Letter "A" represents the reducing agent and letter "B" represents the oxidizing agent because the electrons that "A" loses are transferred to "B."
   b. At the end of the reaction, "A" is the oxidized compound because it has lost electrons and "B" is reduced because it gained electrons.
   c. "A" is glucose, and "B" is O_2.
2. In eukaryotic cells both glycolysis and fermentation take place in the cytoplasm, while the rest of the reactions occur inside the mitochondria. In prokaryotic cells the cytoplasm is the location not only of glycolysis and fermentation but also of the citric acid cycle.
3. See Figures 9.5 and 9.6. Glycolysis begins with a 6-C sugar, and two phosphate groups from two ATP molecules are added to the 6-C sugar to produce the 5-C fructose-1,6-biphosphate (FBP). Then the carbon ring splits into two 3-C phosphorylated molecules: glyceraldehyde 3-phosphate (G3P) and 1,3-bisphosphoglycerate (BPG). The end product of glycolysis is pyruvate. Pyruvate can oxidize to become acetate, which is combined with coenzyme A to form acetyl CoA. The citric acid cycle then begins. CO_2 is produced when the cycle moves from a 6-C to a 5-C molecule, and again at the step from a 5-C to a 4-C molecule.
4. The energy yields are: 6 CO_2, 10 NADH, 2 $FADH_2$, 4 ATP
5. The carbon in the sugar is exhaled in the form of CO_2. The oxidation of one glucose molecule yields 6 CO_2 molecules, therefore all 6 carbon atoms of the sugar are exhaled.
6. Oxygen acts as the terminal electron acceptor in the electron transport pathway. Without it, NADH + H^+ cannot be cycled back to NAD^+. The accumulated NADH + H^+ acts as an inhibitor to the citric acid cycle and effectively shuts it down. Therefore, in the absence of oxygen, a cell can only undergo glycolysis.
7. The proton-motive force results in a concentration and charge gradient across the mitochondrial membrane. For that gradient to equalize, the protons must flow through a channel protein. If the channel protein has an associated ATP synthase, ATP is generated as protons flow through.
8. See also Figure 9.4

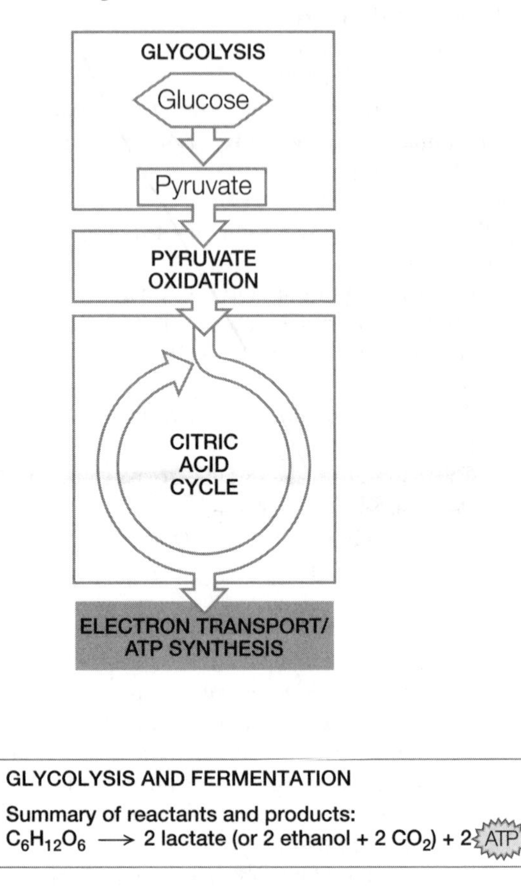

9.

GLYCOLYSIS AND FERMENTATION
Summary of reactants and products:
$C_6H_{12}O_6 \longrightarrow$ 2 lactate (or 2 ethanol + 2 CO_2) + 2 ATP

GLYCOLYSIS AND CELLULAR RESPIRATION
Summary of reactants and products:
$C_6H_{12}O_6 + 6\ O_2 \longrightarrow 6\ CO_2 + 6\ H_2O$ + 32 ATP

10. The reduced efficiency of glycolysis and fermentation for ATP synthesis is one reason that oxygen deprivation is so deadly to humans. (And, as later chapters will show, the resulting lactic acid buildup also causes problems.) In addition, not all cells are capable of carrying out the reactions of fermentation. Brain cells, for example, will die in the absence of oxygen.
11. The bubbles in beer are CO_2 molecules released during the fermentation of pyruvate into ethyl alcohol. See Figure 9.11.
12. If ATP is limited, acetyl CoA enters the citric acid cycle and cellular respiration utilizes it to produce ATP. If ATP is abundant, acetyl CoA is shuttled to fatty acid synthesis, thus storing the energy in chemical bonds.

13.

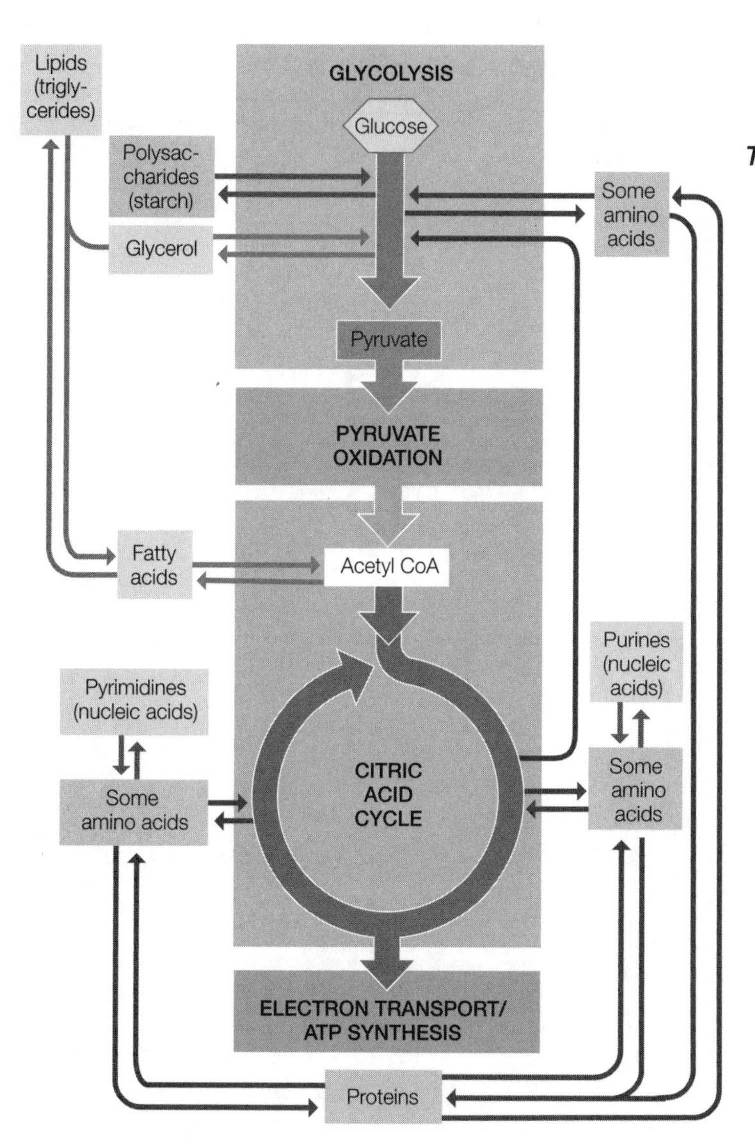

14. Consult Figure 9.14 to see how and where anabolism and catabolism interact in the human body. For example, when a person is exercising, there may be two major organs that most require energy: the leg muscles to power movement, and the heart muscles to circulate blood. This energy can come from the metabolism of glucose during glycolysis, the citric acid cycle, and oxidative phosphorylation. In the leg muscles, there is some stored glycogen, which is hydrolyzed to glucose monomers, while in the liver, glycogen is hydrolyzed and released into the blood as glucose. As glucose is used up by the working muscles, more glucose is made in the liver by anabolism from amino acids and pyruvate. Some of the pyruvate used in this anabolism comes from lactate that was made by fermentation in the leg muscles and then transported back to the liver in the blood.

15. If the level of ATP is high and the citric acid cycle shuts down, the accumulation of citrate activates fatty acid synthase, diverting acetyl CoA to the synthesis of fatty acids for storage. This is one reason why people who overeat accumulate fat.

Test Yourself

1. **c.** Glycolysis proceeds during both fermentation and cellular respiration. Only in cellular respiration is oxygen needed as the terminal electron acceptor of the pathway.
2. **c.** During glycolysis, 6-C glucose is broken down into two 3-C pyruvate molecules. In the process, four total ATP are produced, but two are consumed, leaving a net production of two ATP molecules. No oxygen is required in glycolysis.
3. **c.** Most of the ATP produced during cellular respiration is produced by electron transport and chemiosmosis coupled in oxidative phosphorylation.
4. **a.** The electron transport chain is responsible for oxidizing NADH + H^+ back to NAD^+.
5. **c.** Both ATP and NADH allosterically control metabolism. ATP controls both phosphofructokinase and isocitrate dehydrogenase, which are commitment steps for glycolysis and the citric acid cycle, respectively. NADH controls isocitrate dehydrogenase.
6. **e.** The mitochondrial membrane is necessary for the anchoring of proteins as well as the establishment of a barrier across which a gradient can be established.
7. **a.** A molecule is oxidized when it loses electrons or protons and is reduced when it gains electrons or protons. In this reaction, G3P donates electrons and therefore is oxidized, while NAD^+ accepts them and thus is reduced.
8. **e.** Oxidizing agents accept electrons and cause oxidation of another molecule. Reducing agents donate electrons and cause the reduction of another molecule.
9. **a.** Cyanide stops aerobic cellular respiration because cytochrome *c* oxidase loses the ability to reduce oxygen to water. Those cells that can switch to anaerobic respiration (fermentation) do so and use their remaining glucose more quickly. Cells that cannot make that switch die.
10. **b.** Lipids are broken down into glycerol and fatty acids; polysaccharides are broken down into glucose; proteins are broken down into amino acids. Nucleic acids are converted into some amino acids and fed into the citric acid cycle.
11. **a.** Cellular respiration is the cell's way of converting potential energy in the chemical bonds of glucose to potential energy that the cell ultimately can use.

12. **a.** Substrate level phosphorylation occurs in step 5 of the citric acid cycle in the mitochondria. The intermembrane space is more acidic than the matrix.
13. **a.** Fermentation occurs in the cytosol, whereas all the other processes occur in the mitochondria of eukaryotic cells.
14. **e.** ADP, NAD^+, FAD, and P_i are all recycled and reused in the process of cellular respiration.
15. **a.** Fermentation regenerates NAD^+ so that glycolysis can continue.
16. **c.** Additional malate will increase the carbon compounds cycling through the citric acid cycle, resulting in an increase in the products of the cycle, including ATP and CO_2.

10 Photosynthesis: Energy from Sunlight

The Big Picture

- Photosynthesis allows organisms with the appropriate pigments and metabolic processes to convert light energy from the sun to chemical energy that can be used in the cell or stored. This conversion process forms the basis of all food webs and is vital to life on Earth. Photosynthetic organisms fix CO_2 into sugars. This process requires water and light energy and produces the byproducts O_2 and water.
- The process of photosynthesis occurs in two steps. The first step utilizes light energy to produce ATP and NADPH with the help of the electron transport system. The movement of electrons through the electron transport system generates cellular energy in the form of ATP that can be used in the fixing of CO_2. The fixing of CO_2 occurs in the Calvin cycle and is mediated by an enzyme called rubisco, and the reactions are CO_2 concentration dependent.
- Rubisco can also act as an oxygenase and allow a plant to go through a metabolically expensive process called photorespiration. Plants that grow in conditions favoring photorespiration have evolved a wide variety of systems to avoid it.

Study Strategies

- The process of photosynthesis can be very difficult to understand, and it helps to visualize what occurs in this process. Refer to the figures in the textbook to help you understand where energy and carbon are flowing. Remember that the plant is taking light energy and converting it to chemical energy that can be stored. In the process it uses CO_2 and gives off O_2. Pay attention to where energy is flowing and how this process proceeds.
- The properties of chlorophyll and light greatly affect how well a plant photosynthesizes. Take some time to understand the properties of light and how pigments interact with light.
- Be careful not to confuse the pathways for glycolysis, fermentation, and cellular respiration (see Chapter 9) with the pathways of photosynthesis. There are connections between the two pathways (see Figure 10.18), but they are not interchangeable.
- There is the temptation to learn pathways without understanding why they occur. Be sure to focus on why the plant has a particular pathway before attempting to learn the pathway itself.
- Remember that plants respire as well as photosynthesize and that photorespiration and respiration are different processes. All plant cells go through cellular respiration, which takes the energy stored in photosynthesis and makes it available to drive other cellular processes. Photorespiration is energy expensive and is of little benefit to the plant. Photosynthesis occurs in specialized photosynthetic cells at the same time as respiration.
- Rubisco is the most abundant protein on the planet. Focus on how the relative concentrations of O_2 and CO_2 dictate how the enzyme functions. Also make sure you understand that C_4 and CAM plants have modifications to avoid photorespiration.
- It is very common to think that the light reactions happen in the light and the Calvin cycle occurs in the dark. This is not the case! The light reactions must occur simultaneously with the Calvin cycle to supply the needed energy in the form of ATP and NADPH + H^+.
- Do not simply memorize a process; make sure that you also understand it. It always helps to draw a figure of your own, first by copying a figure from the book and later recalling it and drawing it on your own (with or without all of the details). This will help you create a strong framework that you can build on.
- Go to Web addresses shown to review the following animated tutorials and activities. You can also find them in BioPortal (yourBioPortal.com), along with many additional learning resources.

  Animated Tutorial 10.1 The Source of the Oxygen Produced by Photosynthesis (Life10e.com/at10.1)

  Animated Tutorial 10.2 Photophosphorylation (Life10e.com/at10.2)

  Animated Tutorial 10.3 Tracing the Pathway of CO_2 (Life10e.com/at10.3)

  Activity 10.1 The Calvin Cycle (Life10e.com/ac10.1)

  Activity 10.2 C_3 and C_4 Leaf Anatomy (Life10e.com/ac10.2)

Key Concept Review

10.1 What Is Photosynthesis?

Experiments with isotopes show that O_2 comes from H_2O in oxygenic photosynthesis

Photosynthesis involves two pathways

The photosynthetic production of O_2 by green plants is necessary for most organisms to obtain the energy for life.

Photosynthesis uses CO_2, water, and light to produce carbohydrate and O_2 (see Figure 10.1). The reaction is as follows:

$$6\ CO_2 + 12\ H_2O \rightarrow C_6H_{12}O_6 + 6\ O_2 + 6\ H_2O$$

Light from the sun is required for photosynthesis. In plants, photosynthesis occurs in chloroplasts of the leaves. Water for photosynthesis must be acquired by the roots and transported to the leaves. CO_2, O_2, and water vapor are exchanged through openings, or stomata, in the leaf's surface.

Photosynthesis occurs in a two-pathway process. The light reactions produce ATP from light energy. The light-independent reactions use ATP and NADPH + H^+ produced by the light reactions to trap CO_2 and produce sugars that act as an energy store. Neither pathway operates in the absence of light.

While in oxygenic photosynthesis water is the donor of protons and electrons, in anoxygenic photosynthesis other molecules, such as hydrogen sulfide, sulfide ions, or ferrous ion play this role.

Question 1. What is the general equation of oxygenic photosynthesis?

Question 2. What is the source of the O_2 produced during photosynthesis? Briefly explain the experiment that led to this discovery.

Question 3. Does photosynthesis always produce oxygen?

10.2 How Does Photosynthesis Convert Light Energy into Chemical Energy?

Light energy is absorbed by chlorophyll and other pigments

Light absorption results in photochemical change

Reduction leads to ATP and NADPH formation

Chemiosmosis is the source of the ATP produced in photophosphorylation

The interaction of light and pigment is fundamental to photosynthesis, which involves photochemistry and photobiology. The photochemistry of light depends on its wavelength. Photochemistry is based on the fact that light is a form of electromagnetic radiation that comes in discrete packets called photons and has wavelike properties. Photons can be scattered, transmitted, or absorbed. For light to be biologically available, it must be absorbed by the receptive molecule and contain enough energy to carry out chemical work. When a molecule absorbs a proton, it goes from a ground state to an excited state.

The photobiology of light is based on the use of pigments. Pigments are molecules that absorb wavelengths in the visible spectrum. Compounds have unique absorption spectra, depending on which wavelengths of light are absorbed (see Figure 10.5). The chlorophyll pigment appears green because blue and red light are absorbed and green light is reflected.

An action spectrum can be measured that relates biological activity to particular wavelengths. In plants, photosynthesis relies on chlorophyll *a*, chlorophyll *b*, and accessory pigments. Chlorophylls are more abundant, but they can only absorb photons with blue or orange-red wavelengths. Accessory pigments, such as carotenoids and phycobilins, absorb photons that chlorophylls cannot and transfer energy to the chlorophylls. Light energy is captured by the light-harvesting complexes and transferred to the reaction center, where chlorophyll *a* molecules participate in redox reactions that result in the conversion of the light energy to chemical energy:

$$\text{Chl}^* + \text{acceptor} \rightarrow \text{Chl}^+ + \text{acceptor}^-$$

In photosynthesis, a pigment molecule in the excited state passes the absorbed energy along. Photosynthetic antennae systems function to pass the light energy from one pigment to another until the reaction center is reached, where light energy is converted to chemical energy. In plants, the reaction center is chlorophyll *a*. A ground-state chlorophyll *a* molecule at the reaction center (Chl) absorbs the energy from the adjacent chlorophylls and becomes excited (Chl*). It passes the excited electron to a chemical acceptor, and becomes oxidized (Chl^+). The excitation energy is passed from one pigment that absorbs at a shorter wavelength to another pigment that absorbs at a longer wavelength. Excited chlorophyll acts as a reducing agent, resulting in electron transport.

Noncyclic electron flow requires two photosystems in the thylakoid membrane to continuously absorb light. Photosystem II takes slightly higher energy (680 nm) than photosystem I (700 nm). Water is oxidized to form O_2, H^+, and electrons. Electrons move from water to chlorophyll, through electron carriers, and ultimately to NADP$^+$. Photosystem II passes electrons to photosystem I (see Figure 10.8). Noncyclic electron flow results in the production of ATP and NADPH + H^+. The electron transport pumps protons actively into the thylakoid compartment. This proton gradient powers the formation of ATP. Thus, noncyclic flow produces equal amounts of ATP and NADPH + H^+.

Cyclic electron transport involves only photosystem I and cycles back to the same chlorophyll molecule (see Figure 10.9). The amount of NADPH + H^+ present regulates whether cyclic or noncyclic electron flow occurs. If NADPH + H^+ is abundant, electrons are accepted by plastoquinone, which pumps two protons back across the thylakoid membrane. The light-independent reactions require more ATP than NADPH + H^+. Cyclic electron flow supplies the additional ATP by producing ATP without producing NADPH + H^+.

ATP synthesis in either cyclic or noncyclic electron flow is produced via chemiosmosis. Just as in the mitochondria, a proton-motive force coupled to ATP synthase is established in the chloroplast as electrons are passed along the transport chain. This results in the photophosphorylation of ADP to ATP.

The thylakoid membrane has an electron transport system similar to the respiratory chain of mitochondria, and photophosphorylation looks very similar to the chemiosmotic mechanism for ATP formation in the mitochondrion.

Question 4. Distinguish between cyclic and noncyclic electron flow. Why are both processes necessary?

Question 5. Why are plants green?

Question 6. Label the photosystems, pathways, reactants, and products in the diagram below. Include both electron pathways in your answer.

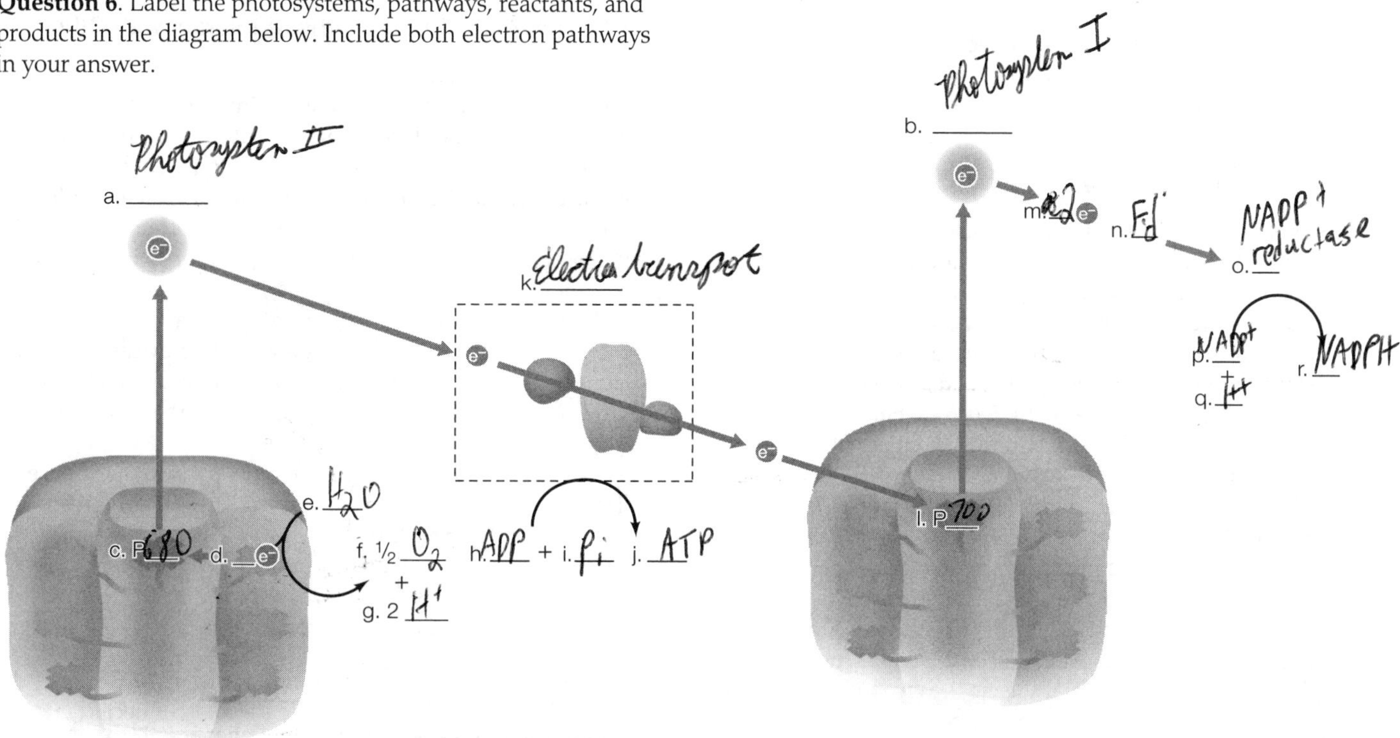

10.3 How Is Chemical Energy Used to Synthesize Carbohydrates?

Radioisotope labeling experiments revealed the steps of the Calvin cycle

The Calvin cycle is made up of three processes

Light stimulates the Calvin cycle

The Calvin cycle occurs in the stroma of the chloroplasts. In the Calvin cycle, ATP and NADPH from the light reactions are used to convert CO_2 to sugars for storage. Because ATP and NADPH are energy-rich coenzymes that cannot be stockpiled, the Calvin cycle reactions occur only in the light when ATP and NADPH can be made.

The cycle was revealed using radiolabeled CO_2 and observing where it was incorporated. Following this process, researchers revealed the pathway by which CO_2 is converted to sugar (see Figure 10.11).

The Calvin cycle is made up of three processes: fixation of CO_2, reduction of 3PG, and regeneration of CO_2. CO_2 is combined with ribulose 1,5-bisphosphate (RuBP), which is then split into two molecules of 3-phosphoglycerate (3PG) by an enzyme called rubisco. Rubisco is the world's most abundant protein. It regulates the Calvin cycle. ATP and NADPH + H^+ from the light reactions are then used to convert the fixed CO_2 in 3PG into carbohydrate (glyceraldehyde 3-phosphate, or G3P). Finally, ATP is used to regenerate RuBP so that additional CO_2 may be fixed. The resulting carbohydrate (G3P) is converted to starch or sucrose, either for storage or for conversion to glucose and fructose for cellular fuel.

Light stimulates the Calvin cycle by inducing pH changes that result in the activation of Calvin cycle enzymes and by reducing disulfide bonds on four of the Calvin cycle enzymes.

Rubisco also mediates the process of photorespiration, since it can act as an oxygenase and fix O_2 at the expense of CO_2 to produce phosphoglycolate. Phosphoglycolate enters membrane-enclosed peroxisomes and is converted to glycine. The glycine is then converted into glycerate and CO_2 in the mitochondrion. This process is known as photorespiration. ATP and NADPH from the light reaction are used in photorespiration, but CO_2 is released instead of being used to make carbohydrate.

Question 7. Why do plants undergo both the light reactions of photosynthesis and the Calvin cycle? Why don't they simply use the ATP produced in the light reactions of photosynthesis to drive cellular processes?

Question 8. What are the three steps of the Calvin cycle? How many ATP molecules are required for each cycle, and which steps use them?

10.4 How Have Plants Adapted Photosynthesis to Environmental Conditions?

Rubisco catalyzes the reaction of RuBP with O_2 or CO_2

C_3 plants undergo photorespiration but C_4 plants do not

CAM plants also use PEP carboxylase

Whether CO_2 or O_2 is fixed depends on relative concentrations of CO_2 and O_2. Excess O_2 (abundant, for example, on hot, dry days) forces photorespiration. Photorespiration is metabolically expensive, so plants have evolved mechanisms to avoid it. To avoid photorespiration, the level of CO_2 around rubisco must be high. This is difficult to achieve because O_2 levels in the air are much higher than CO_2 levels. If hot conditions force the closing of stomata, CO_2 levels are quickly depleted.

Plants such as roses, wheat, and rice that fix CO_2 to 3PG are called C_3 plants and are very sensitive to CO_2/O_2 levels. Photorespiration occurs frequently and limits the range of conditions under which C_3 plants can grow.

Many tropical plants such as corn, sugarcane, and other tropical grasses are C_4 plants. They fix CO_2 to the acceptor phosphoenolpyruvate (PEP) using the enzyme PEP carboxylase to form oxaloacetate. PEP carboxylase has no oxygenase activity and fixes CO_2 at very low levels. The resulting four-carbon compound diffuses to the bundle sheath cells in the interior of the leaf, releases CO_2, and is recycled. The released CO_2 goes through the usual Calvin cycle, including rubisco. The early fixation process is a CO_2-concentrating mechanism around rubisco.

Cacti, pineapples, and other succulents undergo crassulacean acid metabolism (CAM), which separates CO_2 fixation and the Calvin cycle. Plants that carry out CAM open their stomata only at night and store CO_2 for use during the day when light is present. CO_2 is fixed as oxaloacetate and converted to malic acid. CO_2 is stored in malic acid until it is transferred to the chloroplasts or photosynthetic cells when light is available.

Currently, the level of CO_2 is not enough for maximal CO_2 fixation by rubisco, so photorespiration occurs, reducing the growth rates of C_3 plants. But if human-spurred CO_2 levels in the atmosphere continue to rise, C_3 plants will have a comparative advantage. The overall growth rates of crops such as rice and wheat should increase.

Question 9. Rubisco has both carboxylase and oxygenase activities. These processes compete with each other. What determines which function the enzyme has? What conditions favor photorespiration? What conditions favor photosynthesis?

Question 10. Compare and contrast photosynthesis in C_3, C_4, and CAM plants.

Question 11. Humans are responsible for the increase of atmospheric CO_2 levels (e.g., through the burning of fossil fuels). Explain how this may affect the ratio of C_3 and C_4 plants. How is this related to the evolution of these plants?

10.5 How Does Photosynthesis Interact with Other Pathways?

Metabolic pathways in green plants involve photosynthesis and respiration. Because plants are autotrophic, they can synthesize all necessary molecules to survive from CO_2, H_2O, phosphate, sulfate, and NH_4. All cells, whether photosynthetic or not, go through respiration to produce the ATP necessary for cellular processes, and they do so in the light and the dark.

Energy flows from sunlight to reduced carbon in photosynthesis, then to ATP in respiration. Energy can also be stored in polysaccharides, lipids, or proteins. For a plant to grow, overall carbon fixation by photosynthesis must exceed respiration.

The Calvin cycle and the respiratory pathways are closely linked. G3P can be converted to pyruvate and enter respiration, or it can enter the gluconeogenic pathway to form sucrose. In order for plants to grow, the energy stored must exceed the energy used in respiration. Figure 10.18 illustrates how plant metabolism is interconnected.

Question 12. Draw a diagram showing how the Calvin cycle, glycolysis, and the citric acid cycle are interconnected in plants. Label the cycles, pathways, reactants, and products in the diagram. As you work, focus on the connections between the two systems. Note how the products of one system provide the raw materials for the other system.

Question 13. Evaluate how efficient photosynthesis is. What percentage of the sunlight that reaches Earth is converted into plant growth? Where does the energy get lost?

Test Yourself

1. The main function of photosynthesis is the
   a. consumption of CO_2.
   b. production of ATP.
   c. conversion of light energy to chemical energy.
   d. production of starch.
   e. production of O_2.
   Textbook Reference: *10.1 What Is Photosynthesis?*

2. Which of the following best represent the components that are necessary for photosynthesis to take place?
   a. Mitochondria, accessory pigments, visible light, water, and CO_2
   b. Chloroplasts, accessory pigments, visible light, water, and CO_2

c. Mitochondria, chlorophyll, visible light, water, and O_2
d. Chloroplasts, chlorophyll, visible light, water, and CO_2
e. Chlorophyll, accessory pigments, visible light, water, and O_2
Textbook Reference: *10.1 What Is Photosynthesis?*

3. Chlorophyll is suited for the capture of light energy because
a. certain wavelengths of light raise it to an excited state.
b. in its excited state it gives off electrons.
c. its structure allows it to attach to thylakoid membranes.
d. it can transfer absorbed energy to another molecule.
e. All of the above
Textbook Reference: *10.2 How Does Photosynthesis Convert Light Energy into Chemical Energy?*

4. Plants give off O_2 because
a. O_2 results from the incorporation of CO_2 into sugars.
b. they do not respire; they photosynthesize.
c. water is the initial electron donor, leaving O_2 as a photosynthetic by-product.
d. electrons moving down the electron chain bind to water, releasing O_2.
e. O_2 is synthesized in the Calvin cycle.
Textbook Reference: *10.2 How Does Photosynthesis Convert Light Energy into Chemical Energy?*

5. Cyclic and noncyclic electron flow are used in plants to
a. meet the ATP demands of the Calvin cycle.
b. produce excess NADPH + H^+.
c. synthesize proportional amounts of ATP and NADPH + H^+ in the chloroplast.
d. consume the products of the Calvin cycle.
e. produce O_2 for the atmosphere.
Textbook Reference: *10.3 How Is Chemical Energy Used to Synthesize Carbohydrates?*

6. Which of the following statements concerning the light reactions of photosynthesis is true?
a. Photosystem I cannot operate independently of photosystem II.
b. Photosystems I and II are activated by different wavelengths of light.
c. Photosystems I and II transfer electrons and create proton equilibrium across the thylakoid membrane.
d. Photosystem I is more significant than photosystem II.
e. Oxygen gas is a product of photosystem I.
Textbook Reference: *10.2 How Does Photosynthesis Convert Light Energy into Chemical Energy?*

7. ATP is produced during the light reactions via
a. CO_2 fixation.
b. chemiosmosis.
c. reduction of water.
d. glycolysis.
e. noncyclic electron flow from photosystem I.
Textbook Reference: *10.2 How Does Photosynthesis Convert Light Energy into Chemical Energy?*

8. Because of the properties of chlorophyll, plants need adequate _______ light to grow properly.
a. green
b. blue and red
c. infrared
d. ultraviolet
e. blue and blue-green
Textbook Reference: *10.2 How Does Photosynthesis Convert Light Energy into Chemical Energy?*

9. Which of the following statements concerning the Calvin cycle is *false*?
a. Light energy is not required for the cycle to proceed.
b. CO_2 is assimilated into sugars.
c. RuBP is regenerated.
d. It uses energy stored in ATP and NADPH + H^+.
e. All of the above
Textbook Reference: *10.3 How Is Chemical Energy Used to Synthesize Carbohydrates?*

10. Which of the following statements concerning rubisco is true?
a. Rubisco is a carboxylase.
b. Rubisco preferentially binds to O_2 over CO_2.
c. Rubisco is absent from C_4 and CAM plants.
d. Rubisco catalyzes the splitting in water to release O_2.
e. Rubisco is only found in a few plant species.
Textbook Reference: *10.3 How Is Chemical Energy Used to Synthesize Carbohydrates?; 10.4 How Have Plants Adapted Photosynthesis to Environmental Conditions?*

11. Which of the following begins the Calvin cycle that results in the entire pathway being carried out under environmental conditions?
a. 3PG is reduced to G3P using ATP and NADPH + H^+.
b. RuBP is regenerated.
c. CO_2 and RuBP join, forming 3PG.
d. G3P is converted into glucose and fructose.
e. Any of the above can initiate the cycle, since a cycle can start at any point.
Textbook Reference: *10.3 How Is Chemical Energy Used to Synthesize Carbohydrates?*

12. The Calvin cycle results in the production of
a. glucose.
b. starch.
c. rubisco.
d. G3P.
e. ATP.
Textbook Reference: *10.3 How Is Chemical Energy Used to Synthesize Carbohydrates?*

13. Which of the following statements regarding photorespiration is true?
   a. Photorespiration is a metabolically expensive pathway.
   b. Photorespiration is avoided when CO_2 levels are low.
   c. Photorespiration increases the overall CO_2 that is converted to carbohydrates.
   d. Photorespiration increases by 75 percent the net carbon that is fixed.
   e. Photorespiration is most common in C_4 plants.
   Textbook Reference: *10.4 How Have Plants Adapted Photosynthesis to Environmental Conditions?*

14. The fixation of CO_2 by PEP carboxylase functions to
   a. concentrate O_2 for use in photosynthetic cells.
   b. allow plants to close stomata without the occurrence of photorespiration.
   c. allow plants to photosynthesize in the dark.
   d. reduce water loss by the plant.
   e. All of the above
   Textbook Reference: *10.4 How Have Plants Adapted Photosynthesis to Environmental Conditions?*

15. CAM plants differ from C_4 plants in that
   a. photosynthesis can occur at night in CAM plants.
   b. CO_2 is stored in CAM plants as malic acid.
   c. the stomata of CAM plants close during periods that favor photorespiration.
   d. CAM plants use PEP carboxylase to fix CO_2.
   e. the Calvin cycle is only found in C_4 and C_3 plants, not in CAM plants.
   Textbook Reference: *10.4 How Have Plants Adapted Photosynthesis to Environmental Conditions?*

16. Which of the following statements regarding the relationship between photosynthesis and cellular respiration in plants is true?
   a. Photosynthesis occurs in specialized photosynthetic cells.
   b. Cellular respiration occurs in specialized respiratory cells.
   c. Cellular respiration and photosynthesis can occur in the same cell.
   d. Photosynthesis is limited to specialized plant cells and cellular respiration does not occur in plant cells.
   e. Both a and c
   Textbook Reference: *10.5 How Does Photosynthesis Interact with Other Pathways?*

17. Photosynthesis occurs
   a. in all plant cells.
   b. only in photosynthetic plant cells.
   c. only in plant cells lacking mitochondria.
   d. only in the stroma.
   e. only in the thylakoid membrane.
   Textbook Reference: *10.1 What Is Photosynthesis?*

18. Activities such as amino acid synthesis and active transport in plant cells are powered by
   a. the light-dependent and light-independent reactions of photosynthesis.
   b. ATP from the light reactions of photosynthesis.
   c. ATP from fermentation.
   d. ATP from glycolysis and cellular respiration.
   e. All of the above
   Textbook Reference: *10.5 How Does Photosynthesis Interact with Other Pathways?*

Answers

Key Concept Review

1. $6\ CO_2 + 12\ H_2O \rightarrow C_6H_{12}O_6 + 6\ O_2 + 6\ H_2O$
2. In 1941 Samuel Ruben and Martin Kamen performed experiments using the isotopes to identify the source of the oxygen produced during photosynthesis (see Figure 10.2). They labeled the oxygen in these molecules with the isotopes and then tested the oxygen produced by a green plant to find out which molecules contributed the oxygen. Their results showed that all the oxygen gas produced during photosynthesis comes from water.
3. Oxygenic photosynthesis replenishes the oxygen in our atmosphere. But anoxygenic photosynthesis does not produce oxygen because water is not the electron donor. For example, purple sulfur bacteria use hydrogen sulfide (H_2S) as the electron donor. Green sulfur bacteria use sulfide ions, hydrogen, or ferrous iron as electron donors. Another group of bacteria uses compounds derived from arsenic.
4. Noncyclic electron flow involves both photosystems I and II and results in equal amounts of ATP and NADPH being synthesized (see Figure 10.8). However more ATP than NADPH is required for the Calvin cycle. To provide the additional ATP, photosystem I sends electrons to the electron carrier ferredoxin in the electron transport chain, driving ATP synthesis (see Figure 10.9). This cyclic pathway provides the necessary ATP for the Calvin cycle to regenerate RuBP.
5. The primary pigments in plants are chlorophylls. Chlorophylls absorb blue and orange-red wavelengths of light and reflect green light, thus making plants appear green. (See the absorption spectra and action spectra of chlorophyll in Figure 10.5.)
6.

a. Photosystem I	j. ATP
b. Photosystem II	k. Electron transport
c. 680	l. 700
d. 2	m. 2
e. H_20	n. Fd
f. O_2	o. $NADP^+$ reductase
g. H^+	p. $NADP^+$
h. ADP	q. H^+
i. P_i	r. NADPH

7. The light reactions of photosynthesis produce ATP. ATP cannot be stored for use later (such as when light is not available); therefore, there has to be a mechanism for that energy to be stored. The Calvin cycle stores the energy in the chemical bonds of G3P, which can be incorporated into carbohydrates for longer-term storage.
8. The three steps are the (1) fixation of CO_2; (2) the reduction of 3PG to form glyceraldehyde 3-phosphate (G3P); and (3) the regeneration of the CO_2 acceptor, RuBP. All together, 18 ATP molecules are used for each cycle; 12 for phosphorylation during the reduction step and 6 ATP in the conversion of RuMP compound into RuBP during regeneration.
9. Whether rubisco acts as a carboxylase or an oxygenase depends on the relative ratio of O_2 to CO_2 in the leaf. At higher CO_2 levels, it acts as a carboxylase. At low CO_2 levels, it acts as an oxygenase. Photorespiration is favored during hot, dry weather, which forces the closing of stomata and leads to increases in O_2 levels within the leaf. Photosynthesis is favored when stomata can remain open and light intensity is optimal.
10. Refer to Table 10.1 in your textbook.
11. C_3 plants are more ancient than C_4 plants. A possible factor in the emergence of the C_4 pathway was the decline in atmospheric CO_2. If CO_2 levels in the atmosphere continue to rise, the C_3 plants will again have a comparative advantage (just as they did 2.5 billion years ago). The overall growth rates of crops such as rice and wheat should increase.
12. Refer to Figure 10.18 in your textbook.
13. Only 5 percent of the sunlight that reaches Earth is converted into plant growth. The inefficiencies of photosynthesis involve basic chemistry and physics (some light energy is not absorbed by photosynthetic pigments) as well as biology (plant anatomy and leaf exposure, the oxygenase reaction of rubisco, and inefficiencies in metabolic pathways). Around 50 percent of the sunlight is not part of the absorption spectrum, 30 percent is lost because the plant leaves are not properly oriented, 10 percent is lost due to inefficient conversation of light into chemical energy, and 5 percent is lost due to inefficient CO_2 fixation pathways.

Test Yourself

1. **c.** Photosynthetic organisms, including but not limited to plants, are the only life forms capable of trapping light energy and converting it to chemical energy. Because of this they form the basis of many of Earth's food chains.
2. **d.** Chloroplasts are the site of the photosynthetic reactions; chlorophyll is excited by photons of light and serve as reaction centers for the photosystems. Visible light is necessary to excite chlorophyll and accessory pigments. Water is the initial electron donor for the pathway. And CO_2 is necessary to make precursor molecules for energy storage.
3. **e.** The "tails" of chlorophyll molecules are associated with the thylakoid membranes of the chloroplasts. This close membrane association assists with establishing the proton-motive force that will drive ATP synthesis. When excited by light, the chlorophyll moves into an excited state and passes electrons to acceptor molecules. This begins to set up the proton gradient across the membrane that will drive ATP synthesis.
4. **c.** Water is split at photosystem II to donate electrons to the reaction center. The resulting protons are moved across the membrane to establish the proton-motive force, and O_2 is given off as a by-product.
5. **a.** ATP is required at higher levels in the Calvin cycle than NADPH + H^+ is; therefore, there must be a mechanism for producing additional ATP. Cyclic electron flow provides that mechanism. If noncyclic electron flow were to be sped up to meet ATP needs, an excess of NADPH + H^+ would result. Shifting between cyclic and noncyclic flow balances ATP/NADPH + H^+ ratios. Oxygen gas is a by-product of the light reactions, but its production is not the purpose of the reactions.
6. **b.** Photosystem I operates independently of photosystem II during cyclic electron flow. Activity is controlled by the ATP levels in the chloroplast. Photosystem II is activated by light of a higher energy level than photosystem I. Both photosystems transfer electrons and create proton gradients across the thylakoid membranes; photosystem I does this via the cyclic pathway. Water is split by an enzyme embedded in the photosystem II complex.
7. **b.** In the light reactions, ATP synthesis occurs when protons flow through an ATP synthase channel protein in the thylakoid membrane. This is a chemiosmotically driven process. Photosystem II is always involved, whereas photosystem I participates via cyclic electron transport only.
8. **b.** Chlorophyll and accessory pigments absorb light in the blue and red wavelengths of visible light. Green light is reflected; therefore, plants appear green. (Accessory pigments allow energy from additional wavelengths to be absorbed as well.)
9. **a.** Light energy is required for the Calvin cycle to proceed. ATP synthesis is dependent on light energy, and the Calvin cycle is dependent on ATP.
10. **a.** Rubisco, the most abundant enzyme on Earth, has both oxygenase and carboxylase activities. It is present in C_3, C_4, and CAM plants and binds CO_2 with greater affinity than O_2.
11. **c.** The first step of the Calvin cycle is the fixation of CO_2 into 3PG. This is the regulatory step, and it requires ATP and NADPH + H^+. While it is true that the Calvin cycle is a cycle, there is a net consumption of CO_2 for the purpose of building carbohydrates.

12. **d.** The Calvin cycle produces only G3P, which can then be metabolized into storage products like sugars and starch.
13. **a.** Photorespiration uses as much ATP as photosynthesis, but it results in no energy gains for the plant and reduces net carbon fixation by 25 percent compared with the Calvin cycle. If CO_2 is abundant, rubisco acts as a carboxylase rather than an oxygenase.
14. **b.** Plants do not photosynthesize in the dark. PEP carboxylase allows the fixation of CO_2 at low concentrations in the leaf so that it can be sent to rubisco for the Calvin cycle.
15. **b.** CAM plants functionally store CO_2 as malic acid.
16. **e.** Photosynthesis occurs only in plant cells that have the necessary structures, but cellular respiration occurs in every living plant cell that has mitochondria and O_2.
17. **b.** Photosynthesis is limited to photosynthetic plant cells. There are many plant cells that are not exposed to light or that lack chloroplasts; these cells rely on cellular respiration.
18. **d.** Plant cells have mitochondria and rely on the processes of glycolysis and cellular respiration to provide ATP for cellular activities. Photosynthesis converts light energy into potential energy stored in chemical form, but that energy must then be made usable by the cells.

11 The Cell Cycle and Cell Division

The Big Picture

- During the life cycle of a multicellular organism, cells reproduce in a controlled process in both sexually and asexually reproducing species.
- Prokaryotic cells reproduce by a type of cell division known as binary fission.
- In the presence of an internal or external environmental signal, cells undergo a specific set of steps to replicate their DNA, segregate the newly replicated chromosomes, and divide their cytoplasm into two daughter cells. These steps ensure that the daughter cells receive a complete set of genetic instructions.
- The cell cycle is a tightly regulated process involving many regulatory molecules and checkpoints.
- Diploid organisms produce haploid gametes by meiosis, which fuse during fertilization to form new diploid organisms. During the generation of these gametes, crossing over between sister chromatids and random segregation of them creates genetically diverse gametes, leading to increased species diversity and survival.
- Many cells undergo programmed cell death (apoptosis) when they are no longer needed or have accumulated genetic damage that could lead to the development of cancer.
- Cancer cells lose their ability to respond to cell cycle signals.

Study Strategies

- Be sure you understand the differences between mitosis and meiosis. After every DNA replication, there is at least one cell division. In mitosis, the replicated chromosomes are segregated to the two daughter cells, that each have two sets of chromosomes (diploid), like the parent cell. In meiosis, the replicated chromosomes undergo two cell divisions, resulting in four cells that each have only one set of chromosomes (haploid). The homologous chromosomes pair in prophase of meiosis I to allow crossing over to occur.
- It is helpful to work through the processes of mitosis and meiosis with various props. For instance, yarn of different colors, pairs of socks of different colors and design patterns, different flavors of licorice, or Play-Doh "snakes" can represent cells with at least two different chromosome pairs. Build a parent cell and then use additional objects to represent cells at each stage of the cell division process. The hands-on work will allow you to see more clearly what is taking place in the drawings in the book. After you understand the overall process, you can depict it in sequential diagrams. Alternatively, you can photograph your models at different stages, print out the pictures, and then shuffle them. Practice putting the images into the proper order, and explain the purpose of each phase of mitosis and meiosis.
- Once you understand how cell division in mitosis and meiosis works, explore the processes of crossing over and nondisjunction in order to see how these influence genetic composition of daughter cells.
- It also might be helpful to work with a study partner. For instance, you can diagram the starting cell, some of the intermediate steps, and the resulting cells, and then challenge your partner to fill in the missing steps. You and your study partner can also take turns introducing different errors in cell division and challenging each other to explain how these errors will influence the genetic composition of the daughter cells.
- Draw a picture of the cell cycle and indicate the points at which the cyclin–Cdk complexes act during the cycle. Refer to Figure 11.6. Use paper cups to represent Cdk. Replicate the synthesis of cyclin by adding candies to the cups. Draw a large circle and label the stages of the cell cycle. Then place the cyclin–Cdk complexes in their proper places on the circle. What would happen if cyclin could not be degraded (cup stays full) or cyclin could not be synthesized (cup stays empty)? Repeat these activities with a study partner so you can practice breaking the cell cycle in different ways while you challenge each other to explain the outcomes.
- Always keep in mind that the "stages" of the cell cycle, like the stages of mitosis and meiosis, are artificial labels used to help us understand the overall process. The cells do not lurch from stage to stage in

Mitosis: Meiosis:
cross over to produce diff.

incremental jumps, but thinking in terms of phases of the cycle allows us to identify key features of the overall process as it flows from beginning to end.

- Remember that programmed cell death is beneficial. The sacrifice of individual cells aids in the development and/or survival of the multicellular organism.
- This chapter contains a large amount of terminology. Making a terminology/vocabulary list may be helpful.
- You may be confused by the difference between chromosome and chromatid, and about how to count the number of chromosomes in a cell. Remember that a chromosome is an organized package of DNA found in the nucleus of a eukaryotic cell. During the S phase of the cell cycle, the DNA in a chromosome is duplicated, resulting in the formation of a duplicated chromosome, with each DNA copy packaged as a sister chromatid. As a general rule, a chromosome, whether duplicated or unduplicated, contains one centromere region; two sister chromatids attached to a common centromere region are counted as one chromosome. When sister chromatids separate and migrate to opposite poles of the cell, each chromatid now has its own centromere region and is therefore counted as a separate chromosome.
- Go to the Web addresses shown to review the following animated tutorials and activities. You can also find them in BioPortal (yourBioPortal.com), along with many additional learning resources.

  Animated Tutorial 11.1 Mitosis (Life10e.com/at11.1)

  Animated Tutorial 11.2 Meiosis (Life10e.com/at11.2)

  Activity 11.1 Images of Mitosis (Life10e.com/ac11.1)

  Activity 11.2 The Mitotic Spindle (Life10e.com/ac11.2)

  Activity 11.3 Sexual Life Cycle (Life10e.com/ac11.3)

  Activity 11.4 Images of Meiosis (Life10e.com/ac11.4)

Key Concept Review

11.1 How Do Prokaryotic and Eukaryotic Cells Divide?

Prokaryotes divide by binary fission

Eukaryotic cells divide by mitosis or meiosis followed by cytokinesis

Cell division requires a reproductive signal from either inside or outside the cell, subsequent DNA replication, segregation of the newly replicated chromosomes to opposite poles of the cell, and the division of the cytoplasm (cytokinesis) to form two daughter cells.

Prokaryotic cells divide by increasing in size, replicating their DNA (often a single circular chromosome), and dividing into two new cells through the process of binary fission. Environmental signals and nutrient concentrations influence the decision to divide and the rate of division in prokaryotes. DNA replication in prokaryotic cells requires a number of different proteins that form a replication complex through which the DNA is threaded. Special sites on the chromosome facilitate the initiation (the *ori* region) and the termination (the *ter* region) of replication. Segregation of the newly replicated chromosomes occurs with the help of special proteins, ATP hydrolysis, and the prokaryotic cytoskeleton. Cytokinesis is caused by the pinching in of the plasma membrane, and division is completed by the formation of the cell wall between the two cells.

Cell division in multicellular eukaryotes occurs in response to the needs of the entire organism. Eukaryotic cells have more than one chromosome, and these chromosomes are linear. In response to a reproductive signal, eukaryotic cells replicate their chromosomes, segregate the chromosomes (which are closely associated with each other as sister chromatids) during mitosis, and divide their cytoplasm during cytokinesis. Meiosis is the process of nuclear division in cells involved in sexual reproduction. In cells that produce gametes (eggs and sperm), nuclear division occurs via meiosis, which generates diversity.

Question 1. How does the process of cell division differ in prokaryotic cells and eukaryotic cells? Which elements are the same?

Question 2. What is the overall purpose of mitosis? Of meiosis?

11.2 How Is Eukaryotic Cell Division Controlled?

Specific internal signals trigger events in the cell cycle

Growth factors can stimulate cells to divide

The period between cell divisions is called the cell cycle and includes mitosis, cytokinesis, and interphase. The duration of the cell cycle varies considerably in different cell types. While in some cells (e.g., embryonic cells) the cell cycle may be as short as 30 minutes, it can last up to 24 hours in some adult cells.

Interphase is divided into three subphases: G1 (or Gap 1), S (or DNA synthesis), and G2 (or Gap 2). G1 is the most variable phase of the cell cycle. In this phase, the decision and subsequent preparation for DNA synthesis occurs. The G1-to-S transition (called the restrictive point, R) is the point at which the commitment to DNA replication and subsequent cell division occurs. S is the phase of the cell cycle in which all the chromosomes are replicated; each replicated chromosome consists of two identical sister chromatids. During G2, the cell prepares for mitosis (M).

The M phase includes mitosis and cytokinesis. During the M phase, the cell segregates the newly replicated chromosomes to opposite poles of the cell, and nuclear division occurs.

Cdk, a cyclin-*d*ependent *k*inase that phosphorylates other proteins in the cell, is catalytically active when it is bound to another protein, cyclin. Different cyclin–Cdk protein complexes act as checkpoints in the cell cycle: during G1 before the cell enters S, during S phase to stimulate DNA replication, and after S to initiate the transition from G2 to M. Other proteins, such as retinoblastoma protein (RB), are regulated by cyclin–Cdk complexes. RB acts as inhibitor of the cell cycle at the restriction (R) point.

Growth factors are external chemical signals that stimulate cell division, often by activating cyclin-Cdk complexes.

Question 3. In the cell cycle shown in the diagram below, where is the checkpoint that determines whether the cell will proceed from G1 to S? Label G1, S, and G2 phases. Then label Cdk and cyclin in the appropriate locations and indicate if they are present in an intact or degraded form.

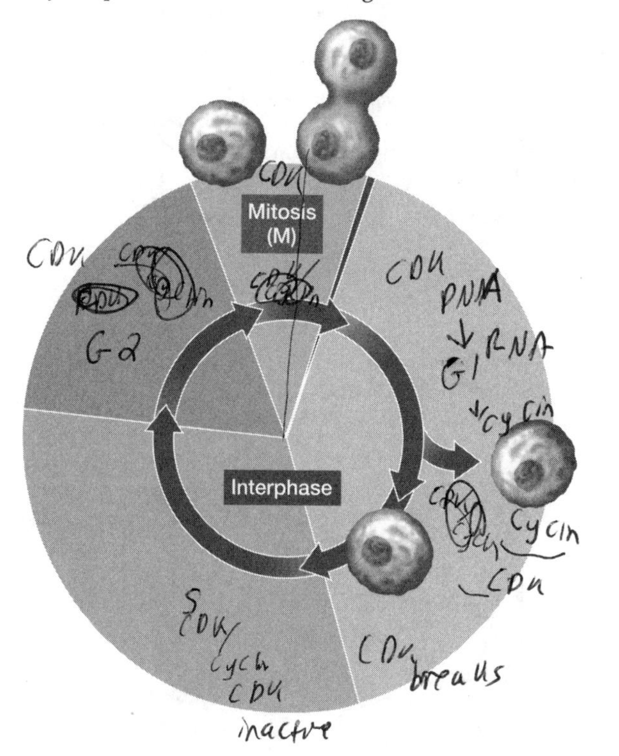

Question 4. What would happen to a cell if it were unable to synthesize cyclin?

11.3 What Happens during Mitosis?

- Prior to mitosis, eukaryotic DNA is packed into very compact chromosomes
- Overview: Mitosis segregates copies of genetic information
- The centrosomes determine the plane of cell division
- The spindle begins to form during prophase
- Chromosome separation and movement are highly organized
- Cytokinesis is the division of the cytoplasm

Eukaryotic chromosomes are composed of protein and DNA, collectively called chromatin. Cohesin is a protein that holds the sister chromatids together along their length. When eukaryotic chromosomes are replicated, the daughter chromosomes, termed sister chromatids, are still joined at the centromere, and condensin proteins make the chromosome more compact. In a eukaryotic nucleus, DNA molecules are wrapped around beadlike particles of protein called histones, thus forming structural units known as nucleosomes. Histones are very basic proteins. The nucleosomes are further condensed into a coil that twists into another larger coil. During mitosis, more compaction occurs, and the sister chromatids become visible (see Figures 11.8 and 11.9). Interphase chromosomes are less densely packed.

During mitosis a single nucleus gives rise to two nuclei that are genetically identical to the parent nucleus. Mitosis is subdivided into five stages: (1) prophase (chromosomes condense and chromatids become visible under the light microscope); (2) prometaphase (nuclear membrane breaks down and sister chromatids attach to spindle apparatus); (3) metaphase (chromosomes line up at the midline of the cell); (4) anaphase (sister chromatids separate and migrate to opposite poles); and (5) telophase (nuclear envelope forms, chromosomes decondense, and spindle apparatus disappears) (see Figure 11.10).

Centrosome duplication occurs during S phase. Centrosomes may contain centrioles, which consist of two hollow tubes of microtubules positioned at right angles to each other. Duplicated centrosomes separate at the beginning of prophase and move to opposite ends of the nuclear envelope, and later, during anaphase, to opposite poles of the cytoplasm. The position of the centrosomes in the cytoplasm determines the plane at which cytokinesis will occur. Plants and fungal cells lack centrosomes, but they possess microtubule-organizing centers that play the same role.

Microtubules grow from a microtubule-organizing center, the centrosome. The microtubules form a spindle along which the chromosomes will move. Spindle fibers develop as tubulin dimers that aggregate around the centrioles and extend toward the middle region of the cell. There are two types of microtubules in the mitotic spindle: polar microtubules and kinetochore microtubules. Polar microtubules form the framework of the spindle and run from one pole to the other. Kinetochore microtubules attach to kinetochores on each chromatid and help separate them to opposite poles of the cell during anaphase.

Prometaphase signals the disappearance of the nuclear envelope and the nucleoli and the attachment of kinetochore microtubules to the kinetochores of each newly replicated chromosome. Metaphase signals the positioning of all of the centromeres at the equatorial plate. At the beginning of anaphase, the cohesion proteins are hydrolyzed by a protease called separase. Separase is activated by an anaphase-promoting complex (APC) which is a cyclin–Cdk complex, and APC is activated by the proper attachment of all the chromosomes to the spindle. Following their separation, sister chromatids are referred to as daughter chromosomes. Sister chromatids share a centromere, while daughter chromosomes have their own centromere. During anaphase, daughter chromosomes move to opposite poles of the spindle, assisted by kinesins and dynein motor proteins acting at the kinetochores and by the shortening of the microtubules at the poles. At telophase, the spindle breaks down, the chromosomes begin to uncoil, and the nuclear envelope forms, resulting in two identical nuclei.

During cytokinesis in animal cells, the cell membrane contracts due to the interaction of actin and myosin microfilaments, forming a contractile ring (see Figure 11.13A). In plant cell cytokinesis, a cell plate forms between the newly segregated chromosomes. This cell plate is derived from

Golgi vesicles located at the equatorial region, which fuse to form a plasma membrane. The plasma membrane secretes plant cell-wall materials into the cell plate to complete cell division (see Figure 11.13B). Upon completion of cytokinesis, there are two distinct cells, each with a full complement of chromosomes.

Question 5. Diagram normal mitosis in a cell with six chromosomes. Label unduplicated chromosomes, duplicated chromosomes, sister chromatids, centromeres, and daughter chromosomes.

Question 6. Describe how 2 meters of DNA in a typical human cell can fit into the nucleus, which is 5 μm in diameter.

Question 7. How does cytokinesis differ in animal and plant cells?

11.4 What Role Does Cell Division Play in a Sexual Life Cycle?

Asexual reproduction by mitosis results in genetic constancy

Sexual reproduction by meiosis results in genetic diversity

There are two kinds of reproduction. The first is asexual (or vegetative reproduction), in which the replicated chromosomes are separated by mitosis. After cytokinesis the offspring are genetically identical (clones) to the parent. The second kind of reproduction is sexual, in which gametes from two different parents fuse to form a zygote.

In sexual reproduction this fusion is called fertilization and produces offspring that are genetically different from both parents. Somatic cells from diploid organisms contain pairs of homologous chromosomes, one set of chromosomes coming from each parent. Homologous chromosomes are similar in size and appearance and bear corresponding genetic information. Gametes each contain a single set of chromosomes (n) resulting from meiosis (and thus are haploid), and the zygote contains two sets of chromosomes ($2n$), one set from each parent, and is diploid.

Mature organisms can exist primarily in the diploid or haploid state or can alternate between haploid and diploid states during their life cycles (see Figure 11.15). In some life cycles, the product of meiosis (spores) undergoes cell division to produce haploid mature organisms (haplontic life cycle). Specialized cells in these haploid organisms differentiate into gametes. In other life cycles, gametes are formed directly from meiosis. Most plants and some protists have mature forms that alternate between diploid and haploid stages. Diplontic organisms are diploid in their mature state; only their gametes are haploid. Fusion of gametes to form a zygote increases diversity because the zygote's genetic makeup is derived from two different parents, each contributing one haploid set of randomly selected chromosomes.

Question 8. How does asexual reproduction differ from sexual reproduction?

Question 9. In the diagrams below, label the sexual life cycle strategy that is represented (diplontic, haplontic, or alternation of generations). For each, provide one example of an organism that demonstrates that life cycle strategy.

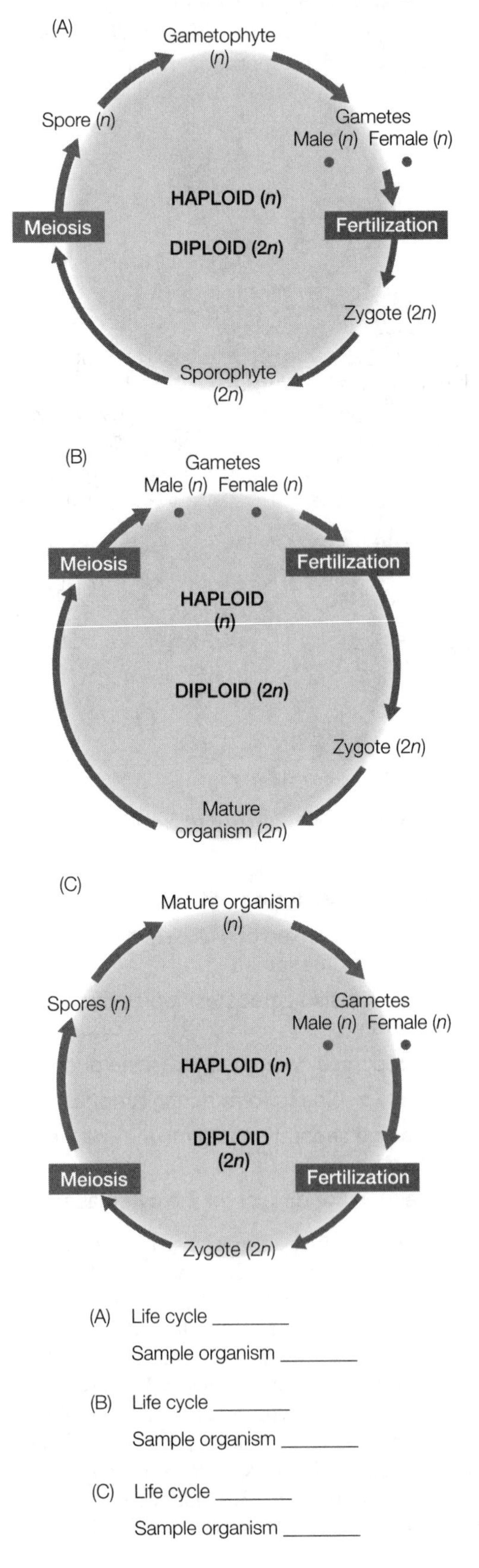

11.5 What Happens during Meiosis?

Meiotic division reduces the chromosome number

Chromatid exchanges during meiosis I generate genetic diversity

During meiosis homologous chromosomes separate by independent assortment

Meiotic errors lead to abnormal chromosome structures and numbers

The number, shapes, and sizes of the metaphase chromosomes constitute the karyotype

Polyploids have more than two complete sets of chromosomes

Meiosis is distinguished from mitosis by *two* nuclear divisions (meiosis I and II), but only *one* round of DNA replication, resulting in a reduction in the number of chromosomes from diploid to haploid (see Figure 11.16). The purpose of meiosis is to reduce the chromosome number, to ensure that haploid products have a complete set of chromosomes, and to promote genetic diversity in gametes and in the species.

In meiosis I, homologous chromosomes, each consisting of two sister chromatids, first pair up along their entire length. This is followed by separation of homologous chromosomes and their segregation into two different nuclei, each with half the original chromosome number of the parent cell. The sister chromatids of each chromosome separate in meiosis II, just as in mitosis. The end product of meiosis is four cells that are genetically different, each containing the haploid number of chromosomes.

In prophase of meiosis I, the newly replicated chromosomes pair with their homologs (synapsis). Because there are four chromatids (two from each homolog), these paired chromosomes are called tetrads or bivalents. Crossing over occurs in tetrads between nonsister chromatids on homologous chromosomes at special sites called chiasmata (see Figure 11.17 and Figure 11.18). The resulting recombinant chromosomes contribute to an increase in the genetic variation in the products of meiosis.

During anaphase of meiosis I, each pole receives one member of the homologous pair. Unlike in mitosis, no DNA replication occurs before meiosis II begins, and the number of chromosomes on the equatorial plate is half the number in the mitotic nucleus (even though each chromosome still consists of two chromatids). Unlike the chromatids in mitosis, each chromatid may be different from its sister chromatid, due to crossing over in meiosis I. As in mitosis, segregation of each chromatid occurs in anaphase II, and each of the resulting four nuclei is haploid. Crossing over in prophase I and independent assortment in meiosis I (of homologs) and in meiosis II (of chromatids) results in genetic diversity in the daughter cells.

Aneuploidy (see Figure 11.20), an abnormal number of chromosomes, occurs in cells whose chromosomes fail to segregate in meiosis I (both homologs fail to separate and go to the same pole) or meiosis II (both chromatids fail to separate and go to one pole. One of the causes of nondisjunction in meiosis I may be the breakdown of cohesins, which orient chromosomes at the equatorial plate during metaphase I. In the absence of cohesins, homologs may line up at random and end up in the wrong daughter cell.

Aneuploid gametes can produce trisomies as in Down syndrome, in which the individual has an extra chromosome 21. Monosomic zygotes have one copy of a homologous chromosome. Translocations, another chromosome abnormality, occur when part of one chromosome breaks away and becomes attached to another chromosome.

Mitotic chromosomes can be characterized by their shape, number, size, and centromere position. Each organism has a distinctive set of chromosomes called a karyotype (see Figure 11.21).

Polyploids are cells that have increased numbers of complete sets of chromosomes. Polyploids with an even number of sets of chromosomes ($4n$, $6n$, $8n$) can undergo meiosis successfully, whereas odd-number polyploids ($3n$, $5n$, etc.) cannot, because each homolog needs to pair with its partner. Polyploid nuclei tend to be larger than their diploid counterparts, a characteristic that is exploited in agriculture. Examples of polyploid crops include bananas, wheat, cotton, and oats. Polyploid wheat is the product of hybridization between species that results in the formation of new allopolyploid species. In this process, haploid gametes fuse to form a diploid zygote. Nondisjunction of the entire chromosome sets of the fertilized egg during mitosis results in a tetraploid. The tetraploid wheat hybridizes with the diploid wild grass, resulting in bread wheat (see Figure 11.22).

Question 10. Starting with a diploid cell ($n = 2$), diagram normal meiosis.

Question 11. A cell at the start of meiosis has two chromosomes, representing the replicated DNA of a diploid cell. If the cell has 1 (μg) of DNA in G1, it would have 2 μg of DNA at the start of meiosis. At the end of meiosis I, there is 1 μg of DNA in a cell, but the cell is not diploid. Why not?

Question 12. Starting with a diploid cell ($n = 2$), diagram meiosis with a nondisjunction in meiosis I, and then with a nondisjunction in meiosis II. Label the gametes that would result from each meiotic event. Describe the types of zygotes that would be produced from these gametes.

Question 13. Describe two ways in which the genetic diversity of organisms is increased during meiosis.

Question 14. Suppose that by using a chemical that inhibits cytokinesis, you have created peaches that are tetraploid. (1) How many sets of chromosomes do these peaches have? (2) What will be the ploidy of the gametes? (3) Are the tetraploid peaches fertile? Would triploid peaches be fertile? Explain.

11.6 In a Living Organism, How Do Cells Die?

Cell death can occur in two ways: by necrosis (cells are damaged by toxins or starved of oxygen or essential nutrients) or by apoptosis (programmed cell death). Apoptosis is a normal process during development and is used to eliminate unneeded tissue. Apoptosis also occurs in older cells in which genetic damage may be more prevalent. Signals for cell death include a lack of mitotic signals and recognition of DNA damage, or changes in a receptor protein that activates signal transduction. Signal transduction can lead to the activation of caspases, which are proteases that hydrolyze target molecules, leading to cell death.

Question 15. Differentiate between necrosis and apoptosis.

Question 16. Describe how a cell exposed to radiation for a prolonged period of time might trigger and proceed with apoptosis.

11.7 How Does Unregulated Cell Division Lead to Cancer?

Cancer cells differ from normal cells

Cancer cells lose control over the cell cycle and apoptosis

Cancer treatments target the cell cycle

Cancer cells differ from normal cells in two ways: unregulated cell division and cell migration to other parts of the body. Cancer cells do not respond to cell division controls and divide continuously, forming tumors. Benign tumors resemble the tissue they came from and can remain localized, but often they need to be removed. Malignant tumors do not look like the parent tissues and metastasize, spreading to other areas of the body (see Figure 11.24).

Oncogenes encode proteins that are positive regulators of cell division (such as HER2). Tumor suppressors (such as RB) prevent the cell cycle from proceeding through cell division. In cancer cells, these regulators do not function normally, leading to uncontrolled cell division (see Figure 11.25). Cell death occurs by necrosis or apoptosis in response to environmental signals. Cancer cells lose their ability to respond to positive regulators of apoptosis. Radiation and cancer drugs target the cell cycle (see Figure 11.26).

Question 17. How are cell cycle regulatory proteins involved in the development of cancer?

Question 18. Explain how mutations in one or more of the cell cycle checkpoints might lead to cancer.

Test Yourself

1. Which of the following statements about mitosis is true?
   a. Cytokinesis follows mitosis.
   b. DNA replication is completed prior to the beginning of mitosis.
   c. The chromosome number of the resulting cells is the same as that of the parent cell.
   d. The daughter cells are genetically identical to the parent cell.
   e. All of the above
   Textbook Reference: *11.3 What Happens during Mitosis?*

2. Which of the following statements about meiosis is true?
   a. The chromosome number in the resulting cells is halved.
   b. DNA replication occurs before meiosis I and meiosis II.
   c. The homologs do not pair during prophase I.
   d. The daughter cells are genetically identical to the parent cell.
   e. The chromosome number of the resulting cells is the same as that of the parent cell.
   Textbook Reference: *11.3 What Happens during Mitosis?*

3. Which of the following statements about kinetochores on mitotic chromosomes is true?
   a. They are located at the centromere of each chromosome.
   b. They are the sites where microtubules attach to separate the chromosomes.
   c. They are organized so that there is one per sister chromatid.
   d. Kinetochore microtubules from opposite poles attach to each sister chromatid.
   e. All of the above
   Textbook Reference: *11.3 What Happens during Mitosis?*

4. Which of the following statements about the mitotic spindle is true?
   a. It is composed of polar and kinetochore microtubules, both of which attach to chromosomes.
   b. It is composed of actin and myosin microfilaments.
   c. It is composed of kinetochores at the metaphase plate.
   d. It is composed of microtubules, which help separate the chromosomes to opposite poles of the cell.
   e. It originates only at the centrioles in the centrosomes.
   Textbook Reference: *11.3 What Happens during Mitosis?*

5. Imagine that there is a mutation in the Cdk gene such that its gene product is nonfunctional. What effect would this mutation have on a mature red blood cell?
   a. The cyclin that bound to this Cdk would not be phosphorylated.
   b. There would be no effect, because mature red blood cells do not enter the cell cycle.
   c. The cell would not be able to replicate its DNA.
   d. The cell would not be able to enter G1.
   e. The cell would not be able to reproduce itself.
   Textbook Reference: *11.2 How Is Eukaryotic Cell Division Controlled?*

6. Imagine that there is a mutation in the Cdk gene such that its gene product is nonfunctional. What effect would this mutation have on a mammalian white blood cell?
   a. The cell would not be able to replicate its DNA.
   b. The cell would not be able to enter mitosis.

c. The cell would not be able to reproduce itself.
d. The cell would not be able to phosphorylate its associated cyclin.
e. All of the above

Textbook Reference: *11.2 How Is Eukaryotic Cell Division Controlled?*

7. Which of the following statements about DNA replication and cytokinesis in *E. coli* is true?
a. DNA replication occurs in the nucleus.
b. Cytokinesis is facilitated by microfilaments of actin and myosin.
c. DNA replication occurs during the S phase of the cell cycle.
d. Cell reproduction is initiated by reproductive signals that result in DNA replication, DNA segregation, and cytokinesis.
e. The *E. coli* chromosome is linear.

Textbook Reference: *11.1 How Do Prokaryotic and Eukaryotic Cells Divide?*

8. Which of the following statements about chromatids is true?
a. They are replicated chromosomes still joined together at the centromere.
b. They are identical in mitotic chromosomes.
c. They undergo recombination in mitosis.
d. They are identical in meiotic chromosomes.
e. Both a and b

Textbook Reference: *11.3 What Happens during Mitosis?*

9. Histones are positively charged because
a. most of the ions in the nucleus of the cell are negatively charged.
b. they interact with acidic residues of proteins found in the nucleus.
c. the basic side chains of histone proteins interact with the negatively charged DNA.
d. they have a majority of acidic residues in their protein sequence.
e. the low pH of the nucleus results in a positive charge.

Textbook Reference: *11.3 What Happens during Mitosis?*

10. Chromosome movement during anaphase is the result of
a. the hydrolysis of ATP by dynein.
b. molecular motors at the kinetochores that move the chromosomes toward the poles.
c. molecular motors at the centrosome that pull the microtubules toward the poles.
d. shortening of the microtubules at the centrosome which pulls the chromosomes toward the poles.
e. a, b, and d

Textbook Reference: *11.3 What Happens during Mitosis?*

11. Programmed cell death (apoptosis)
a. occurs in cells that have been deprived of essential nutrients.
b. occurs only in cells that have damaged DNA.
c. is a natural process during development.
d. is signaled by the initiation of mitosis.
e. is well controlled in cancer cells.

Textbook Reference: *11.6 In a Living Organism, How Do Cells Die?*

12. If the *ori* site on the *E. coli* chromosome is deleted,
a. nothing will happen.
b. replication will start but not be able to continue.
c. replication will not start.
d. replication will initiate at another *ori* site on the chromosome.
e. the chromosome will be replicated but the cell will not be able to divide.

Textbook Reference: *11.1 How Do Prokaryotic and Eukaryotic Cells Divide?*

13. Chiasmata
a. are sites where nonsister chromatids can exchange genetic material during meiosis.
b. are sites where sister chromatids can exchange genetic material during meiosis.
c. increase genetic variation among the products of meiosis.
d. increase genetic variation among the products of mitosis.
e. Both a and c

Textbook Reference: *11.5 What Happens during Meiosis?*

14. The difference between asexual and sexual reproduction is that
a. asexual reproduction occurs only in bacteria, whereas sexual reproduction occurs in plants and animals.
b. asexual reproduction results from meiosis, whereas sexual reproduction results from mitosis.
c. asexual reproduction results in an organism that is identical to the parent, whereas sexual reproduction results in an organism that is not identical to either parent.
d. asexual reproduction results from the fusion of two gametes, whereas sexual reproduction produces clones of the parent organism.
e. asexual reproduction occurs only in haplontic organisms, whereas sexual reproduction occurs only in diplontic organisms.

Textbook Reference: *11.4 What Role Does Cell Division Play in a Sexual Life Cycle?*

15. A chromatid is
a. a chromosome before it has undergone DNA replication.
b. one of the pairs of homologous chromosomes.
c. a homologous chromosome.
d. a newly replicated bacterial chromosome.
e. one-half of a newly replicated eukaryotic chromosome.

Textbook Reference: *11.3 What Happens during Mitosis?*

16. Which of the following has the highest potential for causing cancer?
    a. A mutation in p53 protein that causes it to become inactive, and a mutation in RB protein that renders it overly active
    b. A mutation in p53 protein that causes it to become overly active, and a mutation in RB protein that renders it inactive
    c. Increased production of normal HER2, and a mutation in RB protein that renders it inactive
    d. Increased production of normal HER2, and a mutation in RB protein that renders it overly active
    e. Increased production of normal p53 protein, and a mutation in RB protein that renders it inactive

    Textbook Reference: *11.7 How Does Unregulated Cell Division Lead to Cancer?*

Answers

Key Concept Review

1. All cells divide by replicating their DNA, segregating the DNA, and then splitting the cytoplasm by cytokinesis. In most prokaryotic cells there is only one circular chromosome. As the cell enlarges to prepare for division, the newly replicated daughter chromosomes are separated at opposite sides of the cell. During fission, the cell membrane pinches in, and cell wall components are synthesized between the daughter cells. In eukaryotic cells, there are more chromosomes, and they are linear. The cell undergoes a sequential set of steps called the cell cycle, in which the chromosomes are replicated and then separated to opposite poles of the cell.
2. The purpose of mitosis is to produce daughter cells that are identical to the parent cell. The end product of meiosis is reproductive cells that are genetically different from the parent cell.
3.

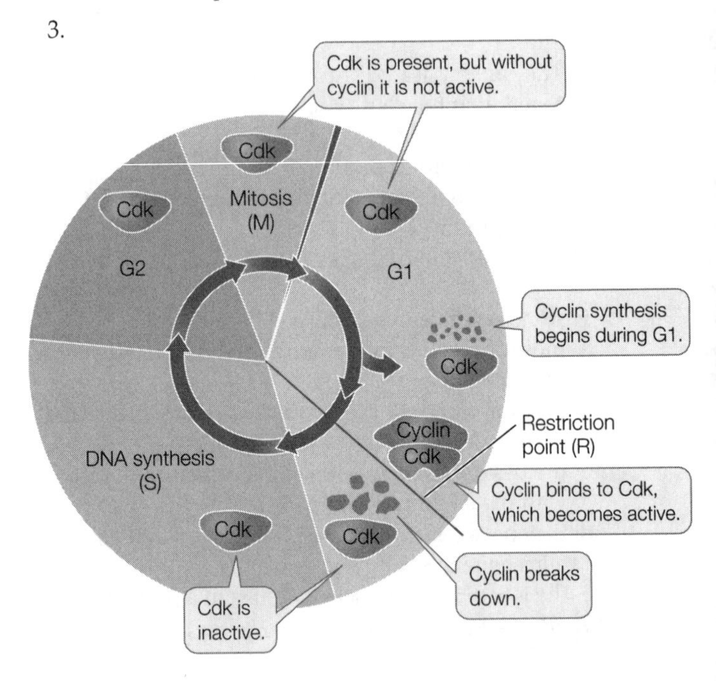

4. The cell would never be able to activate Cdk. Active Cdk is required for the cell to move past the G1-to-S checkpoint.

5.

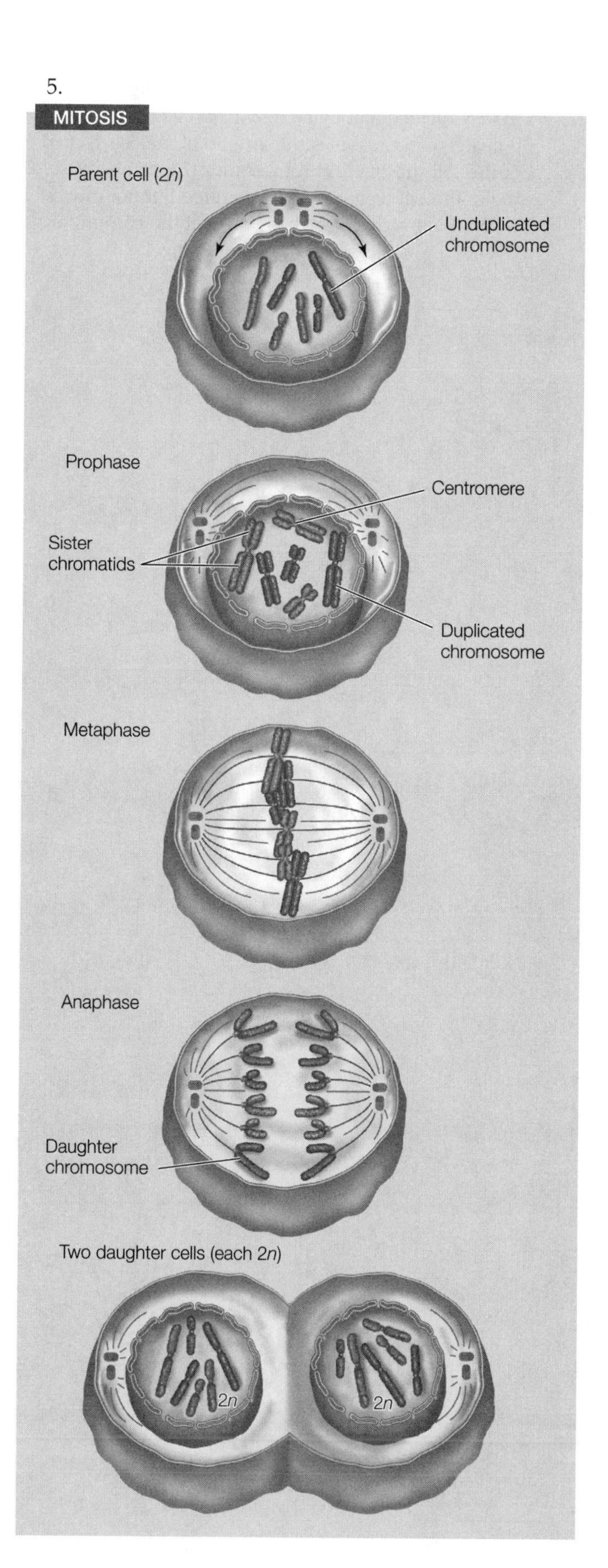

6. The DNA is very tightly condensed because it is packaged with proteins that stabilize the double helix in organized DNA coils. See Figure 11.9 for an illustration of the first level of DNA packaging that results in the condensed chromosome.

7. In animal cells, cytokinesis results from the interaction of actin filaments and myosin, which causes the cell membrane to pinch in and divide the cytoplasm into two cells. In plant cells, a cell plate forms between the newly segregated chromosomes, and Golgi vesicles fuse at that site to form the new cell membranes. Cell wall components are then secreted between the plasma membranes to complete cytokinesis.

8. The difference between mitosis and meiosis is the number and ploidy of the daughter cells. In mitosis, cell division results in two diploid cells, whereas meiosis results in four haploid cells.

9.
   (A) Life cycle: Alternation of generations; Sample organism: Most plants, some fungi
   (B) Life cycle: Diplontic; Sample organism: Animals, brown algae, some fungi
   (C) Life cycle: Haplontic; Sample organism: Most protists, fungi, some green algae

10.

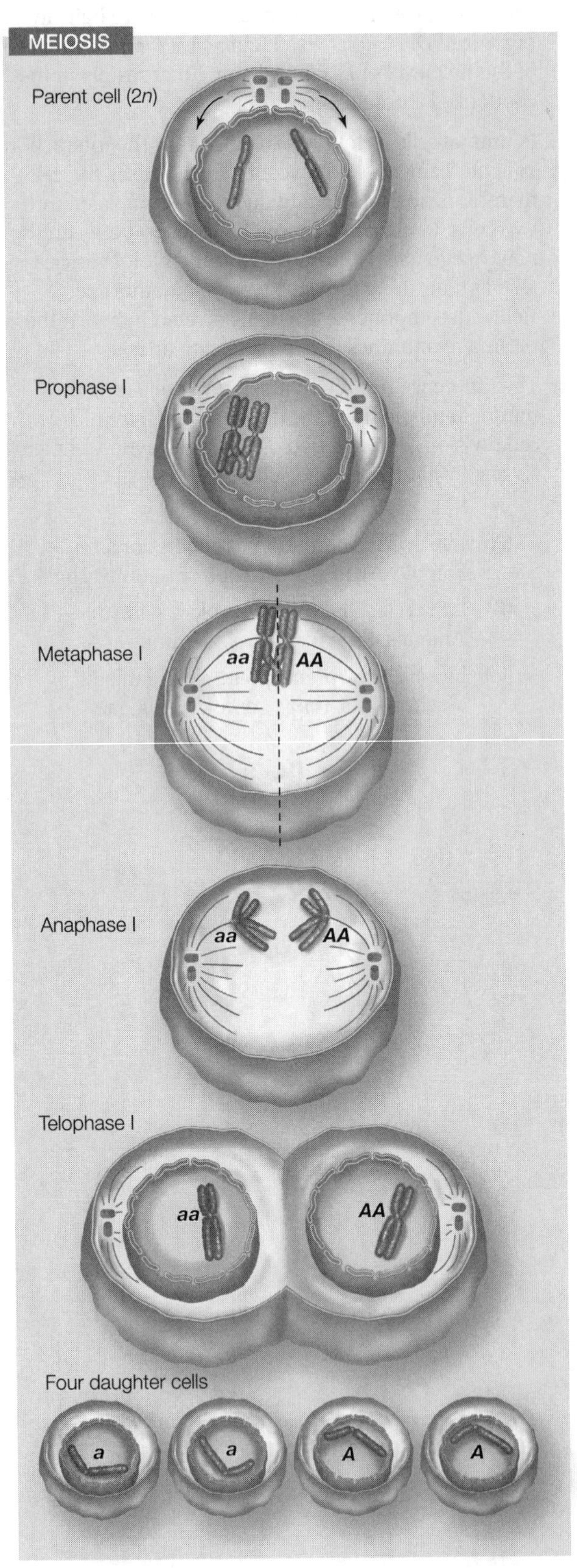

11. At the start of meiosis, the 2 μg of DNA are approximately divided among a total of $2n = 2$ (4 total, or 0.5 μg each) chromosomes. At the end of meiosis I, each of the cells has $n = 2$ chromosomes (2 total) because the homologous pairs have separated but not the sister chromatids, so each cell has one-half the amount of DNA, 1 μg, but is haploid, not diploid.

12. Meiosis with a nondisjunction in meiosis I:

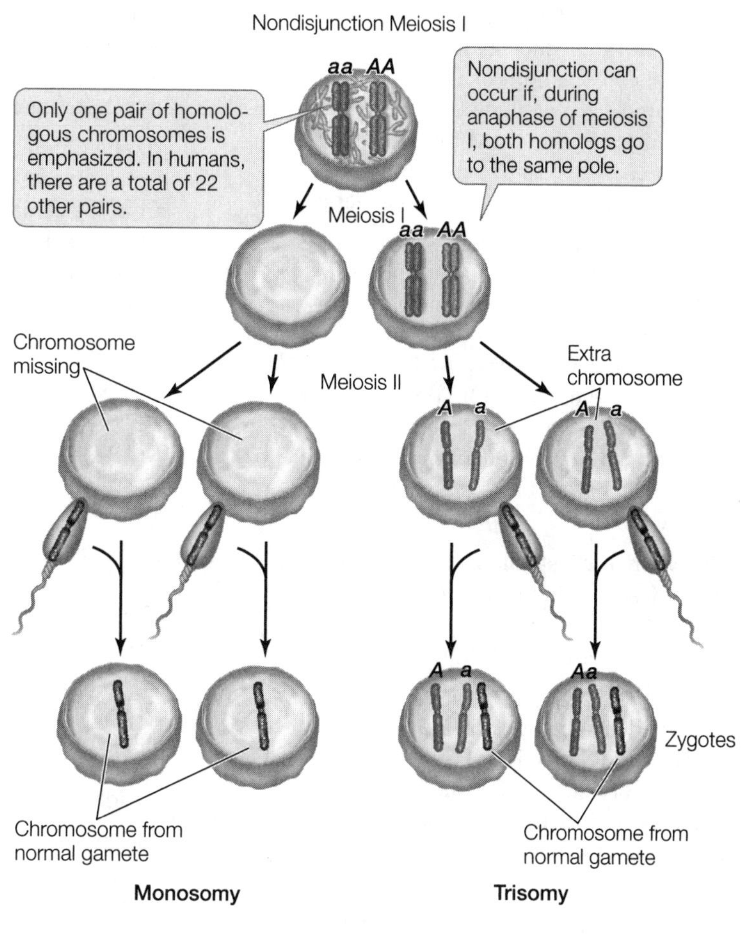

Meiosis with a nondisjunction in meiosis II:

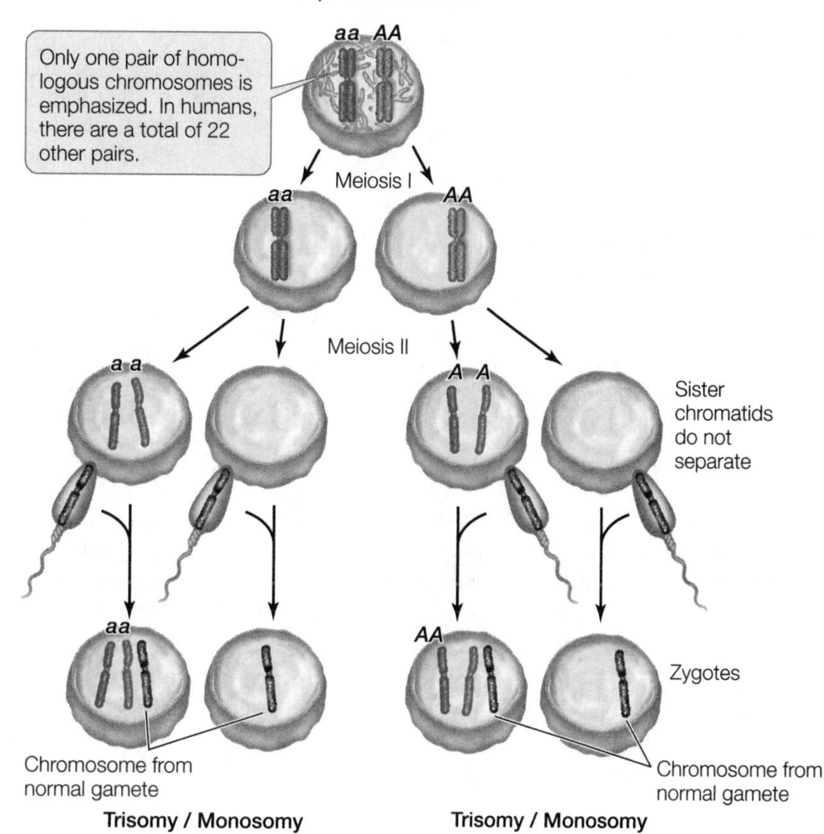

Zygotes would either have one too many chromosomes (trisomy) or would be missing a chromosome (monosomy).

13. Genetic diversity is increased during crossing over of prophase I of meiosis so that each gamete has chromosomes with different combinations of alleles than the parents do. During meiosis, each homologous chromosome is randomly sorted to one of the four daughter cells. This results in 223 different possible combinations of homologous chromosomes per human gamete.

14. (1) Peaches that are tetraploid have four sets of chromosomes. (2) Because there is an even number of chromosomes ($4n$), each replicated homologous chromosome will be able to find a replicated homolog to pair with at meiosis, and fertile gametes will result. These gametes will be diploid ($2n$). (3) Triploid peaches would not be fertile because one of the three homologs would not find its pair during prophase of meiosis I, and the single homologs would be sorted randomly into the daughter cells.

15. Necrosis occurs when a cell is damaged by toxins or starved of oxygen, resulting in bursting of the cell and release of the cytoplasm into the extracellular space. Apoptosis is a programmed and orderly series of events that results in cell death.

16. A cell that has been damaged by radiation likely has DNA damage. This would function as an internal signal for the activation of a signal transduction pathway leading to apoptosis. Activated caspases degrade the nuclear envelope, nucleosomes, and plasma membrane, and the cell dies.

17. There are two regulatory systems that ensure that cells divide only when needed. These systems consist of oncogene proteins (positive regulators that stimulate cells to undergo cell division) and tumor suppressor proteins (negative regulators that inhibit cells from dividing). In cancer cells, these two regulatory systems do not function normally, thus causing loss of control over cell division.

18. The basic role of each checkpoint is to determine whether the cell is functioning normally and should enter into or continue cell division. If these checkpoints malfunction, they may allow a damaged cell to continue through the cell cycle—that is, to divide. Since cancer is uncontrolled cell division, cancer could result from a malfunction of the normal operation of cell cycle checkpoints.

Test Yourself

1. **e.** Mitosis occurs after DNA replication and results in cells with the same number of genetically identical chromosomes as the parent cell. Cytokinesis follows mitosis.

2. **a.** Meiosis occurs after one round of DNA replication. Homologous chromosomes pair during prophase I of meiosis, and after meiosis II the resulting cells have half the number of chromosomes as the parent cell. These chromosomes are not genetically identical to the parent cells.

3. **e.** Kinetochores, one per sister chromatid, are assembled at the centromere of each chromosome and are the sites at which microtubules from opposite poles attach to segregate the chromosomes.

4. **d.** The mitotic spindle is composed of microtubules, not actin and myosin filaments. The spindle originates from the centrosome, which may or may not have centrioles, and only the kinetochore microtubules attach to the chromosomes.

5. **b.** Many cells, such as red blood cells, muscle cells, and nerve cells, lose their ability to divide as they mature.

6. **e.** Cyclin–Cdk complexes affect several events during the cell cycle, including the transition of the cell from G1 phase to the S phase (DNA replication) and the transition of the cell from G2 phase to the M phase (mitosis and therefore cell division). In addition, functional Cdk is required to phosphorylate its associated cyclin.

7. **d.** *E. coli* is a prokaryote. It lacks a nucleus and does not undergo the cell cycle seen in eukaryotes. It has a circular chromosome, and does not synthesize actin or myosin proteins. Cytokinesis in *E. coli* is a result of a reproductive signal that causes the DNA to be replicated and segregated and finally causes the cell to divide.

8. **e.** Chromatids are highly condensed, newly replicated chromosomes that will be segregated to the daughter cells. After DNA replication, chromatids are still attached to each other at the centromere. Meiotic sister chromatids are different from each other due to recombination (crossing over) in prophase of meiosis I. Mitotic sister chromatids are identical.

9. **c.** The positive charges on histone proteins are due to the large number of basic amino acid residues found in these proteins. These positive charges interact with the negatively charged phosphate sugar backbone of DNA during assembly of the DNA on the nucleosome.

10. **e.** Chromosomes are attached to the microtubules at their kinetochores. There are dynein molecular motors at the kinetochores (but not the centrosome) which hydrolyze ATP and help move the chromosomes to opposite poles. Chromosomes are also pulled toward the poles by the shortening of the kinetochore microtubules.

11. **c.** Programmed cell death occurs during the development of many organisms (for instance, tadpoles lose their tails to become adult frogs). One of the stimuli for programmed cell death is DNA damage, but it is not the only cause of death. Necrosis (cell death that is not programmed) occurs when cells have been deprived of essential nutrients. The initiation of mitosis is part of the cell cycle in which cells reproduce and is not a step in programmed cell death. Apoptosis is not well controlled in cancer cells.

12. **c.** Without the origin of replication (there is only one in *E. coli*), there would be no site for the replication proteins to bind to initiate DNA replication, so DNA synthesis would not start.

13. **e.** Chiasmata are sites where nonsister chromatids can exchange genetic material during meiosis, which increases genetic variation in the gametes (the products of meiosis).

14. **c.** Asexual reproduction, which results from mitosis, produces cells that are identical to the parent and can occur in plants. Sexual reproduction can occur in haplontic organisms (such as fungi) and results in an organism that is not genetically identical to either parent.

15. **e.** A chromatid is one-half of a newly replicated eukaryotic chromosome, and is connected to the other (sister) chromatid at the centromere.

16. **c.** Cell division in normal cells is regulated by two systems that ensure that cells divide only when needed, one acting as a positive regulatory system and the other as a negative regulatory system. Cancer can result from malfunctioning in one of these systems. Malfunctioning of both systems further increases the risk of developing cancer, a situation observed when there is simultaneous increase in the activity of the oncogene HER2 (stimulation of cell growth) and inactivation of the tumor suppressor RB.

Inheritance, Genes, and Chromosomes

The Big Picture

- Genes are units of inheritance that are passed down to the progeny by the fusion of gametes to form a new organism.
- Meiosis generates gametes that contain only one allele for each gene, and these alleles are segregated independently (unless they are linked).
- By studying the expression of particular phenotypes in successive generations, one can begin to understand the alleles that gave rise to those phenotypes (i.e., rare recessive, rare dominant, sex-linked).
- Exceptions to Mendel's rules help explain the interactions of gene products and the interaction of alleles that lead to the observed phenotype.
- Bacteria can transfer genes to other recipient cells during conjugation.

Study Strategies

- It is easy to become overwhelmed by all the exceptions to Mendel's rules. Initially, make sure you understand Mendel's laws of segregation and independent assortment. Think about the exceptions to these rules in terms of the molecular role of the alleles. For example, what is the explanation for a codominant phenotype at the molecular level, such as the AB blood type in humans? Both gene products are synthesized by the cell and modify the cell surface glycoproteins, producing both A and B antigens. Or, what is the molecular explanation for epistasis? The enzyme required to produce an initial substrate necessary for pigmentation is missing, even though the other gene products (farther down the biochemical pathway) for producing a particular color are present in the cell.
- Review the figures of meiosis in Chapter 11 to see how Mendel's laws of independent assortment and segregation apply. Draw the chromosomes with different alleles (*S* on one chromosome, *s* on the other) and follow how these alleles are segregated to the gametes during meiosis.
- Use a Punnett square to predict the outcome of different monohybrid and dihybrid crosses. Use the probability rules to predict the genotypes and phenotypes of these same crosses and compare the two methods.
- After reviewing Mendel's laws of independent assortment and allele segregation, make a list of the exceptions to the laws, with specific examples for each exception.
- Go to the Web addresses shown to review the following animated tutorials and activities. You can also find them in BioPortal (yourBioPortal.com), along with many additional learning resources.

  Animated Tutorial 12.1 Independent Assortment of Alleles (Life10e.com/at12.1)

  Animated Tutorial 12.2 Pedigree Analysis (Life10e.com/at12.2)

  Animated Tutorial 12.3 Alleles That Do Not Sort Independently (Life10e.com/at12.3)

  Activity 12.1 Homozygous or Heterozygous? (Life10e.com/ac12.1)

  Activity 12.2 Concept Matching I (Life10e.com/ac12.2)

  Activity 12.3 Concept Matching II (Life10e.com/ac12.3)

Key Concept Review

12.1 What Are the Mendelian Laws of Inheritance?

Mendel used the scientific method to test his hypotheses

Mendel's first experiments involved monohybrid crosses

Mendel's first law states that the two copies of a gene segregate

Mendel verified his hypotheses by performing test crosses

Mendel's second law states that copies of different genes assort independently

Probability can be used to predict inheritance

Mendel's laws can be observed in human pedigrees

Mendel studied the inheritance of characters (such as flower color) and traits (a particular feature, such as a red, purple, or white color) as they were passed (heritable traits) from one generation of pea plants to another. In genetic crosses, the two individuals in the initial cross are the parents (the P generation) and their offspring are the first filial generation (the F_1 generation). The F_2 (second filial generation) are the offspring from a cross of the F_1 generation. Mendel's theory states that discrete particles of inherited traits exist in pea plants in pairs (one from each parent, diploid) and that these particles separated from each other and were assorted independently during gamete formation. Each of the gametes received one of these particles (haploid), whereas the zygote received two particles for a trait, one from each parent. This unit of inheritance is a gene. The totality of all genes in an organism is its genome. Alleles are different forms of the same gene. True-breeding parents are homozygous for a given trait and have two copies of the same allele, one on each homologous chromosome. Heterozygous individuals have two different alleles for a gene, one on each homologous chromosome. Phenotype is the physical appearance of an organism (e.g., having seeds that are spherical or wrinkled).

Mendel's first law, the law of segregation, states that when any individual produces gametes, alleles separate so that each gamete receives only one allele for each gene. Genes are located at particular sites on chromosomes called loci (singular, locus). Punnett squares can be used to analyze the potential outcome of a genetic cross. All of the possible gametes from one parent are lined up on one side of the Punnett square, and all of the gametes from the other parent are lined up on the other side. Genotypes of the progeny are predicted by placing both alleles of the gametes in the square (see Figures 12.3, 12.5, and 12.6). Test crosses are used to determine if a given individual is heterozygous or homozygous for a dominant allele. The individual is crossed with another individual homozygous for the recessive trait (the test cross; see Figure 12.5). Mendel's second law, the law of independent assortment, states that all of the alleles of different genes assort independently of one another during gamete formation. Review the Punnett square in Figure 12.6.

Probability calculations can be used to predict the outcome of genetic crosses. The product rule states that the probability of two independent events happening together (a joint probability) is equal to the product of the probability of each of those individual events (see Figure 12.8). The sum rule is used to predict the probability of an event that can occur in two or more different ways. The probability that the event will occur in one of these ways is equal to the sum of the individual probabilities (see Figure 12.8).

Question 1. (a) Diagram two separate pairs of chromosomes, each bearing a different allele: for instance, *Ss* and *Yy*, as seen in the dihybrid cross with seed shape (*Ss*) and seed color (*Yy*). Show how these alleles assort independently during meiosis to produce haploid gametes (review Figure 12.7). (b) Draw a diagram starting with the same parent (*SsYy*), but assume that *S* and *Y* are linked and are 20 map units apart. For this parent, the *S* and *Y* alleles are on one chromosome and the *s* and *y* alleles are on the other chromosome. Draw the gametes you would expect if there is a crossover between *S* and *Y*.

Question 2. Draw a sample pedigree with three generations in which the paternal grandfather has a rare dominant autosomal trait. What is the probability that one of his children will have the disease? What is the probability that one of his grandchildren will have the disease?

Question 3. Draw a sample pedigree with three generations in which the maternal grandmother and paternal grandfather are carriers of a rare recessive autosomal trait. What is the probability that one of their children will be carriers of this trait? What is the probability that a grandchild will have the disease?

Question 4. Suppose you are a genetics counselor who is working with a 21-year-old pregnant woman who has just discovered that her father has Huntington's chorea, a rare dominant autosomal trait. This disease usually develops in middle age, so people carrying this trait do not find out they have this genetic disorder until midlife. What are the chances that the child she is carrying will develop the disease? (Assume that her husband's family has no history of the disease.) What is the chance that she has Huntington's chorea?

12.2 How Do Alleles Interact?

New alleles arise by mutation

Many genes have multiple alleles

Dominance is not always complete

In codominance, both alleles at a locus are expressed

Some alleles have multiple phenotypic effects

Pedigrees can be used to predict how traits are inherited from generation to generation. If a trait is due to a rare dominant allele, (1) the affected person has an affected parent; (2) on average, one-half of the offspring of an affected parent will be affected; and (3) the phenotype is seen equally in both sexes (see Figure 12.9A). If a trait is due to a rare recessive allele, (1) the affected individual can have unaffected parents; (2) on average, one-fourth of the children from unaffected parents express the trait; and (3) the phenotype occurs equally in both sexes (see Figure 12.9B).

Alleles interact in different ways. New alleles arise by mutation. "Wild type" refers to traits that occur in most individuals in nature, whereas "mutant" refers to traits that are different from wild type. A polymorphic trait is a trait that has many different forms, with each individual phenotype present in less than 99 percent of the population. Because of random mutations, multiple alleles for a given gene may exist in a group of individuals (for example, rabbit fur color; see Figure 12.10).

Incomplete dominance is common in nature. It is a situation in which neither of the two alleles of a gene is dominant and individuals that are heterozygotic for a particular trait appear as intermediate phenotypes of the parents. For example, a cross between an eggplant with purple fruit and an eggplant with white fruit yields F_1 plants with violet fruit (see Figure 12.11).

In codominance, both alleles are expressed in the individual (for example, blood types in humans; see Figure 12.12).

A pleiotropic allele is a single allele that can have more than one effect in an individual. For example, in the Siamese cat, a single allele can affect both pigmentation and crossed eyes.

Question 5. If both parents have AB blood type, what proportion of their children would be expected to have O blood type?

Question 6. For a typical situation of incomplete dominance in which one parent is *BB* and the other is *bb*, what percentage of offspring would be expected to display the heterozygous phenotype in the F_2 generation?

12.3 How Do Genes Interact?

Hybrid vigor results from new gene combinations and interactions

The environment affects gene action

Most complex phenotypes are determined by multiple genes and the environment

In epistasis, the phenotypic expression of one gene is affected or masked by another gene. For example, an allele for yellow coat color in Labrador retrievers can mask another allele for brown or black color (see Figure 12.13). The superiority of individuals that are heterozygous for a particular trait is known as "hybrid vigor," or heterosis. For example, hybrid varieties of corn produce much better yields than their homozygous parents (see Figure 12.14).

Environmental conditions such as light, temperature, or nutrition can affect the expression of a particular trait. Both penetrance and expressivity are related to environmental effects (see Figure 12.15). Penetrance is the proportion of individuals within a group with a particular genotype that actually show the expected phenotype. Expressivity is the degree a particular genotype is expressed in an individual. Variation of phenotypes within a population is called quantitative, or continuous, variation. More than one gene (quantitative trait loci) can affect a particular phenotype, with each allele intensifying or diminishing the phenotype (for example, human height). These traits can be affected by the environment.

Question 7. Compare the closely related concepts of penetrance and expressivity.

Question 8. Explain how a rabbit might have a white body and black legs and ears, assuming only a single gene is involved.

12.4 What Is the Relationship between Genes and Chromosomes?

Genes on the same chromosome are linked

Genes can be exchanged between chromatids and mapped

Linkage is revealed by studies of the sex chromosomes

Different alleles on the same chromosome do not assort independently, but rather are linked, as first noted by Morgan in his studies on *Drosophila* (see Figure 12.17). Genes can be exchanged between chromatids of a homologous pair during prophase I of meiosis. Linkage groups are the full set of loci on a given chromosome. Recombination frequencies between alleles that are linked (due to crossing over during prophase of meiosis I, see Figure 12.18) can be used to map positions of genes on a chromosome. Recombination frequencies are expressed in map units and are equal to the percent of recombinant progeny divided by the total progeny. A map unit (also referred to as a centimorgan, cM) corresponds to a recombination frequency of 0.01 (see In-Text Art, p. 248).

Sex is determined in different ways in different species. Monoecious organisms (earthworms, pea plants, corn, etc.) produce both female and male gametes. Dioecious organisms (some plants and most animals) produce only male or only female gametes, resulting in two separate sexes. In many animals, including humans, sex is determined by one or two sex chromosomes. The other chromosomes are called autosomes. In mammals, there are two X chromosomes in females and one X chromosome and one Y chromosome in males. Males produce two kinds of gametes that carry either an X or a Y chromosome. Nondisjunction of the sex chromosomes produces gametes that can result in Turner syndrome (female, XO) or Klinefelter syndrome (male, XXY).

Genes on sex chromosomes are inherited in special ways. In mammals and other organisms, including some insects, females have two copies of each X-linked gene, whereas males only have one copy of the X-linked genes. Because males are hemizygous for X-linked traits, all their X-linked traits will be expressed. If an allele is located on the X chromosome, reciprocal crosses do not have the same result (see Figure 12.20). X-linked recessive traits show the following characteristics: The phenotype is much more common in males than in females. Males with X-linked traits can pass those traits only to their daughters. Females who are carriers of an X-linked trait (heterozygous) pass that trait on average to one-half of their sons and one-half of their daughters. Because sons always receive the X chromosome from their mother, one-half of those sons, on average, will express the X-linked trait.

Question 9. Draw a pedigree for three generations in which the grandfather has red–green color blindness and his daughter is a carrier. This daughter has four sons. How many of her sons will be color-blind?

Question 10. What is the genetic significance of a hemizygous gene?

12.5. What Are the Effects of Genes Outside the Nucleus?

Cytoplasmic inheritance differs from the Mendelian pattern. Mitochondria and plastids contain genomes that are passed to the zygote from the gamete that makes the largest cytoplasmic contribution (in humans, the egg). These organelles are generally so numerous that genes for cytoplasmic traits are polyploid. Genes in organelles mutate faster than those in the nucleus, leading to multiple alleles for many non-nuclear genes. In plants, mutations in plastid genes affect proteins that assemble chlorophyll molecules into photosystems, resulting in white rather than green leaves (see Figure 12.22). Mutations in mitochondrial genes affect the electron transport chain and the synthesis of ATP. These mutations have a strong effect on tissues that require high ATP concentrations, particularly the nervous and muscular systems and kidneys.

Question 11. Cytoplasmic traits in certain species of trees are passed from the male plant to all of its progeny. Compare this observation to cytoplasmic inheritance in humans.

Question 12. Explain the basis for maternal inheritance of mitochondria in humans.

12.6. How Do Prokaryotes Transmit Genes?

Bacteria exchange genes by conjugation

Bacterial conjugation is controlled by plasmids

Prokaryotes exchange genes by conjugation. *E. coli* has a single circular chromosome that is 1 μM in circumference and carries a few-thousand genes. Some bacteria mate by transferring part of their chromosome through a conjugation tube to a recipient bacterial cell. Those bacteria must contact the recipient cell through a projection called a sex pilus (see Figure 12.23A). Genes that are transferred during the mating process recombine with the recipient cell's chromosome. Many bacterial cells harbor plasmids, which are small circular DNA molecules that replicate independently. Some plasmids harbor genes with metabolic capabilities, genes for conjugation, or genes for antibiotic resistance.

Question 13. When two strains of bacteria with genotypes *ABcd* and *abCD* are grown together in the lab, a small number of bacteria with the genotype *ABCD* eventually arise. How does this likely occur?

Question 14. Many antibiotic resistance genes are located on plasmids. Antibiotics are routinely given to livestock by many farmers to improve productivity and prevent illness. Why are many individuals concerned about this practice as it relates to human health?

Test Yourself

1. Hemophilia is a trait carried by the mother and passed to her sons. The allele for hemophilia, therefore,
   a. is carried on one of the mother's autosomal chromosomes.
   b. is carried on the Y chromosome.
   c. can be carried on the X or Y chromosome.
   d. is on the X chromosome and can be inherited by the son only if the mother is a carrier (heterozygous).
   e. is carried in the mitochondrial genome because sons inherit this allele from their mothers.
   Textbook Reference: *12.4 What Is the Relationship between Genes and Chromosomes?*

2. Genetic inheritance was once thought to be a function of the blending of traits from the two parents. Which exception to Mendel's rules is an example of blending?
   a. X linkage
   b. Polygenic inheritance
   c. Incomplete dominance
   d. Codominance
   e. Pleiotropism
   Textbook Reference: *12.2 How Do Alleles Interact?*

3. True-breeding plants
   a. produce the same offspring when crossed for many generations.
   b have no mutations.
   c. result from a monohybrid cross.
   d. result from a dihybrid cross.
   e. result from crossing over during prophase I of meiosis.
   Textbook Reference: *12.1 What Are the Mendelian Laws of Inheritance?*

4. What is the probability that a cross between a true-breeding pea plant with spherical seeds and a true-breeding pea plant with wrinkled seeds will produce F_1 progeny with spherical seeds?
   a. ½
   b. ¼
   c. 0
   d. ⅛
   e. 1
   Textbook Reference: *12.1 What Are the Mendelian Laws of Inheritance?*

5. What is the pattern of inheritance for a rare recessive allele?
   a. Every affected person has an affected parent.
   b. Unaffected parents can produce children who are affected.
   c. Affected parents do not produce affected children.
   d. Unaffected mothers have affected sons and daughters who are carriers.
   e. None of the above
   Textbook Reference: *12.1 What Are the Mendelian Laws of Inheritance?*

6. What is the pattern of inheritance for a rare dominant allele?
   a. Every affected person has an affected parent.
   b. Unaffected parents can produce children who are affected.
   c. Affected parents do not produce affected children.
   d. Unaffected mothers have affected sons and daughters who are carriers.
   e. None of the above
   Textbook Reference: *12.1 What Are the Mendelian Laws of Inheritance?*

7. What is the pattern of inheritance for a sex-linked allele?
   a. Every affected person has an affected parent.

b. Unaffected parents can produce children who are affected.
c. Affected parents do not produce affected children.
d. Unaffected mothers have affected sons and daughters who are carriers.
e. None of the above

Textbook Reference: *12.4 What Is the Relationship between Genes and Chromosomes?*

8. Penetrance and expressivity are related to
   a. the increased expression of a particular trait when a hybrid species is formed.
   b. quantitative traits that diminish or intensify a particular phenotype.
   c. the influence of environment on the expression of a particular genotype.
   d. the expression of one gene masking the effects of another gene.
   e. the expression of a dominant phenotype in a heterozygote.

   Textbook Reference: *12.3 How Do Genes Interact?*

9. In what way is sex determination similar in humans and *Drosophila*?
   a. Females of both species are hemizygous.
   b. Males of both species have one X chromosome and females have two X chromosomes.
   c. Males from both species have one Y chromosome.
   d. In both species, secondary sex characteristics are determined by genes on the X chromosome.
   e. In both species, the ratio of X chromosomes to sets of autosomes determines maleness or femaleness.

   Textbook Reference: *12.4 What Is the Relationship between Genes and Chromosomes?*

10. Linked genes are genes that
    a. assort independently.
    b. segregate equally in the gametes during meiosis.
    c. always contribute the same trait to the zygote.
    d. are found on the same chromosome.
    e. recombine during mitosis.

    Textbook Reference: *12.4 What Is the Relationship between Genes and Chromosomes?*

11. Cytoplasmic inheritance
    a. results from polygenic nuclear traits.
    b. is determined by nuclear genes.
    c. results from gametes' contributions of equal amounts of cytoplasm to the zygote.
    d. is determined by genes on DNA molecules in mitochondria and chloroplasts.
    e. follows Mendel's law of segregation.

    Textbook Reference: *12.5 What Are the Effects of Genes Outside the Nucleus?*

12. Epistasis is
    a. the degree to which a particular genotype is expressed in an individual.
    b. the proportion of individuals within a group with a particular genotype that show the expected phenotype.
    c. a situation in which a heterozygotic individual expresses an intermediate phenotype of the parents.
    d. a situation in which one gene masks the expression of another gene.
    e. a situation in which both alleles are expressed equally.

    Textbook Reference: *12.3 How Do Genes Interact?*

13. Quantitative traits are traits
    a. that are affected by the environment.
    b. that affect the same physical characteristic.
    c. in which each allele intensifies or diminishes the phenotype.
    d. All of the above
    e. None of the above

    Textbook Reference: *12.3 How Do Genes Interact?*

14. A test cross
    a. is used to determine if an organism that is displaying a dominant trait is heterozygous or homozygous for that trait.
    b. is used to determine if an organism that is displaying a recessive trait is heterozygous or homozygous for that trait.
    c. causes the loss of hybrid vigor.
    d. results in an F_2 generation with a phenotypic ratio of ¾ dominant to ¼ recessive.
    e. results in the same alleles being transferred from generation to generation.

    Textbook Reference: *12.1 What Are the Mendelian Laws of Inheritance?*

15. An individual has a karyotype that is XX but is phenotypically male. What could explain this result?
    a. The Y chromosome was not visible in the karyotype.
    b. Sex is determined by the autosomes and not the X chromosomes.
    c. A translocation has occurred, placing the *SRY* gene on one of the X chromosomes.
    d. The DAX 1 protein is overproduced.
    e. A nondisjunction event has resulted in Kleinfelter syndrome.

    Textbook Reference: *12.4 What Is the Relationship between Genes and Chromosomes?*

16. A bacterial cell's genotype can be altered by
    a. the introduction of a plasmid that carries some of the bacteria's genes.
    b. mating with a bacterial cell with the same genotype.
    c. homologous recombination with human DNA.
    d. the forming of a conjugation tube with another bacterial cell.
    e. the transferring of genetic material from a different strain of bacteria.

    Textbook Reference: *12.6 How Do Prokaryotes Transmit Genes?*

Answers

Key Concept Review

1.

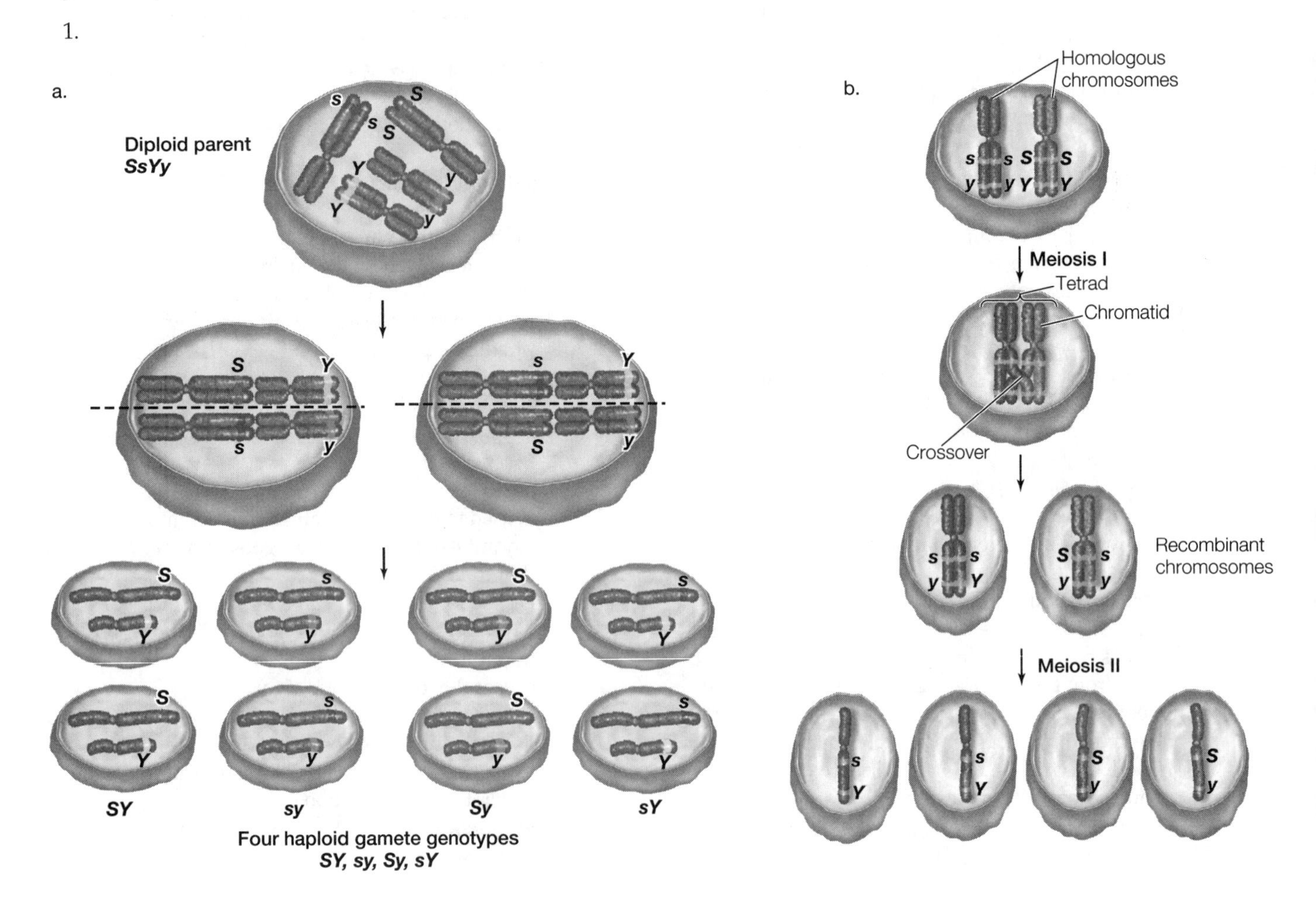

2. See Figure 12.9A. One-half of his children could get the disease; one-fourth of his grandchildren could get the disease.

3. See Figure 12.9B, generations III and IV. One-half of the children of these grandparents could be carriers. One-sixteenth of the children could have the disease.

4. There is a 50 percent chance that she will develop Huntington's chorea. Because the trait is an autosomal dominant allele, one-half of her father's gametes will contain the homologous chromosome carrying that allele and one-half of his gametes will contain the homologous chromosome that carries the wild-type allele. If she received the Huntington's allele, her child has a 50 percent chance of receiving this allele from her. The product rule is used to predict the probability that her child will inherit the Huntington's allele: ½ (the probability that she has the Huntington's allele) × ½ (the probability her child will inherit this allele from her) = ¼ (the probability her child has the allele). Her child has a 25 percent chance of carrying the Huntington's chorea allele and thus of developing the disease.

5. None. The expected ratios would be 25 percent type A, 50 percent type AB, and 25 percent type B.

6. In the F_1 generation, all progeny should be heterozygous with the *Bb* genotype and display an intermediate phenotype. Crossing F_1 individuals then should produce a 1:2:1 ratio of *BB*: *Bb*:*bb*. Therefore, 50 percent of the F_2 generation would be expected to be heterozygotes.

7. Penetrance indicates the percent of a population that displays the expected phenotype for a specified genotype (for some genes, all individuals will not necessarily display the phenotype that one would expect based on the genotype). Expressivity refers to the degree of phenotypic expression among those individuals who do display the phenotype. Some individuals may have slight expression and some a very high degree of expression.

8. This represents an example of how environment can affect gene expression. In this classic case, it was determined that such animals had a mutant allele for black fur. The enzyme responsible for the black color is not functional at core body temperature and the fur remains white for the main mass of the body, but at the slightly cooler temperatures in the legs and ears, the enzyme is functional and the fur is black
9. See Figure 12.21, generations II, III, and IV. One-half of her sons could be color-blind.
10. "Hemizygous" generally refers to X-linked genes in males (XY) and indicates that the individual has only one allele for such genes. It is significant because males will only have the allele from the mother and therefore display the phenotype of that allele, regardless of dominant or recessive, since there is not another allele for the particular gene. This situation for X-linked genes would not occur for females since they would have two copies of the X chromosome.
11. In humans, the gamete with the largest cytoplasmic contribution is the egg, so cytoplasmic inheritance is passed from the female parent to all of her children. In certain tree species, the male gamete contributes the majority of the cytoplasm to the zygote, so all the mitochondria and chloroplasts in the zygote are inherited from the male parent.
12. Upon fertilization of an egg by a sperm cell, the nucleus of the sperm cell enters the egg to create a zygote, but the mitochondria of the sperm cell do not generally enter. Therefore, the zygote and all resulting cells of the new organism have mitochondria from the mother only.
13. Bacterial conjugation would be the likely explanation for the ABCD genotype. DNA can be transferred from one cell to another by passing through the sex pilus and then recombining with or integrating into the recipient cell genome.
14. Antibiotic resistance genes occur on plasmids and plasmids are transferred between bacteria by conjugation. Antibiotics in the livestock feed should select for a population of bacteria that are resistant to those antibiotics. This increased population of bacteria carrying resistance genes makes it more likely that they will transfer the resistance genes to other bacteria that cause disease in humans. If humans have diseases or infections caused by bacteria that are resistant to the antibiotics commonly used for treating the conditions, the success of treatment could be greatly diminished.

Test Yourself

1. **d.** Hemophilia is an X-linked trait and can only be inherited by the son from his mother's X chromosome (and not her mitochondrial chromosome). The father contributes the Y chromosome to his son (not his X chromosome) and thus cannot pass any of his X-linked alleles to his son.
2. **c.** Incomplete dominance results in the progeny's expressing an intermediate form of the two parental alleles. (In a cross between red-flowered plants and white-flowered plants, the expression of pink-flowered plants would be a "blend" of the parental traits.) Codominance is not an example of blending because both alleles are fully expressed in the individual.
3. **a.** Monohybrid and dihybrid crosses produce heterozygous individuals; true-breeding individuals are always homozygous.
4. **e.** This is an example of a monohybrid cross. All of the F_1 progeny would have spherical seeds. (The F_1 generation would all have the genotype *Ss*, producing the phenotype of spherical seeds because the spherical allele, *S*, is dominant to the wrinkled allele, *s*.)
5. **b.** Rare recessive alleles can be carried by both parents but not expressed in the parents. If the parents are heterozygous for this allele (*Aa*), their children will have a one-fourth probability of expressing that recessive allele (*aa*). If both parents are affected (*aa*), their children will also be affected (*aa*).
6. **a.** If an allele is dominant, every affected individual has at least one dominant allele. An affected individual must have received that allele from one of his or her parents. Because the allele is dominant, that parent must also be affected.
7. **d.** The most common sex-linked alleles are X-linked and are passed from a mother to her son (because the mother always donates one of her X chromosomes to her son, and the father always donates the Y chromosome to his son). Daughters can also receive the X-linked allele from their mothers, but the father donates the other X chromosome, so daughters can be carriers.
8. **c.** Penetrance and expressivity are related to the effects of the environment on a particular phenotype. Answer **a** refers to hybrid vigor, answer **b** refers to quantitative traits, answer **d** refers to epistasis, and answer **e** refers to expression of a dominant allele.
9. **b.** In these species, females have two X chromosomes and males have one X chromosome. The alleles for secondary sex characteristics are found on the X chromosome and the autosomes.
10. **d.** Linked genes by definition are on the same chromosome and thus do not sort independently, do not contribute the same trait to the zygote, and do not recombine during mitosis or segregate equally to the gametes during meiosis.
11. **d.** The genes on the mitochondria and chloroplast chromosomes which are cytoplasmically inherited (unlike nuclear genes) are passed on to all of the progeny from the gamete that contributes most of the cytoplasm.
12. **d.** Answer **a** refers to expressivity, answer **b** refers to penetrance, answer **c** refers to incomplete dominance, and answer **e** refers to codominance.

13. **d.** Quantitative traits are traits that are affected by the environment and can either diminish or intensify one phenotype.

14. **a.** A test cross is used to determine if an organism that is expressing a dominant trait is homozygous or heterozygous for that trait. A ratio of ¾ dominant to ¼ recessive in the F_2 generation results from a monohybrid cross. True-breeding individuals continue to express the same alleles generation after generation.

15. **c.** The *SRY* (sex-determining region on the *Y* chromosome) gene has been moved via a translocation to the X chromosome, and in the presence of the SRY protein, the XX individual has developed sperm-producing testes. SRY also inhibits the expression of the *DAX 1* gene, which encodes a male inhibitor. If *DAX 1* was overproduced during embryonic development, a female would be produced. Individuals with Klinefelter syndrome are XXY.

16. **e.** A bacterial cell's genotype changes when new genetic information is introduced either on a plasmid, or by conjugation (which requires a conjugation tube and the transfer of DNA). If conjugation occurs, the genes need to recombine into the recipient cell's chromosome to be maintained. Transferring the same genetic information on a plasmid or mating with a bacterial cell with the same genotype will not change the genotype of the recipient bacteria. Human DNA cannot recombine with a bacterial chromosome unless there are identical (homologous) DNA sequences in both DNA molecules.

DNA and Its Role in Heredity

The Big Picture

- Knowing that DNA is the genetic material has allowed scientists to understand how hereditary information is passed on at the molecular level and how mutations can alter hereditary information.
- Advances in our knowledge of DNA replication provide new technologies for understanding genes, their function, their expression, and genetic relatedness in different organisms.

Study Strategies

- It is easy to get lost in the history of discovery. Do not focus on who and when, but instead on *how* the experiments provided evidence to prove that DNA is the genetic material.
- Understanding DNA replication is a highly visual process. Use the figures in your textbook to "see" what is occurring. Make your own diagrams of these processes.
- 5′ and 3′ ends are often confused when parental (template) strands and daughter (newly synthesized) strands are compared. On both strands, the 3′ end corresponds to the site (or former site) of a hydroxyl group (—OH), and the 5′ end corresponds to the site (or former site) of a phosphate tail.
- Take advantage of laboratory activities that simulate or allow you to experience sequencing or PCR. Nearly all molecular research labs utilize one or both of these techniques.
- Focus on how scientific evidence proved that DNA is the genetic material and how Meselson and Stahl demonstrated that DNA replication is semiconservative.
- Draw a picture of the replication fork. Position the primers on the fork, and then draw in the leading and lagging strands. Label all of the 3′ and 5′ ends. Draw a box to indicate where ligase will seal up the ends.
- Make a list of all the proteins involved in DNA replication. Describe the function of each protein.
- Go to the Web addresses shown to review the following animated tutorials and activity. You can also find them in BioPortal (yourBioPortal.com), along with many additional learning resources.

  Animated Tutorial 13.1 DNA The Hershey–Chase Experiment (Life10e.com/at13.1)

  Animated Tutorial 13.2 Replication and DNA Polymerization (Life10e.com/at13.2)

  Animated Tutorial 13.3 The Meselson–Stahl Experiment (Life10e.com/at13.3)

  Animated Tutorial 13.4 Leading and Lagging Strand Synthesis (Life10e.com/at13.4)

  Animated Tutorial 13.5 Polymerase Chain Reaction (Life10e.com/at13.5)

  Activity 13.1 The Replication Complex (Life10e.com/ac13.1)

Key Concept Review

13.1 What Is the Evidence that the Gene Is DNA?

DNA from one type of bacterium genetically transforms another type

Viral infection experiments confirmed that DNA is the genetic material

Eukaryotic cells can also be genetically transformed by DNA

DNA is the genetic material. By the 1920s, Feulgen's staining techniques showed that DNA is present in the nucleus and the chromosomes, that different organisms show varying amounts of DNA in their nuclei, and that proportional amounts of DNA are found in somatic and germ cells. Griffith's transformation experiments, done around the same time, indicated that some transforming principle (later identified as DNA) could cause a heritable change in living cells (see Figure 13.1). In 1944 Avery, MacLeod, and McCarty used purified DNA to demonstrate that the transforming factor containing the genetic information is DNA (see Figure 13.2). In 1952, Hershey and Chase, using a virus that infected bacteria, demonstrated definitively that DNA carries the hereditary information (see Figures 13.3 and 13.4). The demonstration of transformation in eukaryotes (termed transfection) using a genetic marker was the final proof that DNA is the genetic material.

Question 1. Explain how the Hershey–Chase experiment demonstrated that DNA is the genetic material.

Question 2. Explain how the experiment of Avery et al., ruled out protein and RNA as the transforming substance and suggested that DNA was the genetic material.

13.2 What Is the Structure of DNA?

Watson and Crick used modeling to deduce the structure of DNA

Four key features define DNA structure

The double-helical structure of DNA is essential to its function

A series of experiments elucidated the structure of DNA. In the early 1950s, Franklin and Wilkins prepared X-ray crystallographic images, which were critical to understanding the structure of DNA. In 1950, Chargaff observed that the relative ratios of pyrimidines and purines, the nitrogenous bases of the nucleotides of DNA, are consistent with A = T and G = C (Chargaff's rules). In 1953, Watson and Crick coupled results from previous experimentation and model-building and revealed the three-dimensional, double-helical structure of DNA.

The form of DNA is tied to its function. DNA is a right-handed double-stranded helix formed by two antiparallel DNA strands. The two strands are aligned in the helix, with the 5′ end of one strand pairing with the 3′ end of its complementary strand. The sugar–phosphate backbones of the polynucleotide chains are on the outside of the helix, and the nitrogenous bases point toward the center of the helix. Base pairs are complementary; they consist of a pyrimidine pairing with a purine via hydrogen bonding (A pairs with T, G pairs with C). Hydrophobic interactions between the base pairs help them stack on top of one another in the center of the molecule, stabilizing it and maintaining a constant diameter. The DNA helix has a major and minor groove (see Figure 13.8), and the bases in these grooves provide distinct surfaces for protein recognition. The specific base sequences in DNA are used to store genetic information. Altering the base sequence in the DNA by mutation may cause a change in the genetic information. Precise replication of the DNA molecule is accomplished by means of complementary base pairing. The expression of genetic information in the DNA results in specific phenotypes.

Question 3. Given the following parent strand sequence, what would the daughter strand sequence look like?

5′ – G C T A A C T G T G A T C G T A T A A G C T G A – 3′

Question 4. Diagram the double helix. Be sure to label the properties that make it most suited as the genetic material.

13.3 How Is DNA Replicated?

Three modes of DNA replication appeared possible

An elegant experiment demonstrated that DNA replication is semiconservative

There are two steps in DNA replication

DNA polymerases add nucleotides to the growing chain

Many other proteins assist with DNA polymerization

The two DNA strands grow differently at the replication fork

Telomeres are not fully replicated and are prone to repair

DNA must be able replicate itself in order to be heritable. All of the following must be present for replication to take place: (1) a template strand; (2) all four deoxyribonucleoside triphosphates (dATP, dCTP, dGTP, and dTTP); (3) a primer base-paired to the template DNA; and (4) DNA polymerase.

Meselson and Stahl's experiments demonstrated that DNA replication is semiconservative: Each parental strand of the DNA duplex serves as a template for the newly synthesized (daughter) DNA, and each new DNA molecule is composed of one parental (template) and one newly replicated (daughter) strand (see Figure 13.10). During replication, DNA is unwound, and the new strand is synthesized by complementary base pairing of each incoming nucleotide to the parental strand via hydrogen bonds. New nucleotides are always added at the 3′ (—OH) end of the elongating strand via covalent (phosphodiester) bonds (see Figure 13.11).

Replication complexes assemble at origins (ori) on the DNA to begin replication and are composed of many proteins: Helicases unwind parental DNA, single-stranded binding proteins prevent parental strands from reassociating, primases generate RNA primers, DNA polymerases elongate daughter strands from the 3′ OH end of those primers, and DNA ligases seal the nicks in the sugar–phosphate backbone.

All chromosomes have at least one origin of replication where the replication complex binds. New daughter strands must begin with a short RNA primer, formed with the parental strand as a template. Once a primer is synthesized, DNA polymerase adds deoxyribonucleotides to the 3′ (—OH) end of the primer. There are several different DNA polymerases in cells; all replicate DNA, some remove primers, and some are involved in DNA repair. Replication proceeds in both directions from the replication origin, forming replication forks. Both leading-strand and lagging-strand synthesis occur at replication forks (see Figure 13.18).

Replication proceeds in a specified direction. Nucleotides can be added only to the 3′ (—OH) end of a strand. Thus the parental strand is read in the 3′ -to-5′ direction, and the daughter strand elongates in the 5′-to-3′ direction. The leading strand elongates continuously in the "right" direction (5′ to 3′) as the replication fork opens up. The lagging strands (Okazaki fragments) are synthesized in a direction opposite to the opening of the replication fork (5′ to 3′). This process requires many primers, and the resulting Okazaki fragments that are generated are ligated (covalently joined by DNA ligase) to form a continuous strand. RNA primers are removed prior to this ligation by DNA polymerase I (see Figure 13.17). The sliding clamp increases the efficiency of and rate of DNA replication by keeping DNA polymerase tightly associated with the newly replicated strand (see Figure 13.18).

In eukaryotes, DNA is threaded through a stationary replication complex that is attached to chromatin. Most eukaryotes have many origins of replication for their large linear chromosomes. Telomeres are repetitive sequences found at the ends of eukaryotic chromosomes, which, with the protein telomerase, help maintain the integrity of those ends during each replicative cycle. These telomeres are gradually lost during each replicative cycle, and this loss can eventually lead to cell death. Telomerase is continually expressed in cancer cells.

Question 5. Diagram a replication fork as it would be seen in a replicating segment of DNA. In your diagram, label the 5′ and 3′ ends of each parent strand and daughter strand. Indicate which new strand is the leading strand and which is the lagging strand of the daughter DNA.

Question 6. Based on your diagram in Question 5, construct a flowchart of DNA replication. Divide your chart into three parts: initiation, elongation, and termination of replication. Indicate the roles of helicase, DNA polymerase, single-stranded DNA binding proteins, nucleotides, parental (template) DNA strands, DNA ligase, RNA primase, RNA primers, and the sliding clamp protein.

Question 7. Explain the difference between conservative and semiconservative models of DNA replication. What were the experimental results that supported the semiconservative model? What would the results have looked like if the conservative model of DNA replication had been accurate? Are there any other potential hypotheses?

Question 8. Draw a diagram of the role of Okazaki fragments in the synthesis of the lagging strand.

Question 9. Diagram the replication complex and label the following: helicase, single-stranded DNA binding protein, DNA polymerase, primase, the leading and lagging strands, and the leading and lagging template strands. Be sure to label the 5′ and 3′ ends of the parental and newly replicated DNA. What would happen to DNA replication if the genes encoding each of these proteins were mutated such that the proteins were nonfunctional?

13.4 How Are Errors in DNA Repaired?

Errors can occur in DNA synthesis. DNA polymerase proofreads and repairs base mismatches during replication (see Figure 13.20A). Mismatch repair systems scan for possible errors in newly synthesized DNA (see Figure 13.20B). Excision repair mechanisms replace abnormal bases with functional bases (see Figure 13.20C).

Question 10. Differentiate between proofreading, mismatch repair, and excision repair. Which of these repair mechanisms is responsible for repairing a mutation that occurs in an adult cell from overexposure to the sun? Explain your answer.

Question 11. Explain the ways that DNA polymerase participates in maintaining the fidelity of the DNA.

13.5 How Does the Polymerase Chain Reaction Amplify DNA?

The polymerase chain reaction makes multiple copies of DNA sequences

Advanced techniques in molecular biology have allowed us to make multiple copies of DNA from a single strand and to determine the sequence of bases in DNA. The polymerase chain reaction (PCR) allows the rapid amplification of short DNA sequences into many identical copies (see Figure 13.21). PCR requires primers, dNTPs, a DNA polymerase that is heat-tolerant, the appropriate buffers and salts, and template DNA. The three steps in PCR amplification are denaturation, annealing, and polymerization. DNA sequencing is now a highly automated process in which modified bases are mixed with the normal substrates for DNA replication to generate a mixture of DNA fragments. These fragments are used to determine the base sequence of the DNA.

Question 12. Explain how PCR amplifies a particular sequence of DNA.

Question 13. What is special about DNA polymerase from *T. aquaticus* that makes it valuable in the PCR technique?

Test Yourself

1. Griffith's experiments showing the transformation of R strain pneumococcus bacteria to S strain pneumococcus bacteria in the presence of heat-killed S strain bacteria provided evidence that
   a. an external factor was affecting the R strain bacteria.
   b. DNA was definitely the transforming factor.
   c. S strain bacteria could be reactivated after heat killing.
   d. S strain bacteria required special genes to be pathogenic.
   e. All of the above
   Textbook Reference: *13.1 What Is the Evidence that the Gene Is DNA?*

2. Experiments by Avery, MacLeod, and McCarty supported DNA as the genetic material by showing that
   a. both protein and DNA samples provided the transforming factor.
   b. DNA had to be destroyed by DNase in order to transform the R bacteria.
   c. DNA is not complex enough to be the genetic material.
   d. only samples with DNA provided transforming activity.
   e. even though DNA is molecularly simple, it provides adequate variation to act as the genetic material.
   Textbook Reference: *13.1 What Is the Evidence that the Gene Is DNA?*

3. Hershey and Chase used radioactive ^{35}S and ^{32}P in experiments to provide evidence that DNA is the genetic material. These experiments pointed to DNA because

a. progeny viruses retained ^{32}P but not ^{35}S.
b. the presence of ^{32}P in progeny viruses indicated that DNA was passed on.
c. the absence of ^{35}S in progeny viruses indicated that proteins were not passed on.
d. ^{32}P indicated where the DNA label was localized.
e. All of the above

Textbook Reference: *13.1 What Is the Evidence that the Gene Is DNA?*

4. Chargaff observed that the amount of _______ was roughly equal to the amount of _______ in all tested organisms.
a. purines; pyrimidines
b. A; T
c. A + T; G + C
d. A + G; T + C
e. a, b, and d

Textbook Reference: *13.2 What Is the Structure of DNA?*

5. Watson and Crick's model allowed them to visualize
a. the molecular bonds of DNA.
b. the sugar and phosphate component of the DNA molecule's surface.
c. how the purines and pyrimidines fit together in a double helix.
d. the antiparallel design of two strands of the DNA double helix.
e. All of the above

Textbook Reference: *13.2 What Is the Structure of DNA?*

6. A fundamental requirement for the functioning of genetic material is that it must be
a. conserved among all organisms with very little variation.
b. passed intact from one species to another.
c. accurately replicated.
d. found outside the nucleus.
e. replicated accurately over many millions of years of evolution.

Textbook Reference: *13.2 What Is the Structure of DNA?*

7. Evidence of the semiconservative nature of DNA replication came from
a. DNA staining techniques.
b. X-ray crystallography.
c. DNA sequencing.
d. density gradient studies using "heavy" nucleotides.
e. None of the above

Textbook Reference: *13.3 How Is DNA Replicated?*

8. The primary function of DNA polymerase is to
a. add nucleotides to the growing daughter strand.
b. seal nicks along the sugar–phosphate backbone of the daughter strand.
c. unwind the parent DNA double helix.
d. generate primers to initiate DNA synthesis.
e. prevent reassociation of the denatured parental DNA strands.

Textbook Reference: *13.3 How Is DNA Replicated?*

9. When the lagging daughter strand of DNA is synthesized, unreplicated gaps are formed on the parental DNA. Lagging strand synthesis fills these gaps by
a. synthesizing short Okazaki fragments in a 5′-to-3′ direction.
b. synthesizing multiple short RNA primers to initiate DNA replication.
c. using DNA polymerase I to remove RNA primers from Okazaki fragments.
d. filling in those gaps with new strands of complementary DNA as the replication fork proceeds.
e. All of the above

Textbook Reference: *13.3 How Is DNA Replicated?*

10. RNA primers are necessary in DNA synthesis because
a. DNA polymerase is unable to initiate replication without an origin.
b. the DNA polymerase enzyme can catalyze the addition of deoxyribonucleotides only onto the 3′ (—OH) end of an existing strand.
c. RNA primase is the first enzyme in the replication complex.
d. primers mark the sites where helicase has to unwind the DNA.
e. All of the above

Textbook Reference: *13.3 How Is DNA Replicated?*

11. Proofreading and repair occur
a. at any time during or after synthesis of DNA.
b. only before DNA methylation.
c. only in the presence of DNA polymerase.
d. only in the presence of an excision repair mechanism.
e. only during replication.

Textbook Reference: *13.4 How Are Errors in DNA Repaired?*

12. Thirty percent of the bases in a sample of DNA extracted from eukaryotic cells are adenine. What percentage of cytosine is present in this DNA?
a. 10 percent
b. 20 percent
c. 30 percent
d. 40 percent
e. 50 percent

Textbook Reference: *13.2 What Is the Structure of DNA?*

13. Which of the following represents a bond between a purine and a pyrimidine (in the correct order)?
a. C–T
b. G–A
c. G–C
d. T–A
e. A–G

Textbook Reference: *13.2 What Is the Structure of DNA?*

14. Which of the following statements about DNA replication is *false*?
   a. Okazaki fragments are synthesized as part of the leading strand.
   b. Replication forks represent areas of active DNA synthesis on the chromosomes.
   c. Error rates for DNA replication are reduced by proofreading of the DNA polymerase.
   d. Ligases and polymerases function in the vicinity of replication forks.
   e. The sliding clamp protein increases the rate of DNA synthesis.
   Textbook Reference: *13.3 How Is DNA Replicated? ; 13.4 How Are Errors in DNA Repaired?*

15. The PCR technique
   a. can amplify only very small samples of DNA.
   b. amplifies several random DNA sequences within a genome.
   c. requires synthetic primers to flank the regions of interest.
   d. is accomplished in three sequential steps: annealing, denaturation, and replication.
   e. generates DNA molecules that all have variable sequences.
   Textbook Reference: *13.5 How Does the Polymerase Chain Reaction Amplify DNA?*

16. Which of the following would *not* be found in a DNA molecule?
   a. Purines
   b. Ribose sugars
   c. Phosphates
   d. Sulfur
   e. Nitrogenous bases
   Textbook Reference: *13.2 What Is the Structure of DNA?*

17. If a high concentration of a particular nucleotide lacking a hydroxyl group at the 3′ end is added to a PCR reaction,
   a. no additional nucleotides will be added to a growing strand containing that nucleotide.
   b. strand elongation will proceed as normal.
   c. nucleotides will be added only at the 5′ end.
   d. *T. aquaticus* DNA polymerase will become nonfunctional.
   e. the primer will be unable to anneal with the DNA template.
   Textbook Reference: *13.5 How Does the Polymerase Chain Reaction Amplify DNA?*

18. To determine if DNA replication is semiconservative, conservative, or dispersive, Meselson and Stahl labeled *E. coli* DNA with a regimen of heavy nitrogen (H) for one round of replication and then transferred these cells to light nitrogen (L) for two more rounds of replication. Which of the following statements would *not* have been true within the context of this experiment?
   a. If DNA replication was conservative, no DNA molecules of intermediate density (H-L) would have been seen.
   b. If DNA replication was dispersive, only DNA molecules that were of intermediate density (H-L) would have been seen.
   c. If DNA replication was semiconservative, the DNA molecules that were made would all be heavy density (H-H).
   d. If DNA replication was semiconservative, the DNA molecules would consist of one parental strand base paired to one newly replicated strand.
   e. If DNA replication was semiconservative, a higher proportion of DNA molecules from future divisions would have been low density (L-L).
   Textbook Reference: *13.3 How Is DNA Replicated?*

19. The telomeres at the ends of linear chromosomes allow
   a. the 5′ ends of the chromosomes to undergo recombination.
   b. the gaps left by primer removal of lagging strands to be repaired by telomerase.
   c. DNA repair enzymes to recognize those ends and remove them.
   d. normal cells to divide continuously.
   e. DNA breaks to be examined at cell division checkpoints.
   Textbook Reference: *13.3 How Is DNA Replicated?*

Answers

Key Concept Review

1. By using viral proteins labeled with radioactive sulfur and viral DNA labeled with radioactive phosphate, Hershey and Chase showed that it was the radiolabeled phosphate that was transferred to bacteria, and therefore DNA that was responsible for altering the genetics of the bacterial cell.
2. Even after treating the transforming substance with RNase and protease to destroy RNA and proteins, respectively, bacteria were still transformed from R-type to S-type. When treated with DNase, the transforming ability was lost, thus implying that DNA was responsible.
3. 3′–CGATTGACACTAGCATATTCGACT–5′
4. See Figures 13.7 and In-Text art pp. 265–266 in the textbook.
5. See Figures 13.15 and 13.16 in the textbook.
6. See Figures 13.15 and 13.16 in the textbook.
7. In the conservative model of DNA replication, the parental DNA remains intact, and a newly synthesized molecule consists of two newly replicated daughter strands. If this were indeed DNA's method of replication, Meselson and Stahl would not have seen the intermediate density band in the first generation of

replication; they would have seen a dense band corresponding to the parental DNA (H-H) and a light band corresponding to the daughter DNA (L-L). Because they saw an intermediate band, they knew that one strand was heavy and one was light. Therefore, replication was semiconservative, with each new molecule consisting of one parental strand and one daughter strand. A third hypothesis was the dispersive model, according to which each new molecule contains bits and pieces of both old and new strands. This model was rejected because examination of the second round of replication in light nitrogen showed that the result was an H-L band and an L-L band. The L-L band could not have been generated by dispersive replication; all of the bands would again be of intermediate density (H-L).

8. See Figures 13.16 and 13.17 in the textbook.

9.

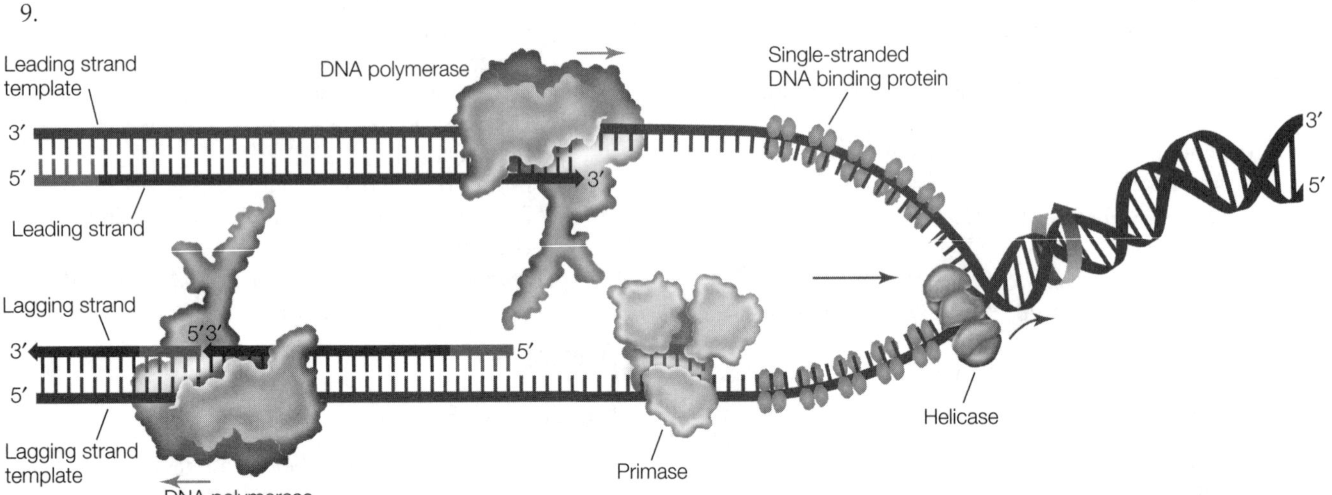

If helicase was nonfunctional, replication would not start because the DNA strands would not be unwound.

If single-stranded DNA binding protein was nonfunctional, the single strands of DNA would not be stabilized, and replication could not start.

If primase was nonfunctional, DNA polymerase could not initiate. DNA polymerase requires a primer with a 3′ OH group base-paired to the template DNA to initiate replication.

If DNA polymerase was nonfunctional, replication could not occur, even though the DNA was unwound and stabilized in an "open" configuration.

10. Proofreading occurs during synthesis of the DNA molecule, mismatch repair occurs after synthesis, and excision repair removes any abnormalities in "mature" DNA. DNA damage due to UV exposure is generally repaired via excision repair mechanisms.

11. DNA polymerase participates in normal proofreading of normal base addition and is also an important component of mismatch and excision repair processes.

12. See Figure 13.21 in the textbook.

13. *T. aquaticus* DNA polymerase is still functional at the 90° denaturation temperature of PCR.

Test Yourself

1. **a.** The experiments showed only that something was causing the R strain pneumococcus to develop the S strain virulence. Additional experiments were necessary to provide evidence that the transforming factor was indeed DNA.

2. **d.** These researchers were able to isolate nearly pure DNA samples. It was only these samples that provided transformation activity.

3. **e.** Only DNA incorporates radioactive ^{32}P. Radioactive ^{35}S is incorporated into proteins. The fact that progeny

viruses retained radioactive ^{32}P and not radioactive ^{35}S indicated that DNA is heritable and protein is not.

4. **e.** Chargaff found that in most DNA sampled, the amount of A equaled the amount of T and the amount of G equaled the amount of C. It follows that the amount of purines (A + G) equals the amount of pyrimidines (T + C). The same does not hold for the amount of A + T versus the amount of G + C.

5. **e.** Model-building by Watson and Crick created a three-dimensional visualization of the size, bond angles, base pairings, and overall structure of the DNA molecule. The information contributing to their model came from a variety of experiments performed by other scientists.

6. **c.** Replication of DNA is a fundamental requirement for its function as the genetic material. DNA must be correctly replicated in each cell of an organism and passed from parent to offspring or from cell to daughter cell. DNA is not passed from one species to another. There is variation in DNA sequences from different organisms. DNA is found outside the nucleus in eukaryotic cells (specifically in the mitochondria, and in the chloroplasts of plants), but this DNA is inherited in a non-Mendelian fashion (see Chapter 12). For evolution to occur, there must be some mistakes in DNA replication over time.

7. **d.** Density gradient labeling allowed Meselson and Stahl to track parental versus daughter strands of DNA. Their research showed that DNA replication is a semiconservative process, with each newly synthesized DNA molecule containing one parental (template) strand base-paired to one daughter (newly synthesized) strand.

8. **a.** DNA polymerase adds nucleotides to an existing nucleotide strand, ligase seals nicks, helicase unwinds the DNA, primase generates primers, and a single-stranded DNA binding protein prevents the reassociation of the parental strands.

9. **e.** Okazaki fragments are small segments of newly synthesized DNA that have been added to short RNA primers. The RNA is removed from these small fragments by DNA polymerase I and replaced with DNA. The remaining DNA fragments are ligated (covalently bound) to form a continuous, newly synthesized DNA strand.

10. **b.** DNA polymerase cannot initiate synthesis of a nucleotide strand; it can only add onto an existing strand of RNA primer or DNA. Primers provide DNA polymerase with the 3′ (—OH) end required for the addition of deoxyribonucleotides. The origin is also required for DNA synthesis, but it is the site where the replication complex is initially assembled. Primase is not the first enzyme in the replication complex.

11. **a.** The mismatch repair mechanism operates before the newly synthesized DNA strand is methylated, but for the integrity of DNA to be maintained, repair mechanisms must be active during synthesis, modification, and utilization of DNA.

12. **b.** If 30 percent of DNA is adenine, then according to Chargaff's rule, 30 percent must be thymine. The remaining 40 percent of the DNA is cytosine and guanine. Because the ratio of cytosine to guanine must be equal, the percentage of cytosine in this DNA must be 20 percent.

13. **c.** Guanine is a purine, and its paired pyrimidine is cytosine.

14. **a.** Okazaki fragments are involved in synthesis of the lagging strand.

15. **c.** PCR amplifies specific DNA sequences from small or large samples of DNA and requires three sequential steps: denaturation, annealing, and replication. The amplified DNA molecules do not have variable DNA sequences.

16. **d.** Sulfur is a constituent of many protein molecules, but it is not found in DNA.

17. **a.** A hydroxyl group at the 3′ position of a nucleotide is necessary for the binding of any additional nucleotides. If this hydroxyl group were absent, no other nucleotides could be added to a growing strand.

18. **c.** DNA replication is semiconservative, and DNA molecules would be of intermediate density (H-L) after one round of replication in heavy nitrogen and one round in light nitrogen.

19. **b.** Telomerase binds the telomeres at the ends of the chromosomes, protects them from being degraded or recombined with other chromosomes, and ensures that these ends will be replicated after primer removal. Telomeres are not associated with cell division checkpoints or continuous cell division.

14 From DNA to Protein: Gene Expression

The Big Picture

- Every protein and RNA in the cell has a DNA blueprint (the gene sequence) that specifies the amino acid or nucleotide sequence of that gene product. These sequences determine how that gene product will fold up three-dimensionally, and ultimately, function in the cell.
- Alterations in these gene products, which are caused by mutations in the genes, help us understand how the proteins and RNAs function in the cell.

Study Strategies

- Students often try to understand every detail of gene expression without comprehending the larger picture: the way genetic information is accessed in the DNA and expressed in the cell so that the cell can function. Familiarize yourself with the details of the central dogma, but then take a step back and think about how all of these details describe the steps required to synthesize gene products.
- It is easy to confuse transcription and translation. Be able to distinguish between the two processes, and for each one, carefully review the template, the product, and the sites of initiation and termination.
- Draw diagrams of the processes of transcription and translation that include initiation, elongation, and termination. Label the components that play a role in each process. Be sure to orient the 5′ and the 3′ ends of the nucleic acid, and the N and the C terminus of the protein.
- Choose a hypothetical gene that encodes a peptide of four amino acids and draw a flowchart that illustrates how the gene in the chromosome is made into protein. Next, follow the details of gene expression: the synthesis of the RNA from the DNA, and the synthesis of the protein from the mRNA. Be sure to include the start and stop signals in each of these processes. Then alter one of those nucleotides in the DNA sequence and transcribe and translate the gene. Is the protein sequence changed? Repeat this process using a different mutation in the DNA.
- Review the experiments that revealed the presence of introns in the primary RNA transcript and describe how these introns are removed during splicing.
- Go to the Web addresses shown to review the following animated tutorials and activities. You can also find them in BioPortal (yourBioPortal.com), along with many additional learning resources.

  Animated Tutorial 14.1 Transcription (Life10e.com/at14.1)

  Animated Tutorial 14.2 Deciphering the Genetic Code (Life10e.com/at14.2)

  Animated Tutorial 14.3 RNA Splicing (Life10e.com/at14.3)

  Animated Tutorial 14.4 Translation (Life10e.com/at14.4)

  Activity 14.1 Eukaryotic Gene Expression (Life10e.com/ac14.1)

  Activity 14.2 The Genetic Code (Life10e.com/ac14.2)

Key Concept Review

14.1 What Is the Evidence that Genes Code for Proteins?

Observations in humans led to the proposal that genes determine enzymes

Experiments on bread mold established that genes determine enzymes

One gene determines one polypeptide

In the early twentieth century, Archibald Garrod studied a rare disease called alkaptonuria, in which patients accumulate homogentisic acid. He concluded that the disease was genetically determined and that these patients lacked an essential enzyme. Beadle and Tatum studied special mutant (auxotroph) and wild-type (prototroph) strains of *Neurospora*, a bread mold. Auxotrophs cannot grow on minimal media without the addition of an external mineral, vitamin, or nutrient, because they are missing the enzymes needed to synthesize these nutrients. Prototrophs can grow on minimal media because all the enzymes needed to synthesize these nutrients are functional in these cells. For the amino acid arginine, for example, each of these enzymes defines a step in the biochemical pathway of arginine synthesis. Supplying mutant strains with chemical intermediates in the arginine pathway allowed Srb and Horowitz to determine which enzymes are nonfunctional in the auxotroph (see Figure 14.1). They also analyzed individual cells for enzyme activity. From their experiments they concluded that one gene encodes one enzyme. Because enzymes can consist of several different polypeptides, this idea has led to the conclusion that one gene encodes one polypeptide. Genes also code for some RNAs (such as ribosomal RNA, transfer RNA, and regulatory RNA) that do not become translated into polypeptides.

Question 1. Suppose that two different mutant strains of a bacterium are unable to grow on a minimal medium without the addition of the amino acid lysine. Explain how this phenotype might be caused by different mutations in each strain, perhaps on the same gene and perhaps in two different genes.

Question 2. Suppose that two auxotrophic mutants that had been isolated are able to grow when fed the same biochemical intermediate. According to the experiments of Beadle and Tatum or Srb and Horowitz, the mutations in each of these auxotrophs should be in the same gene, because they were blocked at the same step in a biochemical pathway. Yet these two auxotrophs had mutations that mapped in different genes. How do you explain this?

14.2 How Does Information Flow from Genes to Proteins?

Three types of RNA have roles in the information flow from DNA to protein

In some cases, RNA determines the sequence of DNA

Information flows from genes to proteins. Genes are transcribed into complementary RNA molecules. RNA sequence is translated into the amino acid sequence of a protein. The central dogma of molecular biology states that information flows from DNA, which is transcribed into RNA, which is translated into protein.

RNA differs from DNA. Unlike DNA, which consists of two polynucleotide strands, RNA is generally a single polynucleotide strand. In RNA, the sugar molecule is ribose rather than deoxyribose, as in DNA. The bases A, G, and C in DNA also occur in RNA, but the T in RNA is replaced by U. In RNA, G still base-pairs with C, but A base-pairs with U. A strand of RNA can base-pair with a single strand of DNA. A strand of RNA may also fold over to base-pair with itself. Three types of RNA are involved in protein synthesis: messenger RNA (mRNA), transfer RNA (tRNA), and ribosomal RNA (rRNA). Messenger RNA is complementary to one DNA strand of the gene, and the information sequence consists of three sequential base units called codons. Transfer RNA (or tRNA) translates the nucleic acid (in the messenger RNA) into proteins, using its anticodon to read the codon in the mRNA. Ribosomal RNA (rRNA) catalyzes peptide bond formation and provides a structural framework for the ribosome.

Some viruses (including the tobacco mosaic virus, the poliovirus, and the influenza virus) use RNA as their genome and during infection make more RNA (using viral RNA as a template). Other viruses, known as retroviruses (like HIV), make a DNA copy of their RNA using reverse transcriptase and use that DNA to synthesize more RNA. Retroviruses can also insert that DNA copy into the host's chromosome. These viruses are exceptions to the central dogma.

Question 3. How have retroviruses caused us to re-evaluate the "central dogma"?

Question 4. What is the functional difference between mRNA and tRNA?

14.3 How Is the Information Content in DNA Transcribed to Produce RNA?

RNA polymerases share common features

Transcription occurs in three steps

The information for protein synthesis lies in the genetic code

DNA is transcribed into RNA. The enzyme RNA polymerase synthesizes RNA using one strand of the DNA as a template. The resulting RNA molecules can be messenger RNAs (which are used to synthesize proteins), or they can become transfer RNA, ribosomal RNA, or regulatory RNA (such as microRNA). RNA polymerase initiates transcription at special sequences on the DNA called promoters. When RNA polymerase binds the promoter, it unwinds the DNA. The promoter tells the RNA polymerase where to start transcription, which strand of DNA to transcribe, and in which direction to move on the DNA to begin synthesizing RNA. RNA polymerase unwinds about 10 base pairs in the DNA as it reads and synthesizes the complementary RNA strand. New RNA is synthesized in a 5′-to-3′ direction, antiparallel to the template DNA strand. New nucleotides are added to the 3′ end of the growing strand. No primer is required for initiation of RNA synthesis. RNA polymerase does not proofread, resulting in transcription errors of one for every 10^4 to 10^5 bases. RNA polymerase terminates transcription at special sequences in the DNA.

The information for translation is in the genetic code, which provides the key to "reading" mRNA sequences. Genetic information in the messenger RNA is read one codon (three bases) at a time. There are four possible bases that can be used in a codon, so there are 4^3, or 64, different codons. Sixty-one codons specify specific amino acids, so the genetic code is redundant, meaning that there is more than one codon for many of the 20 amino acids. Translation starts with the AUG codon, which also specifies the amino acid methionine. Three codons—the stop codons UAG, UAA, and UGA—are used to terminate translation. The genetic code for nuclear genes is almost identical in all living cells. There are slight differences in the genetic code of one group of protists and in mitochondria and chloroplasts. The genetic code was first determined by experiments in which artificial RNAs were added to reaction mixtures containing all of the necessary components needed for protein translation (see Figure 14.5). The resulting polypeptides were analyzed to determine the protein-synthesizing instructions for these RNA sequences. A simple UUU mRNA caused the tRNA carrying phenylalanine to bind to the ribosome.

Question 5. What are two important general differences between DNA polymerase and RNA polymerase?

Question 6. Why is it necessary for each codon to be composed of at least three nucleotides?

14.4 How Is Eukaryotic DNA Transcribed and the RNA Processed?

Many eukaryotic genes are interrupted by noncoding sequences

Eukaryotic gene transcripts are processed before translation

Each eukaryotic gene has its own promoter. Eukaryotic RNA polymerase requires other molecules to help it recognize promoters. Terminators are special sequences in eukaryotic DNA that signal the end of transcription. Eukaryotic genes contain noncoding regions called introns interspersed with coding regions (exons). Introns must be removed from the pre-mRNA transcripts by RNA processing to ensure that all the exons are adjacent to each other for translation. Introns were discovered by means of nucleic acid hybridization. Mature mRNAs were hybridized to denatured DNA containing the gene for that mRNA (see Figure 14.7). Introns in the DNA were observed to be single-stranded loops that did not hybridize to the mature mRNA. When unprocessed pre-mRNA was used in these experiments, no DNA loop was observed, indicating that intron removal occurred during mRNA maturation. Some exons encode functional domains of a protein.

Modifications of pre-mRNA occur in the nucleus of eukaryotic cells. A cap of modified GTP (a G cap) is added to the 5′ end of eukaryotic mRNA to facilitate binding of mRNA to the ribosome in the cytoplasm and to protect the mRNA from ribonuclease degradation in the cytoplasm. The pre-mRNA is cut by an enzyme that recognizes the sequence A A U A A A at the 3′ end of the mRNA, and a poly A tail (100–300 nucleotides) is added to the 3′ end. The poly A tail assists in the transport of the mRNA from the nucleus to the cytoplasm and increases the mRNA stability in the cytoplasm. Introns are "spliced," or removed, from pre-mRNAs by small nuclear ribonucleoprotein particles (snRNPs). snRNPs bind consensus sequences at the 3′ and 5′ boundaries of the introns by means of complementary base pairing. Other proteins bind the snRNPs to form the spliceosome, which uses the energy of ATP hydrolysis to cut the RNA, release the intron, and rejoin the RNA to form the mature mRNA (see Figure 14.10). Some forms of β-thalassemia (in which there is an inadequate amount of β-globin, resulting in severe anemia) are the result of a mutation in the consensus sequence, causing inappropriate splicing of β-globin RNA and a nonfunctional gene product.

Mature RNA is transported from the nucleus to the cytoplasm through the nuclear pores.

Question 7. Since introns are normally excised before translation, would a mutation in an intron necessarily result in no effect or a lesser effect than a mutation in an exon?

Question 8. What characteristic of newly synthesized mRNA could be used in the lab to separate it from other RNAs?

14.5 How Is RNA Translated into Proteins?

Transfer RNAs carry specific amino acids and bind to specific codons

Each tRNA is specifically attached to an amino acid

The ribosome is the workbench for translation

Translation takes place in three steps

Polysome formation increases the rate of protein synthesis

RNA is translated into proteins. Transfer RNA (tRNA) is the link between a codon and its amino acid. Each tRNA binds (is "charged") to a particular amino acid; they base-pair with codons in the mRNA and interact with ribosomes. tRNA is a small RNA molecule (75–80 nucleotides) that forms a characteristic shape found in all tRNAs (see Figure 14.11). The anticodon of the tRNA (found at about the mid-point in the molecule) base-pairs with the codon of the mRNA in an antiparallel fashion. This base pairing occurs on the ribosome during translation. Base pairing between the anticodon and the codon is always complementary (A pairs with U, C pairs with G) in the first two positions. At the third position of the codon (and the first position of the anticodon), base pairing does not have to be exact. This "wobble" position allows the cell to use one tRNA to base-pair with up to four different codons. An amino acid is covalently attached to the 3′ end of the tRNA; thus the tRNA (with an anticodon of AAA) has arginine covalently attached to its 3′ end. The enzymes that catalyze the covalent attachment of the amino acids to the correct tRNAs are tRNA synthase (see Figure 14.12).

The ribosome is the site in the cell where protein translation occurs. Ribosomes are composed of a large subunit and a small subunit; both subunits consist of different ribosomal proteins and RNA molecules (ribosomal RNA) (see Figure 14.13). The important sites on a ribosome are the A site (where the anticodon of the charged tRNA binds the codon of the mRNA), the P site (where the tRNA with the growing polypeptide chain is located), and the E site (where the tRNA exits from the ribosome).

Translation takes place in three steps: initiation, elongation, and termination. Initiation of translation requires: (1) the small subunit of the ribosome; (2) the initiator tRNA charged with methionine; (3) the mRNA with the start codon AUG positioned correctly on the ribosome; and (4) the initiation factors (see Figure 14.14). In prokaryotes, the rRNA of the small subunit of the ribosome base-pairs with a complementary sequence (the Shine–Dalgarno sequence) upstream from the start codon (AUG). In eukaryotes, the small subunit of the ribosome loads onto the 5′ cap of the mRNA and moves along the RNA until it finds the first AUG. When all of these components are in place, the large subunit of the ribosome joins the small subunit of the ribosome, and the incoming charged tRNA is positioned at the A site. The anticodon of this tRNA base-pairs with the codon on the mRNA that is at the A site of the ribosome. The large subunit of the ribosome catalyzes the formation of a peptide bond between the amino acid on the tRNA at the P site and the amino acid on the tRNA at the A site. This peptidyl transferase activity can be attributed to the RNA (not to any of the proteins) in the large ribosomal subunit. Once the peptide bond is formed, the ribosome shifts down the messenger RNA so that the next codon is placed at the A site. The tRNA that was in the P site is moved to the E site and will exit the ribosome. The tRNA that was at the A site is shifted to the P site. During elongation, the messenger RNA is read in the 5′-to-3′ direction, with each new codon being moved to the A site during elongation so that the next incoming charged tRNA can base-pair with the codon in the mRNA (see Figure 14.15). Termination occurs when a stop codon is moved to the A site of the ribosome. Release factor binds the stop codon, and peptidyl transferase hydrolyzes the bond between the amino acid (which will be the C terminus) that is covalently attached to the tRNA in the P site. The polypeptide is released from the ribosome, and the small and large subunits of the ribosome dissociate (see Figure 14.16). In prokaryotes, one messenger RNA is usually bound by many translating ribosomes, forming a polysome (see Figure 14.17).

Question 9. What would happen if the tRNA synthase for tryptophan added a phenylalanine to the tryptophan tRNAs instead of tryptophan?

Question 10. Draw a diagram showing ribosome structure and initiation, elongation, and termination of protein synthesis (at a minimum, include A, P, and E sites, codons and anticodons, tRNA, rRNA, and small and large subunits).

14.6 What Happens to Polypeptides after Translation?

Signal sequences in proteins direct them to their cellular destinations

Many proteins are modified after translation

Changes and modifications alter translation. Signal sequences within the amino acid sequence of some proteins in eukaryotic cells serve as addresses for the protein's final cellular destination. In the case of nuclear, mitochondrial, plastid, or peroxisomal proteins, these signal sequences can be recognized by docking proteins, which form pores in the membrane of their destination. Alternatively, they can be recognized by signal recognition particles in the cytoplasm that direct the partially translated protein to the endoplasmic reticulum for further addressing and completion of translation (see Figure 14.18). The signal recognition particle binds the docking protein in the endoplasmic reticulum and protein translation continues into the ER through a receptor pore. Proteins synthesized in the endoplasmic reticulum may: (1) be sent to the Golgi, lysosomes, or the plasma membrane; (2) be secreted; or (3) stay in the endoplasmic reticulum and be modified further. Proteins that lack signal sequences remain in the cytoplasm. Protein modification after translation includes proteolysis, glycosylation, and phosphorylation. Proteolysis is the process of cutting proteins into smaller pieces. The removal of the signal sequence from a protein in the ER is an example of proteolysis. Glycosylation is the addition of sugars to proteins. Glycosylation in the Golgi apparatus marks proteins as destined for lysosomes, the plasma membrane, or the vacuole (in plants). Phosphorylation is the addition of phosphate groups to proteins and is catalyzed by protein kinases. Phosphorylation can alter the tertiary structure of a protein.

Question 11. Draw a diagram of a eukaryotic cell and show where in the cell the gene is transcribed and translated. In your diagram, indicate how this particular gene product is targeted to a compartment in the cell such as the endoplasmic reticulum. Create a stepwise list of all the important proteins and enzymes required for this process. Imagine that the signal sequence has been deleted from this gene without deleting the start codon. How will this deletion affect gene expression? How will this deletion affect the targeting of the protein to the endoplasmic reticulum?

Question 12. What would be the effect of a deletion of the DNA encoding the targeting sequence for that gene product? (Imagine that this protein was targeted to go to the endoplasmic reticulum and the signal sequence was removed as a result of this deletion.)

Test Yourself

1. Which of the following statements about transcription in prokaryotic cells is true?
   a. It occurs in the nucleus, whereas translation occurs in the cytoplasm.
   b. It is initiated at a start codon with the help of initiation factors and the small subunit of the ribosome.
   c. It is initiated at a promoter and uses only one strand of DNA (the template strand) to synthesize a complementary RNA strand.
   d. It is terminated at stop codons.
   e. It is initiated at an *ori* site on the chromosome.

   Textbook Reference: *14.3 How Is the Information Content in DNA Transcribed to Produce RNA?*

2. Which of the following statements about RNA polymerase is *false*?
   a. It synthesizes mRNA in a 5′-to-3′ direction, reading the DNA strand 3′-to-5′.
   b. It synthesizes mRNA in a 3′-to-5′ direction, reading the DNA strand 5′-to-3′.
   c. It binds at the promoter and unwinds the DNA.
   d. It does not require a primer to initiate transcription.
   e. It uses only one strand of DNA as a template for synthesizing RNA.

   Textbook Reference: *14.3 How Is the Information Content in DNA Transcribed to Produce RNA?*

3. Translation of messenger RNA into protein occurs in a _______ direction, and from _______ terminus to _______ terminus.
   a. 3′-to-5′; N; C
   b. 5′-to-3′; N; C
   c. 3′-to-5′; C; N
   d. 5′-to-3′; C; N
   e. 3′-to-5′; C; C

   Textbook Reference: *14.5 How Is RNA Translated into Proteins?*

4. If codons were read two bases at a time instead of three bases at a time, how many different possible amino acids could be specified?
   a. 16
   b. 64
   c. 8
   d. 32
   e. 128

   Textbook Reference: *14.3 How Is the Information Content in DNA Transcribed to Produce RNA?*

5. Translate the following mRNA:

   3′ – G A U G G U U U U A A A G U A – 5′

   a. NH_2 met—lys—phe—leu—stop COOH
   b. NH_2 met—lys—phe—trp—stop COOH
   c. NH_2 asp—gly—phe—lys—val COOH
   d. NH_2 met—gly—phe—lys—val COOH
   e. NH_2 asp—gly—phe—lys—stop COOH

   Textbook Reference: *14.3 How Is the Information Content in DNA Transcribed to Produce RNA?*

6. What would happen if a mutation occurred in DNA such that the second codon of the resulting mRNA was changed from UGG to UAG?
   a. Translation would continue and the second amino acid would be the same.
   b. Nothing. The ribosome would skip that codon and translation would continue.
   c. Translation would continue, but the reading frame of the ribosome would be shifted.
   d. Translation would stop at the second codon, and no functional protein would be made.
   e. Translation would continue, but the second amino acid in the protein would be different.

   Textbook Reference: *14.3 How Is the Information Content in DNA Transcribed to Produce RNA?*

7. If the following synthetic RNA were added to a test tube containing all the components necessary for protein translation to occur, what would the amino acid sequence be?

   5′ – A U A U A U A U A U A U – 3′

   a. Polyphenylalanine
   b. Isoleucine–tyrosine–isoleucine–tyrosine
   c. Isoleucine–isoleucine–isoleucine–isoleucine
   d. Tyrosine–tyrosine–tyrosine–tyrosine
   e. Aspargine–aspargine–aspargine–aspargine

   Textbook Reference: *14.3 How Is the Information Content in DNA Transcribed to Produce RNA?*

8. Which part of the tRNA base-pairs with the codon in the mRNA?
   a. The 3′ end, where the amino acid is covalently attached
   b. The 5′ end
   c. The anticodon
   d. The start codon
   e. The promoter

   Textbook Reference: *14.5 How Is RNA Translated into Proteins?*

9. Peptidyl transferase is an
   a. enzyme found in the nucleus of the cell that assists in the transfer of mRNA to the cytoplasm.
   b. enzyme that adds the amino acid to the 3′ end of the tRNA.
   c. enzyme found in the large subunit of the ribosome that catalyzes the formation of the peptide bond in the growing polypeptide.
   d. RNA molecule that is catalytic.
   e. Both c and d

   Textbook Reference: *14.5 How Is RNA Translated into Proteins?*

10. Termination of translation requires
    a. a termination signal, RNA polymerase, and a release factor.
    b. a release factor, initiator tRNA, and ribosomes.
    c. initiation factors, the small subunit of the ribosome, and mRNA.
    d. elongation factors and charged tRNAs.
    e. a stop codon positioned at the A site of the ribosome, peptidyl transferase, and a release factor.

    Textbook Reference: *14.5 How Is RNA Translated into Proteins?*

11. If the DNA encoding a nuclear signal sequence were placed in the gene for a cytoplasmic protein, the protein would
    a. be modified in the Golgi.
    b. be directed to the lysosomes.
    c. be directed to the nucleus.
    d. be directed to the cytoplasm.
    e. stay in the endoplasmic reticulum.

    Textbook Reference: *14.6 What Happens to Polypeptides after Translation?*

12. The central dogma of molecular biology states that _______ is transcribed into _______, which is translated into _______.
    a. a gene; polypeptides; a gene product
    b. protein; DNA; RNA
    c. DNA; mRNA; tRNA

d. DNA; RNA; protein
e. RNA; DNA; protein
Textbook Reference: *14.2 How Does Information Flow from Genes to Proteins?*

13. A gene product can be
    a. an enzyme.
    b. a polypeptide.
    c. RNA.
    d. microRNA.
    e. All of the above
    Textbook Reference: *14.1 What Is the Evidence that Genes Code for Proteins?*

14. The enzyme that catalyzes the synthesis of RNA is
    a. peptidyl transferase.
    b. DNA polymerase.
    c. tRNA synthase.
    d. ribosomal RNA.
    e. RNA polymerase.
    Textbook Reference: *14.3 How Is RNA Translated into Proteins?*

15. If a mutation occurs such that a spliceosome cannot remove one of the introns in a gene, what effect will this have on that gene?
    a. It will have no effect; the gene will be transcribed and translated into protein.
    b. Transcription will terminate early and the protein will not be made.
    c. Transcription will proceed, but translation will stop at the site where the intron remains.
    d. Translation will continue, but it is likely that a nonfunctional or aberrant protein will be made.
    e. Translation will continue and will skip the intron sequence.
    Textbook Reference: *14.3 How Is the Information Content in DNA Transcribed to Produce RNA?*

Answers

Key Concept Review

1. The mutations in these two strains of bacteria apparently interfere with lysine synthesis. Both mutations might be in the same gene coding for an enzyme necessary for lysine synthesis, but one, for example, could be a nonsense mutation in the fifth codon and the other a frame-shift mutation in the twenty-third (the number of mutations that can disable a gene is enormous). If lysine synthesis in this bacterium requires more than one enzyme (as is likely), the two mutations could be in different genes coding for different enzymes. In this case, the phenotypes would not be strictly identical; it should be possible to distinguish the two by trying to grow them on minimal media to which different intermediates in the synthesis of lysine have been added (see Figure 14.1).
2. These two genes must encode different polypeptides that are both subunits for the same enzyme in this biochemical pathway.
3. Retroviruses such as HIV use RNA as their genetic material. Upon entering a host cell, the virus produces a complementary DNA from its RNA. This DNA then produces more RNA for translation or for more viral genomes. This runs contrary to the central dogma of "DNA makes RNA which makes proteins."
4. mRNA contains a particular nucleotide sequence that determines the amino acid sequence of a protein; after synthesis and modification, it travels from the nucleus to the cytosol to be translated into that protein. tRNA is synthesized and transported from the nucleus to the cytosol. There, a specific tRNA binds both a specific amino acid and a specific nucleotide sequence on the mRNA to facilitate its addition to a nascent polypeptide chain.
5. DNA polymerase requires a primer and also proofreads as it moves along the DNA. RNA polymerase does not require a primer and does not proofread the RNA.
6. Three nucleotides are the minimum necessary to code for 20 amino acids. Each codon position can be only 1 of 4 nucleotides, and with a codon length of 3 nucleotides, a total of 64 different codons can be assembled ($4 \times 4 \times 4$), or more than enough to produce the 20 amino acids.
7. No, a mutation in an intron could result in a splicing error that produces a nonfunctional protein.
8. A poly A tail (100–300 adenine nucleotides) is added to newly synthesized mRNA (this seems to have a role in mRNA export from the nucleus and for stability). A variety of techniques could be used to bind or identify the poly A tail to isolate mRNA.
9. If the tRNA synthetase for tryptophan added phenylalanine to the tryptophan tRNAs, whenever a tryptophan codon was read by these tryptophan tRNAs,

phenylalanine would be added to the polypeptide. This would create proteins that were nonfunctional, and the cell would die.

10. See Figures 14.13–14.16 in the textbook.

11.

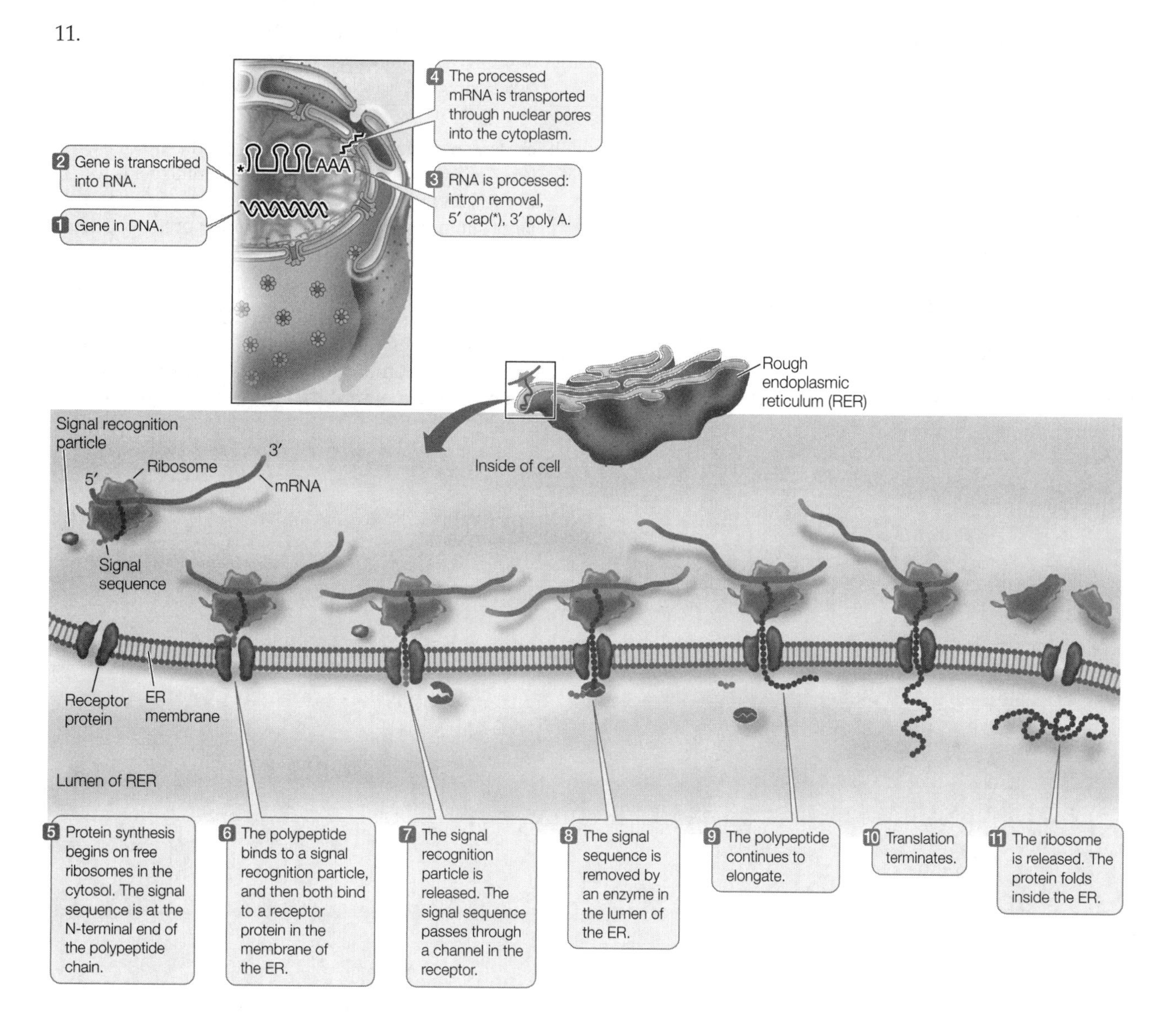

Deleting the signal sequence will not affect the transcription or translation of this gene, but it will affect the targeting of this gene product. During translation of this mRNA, there will be no signal peptide made and the signal recognition particle will be unable to bind. The mRNA will continue to be translated in the cytoplasm and remain there.

12. Because the protein lacked its targeting sequence, it would no longer be moved to the mitochondria and would remain in the cytoplasm after it had been translated.

Test Yourself

1. **c.** Answer **a** describes transcription and translation in eukaryotic cells; answers **b** and **d** describe translation. Answer **e** refers to the *ori* site, where replication of the circular chromosome starts.

2. **b.** RNA polymerase binds at a promoter, unwinds the DNA, synthesizes mRNA in a 5′-to-3′ (*not* 3′-to-5′) direction, and does not require a primer to synthesize the RNA.

3. **b.** Translation of messenger RNA occurs 5′-to-3′, and the polypeptide is synthesized from the N terminus to the C terminus.

4. **a.** Four possible bases read two at a time would yield 4^2, or 16, different codons.

5. **b.** See the codon table (Figure 14.6). Recall that translation occurs in the 5′-to-3′ direction.

6. **d.** UAG is a stop codon, so translation would terminate at that site.
7. **b.** See the codon table (Figure 14.6).
8. **c.** Neither the 3′ end nor the 5′ end of the tRNA is part of the anticodon. The promoter is a DNA sequence, to which RNA polymerase binds to initiate transcription. The start codon is found in the mRNA.
9. **e.** Peptidyl transferase is the enzyme that catalyzes the formation of the peptide bond, and it is located in the large subunit of the ribosome. Its catalytic activity is due to ribosomal RNA found in the large subunit of the ribosome.
10. **e.** Termination of translation requires a stop codon positioned at the A site of the ribosome, peptidyl transferase, and a release factor. Peptidyl transferase hydrolyzes the last amino acid attached to the tRNA in the P site, creating the C terminus.
11. **c.** The nuclear sequence would direct this protein to the nucleus.
12. **d.** Genes are not transcribed into polypeptides, protein is not used to synthesize DNA, and messenger RNAs are not translated into tRNAs. RNA can be used to synthesize DNA using reverse transcriptase, but DNA cannot be utilized to make protein.
13. **e.** Gene products can be RNAs (such as rRNA, tRNA, and microRNA) as well as enzymes and other polypeptides. Messenger RNA is translated into a gene product, protein.
14. **e.** DNA polymerase catalyzes the synthesis of DNA, tRNA synthase covalently attaches amino acids to tRNAs, and ribosomal RNA (peptidyl transferase in the large subunit) catalyzes the formation of the peptide bond during translation.
15. **d.** When an intron fails to be removed, that noncoding sequence is retained in the RNA within the coding sequence. When this RNA is translated, the protein will likely be nonfunctional due to the insertion of a noncoding sequence within the coding sequence.

Gene Mutation and Molecular Medicine

The Big Picture

- Some diseases in humans can be directly related to mutations in the genome, causing the production of an abnormal gene product. Other molecular diseases are affected by the individual's environment. Understanding the molecular alterations in a particular gene product and its effects can lead to therapeutic approaches, new diagnostic tools, and preventive measures in health care.

Study Strategies

- Allele-specific oligonucleotide hybridization and similar techniques are used in genetic screening to detect mutant alleles. Individual differences in DNA sequences (including simple base changes and DNA deletions, as well as a difference in the number of repeated sequences) can be utilized to distinguish individuals and to map genes (review Figure 15.18).
- Many different mutations can occur in a gene that will alter the function of the gene product. Some of those mutations may have more deleterious effects than others. Review the different kinds of mutations and how they affect gene function and the health of the individual.
- Make a list of all the kinds of mutations discussed in this chapter. Describe how these mutations can result in a nonfunctional gene product and give examples of diseases that are caused by those mutations.
- Make a list of the molecular techniques that are used to screen for abnormal genes. Select a few of the human diseases that are discussed in this chapter and describe the techniques that would be useful to screen for these diseases.
- Describe some of the techniques that are used to treat genetic diseases.
- Go to the Web addresses shown to review the following animated tutorials and activity. You can also find them in BioPortal (yourBioPortal.com), along with many additional learning resources.

  Animated Tutorial 15.1 Gene Mutations (Life10e.com/at15.1)

  Animated Tutorial 15.2 Gel Electrophoresis (Life10e.com/at15.2)

  Animated Tutorial 15.3 DNA Testing (Life10e.com/at15.3)

  Activity 15.1 Allele-Specific Cleavage (Life10e.com/ac15.1)

Key Concept Review

15.1 What Are Mutations?

Mutations have different phenotypic effects

Point mutations are changes in single nucleotides

Chromosomal mutations are extensive changes in the genetic material

Retroviruses and transposons can cause loss of function mutations or duplications

Mutations can be spontaneous or induced

Mutagens can be natural or artificial

Some base pairs are more vulnerable than others to mutation

Mutations have both benefits and costs

Mutations are heritable changes in genetic information. They can occur in germline cells and thus be passed on to progeny, or they can occur in somatic cells. Silent mutations are mutations that do not affect protein function. They can occur in noncoding DNA or in coding DNA where, because of the redundancy of the codons, the meaning of the codon is not altered by the mutation. Loss of function mutations lead to a loss of protein function and are almost always recessive. Gain of function mutations lead to altered protein function and usually show dominant inheritance to the wild-type allele. Conditional mutants are a class of mutations that are observable only under restrictive conditions, such as high temperature. The inability of certain mutants to grow at high temperatures is related to the unstable tertiary structure of the protein whose gene sequence has been altered.

All mutations are alterations in DNA sequence and can be classified into two categories: point mutations and chromosomal mutations. Point mutations are changes of one or two base pairs and include silent mutations, missense mutations, nonsense mutations, and frame-shift mutations. Silent mutations are base changes that do not alter the protein function because they do not change the meaning of the codon. For example, the change of an A to a C in the third position of a codon would change a CCA to a CCC, but both codons specify proline (see Figure 15.2). Missense mutations are single base changes that change one codon to another (e.g., GAU, which codes for asparagine, to GUU, which codes for valine; see Figure 15.2). Nonsense mutations are single base changes that change a codon to a stop codon (e.g., UGG, which codes for tryptophan, to UAG, stop). Nonsense mutations result in a shortened polypeptide (see Figure 15.2). Frame-shift mutations are deletions or insertions of one or two bases and alter the reading frame of the genetic message (see Figure 15.2). Chromosomal mutations are extensive changes in the chromosome and include larger deletions, duplications, inversions, or translocations (see Figure 15.4).

Mutations can be spontaneous or induced. Spontaneous mutations occur without any outside influence and can be caused by tautomerization of the bases, deamination of the bases, errors during DNA replication, and errors during meiosis (leading to unequal crossovers or unequal separation of the chromosomes). Induced mutations are caused by chemicals and radiation, which alter the bases, add groups to the bases, or break the backbone of the DNA. Hot spots, which are usually sites on the DNA containing methylated cytosines, are more prone to mutation than average. Loss of the amino group on a cytosine converts that base to uracil, which is removed from the DNA by repair enzymes. Loss of an amino group on a methylated cytosine converts that base to thymine, which is retained in the DNA, resulting in a base change at that site. Repair of the GT base pair may result in a correction to the original GC base pair or the introduction of a new AT mutation in the DNA at that site.

Mutagens can be natural (e.g., aflatoxins in plants) or artificial (e.g., isotopes made in nuclear reactors). Mutations for the most part are deleterious, but some mutations may benefit an organism by creating an altered gene product that improves survival. Mutations in germ line cells are carried on to the next generation. Somatic mutations can lead to cancer (see Chapter 11).

Question 1. Put the following sequence on one of the DNA strands:

5′ – ATGAAATTTTTG – 3′

Consider ATG the start codon for the peptide that is made from this gene. Then, for each of the following, write the new DNA sequence and translate the sequence into amino acids:

a. For the second codon, alter one of the bases to make it a silent mutation.
b. For the fourth codon, alter one of the bases to make it a nonsense mutation.
c. Pick any codon and change it to a missense codon.
d. Insert an A after the first A in the sequence and translate the resulting peptide.
e. Invert the sequence AATTT and translate the resulting peptide.

Question 2. Two individuals have a mutation in gene *X* but at different sites. The mutation affects the first individual adversely, while the second individual experiences no effect. How can this difference be explained?

15.2 What Kinds of Mutations Lead to Genetic Diseases?

- Genetic mutations may make proteins dysfunctional
- Disease-causing mutations may involve any number of base pairs
- Expanding triplet repeats demonstrate the fragility of some human genes
- Cancer often involves somatic mutations
- Most diseases are caused by multiple genes and environment

Many diseases in humans are caused by a specific genetic defect that leads to defective proteins. Proteins function in cells in many ways: as enzymes, receptors, transporters, and structures. Mutations in genes encoding these proteins can cause many diseases. Phenylketonuria (PKU) is caused by a mutation in the gene encoding an enzyme (phenylalanine hydroxylase) in the biochemical pathway for the synthesis of tyrosine (see Figure 15.7). When this enzyme is nonfunctional, a buildup of phenylalanine in the blood occurs. As phenylalanine accumulates, it is converted to phenylpyruvic acid, which in high concentrations prevents normal brain development in infants. Patients with PKU also have light-colored skin and hair because the skin pigment melanin is made from tyrosine (see Figure 15.7). The molecular alterations in many patients with PKU is a change in the 408th amino acid (from arginine to tryptophan) or a change in the 280th amino acid (from glutamate to lysine) of phenylalanine hydroxylase. Many proteins are polymorphic. Protein sequencing has revealed a 30 percent variation in amino acid sequences for the same protein in different individuals. Some mutations in these genes have no effect, producing functional gene products.

In sickle-cell disease, which is homozygous in 1 in 655 African-Americans, abnormal alleles produce an abnormal protein (β-globin). The abnormal hemoglobin protein aggregates in the red blood cells, causing the cells to sickle. Sickle cells block narrow capillaries and damage highly vascularized tissues, resulting in organ damage, especially at low blood oxygen concentrations. The abnormal allele in sickle-cell disease is caused by a mutation that results in a change at amino acid position 6 (from a glutamic acid to a valine) in the β-globin gene. This changes the structure of hemoglobin, reducing its oxygen-carrying capacity and resulting in anemia. Hemoglobin is easy to isolate and to study. Hundreds of β-globins with alterations in their primary amino acid sequence have been isolated. Some alterations have no effect, whereas others are quite severe. Hemoglobin C disease results from a change at position 6 of β-globin, which changes the glutamic acid to a lysine. The result is anemia that is not as severe as in sickle-cell disease (see Figure 15.8).

Most common genetic diseases result from altered proteins. Familial hypercholesterolemia (FH), cystic fibrosis,

Duchenne muscular dystrophy, and hemophilia are all examples of human disease in which abnormal proteins are formed due to specific mutations in the genes encoding those proteins.

Disease-causing mutations may involve a single base pair, a long stretch of DNA, multiple segments of DNA, or even an entire chromosome. Some human diseases have the same point mutation (as seen in sickle-cell anemia), while others can have a wide variety of different mutations. For example, there are 500 different mutations that have been isolated in different patients with PKU. Larger mutations include deletions, which can vary in size. Small deletions generally have less severe consequences than large deletions, which can remove more than one gene. Duchenne muscular dystrophy is due to deletions in the X chromosome that include the gene for the protein dystrophin. Chromosomal abnormalities include the gain or loss of one or more chromosomes, loss of a piece of a chromosome, translocations, and abnormal constrictions of chromosomes. Triplet repeats can be found within or outside the coding sequence of a gene. Expanding triplet repeats are found in fragile-X syndrome (a repeat of CGG in the *FMR1* gene), in myotonic dystrophy (a repeat of CTG), and in Huntington's disease (a repeat of CAG). For fragile-X syndrome, triplet repeats are more extensive in affected individuals (200–2,000) and are thought to expand during DNA replication due to slippage of DNA polymerase in those regions. Premutated individuals carry a moderate number of repeats (55–200) and show no symptoms, while normal individuals carry a small number of repeats (6–54) (see Figure 15.10).

Mutations in somatic cells can result in cancers. Generally, a cell must undergo mutations of two or more genes to produce cancer. In the case of colon cancer, at least three tumor suppressor genes and one oncogene are required to be mutated in order to form a malignant tumor (see Figure 15.11).

Most diseases are multifactorial. Although some diseases are the result of a mutant allele that produced a single dysfunctional protein, most diseases are caused by many genes and proteins interacting with the environment. Up to 60 percent of people may be affected by diseases that are genetically influenced.

Question 3. Describe the expected outcome of a somatic cell mutation that results in the gain of function of an oncogene.

Question 4. Describe the expected outcome of a somatic cell mutation that results in the loss of function of a tumor suppressor gene.

15.3 How Are Mutations Detected and Analyzed?

Restriction enzymes cleave DNA at specific sequences

Gel electrophoresis separates DNA fragments

DNA fingerprinting combines PCR with restriction analysis and electrophoresis

Reverse genetics can be used to identify mutations that lead to disease

Genetic markers can be used to find disease-causing genes

The DNA barcode project aims to identify all organisms on Earth

DNA molecules and mutations can be analyzed. Bacteria defend themselves against viral infection by synthesizing enzymes known as restriction endonucleases. Restriction endonucleases cut viral DNA but leave their own DNA unharmed. Bacteria use their own enzymes (methylases) to add a methyl group to the bases of the DNA, making their DNA unable to be cut by their own restriction enzymes. Restriction endonucleases cut the backbone of the DNA at specific recognition sites (restriction sites) between the 3′ hydroxyl group of one nucleotide and the 5′ phosphate group of the next nucleotide. For example, *Eco*RI is a restriction endonuclease that cuts the sequence 5′ – GAATTC – 3′ and its complementary strand, 3′ – CTTAAG – 5′, between the G and the A on each strand. This sequence is a palindrome; it reads identically in both directions. *Eco*RI cuts, on average, about 1 in every 4,000 base pairs in prokaryotic genomes, but sizes of the DNA fragments vary, depending on the location of these *Eco*RI sites in the chromosome. Hundreds of purified restriction enzymes are used in research laboratories to manipulate DNA.

DNA fragments can be separated by means of gel electrophoresis. A gel is a porous molecular sieve that allows smaller DNA molecules to move through it more quickly than larger DNA molecules. When an electric field is applied to the gel, the negatively charged phosphates of the DNA are pulled toward the positively charged electrode (see Figure 15.13). Restriction digests of a particular DNA sample will reveal the number of DNA fragments, their sizes, and their abundance.

DNA fingerprinting uses electrophoresis and restriction digestion. Genes that are used for DNA fingerprinting are highly polymorphic and include single nucleotide polymorphisms (SNPs) or short tandem repeats (STRs) (see Figure 15.14A). DNA fingerprinting is used in forensic cases to identify suspects and to determine relatedness between skeletal remains and living relatives (e.g., the identification of Saddam Hussein; see Figure 15.14B). Scientists have proposed a DNA barcode using a short sequence from a highly conserved gene (cytochrome oxidase) to classify all organisms on Earth (see Figure 15.16). This project has the potential to help us better understand evolution and species diversity, identify new species, and detect undesirable microbes and bioterrorism agents.

Question 5. Bacterial restriction endonucleases provide a type of "immunity" to viral infections. Describe how these endonucleases protect a bacterial cell from viral infections.

Question 6. You have characterized over 100 mutations in a particular gene that are linked to a certain disease. The mutations you have identified are all due to a single base change in one codon of the gene encoding protein X. Which DNA technique would you use to analyze patients who might have this disease?

15.4 How Is Genetic Screening Used to Detect Diseases?

Screening for disease phenotypes involves analysis of proteins and other chemicals

DNA testing is the most accurate way to detect abnormal genes

Allele-specific oligonucleotide hybridization can detect mutations

Genetic screening is used to identify people who might be carriers of a particular disease. Genetic screening can be done prenatally, on newborns, or on asymptomatic individuals who have close relatives with a genetic disease. DNA testing can be used to look for mutations responsible for human disease. DNA extracted from several cells can be PCR amplified, and the resulting sequences are analyzed. In PKU screening, phenylalanine concentrations are assayed in blood samples from newborns (see Figure 15.17). Individuals who test positive for PKU are fed special diets low in phenylalanine to prevent abnormal brain development. Preimplantation screening for mutant alleles for *in vitro* fertilized human embryos can be done at the eight-cell stage. One of the embryonic cells is removed, and its DNA is analyzed for the disease gene. Once a normal genotype has been confirmed, the seven-cell embryo can be implanted in the mother and will develop normally. Fetuses can also be screened with chorionic villus sampling or by amniocentesis. Once the DNA has been isolated, different types of genetic screens can be used. Allele-specific cleavage differences can detect changes in restriction endonuclease sites in the DNA containing the gene of interest (see Figure 15.18). Oligonucleotides that are specific for the mutant allele and the normal allele can hybridized with DNA isolated from individuals who are being tested.

Question 7. What is the anticipated future method of screening fetal cells for genetic diseases and how does it work? What are its advantages compared to current methods?

Question 8. What probe is used in allele-specific oligonucleotide hybridization?

15.5 How Are Genetic Diseases Treated?

Genetic diseases can be treated by modifying the phenotype

Gene therapy offers the hope of specific treatments

Treating genetic diseases includes modifying the phenotype and gene therapy. There are three ways to treat genetic disease: restrict the substrate of the deficient enzyme, inhibit a harmful metabolic reaction, or supply the normal version of the protein. For individuals with PKU, treatment involves restricting the substrate (phenylalanine) for the missing enzyme (phenylalanine hydroxylase), which prevents the buildup of phenylpyruvic acid. For individuals with chronic myelogenous leukemia, drugs have been developed that target the protein responsible for cell proliferation. Patients with hemophilia A are given the missing gene product, which is a clotting factor, in purified form. In gene therapy, the new gene is inserted into the affected individual. The gene must be taken up efficiently, precisely inserted into host DNA, and appropriately expressed (see Figure 15.20). This technique has been used to replace the γ-aminobutyric acid (GABA) gene using viral DNA to infect the deficient cells in patients with Parkinson's disease (see Figure 15.20). *Ex vivo* gene therapy is a technique in which target cells are removed from the patient, given the new gene, and then reinserted into the patient. Another approach to gene therapy involves inserting the gene directly into the cells in which the defect exists.

Question 9. Phenylketonuria (PKU) and Duchenne muscular dystrophy are both molecular diseases that are caused by mutations in particular genes, yet PKU is easier to treat than muscular dystrophy. Why?

Question 10. What are the obstacles to the success of gene therapy, even in cases in which the defective disease-causing gene has been identified?

Test Yourself

1. Fragile-X syndrome is caused by
   a. a single base change in the DNA.
   b. the changing of a valine into a glutamic acid.
   c. triplet expansion.
   d. a chromosomal translocation.
   e. nondisjunction of the X chromosome in meiosis.
   Textbook Reference: *15.2 What Kinds of Mutations Lead to Genetic Diseases?*

2. A nonfunctional membrane protein is responsible for
   a. hemophilia.
   b. sickle-cell disease.
   c. Duchenne muscular dystrophy.
   d. cystic fibrosis.
   e. PKU.
   Textbook Reference: *15.2 What Kinds of Mutations Lead to Genetic Diseases?*

3. Expanding triplet repeats
   a. occur during DNA replication due to slippage of DNA polymerase.
   b. are caused by errors in DNA synthesis with reverse transcriptase.
   c. have been observed in more than a dozen diseases.
   d. occur in individuals with a normal number of triplet codons.
   e. Both a and c
   Textbook Reference: *15.2 What Kinds of Mutations Lead to Genetic Diseases?*

4. In sickle-cell disease,
   a. a clotting factor in the blood is nonfunctional.
   b. the sixth amino acid is changed from a valine to a glutamic acid.
   c. the sixth amino acid is changed to a stop codon.
   d. hemoglobin builds up in the red blood cells.

e. the structure of β-globin is altered, and the hemoglobin protein forms aggregates in the red blood cells.

Textbook Reference: *15.2 What Kinds of Mutations Lead to Genetic Diseases?*

5. Metabolic disorders are
   a. caused by mutations in centromeres.
   b. caused by mutations in genes encoding enzymes that are required to synthesize particular compounds in the cell (such as an amino acid).
   c. caused by abnormal membrane proteins that transport chloride ions.
   d. caused by prions.
   e. All of the above

   Textbook Reference: *15.2 What Kinds of Mutations Lead to Genetic Diseases?*

6. Human genetic disease can be caused by
   a. an autosomal recessive trait.
   b. translocations.
   c. a deletion in a chromosome.
   d. a mutant allele that is dominant to the wild-type allele.
   e. All of the above

   Textbook Reference: *15.2 What Kinds of Mutations Lead to Genetic Diseases?*

7. Mutations that result in human disease include
   a. expanding triplet repeats that occur during DNA synthesis.
   b. point mutations that do not change the amino acid sequence of the gene.
   c. silent mutations.
   d. mutations in restriction sites in the DNA.
   e. Both a and d

   Textbook Reference: *15.1 What Are Mutations?; 15.2 What Kinds of Mutations Lead to Genetic Diseases?; 15.3 How Are Mutations Detected and Analyzed?*

8. Genetic screening
   a. has been used to treat embryos carrying a mutant allele.
   b. requires that gene sequences are the same in affected and unaffected individuals.
   c. has been used to identify fathers who are carriers for X-linked alleles.
   d. has been used to identify mutant alleles in embryos.
   e. requires dissimilarity between the alleles being tested and those that are linked to a particular disease.

   Textbook Reference: *15.4 How Is Genetic Screening Used to Detect Diseases?*

9. For gene therapy to be most effective, genes should
   a. be delivered to the germ-line cells using a viral vector.
   b. be inserted into the host DNA chromosome at random sites.
   c. be expressed in the host cells in response to the appropriate environmental signals.
   d. be similar to the defective gene.
   e. have antibiotic resistance markers.

   Textbook Reference: *15.5 How Are Genetic Diseases Treated?*

10. Allele-specific oligonucleotides
    a. should be tightly linked to the disease allele.
    b. can be used to induce spontaneous mutations.
    c. can detect a change in a restriction endonuclease pattern.
    d. are used to detect mutations in protein.
    e. a, c, and d

    Textbook Reference: *15.4 How Is Genetic Screening Used to Detect Diseases?*

11. Conditional mutations
    a. are mutations that are expressed in wild-type cells.
    b. are deletion mutations.
    c. are analyzed under permissive and restrictive conditions.
    d. produce gene products that are nonfunctional under all conditions.
    e. can be analyzed only under restrictive conditions.

    Textbook Reference: *15.1 What Are Mutations?*

12. Gel electrophoresis
    a. causes DNA to be pulled through the gel toward the negative end of the field.
    b. causes larger DNA fragments to move more quickly through the gel than smaller DNA fragments.
    c. is used to identify and isolate DNA fragments.
    d. is required for PCR reactions.
    e. is used in allele-specific oligonucleotide hybridization.

    Textbook Reference: *15.3 How Are Mutations Detected and Analyzed?*

13. Gain of function mutations
    a. are dominant mutations that are expressed in wild-type cells.
    b. are dominant mutations that are expressed in mutant cells.
    c. are the cause of continuous division in cancer cells.
    d. can be analyzed only under restrictive conditions.
    e. are expressed only with the appropriate environmental signals.

    Textbook Reference: *15.1 What Are Mutations?*

14. Which of the following mutations would still allow protein X to be functional?
    a. A missense mutation in the third codon of gene *X*
    b. A silent mutation in the third codon of gene *X*
    c. A deletion of the first three codons of gene *X*
    d. A frame-shift mutation in the third codon of gene *X*
    e. A nonsense mutation in the third codon of gene *X*

    Textbook Reference: *15.1 What Are Mutations?*

15. Which of the following chromosomal mutations would still allow protein X to be functional?
    a. Deletion of the last 100 codons of gene *X*
    b. A duplication of gene *X*
    c. An inversion of the last 100 codons of gene *X*
    d. A translocation of the last 100 codons of gene *X* to another chromosome
    e. None of the above
    Textbook Reference: *15.1 What Are Mutations?*

16. Which of the following mutations would probably be the most deleterious?
    a. A missense mutation in the second codon
    b. A frame-shift mutation in the second codon
    c. A nonsense mutation in the last codon
    d. A silent mutation in the second codon
    e. A missense mutation in the last codon
    Textbook Reference: *15.1 What Are Mutations?*

Answers

Key Concept Review

1. 5′ – ATGAAATTTTGG – 3′ met-lys-phe-trp
   a. 5′ – ATGAA**G**TTTTGG – 3′ met-lys-phe-trp
   b. 5′ – ATGAAATTTT**A**G – 3′ met-lys-phe-stop
   c. 5′ – ATG**C**AATTTTGG – 3′ met-gln-phe-trp
   d. 5′ – **A**ATGAAATTTTGG – 3′ asn-glu-ile-leu
   e. 5′ – ATGA**TTTAA**TGG – 3′ met-ile-stop
2. The mutation in gene *X* in the first individual must have occurred in an essential region of the gene that is required for its function. The mutation in gene *X* in the second individual may be a silent mutation, or it may be in a region that is nonessential for the function of that protein.
3. A gain of function for an oncogene would be the same as activation of the oncogene and would be expected to enhance the likelihood that the cell might be transformed to a tumor cell.
4. Since a tumor suppressor gene produces a protein that prevents a normal cell from becoming a tumor cell, the loss of function of a tumor suppressor gene would remove that inhibition and allow the cell to progress toward transformation into a tumor cell.
5. Restriction endonucleases cut DNA at specific sites and can degrade the viral DNA once it has infected the bacterial cell. Bacteria modify their own DNA, so the restriction endonucleases cannot cut their own DNA.
6. Since all of the mutations you isolated occur in the same codon, you could use a set of SNPs for that codon. SNPs identify single base changes in the DNA.
7. Typically, small numbers of fetal cells are released into the mother's blood. Recent techniques permit isolation of the fetal cells from the mother's blood. DNA can then be isolated from the fetal cells, amplified, and assayed for various mutations. Unlike amniocentesis and chorionic villus sampling, this is a relatively noninvasive procedure that does not carry the risk of miscarriage.
8. Generally, the probe in allele-specific oligonucleotide hybridization would be a single-stranded segment of a nucleic acid that is complementary to the gene of interest.
9. Phenylketonuria is a metabolic disease that results from a deficient enzyme, phenylalanine hydroxylase. Patients with PKU can build up harmful levels of phenylalanine in their systems and can be treated by restricting their intake of phenylalanine. Muscular dystrophy is caused by a defect in a structural protein, dystrophin, and results in dysfunctional muscle tissue. This structural dysfunction cannot be reversed with the application of any drug.
10. The problem lies in getting the corrective gene into the deficient cells. Both *in vivo* and *ex vivo* methods have been attempted. Using adeno-associated virus as a delivery mechanism for the corrective gene has provided some success.

Test Yourself

1. **c.** Fragile-X syndrome is caused by a triplet expansion on the X chromosome, not a single base change, a translocation, or an alteration of protein sequence. Nondisjunction of the X chromosome would result in XO, also known as Turner's syndrome.
2. **d.** The chloride transporter is a membrane protein that, when nonfunctional, causes cystic fibrosis. Hemophilia is caused by a defect in a clotting factor. Sickle-cell disease is caused by a defect in β-globin. PKU is caused by a defect in an enzyme in a biochemical pathway. Duchenne muscular dystrophy is caused by a defect in the muscle protein dystrophin.
3. **e.** Expanding triplet repeats are thought to be due to slippage of DNA polymerase during DNA replication, and result in an increasing number of triplet repeats in those cells. The triplet repeats have been observed in more than a dozen diseases. Reverse transcriptase is an enzyme from HIV that copies RNA into DNA.
4. **e.** In sickle-cell disease, the sixth amino acid is changed from a glutamic acid to a valine (not from a valine to a glutamic acid, and not to a stop codon). The resulting change causes the hemoglobin protein to form aggregates in the red blood cell. The concentration of hemoglobin in the red blood cell is not altered in sickle-cell disease.
5. **b.** Some metabolic disorders are caused by mutations in genes that encode enzymes in a biochemical pathway (such as the pathway for the synthesis of phenylalanine). Mutations in centromeres and genes for chloride transporters do not result in metabolic disorders. Prions are not the result of a mutation; they are the result of the abnormal folding of a protein.

6. **e.** All these mutations and chromosome alterations can cause human disease.

7. **e.** Expanding triplet repeats cause a variety of diseases, including fragile-X syndrome, Huntington's chorea, and myotonic dystrophy. Point mutations that do not alter the amino acid sequence of a protein are silent mutations and do not cause disease. Prion diseases are not the result of a mutation in a gene, but are due to the abnormal folding of a particular protein. Mutations in a gene coding sequence or regulatory sequence could alter a restriction site.

8. **d.** Genetic screening has been used to analyze *in vitro* fertilized embryos (at the eight-cell stage) but cannot be used to treat those embryos. To be useful, the alleles being tested have to be different in affected versus unaffected individuals, and the specific alleles that are linked to a disease trait should be the same. Fathers are not carriers for X-linked diseases; they express all their X-linked alleles.

9. **c.** Gene therapy is most effective for genes that reside permanently at the appropriate site in the host chromosome in the individual's cells and are expressed under the correct environmental signals. The gene should be identical to the wild-type copy of the gene and does not require antibiotic resistance markers. Viral delivery of the gene to germ line cells will not affect the diseased somatic cells.

10. **e.** Allele-specific oligonucleotides can detect specific changes in the DNA, including mutations, disease alleles, and changes in restriction sites. To be useful, they should be tightly linked to the disease allele. They cannot induce spontaneous mutations.

11. **c.** Conditional mutations are used to compare gene expression under permissive and restrictive conditions. Conditional mutations are not expressed in mutant cells under restrictive conditions but are expressed under permissive conditions. Under restrictive conditions they produce nonfunctional gene products. They cannot be deletions, because the gene product has to be functional at the permissive temperatures.

12. **c.** DNA fragments migrate toward the positive end of the electric field, with the smallest fragments migrating the fastest. PCR reactions do not require gel electrophoresis, although their products are analyzed by gel electrophoresis. Allele-specific oligonucleotide hybridization does not require gel electrophoresis.

13. **b.** Gain of function mutations are dominant and are expressed in mutant cells. They do not respond to the appropriate environmental signals and do not have a restrictive condition under which they are expressed. Some gain of function mutations are seen in cancer cells, but mutations in tumor suppressors are also responsible for unregulated cell division in cancer cells.

14. **b.** A silent mutation (no change in the amino acid sequence) would result in no change in protein function. All of the other mutations would alter the amino acid sequence. A missense mutation would change the amino acid sequence at that codon and could result in a nonfunctional protein. A deletion would remove the first three amino acids of the protein. A frame shift would cause all of the codons after that mutation to be out of frame. A nonsense mutation would terminate translation at that codon.

15. **b.** A duplication of gene *X* might allow the protein to be functional. All of the rest of these chromosomal mutations would remove large regions of the coding sequence or would position part of that sequence in the opposite direction (inversion), which would produce a nonfunctional protein X.

16. **b.** The mutation that is potentially the most dangerous is the frame shift in the second codon. All of the other codons would be out of register and no functional protein could be made. A missense mutation in the second codon could be disastrous or minimal, depending on how crucial the second amino acid was to the folding and functioning of the protein. A nonsense or missense mutation at the end of the coding sequence would have less of an effect on protein function because almost every amino acid in the protein would have been synthesized correctly. Again, this would depend on how important that last codon was to the proper folding and functioning of the protein. A silent mutation would have no effect on protein function because the amino acid inserted at that codon would be the same.

Regulation of Gene Expression

The Big Picture

- Prokaryotes regulate gene expression transcriptionally at promoters and posttranscriptionally by mRNA degradation, translational regulation, protein hydrolysis, and the inhibition of protein function.
- Prokaryotic cells use operons to coordinately regulate the expression of several genes. In response to environmental signals (the presence of lactose or the absence of tryptophan, for example) the repressor protein releases the operator site, leaving the promoter accessible to RNA polymerase to initiate transcription. In bacteriophage λ, other proteins bind promoters and control whether lysis or lysogeny will occur.
- Eukaryotes regulate gene expression with transcription factors, repressors, and activators, and they coordinate that regulation by placing regulatory sequences in front of different genes. Histone modifications also affect gene expression. Posttranscriptional regulation includes alternative splicing, modification of the 5′ cap, control of translational initiation, and protein degradation.

Study Strategies

- It is easy to confuse inducible operons (the *lac* operon) with repressible operons (the *trp* operon) because they both use repressors to regulate gene expression. Review the environmental conditions that must exist for each repressor to bind its operator site and what causes each repressor to release the operator site. (In the presence of lactose, the *lac* repressor cannot bind the *lac* operator site; in the presence of tryptophan, the *trp* repressor binds the *trp* operator site.) Review gene expression (Chapter 14) to see how, once transcription is initiated, proteins are produced that will act on those environmental signals.
- It may seem puzzling at first that some viruses insert their chromosome into the host chromosome, because the goal of viral infection seems to be rapid multiplication and infection of adjacent cells. However, lysogeny allows the viral genome to persist while environmental resources are abundant. When resources are depleted or cell damage occurs, the virus can enter the lytic cycle to infect other cells.
- Gene expression in eukaryotes can occur at transcriptional and posttranscriptional levels but can be regulated epigenetically. Eukaryotic gene expression requires that the DNA sequences be made accessible through chromatin remodeling before transcription can begin.
- Gene expression often occurs in response to environmental signals. Make a list of the environmental signals that a prokaryote receives, then outline the steps initiated by the cell in response to those signals. Include the following signals: lactose in the cell, glucose and lactose in the cell, high levels of glucose in the cell, high or low levels of tryptophan in the cell, bacteriophage λ infection under poor or rich environmental conditions.
- Outline the different animal virus life cycles, from the start of infection through multiplication of virus to viral release (review Figures 16.15 and 16.16). Compare these life cycles to the lytic and lysogenic life cycles of bacteriophages (see Figure 16.14).
- Gene regulation can occur at many steps in a eukaryotic cell. Starting with chromatin remodeling, list the steps that must occur for a gene to be made into functional protein in the cytoplasm of the cell.
- Go to the Web addresses shown to review the following animated tutorials and activities. You can also find them in BioPortal (yourBioPortal.com), along with many additional learning resources.

Animated Tutorial 16.1 The *lac* Operon (Life10e.com/at16.1)

Animated Tutorial 16.2 The *trp* Operon (Life10e.com/at16.2)

Animated Tutorial 16.3 Initiation of Transcription (Life10e.com/at16.3)

Activity 16.1 Eukaryotic Gene Expression Control Points (Life10e.com/ac16.1)

Activity 16.2 Concept Matching (Life10e.com/ac16.2)

Key Concept Review

16.1 How Is Gene Expression Regulated in Prokaryotes?

Regulating gene transcription conserves energy

Operons are units of transcriptional regulation in prokaryotes

Operator–repressor interactions control transcription in the *lac* and *trp* operons

Protein synthesis can be controlled by increasing promoter efficiency

RNA polymerases can be directed to particular classes of promoters

Genes can be negatively or positively regulated. Gene expression can be negatively regulated by a repressor that inhibits transcription or positively regulated by an activator protein that stimulates transcription (see Figure 16.1).

Prokaryotes conserve energy by making proteins only when they need them. They regulate gene expression by (1) downregulating transcription of mRNAs; (2) hydrolyzing mRNA before its translation; (3) preventing mRNA from being translated by the ribosome; (4) hydrolyzing the protein after it is made; and (5) inhibiting the function of the protein. There are three proteins involved in lactose uptake: β-galactoside permease, β-galactosidase, and β-galactoside transacetylase. β-galactoside permease transports lactose into the cell. β-galactosidase breaks down lactose into glucose and galactose. β-galactoside transacetylase transfers acetyl groups to certain β-galactosides.These structural genes are expressed in the presence of the inducer lactose (see Figure 16.5). Genes can be expressed in response to specific environmental conditions (inducible) or constitutively (all the time). Enzyme activity can be repressed by end-product feedback (also known as feedback inhibition) or by regulating gene expression (see Figure 16.3).

In prokaryotic cells, one promoter can control more than one gene. Transcription in prokaryotes can be regulated by proteins that bind operator sites on the DNA. The structural genes for the metabolism of lactose are adjacent to one another in the *lac* operon and are all transcribed from one promoter (see Chapter 14 for a review of transcription). There are three structural genes which code for three enzymes necessary for the cell to use lactose (see Figures 16.4 and 16.5). The *lac* operon is an inducible operon.

Operons are units of transcription in prokaryotes that consist of structural genes and the DNA sequences that control their transcription. The control sequences include the promoter (which is bound by RNA polymerase) and the operator (which is bound by the *lac* repressor). When RNA polymerase binds the promoter of the operon, all three structural genes are transcribed. The operator is a DNA sequence immediately adjacent to the promoter in the operon. If a repressor molecule is bound to the operator, it prevents RNA polymerase from binding to the promoter. Operons can be inducible or repressible and can be regulated by a repressor. Operons can also be regulated by an activator protein. The repressor for the *lac* operon can bind the operator site of the *lac* operon and the inducer, lactose. When the repressor binds the inducer, it cannot bind the operator, and transcription of the *lac* genes will occur (see Figure 16.5). The repressor regulates the expression of the lactose genes by binding the operator site on the DNA.

In summary, inducible operons have the following features: the operator and promoter are bound by a regulatory protein (the repressor) to control transcription; in the absence of the inducer, the operon is off; the repressor prevents transcription of the operon; and adding the inducer changes the tertiary structure of the repressor so transcription of the operon occurs.

The *trp* operon is a repressible operon. This means that its repressor is not bound to the operator unless another molecule, the corepressor (tryptophan, in this case), first binds to the repressor. When the corepressor binds to the repressor, the repressor becomes active and binds to the operator, preventing transcription of the structural genes. In inducible systems (like that of the *lac* operon), the substrates of these pathways bind the repressor, allowing transcription to occur. The presence of the substrate thus turns on production of the machinery necessary to catabolize the substrate. In repressible systems (like that of the *trp* operon), the product of the metabolic pathway (the corepressor) interacts with the regulatory protein, enabling it to bind the promoter and block transcription. The presence of the product thus turns off the transcriptional machinery necessary to produce more of that product.

Transcription can be positively regulated by increasing promoter efficiency. The cAMP receptor protein (CRP), when bound to cAMP, will cause the transcription of certain operons to increase by helping RNA polymerase bind the promoter more efficiently (see Table 16.1 for an outline of positive and negative regulation of transcription). Glucose reduces the levels of cellular cAMP, resulting in lower transcription efficiency of the *lac* operon, which is termed catabolite repression. This type of gene regulation occurs when the presence of a preferred energy source (glucose) represses other catabolic pathways.

Question 1. Suppose that a cell has a mutation that deletes the gene encoding the repressor for a certain operon, and a plasmid is introduced into the host cell that carries a wild-type copy of the gene for the repressor. Is normal regulation of this operon restored in the presence of this plasmid?

Question 2. Suppose that a cell has a mutation that deletes the gene encoding the operator for a certain operon, and a plasmid is introduced into the host cell that carries a wild-type copy of the operator. Is normal regulation of this operon restored in the presence of this plasmid?

16.2 How Is Eukaryotic Gene Expression Regulated?

General transcription factors act at eukaryotic promoters

Specific proteins can recognize and bind to DNA sequences and regulate transcription

Specific protein–DNA interactions underlie binding

The expression of transcription factors underlies cell differentiation

The expression of sets of genes can be coordinately regulated by transcription factors

The expression of eukaryotic genes is precisely regulated. Eukaryotic promoters have two important sequences: a recognition sequence and a TATA box. Initiation of transcription in eukaryotic cells requires regulatory proteins called transcription factors. These are proteins that bind to the promoter and form a transcription complex to which RNA polymerase II can bind in order to initiate transcription. TFIID, a transcription factor, binds the TATA box, changing the shape of the DNA and the transcription factor. Other transcription factors then bind the promoter, and RNA polymerase binds to initiate transcription. Other short sequences bind regulatory proteins that affect transcription positively (enhancers, which bind activator proteins) and negatively (silencers, which bind repressors). Regulatory sequences can be located close to or far away from the gene they affect. The transcription of any eukaryotic gene is determined by the combination of transcription factors, repressors, and activators.

Proteins that bind DNA have characteristic structural motifs. Activators, transcription factors, regulatory proteins, and repressors have characteristic protein motifs. One of the common DNA protein-binding motifs is the helix-turn-helix (see In-Text Art, p. 335). These proteins can interact with the DNA by fitting into the major or minor groove, using amino acid side chains that can project into the interior of the helix. Some of these may form hydrogen bonds with the bases.

Gene regulation is a coordinated process. Coordinated gene regulation is achieved by the positioning of regulatory sequences that bind the same transcription factors near the promoters of each of those genes. Proteins bind to these regulatory sequences and stimulate RNA synthesis. In plants, regulatory elements known as stress response elements (SREs) are found near the promoters of genes that respond to drought (see Figure 16.11).

Question 3. Suppose you are engineering gene *Y*, such that when it is inserted into a eukaryotic chromosome it will be expressed continuously. Which specific sequences must be part of this gene so that it will be expressed?

Question 4. Suppose you are engineering a new plant in which gene *Y* will be activated under drought conditions. What kinds of DNA sequences need to be present to ensure activation of the gene under these conditions?

Question 5. Diagram a gene in both a prokaryotic cell and a eukaryotic cell. Include the types of sequences that are important for transcriptional regulation, the location of those sequences, and the proteins that bind these regions.

16.3 How Do Viruses Regulate Their Gene Expression?

Many bacteriophages undergo a lytic cycle

Some bacteriophages can undergo a lysogenic cycle

Eukaryotic viruses can have complex life cycles

HIV gene regulation occurs at the level of transcription elongation

Some bacteriophages are lytic, and some alternate between lytic and lysogenic life cycles. Outside the host cell, viruses exist as individual particles called virions, which consist of nucleic acid and proteins. Viruses are acellular, do not regulate transport in and out of cell membranes, and perform no metabolic function. They develop and reproduce only inside the cells of specific hosts. Some bacteriophages (the virulent viruses) are lytic. During infection they take over the host machinery to synthesize new viral particles and lyse the host cell. During the early phase of the lytic cycle, the phage injects its nucleic acid into the host cytoplasm, and viral genes are transcribed and translated. These early gene products shut down host transcription, degrade host DNA, and stimulate viral genome replication and viral gene transcription. Late gene products include viral capsid proteins and proteins that lyse the cell at the end of the lytic cycle.

The lytic cycle takes about 30 minutes and produces hundreds of bacteriophages per cell (see Figure 16.14). During the lytic cycle, some bacteriophages can package host DNA and transfer it to the next bacterial cell. This process is called transduction. Other bacteriophages (the temperate viruses) can alternate between lytic and lysogenic life cycles.

Lysogeny occurs when the viral genome inserts itself into the host genome and the viral genome is replicated as part of the host chromosome. The virus is called a prophage at this stage. When the cell encounters stressful environmental conditions, lysogenic viruses respond with a change in gene expression, resulting in excision from the host chromosome and completion of the lytic cycle. Lysogenic phage have a genetic switch that senses conditions in the host. Two proteins (Cro and cI) compete for two promoters in lambda (λ), a lysogenic phage. When the Cro protein binds those promoters, lytic genes are activated. When the cI protein binds those promoters, genes involved in lysogeny are activated (see Figure 16.15).

Eukaryotic viruses are diverse. Animal viruses include RNA and DNA viruses and retroviruses. DNA viruses can be double-stranded or single-stranded and can be both lytic and lysogenic (e.g., herpes and papillomaviruses). RNA viruses are usually single-stranded (like influenza). Their genome is translated by the host machinery, and some of those gene products are used to replicate the RNA genome. Retroviruses are viruses whose genome is RNA (e.g., HIV). Once the host is infected, that RNA is used as a template to synthesize double-stranded DNA, which is integrated into the host genome. Human immunodeficiency virus (HIV) is an enveloped virus. It is enclosed in a plasma membrane derived from the host cell. HIV is a retrovirus that copies its RNA genome into DNA using reverse transcriptase. The DNA copy of the viral genome is then integrated into the host cell's chromosome, forming a provirus (see Figure 16.16). The provirus resides permanently in the host chromosome. When activated, the provirus is transcribed as mRNA and translated into protein by the host cell's translation machinery. The viral protein tat (*trans*activation of *t*ranscription) prevents termination, allowing viral gene transcription by the host RNA polymerase. Viruses are assembled at the host cell membrane, and mature virions bud from the cell membrane. Viral proteins activate a switch in response to environmental factors, which can determine if temperate phage (such as lambda, λ) will make a "decision" to lyse or lysogenize.

Question 6. Contrast the key difference between lytic and lysogenic cycles of bacteriophages.

Question 7. Describe the mechanisms of action of current drugs used to treat HIV.

16.4 How Do Epigenetic Changes Regulate Gene Expression?

DNA methylation occurs at promoters and silences transcription

Histone protein modifications affect transcription

Epigenetic changes can be induced by the environment

DNA methylation can result in genomic imprinting

Global chromosome changes involve DNA methylation

Histone protein modifications affect transcription. The term "epigenetics" refers to changes in the expression of a gene or genes that occur without any change in the gene sequence. This process includes DNA methylation and chromosomal protein alterations. In methylation, cytosine residues are methlyated by DNA methyl transferase, usually at CpG islands. Maintenance methylases catalyze the formation of 5′-methylcytosine on the replicated DNA strand; demethylases remove the methyl group from the cytosine. Repressor proteins bind methylated regions on DNA. Demethylation occurs in male and female genomes upon fertilization and in tumor suppressor genes of cancer cells. The DNA in eukaryotic chromosomes is associated with nucleosomes and other chromosomal proteins, which can inhibit the initiation and elongation steps of transcription. Chromatin remodeling must occur to make the DNA accessible to transcription complexes. Nucleosomes become disassembled and then reassembled with the help of acetylation, deacetylation, methylation, and phosphorylation of histones (see Figure 16.19). Epigenetic changes induced by the environment can be inherited. DNA methylation patterns differ in male and female gametes, leading to genomic imprinting (see Figure 16.20).

Examples of genomic imprinting include Angelman syndrome (male imprinting for that gene is inherited and the female region for that gene is deleted) and Prader-Willi syndrome (female imprinting for that gene is inherited and the male region for that gene is deleted). The genotype for the sex chromosomes in mammals is XY in males and XX in females. However, the expression of X-linked genes (the gene dosage) is the same in both sexes due to X-inactivation in females. Early in embryonic development, one of the two X chromosomes is randomly inactivated, producing a highly condensed heterochromatic chromosome called a Barr body (see Figure 16.21). The number of Barr bodies is equal to the number of X chromosomes minus one; an XXX female has two Barr bodies and normal females (XX) have one Barr body. Most of the DNA of the inactive X is heavily methylated, with methyl groups added to the 5′ carbon of cytosine. However, transcription of the *Xist* gene occurs on the inactive X chromosome. The *Xist* RNA (also known as an interference RNA) binds the X chromosome from which it was transcribed and inactivates it (see Figure 16.21).

Question 8. What are CpG islands and what is their significance?

Question 9. What is a consequence of acetylation of histone tails?

16.5 How Is Eukaryotic Gene Expression Regulated after Transcription?

Different mRNAs can be made from the same gene by alternative splicing

Small RNAs are important regulators of gene expression

Translation of mRNA can be regulated by proteins and riboswitches

Different RNAs can be made from the same eukaryotic gene after transcription. Different RNAs can be made from the same gene by alternative splicing (see Figure 16.22). MicroRNAs are small RNAs (about 22 bp) that bind specific mRNAs and block their translation. MicroRNAs are transcribed as longer precursors in the cell and are then cleaved by dicer protein to produce small double-stranded microRNAs. Those RNAs are converted to single-stranded RNA by a protein complex and inhibit translation of mRNAs, targeting them for degradation.

Translation of mRNA can be regulated. Messenger RNAs in the cytoplasm are not always translated, because other factors regulate the accessibility of these mRNAs to the translational machinery. If the G of the 5′ cap is not modified (as in the egg cells of the tobacco hornworm moth), the mRNA is not translated. After fertilization, this G cap is modified, and the mRNA is translated.

Ferritin is a storage protein for iron. The mRNA necessary for making ferritin is present in cells at steady levels, regardless of the concentration of iron ions in the cell. When the iron level in the cell is low, a repressor protein binds to ferritin mRNA, preventing its translation into ferritin. As the iron level rises in the cell, some iron ions bind to the repressor protein, altering its shape so that it can no longer bind the ferritin mRNA. This permits translation of ferritin from the ferritin mRNA; the new ferritin is then available for storage of iron ions.

Protein degradation is regulated. Proteins that are targeted for degradation are first linked to ubiquitin through a lysine residue. Subsequently, more ubiquitin chains attach, forming a polyubiquitin chain. This polyubiquitin complex binds to a proteasome, which has ATPase activity. When a protein–polyubiquitin complex enters the proteasome, ubiquitin is removed, the protein is unfolded, and three proteases digest the protein, using ATP hydrolysis (see Figure 16.25).

Question 10. What is a general overall similarity in the actions of miRNA and siRNA and what is a general difference?

Question 11. Explain the sequence of events that lead to proteasomal degradation of proteins.

Test Yourself

1. A virus consists of
   a. a protein core and a nucleic acid capsid.
   b. a cell wall surrounding nucleic acid.
   c. RNA and DNA enclosed in a membrane.
   d. a nucleic acid core surrounded by a protein capsid, and in some cases, a membrane.
   e. a nucleic acid core surrounded by a cell membrane.
   Textbook Reference: *16.3 How Do Viruses Regulate Their Gene Expression?*

2. Lytic bacterial viruses
   a. infect the cell, replicate their genomes, and lyse the cell.
   b. infect the cell, replicate their genomes, transcribe and translate their genes, and lyse the cell.
   c. infect the cell, replicate their genomes, transcribe and translate their genes, package those genomes into viral capsids, and lyse the cell.
   d. infect the cell, transcribe and translate their RNA, replicate their genomes, package those genomes into viral capsids, and lyse the cell.
   e. insert their chromosome into the host chromosome.
   Textbook Reference: *16.3 How Do Viruses Regulate Their Gene Expression?*

3. Animal viruses that integrate their DNA into the host chromosome
   a. have DNA as their genome.
   b. are prophages.
   c. copy their RNA genome into DNA using reverse transcriptase.
   d. replicate their genome using RNA polymerase.
   e. can undergo only a lytic infection cycle.
   Textbook Reference: *16.3 How Do Viruses Regulate Their Gene Expression?*

4. An operon
   a. is regulated by a repressor binding at the promoter.
   b. has structural genes that are all transcribed from the same promoter.
   c. has several promoters, but all of the structural genes are related biochemically.
   d. is a set of structural genes that are all under the same translational regulation.
   e. is transcribed when RNA polymerase binds the operator.
   Textbook Reference: *16.1 How Is Gene Expression Regulated in Prokaryotes?*

5. If the gene encoding the *lac* repressor is mutated so that the repressor can no longer bind the operator, will transcription of that operon occur?
   a. Yes, because the repressor transcriptionally activates the *lac* genes.
   b. Yes, but only when lactose is present.
   c. No, because RNA polymerase is needed to transcribe the genes.
   d. Yes, because RNA polymerase will be able to bind the promoter and transcribe the operon.
   e. No, because cAMP levels are low when the repressor is nonfunctional.
   Textbook Reference: *16.1 How Is Gene Expression Regulated in Prokaryotes?*

6. If the gene encoding the *trp* repressor is mutated such that it can no longer bind tryptophan but can still bind the operon, will transcription of the *trp* operon occur?
   a. Yes, because the *trp* repressor can bind the *trp* operon and block transcription only when it is bound to tryptophan.
   b. No, because this mutation does not affect the part of the repressor that can bind the operator.
   c. No, because the *trp* operon is repressed only when tryptophan levels are high.
   d. Yes, because the *trp* operon can allosterically regulate the enzymes needed to synthesize the amino acid tryptophan.
   e. No, because the repressor will be continuously bound to the operator.
   Textbook Reference: *16.1 How Is Gene Expression Regulated in Prokaryotes?*

7. Transcriptional regulation in prokaryotes can occur by
   a. a repressor binding an operator and preventing transcription.
   b. an activator binding upstream from a promoter and positively affecting transcription.
   c. different promoter sequences binding RNA polymerase more tightly, resulting in more effective transcriptional initiation.
   d. the control of promoter efficiency.
   e. All of the above
   Textbook Reference: *16.1 How Is Gene Expression Regulated in Prokaryotes?*

8. For the bacteriophage λ, the "decision" to become a prophage is made
   a. if environmental resources (e.g., food) for the host are limited.
   b. when the Cro protein binds the promoter.
   c. when the cI protein binds the promoter.
   d. when bacterial lysis occurs.
   e. at the end of the lytic infection.
   Textbook Reference: *16.3 How Do Viruses Regulate Their Gene Expression?*

9. Imagine that the TATA box for gene *X* becomes highly methylated. How will this affect the expression of gene *X*?
   a. There will be no effect.
   b. Gene *X* will be transcribed but not translated.
   c. Gene *X* will be transcribed if the transcription factors receive the appropriate environmental signal.
   d. Gene *X* will not be transcribed or translated.
   e. Gene *X* will be transcribed if the histones become acetylated.
   Textbook Reference: *16.4 How Do Epigenetic Changes Regulate Gene Expression?*

10. Imagine that gene *X* is moved to a part of the chromosome where the histones are highly acetylated. How will this affect its expression?
   a. There will be no effect.
   b. It will be transcribed but not translated.
   c. It will be transcribed if the transcription factors receive the appropriate environmental signal.
   d. It will not be transcribed or translated.
   e. It will be transcribed if the histones become deacetylated.
   Textbook Reference: *16.4 How Do Epigenetic Changes Regulate Gene Expression?*

11. Which of the following is an example of regulation of eukaryotic transcription?
   a. Iron binding the repressor protein for the ferritin mRNA and increasing ferritin expression
   b. Proteasome breakdown of protein–polyubiquitin complexes
   c. MicroRNAs binding their target mRNA and causing its degradation
   d. Alternative splicing of an mRNA transcript
   e. Activator proteins binding an enhancer
   Textbook Reference: *16.2 How Is Eukaryotic Gene Transcription Regulated?*

12. Which of the following statements about histone modifications is *false*?
   a. They cause some genes to be transcriptionally activated.
   b. They can result in the repression of gene transcription.
   c. They are inherited from parental cells in a Mendelian fashion.
   d. They cause Barr bodies to form.
   e. All of the above are false.
   Textbook Reference: *16.4 How Do Epigenetic Changes Regulate Gene Expression?*

13. What would happen initially to cells that lack a functional ubiquitin?
   a. Nothing would happen.
   b. Transcriptional initiation would increase.
   c. Protein degradation would decrease.
   d. Histone modifications would increase.
   e. Translation of proteins would be more efficient.
   Textbook Reference: *16.5 How Is Eukaryotic Gene Expression Regulated after Transcription?*

14. Which of the following would *not* affect gene expression in a eukaryotic cell?
   a. Deletion of a promoter
   b. Deletion of an enhancer
   c. Lack of modification of the cap structure on mRNA.
   d. Inability of transcription factor to bind the promoter
   e. Deletion of a DNA *ori* site
   Textbook Reference: *16.2 How Is Eukaryotic Gene Transcription Regulated?*

15. Transcription factors
   a. have a particular motif that allows them to interact with mRNA.
   b. assemble at the promoter to help translation begin.
   c. mark particular proteins for degradation.
   d. interact with RNA polymerase to initiate transcription.
   e. help stabilize the mRNAs in the cytoplasm.
   Textbook Reference: *16.2 How Is Eukaryotic Gene Transcription Regulated?*

Answers

Key Concept Review

1. Yes. The repressor gene can be transcribed and translated from the plasmid DNA, and normal regulation will be restored.
2. No. The operator site on the plasmid cannot restore regulation unless it recombines with the host operator site in such a way that it replaces the mutant operator on the host chromosome. The DNA site on the plasmid would bind the repressor, but because that site is not adjacent to the promoter or the structural genes on the chromosome, normal regulation of those genes cannot occur.
3. You will need a promoter that binds transcription factors (such as a TATA box), an RNA polymerase, and regulatory binding sites that bind activator proteins. You will also need to put it into a chromosomal region that has not been silenced by condensed nucleosomes.
4. The plant will need stress response elements (SREs) in front of the promoter for gene Y. SREs are bound by transcription factors that are sensitive to drought, and genes with SRE sequences in front of their promoters can be coordinately regulated.

5.

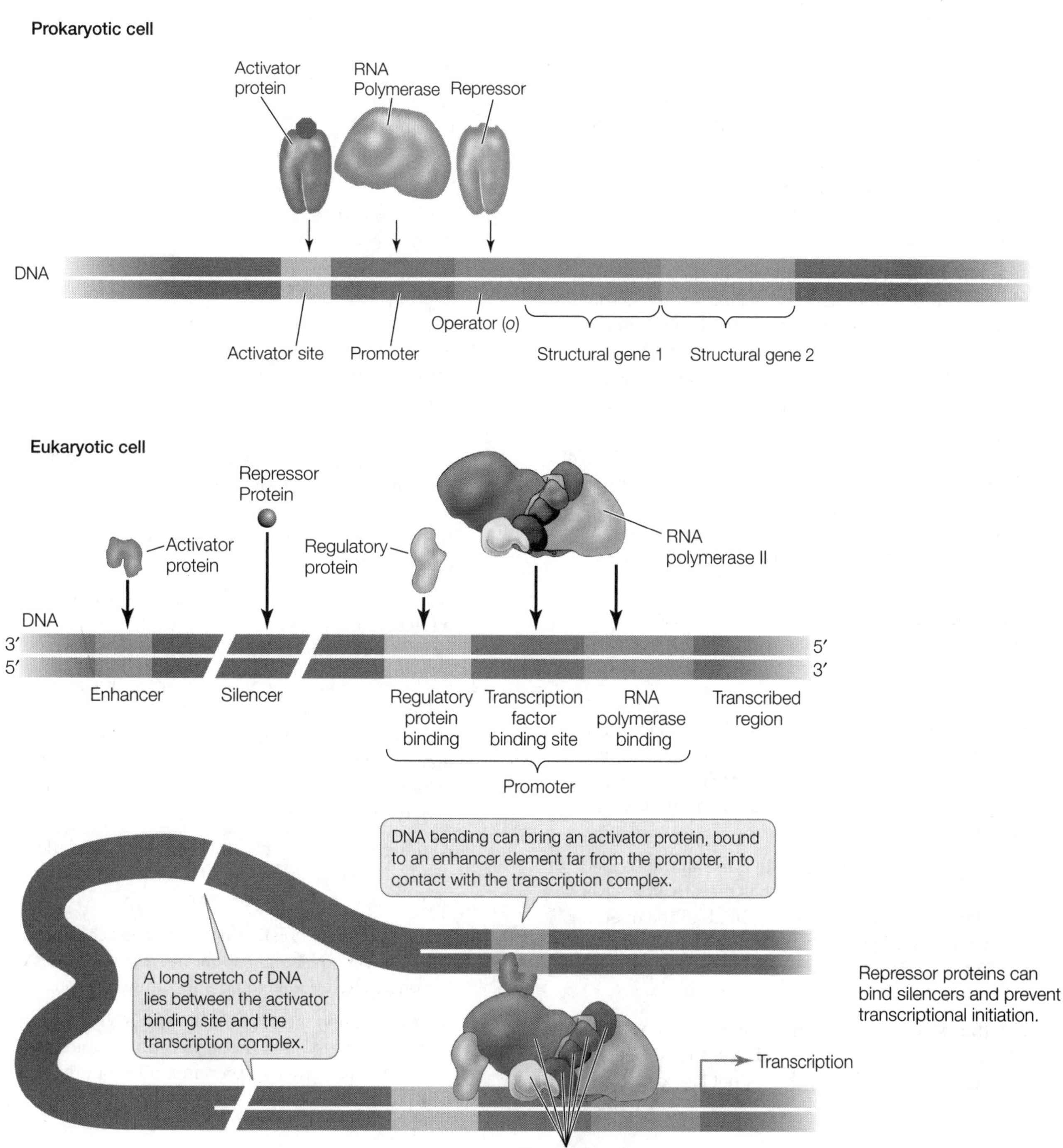

6. In the lytic cycle, the viral DNA infects the cell, multiplies, lyses the cell, and releases the newly made virus. In the lysogenic cycle, the viral DNA incorporates into the host DNA (called a prophage at this point) and replicates when the host DNA is replicated. Under the appropriate environmental conditions, the prophage DNA can excise itself from the host DNA and enter the lytic cycle.

7. Current HIV drugs focus primarily on inhibiting reverse transcriptase to block viral DNA synthesis, integrase to block incorporation of the viral DNA into the chromosome, and proteases that block posttranslational processing of viral proteins.

8. Promoter regions often have an abundance of C residues beside G residues. The C residues are prone to addition of a methyl group to form 5′–methylcytosine.

These regions of the promoter are referred to as CpG islands and are associated with repression of gene transcription.

9. Acetylation of histone tails allows the DNA in that region to be accessible to RNA polymerase, transcription factors, and the rest of the transcription apparatus to enable transcription.
10. They both inhibit mRNA translation, but miRNA binds the mRNA in a complementary manner to inhibit, while siRNA degrades the mRNA.
11. Ubiquitin binds a lysine on the protein to be destroyed and many more ubiquitins bind to form a polyubiquitin, chain on the protein. This targets the protein to a proteasome where the polyubiquitin chain is removed and recycled, the protein is unfolded, and proteases digest the protein into small peptides and amino acids.

Test Yourself

1. **d.** Nucleic acids do not form capsids; cell walls are found in bacterial and plant cells, not viruses. Viruses are organized so that the nucleic acid is surrounded by protein (not membranes), and a membrane may surround the protein capsid.
2. **d.** This sequence includes the most complete details of the viral life cycle. Viral transcription and translation have to occur first so that viral gene products needed for viral replication will be synthesized.
3. **c.** Answer **b** describes a provirus, which is a bacterial virus that has inserted its genome into a host chromosome. Animal viruses replicate their RNA genomes using reverse transcriptase and do not lyse the cells they infect.
4. **b.** An operon is a set of genes that are all transcribed from the same promoter, which is the site where RNA polymerase binds. The repressor binds at the operator site, which overlaps the promoter. Answer **d** is not correct because the operon is regulated transcriptionally, not translationally.
5. **d.** If the *lac* repressor is nonfunctional, it cannot bind the operator site, and transcription of the *lac* operon will occur at all times, whether or not lactose is present.
6. **a.** If the repressor can no longer bind tryptophan, then it cannot bind the operator, and transcription of the *trp* operon will always be on, whether tryptophan levels in the cell are high or low.
7. **e.** Answer **a** refers to the *lac* and *trp* repressors, answer **b** to the CRP protein, and answer **c** to promoters that have different transcriptional efficiencies. Answer **d** refers to the *lac* operon.
8. **c.** The "decision" to become a prophage by bacteriophage λ is made when cI binds the promoter early in infection. This decision is made if the host is experiencing rich nutrient conditions and results in lysogeny, not cell lysis.
9. **d.** Gene *X* will not be transcribed or translated, since methylation sites on DNA are transcriptionally inactive.
10. **c.** Acetylation leads to loosening of the nucleosomes, resulting in more DNA sequence being accessible for transcription. If transcription factors receive the appropriate environmental signal, transcription will occur. Deacetylation leads to tighter packing of the nucleosomes.
11. **e.** Answer **a** refers to translational regulation; answer **b** refers to the regulation of protein longevity; answers **c** and **d** refer to posttranscriptional regulation.
12. **d.** Histone modifications are inherited in a non-Mendelian fashion. Depending on the parent in which the modification (genomic imprinting) occurs (i.e., Angelman syndrome versus Prader-Willi syndrome), the resulting phenotype can be different, even though the genotype is the same.
13. **c.** In the absence of ubiquitin, protein degradation would decrease, since ubiquitin targets proteins for degradation in the proteasome.
14. **e.** Deletion of an *ori* site on a eukaryotic chromosome would affect DNA replication, not transcription. All of the other alterations would affect transcription.
15. **d.** Transcription factors interact with RNA polymerase to help transcriptional (not translational) initiation. They do have particular protein motifs, but they bind DNA, not RNA. They do not stabilize mRNAs in the cytoplasm or mark proteins for degradation.

Genomes

The Big Picture

- Variations within a single species can be caused by allelic differences within a single gene. Some diseases in humans can be directly related to single nucleotide changes in the genome, causing the production of an abnormal gene product. Other molecular diseases are affected by multiple genes interacting with the individual's environment. Understanding the molecular alterations in a particular gene product and their effects can lead to therapeutic approaches, new diagnostic tools, and preventative measures in health care.
- The development of high-throughput sequencing methods has permitted the sequencing of whole genomes. The science of genomics is the study and comparison of genomes from different organisms and is used to identify genes, potential genes, and their regulatory sequences. Bioinformatics is used to analyze the sequences to trace evolutionary relationships between different species; it is also used to identify genes for particular traits, such as genes associated with phenotypic variations or diseases.
- The science of proteomics seeks to identify and characterize all of the proteins expressed by a genome. Due to alternative splicing of mature mRNAs and posttranslational modifications, the proteome of an organism is more complex than its genome and varies in response to environmental and developmental stimuli. Metabolomics is the study of the active metabolites in the cell (primary and secondary metabolites) that control cell functions, and how metabolite pools change under different environmental and developmental conditions.
- These subdisciplines have led to a better understanding of diseases and disabilities as well as improvements in agriculture, pharmacology, and environmental reclamation. In agriculture, important areas of research are related to the production of disease-resistant compounds and methods of enhancing the expression of inserted genes. Information from these studies has also been used to design new medicines to combat pathogens and to understand the minimal requirements for a cell to sustain life.

Study Strategies

- There are many methods for analyzing genomes and proteomes, including those of pharmagenomics and metagenomics. Review what kinds of sequences are analyzed by each of these methods, and give examples of how those analyses have furthered our understanding of the functions of genes, proteins, and cells in different organisms.
- Describe the steps that must be followed in order to generate a genome sequence from a particular organism.
- Comparing genomes among different organisms has allowed us to determine which genes are essential for all cells and which genes are designated for specific adaptive functions. Compare prokaryotic and eukaryotic genomes; list genes that are essential for both types of organisms and genes that are specialized for each type of organism.
- Compare the information and applications of genome and proteome data.
- List the agricultural, medical, and environmental benefits of genome sequencing.
- Go to the Web addresses shown to review the following animated tutorials and activity. You can also find them in BioPortal (yourBioPortal.com), along with many additional learning resources.

  Animated Tutorial 17.1 Sequencing the Genome (Life10e.com/at17.1)

  Animated Tutorial 17.2 High-Throughput Sequencing (Life10e.com/at17.2)

  Activity 17.1 Concept Matching (Life10e.com/ac17.1)

Key Concept Review

17.1 How Are Genomes Sequenced?

New methods have been developed to rapidly sequence DNA

Genome sequences yield several kinds of information

The Human Genome Project, completed in 2003, sequenced all of the human genome's nearly 3.2 billion base pairs (in haploid cells). The effort resulted in the development of more rapid and less costly sequencing methods. These high-throughput sequencing methods have enabled the sequencing of prokaryotic, viral, and much larger eukaryotic genomes (see Figure 17.1 and Table 17.2).

A current approach to genomic sequencing involves cutting the DNA into smaller (100-bp) fragments. DNA fragments are denatured using heat, creating single-strand DNA templates. These fragments are then attached to short adapter sequences that are mounted on a solid surface and amplified by PCR. Synthesis of DNA complementary to each tethered, amplified fragment is carried out one nucleotide at a time using universal primers, DNA polymerase, and the four nucleotides. Each nucleotide is labeled with a different colored fluorescent tag. Unused nucleotides are removed after each nucleotide addition. A camera records the color of the fluorescent tag identifying the nucleotide after each addition, building a color-coded sequence of the DNA fragment (see Figure 17.1). This powerful method is fully automated, can process millions of fragments simultaneously, and is fast and relatively inexpensive. Determining the entire genome sequence from these fragments is possible because the fragments are overlapping and computers can be used to align the overlapping sequences (see Figure 17.2).

The analysis of genomic data is facilitated by the development of increasingly powerful computers and the development of the field of bioinformatics, which utilizes complex mathematics and computer programs to analyze DNA sequences.

Genome sequences yield several kinds of information: open reading frames, amino acid sequences of proteins, regulatory sequences, RNA genes, and other noncoding sequences (see Figure 17.3). In functional genomics, biologists use sequence information to identify the functions of various parts of the genome. In comparative genomics, sequences or entire genomes of different organisms are compared. This can provide further information about the functions of sequences and help establish evolutionary relationships.

Question 1. List the steps required to sequence a DNA molecule using the high-throughput sequencing method. Be sure to include all of the components of DNA sequencing, along with the technologies involved.

Question 2. Suppose you possess the sequenced genome for a prokaryote and you are looking for the gene(s) that code for a given protein. How would you begin your search?

17.2 What Have We Learned from Sequencing Prokaryotic Genomes?

Prokaryotic genomes are compact

The sequencing of prokaryotic and viral genomes has many potential benefits

Metagenomics allows us to describe new organisms and ecosystems

Some sequences of DNA can move about the genome

Will defining the genes required for cellular life lead to artificial life?

Genomic sequencing of prokaryotes has revealed the genes that are used for different cellular functions and how those specialized functions are carried out. Prokaryotic genomes are relatively small compared with eukaryotic genomes, and are compact, with little non-coding sequence, and usually lacking intronic sequences. In addition to the single chromosome, prokaryotes often have plasmids that can be transferred between cells.

Functional genomic analysis has been used to assign functions to gene products. Genes can be identified that are involved in the prokaryote's metabolism, transport, and infectious properties. Comparative genome analysis has been used to compare genomes from different organisms to understand their physiology.

Genomic analysis of prokaryotes has revealed useful information about unique genes in microorganisms that are important for agriculture and medicine. Genomic analysis has identified genes in the nitrogen-fixing *Rhizobium* family. These species of bacteria are important for their symbiotic relationship with plants, and this information may help increase crop yield. Genes unique to pathogenic strains of *E. coli*, *Salmonella*, and *Shigella* have also been identified in this way, as have several novel proteins in the virus responsible for severe acute respiratory syndrome (SARS). Some of the genome information will be potentially useful in developing new vaccines against pathogens.

The sequencing approach, metagenomics, allows us to describe new organisms and ecosystems. Metagenomics analyzes genes without isolating the intact organism. PCR can be used to amplify specific sequences from an environmental sample without the need to culture those organisms. Using this method, thousands of new viruses and bacteria have been found in seawater, marine sediment, and mine water runoff (see Figure 17.4).

Genome sequencing has enabled scientists to understand more about transposable elements (or transposons), mobile genetic elements that can move from one site in the genome to another. Sometimes two nearby transposable elements on the same DNA strand, called composite transposons, can carry the intervening DNA sequence with them when they move (see Figure 17.5).

Genome sequencing and transposon mutagenesis have been used to determine the minimal genome of the prokaryote *Mycoplasma genitalium* (see Figure 17.6). Through the inactivation of all of *M. genitalium*'s genes one by one, the smallest number of genes needed for survival (under laboratory conditions) was determined to be 382 genes. This knowledge has the potential to help us design new microbes with specific purposes, such as bacteria that will degrade oil or make plastics. Experiments are now under way to make synthetic genomes based on the genome of *M. genitalium* (see Figure 17.7).

Question 3. Genomics has revealed that the bacterium that causes tuberculosis has more than 250 genes that metabolize lipids. What does this finding suggest about the bacterium and about medicinal approaches to this disease?

Question 4. The *E. coli* strain O157:H7 causes illness, yet most *E. coli* strains in the human gut do not. What method could be used to identify genes that are potentially involved in *E. coli* O157:H7 pathogenesis?

17.3 What Have We Learned from Sequencing Eukaryotic Genomes?

- Model organisms reveal many characteristics of eukaryotic genomes
- Eukaryotes have gene families
- Eukaryotic genomes contain many repetitive sequences

Eukaryotic genomes are larger than those of prokaryotes, and have more protein-coding genes and regulatory sequences. These differences reflect the fact that multicellular organisms have many cell types with specialized functions (see Figure 17.8). Much of eukaryotic DNA is noncoding. Eukaryotes have multiple chromosomes.

Model organisms reveal many characteristics of eukaryotic genomes and include the yeast *Saccharomyces cerevisiae*, the nematode *Caenorhabditis elegans*, the fruit fly *Drosophila melanogaster*, and the plant *Arabidopsis thaliana* (see Table 17.2).

The prokaryote *E. coli* has 4,288 protein-encoding genes on a single circular chromosome of 4.6 million bp, whereas the simple eukaryote *Saccharomyces cerevisiae* (budding yeast) has 6,275 protein-coding genes, with a haploid content of 12.2 million bp on16 linear chromosomes. The genomes of *E. coli* and *S. cerevisiae* have the same number of genes for the basic functions of cell survival, but the more complex yeast requires additional genes that are involved in protein targeting and organelle function (see Table 17.3).

The nematode *Caenorhabditis elegans* is a simple multicellular organism used to study development. The worm's body is transparent, and its growth from a fertilized egg to a 1,000-celled differentiated adult organism takes just 3 days. The genome of *C. elegans* (100 million bp) is eight times larger than that of yeast and has 3.3 times the number of protein-coding genes (20,470 genes versus 6,275 genes, respectively). Gene inactivation studies have revealed that the worm can survive with only 10 percent of those genes in laboratory cultures. Other gene products in *C. elegans* include proteins used for holding cells together to form tissues, for cellular differentiation, and for intracellular communication (see Table 17.4).

The fruit fly *Drosophila melanogaster* has ten times more cells than *C. elegans* and its genome is larger (140 million bp versus 100 million bp, respectively). However, *D. melanogaster*'s 15,016 genes are fewer in number than those of *C. elegans*, which has 20,470 genes, thereby illustrating that genome size does not necessarily correlate with the number of genes encoded (see Figure 17.8 for distribution of gene functions in *D. melanogaster*).

Compared with the genomes of many plants used for food and fiber, the 125 million bp genome of the *Arabidopsis thaliana* plant, with 25,498 protein-coding genes, is relatively small (wheat, for example, has a genome of 16 billion bp). Furthermore, many of these genes are duplicates of each other and only 15,000 genes remain when the duplicates are removed. *Arabidopsis* contains genes unique to plants, including genes involved in photosynthesis, water transport into the root, cell-wall synthesis, uptake and metabolism of inorganic substances, and defense against herbivores (see Table 17.5). The plant-specific genes in *Arabidopsis* can also be found in rice, *Oryza sativa* (see Figure 17.9).

In eukaryotes, approximately half of the genes exist as one copy per haploid genome, but the rest are present in multiple copies, which arose from gene duplications. These sets of duplicate or closely related genes, such as those encoding the globin and immunoglobulin genes, form gene families. Gene families provide the organism with a functional gene while allowing mutations in other members of the gene family. Some of these mutations may create new genes that are advantageous to the organism.

In humans, the globin gene family consists of three functional members of the α-globin genes and five β-globin genes. During human development, different globins are expressed at different times and in different tissues. In addition to genes that encode functional products, many gene families include pseudogenes, which are copies of genes that result from mutations that cause a loss of function. Pseudogenes may be nonfunctional because they lack promoters and/or recognition sites for intron removal (see Figure 17.10).

Eukaryotic genomes contain many repetitive sequences. Highly repetitive sequences are short (less than 100 bp) and are repeated thousands of times in tandem in the genome. They can be densely packed in heterochromatic regions or scattered around the chromosome (as seen for short tandem repeats, STRs). The number of repeats varies in individual organisms and provides unique molecular markers that can be used to identify individuals.

Moderately repetitive DNA sequences include sequences that are repeated 10 to 1,000 times in the genome and include tRNA and rRNA genes. Multiple copies of the tRNA and rRNA genes are needed to provide the cell with high enough concentrations of components needed for protein translation (see Figure 17.11).

Transposons, segments of DNA that can move from place to place in the genome, make up more than 40 percent of the human genome. There are two major types of transposons in eukaryotes: retrotransposons and DNA transposons (see Table 17.6).

Retrotransposons are transposons that utilize an RNA intermediate when they move to a different site in the genome; the RNA serves as a template for new DNA, which is then inserted into the genome. This "copy and paste" method results in two copies of the transposon. There are three types of retrotransposons: LTRs, LINEs, and SINEs. LTRs have *l*ong *t*erminal *r*epeats and make up about 8 percent of the human genome. SINEs are *s*hort *in*terspersed *e*lements <300 bp; they are transcribed but not translated. LINEs are *long interspersed elements* of 6,000–8,000 bp; some are transcribed and translated. LINEs make up about 17 percent of the human genome. They include the 300-bp *Alu I* element (see Table 17.6).

DNA transposons do not replicate when they move to a new site on the chromosome, nor do they use RNA as an intermediate. Instead, they use a "cut and paste" method—they are excised from the original location and become inserted at a new location. DNA transposons make up about 3 percent of the human genome (see Table 17.6).

Transposons can adversely affect the cell by inserting their DNA into different genes and inactivating them. Transposons can replicate themselves and adjacent chromosomal genes, resulting in gene duplication. Transposons can also pick up adjacent genes during transposition and move them to a new site on the chromosome, potentially creating a new gene that will be advantageous to the organism's survival. Transposons help explain why some mitochondrial and chloroplast genes are located in the nucleus, whereas other organelle genes remain inside the organelle.

Question 5. Name a model organism that you would use to study the functional genomics of neuron signaling, and explain this choice.

Question 6. Compare and contrast gene families with repetitive sequences in eukaryote genomes.

Question 7. Some transposons carry the genes for antibiotic resistance when they move. By what mechanism could this happen?

17.4 What Are the Characteristics of the Human Genome?

The human genome sequence held some surprises

Comparative genomics reveals the evolution of the human genome

Human genomics has potential benefits in medicine

Of the nearly 3.2 billion bp in the human genome, only an estimated 1.2 percent of the DNA (about 21,000 genes) makes up protein-coding regions. Human cells have a greater number of different proteins than the actual number of protein-coding genes. Thus posttranscriptional mechanisms (such as alternative splicing) must account for the greater number of types of proteins.

Gene sizes vary greatly, from about 1,000 bp to 2.4 million bp, with the average gene having 27,000 bp. Virtually all human genes have many introns. About half of the genome contains transposons and other highly repetitive sequences.

Although the genomes of unrelated individuals are about 99.5 percent identical, it is estimated that each haploid genome contains about 3.3 million single nucleotide polymorphisms (SNPs). SNPs account for one-fifth of the variation between two individuals; the remaining four-fifths are due to copy number variation. Genes are not evenly distributed over the genome. Some chromosomes are densely packed with genes, whereas others have long stretches without coding regions.

Comparisons of genes from different organisms have revealed evolutionary relationships (see Figure 17.12). Ninety-nine percent of the human genome is shared with the chimpanzee. Genome sequencing and analyses of an ancient human relative, the Neanderthals, has revealed that their DNA is 99 percent identical to that of modern humans (see Figure 17.13).

A DNA segment that contains two or more SNPs that are usually inherited together is called a haplotype. Haplotype mapping determines an individual's set of haplotypes. Complex phenotypes are determined by multiple genes interacting with the environment. Comparison of the haplotypes of healthy individuals with those of individuals with particular diseases has revealed correlations between genetic loci—sometimes specific SNPs—and complex diseases (see Figure 17.14).

Currently, more than 500,000 SNPs can be placed on a chip and used to analyze disease states. Statistical measures of association of SNP data can be used to determine increased risk for particular diseases (see Table 17.7). Due to continually improving technologies that increase the rate but decrease its cost, genome sequencing is expected eventually to replace SNP testing.

Pharmagenomics is the study of how an individual's genome can affect his or her response to drugs or other outside agents. Correlations between genotypes and responses to drugs will help physicians develop personalized medical care (see Figure 17.15).

Question 8. Suppose you use a radioactively labeled DNA probe to identify a DNA fragment from an *Eco*RI digestion of human DNA samples from multiple individuals. When you complete your hybridization, you notice that the probe has hybridized to one DNA fragment but not in all of the samples. How would you explain this result?

Question 9. Analysis of the human genome has revealed many genes that cause disease when mutated. Now that we know the genes that are involved, why haven't these diseases been eliminated?

17.5 What Do the New Disciplines of Proteomics and Metabolomics Reveal?

The proteome is more complex than the genome

Metabolomics is the study of chemical phenotype

Because of alternative splicing and posttranslational modifications, the sum total of proteins produced by a cell, tissue, or organism at a given time under defined conditions (the proteome) is more complex than the genome (see Figure 17.16A). Proteome analysis includes two-dimensional gel electrophoresis (see Figure 17.16B) and mass spectrophotometry. Comparative proteome analysis and subsequent functional analysis have revealed a common set of proteins that provide the basic metabolic functions of a eukaryotic cell in humans, worms, flies, and yeast (see Figure 17.17). The unique proteins in each organism may result from a reshuffling of the same domains that exist in other organisms. This reshuffling of the genetic deck is a key to evolution.

Metabolomics is the quantitative description of all the small molecules in a cell or organism under defined conditions. Primary metabolites are involved in normal processes such as glycolysis. Secondary metabolites unique to particular organisms are often involved in special responses to the environment. Patterns of metabolites may be helpful in diagnosing diseases or may provide insights into how plants cope with stress.

Question 10. Why do different organisms have more similarities in their proteomes than in their genomes?

Question 11. Describe the set of small molecules present in the cell that make up the metabolome. What type of metabolite would you study if you were interested in a process that occurs in response to an infection?

Test Yourself

1. Functional genomics
   a. assigns functions to the products of genes.
   b. assigns functions to regulatory sequences.
   c. compares genes in different organisms to see how those organisms are related physiologically.
   d. Both a and c
   e. All of the above
   Textbook Reference: *17.2 What Have We Learned from Sequencing Prokaryotic Genomes?*

2. Comparative genomics
   a. assigns functions to the products of genes.
   b. assigns functions to regulatory sequences.
   c. compares genes in different organisms to see how those organisms are related physiologically.
   d. Both a and c
   e. All of the above
   Textbook Reference: *17.2 What Have We Learned from Sequencing Prokaryotic Genomes?*

3. Which of the following statements is *false*?
   a. Cancer can be caused by a transposon inserting into the coding region of a gene in a somatic cell.
   b. Cancer can be caused by genes that are inherited from one or both parents.
   c. Cancer can be caused by environmental conditions but not by genetic factors.
   d. Cancer can be caused by multiple genes, epigenetic effects, and environmental conditions.
   e. Cancer can be linked to several key SNPs.
   Textbook Reference: *17.4 What Are the Characteristics of the Human Genome?*

4. Genes that cause cancer when mutated
   a. can be analyzed by means of linked markers such as SNPs.
   b. can be analyzed to determine which cancer treatment will work the best for an individual.
   c. do not have any homologs in other organisms.
   d. can help us predict which individuals are more likely to develop cancer.
   e. a, b, and d only
   Textbook Reference: *17.4 What Are the Characteristics of the Human Genome?*

5. The sequencing of the human genome has allowed scientists to
   a. understand regulatory sequences that are important for gene expression.
   b. locate genes that cause disease.
   c. understand evolutionary relationships by comparing human genes to genes in other organisms.
   d. investigate gene families and their origins.
   e. All of the above
   Textbook Reference: *17.4 What Are the Characteristics of the Human Genome?*

6. Proteomics has been used to compare
   a. DNA sequences between closely related species.
   b. gene expression during embryonic development.
   c. protein sequences between closely related species.
   d. shotgun cloned sequences.
   e. prokaryotic genomes.
   Textbook Reference: *17.5 What Do the New Disciplines of Proteomics and Metabolomics Reveal?*

7. Which of the following was *not* a component of the sequencing of the human genome?
   a. Large segments of DNA cloned into bacterial artificial chromosomes (BACs)
   b. cDNA cloning
   c. Computer alignment of overlapping pieces of chromosomes
   d. Sequencing of DNA with deoxyribonucleotides triphosphates
   e. Bioinformatics
   Textbook Reference: *17.1 How Are Genomes Sequenced?*

8. Which of the following statements about transposable elements is *false*?
   a. They can inactivate genes into which they are inserted.
   b. They can contain gene sequences.
   c. They are mobile genetic elements that move from RNA molecule to RNA molecule.
   d. They may be spliced out of one region of the genome and inserted into another.
   e. They replicate themselves before moving to another site on the genome.
   Textbook Reference: *17.2 What Have We Learned from Sequencing Prokaryotic Genomes?*

9. Gene inactivation studies have allowed us to
   a. determine the minimum number of genes humans need to survive.
   b. demonstrate that *C. elegans* needs most of its genes.
   c. investigate the minimum number of genes needed to sustain life.

d. create artificial life in a test tube.
e. All of the above
Textbook Reference: *17.2 What Have We Learned from Sequencing Prokaryotic Genomes?; 17.3 What Have We Learned from Sequencing Eukaryotic Genomes?*

10. Comparisons of yeast and bacterial cell genomes have revealed that
a. yeast cells have more genes devoted to the basic functions of survival than bacteria do.
b. eukaryotic cells are structurally similar to bacterial cells in terms of complexity.
c. there are more genes for targeting proteins to organelles in yeast than in bacteria.
d. the histones of bacteria are very similar to those of yeast.
e. bacteria and yeast are both haploid.
Textbook Reference: *17.3 What Have We Learned from Sequencing Eukaryotic Genomes?*

11. Which one of the following does *not* represent information that we have gained from genome sequencing?
a. Mycobacteria have many genes devoted to metabolizing fats.
b. There is extensive genetic exchange between different kinds of bacteria.
c. Genomes can be sequenced even when organisms cannot be cultured.
d. Much of the eukaryotic genome contains coding sequences.
e. The majority of genes in *C. elegans* are cell signaling genes.
Textbook Reference: *17.2 What Have We Learned from Sequencing Prokaryotic Genomes?; 17.3 What Have We Learned from Sequencing Eukaryotic Genomes?; 17.4 What Are the Characteristics of the Human Genome?*

12. Which of the following was *not* one of the discoveries that resulted from the genome sequencing of plants?
a. There are more protein-coding genes in animals than in plants.
b. Many *Arabidopsis* genes are duplicated due to chromosomal rearrangements.
c. There are more genes in plants that are similar to each other than there are genes that are unique.
d. Plants have many genes whose products are used for defense against microbes and herbivores.
e. All of the above were discoveries resulting from the genome sequencing of plants.
Textbook Reference: *17.3 What Have We Learned from Sequencing Eukaryotic Genomes?*

13. Which of the following statements about eukaryote retrotransposons is *false*?
a. They are translated before they are transcribed.
b. They use an RNA intermediate to move from one region of the genome to another.
c. They are highly repetitive sequences found throughout the genome.
d. They can encode gene products that are required for their own transposition.
e. All of the above are true; none is false.
Textbook Reference: *17.3 What Have We Learned from Sequencing Eukaryotic Genomes?*

14. Which of the following statements about the human genome is *false*?
a. Over 50 percent of the genome contains transposons.
b. Almost every gene has introns.
c. Genes are evenly distributed over the genome.
d. About 2 percent of the genome codes for genes.
e. Humans have about the same number of genes as fruit flies have.
Textbook Reference: *17.4 What Are the Characteristics of the Human Genome?*

15. Single nucleotide polymorphisms (SNPs)
a. can be used to map unlinked genes in order to follow the inheritance of disease traits.
b. can be used to predict if a patient is at risk for a particular disease.
c. can be linked in order to generate gene sequences.
d. have limited sequence variations.
e. All of the above
Textbook Reference: *17.4 What Are the Characteristics of the Human Genome?*

16. The study of proteomes allows scientists to compare
a. the proteome with the genome to see if the gene sequences are correct.
b. proteome sequences between species to see if similar proteins are expressed in all species.
c. transcriptional patterns in different organisms.
d. highly repetitive DNA sequences in different organisms.
e. how noncoding regions of the genome differ in different organisms.
Textbook Reference: *17.5 What Do the New Disciplines of Proteomics and Metabolomics Reveal?*

17. Which of the following is *not* a characteristic you would look for when choosing the next "model" plant genome to study agriculturally important traits?
a. A small genome
b. A genome with little redundancy
c. Ease of laboratory culture
d. Rapid life cycle
e. Large numbers of transposons
Textbook Reference: *17.3 What Have We Learned from Sequencing Eukaryotic Genomes?*

18. Which of the following is *not* something you would expect about susceptibility to a disease revealed by haplotype mapping?
a. The SNP patterns would have a good chance of being altered by transposable elements.

b. The SNP patterns might involve interactions with environmental cues.
c. The susceptibility would be shared by genetically related individuals.
d. Some of the SNP patterns would affect posttranscriptional events.
e. The susceptibility might be greater if the individual were homozygous for the haplotype.

Textbook Reference: *17.4 What Are the Characteristics of the Human Genome?*

Answers

Key Concept Review

1. Your list should include the following steps:
   1. Cut the DNA into 100-bp fragments physically or using enzymes to hydrolyze phosphodiester bonds.
   2. Heat the DNA to break the hydrogen bonds holding the two strands together.
   3. Attach each end to a short adapter sequence that is anchored to a bead or flat surface.
   4. Amplify using PCR.
   5. Heat the fragments to denature them.
   6. Add a universal primer complementary to one of the adaptor sequences, DNA polymerase, and the four nucleotides, each with an identifying fluorescent dye.
   7. Once a nucleotide is incorporated into the replicating strand, remove all unincorporated nucleotides.
   8. Use a camera to record the color of the added nucleotide.
   9. Remove the fluorescent tag from the added nucleotide and repeat the process from step 5.
   10. Run a computer analysis of the captured images to determine the sequence of nucleotides.
   11. Use a computer program to determine how the sequences of the DNA fragments overlap and to reconstruct the genome sequence.
2. The search would begin with an examination of the genome sequence for start and stop codons for translation, which would indicate the location of genes. Then each gene sequence would be checked for correct codons to produce the protein's sequence of amino acids.
3. The tuberculosis bacterium must use lipids as a source of energy-rich compounds; a drug that inhibits lipid uptake or metabolism in this bacterium could inhibit its growth.
4. Comparative genomics could be used to identify sequences that are unique to *E. coli* O157:H7 and not found in nonpathogenic *E. coli* strains.
5. The simplest model organism that includes genes for a nervous system and intercellular communication is the nematode *Caenorhabditis elegans,* so this would be a good choice.
6. Gene families are duplicated genes that code for proteins. They can include a few to hundreds of copies of the genes in a single genome. Usually different copies contain slight mutations that code for a functional variant of the gene product. Pseudogenes are nonfunctional duplicate genes. Repetitive sequences may be short, nontranscribed sequences that exist in thousands of tandem copies, or they may be moderately repetitive (10–1,000 repeats) that code for tRNAs and rRNAs.
7. Some transposons become duplicated, with two copies flanking one or more genes. The duplicated transposons together with the genes form a single transposable element. If the flanked genes contain antibiotic resistance factors, they could be transported along with the new large transposable element to different parts of the genome or onto plasmids.
8. The DNA sequence of the probe is specific for a unique gene in humans that has two or more alleles, of which one or more do not bind the probe. Some individuals may have different alleles due to one or more SNPs, or other sequence modifications, such as deletions or insertions.
9. We have not been able to eliminate genetic diseases because, except in a few rare cases, we are not able to replace the mutated genes in the diseased individual.
10. More sequence similarities are revealed when proteomes are compared than when genomes are compared because the genetic code is redundant. There can be more than one codon for a particular amino acid, so the DNA sequences can vary and still generate the same amino acid sequences.
11. The metabolome is the complete set of small molecules present in a cell, tissue, or organism under defined conditions. Primary metabolites are involved in normal processes. If one were interested in small molecules present only in response to a form of stress, then one would study secondary metabolites.

Test Yourself

1. **a.** Functional genomics assigns functions to gene products, not to the regulatory sequences. Comparing the genes of different organisms is the work of comparative genomics.
2. **c.** Comparative genomics compares genes between different organisms to see which genes are possessed by one organism but not another. These comparisons can be related to the physiology of the organisms being compared.
3. **c.** Multiple genes, environmental influences, and epigenetic effects all contribute to the development of cancer. Transposon insertion into a somatic gene may also result in cancer, and a small percentage of cancers are inherited. Cancer genes have been linked to many different SNPs.

4. **e.** Mutated genes that are linked to cancer can be linked to particular SNPs, can be used to predict if a person might be susceptible to developing cancer, and are being used to try to determine the best treatment. Those genes do have homologs in other organisms.

5. **e.** Human genome sequencing provides all these advantages.

6. **c.** Proteomics is used to compare protein sequences between different organisms. Answers **a**, **b**, and **d** all refer to techniques used in genomics.

7. **b.** cDNA cloning is used to determine what mRNAs are being expressed in cells. All of the other techniques were utilized to sequence the human genome and order those sequences.

8. **c.** Some transposons do contain gene sequences. They can be spliced out of one region and inserted into another and also replicate themselves and then move to another site. They do not move from one RNA molecule to the next.

9. **c.** Gene inactivation studies have been used in some organisms (not humans) to determine the minimum number of genes needed by cells in order to survive. *C. elegans* needs only about 10 percent of its genes to survive. Artificial life has not yet been created in a test tube, but scientists are actively pursuing this possibility.

10. **c.** A comparison of yeast and bacterial cell genomes revealed that the number of genes needed for survival in these two organisms was roughly the same but that yeast cells had many genes for targeting proteins to organelles. Yeast cells are structurally more complex than bacterial cells. Bacterial cells do not have histones. Yeast cells can be haploid or diploid; bacteria are haploid (which was known long before comparative genomics was developed).

11. **d.** Only 2 percent of the human genome codes for protein-coding genes.

12. **a.** The reverse is true: There are more protein-coding genes in plant cells than in animal cells.

13. **a.** Retrotransposons can move from one region of the genome to another using an RNA intermediate to make a DNA copy of themselves. They are highly repetitive sequences found throughout the genome. Some transposons encode gene products needed for their transposition. Retrotransposons are not translated before they are transcribed.

14. **c.** Genes are not evenly distributed over the genome; some chromosomes have many more genes than others have.

15. **b.** SNPs that are linked to a disease gene can be used to analyze patients' DNA samples to determine if they are at risk for that disease. SNPs are quite variable. They are small sequences and are not used to generate gene sequences.

16. **b.** Proteomic studies can be used to compare proteins in different organisms. The gene sequence will determine the protein sequence. Proteomic studies cannot be used to study transcriptional patterns, which are cellular processes involving RNA. Proteins are coding regions of the genome and highly repetitive sequences do not code for protein.

17. **e.** A plant with large numbers of transposons would be subject to gene interruptions and movement, which could complicate studies. A small genome is more manageable and means less redundancy and noncoding DNA, thus simplifying the search for genes of interest. The ability to cultivate many generations of a research organism minimizes the time and effort needed to observe expressed phenotypes.

18. **a.** Because the haplotype is a set of closely linked SNPs on a chromosome, they would not be likely to be interrupted by a transposable element.

Recombinant DNA and Biotechnology

The Big Picture

- The ability to isolate DNA from any organism, ligate it to vector DNA, introduce that DNA into host cells, and propagate those cells has had an enormous impact on our understanding of genetics, molecular biology, and cell function and development. These techniques have been used to elucidate evolutionary relationships among different organisms and to understand gene regulation and function in greater depth. Applications of these techniques have been used to develop new medicines and diagnostics, agricultural products, and powerful forensic tools.

Study Strategies

- There are a variety of ways to clone DNA fragments, and trying to remember all the different cloning methods can be challenging. Consider the following as you review the different cloning procedures: the size of the cloned DNA, the host cell in which the DNA could be cloned, and the expression of the cloned DNA in the host cell. Different vectors can be used in different host cells to address each of these considerations.
- Many different experimental questions can be answered using cloning. Ask yourself what cloning strategies could be used to answer the following questions: What is the sequence of a gene? What sequences are important for regulation of that gene? What sequences are important for targeting that gene to a particular site in a eukaryotic cell? What is the difference in function between a mutant gene product and a wild-type gene? What kinds of genes are expressed during the development of an organism? How can cloned genes be expressed in plants or in the milk of mammals? Review the textbook for answers.
- Outline the specific steps needed to clone a gene in a bacterial cell. Include how the gene is initially isolated, what kind of vectors can be used, how the recombinant DNA is introduced into the cell, and how the placement of the recombinant molecule in the host cell can be confirmed. Then outline the steps needed to clone a gene in a eukaryotic cell.
- Complementary base pairing is important for many aspects of biotechnology. Describe each technique in gene cloning that uses complementary base pairing, detailing specifically how base pairing is involved.
- Go to the Web addresses shown to review the following animated tutorial and activity. You can also find them in BioPortal (yourBioPortal.com), along with many additional learning resources.

  Animated Tutorial 18.1 DNA Microarray Technology (Life10e.com/at18.1)

  Activity 18.1 Expression Vectors (Life10e.com/ac18.1)

Key Concept Review

18.1 What Is Recombinant DNA?

Recombinant DNA, which is a single DNA molecule made from at least two different sources of DNA, is used for genetic modification of organisms. Restriction endonucleases are used to cleave DNA. These enzymes recognize palindromic sequence in DNA and cut at or near those sites. DNA ligase (the enzyme that covalently joins the Okazaki fragments during DNA replication and mends broken DNA; see Chapter 13) is used to form a covalent bond on each DNA strand of the recombinant molecule, even if they are from different species (see Figure 18.1).

Some restriction endonucleases cut between the same bases on both DNA strands, producing blunt ends. When restriction endonucleases cut DNA, they often leave ends that have 5′ or 3′ overhangs of single-stranded DNA. These ends are called "sticky ends" and can form complementary base pairs with other DNA molecules that have the same sticky ends (see Figure 18.2).

Question 1. Explain and diagram the steps required to make a recombinant plasmid using two different *E. coli* plasmids: plasmid A^r expresses the resistant gene for ampicillin and plasmid K^r contains the resistance gene to kanamycin. Include how you would prepare the plasmids for recombination and how you would recombine the plasmids. Explain how, after transforming bacteria with your plasmid mixture, you would select for bacteria that contain the recombinant molecule.

Question 2. You need to insert a fragment into a vector in a specific direction. Explain which type of restriction cut (i.e., blunt-ended or sticky-ended) would be the best choice for preparing the ends of the fragment and cutting the plasmid.

18.2 How Are New Genes Inserted into Cells?

Genes can be inserted into prokaryotic or eukaryotic cells

A variety of methods are used to insert recombinant DNA into host cells

Reporter genes help select or identify host cells containing recombinant DNA

Recombinant DNA is inserted into a host cell by transformation (or transfection, if the host cell is an animal cell), creating a transgenic cell or organism. Because only a few of the cells exposed to the recombinant DNA are transformed, selectable markers, such as genes that confer resistance to antibiotics, are often included on the recombinant DNA molecule (see Figure 18.3).

One goal of recombinant DNA technology is to clone genes. Cloned genes are useful for sequence analysis, to produce quantities of their protein products for medicinal or agricultural use, or to create a transgenic cell or organism. In theory, any cell type can be used for cloning, but most cloning is carried out using model organisms. In addition to their chromosomal DNA bacteria contain plasmid DNA, circular DNA molecules which are easily manipulated to carry recombinant DNA into the cell. Yeast cells have rapid cell division cycle (2–8 hours), are easy to grow, have a small genome size (12 million base pairs), and have been used to clone many eukaryotic genes. Plant cells are also good hosts for cloning genes. Many plant cells are totipotent. They can be grown in culture, transformed with recombinant DNA, and manipulated to form an entire new transgenic plant containing the recombinant DNA molecule.

A variety of methods are used to insert recombinant DNA into host cells. For example, cells can be treated chemically to make their outer membranes more permeable, or electroporation can be used to create temporary pores. Viruses can carry recombinant DNA into cells. Once within the cell, the new DNA must be replicated when the cell divides. Recombinant DNA can be inserted into the host chromosome, where it is replicated when the chromosome is replicated. Such insertion might happen randomly after the DNA is introduced into the cell. Alternatively, recombinant DNA molecules can enter the host cell as part of a vector that already has an origin of replication.

Plasmids are useful vectors for recombinant DNA cloning. Plasmids synthesize their DNA independently from their own origins of replication, making many copies of plasmid DNA per cell. A typical plasmid vector has been engineered to have one or more unique restriction sites to allow the insertion of the recombinant DNA, and to carry genes that confer resistant against selectable markers, such as antibiotics. *E. coli* plasmids are small (from 2,000 to 6,000 base pairs) and are commonly used as vectors. They have a single set of unique restriction sites where the cut DNA is inserted, and a drug resistance marker allows the investigator to confirm the presence of the plasmid in the *E. coli* cell.

Plasmid vectors can also be used to insert recombinant DNA into plant cells. The plant-specific bacterium *Agrobacterium tumefaciens*, which causes the disease crown gall in plants, harbors a Ti plasmid. This plasmid contains a transposon, T DNA, which inserts copies of itself into the host plant cell's DNA when the bacteria infect the plant. This plasmid has been modified to be nonpathogenic-inducing and is widely used to insert desirable genes into plant genomes.

Due to replication restraints, the size of DNA that can be inserted into plasmids is restricted to about 10,000 base pairs. Most eukaryotic genes—with introns and flanking regulatory regions—are larger than this. Consequently, both prokaryotic and eukaryotic virus vectors are usually used to clone larger DNA sequences (up to 20,000 base pairs of inserted DNA). Viruses infect cells naturally, allowing easy entry of cloned sequences into the cytoplasm of the cell.

Reporter genes help select or identify host cells containing recombinant DNA. Selectable markers such as markers for antibiotic resistance can be used to determine if the host cell contains the recombinant DNA molecule by inhibiting the growth of non-carriers (see Figure 18.3). Reporter genes such as β-galactosidase and green fluorescent protein (GFP) are selectable markers that can be used to visually detect the expression of recombinant DNA molecules in host cells. These reporter genes can be attached to promoters of gene coding regions to visualize transcription of the gene of interest (β-galactosidase) or localization of its protein product in cells (GFP) (see Figure 18.4).

Question 3. What are the critical elements needed for a useful plasmid vector for recombinant DNA?

Question 4. What are the constraints on the use of plasmids as vectors for inserting recombinant DNA into cells and organisms?

Question 5. You have inserted gene *X* into a plasmid vector and transfected a cell line. However, upon looking at the cells through a microscope, you find that you are unable to determine if protein X is expressed and localized to the correct cell compartment. What do you have to change in your procedure to clone gene *X* so that you can visualize protein expression and localization?

18.3 What Sources of DNA Are Used in Cloning?

Libraries provide collections of DNA fragments

cDNA is made from mRNA transcripts

Synthetic DNA can be made by PCR or by organic chemistry

One of the reasons for cloning DNA is to study its function, including the proteins it codes for and their regulatory sequences. Sources of DNA for cloning include genomic libraries, cDNA libraries, and synthetic data.

Genomic libraries are a collection of DNA fragments from the entire genome of an organism. After the DNA is cut or broken into fragments, each fragment is inserted into a vector, usually a bacteriophage λ, which is used to insert the gene into bacteria. DNA hybridization with colonies of transformed bacteria can identify fragments that contain particular coding sequences (see Figure 18.5A).

Complementary DNA (cDNA) libraries consist of all the genes transcribed in a particular tissue. To make cDNA, the messenger RNA (mRNA) must first be isolated from the cell. A DNA molecule complementary to that mRNA is synthesized using reverse transcriptase, and the new molecule can then be cloned (see Figure 18.5B). cDNA clones are used to compare gene expression in different tissues at different stages of development.

Synthetic DNA can be made using PCR to amplify a specific sequence. Artificial genes can be made if the amino acid sequence of the gene is known. DNA sequences can be created with desirable characteristics, such as restriction sites or specific mutations. Sequences for the initiation of transcription and translation and for termination can be added to the gene sequence.

Question 6. DNA libraries are useful tools in biotechnology. Explain which type of DNA library would work best for studying gene expression at a particular stage of cell development.

Question 7. What two methods of DNA synthesis are used for making synthetic DNA?

18.4 What Other Tools Are Used to Study DNA Function?

Genes can be expressed in different biological systems

DNA mutations can be created in the laboratory

Genes can be inactivated by homologous recombination

Complementary RNA can prevent the expression of specific genes

DNA microarrays reveal RNA expression patterns

Genes and their products can be studied in different biological systems; for example, human genes can be studied in a mouse, or a bacterial gene can be studied in a plant. For cells to produce a cloned gene product, the vector must have appropriate DNA sequences that allow the cloned gene to be expressed in the host organism.

Gene function can be studied by overexpressing it. To achieve this, the gene is inserted into a vector downstream of a stronger promoter.

Recombinant technology permits generation of mutant genes in the lab. Mutants can be synthesized and compared to wild-type genes to analyze gene function.

Another method of studying gene function is to eliminate gene expression by creating knockout genes. This can be done in a variety of ways. One of these is called homologous recombination, which involves an exchange of DNA between molecules with identical, or nearly identical, sequences. In constructing a knockout mouse, the gene of interest is disrupted by the insertion of a reporter or marker gene into the middle of the native gene. A plasmid containing the inactivated gene with the marker is transfected into a mouse stem cell. If recombination occurs, the marker gene will be expressed. The transfected stem cell is then transplanted into an early mouse embryo, and the consequences of carrying an inactivated gene are analyzed (see Figure 18.6). To obtain mice with both alleles knocked out in all tissues, the knockout mice are inbred.

Antisense messenger RNA and RNAi can prevent the translation of specific genes. Antisense RNA will base pair with mRNA in the cytoplasm and form a double-stranded RNA molecule, which cannot be translated and will be degraded by the cell. Interference RNAs, known as small interfering RNAs (siRNAs), are short (about 20 nucleotides) double-stranded RNA molecules that bind specific mRNAs and target them for degradation. Because siRNAs are more stable than antisense RNA, they are the preferred method in medicine and in the laboratory for inhibiting translation (see Figure 18.7).

DNA microarrays reveal RNA expression patterns using hybridization in a large number of sequences simultaneously. It is possible to examine patterns of expression in different tissues, under different conditions, and in individuals with known mutations. Microarrays are small glass chips containing thousands of copies of each DNA sequence per chip. These sequences (greater than 20 bp) are attached to the chip in spots in a precise order. Each spot on the chip contains a unique sequence. cDNAs from a tissue or cell of interest is then hybridized to the chip, and the level of hybridization is analyzed. Chip technology has been used to look at gene expression from different breast cancers to predict the prognosis for patients and determine treatments (see Figure 18.8).

Question 8. A researcher would like to study the effects of a specific mutation in gene *Y*. How would the researcher create a mouse that expresses only this mutated gene?

Question 9. A researcher would like to study the effect of inhibiting the expression of a specific gene product (protein) in a cell line. What methodology could the researcher use?

18.5 What Is Biotechnology?

Expression vectors can turn cells into protein factories

Biotechnology is the use of living cells to produce useful materials, including food, medicines, and chemicals. Historically, humans have manipulated living organisms without being aware of the genes involved, and sometimes even the organisms. Today, microbes are manipulated to express cloned genes at high levels and can be used to produce antibiotics and other useful chemicals.

For cells to produce a cloned gene product, the vector (an expression vector) must have DNA sequences that allow the cloned gene to be expressed. Prokaryotic expression vectors require a promoter, a termination site for transcription, and a ribosome-binding site. Eukaryotic expression vectors require the same elements as well as a poly A–addition site, transcription factor binding sites, and enhancers. Modifications of expression vectors include the addition of inducible promoters (which respond to a specific signal), tissue-specific promoters, and signal sequences (see Figure 18.9).

Question 10. Describe three yeast-based biotechnologies that people employed long before microbes and genes were identified.

Question 11. Suppose that your lab assistant has cloned gene *X* into yeast and confirmed that the recombinant DNA molecule is present in the yeast cells. However, the yeast cells are unable to synthesize protein X. Suggest why this part of the experiment is not working and what modifications are needed in the cloning procedure.

18.6 How Is Biotechnology Changing Medicine and Agriculture?

Medically useful proteins can be made using biotechnology

DNA manipulation is changing agriculture

There is public concern about biotechnology

Medically useful products that have been cloned in expression vectors include tissue plasminogen activator, human insulin, and vaccine proteins (see Figure 18.10 and Table 18.1).

Recombinant DNA offers breeders the opportunity to choose specific genes that will be incorporated into an organism, to introduce any gene into a plant or animal species, and to generate new organisms quickly (see Figure 18.11 and Figure 18.12).

Transgenic plants have been created that express toxins for insect larva (using the gene for the toxin naturally produced by the bacterium *Bacillus thuringiensis*), that produce extra nutrients (adding genes for β-carotene to rice to reduce vitamin A deficiencies in human populations), or that can tolerate drought or high-salt conditions (adding salt-tolerance genes to tomatoes) (see Figure 18.13, Figure 18.14, and Table 18.2).

The creation of transgenic plants has raised some concerns that these crops could be unsafe for human consumption, that it is unnatural to interfere with nature, and that transgenes could be transferred into other crops or noxious plants. Transgenic plants are extensively field tested, and scientists are examining these issues and proceeding cautiously.

Recombinant bacteria have been developed for environmental clean-up in composting programs, wastewater treatments, oil spills, and other such efforts. As with transgenic plants, such organisms have not been released into the environment because of the unknown effects such organisms might have on a natural ecosystem.

Question 12. Describe three useful products that have been produced using biotechnology.

Question 13. Outline two potential dangers of producing organisms that contain foreign genes.

Test Yourself

1. Cloning a gene may involve
   a. restriction endonucleases and ligase.
   b. plasmids and bacteriophage λ.
   c. transformation or transfection.
   d. selectable markers and/or reporter genes.
   e. All of the above
   Textbook Reference: *18.1 What Is Recombinant DNA?*

2. Complementary base pairing is important for
   a. ligation reactions with blunt-end DNA molecules.
   b. hybridization between DNA and transcription factors.
   c. restriction endonucleases for cutting cell walls.
   d. synthesizing cDNA molecules from mRNA templates.
   e. the transcriptional activation of expression vectors.
   Textbook Reference: *18.3 What Sources of DNA Are Used in Cloning?*

3. For a prokaryotic vector to be propagated in a host bacterial cell, the vector needs
   a. an origin of replication.
   b. telomeres.
   c. centromeres.
   d. drug-resistance genes.
   e. reporter genes.
   Textbook Reference: *18.2 How Are New Genes Inserted into Cells?*

4. For a Ti plasmid to be propagated in a host plant cell, the vector needs
   a. telomeres.
   b. centromeres.
   c. an origin of replication.
   d. a reporter gene.
   e. a, b, and c
   Textbook Reference: *18.2 How Are New Genes Inserted into Cells?*

5. Reporter genes include genes for
   a. drug resistance.
   b. bioluminescence.
   c. DNA origins.
   d. restriction endonucleases.
   e. Both a and b
   Textbook Reference: *18.2 How Are New Genes Inserted into Cells?*

6. Vectors include
   a. bacterial plasmids.
   b. viruses.
   c. plant plasmids.
   d. bacteriophage λ.
   e. All of the above
   Textbook Reference: *18.2 How Are New Genes Inserted into Cells?*

7. A cDNA clone is
   a. mostly cytosine.
   b. a copy of the DNA identical to the nuclear gene.
   c. a copy of noncoding DNA.
   d. a DNA molecule complementary to an mRNA molecule.
   e. a fragment of DNA inserted into the host chromosome.

   Textbook Reference: *18.3 What Sources of DNA Are Used in Cloning?*

8. Gene expression can be inhibited by
   a. antisense RNA.
   b. knockout genes.
   c. DNA microarrays.
   d. microRNA.
   e. a, b, and d

   Textbook Reference: *18.4 What Other Tools Are Used to Study DNA Function?*

9. Expression vectors are different from other vectors because they contain
   a. drug-resistance markers.
   b. telomeres.
   c. regulatory regions that permit the cloned DNA to produce a gene product.
   d. DNA origins.
   e. reporter genes that are expressed in the host.

   Textbook Reference: *18.5 What Is Biotechnology?*

10. RNAi
    a. is more effective than antisense RNA in inhibiting translation.
    b. inhibits transcription in eukaryotes.
    c. is produced only by viruses.
    d. requires other proteins to modify the RNAi.
    e. Both a and d

    Textbook Reference: *18.4 What Other Tools Are Used to Study DNA Function?*

11. DNA chip technologies can be used to
    a. predict who will get cancer.
    b. show transcriptional patterns in an organism during different times of development.
    c. clone DNA.
    d. make transgenic plants.
    e. inhibit transcription of disease genes.

    Textbook Reference: *18.4 What Other Tools Are Used to Study DNA Function?*

12. Which of the following is *not* an application of recombinant DNA technology?
    a. Generating large amounts of tissue plasmonigen factor to help dissolve blood clots
    b. Making plants more resistant to insect larva
    c. Creating vaccines for pathogens
    d. Reducing the salt in environmental soil so that plants can grow
    e. Creating bacteria that can accelerate the breakdown of wood chips and paper

    Textbook Reference: *18.6 How Is Biotechnology Changing Medicine and Agriculture?*

13. A disadvantage of recombinant DNA technology is that it
    a. has the capacity to spread transgenes from crops to other species.
    b. makes weeds herbicide-resistant.
    c. creates transgenic plants that kill beneficial insects.
    d. creates transgenic plants that could be harmful for human consumption.
    e. All of the above

    Textbook Reference: *18.6 How Is Biotechnology Changing Medicine and Agriculture?*

14. Expression vectors
    a. are useful for analyzing RNA transcription patterns.
    b. are useful for isolating large amounts of DNA.
    c. can be used to make large amounts of protein in *E. coli* cells.
    d. are useful only in prokaryotic cells and cannot be expressed in eukaryotic cells.
    e. are used to create SNPs for DNA microarrays.

    Textbook Reference: *18.5 What Is Biotechnology?*

15. Which of the following is minimally required for a transgene to be expressed in a eukaryotic host?
    a. A promoter, a transcriptional termination site, and a ribosome binding site
    b. A DNA origin and an antibiotic resistance marker
    c. Transcription factor binding sites, enhancers, and a poly A–recognition sequence
    d. An inducible promoter and repressor proteins
    e. a, c, and d

    Textbook Reference: *18.5 What Is Biotechnology?*

Answers

Key Concept Review

1.

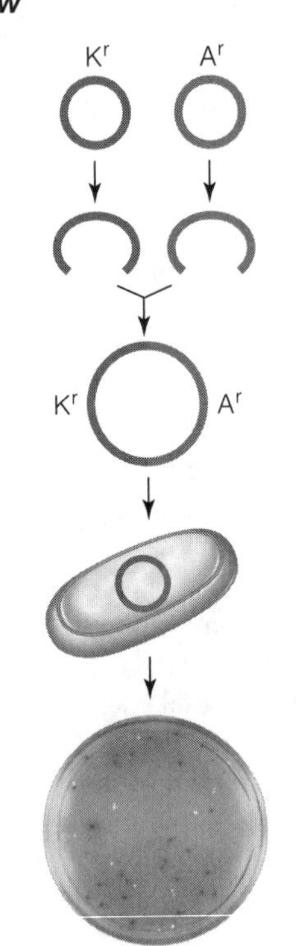

1. Cut plasmids A^r and K^r with a unique restriction enzyme (RE) that produces a sticky end, such as *Eco*R1. (An RE that produces a blunt end can also be used, but the number of successful recombinations after the ligation step will be significantly lower.)
2. Combine RE-digested plasmid A^r and plasmid K^r mixtures in a reaction mixture that contains DNA ligase.
3. Transform *E. coli* with ligated plasmids (see section 18.2)
4. To isolate the bacteria that contain the recombined plasmids, spread the transformed bacteria on agar plates that contain both ampicillin and kanamycin. Only the bacteria that contain a recombinant molecule containing both resistance genes will survive and grow to form colonies.

Note: A bacterium that takes up copies of both plasmid A^r and K^r will also be able to grow and form a colony. To distinguish between these bacteria and the bacteria with the recombined plasmids, you would isolate and analyze the plasmids from samples of the bacterial colonies.

2. Using restriction enzymes (REs) that produce sticky-end cuts would yield a greater number of recombinant molecules with the DNA insert in the correct orientation. Blunt-ended DNA fragments can be ligated with any other blunt-ended fragment. Thus, a fragment prepared with a blunt end–producing RE could insert in either direction into a vector that has been cut with a blunt end–producing RE. Sticky-end overhangs, in contrast, will only reanneal with a complementary strand, such as that produced by the same restriction enzyme or another RE that generates the same sequence overhang. To ensure a correct orientation, two different unique sticky-end REs (or one sticky-end and one blunt-end RE) should be used. The vector used would need to have the same unique RE sites in the proper 5′ to 3′ order.
3. A successful plasmid vector must include 20 or more unique restriction sites, origins of replication (*ori*) sites, and reporter genes and selectable marker genes.
4. Constraints on plasmid replication limit the size of the new DNA that can be inserted to 10,000 bp, while most eukaryotic genes together with their introns and flanking sequences are larger. In addition, plasmids do not naturally infect cells, and sometimes it is difficult to get them inside the desired host cells.
5. To visualize protein X you need to use a reporter gene, such as the gene that encodes green fluorescent protein (GFP). Insert gene X into an expression vector that will fuse the GFP gene with the inserted DNA at either the 5′ end (with the GFP gene sequence providing the START codon) or the 3′ end (with GFP sequence providing the STOP codon) of the inserted gene X. When the gene X GFP mRNA transcript is translated, protein X will be fused with GFP. After transfecting the cell line with the recombinant DNA, expose the cells to ultraviolet light to visualize the production of protein X GFP and to determine its localization within the cell.
6. cDNA libraries can be made from different developing tissues in order to see which genes are being expressed. DNA microarrays can also be used to analyze transcriptional patterns during development.
7. Synthetic DNA can be created using the polymerase chain reaction (PCR) or synthesized using organic chemistry. PCR requires the use of template DNA. Organic chemistry can link individual nucleotides to form any specified sequence.
8. Homologous recombination can be used to replace the normal mouse gene Y with a copy of Y that has a specific mutation. A vector containing the mutated gene is inserted into a mouse embryonic stem cell (derived from a blastocyst), where the nearly homologous (one nucleotide difference) gene Y in the vector lines up with the normal gene Y in the mouse genome. Homologous recombination occurs, and the mutated gene Y is now in the chromosome. The stem is then transplanted into an early mouse blastocyst, which is then implanted into a recipient female mouse. To obtain mice that are homozygous for mutant gene Y, the mice are inbred.
9. The researcher could use antisense RNA or RNA interference (RNAi) to inhibit the expression of the specific gene. For antisense RNA, the researcher would produce short, single-stranded microRNAs (miRNAs) that are complementary to mRNA sequences of the gene.

Upon hybridization, these would inhibit translation. For RNAi, the researcher would synthesize small interfering RNA (siRNA) which, upon binding to the target mRNA, induces its degradation.

10. Thousands of years ago people started using yeast, without realizing what they were using, to brew beer and wine and to make bread.
11. The *X* gene can be cloned into a yeast cell, but unless the vector has the appropriate regulatory signals (promoters, poly A–addition sites, translational initiation, and termination signals), no expression of gene *X* will occur. Recloning gene *X* in a yeast expression vector will result in the expression of the *X* gene.
12. Useful products include rice grains that produce β-carotene, plants that are resistant to herbicides and insect larvae, cow's milk containing human growth hormone, and others (see Tables 18.1 and 18.2).
13. Potential dangers include the creation of genetically engineered foods that could adversely affect human nutrition, the transfer of herbicide- and insect-resistant genes from crop plants to noxious weeds, and the introduction into the wild of new organisms that might have unforeseen ecological consequences.

Test Yourself

1. **e.** Cloning a gene requires restriction endonucleases to cut the gene of interest; vectors, including plasmids and bacteriophage λ; and ligase to covalently join the DNA to the vector. Transformation or transfection is needed to introduce the vector into the host cells, and reporter genes and selectable markers can be used to detect the presence of the gene in the host cells.
2. **d.** No complementary base pairing can occur between blunt-end cut DNA molecules. Transcription factors are proteins that bind DNA through interactions with their side chains (which are amino acids) and the nucleotides of the DNA. Restriction endonucleases cut double-stranded DNA, not cell walls. The activation of transcription vectors does not require complementary base pairing.
3. **a.** A prokaryotic vector needs an origin of replication to be propagated in a prokaryotic cell.
4. **c.** The Ti plasmid requires an origin of replication but not telomeres or centromeres to propagate itself inside the plant cell.
5. **e.** Reporter genes include genes for drug resistance and bioluminescence.
6. **e.** All of these molecules can serve as vectors for cloned DNA.
7. **d.** cDNA clones do not contain mostly cytosine, nor are they copies of noncoding genes. A cDNA clone is generated by making a DNA copy of a particular messenger RNA using reverse transcriptase. The cDNA clone is not identical to the nuclear gene because in the messenger RNA (which served as a template for the cDNA) the introns have been removed, leaving coding sequence and 5′ and 3′ flanking sequences.
8. **e.** Gene expression can be inhibited by antisense RNA, which complementarily base pairs with the target messenger RNA, making it inaccessible to the translation machinery in the cell. MicroRNAs bind mRNAs and target them for degradation. Knockout genes are genes that have been inactivated by the insertion of DNA into their coding sequences. DNA microarrays are used in hybridization experiments and do not inhibit gene expression.
9. **c.** Expression vectors may contain drug-resistance markers and reporter genes, and they must contain DNA origins, but none of these features distinguishes them from other vectors. Expression vectors are unique because they contain regulatory sequences that allow the cloned gene to be expressed in the host cell.
10. **e.** RNAi is more effective than antisense RNA at inhibiting translation (not transcription). It is produced by viruses and by eukaryotic cells in small amounts and requires other modifying proteins.
11. **b.** DNA chips can be used to analyze gene expression at different times in development and to predict if an individual is at risk of developing cancer. Other factors (e.g., environmental) also determine if a person will get cancer. They are not used to make transgenic plants, to clone DNA, or to inhibit transcription of disease genes.
12. **d.** Soils have not been made less salty using biotechnology; plants have been made more salt-tolerant by means of recombinant DNA techniques.
13. **e.** All of the problems listed are disadvantages of recombinant DNA technology.
14. **c.** Expression vectors have the appropriate control regions that allow them to be expressed (make protein) in both prokaryotic and eukaryotic host cells. They are not used to create SNPs, nor are they used to isolate DNA or study transcription.
15. **e.** A promoter, a site for transcription factors, enhancers, a transcriptional termination sequence including a poly A–addition sequence, and a ribosome site are minimally required for an expression vector to produce its cloned gene product in a host eukaryotic cell. Inducible promoters and repressors will regulate gene expression but are not required for expression. A tissue-specific promoter will cause the gene to be expressed in particular cells (and would be required for the expression vector to work in a particular tissue), and signal sequences will target the protein to particular compartments in the cell, but they are not part of the minimal requirements for expression.

Differential Gene Expression in Development

The Big Picture

- The development of a mature differentiated organism from a fertilized egg involves the differential activation of genes in response to environmental signals. In many organisms, positional determinants may be present in different regions of the egg due to maternal factors. The asymmetric distribution of these factors will activate genes differentially, and as cell division proceeds, that asymmetric gene expression continues. As a result, different cells experience different environmental signals based on their position in the developing embryo, and they respond by activating different genes. This differential gene activation determines the fate of the cell.
- The zygote and early embryonic cells are totipotent; they can develop into any structure in the adult organism. Cells of the inner cell mass of the blastocyst are pluripotent, and adult stem cells are multipotent. As development proceeds, the developmental potential of embryonic cells narrows. In the adult organism, differentiated cells express particular genes that give those tissues a particular structure and function, even though all the genes are still present in the nucleus of those cells.
- Positional information, often coming from inducers called morphogens, leads to changes in the expression of key developmental genes, which in turn control morphogenesis, the creation of body form.

Study Strategies

- It may be difficult to visualize how a single cell can divide and grow to produce a mature functional organism with highly differentiated tissues. At some point very early in development (either before fertilization or in one of the first set of divisions), different genes begin to be expressed in different cells. This gene expression can occur in response to cytoplasmic factors or signals from other cells. Early gene expression sets up positional determinants that activate another wave of genes, and this wave further divides regions of the embryo into different developmental areas. As cell division proceeds in the embryo, these developmental genes continue to be differentially expressed, resulting in differentiation and morphogenesis in the organism.
- Transplantation experiments help to illustrate totipotency. Make sure that you can explain how the transplantation experiments in frogs using early embryonic tissues and adult somatic cells addressed the totipotency of these tissues. Make sure that you understand the different potentials of totipotent, pluripotent, and multipotent stem cells.
- Review the experiments that demonstrated how tissue induction causes the lens of the vertebrate eye to form.
- Review some of the important genes whose products direct development in the organisms discussed in this chapter. Identify homologs from different organisms.
- Go to the Web addresses shown to review the following animated tutorials and activity. You can also find them in BioPortal (yourBioPortal.com), along with many additional learning resources.

  Animated Tutorial 19.1 Cell Fates (Life10e.com/at19.1)

  Animated Tutorial 19.2 Early Asymmetry in the Embryo (Life10e.com/at19.2)

  Animated Tutorial 19.3 Pattern Formation in the *Drosophila* Embryo (Life10e.com/at19.3)

  Animated Tutorial 19.4 Embryonic Stem Cells (Life10e.com/at19.4)

  Activity 19.1 Stages of Development (Life10e.com/ac19.1)

Key Concept Review

19.1 What Are the Processes of Development?

Development involves distinct but overlapping processes

Cell fates become progressively more restricted during development

During development, an organism progresses through successive forms as it moves through its life cycle (see Figure 19.1). The embryo is an early stage of development that is nourished directly (via a placenta) or indirectly (via nutrients stored in the egg or seed).

Development involves four distinct but overlapping processes: determination, differentiation, morphogenesis, and growth. The developmental fate of a cell is set during determination and is influenced by internal and external conditions. During differentiation, cells become specialized to contain specific structures and to perform particular functions. Cell fate becomes apparent as cells differentiate. Morphogenesis is the organization and spatial distribution of differentiated cells to create the form and shape of the body and its organs. It is a result of pattern formation during development, which helps direct differentiated tissues to form specific structures. Morphogenesis can occur in several ways—through cell division, cell expansion, cell movement, and cell death. Growth is an increase in size due to cell division (an increase in the number of cells) or cell expansion (an increase in the size of existing cells). Cell division is important in the development of both plants and animals; cell expansion is particularly important in the development of plants.

Cell fates become progressively more restricted during development. Embryonic cells eventually become committed to a particular developmental fate, even though they are not yet differentiated into a particular cell or tissue type. Transplantation experiments have helped define the timing of cell fate determination. When early embryonic cells (at the blastula stage) are transplanted from their environment to another region of the embryo, the exposure to the new environment can redirect them along different developmental paths. In contrast, when older embryonic cells (at the gastrula stage) are transplanted to other locations in the embryo, they continue to develop into the original differentiated tissue, regardless of their new environment (see Figure 19.2).

During animal development the potential for a cell to differentiate into other cell types becomes more restricted. The cells of an early embryo are totipotent and have the ability to form all types of cells, including embryonic cells. Later stage cells are pluripotent, able to form many different cell types but no longer able to form embryos. In later developmental stages, certain cells are multipotent. These stem cells can differentiate into several different cell types. Unipotent cells can only produce copies of themselves.

Researchers have developed ways to manipulate differentiated cells, such as dedifferentiating plant cells to become totipotent and animal cells to become pluripotent.

Question 1. Describe three ways in which plant and animal development differ.

Question 2. Create a flow chart detailing the three different types of stem cells present in a developing animal. Name and define each type of stem cell, and include the developmental stage (and location, when possible) at which each type of stem cell is present.

19.2 How Is Cell Fate Determined?

Cytoplasmic segregation can determine polarity and cell fate

Inducers passing from one cell to another can determine cell fates

Cell determination is determined by cytoplasmic segregation and induction. Initially, in the fertilized egg unequal distribution of cytoplasmic determinants directs embryonic development and controls the polarity of the organism (see Figure 19.3). By the 8-cell stage of embryonic development, the different sets of cells have already developed distinct fates. For example, if a sea urchin embryo is cut in half at the 8-cell stage and split from left to right, the resulting larvae are normal, but small. If the sea urchin is cut so that its upper half is split from its lower half, the cells from the upper half do not develop at all, and the cells from the lower half develop into abnormal larvae.

Cytoplasmic determinants include specific proteins, small regulatory RNAs, and mRNAs. The cytoskeleton contributes to the asymmetric distribution of cytoplasmic determinants through the functioning of microtubules and microfilaments. For example, during sea urchin development, an mRNA determinant is carried by growing microfilaments to one end of the embryonic cell.

Cells or tissues in developing embryos can induce other cells or tissues to follow a particular developmental path. Cells can induce other cells to differentiate by secreting induction factors. The lens of the vertebrate eye forms when a portion of the forebrain bulges outward, forming the optic vesicle, which then contacts cells at the surface of the head. The optic vesicle induces the surface cells to form a lens placode, which eventually develops into the lens. If a barrier is placed between the optic vesicle and surface cells in a developing embryo, or if the optic vesicle is removed before it can contact the surface cells, no lens develops. Under normal conditions, the developing lens moves away from the surface tissue and induces it to form the cornea (see Figure 19.4).

The nematode *Caenorhabditis elegans* is a model organism used in developmental studies. The egg of *C. elegans* develops into a larva in 8 hours and into an adult in 3.5 days. The animal has a transparent body, making it easy to follow the developmental fate of its 959 somatic cells (see Figure 19.5A). The adult is hermaphroditic (i.e., it has female and male reproductive organs). Eggs are laid through a pore called the vulva, which is induced to develop from a single cell called an anchor cell. If this cell is destroyed, no vulva develops. The anchor cell determines the fates of six cells on the ventral surface of *C. elegans* by producing a primary inducer (LIN-3 protein) that diffuses toward adjacent cells, establishing a concentration gradient. The closest cell is exposed to the highest concentration of LIN-3 and becomes the primary precursor cell. It then produces a secondary inducer (known as the lateral signal) that causes the next closest cells to become secondary precursors. Descendants of primary and secondary precursor cells form the vulva. The final three cells (which are farthest from the anchor cell) become epidermal cells (see Figure 19.5B).

Question 3. What is meant by the phrase "cytoplasmic determinants"? Describe two mechanisms that ensure the proper distribution of the cytoplasmic determinants within a fertilized egg.

Question 4. A researcher removes the anchor cell in a developing *C. elegans* and it develops into an adult lacking a vulva. What developmental process did the researcher disrupt by removing the anchor cell? Explain why the nematode failed to develop a vulva.

19.3 What Is the Role of Gene Expression in Development?

Cell fate determination involves signal transduction pathways that lead to differential gene expression

Differential gene transcription is a hallmark of cell differentiation

Differentiated cells express only a very small subset of genes. Transcription factors are proteins that bind DNA and regulate the expression of genes. The binding of an inducer molecule to its receptor on the surface of a cell can lead to the activation of one or more transcription factors. Molecular switches, such as inducers, allow a developing cell to proceed down one of two alternative pathways, since different concentrations of inducer molecules can result in differential gene expression (see Figure 19.6).

LIN-3, the primary inducer for the development of the vulva in *C. elegans*, is homologous to mammalian epidermal growth factor (EGF). When LIN-3 binds to a receptor on the surface of the closest cell, it causes a signal transduction cascade involving the Ras protein and MAP kinases. The end result of this cascade is the differentiation of vulval cells.

Researchers can detect gene expression by using probes for RNA. For example, a DNA probe for the β-globin gene will bind DNA in both brain cells and immature red blood cells (RBCs), whereas only immature RBCs will express β-globin mRNA and, therefore, bind a β-globin RNA probe.

Sometimes a single transcription factor causes a cell to differentiate. In other cases, several different transcription factors promote differentiation. One well-studied example of cell differentiation is the conversion of undifferentiated muscle precursor cells into cells that are destined to form muscle. The transcription factor myoblast-determining gene (myoD) induces the expression of the p21 gene, whose protein product is an inhibitor of cyclin-dependent kinases (Cdk's). In response, undifferentiated muscle precursor cells stop dividing; this step is necessary for the eventual differentiation of these cells into mature muscle cells (see Figure 19.7).

Question 5. Activated MyoD has been found in muscle stem cells of adult vertebrates. What might its role be in this location?

Question 6. Describe the hybridization experiments that showed that fully differentiated cells still contain all their genes, even though they express only a small set of those genes.

19.4 How Does Gene Expression Determine Pattern Formation?

Multiple proteins interact to determine developmental programmed cell death

Plants have organ identity genes

Morphogen gradients provide positional information

A cascade of transcription factors establishes body segmentation in the fruit fly

Pattern formation is the spatial organization of tissue that results in the appearance of body form (morphogenesis). Programmed cell death (apoptosis) is an important part of development in both plants and animals. During morphogenesis, some cells are programmed to die by death genes through the process of apoptosis.

As *C. elegans* develops from a fertilized egg into an adult, 1,090 cells are produced. However, 131 of these cells are programmed to die due to the sequential expression of the *ced-4* and *ced-3* genes. The protein CED-9 inhibits the actions of the two proteins, CED-3 and CED-4 during development (see Figure 19.8A). In human embryos, apoptosis occurs in the webs of skin that initially form between fingers and toes. Caspases, a group of enzymes homologous to CED-3 in *C. elegans*, cause this apoptosis. The human homologs to CED-9 and CED-4 are Bcl-2 and Apaf 1, respectively (see Figure 19.8B). The genes controlling apoptosis are crucial for proper development in many organisms. These genes have been conserved, for example, in nematodes and humans—organisms separated by 600 million years of evolution.

The areas of undifferentiated, rapidly dividing cells in the growing tips of plants (such as the shoot apex and root tip) are known as meristems. During plant development, organs such as leaves, roots, and flowers develop from meristems. Flowers have four types of organs: sepals, petals, stamens (male reproductive organs), and carpels (female reproductive organs). Flower organs occur in whorls organized around a central axis and are derived from floral meristems that contain about 700 undifferentiated cells (see Figure 19.9A).

Plant geneticists have studied the development of flower organs in *Arabidopsis* and determined that plants have organ identity genes. For example, there are four whorls of organs in *Arabidopsis.* Three genes expressed in the whorls act as organ identity genes to guide differentiation in each whorl (see Figure 19.9B). Gene A is expressed in whorls 1 and 2, which form the sepals and petals, respectively; gene B is expressed in whorls 2 and 3, which form petals and stamens, respectively; and gene C is expressed in whorls 3 and 4, which form stamens and carpels, respectively.

The three *Arabidopsis* genes (A, B, and C) all encode transcription factors that are active as dimers (proteins with two polypeptide subunits). Gene regulation is combinatorial, and the combination of different dimers determines which genes are activated. A dimer of the transcription factor encoded by gene A will activate genes that make sepals. A dimer of transcription factor A with transcription factor B will result in petals (see Figure 19.9C).

LEAFY is a transcription factor that regulates the transcription of the genes A, B, and C. Plants bearing a mutation in *LEAFY* produce leaves but no flowers. In plants, cell fate is often determined by genes that contain a DNA-binding domain called the MADS box. Loss-of-function and gain-of-function mutations in organ identity genes cause the loss or replacement of one organ for another, respectively. The replacement of one organ for another is known as homeosis.

The homeotic genes A, B, and C, as well as *LEAFY*, may be able to be genetically modified so that the flower organs can produce more fruit and seeds (which come from the carpels).

Positional information allows developing cells to determine where they are in the developing organism. Morphogens are signals that establish positional information. They act directly on the target cell, and differential concentrations within the embryo cause different effects. Concentration gradients of the morphogen Sonic hedgehog, secreted by the zone of polarizing activity (ZPA) in the limb bud, determine the anterior–posterior axis of the developing vertebrate limb. According to the "French flag" model (see Figure 19.10), cells in the limb bud form different digits depending on the concentration of the morphogen (see Figure 19.11).

A cascade of transcription factors controls development of the body of *Drosophila*. The body of the embryo is segmented, consisting of a head (which is formed from fused segments), three thoracic segments, and eight abdominal segments. Although the segments look similar in the embryo, the fates of these cells are already determined, and during the transformation of the embryo into an adult fly each segment develops into different body parts: for example, antennae and eyes develop from head segments (see Figure 19.14).

The first step of development in *Drosophila* is the establishment of anterior–posterior and dorsal–ventral polarity. Polarity is based on the cytoplasmic distribution of mRNA and proteins produced by maternal effect genes. These morphogens are able to diffuse easily within the early embryo, a multi-nuclear single cell created by 12 nuclear divisions unaccompanied by cytokinesis. Mutations in maternal effect genes (*Bicoid* and *Nanos*) create larvae lacking anterior structures (*Bicoid*) or abdominal segments (*Nanos*). Bicoid encodes a transcription factor that positively affects some genes (*Hunchback*) and negatively affects others. Nanos protein inhibits the translation of *Hunchback* (see Figure 19.12).

Segmentation genes are expressed when the embryo is at the 6,000-nuclei stage and determine the number, boundaries, and polarity of the segments. Three types of segmentation genes are involved in regulation of segmental development: gap genes, pair rule genes, and segment polarity genes. Gap genes organize large areas along the anterior–posterior axis of the developing embryo. Gap mutants produce larvae that are missing consecutive larval segments. Pair rule genes divide the larva into units of two segments each. Mutations in the pair rule genes produce larvae that are missing every other segment. Segment polarity genes determine the boundary and the anterior–posterior organization of each segment. Mutations in segment polarity genes produce larvae in which the posterior structures in the segments have been replaced by reversed anterior structures.

The expression of segmentation genes is sequential (see Figure 19.13). Gap genes control the expression of pair rule genes, some of which are transcription factors that control segment polarity. The pair rule genes activate segment polarity genes.

Differences between segments are encoded by Hox genes, which give each segment an identity. For instance, in response to Hox gene expression, cells in a thoracic segment produce legs and cells in the head segment produce eyes (see Figure 19.14). Hox genes are shared by all animals and are homeotic genes. Homeotic mutants include *antennapedia* (legs in place of antennae; see Figure 19.15) and *bithorax* (produces a fly with an extra set of wings). Homeotic genes are clustered on the chromosome in the order of the segments they determine. The first set of genes in the homeotic cluster specifies anterior segments (head and then thoracic segments). The second cluster begins with a gene specifying the last thoracic segment and has genes for anterior and posterior abdominal segments.

Nucleic acid hybridization experiments confirm that Hox genes have arisen from duplications of a single gene in an ancestral unsegmented organism. A DNA sequence, the homeobox, encodes a 60-amino acid sequence, the homeodomain, which includes a DNA-binding motif. This domain is found in transcription factors that regulate anterior–posterior axis development in many other animals as well.

Question 7. Create a flow chart showing the gene cascade that controls body segmentation in the fruit fly. For each class of genes, include a brief description of general function.

Question 8. During the first 24 hours after fertilization of the *Drosophila* egg, morphogens can easily diffuse within the embryo, thereby establishing a concentration gradient. Explain how this is possible.

19.5 Is Cell Differentiation Reversible?

- Plant cells can be totipotent
- Nuclear transfer allows the cloning of animals
- Multipotent stem cells differentiate in response to environmental signals
- Pluripotent stem cells can be obtained in two ways

Early embryonic cells are totipotent: they can develop into any of the different kinds of cells in the mature organism. Developmental possibilities narrow as cell determination and differentiation occur.

In plants, differentiation is reversible in some cells. Under certain conditions, plant cells in culture can give rise to a new, genetically identical plant (a clone). Plant cells first dedifferentiate, and then produce a callus (mass of cells), which develops into a plant embryo when placed in the appropriate medium. Thus the original differentiated cells contained all the genetic information needed to express genes in the correct sequence to generate a whole plant (see Figure 19.16).

The cells of the early animal embryo are totipotent and can give rise to every type of cell in the adult body. A few cells removed from the early embryo can be used for genetic screening. The embryo, with its remaining cells, can be implanted into a mother's uterus, where it can develop into a fetus.

The totipotency of early embryonic animal cells has been shown in experiments in which nuclei isolated from frog embryos were fused to enucleated eggs, resulting in the development of tadpoles and mature frogs. Such experiments show that cytoplasm has a large influence on the fate of a cell. These experiments also established the principle of genomic equivalence, which states that no information is lost from the nuclei of cells as they pass through the early stages of development.

Under the right environmental conditions, cells can be induced to dedifferentiate and give rise to a new, cloned individual. This method, called somatic cell nuclear transfer, involves fusing somatic cells with enucleated eggs and implanting the resulting embryos in surrogate mothers. The donor somatic cell is first induced to halt in G1 of the cell cycle. A G1-arrested cell is then fused with an enucleated egg. The fused cell is naturally stimulated by cytoplasmic factors in the egg to enter the S phase. An early embryo develops and is then implanted in the uterus of a surrogate mother. Dolly, a Finn Dorset sheep, was the first mammal cloned by the technique of nuclear transfer (see Figure 19.17), and several mammals, including sheep, mice, and cattle, have also been cloned in this way. Practical uses of cloning include increasing the number of valuable animals (for example, genetically engineered animals) and the preservation of endangered species and pets.

Undifferentiated dividing cells in animals are known as stem cells. Stem cells in adult animals are specific for the kinds of tissue they replace—typically skin, the lining of the intestines, and blood cells. These stem cells are multipotent, meaning that they can differentiate into a limited number of cell types. Stem cell proliferation is "on demand": the cells proliferate in response to signals, such as growth factors.

Bone marrow contains two types of multipotent stem cells: hematopoietic and mesenchymal. Hematopoietic stem cells produce red and white blood cells, while mesenchymal stem cells produce bone and muscle cells. Hematopoietic stem cells proliferate in the bone marrow in response to growth factors, and the extra stem cells are released into the blood. This is the basis for hematopoietic stem cell transplantation to counteract the effects of cancer treatment (which can kill stem cells). Stem cells are removed from the patient's blood, stored, and then reinfused into the patient once the treatment has been completed (see Figure 19.18).

There is evidence that the injection of stem cells into damaged tissue can aid in tissue repair, but the mechanism is not clear. Growth factors and adjacent cells influence the differentiation of stem cells. Alternatively, stem cells may secrete growth factors and other factors that induce the surrounding tissue to regenerate into healthy tissue.

In embryonic mammals, cells from a part of the blastocyst (the inner cell mass) are pluripotent, which means they are capable of forming nearly every type of cell. These pluripotent cells are more restricted than totipotent cells because they cannot form cells of the placenta. In mice, these embryonic stem cells (ESCs) can be harvested and grown in the laboratory. ESCs injected back into blastocysts will mix with the resident cells and form all the different cell types of the mouse. ESCs can be induced to differentiate into specific cell types. For instance, a derivative of vitamin A causes ESCs to form neurons. Because of this, there is the possibility that ESCs can be used to repair specific tissues.

Human ESCs can be harvested from embryos that had been prepared for in vitro fertilization procedures but not used (see Figure 19.19A). These ESCs can be grown in the laboratory and have the potential to be used to repair damaged tissues. However, some people object to using human embryos for this purpose, and ESCs transplanted into patients can provoke an immune response. A new method for obtaining stem cells involves inducing adult cells to dedifferentiate. These induced pluripotent stem (IPS) cells are obtained through the transfection of adult cells, such as skin cells, with a vector carrying several genes believed to be essential to the undifferentiated state and function of stem cells (see Figure 19.19B). This technique does not destroy human embryos nor provoke an immune response in recipients.

Question 9. Describe one application for human health of multipotent stem cells and one benefit for human health of cloned mammals.

Question 10. When mammals are cloned by the fusing of whole cells with enucleated eggs, why is the original nucleus removed from the egg?

Test Yourself

1. Cell differentiation
   a. results from the loss of particular genes from the nucleus of the differentiated cell.
   b. results from the differential expression of genes that are responsive to environmental signals.
   c. involves the persisting totipotency of early embryonic cells in the mature organism.
   d. results from mutations in genes that control the synthesis of DNA.
   e. precedes cell determination.

   Textbook Reference: *19.1 What Are the Processes of Development?*

2. A totipotent cell is a cell
   a. whose developmental fate has been decided.
   b. that has differentiated into a specialized tissue.
   c. that is fated to form a particular structure.
   d. whose developmental potential is extremely broad.
   e. whose developmental potential is narrower than that of a pluripotent cell.

   Textbook Reference: *19.1 What Are the Processes of Development?*

3. Apoptosis
   a. occurs during embryonic development of invertebrates, but not vertebrates.
   b. prompts cell differentiation.
   c. is controlled by similar genes in humans and nematodes.
   d. does not occur during morphogenesis.
   e. involves caspase enzymes in the nematode *C. elegans*.

   Textbook Reference: *19.4 How Does Gene Expression Determine Pattern Formation?*

4. The fate of a cell
   a. refers to the cell's original type.
   b. describes its genes.
   c. refers to the type of cell into which it will differentiate.
   d. describes its death.
   e. cannot by modified by its cytoplasmic environment.

   Textbook Reference: *19.1 What Are the Processes of Development?*

5. Mammals can be cloned using donor somatic cells fused to enucleated eggs if the
   a. donor cell is in the S phase of the cell cycle.
   b. donor cell is in the G1 phase of the cell cycle.
   c. mammalian embryo can be propagated in culture.
   d. donor cell is a mature red blood cell.
   e. donor cell and enucleated egg are derived from individuals of the same species.

   Textbook Reference: *19.5 Is Cell Differentiation Reversible?*

6. The protein MyoD
   a. is a transcription factor that controls the expression of genes involved in the differentiation of muscle cells.
   b. controls segment identity in mice.
   c. is a transcription factor that activates a gene that causes the cell cycle in muscle precursor cells to start.
   d. is absent in adult muscle.
   e. is the only transcription factor involved in differentiation of mesoderm cells into mature muscle cells.

   Textbook Reference: *19.3 What Is the Role of Gene Expression in Development?*

7. Induction occurs when
   a. nuclear genes are lost in some tissues.
   b. one cell contacts another cell and fails to alter the second cell's developmental fate.
   c. a cell or tissue sends a chemical signal to another, causing differentiation.
   d. two tissues come into contact, causing the release of transcription factors.
   e. factors in the cytoplasm of the cell are unequally distributed.

   Textbook Reference: *19.2 How Is Cell Fate Determined?*

8. Which of the following statements about plant development is *false*?
   a. Both cell division and cell expansion contribute to growth in early plant embryos.
   b. Differentiation in plant cells is irreversible.
   c. Meristems at the tips of the roots and stems consist of undifferentiated cells.
   d. Flower development involves organ identity genes.
   e. The fertilized egg in plants is called a zygote.

   Textbook Reference: *19.1 What Are the Processes of Development?; 19.4 How Does Gene Expression Determine Pattern Formation?; 19.5 Is Cell Differentiation Reversible?*

9. It may be possible to genetically engineer plants to produce more seeds and fruits by
   a. fertilizing them aggressively in the spring.
   b. manipulating their gap genes.
   c. altering the homeotic genes that control carpel development.
   d. inducing a mutation that eliminates *LEAFY* gene function.
   e. altering the homeotic genes that control stamen development.

   Textbook Reference: *19.4 How Does Gene Expression Determine Pattern Formation?*

10. Morphogens
    a. diffuse within the embryo to set up a concentration gradient.
    b. are expressed as positional signals in developing embryos.
    c. direct differentiated cells to form organs.
    d. act directly on target cells.
    e. All of the above

    Textbook Reference: *19.4 How Does Gene Expression Determine Pattern Formation?*

11. Maternal effect genes
    a. begin setting up positional axes in the egg prior to fertilization.
    b. are the same as segmentation genes.
    c. are masked by paternal genes.
    d. operate late in the gene cascade regulating development of *Drosophila* embryos.
    e. determine cell fate within each segment.

    Textbook Reference: *19.4 How Does Gene Expression Determine Pattern Formation?*

12. Hox genes
    a. encode protein domains that are important in development and have been highly conserved over evolutionary time.
    b. are found in diverse organisms.
    c. can produce the wrong structure in the wrong place when mutated.
    d. help determine cell fate within each segment of a developing *Drosophila* embryo.
    e. All of the above

    Textbook Reference: *19.4 How Does Gene Expression Determine Pattern Formation?*

13. The experiments of Briggs and King established that the nuclei of cells do not lose any genetic information during the early stages of development. This characteristic of cells became known as
    a. genomic equivalence.
    b. apoptosis.
    c. totipotency.
    d. transcription.
    e. fate mapping.

    Textbook Reference: *19.5 Is Cell Differentiation Reversible?*

14. What is the role of cytoplasmic segregation in determining the fate of a cell?
    a. It keeps cells totipotent.
    b. It determines polarity within the organism.
    c. It ensures that gradients do not develop in the developing organism.
    d. It stops cell division.
    e. It has no role in determining cell fate.
    Textbook Reference: *19.2 How Is Cell Fate Determined?*

15. Organ identity genes are found in
    a. bacteria.
    b. animals.
    c. plants.
    d. fungi.
    e. None of the above
    Textbook Reference: *19.4 How Does Gene Expression Determine Pattern Formation?*

Answers

Key Concept Review

1. The MADS box gene family plays a role in regulating plant development; these genes are not involved in animal development, in which homeobox genes are important. With regard to morphogenesis and growth, cell division is important in plants and animals, whereas cell expansion is particularly important in plants. Also, whereas cell movement is critical to animal development, plant cells exist within rigid cell walls, making cell movement during development impossible.

2.

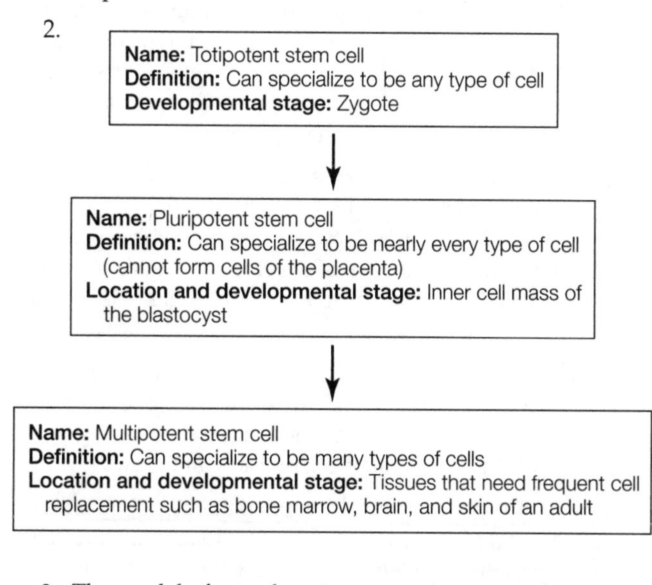

3. The model of cytoplasmic segregation states that certain materials, called cytoplasmic determinants, are distributed unequally in the egg cytoplasm. These materials can be proteins, small regulatory RNAs, or mRNAs, and they play roles in directing the embryonic development of many organisms. One of the mechanisms by which the cytoplasmic determinants are distributed within the fertilized egg is simple diffusion, which creates a concentration gradient. Some cytoplasmic determinants can bind to motor proteins that transport the material within the cell via the cytoskeleton, thereby creating asymmetric distribution of the determinants.

4. The process that the researcher has disrupted in a developing *C. elegans* is called induction. The anchor cell in the developing nematode secretes a molecule, LIN-3 protein, which induces neighboring cells to acquire specific fates. The most proximal cell, the center cell, will be induced to proliferate and differentiate into vulval cells. This cell is also induced by exposure to high LIN-3 levels to produce a secondary inducer molecule. This secondary inducer molecule, along with low levels of the primary inducer molecule (LIN-3), will induce two cells proximal to the center cell to also proliferate and develop into vulval cells. Thus the removal of the anchor cell abolished expression of the inducer LIN-3, which was required for the proper fate determination of the neighboring four cells.

5. The discovery of activated MyoD in muscle stem cells of adult vertebrates suggests that this transcription factor may also be involved in muscle repair. In other words, its role is not restricted to stimulating muscle cell differentiation from mesoderm in vertebrate embryos.

6. Hybridizations were carried out using a specific DNA probe to nuclear DNA and mRNA. If the DNA probe was complementary to a gene that was not being expressed in the cell, it would not hybridize to the mRNA isolated from that cell, but it still would hybridize to the nuclear DNA. This indicated that the gene was still present in the nucleus of the cell, even though it was not being expressed in the cytoplasm.

7.

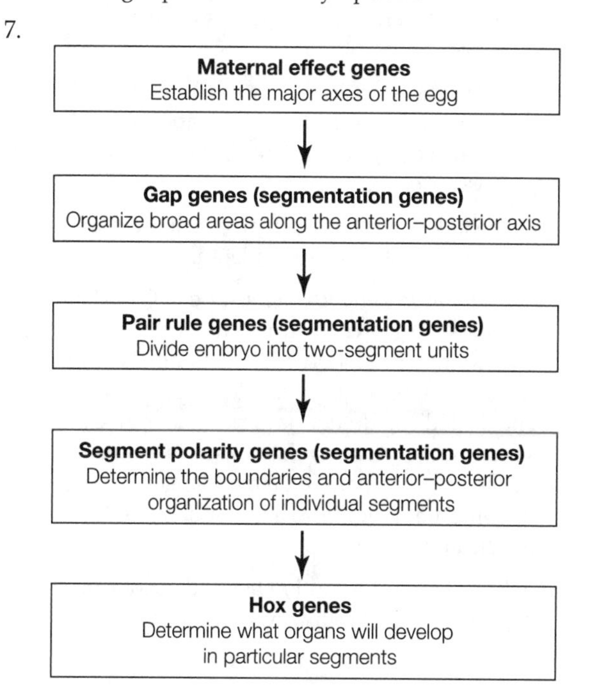

8. Although there are 12 nuclear divisions within the *Drosophila* embryo during the first 24 hours after fertilization, these events are not accompanied by cytokinesis. Therefore, there are no cell membranes to inhibit the free diffusion of the morphogens within the *Drosophila* early embryo.
9. One application for human health of multipotent stem cells is hematopoietic stem cell transplantation. Hematopoietic stem cells are multipotent stem cells that produce red and white blood cells. If left in the body, these dividing cells are killed during cancer treatment along with cancer cells. In hematopoietic stem cell transplantation, stem cells are removed prior to cancer treatment, stored, and then reinfused into the patient once treatment has been completed. One benefit for human health of cloned animals is that it can increase the number of particularly valuable animals, such as transgenic cows. Transgenic cows produce human growth hormone, and cloning thereby increases the supply of growth hormone needed for treating human hormone deficiencies.
10. The nucleus of the egg contains a haploid set of chromosomes. Fusing a diploid somatic cell with an egg cell that still contained a nucleus would result in a cell with three copies of each chromosome and little chance of developing normally.

Test Yourself

1. **b.** Genes are not lost from differentiated cells. There are no embryonic cells in a mature organism. Mutations in cells that affect DNA replication would result in cells that were unable to replicate their DNA. Cell differentiation occurs after cell determination.
2. **d.** A totipotent cell is a cell that can develop in any number of ways. Its fate has not been determined, it has not differentiated, and it is not expressing morphogens.
3. **c.** Apoptosis is controlled by similar genes in humans and nematodes, suggesting the importance of this pathway (i.e., mutations are extremely harmful, so the pathway is conserved).
4. **c.** The fate of a cell is the type of cell it will eventually become once it has differentiated.
5. **b.** The donor somatic cell must be in G1. Mammalian embryos are grown in surrogate mothers, not in culture. Mature red blood cells in mammals have no nuclei and cannot be used for cloning. The donor cell and enucleated egg do not have to come from exactly the same species; closely related species can be used, as in the cloning of a banteng (an endangered close relative of domestic cattle) from a donor banteng cell and an enucleated cow egg.
6. **a.** MyoD is a transcription factor that controls the expression of genes involved in the differentiation of mesoderm cells into mature muscle cells. It activates a gene that causes the cell cycle to stop, thereby allowing differentiation to begin. The transcription factor MyoD is one of several involved in the differentiation of muscle cells. It is found in adult muscle cells, where it may play a role in repair.
7. **c.** Induction occurs when one cell or tissue secretes factors that influence the developmental fate of other cells or tissues. Nuclear genes are not lost in developing cells (red blood cells are an exception). Transcription factors are found in the nucleus and are not released to adjacent tissues.
8. **b.** In general, it is easier to reverse differentiation in plant cells than in animal cells, as demonstrated by the cloning of a carrot from a differentiated storage cell in its root.
9. **c.** Fertilization is not genetic engineering. Gap genes are found in *Drosophila,* not plants. The LEAFY protein transcriptionally activates the homeotic genes that control organ development (and subsequent seed and fruit formation). In its absence, no fruit or seeds would develop. Altering the homeotic genes that control carpel (not stamen) development could lead to more fruit and seeds.
10. **e.** Morphogens diffuse in the embryo to form concentration gradients that act as positional signals to direct differentiated tissues to form organs. They act directly on target cells rather than triggering a secondary signal that acts on target cells.
11. **a.** Beginning before fertilization, maternal effect genes determine the anterior–posterior axis. These genes, which operate early in the gene cascade, also induce segmentation genes. The paternal genotype does not affect the expression of maternal gene products in the egg. Hox genes, and not maternal effect genes, determine cell fate within each segment.
12. **e.** Hox genes encode proteins that are highly conserved and important for directing development in many different organisms. Mutations in these genes can produce an abnormal developmental event in the organism. In *Drosophila* embryos, Hox genes determine segment identity.
13. **a.** Genomic equivalence is a fundamental principle of developmental biology. It states that none of the genetic information contained in the original zygote is lost during subsequent cell division and growth.
14. **b.** Cytoplasmic segregation refers to the unequal distribution of factors within the developing embryo. This asymmetry helps set up, for example, the polarity (distinct top and bottom) of the developing organism.
15. **c.** Organ identity genes are found in plants, where they help to direct the development of structures such as flowers.

Genes, Development, and Evolution

The Big Picture

- Evolutionary developmental biology (evo-devo) is the study of how organisms evolved. Evo-devo biologists study evolution by examining the differences in developmental gene expression and regulation among species.
- Animals have highly conserved genes, such as homeobox genes, that control development. Genetic switches determine where and when genes will be expressed. Changes in these switches can result in the evolution of species differences.
- Differences in organisms have arisen because of differences in the timing of different developmental processes, which is called heterochrony. The timing of gene expression can differ between organisms because of differences in modular makeup.
- The ability of an organism to modify its development in response to environmental conditions is called developmental (or phenotypic) plasticity. These environmental signals can provide an accurate prediction of the future, but sometimes they are poorly correlated with future conditions.

Study Strategies

- This chapter introduces many concepts. Learn the concepts involved in evolutionary developmental biology first. Then go back and see how the examples fit with the concepts.
- The concept of heterochrony can be difficult to grasp. Try to remember that the concept refers to differences among species, not within species.
- Go to the Web addresses shown to review the following animated tutorial and activity. You can also find them in BioPortal (yourBioPortal.com), along with many additional learning resources.

  Animated Tutorial 20.1 Modularity (Life10e.com/at20.1)

  Activity 20.1 Concept Matching (Life10e.com/ac20.1)

Key Concept Review

20.1 How Can Small Genetic Changes Result in Large Changes in Phenotype?

Developmental genes in distantly related organisms are similar

Many of the genes involved in controlling development are conserved across different organisms, yet diverse phenotypes can result from differences in the expression of these genes during development. Evolutionary developmental biology is the study of the relationship between evolution and development. A major focus of evo-devo studies is the comparison of genes that regulate development and how their expression is controlled in time, space, and amount in different multicellular organisms.

Evo-devo biologists have found that organisms share a "toolkit" of signaling molecules that control the expression of genes. Independent activity of these signaling molecules in different tissues and regions of the body enables modular evolutionary change. The major differences in body form arise because certain genes are turned on and off at different times and locations during development. Developmental differences can also arise from changes in location of expression, from the quantity of expression, and from alterations in the pattern of expression. Environmental signals have also been found to influence development.

Genes involved in regulating development in insects and vertebrates are very similar. For example, the gene *eyeless*, which encodes a transcription factor necessary for the development of the *Drosophila* eye, contains regulatory sequence that is identical to *Pax6*, a gene that is necessary for the development of the mammalian eye. Genes such as *eyeless* and *Pax6* that share high sequence identity and function appear to have evolved from a gene present in a common ancestor; these are called homologous genes. *Eyeless* homologs control eye development in a wide range of species with phenotypically dissimilar eyes (see Figure 20.1).

Another example of gene homology is exemplified by the Hox gene cluster. This set of genes, which encodes transcription factors and provides positional information during the development of the anterior–posterior axis in insects and mammals, shares a homologous sequence called the homeobox domain. The high similarity of the genes suggests a common ancestral gene that underwent duplications and divergence to produce the various modern Hox genes. This duplication-and-divergence theory is also supported by the correlation of increasing numbers of Hox genes with increasing complexity in the patterning of the body axis. Nevertheless, a comparison of the expression patterns of Hox genes in *Drosophila* and mouse embryos shows that the genes are arranged in similar clusters and are expressed in similar patterns along the axis of both species (see Figure 20.2).

The similarities between genes in organisms that last shared a common ancestor millions of years ago has led evo-devo biologists to develop the concept of the genetic toolkit. According to this fundamental principle of evo-devo, the remarkable diversity of organisms today is largely due to small changes in terms of when, where, and to what extent these transcription factor genes are expressed.

Question 1. Why do scientists consider the gene *eyeless* in fruit flies and the gene *Pax6* in mice to be homologous?

Question 2. Biologists found that if they substituted *Pax6* for *eyeless* in *Drosophila*, a *Drosophila* eye still developed, (not a mammalian eye). How does the genetic toolkit theory explain this result?

20.2 How Can Mutations with Large Effects Change Only One Part of the Body?

Genetic switches govern how the genetic toolkit is used

Modularity allows for differences in the patterns of gene expression

The many parts of the developing embryo can change independently of one another because an organism consists of developmental modules that are based on a common set of genetic instructions. A module, such as the genes and signaling pathways that lead to development of the eyes, can change independently of the other modules because developmental genes can be controlled separately in the different modules. Developmental modules can evolve separately within a species because genetic switches, which consist of gene promoters and the transcription factors that bind them, as well as enhancers and repressors, control the way in which a gene is used during development. Each gene is controlled by multiple switches that influence where and when it is expressed. For example, genetic switches control spatial development of the embryo. The pattern of development depends on the unique Hox gene or combination of Hox genes that are expressed within the module. In most insect groups, for example, the expression of the Hox gene *antennapedia* (*Antp*) in the second and third thoracic segments results in the development of wings from both segments. In *Drosophila* and other members of the group *Diptera*, a second Hox gene, *ultrabithorax* (*Ubx*), is co-expressed in the third segment. The protein product of *Ubx* suppresses the expression of *Antp*, thereby suppressing the development of wings from the third segment (see Figure 20.3).

The modularity of embryos allows different developmental processes to shift independently of one another. Heterometry, the expression of structural genes in different amounts, plays a role in the development of the various beak sizes and shapes of the Galápagos finches (see Figure 20.4).

Heterochrony occurs when the relative timing of developmental processes independently shifts in different species. It is the basis for the elongated cervical vertebrae of the giraffe compared with humans (see Figure 20.5).

The retention of webbing between the digits of a duck's foot but the absence of webbing in the foot of a chicken is an example of heterotopy, the occurrence of spatial differences in the expression of a developmental gene. The signaling protein bone morphogenetic protein 4 (BMP4) causes the cells of the webbing to undergo apoptosis. Although BMP4 is present in the webbing of both embryonic ducks and chickens, these species differ in whether the gene *Gremlin* is expressed. *Gremlin* encodes a protein that inhibits BMP4 activity. In duck embryos, but not chicken embryos, *Gremlin* is expressed in the webbing, thereby preserving the webbing in the duck's feet (see Figure 20.6).

Question 3. A researcher has discovered a *Drosophila* mutant that expresses two sets of wings. Based on what you know about Hox genes and insect wing development, what is the most likely genetically-based explanation for the two sets of wings?

Question 4. There are morphological differences between adult chickens and ducks that relate to the duck's aquatic lifestyle. Why were these adaptations able to occur without influencing the development of the rest of the duck's morphology?

20.3 How Can Developmental Changes Result in Differences among Species?

Differences in Hox gene expression patterns result in major differences in body plans

Mutations in developmental genes can produce major morphological changes

Genetic switches can explain the evolution of major morphological differences among species. Evolutionary changes in morphology have occurred due to the change in the timing or spatial expression of regulatory genes during development. In vertebrates, the spatial pattern of Hox gene expression regulates the number of each kind of vertebra (cervical, thoracic, lumbar, sacral, and caudal). In an embryonic mouse, the transition from cervical to thoracic vertebrae occurs at the anterior limit of *Hoxc6* expression, resulting in seven cervical vertebrae. In the embryonic chicken, the anterior limit of *Hoxc6* expression is farther down the vertebral column, resulting in more than seven cervical vertebrae (see Figure 20.8).

Sometimes a major developmental change is due to heterotopy, an alteration in a single gene. The insect *ultrabithorax* (*Ubx*) homeotic gene has a mutation that results in a modified Ubx protein that represses the expression of a gene involved in limb formation, the *Distalless* (*Dll*) gene. The modified protein product of the *Ubx* gene is expressed in the abdomen during development, leading to the absence of limbs on the abdomen in insects. Other arthropods, such as centipedes, do not have this mutation and expression of the *Ubx* gene promotes the development of abdominal legs. Thus, a mutation in a Hox gene changed the number of legs in insects relative to other arthropods (see Figure 20.9).

Heterotopy is also partly responsible for the easily accessible kernels of domesticated corn. The kernels of its wild relative, the plant teosinte, are encased in tough shells (glumes). A mutation in a gene called *Tga1* in domesticated corn appears to be partially responsible for the liberation of domesticated corn kernels from glumes. Expression of the mutated form in teosinte allows its kernels to break partially free of glumes, whereas expression of the teosinte form of *Tga1* in domesticated corn induces the formation of glumes over the kernels (see Figure 20.10).

Question 5. Evaluate the following statement: "Most changes in morphology over evolutionary time occur as a result of the introduction of radically new developmental mechanisms."

Question 6. Explain in terms of genetics and developmental expression why insects have legs only on thoracic segments, whereas centipedes have legs on thoracic and abdominal segments?

20.4 How Can the Environment Modulate Development?

Temperature can determine sex

Dietary information can be a predictor of future conditions

A variety of environmental signals influence development

The environment can greatly influence the development of plants and animals. Organisms have the ability to exhibit different phenotypes based on a single genotype. When the phenotype is dependent on the environment in which the organism develops and lives this response is known as developmental (or phenotypic) plasticity.

Sex determination in mammals is based on sex chromosomes: XX individuals are female and XY individuals are male. In contrast to the chromosomal sex determination of mammals, some reptiles exhibit temperature-dependent sex determination, whereby the temperature at which an embryo is incubated determines its gender. The specific effects of incubation temperature can vary among species, but the general phenomenon is an example of developmental plasticity (see Figure 20.11).

In vertebrates, the biosynthesis of sex steroid hormones begins with cholesterol, which first forms testosterone. Testosterone may be converted via the enzyme aromatase to estrogens (female sex steroids) or via the enzyme reductase to other androgens (male sex steroids). In animals with temperature-determined sex, incubation temperature influences gender by controlling the amount of aromatase expressed. When aromatase is expressed in large quantities, estrogens dominate and female sex organs develop. When aromatase is not expressed, testosterone dominates and male sex organs develop. Some experiments have demonstrated the fitness value of temperature-determined sex by showing that this form of sex determination is associated with sex-specific differences in fitness (see Figure 20.12).

An example of developmental plasticity in invertebrates is the spring and summer caterpillar forms of the moth *Nemoria arizonaria*. Whereas spring caterpillars resemble the oak tree flowers (catkins) on which they feed, summer caterpillars resemble oak twigs and feed on oak leaves. Spring and summer caterpillars look alike at hatching; the different appearances are triggered by their different diets. Thus, developmental plasticity allows both forms of caterpillars to avoid predation.

Environmental signals can either be accurate predictors of the future environment or poorly correlated with the future environment. Day length is an accurate predictor of future conditions. Many insects use day length to time periods of developmental arrest called diapause, and in mammals such as deer and elk, antlers are lost in response to changes in day length. Day length is critical to the timing of reproduction in many animals.

Environmental influences can affect the development of an organism for the duration of its life, beyond the point at which maturity is reached. For example, a thousand-year-old redwood tree retains meristems and can, therefore, produce new differentiated tissues throughout its life. Light is an important environmental signal in plant development. In plants, low light levels result in elongation of cells in the stem. This allows a plant to grow tall and reach a sunny patch more quickly than a compact plant could. When light is abundant, plants invest their energy in growing leaves rather than in lengthening their stems (see Figure 20.14). Thus, developmental plasticity can also result in adaptive changes in the forms and functions of adult organisms.

Question 7. Day length and temperature are both potential environmental signals that may influence development. What do changes in these two environmental factors tell an organism about future conditions?

Question 8. Conservation efforts to restore sea turtle populations often rely on collecting eggs, incubating them in captivity, and then releasing hatchlings. Why would knowledge of temperature-determined sex be critical to such efforts?

Question 9. In this chapter you learned about some environmental signals that influence development. Write "Developing organism" on a piece of paper and then indicate, using arrows pointing toward this term, at least six environmental signals that could potentially influence development. This will require you to go beyond the number of environmental signals covered in the textbook.

20.5 How Do Developmental Genes Constrain Evolution?

Evolution usually proceeds by changing what's already there

Conserved developmental genes can lead to parallel evolution

Evolution is constrained by existing structures and highly conserved genes. Evolution works on existing genes and their expression and can occur through changes in Hox gene expression in different modules. Thus, most evolutionary innovations are modifications of previously existing structures. For example, wings evolved three times in vertebrates (birds, bats, and extinct pterosaurs), but rather than being new structures, they are modified limbs. Evolutionary losses are also controlled by Hox genes, such as the loss of forelimbs in the ancestors of present-day snakes.

Similar traits evolve repeatedly because of the highly conserved nature of the genetic code. This can lead to parallel phenotypic evolution of a trait, such as the evolutionary loss of body armor in numerous species of freshwater stickleback fish. Freshwater populations of sticklebacks are descended from marine populations. In marine sticklebacks, the gene *Pitx1* codes for a transcription factor that stimulates the production of protective plates and spines. The *Pitx1* gene has changed in various freshwater populations of sticklebacks, leading to the loss of body armor in these populations.

Question 10. Amphipods are small aquatic crustaceans. Researchers have discovered the independent reduction of eyes in different populations of a species of cave-dwelling amphipod. What phenomenon does this represent?

Question 11. How does a comparison of the wings of bats, the fore flippers of a seal, and the human hand support the notion that evolution works by modifying preexisting structures?

Test Yourself

1. Genes that regulate development are highly conserved, meaning that
   a. large differences have evolved among multicellular organisms.
   b. they have changed very little over the course of evolution.
   c. they are always turned on.
   d. they have undergone mutation.
   e. they have been lost in some lineages.
   Textbook Reference: *20.1 How Can Small Genetic Changes Result in Large Changes in Phenotype?*

2. _______ genes are involved in the development of the anterior–posterior axis in mammals and insects.
   a. Developmental plasticity
   b. Heterochrony
   c. Hox
   d. *Pax6*
   e. *Gremlin*
   Textbook Reference: *20.1 How Can Small Genetic Changes Result in Large Changes in Phenotype?*

3. Which of the following is *not* true of heterochrony?
   a. It explains the long neck of the giraffe.
   b. It involves shifts in the relative timing of developmental processes.
   c. It can lead to major morphological changes and new lifestyles.
   d. It explains whether webbing remains between the digits of adult salamanders and birds or not.
   e. All of the above are true of heterochrony.
   Textbook Reference: *20.2 How Can Mutations with Large Effects Change Only One Part of the Body*

4. Module development has allowed for evolutionary changes to occur while still resulting in a viable organism because
   a. all modules must change together.
   b. modules can change independently of one another.
   c. modules are not important in development.
   d. modules prevent heterochrony.
   e. developmental genes always exert their effects on more than one module.
   Textbook Reference: *20.2 How Can Mutations with Large Effects Change Only One Part of the Body?*

5. Homologous genes
   a. are adjacent to one another on the same chromosome.
   b. never involve whole sets (clusters) of genes.
   c. are found exclusively in mammals.
   d. evolved from a gene present in a common ancestor.
   e. are expressed at the same time in an organism.
   Textbook Reference: *20.1 How Can Small Genetic Changes Result in Large Changes in Phenotype?*

6. Developmental plasticity
   a. is the ability of an organism to change its development in response to environmental conditions.
   b. is limited to plants.
   c. is the ability of an organism to express only one phenotype under different environmental conditions.
   d. means that an organism can begin as one genotype and later change genotype in response to environmental conditions.
   e. is also called genotypic plasticity.
   Textbook Reference: *20.4 How Can the Environment Modulate Development?*

7. Which of the following statements is *false*?
   a. Diet can initiate developmental changes.
   b. Day length is a reliable environmental signal.
   c. Temperature can initiate developmental changes.
   d. Cells of plant stems elongate in response to low light levels.
   e. The cycle by which deer form and lose their antlers occurs in response to temperature.
   Textbook Reference: *20.4 How Can the Environment Modulate Development?*

8. Genetic switches
   a. stop development.
   b. help various modules develop differently.
   c. ensure that all modules develop into the same type.
   d. do not involve promoters and transcription factors.

e. play no role in integrating positional information in the embryo.

Textbook Reference: *20.2 How Can Mutations with Large Effects Change Only One Part of the Body?*

9. The difference in appearance of *Nemoria arizonaria* caterpillars in spring and summer is triggered by
   a. temperature.
   b. day length.
   c. the relative availability of food.
   d. different genotypes.
   e. light intensity.

   Textbook Reference: *20.4 How cAn the Environment Modulate Development?*

10. Which of the following statements is true?
    a. Wings have evolved independently four times in vertebrates.
    b. Wings arose as a result of new "wing genes."
    c. Wings arose as modifications to already-existing structures.
    d. Wings have different skeletal components in birds and bats.
    e. Wings have similar structural components in pterosaurs and insects.

    Textbook Reference: *20.5 How Do Developmental Genes Constrain Evolution?*

11. Which of the following statements is *false*?
    a. All bird embryos have webbed feet, although some adult birds do not.
    b. Bone morphogenetic protein 4 (BMP4) directs cells in the foot webbing of bird embryos to undergo apoptosis.
    c. The protein Gremlin inhibits BMP4 protein from signaling for apoptosis, causing webbed feet.
    d. The gene *Gremlin* is expressed in the webbing of both chicken and duck embryos.
    e. Adding the protein Gremlin to a developing chicken foot will cause the foot to resemble that of a duck.

    Textbook Reference: *20.2 How Can Mutations with Large Effects Change Only One Part of the Body?*

12. Which of the following statements is *false*?
    a. In arthropods, the gene *Distalless* causes legs to form from segments.
    b. In centipedes, the Hox gene *Ubx* activates expression of the gene *Distalless*, causing legs to form from segments.
    c. A mutation in the Hox gene *Ubx* in the ancestor of insects resulted in legs forming from thoracic but not abdominal segments in this lineage.
    d. Changes in the spatial pattern of Hox gene expression can account for the different number of cervical vertebrae in mouse and chicken embryos.
    e. Morphological differences among species do not evolve through mutations in Hox genes.

    Textbook Reference: *20.3 How Can Developmental Changes Result in Differences among Species?*

13. Temperature-determined sex
    a. occurs in mammals.
    b. is an example of developmental plasticity.
    c. is not associated with sex-specific fitness differences.
    d. relies on a mechanism whereby high levels of the enzyme aromatase make testosterone dominant and lead to development of male sex organs.
    e. relies on a mechanism that makes high temperatures produce males.

    Textbook Reference: *20.4 How Can the Environment Modulate Development?*

14. Which of the following statements is true?
    a. Development ceases when an organism reaches maturity.
    b. Development continues until an organism dies.
    c. Development is fixed and cannot change in response to environmental conditions.
    d. In every organism, the genes governing development are unique to that organism.
    e. Development cannot constrain evolution.

    Textbook Reference: *20.1 How Can Small Genetic Changes Result in Large Changes in Phenotype?*

15. Which of the following statements is *false*?
    a. Body armor is absent from sticklebacks living in freshwater environments due to a genetic mutation.
    b. The *Pitx1* gene promotes the production of protective plates and spines in marine populations of sticklebacks.
    c. Similar traits are likely to evolve repeatedly because of the highly conserved nature of developmental genes.
    d. The *Pitx1* gene is inactive in freshwater stickleback populations.
    e. Environmental conditions induce the loss of body armor in freshwater sticklebacks.

    Textbook Reference: *20.5 How Do Developmental Genes Constrain Evolution?*

Answers

Key Concept Review

1. The gene *eyeless* in fruit flies and the gene *Pax6* in mice each control the developmental switch that turns on eye development. *Eyeless* and *Pax6* exhibit DNA sequence similarity even though fruit flies and mice have very different eyes. These sequence similarities suggest that *eyeless* and *Pax6* evolved from a gene present in a common ancestor of fruit flies and mice.

2. *Eyeless* and *Pax6* are homologous genes and share such a close identity that *Pax6* can substitute for *eyeless* during *Drosophila* eye development. However, although the expression of *eyeless*, or a homolog, is necessary for *Drosophila* eye development, it is not sufficient for normal *Drosophila* eye development. The expression of eyeless homologs, and other genes necessary for

each type of organism's eye development, is controlled by each organism's specific application of its genetic toolkit. This determines when, where, and to what extent the development-controlling transcriptions are expressed—and this ultimately determines the phenotype of the eye.

3. In most insect groups, the expression of the Hox gene *antennapedia* (*Antp*) in the second and third thoracic segments results in the development of wings from both segments. In *Drosophila*, the Hox gene *ultrabithorax* (*Ubx*) is expressed in the third segment and its protein product suppresses the expression of *Antp*, thereby suppressing the development of wings from the third segment. In the mutant fly with wings from both second and third segments, the *Ubx* gene has likely acquired an inactivating mutation, and *Antp* expression is no longer suppressed in the third thoracic segment.
4. Organisms are made up of modules. Change can occur in a module independent of other modules. The webbing in duck feet is due to expression of the *Gremlin* gene in the foot module. Expression of *Gremlin* in the webbing cells of the feet results in a protein that inhibits the BMP4 protein responsible for prompting apoptosis.
5. The statement is incorrect. In fact, most evolutionary changes in morphology occur by modification of existing development genes and developmental pathways.
6. Leg development in arthropods involves the homeotic gene *ultrabithorax* (*Ubx*) and the gene *Distalless* (*Dll*). Expression of *Dll* in a segment results in the development of a pair of legs in that segment. In insects, the *Ubx* gene has a mutation that suppresses the expression of *Dll* in the abdomen. As a result, no legs will develop in the abdomen. Centipedes do not have this mutation, therefore legs develop in thoracic and abdominal segments.
7. Day length and temperature are accurate predictors of the future environment. Changes in day length act as an environmental signal that accurately predicts when future conditions will change. Shorter days signify that fall and winter are approaching. Similarly, changes in temperature occur with changes in the seasons and can signal to an organism that a change in the season will soon occur.
8. The knowledge that sea turtles have temperature-determined sex would be critical to conservation efforts because if all eggs were incubated at the same temperature, then hatchlings would all be the same sex, a situation unlikely to be helpful to such efforts. Knowledge of which temperatures produce males and which produce females also would allow scientists to manipulate incubation temperature to produce the sex ratio best conducive to the restoration of sea turtle populations.
9.

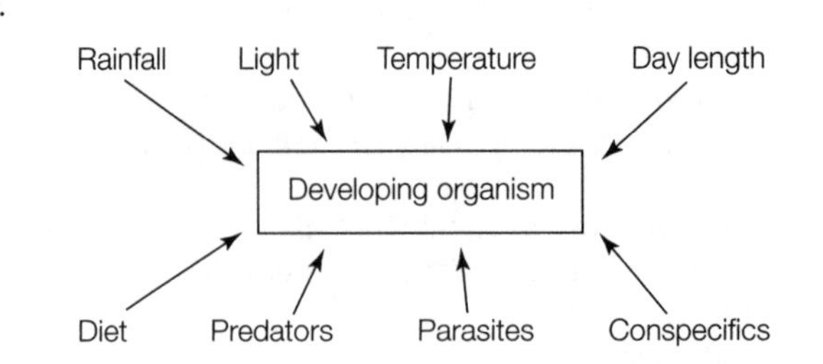

10. The independent reduction of eyes in several populations of a cave-dwelling amphipod is an example of parallel phenotypic evolution, the phenomenon whereby similar traits evolve repeatedly across multiple populations of a species.
11. The framework of the bat's wing is created by skeletal components that are analogous to the skeletal components that make up the fore flipper of the seal. These skeletal components are analogous to the metacarpals and phalanges of the human hand. These features suggest that all three organisms share a common ancestor whose descendants evolved structural modifications to better fit their environmental niches.

Test Yourself

1. **b.** Conserved genes are genes that are found in many organisms and have undergone very little change.
2. **c.** Hox genes regulate development along the anterior–posterior axis in animals.
3. **d.** Heterochrony is the process by which the timing of gene expression differs between two or more species. Heterochrony can explain the long neck of the giraffe (delays occur in the signaling process and stops bone growth, so the neck vertebrae grow longer) and whether webbing remains between the digits of adult salamanders and ducks.
4. **b.** Embryos are composed of self-contained units called modules that can change independently of one another. This allows for one module to evolve without disrupting the other modules of the animal.
5. **d.** Homologous genes evolved from a gene present in a common ancestor. They can be clusters of genes (such as Hox genes) and they are not restricted to mammals.
6. **a.** Developmental plasticity is found in both plants and animals. It allows a developing organism to express different phenotypes, depending on the environment.
7. **e.** Day length, not temperature, influences antler development in deer.
8. **b.** Genetic switches help each module develop into its appropriate form. They work by turning Hox genes on and off. They are involved in integrating positional information in the embryo, and they do involve promoters and transcription factors.

9. **c.** The difference in appearance of the *Nemoria arizonaria* spring caterpillar and summer caterpillar results from differences in diet.

10. **c.** Wings arose as modifications of preexisting structures. In vertebrates, for example, wings are modified forelimbs.

11. **d.** The *Gremlin* gene is expressed in the foot webbing cells of ducks, but not chickens; the Gremlin protein inhibits the BMP4 protein from signaling for apoptosis, so the feet of adult ducks are webbed.

12. **e.** Morphological differences between species can arise through mutations in Hox genes. This is exemplified by the mutation in the *Ubx* gene of insects, which causes leg formation in abdominal segments to be inhibited rather than prompted.

13. **b.** Temperature-determined sex is an example of developmental plasticity. The sex of some reptiles, for example, is determined by the temperature at which eggs incubate rather than by sex chromosomes (as in mammals).

14. **b.** Development occurs throughout an organism's lifespan and ends when it dies.

15. **e.** Absence of body armor in freshwater sticklebacks results from a mutation in the *Pitx1* gene; the absence of armor is not induced by environmental conditions.

Mechanisms of Evolution

The Big Picture

- Most biologists rank Darwin's *On The Origin of Species* as the most significant book ever written about their subject. Its importance is twofold. First, it presents a vast amount of evidence for the fact of evolution. Second, it describes a mechanism—natural selection—to explain evolutionary change. Evolution through natural selection remains the most important unifying concept in biology.
- Population genetics unites the concepts of Darwin with the insights into the mechanisms of heredity provided by Mendel. It enables biologists to study evolution with quantitative rigor.
- The loss of genetic variation that occurs during a population bottleneck is a major concern for biologists working to preserve endangered species. As the discussion of the greater prairie chicken makes clear, genetic uniformity in a population renders it more vulnerable to extinction.

Study Strategies

- A common mistake made by students when attempting to solve Hardy–Weinberg problems is the use of the wrong equation. For example, if you are given phenotypic frequencies for a trait that is at genetic equilibrium and shows dominance, remember that the frequency of the recessive phenotype (and genotype) is equal to q^2, not q. By taking the square root of q^2, you can obtain the frequency of the recessive allele and then determine the frequency of the dominant allele by subtraction ($p = 1 - q$).
- It is important to bear in mind that for any gene locus with two alleles, the allele frequencies must total 1 (i.e., it is always the case that $p + q = 1$). The equation that specifies genotype frequencies ($p^2 + 2pq + q^2 = 1$) is valid only for a population at genetic equilibrium.
- Population genetics is learned best by doing problems, so be sure you can solve problems similar to those presented in the "Test Yourself" section of this chapter of the Study Guide. Your instructor may provide a worksheet with additional examples.
- Go to the Web addresses shown to review the following animated tutorials and activity. You can also find them in BioPortal (yourBioPortal.com), along with many additional learning resources.

  Animated Tutorial 21.1 Natural Selection (Life10e.com/at21.1)

  Animated Tutorial 21.2 Genetic Drift (Life10e.com/at21.2)

  Animated Tutorial 21.3 Hardy–Weinberg Equilibrium (Life10e.com/at21.3)

  Animated Tutorial 21.4 Assessing the Costs of Adaptation (Life10e.com/at21.4)

  Activity 21.1 Darwin's Voyage (Life10e.com/ac21.1)

Key Concept Review

21.1 What Is the Relationship between Fact and Theory in Evolution?

Darwin and Wallace introduced the idea of evolution by natural selection

Evolutionary theory has continued to develop over the past century

Genetic variation contributes to phenotypic variation

Evolutionary theory encompasses our understanding of the mechanisms that lead to biological changes in populations over time. The theory of evolution by natural selection rests on two facts. First, all populations have the capacity to increase in number exponentially. Because populations rarely increase rapidly, high death rates must balance potential growth rates. Second, all populations show variations among individuals with regard to numerous inherited characteristics, some of which may affect their chances of surviving and reproducing. From these two facts, Darwin made the following inference: Those individuals whose heritable traits enable them to survive and reproduce more successfully than others will pass those traits on to greater numbers of offspring. In short, natural selection is the differential contribution of offspring to the next generation by various genetic types belonging to the same population. The rediscovery of Mendel's laws of inheritance led to the development of population genetics, which applies these laws to entire populations in order to understand the genetic basis of evolutionary change.

The discovery of the role of chromosomes in inheritance and the structure and function of DNA over the past century has greatly advanced our understanding of the theory of evolution and how it operates in populations.

Genetic variation contributes to phenotypic variation. A genotype is the genetic constitution of a particular trait of an individual. Evolution occurs when individuals with different genotypes survive or reproduce at different rates. The features of a phenotype are its characters. A trait is a specific form of a character. The gene pool of a population is the sum of all the alleles (alternative versions of a gene) found within it. Although a single diploid individual can have at most two different alleles for any gene, a population may contain many alleles for that gene. The gene pool contains all the variations that result in individuals with differing phenotypes on which agents of evolution act. Genotypes do not uniquely determine phenotypes. For example, if one allele is dominant to another, two genotypes (e.g., *AA* and *Aa*) can determine the same phenotype. Conversely, one genotype can produce different phenotypes because of the interaction of genetic and environmental factors during development.

Question 1. What is the meaning of the phrase "evolutionary theory"?

Question 2. How are characters, traits, and alleles related to one another?

21.2 What Are the Mechanisms of Evolutionary Change?

Mutation generates genetic variation

Selection acting on genetic variation leads to new phenotypes

Gene flow may change allele frequencies

Genetic drift may cause large changes in small populations

Nonrandom mating can change genotype or allele frequencies

A population is a group of individuals of a single species that live and interbreed in a particular area at the same time. Populations evolve, whereas individuals do not. The ultimate source of genetic variation in a population is mutation. Because mutations are random changes in genetic material, most are harmful or neutral, but it is the environment that determines whether a particular mutation is disadvantageous or adaptive. Allele frequencies measure the amount of genetic variation in a population, whereas genotype frequencies show how this variation is distributed among its members. Populations that have the same allele frequencies may nevertheless differ in their genotype frequencies.

When selection of individuals with desirable traits is carried out by humans, such as plant and animal breeders, it is referred to as artificial selection. An adaptation is a characteristic that helps its bearer survive and reproduce; the term also refers to the evolutionary process that produces such characteristics. Agents of evolution actually act on the phenotype, which is the physical expression of the genotype.

Gene flow occurs when individuals migrate from one population to another and breed in their new location. Gene flow can add new alleles to a population's gene pool, change the frequencies of alleles already present, or both.

Genetic drift is caused by chance events that alter allele frequencies in a population. It has its greatest impact on small populations, in which it may even cause harmful alleles to increase in frequency. During a population bottleneck, when a large population is severely reduced in size, allele frequencies may shift drastically, and genetic variation may be reduced as a result of genetic drift. The change in genetic variation that occurs when a few individuals originate a new population is called the founder effect. As in the case of a population bottleneck, some alleles found in the source population will be missing from the founding population, and others will occur with altered frequencies.

The preferential mating of individuals with others either of the same genotype or of a different genotype is called nonrandom mating. The effect of nonrandom mating is a population with either fewer or more heterozygous individuals than would be expected in a population in Hardy–Weinberg equilibrium. The effect of self-fertilization, another form of nonrandom mating, is a reduction in the frequency of heterozygotes. In most types of nonrandom mating (except for sexual selection), the allele frequencies remain the same despite the changes in genotype frequencies.

Sexual selection favors traits that benefit their bearers (generally males) in the competition for access to members of the opposite sex, or make their bearers more attractive to members of the opposite sex. Sexual selection often results in sexually dimorphic species, in which males and females differ in appearance.

Question 3. Describe what is meant by sexual selection and how the sexual selection of tail length in male widowbirds has been studied by behavioral ecologists.

Question 4. Describe some of the effects that genetic drift can have on allele frequencies in a population.

21.3 How Do Biologists Measure Evolutionary Change?

Evolutionary change can be measured by allele and genotype frequencies

Evolution will occur unless certain restrictive conditions exist

Deviations from Hardy–Weinberg equilibrium show that evolution is occurring

Natural selection acts directly on phenotypes

Natural selection can change or stabilize populations

Allele frequencies at a given locus can be estimated by counting alleles in a sample from the population. In a population, the frequency of all alleles at a particular locus is equal to 1. If there is only one allele at a locus, the population is monomorphic, and the allele is fixed. If two or more alleles exist, the population is polymorphic at that locus. The genetic structure of a population is described by the frequencies of different alleles at each locus and the frequencies of different genotypes.

To be at Hardy–Weinberg equilibrium, a population must meet five conditions: (1) mating must be random; (2) the population must be infinite; (3) there must be no migration into or out of the population; (4) there must be no mutation; (5) natural selection must not be affecting the survival of particular genotypes. If these conditions are met, allele frequencies at a locus remain the same from generation to generation. Moreover, after one generation of random mating, the genotype frequencies will remain in the proportions $p^2 + 2pq + q^2 = 1$, where p^2, $2pq$, and q^2 represent the frequencies of the homozygous dominant, heterozygous, and homozygous recessive genotypes, respectively.

For a locus with two alleles, p and q are typically used to represent the frequencies of the dominant and recessive alleles, respectively, and thus $p + q = 1$. Biologists estimate allele frequencies by measuring numbers of alleles in a sample of individuals from the population.

Deviations from Hardy–Weinberg equilibrium show that evolution is occurring. Though populations in nature can never fully meet the conditions of the Hardy–Weinberg equilibrium, this equation is often useful for predicting the approximate genotype frequencies in a population. It is also important because deviations from it may show that evolution is occurring. Moreover, the pattern of deviations is useful in identifying the agents of evolutionary change operating on the population.

Natural selection acts on the phenotype (the physical features expressed by an organism) rather than directly on the genotype. Fitness is the contribution of a phenotype to the composition of later generations, relative to the contribution of alternative phenotypes. The fitness of a phenotype is determined by the average rates of survival and reproduction of individuals with that phenotype, as compared to individuals with other phenotypes. Differences in fitness among phenotypes (and hence in the controlling genotypes) lead to changes in allele frequencies in the gene pool of later generations. Changes in allele frequencies due to differences in fitness result in adaptation. Natural selection, unlike other agents of evolution, adapts organisms to their environment. In cases in which the distribution of a phenotype approximates a bell-shaped curve because it is controlled by many gene loci, selection can produce any one of three results: (1) stabilizing selection, which reduces variation in the population by favoring average individuals; (2) directional selection, which changes the mean value for a character by favoring individuals that vary in one direction; or (3) disruptive selection which, by favoring both extremes, leads to a population with two peaks in the distribution of the character. Genetic drift, stabilizing selection, and directional selection tend to lessen the genetic variation within a population.

Question 5. Construct a concept map with the theme of "Evolutionary Agents." Include in your map the following terms: evolutionary agents, allele frequencies, directional selection, disruptive selection, founder effects, gene flow, genetic drift, genetic structure of a population, genotype frequencies, mutation, natural selection, nonrandom mating, phenotypic variation, population bottlenecks, and stabilizing selection. Connect these terms with verbs or short phrases to indicate the relationships among them.

Question 6. In a population with 600 members, the numbers of individuals of three different genotypes are $AA = 350$, $Aa = 100$, $aa = 150$.

a. What are the genotype frequencies in this population?
   1. $AA =$
   2. $Aa =$
   3. $aa =$
b. What are the allele frequencies in this population?
   1. $A =$
   2. $a =$
c. What would be the expected genotype frequencies if this population were in genetic equilibrium?
   1. $AA =$
   2. $Aa =$
   3. $aa =$
d. Is this population in genetic equilibrium? Explain.

21.4 How Is Genetic Variation Distributed and Maintained within Populations?

Neutral mutations accumulate in populations

Sexual recombination amplifies the number of possible genotypes

Frequency-dependent selection maintains genetic variation within populations

Heterozygote advantage maintains polymorphic loci

Genetic variation within species is maintained in geographically distinct populations

Neutral mutations—that is, mutations that do not affect the fitness of an individual—tend to accumulate and thereby increase the genetic variation in a population. Sexual recombination amplifies the number of possible genotypes by generating new combinations of alleles. Sexual reproduction has short-term disadvantages, including disruption of adaptive combinations of genes and reduction both of the rate at which females pass genes on to their offspring and of the overall reproductive rate. Possible advantages of sexual reproduction that may account for its evolution include facilitation of DNA repair, elimination of harmful mutations, and defense against pathogens and parasites.

Frequency-dependent selection, in which the less-common genotype or phenotype is favored by natural selection, preserves variation as a polymorphism. Heterozygote advantage maintains polymorphic loci. If heterozygous individuals have a selective advantage over homozygotes for either allele, both alleles will persist in a population. Geographically distinct subpopulations of a population often vary genetically because they are adapted to different environments. Clinal variation is gradual change in phenotype across a geographic gradient.

Question 7. Describe the advantages and disadvantages of asexual and sexual reproduction.

Question 8. Provide an example of heterozygote advantage in a population. What are the benefits that heterozygous individuals have compared to homozygous individuals?

21.5 What Are the Constraints on Evolution?

Developmental processes constrain evolution

Trade-offs constrain evolution

Short-term and long-term evolutionary outcomes sometimes differ

Evolutionary changes must be based on modifications of previously existing traits. The evolution of adaptations depends on the trade-off between costs and benefits. For an adaptation to be favored, the fitness benefit it confers must outweigh the fitness cost.

Long-term evolutionary changes are strongly influenced by events that occur so infrequently or so slowly that they are rarely observed during short-term evolutionary studies. To understand the long-term course of evolution, biologists seek evidence of rare events and search for trends in the fossil record.

Question 9. Give an example of how developmental processes constrain evolution.

Question 10. Male whitetail deer produce antlers every year and then shed them after the breeding season. Discuss the trade-offs incurred by males in producing antlers.

Test Yourself

1. Evolution occurs at the level of
   a. the individual genotype.
   b. the individual phenotype.
   c. environmentally based phenotypic variation.
   d. the population.
   e. the species.
   Textbook Reference: *21.1 What Is the Relationship between Fact and Theory in Evolution?*

2. Natural selection acts on
   a. the gene pool of the species.
   b. the genotype.
   c. the phenotype.
   d. multiple gene inheritance systems.
   e. the environment.
   Textbook Reference: *21.3 How Do Biologists Measure Evolutionary Change?*

3. In comparing several populations of the same species, the population with the greatest genetic variation would have the
   a. greatest number of genes.
   b. greatest number of alleles per gene.
   c. greatest number of population members.
   d. largest gene pool.
   e. None of the above
   Textbook Reference: *21.4 How Is Genetic Variation Maintained within Populations?*

4. The ability to taste the chemical PTC (phenylthiocarbamide) is determined in humans by a dominant allele T, with tasters having the genotypes Tt or TT and nontasters having tt. If 36 percent of the members of a population cannot taste PTC, then according to the Hardy–Weinberg rule, the frequency of the T allele should be
   a. 0.36.
   b. 0.4.
   c. 0.6.
   d. 0.64.
   e. 0.8.
   Textbook Reference: *21.3 How Do Biologists Measure Evolutionary Change?*

5. A gene in humans has two alleles, M and N, that code for different surface proteins on red blood cells. If you know that the frequency of allele M is 0.2, according to the Hardy–Weinberg rule, the frequency of the genotype MN in the population should be
   a. 0.16.
   b. 0.2.
   c. 0.32.
   d. 0.64.
   e. 0.8.
   Textbook Reference: *21.3 How Do Biologists Measure Evolutionary Change?*

6. If the frequency of allele b in a gene pool is 0.2, and the population is at Hardy–Weinberg equilibrium, the expected frequency of the genotype $bbbb$ in a tetraploid ($4n$) plant species would be
   a. 0.0016.
   b. 0.04.
   c. 0.08.
   d. 0.2.
   e. The answer cannot be determined from this information.
   Textbook Reference: *21.3 How Do Biologists Measure Evolutionary Change?*

7. Random genetic drift would probably have its greatest effect on a
   a. small, isolated population.
   b. large population in which mating is nonrandom.
   c. large population in which mating is random.
   d. large population with regular immigration from a neighboring population.
   e. large population with a high mutation rate.
   Textbook Reference: *21.2 What Are the Mechanisms of Evolutionary Change?*

8. Allele frequencies for a gene locus are *least* likely to be significantly changed by
   a. mutation.
   b. the founder effect.
   c. self-fertilization.
   d. gene flow.
   e. natural selection.
   Textbook Reference: *21.2 What Are the Mechanisms of Evolutionary Change?*

9. Which of the following evolutionary agents would produce nonrandom changes in the genetic structure of a population?
   a. Self-fertilization
   b. Population bottlenecks
   c. Mutation
   d. Natural selection
   e. Both a and d
   Textbook Reference: *21.2 What Are the Mechanisms of Evolutionary Change?*

10. Suppose that a particular species of flowering plant that lives only one year can produce red, white, or pink blossoms, depending on its genotype. Biologists studying a population of this species count 300 red-flowering, 500 white-flowering, and 800 pink-flowering plants in a population. When a census of the population is taken the following year, 600 red-flowering, 900 white-flowering, and 1,000 pink-flowering plants are observed. Which color has the highest fitness?
    a. Red
    b. White
    c. Pink
    d. All colors are equally fit.
    e. The answer cannot be determined from this information.
    Textbook Reference: *21.3 How Do Biologists Measure Evolutionary Change?*

11. The graph below shows the range of variation among population members for a trait determined by multiple genes.

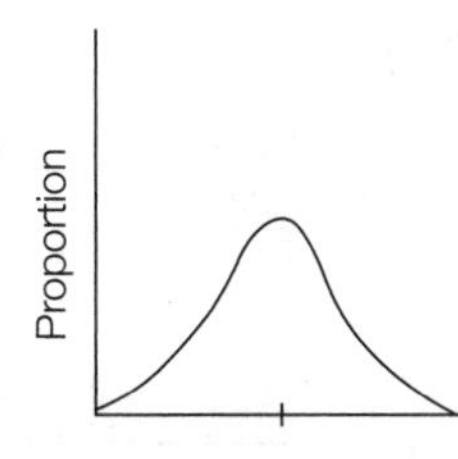

If this population is subject to *stabilizing selection* for several generations, which of the distributions would be most likely to result?

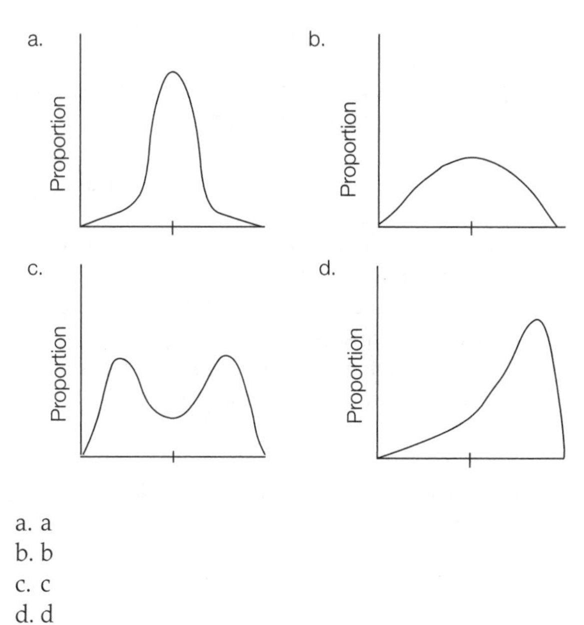

    a. a
    b. b
    c. c
    d. d
    e. Both a and b
    Textbook Reference: *21.3 How Do Biologists Measure Evolutionary Change?*

12. In areas of Africa in which malaria is prevalent, many human populations exist in which the allele that produces sickle-cell disease and the allele for normal red blood cells occur at constant frequencies, despite the fact that sickle-cell disease frequently causes death at an early age. This phenomenon is an example of
    a. the founder effect.
    b. a stable polymorphism.
    c. mutation.
    d. nonrandom mating.
    e. Both b and c
    Textbook Reference: *21.4 How Is Genetic Variation Maintained within Populations?*

13. Which of the following is *not* a disadvantage of sexual reproduction?
    a. When it involves separate genders, it reduces the overall reproductive rate.
    b. It breaks up adaptive combinations of genes.
    c. It reduces the rate at which females pass genes on to their offspring.
    d. It increases the difficulty of eliminating harmful mutations from the population.
    e. All of the above are disadvantages of sexual reproduction.
    Textbook Reference: *21.4 How Is Genetic Variation Maintained within Populations?*

14. Genetic variation within a population may be maintained by
    a. frequency-dependent selection.
    b. the accumulation of neutral alleles.
    c. sexual recombination.

d. heterozygote advantage.
e. All of the above
Textbook Reference: *21.4 How Is Genetic Variation Maintained within Populations?*

15. Which of the following can act as a constraint on the evolutionary process?
    a. The trade-off between the cost and benefit of an adaptation
    b. The occurrence of rare catastrophic events, such as meteorite impacts
    c. The fact that all evolutionary innovations are modifications of previously existing structures
    d. Both a and c
    e. All of the above
    Textbook Reference: *21.5 What Are the Constraints on Evolution?*

Answers

Key Concept Review

1. "Evolutionary theory" refers to a large body of scientific evidence that shows how physical changes have occurred in living organisms and what changes have occurred in the past. It provides an understanding of the mechanisms of evolutionary change.
2. Characters are features of the phenotype, such as eye color or hair color. Traits are specific forms of a character, such as brown eyes or blonde hair. Alleles are different forms of a gene. An allele (genotype) is the genetic information in the DNA that will be manifested as a particular trait (phenotype).
3. Sexual selection is the spread of a trait that improves the reproductive success of an individual. The trait may improve the ability of its bearer to compete with other members of its sex for access to mates, or it may make its bearer more attractive to members of the opposite sex. By artificially lengthening or shortening the tails of male widowbirds, behavioral ecologists were able to show that increased tail length attracted more females but did not confer an advantage in males' interactions with other males.
4. Genetic drift can result in large changes in allele frequencies in a population over time. Harmful alleles could increase in the population while rare advantageous alleles may be lost. Its effects can be pronounced when there is a dramatic reduction in population size, such as when a population bottleneck occurs. Similarly, when the founder effect occurs, the colonizing individuals likely represent a small proportion of the alleles from the source population.
5.

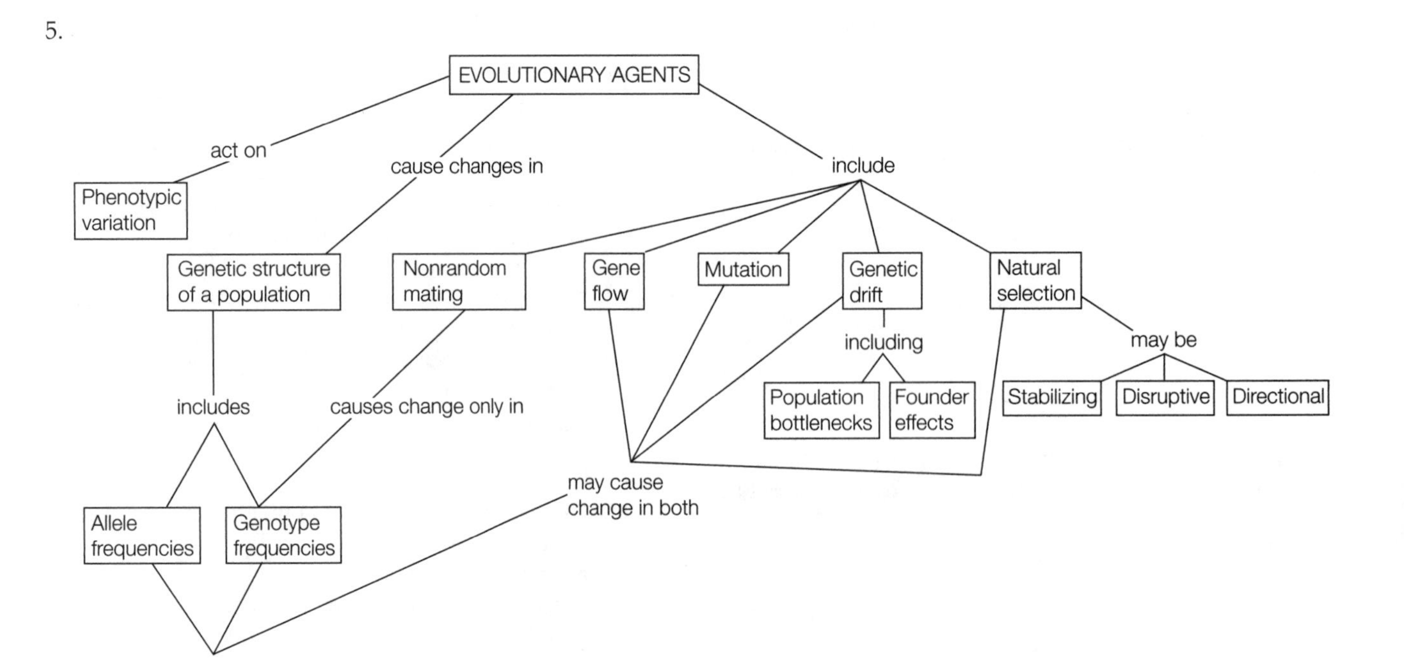

6. a.
   1. $AA = 350/600 = 0.58$
   2. $Aa = 100/600 = 0.17$
   3. $aa = 150/600 = 0.25$

   b.
   1. $A = (700 + 100)/1200 = 0.67$
   2. $a = (300 + 100)/1200 = 0.33$

   c.
   1. $AA = p^2 = (0.67)^2 = 0.45$
   2. $Aa = 2pq = 2 \times 0.67 \times 0.33 = 0.44$
   3. $aa = q^2 = (0.33)^2 = 0.11$

   d. No. The observed genotypic frequencies differ from those predicted by the Hardy–Weinberg rule using the observed allele frequencies for p and q.

7. In asexual reproduction, an organism's offspring is genetically identical to the parent. One advantage is the rapid rate of reproduction possible in asexual species. The major disadvantage is that deleterious mutations cannot be easily eliminated from the population. Sexual reproduction has the disadvantage of being slower and potentially breaking up advantageous combinations of genes. It has the advantage of creating greater genetic variation and a means to eliminate deleterious alleles from the population.

8. One example is provided by butterflies of the genus *Colias*, which are polymorphic for a gene that encodes phosphoglucose isomerase (PGI), which affects flight at different temperatures. Heterozygous individuals can fly over a greater range of temperatures, which gives them an advantage in both finding mates and foraging.

9. The evolution of ventrally flattened fishes that spend most of their time on the sea floor provides a good example of developmental processes constraining evolution. Cartilaginous fishes have somewhat ventrally flattened bodies. Evolution in skates and rays led to further flattening, allowing them to swim along the ocean floor. Bony fish relatives (plaice, sole, and flounder), which evolved from laterally flattened relatives, are constrained by this evolutionary history. They are able to lie on their sides on the bottom of the ocean floor, and during development the position of the eyes moves so that both eyes are on the same side of the body. While evolution has resulted in both types of fishes being able to exploit the ocean floor, the less-than-optimal body form of the bony fishes is constrained by development.

10. The benefit of producing antlers for whitetail deer is that it allows them to compete with other males for access to females during the breeding season, with a larger set of antlers potentially giving a male an advantage over other males. However, there is a metabolic cost to growing antlers. The persistence of this feature in male whitetail deer suggests that the benefits of producing antlers outweigh the costs.

Test Yourself

1. **d.** Evolution is defined as changes in the genetic structure of a population over time, so evolution occurs at the level of the population.

2. **c.** Natural selection acts on phenotypes, not genotypes. For example, a harmful recessive allele is "invisible" to natural selection when it occurs in a heterozygote, where its harmful effect is masked by the dominant allele.

3. **b.** Genetic variation is related to the number of different alleles per gene. Regarding choice **a**, recall that all species members have the same number of genes. Population size in itself has little to do with genetic variation, so choices **c** and **d** are also incorrect.

4. **b.** Recall that $q^2 = 0.36$ is the frequency of the *tt* genotype, so q (0.6, the square root of 0.36) is the frequency of the t allele. If there are only two alleles for this trait, then $T + t = 1$. The frequency of the T allele is therefore $1 - t = 1 - 0.6 = 0.4$.

5. **c.** Because $p = 0.2$ (and therefore $q = 1 - 0.2 = 0.8$), the frequency of the *MN* genotype is $2pq$, or $2 \times 0.2 \times 0.8 = 0.32$.

6. **a.** The probability of one allele b in a genotype is equal to its frequency, or 0.2, so the probability that all four of the alleles in a tetraploid organism will be b would be $(0.2)^4$, or 0.0016.

7. **a.** Genetic drift is most significant in small populations.

8. **c.** Unlike the founder effect, mutation, gene flow, and natural selection, all of which may change allele frequencies in a population, self-fertilization (like other types of nonrandom mating, with the exception of sexual selection) only causes a deviation from the frequency of heterozygotes predicted by Hardy–Weinberg equilibrium.

9. **e.** Self-fertilization (**a**) leads to increased numbers of homozygous individuals, and natural selection (**d**) is also nonrandom.

10. **a.** Fitness measures the relative contribution of a genotype or phenotype to subsequent generations. The red-flowering plants, which doubled in number, had the greatest percentage increase of any of the plants and thus had the highest fitness.

11. **a.** Stabilizing selection results when individuals that are intermediate in phenotype make a larger contribution to future generations than individuals of more extreme phenotype. This leads to reduced variation for the trait and causes the curve to be higher and narrower. Curve *b* shows greater variation, curve *c* would result from disruptive selection, and curve *d* would result from directional selection.

12. **b.** Polymorphism in a population is the existence of two or more alleles at a particular gene locus. If phenotypes are stable through time, then the underlying

alleles will also be constant. In this instance, the polymorphism is stable because malaria is a significant cause of mortality in some parts of Africa, and heterozygotes have greater resistance to this disease than individuals with a "normal" phenotype. Thus, the superior fitness of the heterozygotes maintains both alleles in the population.

13. **d.** Sexual recombination produces some individuals in a population that are less fit than others because they carry a greater-than-average number of deleterious mutations. Because these individuals are selected against, sexual reproduction is able to reduce the number of deleterious mutations in the population over time.

14. **e.** Frequency-dependent selection, the accumulation of neutral alleles, sexual recombination, and heterozygote advantage are the four major forces that maintain genetic variation in a population.

15. **e.** Cost–benefit trade-offs, developmental constraints, and major environmental disruptions can all act as constraints on natural selection and thus on the evolution of adaptive traits.

Reconstructing and Using Phylogenies

The Big Picture

- It is fair to say that several decades ago systematics was widely regarded as one of the less dynamic arenas of research within the biological sciences. Several developments have, in recent years, brought new excitement to the field. First was the development of a new, rigorous approach to systematics, known as cladistics. More recently, new methods of sequencing DNA and RNA, coupled with enormous increases in computer power, have enabled biologists to apply sophisticated mathematical techniques to phylogenetic studies, such as maximum likelihood analyses. As a result, systematists are now contributing to a wide array of biological studies, including subjects of medical importance such as the origins and types of human immunodeficiency virus (HIV) and other infectious organisms.
- The parsimony principle, which holds that one should prefer the simplest hypothesis capable of explaining the known facts, is fundamental not only to the reconstruction of phylogenies, but also to every other field of scientific research.
- The concept of evolution revolutionized taxonomy. Before Darwin's time, Linnaeus and others had developed "natural" systems of classification based primarily on similarity of morphology. Modern biologists realize that similarities of organisms (other than homoplasies) are the result of descent from a common ancestor and believe that a truly natural system of classification should reflect evolutionary relationships.
- Knowing that organisms are evolutionarily related enables biologists to make predictions about their characteristics. This knowledge can provide important hints in the search for organisms with valuable properties, such as the ability to produce medically useful drugs.

Study Strategies

- The concept of homology can be confusing. Figure 22.4 provides a helpful image of the difference between homologous and homoplastic traits.
- Distinguishing monophyletic, paraphyletic, and polyphyletic groups can be quite difficult. Bear in mind that a monophyletic group is analogous to a branch (or twig) of a tree: a single "cut" can remove it from a phylogenetic tree. This analogy should help you work out which kinds of "cuts" would result in paraphyletic and polyphyletic groups.
- Here is a mnemonic for the hierarchy of taxa (kingdom, phylum, class, order, family, genus, species) in the Linnaean system of classification: "Kindly Professors Cannot Often Fail Good Students." Note that the plural of genus is genera; the words general and generic come from the same root; this can help you remember that the genus is the more general taxon in the genus/species binomial. Species and specific also come from the same root, and the species is, of course, the most specific taxon.
- Go to the Web addresses shown to review the following animated tutorials and activities. You can also find them in BioPortal (yourBioPortal.com), along with many additional learning resources.

  Animated Tutorial 22.1 Using Phylogenetic Analysis to Reconstruct Evolutionary History (Life10e.com/at22.1)

  Animated Tutorial 22.2 Phylogeny and Molecular Evolution (Life10e.com/at22.2)

  Activity 22.1 Constructing a Phylogenetic Tree (Life10e.com/ac22.1)

  Activity 22.2 Types of Taxa (Life10e.com/ac22.2)

Key Concept Review

22.1 What Is Phylogeny?

All of life is connected through evolutionary history

Comparisons among species require an evolutionary perspective

A phylogeny is a description of the evolutionary history of relationships among organisms or their genes. Phylogenetic trees display the order in which lineages are hypothesized to have split. Each split (or node) in a phylogenetic tree represents a point at which lineages diverged in the past. The common ancestor of all the organisms in the tree forms the root of the tree. The timing of separations between lineages of organisms is shown by the positions of nodes on a time or divergence axis. A taxon is any named group of species. A taxon consisting of all the evolutionary descendants of a common ancestor is called a clade. Just as species that are each other's closest relatives are called sister species, so clades that are each other's closest relatives are called sister clades. Systematics is the study and the classification of the diversity of life.

All of life is connected through its evolutionary history, known as the tree of life. The evolutionary relationships among species, as shown by the tree of life, form the basis for biological classification.

Homologous traits are keys to reconstructing phylogenetic trees because they are features that are shared by members of a lineage owing to their descent from a common ancestral trait. A trait that differs from its ancestral form is called a derived trait. Conversely, a trait that was present in the ancestor of a group is known as an ancestral trait for that group. Synapomorphies are derived traits that are shared among a group of organisms and are viewed as evidence of the common ancestry of the group. Homoplasies (homoplastic traits) create confusion in reconstructing the evolutionary history of a lineage because they are features that are similar for some reason other than descent from a common ancestral trait. Two processes generate homoplasies: convergent evolution and evolutionary reversals. In convergent evolution, features that evolved independently become superficially similar. In an evolutionary reversal, a character reverts from a derived state to an ancestral one.

Question 1. All but one of the trees shown in the diagram below portray the same phylogenetic relationships among taxa A, B, C, D, E, F, and G. Which one depicts a different phylogeny?

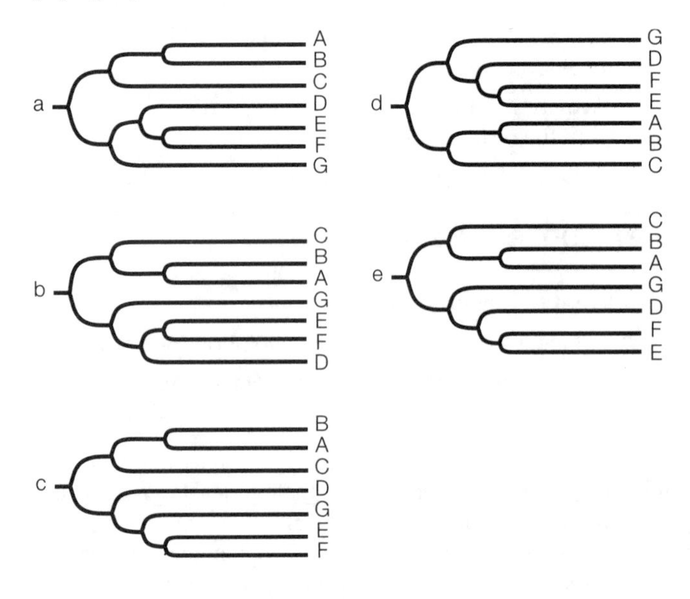

Question 2. In the phylogeny shown below, which group of taxa would *not* constitute a clade?

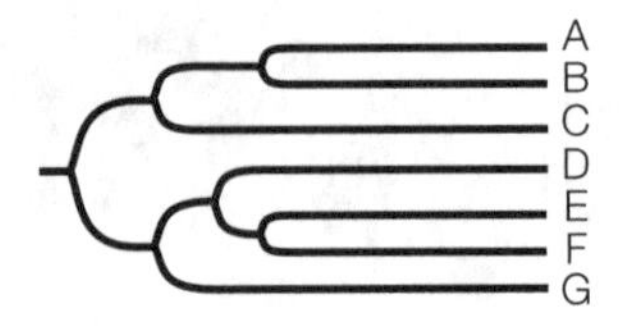

a. A, B, C, D, E, F, and G and their common ancestor
b. E, F, and G and their common ancestor
c. A, B, and C and their common ancestor
d. C, D, E, and F
e. Both b and d

Question 3. Discuss the implications of the following statement for the field of systematics: "DNA is the genetic material for all prokaryotes and eukaryotes."

22.2 How Are Phylogenetic Trees Constructed?

Parsimony provides the simplest explanation for phylogenetic data

Phylogenies are constructed from many sources of data

Mathematical models expand the power of phylogenetic reconstruction

The accuracy of phylogenetic methods can be tested

The first step in reconstructing a phylogeny is the choice of the ingroup and the appropriate outgroup. The ingroup is an assemblage of organisms whose phylogeny is to be determined. One method of distinguishing ancestral and derived traits is to compare the ingroup to an outgroup, which can be any species or group of species outside the group of interest. Ancestral traits should be present in both the ingroup and the outgroup, whereas derived traits should occur only in the ingroup. The root of the tree is determined by the relationship of the ingroup to the outgroup.

In reconstructing phylogenies, systematists are guided by the parsimony principle, which states that the preferred explanation of the observed data is the simplest explanation. In practice, this means that the best phylogenetic reconstruction is the one that minimizes the number of evolutionary changes that need to be assumed over all characters in all groups in the tree. In other words, the best hypothesis is one that requires the fewest homoplasies.

Morphology—the presence, size, shape, and other attributes of body parts—is an important source of traits for phylogenetic analysis. Similarities in developmental pattern may reveal evolutionary relationships. Structures in early developmental stages sometimes reveal evolutionary relationships that are not evident in adults. The morphology of fossils is particularly useful in helping to distinguish ancestral and derived traits. The fossil record also reveals when lineages diverged. In groups with few living representatives, information on extinct species may be critical to an understanding of the large divergences between the surviving species.

Behavior, if genetically determined, also is a useful source of information about evolutionary relationships.

Like morphological characters, molecules are heritable characteristics that may diverge among lineages. The complete genome of an organism contains an enormous set of traits (the individual nucleotide bases of DNA) that can be used to analyze phylogenies. Both nuclear and organelle DNA sequences are used in phylogenetic studies, as are sequences in gene products (such as the amino acid sequences of proteins). Because the chloroplast genome has changed slowly over evolutionary time, it is often used for the study of relatively ancient phylogenetic relationships among plants. Animal mitochondrial DNA has changed more rapidly, making it useful for studies of evolutionary relationships among closely related animal species.

Biologists have conducted experiments in living organisms and with computer simulations; both have demonstrated the effectiveness and accuracy of phylogenetic methods. The maximum likelihood method uses computer analysis for the reconstruction of phylogenies. A likelihood score of a tree is based on the probability that the observed data evolved on the specified tree, given an explicit mathematical model of evolution for the characters. An advantage of maximum likelihood analyses is that they incorporate more information about evolutionary change than parsimony methods do.

Question 4. The phylogenetic tree below shows the evolutionary relationships of five species (A–E) relative to five traits (1–5). Based on this tree, fill in the table below it, using 1 to indicate the presence of a derived trait and 0 to indicate the presence of an ancestral trait.

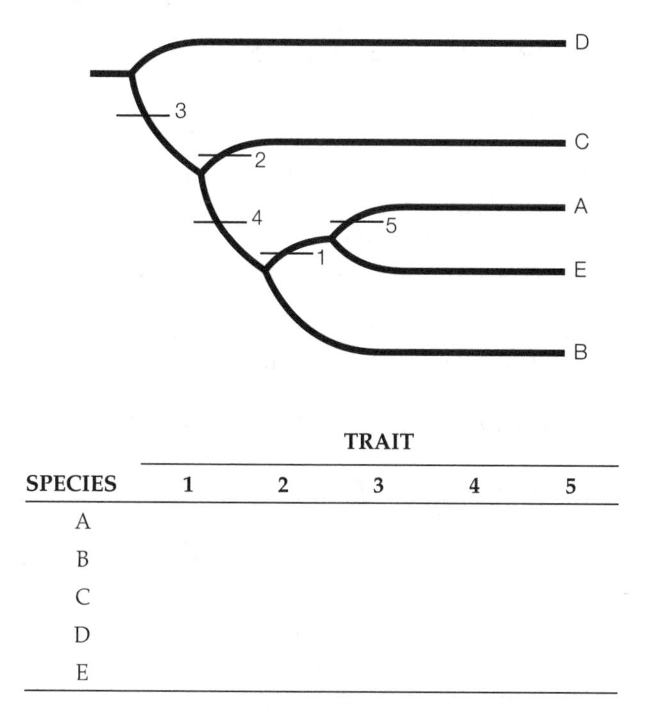

SPECIES	TRAIT 1	2	3	4	5
A					
B					
C					
D					
E					

Question 5. In the phylogenetic tree shown in Question 4, which species is considered the outgroup?

Question 6. The following table shows the ancestral and derived traits of five species (A–E). Based on the table, and following conventions presented in the textbook, construct a phylogenetic tree that represents the evolutionary relationships of this group. In this table, the ancestral state of each trait is indicated by 0 and the derived state is indicated by 1.

SPECIES	TRAIT 1	2	3	4	5
A	1	1	0	0	0
B	0	0	0	0	0
C	1	1	0	1	0
D	0	1	0	0	0
E	1	1	0	1	1

Question 7. Discuss the application of the parsimony principle in the construction of phylogenetic trees.

22.3 How Do Biologists Use Phylogenetic Trees?

Phylogenetic trees can be used to reconstruct past events

Phylogenies allow us to compare and contrast living organisms

Phylogenies can reveal convergent evolution

Ancestral states can be reconstructed

Molecular clocks help date evolutionary events

Phylogenetic trees are used to determine how many times a particular trait has evolved within a lineage. They can also be used to determine when, where, and how zoonotic diseases (diseases caused by infectious organisms that have been transferred to humans from another animal host) first entered human populations. Phylogenetic analysis is used by biologists to help them make relevant biological comparisons among genes, populations, and species. Phylogenetic methods are used not only to discover the evolutionary relationships among lineages of living organisms, but also to reconstruct morphological, behavioral, and molecular characteristics of ancestral species.

Phylogenies can be used to test hypotheses about evolution of particular traits by examining closely related species. Sister species on the same branch of the phylogeny may provide explanations for observed traits. Using molecular data, researchers can reconstruct phylogenies beyond morphological traits to infer the evolutionary relationships among lineages. Phylogenies can be used to reconstruct the morphology, behavior, or nucleotide and amino acid sequences of ancestral species. They can also provide information about the biology of long-extinct organisms.

In a molecular clock analysis, biologists assume that particular DNA sequences evolve at a reasonably constant rate and can be used as a metric to gauge the time of divergence for a particular split in a phylogeny. Molecular clocks must be calibrated with independent data such as the fossil record, known times of divergence, or biogeographic dates.

Question 8. How can molecular clocks be used to construct a phylogenetic tree?

Question 9. Provide an example of how phylogenies allow scientists to compare living organisms.

22.4 How Does Phylogeny Relate to Classification?

Evolutionary history is the basis for modern biological classification

Several codes of biological nomenclature govern the use of scientific names

The system of binomial nomenclature developed by Linnaeus in 1758 assigns each species two names, one identifying the species itself and the other the genus to which it belongs. The generic name (always capitalized) is followed by the species name (lower case), and both are italicized. In the Linnaean system of classification, species are grouped into higher-order taxa. The hierarchy of taxa, ranked from most to least inclusive, is kingdom, phylum, class, order, family, genus, species. Thus, a genus includes one or more species, a family includes one or more genera, and so forth. Biologists today recognize the tree of life as the basis for classification and often name clades without placing them into any Linnaean rank.

Taxa in biological classifications are expected to be monophyletic. A monophyletic taxonomic group (a clade) contains an ancestor and all descendants of that ancestor and no other organisms. A polyphyletic group does not include its common ancestor, whereas a paraphyletic group includes some, but not all, descendants of a particular ancestor. Virtually all taxonomists agree that polyphyletic and paraphyletic groups are inappropriate as taxonomic units.

Several sets of rules govern the use of scientific names, with the goal of providing unique and universal names for biological taxa. One such rule is that if a species is named more than once, the valid name is the first name proposed. In the past, different sets of taxonomic rules were developed by zoologists, botanists, and microbiologists, with the result that there are many duplicated names. Today, taxonomists are developing rules to ensure that every taxon has a unique name.

Question 10. Why are polyphyletic and paraphyletic groups considered to be inappropriate as taxonomic units?

Question 11. Why do scientists use explicit rules when assigning scientific names to organisms?

Test Yourself

1. A group that consists of all the evolutionary descendants of a common ancestor is called a(n)
   a. grade.
   b. taxon.
   c. homology.
   d. ingroup.
   e. clade.
   Textbook Reference: *22.1 What Is Phylogeny?*

2. A synapomorphy is
   a. the product of convergent evolution.
   b. the result of an evolutionary reversal.
   c. a shared derived characteristic.
   d. a trait that was present in the ancestor of a group.
   e. a phylogenetic tree.
   Textbook Reference: *22.1 What Is Phylogeny?*

3. Members of genus X, a hypothetical taxon of invertebrates, have antennae with a variable number of segments. Species A and B have 10 segments; species C and D have 9 segments; species E has 8 segments. In all other genera in this family (including genus Y), all species have antennae with 10 segments. Which of the following character states is a synapomorphy that would be useful for determining evolutionary relationships within genus X?
   a. 10-segment antennae in species A and B
   b. 10-segment antennae in genus Y and in two species of genus X
   c. Antennae with fewer than 10 segments in species C, D, and E
   d. 8-segment antennae in species E
   e. Both c and d
   Textbook Reference: *22.1 What Is Phylogeny?*

4. Which of the following would *not* be expected to result in homoplasy?
   a. Convergent evolution
   b. The independent evolution of similar structures in different lineages
   c. Selection for traits that perform similar functions
   d. The inheritance of ancestral traits
   e. An evolutionary reversal
   Textbook Reference: *22.1 What Is Phylogeny?*

5. A derived trait is one that
   a. differs from its ancestral form.
   b. is homologous with another trait found in a related species.
   c. is the product of an evolutionary reversal.
   d. has the same function, but not the same evolutionary origin, as a trait found in another species.
   e. is found only in members of the outgroup.
   Textbook Reference: *22.1 What Is Phylogeny?*

6. Which of the following statements about reconstructing phylogenies is *false*?
   a. Traits found in the outgroup as well as in the ingroup are likely to be ancestral traits.
   b. Shared traits are generally assumed to be homoplastic until they can be proven to be homologous.
   c. Phylogenetic trees do not always provide an explicit time scale by which to date the splits between lineages.
   d. In a phylogenetic tree, branches can be rotated around any node without changing the meaning of the tree.
   e. A particular trait may be either ancestral or derived depending on the point of reference of the phylogeny.

   Textbook Reference: *22.2 How Are Phylogenetic Trees Constructed?*

7. Which of the following is the most significant limitation of fossils as a source of information about evolutionary history?
   a. It is sometimes impossible to determine when a fossil organism lived.
   b. The fossil record for many groups is fragmentary or even nonexistent.
   c. Most fossils contain no nucleic acids or proteins and therefore are not useful for studies of molecular evolution.
   d. It is impossible to determine if morphologically similar fossils belong to the same species, because one cannot know if the fossil species interbred.
   e. Most fossils provide no information about the morphology of soft anatomical structures or about external characteristics such as color.

   Textbook Reference: *22.2 How Are Phylogenetic Trees Constructed?*

8. Which of the following sources of molecular data would be most helpful for a study of the evolutionary relationships of closely related animal species?
   a. Chloroplast DNA
   b. Mitochondrial DNA
   c. The amino acid sequences of a protein found in all animals, such as cytochrome *c*
   d. Ribosomal RNA sequences
   e Both b and d

   Textbook Reference: *22.2 How Are Phylogenetic Trees Constructed?*

9. Which of the following statements about the use of molecular clocks in phylogenetic analyses is true?
   a. A given gene usually evolves at the same rate in two different species regardless of differences in generation time of the species.
   b. Because changes in DNA sequences occur very slowly, molecular clocks can be used only to date evolutionary divergences that occurred millions of years ago.
   c. Molecular clocks must be calibrated with independent data, such as the fossil record.
   d. Even in a group of closely related species, different genes have been found to evolve at different rates.
   e. Both c and d

   Textbook Reference: *22.3 How Do Biologists Use Phylogenetic Trees?*

10. Which of the following statements describes a purpose for which biologists use phylogenetic trees?
    a. For human diseases once found only in other animals, phylogenetic trees are helpful in determining when and where the infectious organisms first entered human populations.
    b. Phylogenetic trees are useful for determining how many times a particular trait may have evolved independently within a lineage.
    c. Phylogenetic trees can be used to reconstruct ancestral traits.
    d. Phylogenetic trees can be used in conjunction with molecular clocks to estimate the timing of evolutionary events.
    e. All of the above

    Textbook Reference: *22.3 How Do Biologists Use Phylogenetic Trees?*

11. The *most* important attribute of a biological classification scheme is that it
    a. avoids the ambiguity created by the use of common names.
    b. reflects the evolutionary relationships among organisms.
    c. helps us remember organisms and their traits.
    d. improves our ability to make predictions about the morphology and behavior of organisms.
    e. groups together organisms with similar traits.

    Textbook Reference: *22.4 How Does Phylogeny Relate to Classification?*

12. Suppose you are writing a scientific paper about a unicellular green alga called *Chlamydomonas reinhardtii.* What is the proper way to refer to this species after the full binomial has been used once?
    a. *Chlamydomonas reinhardtii*
    b. *Chlamydomonas* spp.
    c. *Chlamydomonas* sp.
    d. *C. reinhardtii*
    e. *Chlamydomonas r.*

    Textbook Reference: *22.4 How Does Phylogeny Relate to Classification?*

13. The organisms that make up a class are _______ diverse and _______ numerous than those in a family within that class. The organisms that make up a phylum all diverged from a common ancestor _______ recently than did the organisms in an order within that phylum.
    a. more; more; less

b. more; more; more
c. more; less; less
d. less; less; more
e. less; less; less
Textbook Reference: *22.4 How Does Phylogeny Relate to Classification?*

14. Which of the following lists of taxonomic categories ranks them properly (from most inclusive to least inclusive)?
a. Phylum, order, family, genus
b. Class, phylum, order, species
c. Order, class, family, genus
d. Family, order, class, kingdom
e. Kingdom, class, species, genus
Textbook Reference: *22.4 How Does Phylogeny Relate to Classification?*

15. The ratites are a group of flightless birds comprising the ostrich, emu, cassowaries, rheas, and kiwis. All share certain morphological similarities (such as a breastbone without a keel) not found in other birds, but they live on different continents. In the past, some ornithologists regarded their similarities as homoplasies, but they are now thought to be synapomorphies. Based on this information, you would conclude that the ratites were once regarded as a _______ group but are now believed to be _______.
a. polyphyletic; paraphyletic
b. paraphyletic; monophyletic
c. polyphyletic; monophyletic
d. monophyletic; polyphyletic
e. monophyletic; paraphyletic
Textbook Reference: *22.4 How Does Phylogeny Relate to Classification?*

Answers

Key Concept Review

1. **c.** Recall that branches of a phylogenetic tree can be rotated around any node without changing the meaning of the tree. Tree "c" is different from all the others because it shows G (rather than D) as the sister taxon to taxa E and F.

2. **e.** E, F, and G and their common ancestor constitute a paraphyletic group; to be a clade, D would have to be included in the group. C, D, E, and F represent a polyphyletic group because the common ancestor of these taxa is not included in the group. (If the common ancestor of these four taxa were included, they would make up a paraphyletic group.)

3. Some of the implications of this statement are that DNA evolved as the genetic material before eukaryotes had diverged from prokaryotes, that DNA is an ancestral and general homologous trait, and that all surviving eukaryotes have DNA as their genetic material.

4.

	TRAIT				
SPECIES	**1**	**2**	**3**	**4**	**5**
A	1	0	1	1	1
B	0	0	1	1	0
C	0	1	1	0	0
D	0	0	0	0	0
E	1	0	1	1	0

5. Species D. The outgroup has the ancestral form for all traits.

6.

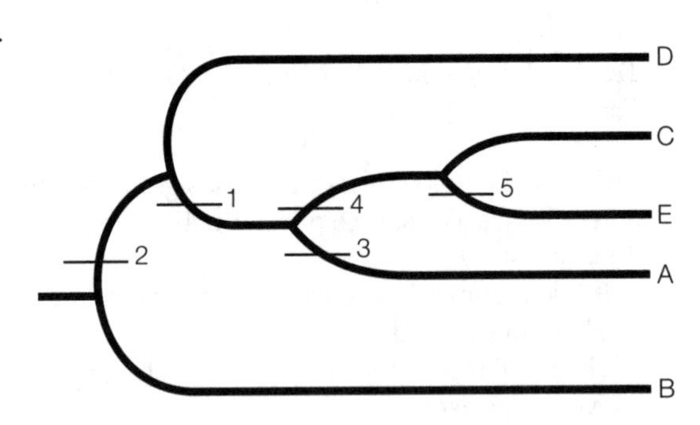

7. In the construction of a phylogenetic tree, the initial assumption is that derived traits appear only once and never disappear. Given a set of traits for a group of species, these restrictions sometimes must be relaxed to produce a phylogenetic tree for the group. Parsimony involves arranging the species so that the number of required reversals and multiple origins is minimized. Generally, the simplest explanation is most likely to be the most accurate.

8. A molecular clock is the average rate at which a gene or protein accumulates change. By knowing the rate of change, scientists can gauge the time of divergence of closely related species and estimate the time of a particular split in a phylogeny.

9. Scientists wanted to know, for example, why male swordtails evolved their characteristic appendage. One hypothesis was that male sword length exploited a preexisting bias in female sensory systems. By means of phylogenetic reconstruction, they were able to determine the nearest species that had most recently split from the lineage before the evolution of the sword. They found that the platyfishes, the closest relative, did not have swords, but when swords were attached to the platyfishes, female platyfishes preferred the experimental males over those without the sword.

10. Polyphyletic and paraphyletic groups are considered inappropriate as taxonomic units because they do not correctly reflect evolutionary history. Polyphyletic groups do not include the common ancestor, and paraphyletic groups do not include all the descendants of a common ancestor, and thus do not give a complete picture of the phylogenetic tree.

11. Scientists use specific rules in assigning scientific names to organisms to facilitate communication and dialogue. Typically there are many common names, using different languages, for an organism. The use of binomial nomenclature to assign one name to a species eliminates confusion.

Test Yourself

1. **e.** A clade can be thought of as a complete branch on the tree of life. It includes the ancestor of a group, all the ancestor's descendants, and no other organisms.
2. **c.** Synapomorphies are traits that are not found in the ancestor of a group (hence they are derived) and that are found in more than one member of a group (hence they are shared).
3. **c.** Because antennae with 10 segments are found in all genera in this family except for genus X, the trait of 10-segment antennae is best regarded as ancestral; hence, having fewer than 10 segments in the antennae is a derived trait. The presence of 8-segment antennae in species E is not a synapomorphy because it is a trait found only in that species.
4. **d.** Homoplasy is the appearance of similar structures in different lineages that were not present in the common ancestor.
5. **a.** Derived traits are those that have undergone a change during evolution from the ancestral (original) character state.
6. **b.** Most shared traits, especially in species with a recent common ancestor, are likely to be homologous, not homoplastic. Therefore, the assumption that traits are homologous until proven homoplastic is more consistent with the parsimony principle than the reverse assumption is.
7. **b.** The incompleteness of the fossil record is by far the greatest limitation of its usefulness in determining phylogenies.
8. **b.** Mitochondrial DNA in animals changes rapidly over evolutionary time and hence would be most useful in determining evolutionary relationships among species that have diverged from one another only recently.
9. **e.** It is true that molecular clocks must be calibrated with independent data, and it is true that different genes and other DNA sequences evolve at different rates (see Chapter 24). But it is not true that the rate of gene evolution is independent of generation time or several other biological factors, or that molecular clocks cannot be used to date comparatively recent events (e.g., the evolution of HIV-1).
10. **e.** The textbook describes specific examples of all four of the purposes listed as answers.
11. **b.** Although all of the statements listed are important attributes of biological classification schemes, the most important attribute is that biological classification reflects evolutionary relationships.
12. **d.** After a scientific name is referenced once in a text, the genus typically is abbreviated but the species name is given in full.
13. **a.** Organisms in a higher taxon are *more* diverse, include *more* species than organisms in a lower included taxon, and have diverged from a common ancestor *less* recently.
14. **a.** The complete hierarchy of taxonomic categories, from most to least inclusive, is: kingdom, phylum, class, order, family, genus, species.
15. **c.** If the shared characteristics of the ratites are homoplasies (meaning that they evolved independently by convergent evolution), then the group did not have a single common ancestor and is polyphyletic. If (as is now accepted) the shared traits are synapomorphies shared among all ratites and not found in other birds, then the group is monophyletic.

Speciation

The Big Picture

- Speciation is the process that has produced the millions of life forms on Earth, each adapted to a particular environment and way of life.
- Archipelagos such as the Galápagos and Hawaiian Islands have been called natural laboratories of evolution. Studies of the evolutionary radiations of the Galápagos finches and of several Hawaiian groups, such as *Drosophila* and the silverswords, have provided evolutionary biologists with crucial insights since the time of Darwin.
- Chapter 59, "Biodiversity and Conservation Biology," discusses the reasons for the rapid pace of human-caused species extinctions, including the disappearance of many members of groups that have undergone evolutionary radiations, such as Australian marsupials and Hawaiian honeycreepers. It also describes some strategies for the preservation of Earth's precious biological diversity.

Study Strategies

- The concept that hybrids between species with different numbers of chromosomes are inevitably sterile unless the hybrid is an allopolyploid is a difficult concept. Recall that in meiosis I, homologous chromosomes undergo synapsis. This process cannot occur properly if the haploid sets of chromosomes inherited from the parents contain different numbers of chromosomes, because it is then impossible for every chromosome to have a homolog. As a consequence, meiosis does not proceed normally and few if any normal gametes will be produced. Allopolyploids possess four sets of chromosomes (two from each parent), so their chromosomes can synapse normally and they can produce viable gametes.
- The heart of this chapter is the discussions of allopatric (geographic) speciation and reproductive isolating mechanisms, so you should pay particular attention to these topics.
- Go to the Web addresses shown to review the following animated tutorials and activity. You can also find them in BioPortal (yourBioPortal.com), along with many additional learning resources.

  Animated Tutorial 23.1 Speciation Simulation (Life10e.com/at23.1)

  Animated Tutorial 23.2 Speciation Mechanisms (Life10e.com/at23.2)

  Animated Tutorial 23.3 Founder Events and Allopatric Speciation (Life10e.com/at23.3)

  Activity 23.1 Concept Matching (Life10e.com/ac23.1)

Key Concept Review

23.1 What Are Species?

We can recognize many species by their appearance

Reproductive isolation is key

The lineage approach takes a long-term view

The different species concepts are not mutually exclusive

A species is a group of organisms that can mate with one another, producing fertile offspring. Speciation is the process by which one species splits into two species. Determining whether two populations constitute different species may be difficult because speciation is frequently a gradual process. A number of species concepts exist. The morphological concept of species used by early biologists grouped organisms into species on the basis of their appearance. This concept has limitations; in some instances not all members of a species look alike, and in other instances two or more cryptic species are morphologically indistinguishable but do not interbreed. The biological species concept defines a species as a group of actually or potentially interbreeding natural populations that are reproductively isolated from other such groups. It emphasizes that reproductive isolation is essential to keep sexual lineages on the tree of life separated from one another. This definition cannot be applied to organisms that reproduce asexually, and it is limited to a single point in evolutionary time. The lineage species concept regards species as the smallest branches on the tree of life. The lineage splitting may be sudden or gradual, but in either case the lineages are thereafter independent of each other, allowing biologists to consider species over evolutionary time.

Question 1. Discuss the relationship between the three major species concepts. How are they similar? How do they differ?

Question 2. In the eastern United States, populations of the white-footed mouse, *Peromyscus leucopus*, can be found almost continuously from Maine to Georgia. Based on the biological species concept, why would a population of this species from Maine *not* be considered a separate species from a population in Georgia?

23.2 What Is the Genetic Basis of Speciation?

Incompatabilities between genes can produce reproductive isolation

Reproductive isolation develops with increasing genetic divergence

Evolutionary change can occur without speciation. A single lineage may change through time without diverging into two species. Speciation requires that the gene pool of the original species divide into two isolated gene pools. After separation, according to the Dobzhansky–Muller model, the isolated populations will accumulate allelic differences at gene loci (or chromosomal differences) that eventually will make it impossible for members of the two populations to interbreed successfully if they come together again. In some cases, complete reproductive isolation may take millions of years to develop, whereas in other cases it may take only a few generations.

Question 3. Describe how centric fusion can play a role in speciation.

Question 4. Explain how incompatibilities between genes can produce reproductive isolation in a population.

23.3 What Barriers to Gene Flow Result in Speciation?

Physical barriers give rise to allopatric speciation

Sympatric speciation occurs without physical barriers

In allopatric speciation, the population is initially divided by a geographic barrier. If the barrier results from a geological or climatic change, the two isolated populations are often large and genetically similar. These populations diverge not only because of genetic drift, but especially because the environments in which they live are, or become, different. Alternatively, separation may occur when some members of a population cross a barrier and found a new, isolated population. Evidence suggests that allopatric speciation is the most common mechanism of speciation in most groups of organisms.

Sympatric speciation occurs without geographic subdivision of the gene pool of the original species. Disruptive selection, in which different genotypes have high fitness on one of two different food resources, may be a widespread mechanism of sympatric speciation among insects. Sympatric speciation by polyploidy, the production of duplicate sets of chromosomes within an individual, is common in plants. Polyploidy produces new species because the polyploid organisms cannot interbreed with members of the parent species. Polyploid species that have a single ancestor are called autopolyploids, whereas those that have resulted from the hybridization of two species are referred to as allopolyploids. New species may arise by polyploidy much more easily among plants than among animals because plants of many species can reproduce by self-fertilization.

Question 5. Discuss the conditions on the Galápagos Islands that led to the evolution of the birds known as Darwin's finches.

Question 6. In autopolyploidy, a new species of plant can arise by the doubling of chromosome numbers in a single individual of one species (provided that the individual is capable of self-fertilization). Why is it virtually impossible for such a tetraploid plant to interbreed successfully with diploid individuals of the "same" species?

23.4 What Happens When Newly Formed Species Come into Contact?

Prezygotic isolating mechanisms prevent hybridization

Postzygotic isolating mechanisms result in selection against hybridization

Hybrid zones may form if reproductive isolation is incomplete

Prezygotic barriers prevent members of different species from mating. Differences in reproductive organs may prevent interbreeding (mechanical isolation). Species may not be able to interbreed because they are fertile at different times (temporal isolation). The two species may not recognize or respond to each other's mating behaviors, or in flowering plants, the behavioral preferences of the pollinating animals may prevent interbreeding (behavioral isolation). Species may simply mate in different areas or different parts of a habitat (habitat isolation). The sperm and egg may be chemically incompatible (gametic isolation).

Postzygotic barriers can prevent effective gene flow between species, even if mating occurs. Hybrid zygotes may not mature normally (low hybrid zygote viability). Hybrids may survive less well than either parent species (low hybrid adult viability). Hybrids may be infertile (hybrid infertility). The evolution of more effective prezygotic reproductive barriers is known as reinforcement. It may occur if the hybrid offspring of two species survive poorly.

If two populations reestablish contact before reproductive isolation is complete, several results are possible. If hybrid offspring are not at a selective disadvantage, they may spread through both populations with the result that the gene pools of the populations combine. Thus no new species result from the period of isolation. If hybrid offspring are less successful, reinforcement may strengthen prezygotic reproductive barriers. If hybrid offspring are at a disadvantage but reinforcement fails to occur, a stable, narrow hybrid zone may form.

Question 7. Construct a concept map whose theme is "Species." Include in your map the following terms: species, allopatric, allopolyploidy, autopolyploidy, biological, concepts, founder events, independent evolution, interruption of gene flow, lineage, morphological, physical similarity, postzygotic barriers, prezygotic barriers, reproductive isolation, speciation, and sympatric. Connect these terms by verbs or short phrases to indicate the relationships among them.

Question 8. Suppose that members of two populations are separated by a geographic barrier and begin to diverge genetically. Many generations later, when the barrier is removed, the two populations can interbreed, but the hybrid offspring do not survive or reproduce well. Explain how natural selection might lead to the evolution of more effective prezygotic barriers in these species.

Question 9. The yellow-rumped warbler was formerly split into two species (myrtle and Audubon's warblers), but in 1973 it was reclassified as a single species. Myrtle warblers and Audubon's warblers have largely allopatric ranges but hybridize where they are sympatric, in the Canadian Rockies. They are similar in appearance but are readily distinguished by experienced birders. What further data about these two forms should ornithologists collect and analyze in order to decide whether they should continue to be classified as a single species?

23.5 Why Do Rates of Speciation Vary?

Several ecological and behavioral factors influence speciation rates

Rapid speciation can lead to adaptive radiation

Rates of speciation vary greatly among groups of organisms. Characteristics of a species that make it prone to speciation include membership in a large evolutionary lineage and poor dispersal ability. Among plants, high rates of speciation are found in groups with specialized animal pollinators. Among animals, dietary specialization and sexual selection typically stimulate speciation. In an evolutionary radiation, many daughter species arise from a single ancestor. An adaptive radiation results in an array of species that differ significantly in their habitats and resource utilization. The native biota of the Hawaiian Islands illustrates that populations colonizing environments that have underutilized resources are particularly likely to produce adaptive radiations.

Question 10. Discuss what is known about the evolutionary radiations on islands based on studies of Hawaiian silverswords and tarweeds.

Question 11. Why would animals with complex sexually selected behaviors be more likely to form new species at a higher rate than those without such behaviors?

Test Yourself

1. It is difficult to apply the biological species concept to groups of organisms that
   a. are asexual.
   b. produce hybrids only in captivity.
   c. show little morphological diversity.
   d. exist only in the fossil record.
   e. Both a and d

   Textbook Reference: *23.1 What Are Species?*

2. Which of the following statements about allopatric speciation is *false?*
   a. It can sometimes involve small populations.
   b. It occurs only in species that are widely distributed.
   c. It always involves a physical barrier that interrupts gene flow.
   d. It sometimes can involve chance events.
   e. It is the dominant mode of speciation in most groups of organisms.

   Textbook Reference: *23.3 What Barriers to Gene Flow Result in Speciation?*

3. A long, narrow hybrid zone exists in Europe between the ranges of the fire-bellied toad and the yellow-bellied toad. The persistence of this zone can be attributed to which of the following factors?
   a. Reinforcement strengthens the prezygotic barriers between the two species.
   b. Hybrid offspring have the same fitness as nonhybrid offspring.
   c. Both species travel long distances over the course of their lives.
   d. Individuals from outside the hybrid zone regularly move into the hybrid zone.
   e. None of the above

   Textbook Reference: *23.4 What Happens When Newly Formed Species Come into Contact?*

4. Which type of speciation is most common among flowering plants?
   a. Geographic
   b. Sympatric
   c. Allopatric
   d. Disruptive
   e. None of the above

   Textbook Reference: *23.3 What Barriers to Gene Flow Result in Speciation?*

5. Which of the following would *not* be considered an example of a prezygotic reproductive isolating mechanism?
   a. One bird species forages in the tops of trees for flying insects, whereas another forages on the ground for worms and grubs.
   b. The males of one species of moth cannot detect and respond to the sex attractant chemicals produced by the females of another species.
   c. Sperm of one species of sea urchin are unable to penetrate the egg plasma membrane of another species.
   d. Mosquitoes of one species are active in foraging and searching for mates at dusk, whereas those of another species are active at dawn.

e. Flowers of one orchid species mimic female bees of species A, whereas flowers of another orchid species mimic female bees of species B.

Textbook Reference: *23.4 What Happens When Newly Formed Species Come into Contact?*

6. Which of the following factors would *not* be expected to increase the rate of speciation in a group of organisms?
   a. Fragmentation of populations
   b. Poor dispersal ability
   c. High birthrates
   d. Dietary specialization
   e. Pollination by animals

   Textbook Reference: *23.5 Why Do Rates of Speciation Vary?*

7. Which of the following is *not* a suggested reason for the adaptive radiation of silverswords on the Hawaiian archipelago?
   a. Water is an effective barrier for many organisms.
   b. Because islands are small compared with mainland areas, more species would be expected to develop there.
   c. Competition is frequently reduced on islands.
   d. More ecological opportunities exist on islands that have not been colonized by many species.
   e. Neither b nor d is a suggested reason.

   Textbook Reference: *23.5 Why Do Rates of Speciation Vary?*

8. More than 800 species of *Drosophila* occur in the Hawaiian Islands, representing 30 to 40 percent of all the species in this genus. The occurrence of so many *Drosophila* species in this island chain is
   a. the result of many founder events followed by genetic divergence.
   b. an example of an evolutionary radiation.
   c. largely the result of sympatric speciation.
   d. evidence that the genus *Drosophila* first evolved in the Hawaiian Islands.
   e. Both a and b

   Textbook Reference: *23.3 What Barriers to Gene Flow Result in Speciation?*

9. Which of the following statements about speciation is *false?*
   a. A small founding population can be involved in speciation.
   b. Speciation always involves interruption of gene flow between different groups of organisms.
   c. The rate of speciation can vary for different groups of organisms.
   d. Speciation always requires many generations.
   e. Speciation may occur because certain genotypes within a population prefer distinct microhabitats where mating takes place.

   Textbook Reference: *23.3 What Barriers to Gene Flow Result in Speciation?*

10. Which of the following observations would constitute conclusive evidence that two overlapping populations that had been geographically separated have *not* diverged into distinct species?
    a. Matings between members of the two populations produce viable hybrids.
    b. A stable hybrid zone exists where their ranges overlap.
    c. Interbreeding is common between members of the two populations.
    d. All of the above
    e. None of the above

    Textbook Reference: *23.4 What Happens When Newly Formed Species Come into Contact?*

11. Which of the following processes is *least* likely to be important in allopatric speciation?
    a. A founder event
    b. Allopolyploidy
    c. Behavioral isolation
    d. Genetic drift
    e. Both b and d

    Textbook Reference: *23.3 What Barriers to Gene Flow Result in Speciation?*

12. A field contains two related species of flowering plants. Species A has a diploid chromosome number of 16, and species B has a diploid number of 18. If a third species arises as a result of hybridization between A and B, how many chromosomes will it have?
    a. 17
    b. 32
    c. 34
    d. 36
    e. 68

    Textbook Reference: *23.3 What Barriers to Gene Flow Result in Speciation?*

13. Speciation by polyploidy occurs far more often in plants than in animals because
    a. plants are more likely to be capable of self-fertilization than animals.
    b. plant cells can tolerate extra sets of chromosomes, whereas animal cells cannot.
    c. plants as a rule have higher reproductive rates than animals.
    d. many plants are specialized with respect to their pollinating agent.
    e. All of the above

    Textbook Reference: *23.3 What Barriers to Gene Flow Result in Speciation?*

14. Two species of narrowmouth frogs in the United States have mating calls that differ more in their region of sympatry than in those parts of their ranges that do not overlap. If this difference in their vocalizations has the function of preventing hybridization between the two species, it is an example of
    a. a hybrid zone.

b. reinforcement.
c. sympatric speciation.
d. a postzygotic reproductive barrier.
e. allopatric speciation.
Textbook Reference: *23.4 What Happens When Newly Formed Species Come into Contact?*

15. Four hypothetical families of birds are (1) endemic either to a single large island or to an island group (archipelago) far from the nearest continental land mass, and (2) have either monogamous or promiscuous mating systems in their species. In which of the four families would you predict the highest rate of speciation?
a. The single-island family whose species are promiscuous
b. The archipelago family whose species are monogamous
c. The single-island family whose species are monogamous
d. The archipelago family whose species are promiscuous
e. The rate of speciation cannot be predicted from the information given.
Textbook Reference: *23.3 What Barriers to Gene Flow Result in Speciation?; 23.5 Why Do Rates of Speciation Vary?*

Answers

Key Concept Review

1. The three major species concepts emphasize different aspects of species or speciation. The morphological species concept emphasizes similarity in appearance, but sometimes results in under- or overestimation of actual number of species. The biological species concept emphasizes reproductive isolation. The lineage species concept emphasizes reproductive isolation over evolutionary time. This concept also accommodates asexually reproducing species. Significant reproductive isolation between species is necessary to maintain distinct lineages over evolutionary time. Reproductive isolation is also responsible for morphological distinctness of most species. The biological and lineage concepts both address reproductive isolation, but differ in the evolutionary time scale each addresses. They both differ from the morphological species concept, which does not address reproductive isolation, only similarity in appearance.

2. The biological species concept includes the key phrase "actually or potentially" interbreeding individuals. While individual mice from populations in Maine are not actually breeding with individuals from populations in Georgia, there is the potential for individuals from both populations, if brought together, to interbreed and produce viable offspring.

3. Centric fusion is the fusion of two acrocentric (one-armed) chromosomes to form a metacentric chromosome. If centric fusion becomes fixed at one chromosome in one population of organisms but at a different chromosome in another population, individuals from the two populations will not be able to produce viable offspring because the offspring will not be able to produce normal gametes in meiosis.

4. If an initial population became subdivided into two populations by some barrier to gene flow, new alleles at different loci could arise and become fixed in these separated populations. Either new allele might cause reproductive incompatibility. If these two new alleles from the isolated subpopulations were to come together, they might not be compatible with each other, leading to functionally inferior offspring or lethality.

5. The relatively great distance between the Galápagos Islands and the South American mainland and also between each of the islands in the archipelago ensured that once immigrants had arrived on an island, they would be genetically isolated for a substantial period of time. Also, because the islands differ greatly in climate and vegetation, the resident birds were subject to different selection pressures. This, in combination with reduced gene flow between the islands, led to a rapid evolutionary radiation of finches.

6. Recall from Chapter 11 that pairs of homologous chromosomes synapse during prophase and metaphase of the first division of meiosis. The homologs then separate, so that each cell resulting from meiosis I is haploid, as are the products of meiosis II. In tetraploids as in diploids, meiosis is normal because pairing of homologs can occur. Any offspring of a cross between tetraploid and diploid individuals would be triploid, however, and therefore sterile because correct synapsis of homologs could not occur. Because the tetraploid product of autopolyploidy is reproductively isolated from its diploid relatives, it is a new species.

7.

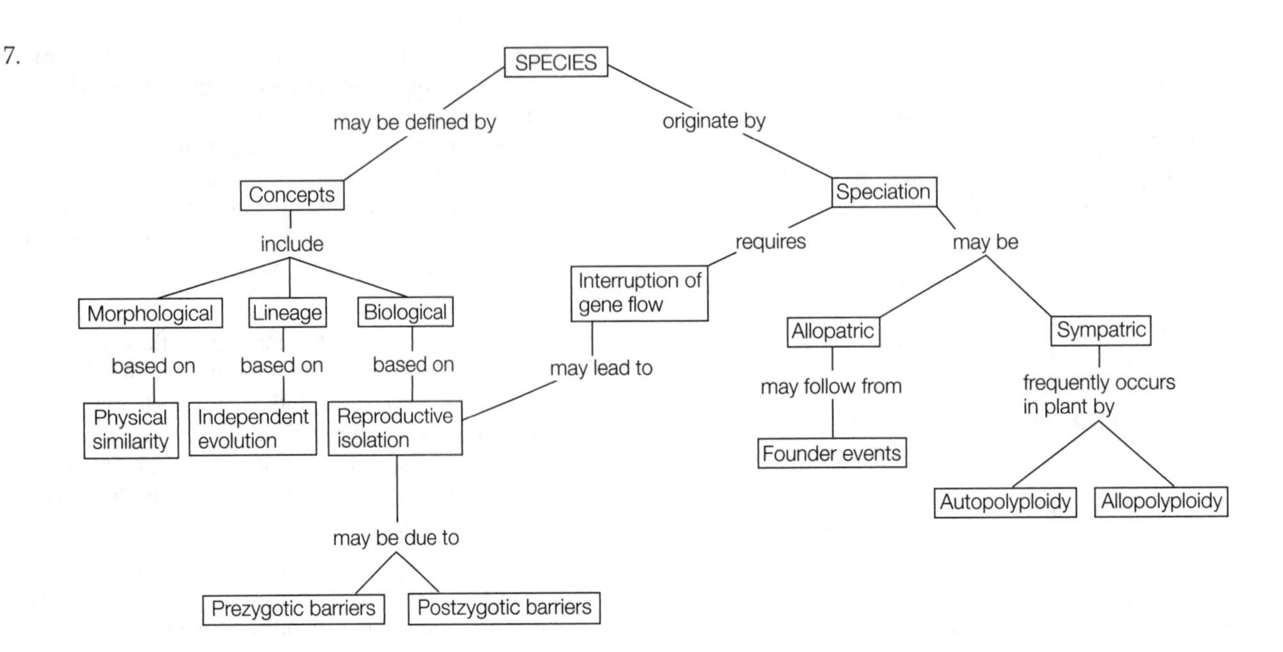

8. Recall that natural selection tends to remove from a population traits that reduce survival or reproductive success. Individuals that interbreed between populations will have lower fitness (they will contribute fewer offspring to future generations) than those who breed within their own population. If the tendency to avoid interbreeding is heritable (and not just the result of chance), the frequency of alleles that prevent interbreeding will increase in each population. How might such a trait be heritable? Any of the prezygotic barriers to interbreeding might be heritable traits. For example, if the species in question are frogs, and if their mating calls started to diverge while the populations were separated, the following traits might be heritable: a tendency to make a call that is more distinct from that of the other population, or the ability to distinguish between the existing calls of the two populations (coupled with a preference for the call of one's own population). As alleles for these traits increased in frequency, they would contribute to the behavioral isolation of the two populations and perhaps eventually to more complete speciation.

9. During allopatric speciation, divergence of two populations often occurs gradually. In such cases it is inevitable that intermediate stages of speciation occur, and it may be a matter of opinion whether two forms have diverged sufficiently to be considered separate species. With respect to these two warbler populations, biologists would seek answers to these questions: Are hybrid offspring as fit as those resulting from mating of individuals of the same population? Is there evidence that the zone of hybridization is expanding, indicating that the gene pools of the populations are combining? Is there evidence of reinforcement of prezygotic barriers to interbreeding (e.g., a greater difference in the songs of the two forms in the area of hybridization than in allopatric parts of their ranges)? Lesser fitness of hybrid offspring, a stable, narrow zone of hybridization, and evidence of reinforcement would favor the conclusion that the populations are best regarded as separate species.

10. Studies of the silverswords of the Hawaiian archipelago show that taxa that evolve on islands frequently show great morphological diversity because of the reduced competition that immigrants encounter on islands. Thus Hawaiian silverswords, unlike their mainland tarweed relatives, have evolved tree- and shrublike species because there were few resident tree and shrub species with which they had to compete.

11. Animals with complex sexually selected behaviors are likely to form new species at higher rates because they are more discriminating among potential partners. They make subtle discriminations among members of their own species based on size, shape, appearance, and behavior. Such choices can influence mating success, so may lead to rapid evolution of behavioral isolating mechanisms among populations.

Test Yourself

1. **e.** The key criterion of a biological species is that its members are reproductively isolated from other such groups. This criterion is impossible to evaluate in asexual and fossil species.
2. **b.** A wide distribution is not a requisite for allopatric speciation.
3. **d.** These toads are an example of related species in which reinforcement does not strengthen prezygotic barriers, even though hybrid offspring are only half as fit as nonhybrid offspring. The reason for this is that toads from outside the hybrid zone (not subject to the selective pressure against hybridizing) regularly move into the hybrid zone and mate with members of the other species. The hybrid zone remains narrow because toads do not travel long distances; natural

selection removes hybrids from the population before they can disperse very far.

4. **b.** Sympatric speciation is most common among flowering plants. It has been estimated that about 70 percent of all flowering plant species are polyploid.
5. **a.** Provided that the two species are active in the same locality at the same time, a difference in the habitat in which they forage would not in itself be a barrier to interbreeding (though seeking mates in different habitats might well be a barrier to interbreeding). All the other choices describe reproductive barriers that would act prior to fertilization.
6. **c.** Birthrates per se do not seem to affect the rate of speciation in organisms. All other factors have been shown to increase speciation rates in the lineages of some organisms.
7. **b.** Actually, biogeographers have found that larger land masses tend to have more species than smaller land masses, so you might expect the reverse effect.
8. **e.** Island groups are frequently sites of evolutionary radiations through repeated allopatric speciation events that are initiated by individuals (or groups) dispersing from one island to another. The numerous species of *Drosophila* in the Hawaiian Islands are believed to have originated in this way.
9. **d.** New species formed by polyploidy can arise in only two generations.
10. **e.** Interbreeding, production of viable hybrids, and establishment of a hybrid zone do not necessarily mean that speciation is incomplete. If, however, the hybrids were successful, fertile, and bred freely with members of both original populations, their gene pools would merge, and you would conclude that speciation had not taken place.
11. **b.** Genetic drift, founder events, and behavioral isolation may all play a role in allopatric speciation. Allopolyploidy, however, can occur only as the result of hybridization between individuals of different species; hence the two parent species cannot be allopatric.
12. **c.** If haploid gametes of species A and B joined, the result would be a zygote with 17 chromosomes. A mature plant with 17 chromosomes would be sterile, because chromosomes would be unable to pair properly during prophase and metaphase of meiosis I. A fertile allopolyploid would therefore have to have 34 chromosomes so that each chromosome would have a homolog with which to pair. Autopolyploids of species A and B would have 32 and 36 chromosomes, respectively.
13. **a.** If a polyploid plant or animal is capable of self-fertilization, then a new species can arise from a single individual. The ability to self-fertilize is far more common among plants than animals.
14. **b.** Reinforcement is defined as the evolutionary strengthening of prezygotic barriers to interbreeding within the zone of sympatry of two closely related species.
15. **d.** Speciation occurs more readily in archipelagos than on isolated islands because the establishment of geographical isolation through founder events involving dispersing individuals occurs much more readily within archipelagos than on a single island. Speciation would be more rapid in a family with promiscuous mating systems because species with this system are frequently characterized by a high degree of sexual dimorphism and the capacity of individuals to make subtle discriminations between members of their own species and between members of their own species and other species. As a consequence, even slight differences in the appearance or behavior of members of different populations may lead to the evolution of reproductive barriers based on the mating preferences of individuals in the different populations.

Evolution of Genes and Genomes

The Big Picture

- The growing importance of macromolecules in the reconstruction of phylogenies is an example of the far-reaching impact that advances in molecular biology have had in the decades since the discovery of the structure of DNA by Watson and Crick in 1953.
- The principles of molecular evolution are critical for our understanding of how new diseases—especially those caused by viruses—originate and evolve. This understanding in turn may lead to more effective strategies for combating these diseases.

Study Strategies

- Figure 24.1 shows how a similarity matrix is constructed. Similarity matrices are essential to reconstructing phylogenies based on molecular data.
- The concept of concerted evolution and the two mechanisms by which it can occur may seem confusing. Study Figure 24.12 to understand the difference between unequal crossing over and biased gene conversion.
- The difference between orthologs and paralogs may also be confusing. Figure 24.13 may help you understand the distinction.
- Go to the Web addresses shown to review the following animated tutorial and activities. You can also find them in BioPortal (yourBioPortal.com), along with many additional learning resources.

  Animated Tutorial 24.1 Concerted Evolution (Life10e.com/at24.1)

  Activity 24.1 Amino Acid Sequence Alignment (Life10e.com/ac24.1)

  Activity 24.2 Similarity Matrix Construction (Life10e.com/ac24.2)

  Activity 24.3 Gene Tree Construction (Life10e.com/ac24.3)

Key Concept Review

24.1 How Are Genomes Used to Study Evolution?

Evolution of genomes results in biological diversity

Genes and proteins are compared through sequence alignment

Models of sequence evolution are used to calculate evolutionary divergence

Experimental studies examine molecular evolution directly

The genome of eukaryotes includes both the nuclear genes located on chromosomes and genes present in mitochondria and chloroplasts. Scientists study genomes and particular nucleic acids and proteins to determine how rapidly and why they have changed. The answers to these questions are crucial to an understanding of the evolutionary history of genes and of the organisms that carry them. The evolution of nucleic acids and proteins depends on variation introduced by mutation. Nucleotide substitution mutations may result in amino acid replacements on the encoded proteins. Evolutionary changes in genes and proteins are detected by comparing the nucleotide and amino acid sequences among different organisms.

With the use of the sequence alignment technique, biologists identify homologous sequences within nucleic acids or proteins. The concept of homology (similarity that results from common ancestry) extends down to particular positions in nucleotide or amino acid sequences. Having identified homologous regions of a nucleic acid or protein, biologists construct a similarity matrix to measure the minimum number of changes that have occurred during the divergence between pairs of organisms. The assumption is that the longer the molecules have been evolving separately, the more differences they will have. Differences in homologous DNA sequences in two species underestimate the number of substitutions that actually have occurred since the sequences diverged from a common ancestor. Molecular evolutionists use mathematical models to correct for this undercounting. Because viruses, bacteria, and unicellular eukaryotes have short generation times, biologists use them to study molecular evolution in the laboratory.

Question 1. The table below shows the amino acid sequences for an eight-residue section of a small protein for five different species (1–5).

Position	1	2	3	4	5	6	7	8
Species 1:	Arg	Cys	Leu	Leu	Ser	Thr	Asn	Met
Species 2:	Arg	Cys	Phe	Leu	Leu	Ser	Thr	Asn
Species 3:	Arg	His	Leu	Leu	Ser	Thr	Asn	Met
Species 4:	Arg	Cys	Leu	Ser	Ser	Thr	Asn	Met
Species 5:	Arg	His	Leu	Leu	Ser	Gln	Asn	Met

Complete the following similarity matrix using these sequences.

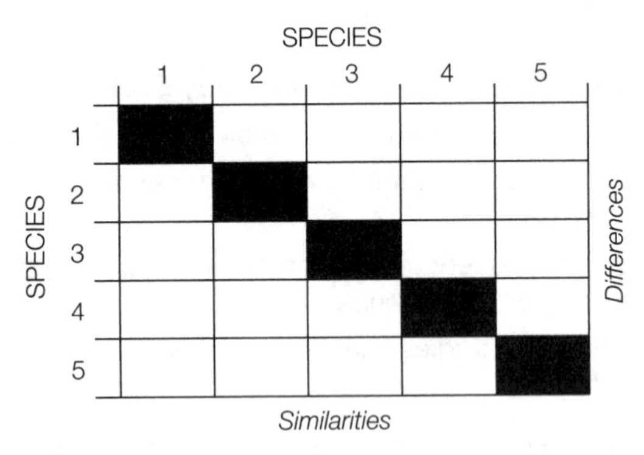

Which species differs from Species 1 because of an amino acid insertion?

Which species differs from Species 1 because of amino acid substitutions?

Question 2. What does sequence alignment tell us about evolutionary change? What are the drawbacks of using this technique?

24.2 What Do Genomes Reveal about Evolutionary Processes?

Much of evolution is neutral

Positive and purifying selection can be detected in the genome

Genome size also evolves

A synonymous (silent) mutation replaces a nucleotide base in a codon but does not change the amino acid specified by the codon. Synonymous mutations do not affect the functioning of a protein and are unlikely to be affected by natural selection. A nonsynonymous mutation changes the amino acid specified by the codon. Though such mutations are likely to be harmful, they are sometimes selectively neutral and are occasionally advantageous. Within functional genes, nucleotide substitution rates are highest at nucleotide positions that do not change the amino acid being expressed. The rate of substitution is higher in pseudogenes—duplicate copies of genes that are never expressed—than in functional genes. The neutral theory of molecular evolution postulates that most evolutionary change in macromolecules and much of the genetic variation within species is the result of random genetic drift rather than natural selection. According to the neutral theory of molecular evolution, it is possible to distinguish among evolutionary processes by comparing the rates of synonymous and nonsynonymous substitutions in a protein-coding gene. If an amino acid substitution is neutral in its effect on fitness, then the rates of synonymous and nonsynonymous substitutions in the corresponding DNA sequences are expected to be very similar. If an amino acid position is under strong selection for change, then the rate of nonsynonymous substitutions in the corresponding DNA is expected to exceed the rate of synonymous substitutions. If an amino acid position is under purifying selection, then the rate of synonymous substitutions in the corresponding DNA is expected to be much higher than the rate of nonsynonymous substitutions.

The enzyme lysozyme, while serving in almost all animals as an important first line of defense against invading bacteria, also has evolved to take on an essential role in the digestive process of several groups of foregut fermenters. By comparing the lysozyme-coding sequences in foregut fermenters with several of their nonfermenting relatives, molecular evolutionists have discovered that neutral evolution, purifying selection, and selection for change have all occurred as lysozyme evolved to take on its new function. The independent evolution in several groups of foregut fermenters of a type of lysozyme adapted to its new environment and function shows that convergent evolution occurs at the molecular level. Genome size and organization also evolve. The size of the coding portion of the genome is larger in more complex organisms. Thus, eukaryotes have many times more genes than prokaryotes, and multicellular eukaryotes with tissue organization have more genes than single-celled eukaryotes. Most of the variation in genome size of various organisms is due not to differences in the number of functional genes, but in the amount of noncoding DNA. Although much of the noncoding DNA appears to be nonfunctional, it may alter the expression of surrounding genes. Important categories of noncoding DNA include pseudogenes and parasitic transposable elements. Studies of the rate of retrotransposon loss show that species differ greatly in the rate at which they gain or lose apparently functionless DNA. The reason for the differences between species is unclear. It may be related to the rate at which the organism develops or to its population size.

Question 3. How might the proportion of coding to noncoding DNA in the genome of a species be related to the relative importance of natural selection and genetic drift in the evolution of the species?

Question 4. Why is much of the genetic variation observed in populations the result of neutral evolution?

24.3 How Do Genomes Gain and Maintain Functions?

Lateral gene transfer can result in the gain of new functions

Most new functions arise following gene duplication

Some gene families evolve through concerted evolution

Lateral gene transfer occurs when a species picks up fragments of foreign DNA directly from the environment, either from a virus or through hybridization with another species. This process increases the genetic variability of the species and thus provides additional raw material on which natural selection can act. It can also result in the spread of genetic functions between distantly related species. The endosymbiotic events that gave rise to mitochondria and chloroplasts in the eukaryotic lineage can be viewed as lateral transfers of entire bacterial genomes. Lateral transfer appears to be much more common among species of bacteria than among most eukaryotic lineages.

When a gene is duplicated, four evolutionary outcomes are possible: both copies can retain the gene's original function; both copies can retain the ability to produce the original gene product but gene expression diverges in different tissues or at different times of development; one copy can become a functionless pseudogene; or one copy can retain its original function while the second mutates so that it performs a different function. When an entire genome is duplicated, there are opportunities for new gene functions to evolve. Genome duplication events that occurred in the ancestor of the jawed vertebrates have permitted many individual vertebrate genes to become highly tissue-specific in their expression.

Successive rounds of gene duplication and mutation can result in a gene family (e.g., the globin gene family)—a group of homologous genes with related functions. Concerted evolution results in similar DNA sequence changes in all copies of highly repeated genes, such as those that code for ribosomal RNA. Concerted evolution can occur if homologous chromosomes align imprecisely during meiosis. Unequal crossing over may occur, resulting in one chromosome with extra copies of a highly repeated gene and the other chromosome with fewer copies. If the favored copy of a repeated gene on one homolog is used as the template for the repair of damage to the copies of the gene on the other homolog, the result is biased gene conversion, the rapid spread of the favored sequence across all the copies of the gene.

Question 5. How is one of the mechanisms of concerted evolution related to the pairing of homologous chromosomes that occurs during meiosis?

Question 6. Why can lateral gene transfer be advantageous to a species?

24.4 What Are Some Applications of Molecular Evolution?

Molecular sequence data are used to determine the evolutionary history of genes

Gene evolution is used to study protein function

In vitro evolution is used to produce new molecules

Molecular evolution is used to study and combat diseases

A gene tree depicts the evolutionary history of a particular gene or of the members of a gene family. Orthologs are genes found in different organisms that arose from a single gene in their common ancestor. Paralogs are related genes that have resulted from gene duplication in a single lineage.

The principles of molecular evolution help us understand function and diversification of function in many proteins. For example, detection of strong selection for change in a nucleotide sequence can help us identify molecular changes that have resulted in functional changes. Molecular evolutionary principles underlie the field of in vitro evolution, in which new molecules are produced in the laboratory to perform particular desired functions. The basis of in vitro evolution is the creation of random molecular variation followed by selection by the experimenter. Biomedical scientists are using principles of molecular evolution to identify and combat human diseases.

Question 7. Construct a concept map whose theme is "Genome." Include in your map the following terms: genome, duplication, endosymbiosis, eukaryotes, genes, gene families, hybridization, lateral gene transfer, noncoding DNA, organelle DNA, paralogs, prokaryotes, pseudogenes, and size. Connect these terms by verbs or short phrases to indicate the relationships among them.

Question 8. Compare in vitro evolution with molecular evolution in organisms.

Question 9. How is the study of molecular evolution important in efforts to combat HIV and other viral pathogens that have emerged recently?

Test Yourself

1. The genome of a eukaryotic organism is best defined as
   a. all of the organism's protein-coding genes.
   b. all of the DNA contained in the organism's nucleus.
   c. all of the organism's genetic material.
   d. a haploid set of the organism's chromosomes.
   e. all of the organism's DNA that is transcribed.
   Textbook Reference: *24.1 How Are Genomes Used to Study Evolution?*

2. The sequence alignment technique
   a. permits comparison of sequences of amino acids in proteins or sequences of nucleotides in DNA.
   b. enables the detection of deletions and insertions in sequences that are being compared.
   c. enables the detection of back substitutions and parallel substitutions in sequences that are being compared.
   d. Both a and b
   e. All of the above
   Textbook Reference: *24.1 How Are Genomes Used to Study Evolution?*

3. Experimental molecular evolutionary studies have shown that
   a. a heterogeneous environment favors adaptive radiation.
   b. a heterogeneous environment induces an increase in the mutation rate.
   c. even in bacteria, substantial molecular evolution cannot be observed in less than a year.
   d. Both a and b
   e. None of the above

   Textbook Reference: *24.1 How Are Genomes Used to Study Evolution?*

4. In a eukaryote, one would expect to find the lowest rate of nonsynonymous nucleotide substitutions in an
   a. intron of a protein-coding gene.
   b. exon of a protein-coding gene.
   c. intron of a pseudogene.
   d. exon of a pseudogene.
   e. intron or exon of a protein-coding gene.

   Textbook Reference: *24.2 What Do Genomes Reveal about Evolutionary Processes?*

5. Which of the following statements about mutations is *false*?
   a. A silent mutation results in no change in the amino acid sequence of a protein.
   b. According to the neutral theory of molecular evolution, most substitution mutations are selectively neutral and accumulate through genetic drift.
   c. A base substitution mutation in the third codon position is more likely to be neutral than a substitution at the first or second codon position.
   d. Nonsynonymous mutations are virtually always deleterious to the organism.
   e. The rate of fixation of neutral mutations is equal to the mutation rate and is independent of population size.

   Textbook Reference: *24.2 What Do Genomes Reveal about Evolutionary Processes?*

6. Studies of the structure of proteins such as cytochrome *c* and lysozyme in different species have shown that
   a. the rate of evolution of particular proteins is often relatively constant over time.
   b. many nucleotide substitutions result either in no change in the amino acid sequence of the protein or in a change to a functionally equivalent amino acid.
   c. there are fewer differences in the amino acid sequences of proteins whose organismal sources were closely related than in those whose sources were distantly related.
   d. functionally important regions of a protein can be discovered by identifying the regions with the most amino acid substitutions.
   e. All of the above

   Textbook Reference: *24.1 How Are Genomes Used to Study Evolution?; 24.2 What Do Genomes Reveal about Evolutionary Processes?*

7. Which of the following statements about the enzyme lysozyme is true?
   a. A small group of closely related mammals has evolved a special form of lysozyme that functions in digestion.
   b. The lysozymes found in the foregut fermenters resulted from convergent evolution.
   c. Lysozyme could not have evolved a secondary function if it had been an enzyme with a vital primary function.
   d. A high mutation rate in foregut fermenters allows their lysozymes to evolve rapidly.
   e. Lysozymes first evolved as a defense against bacteria in the common ancestor of mammals.

   Textbook Reference: *24.2 What Do Genomes Reveal about Evolutionary Processes?*

8. Which of the following ranks the organisms correctly in terms of the expected total amount of coding DNA in their genomes (from least coding DNA to most coding DNA)?
   a. Bacterium, single-celled eukaryote, *Drosophila*, bird
   b. Bacterium, *Drosophila*, bird, single-celled eukaryote
   c. Single-celled eukaryote, bacterium, *Drosophila*, bird
   d. *Drosophila*, single-celled eukaryote, bird, bacterium
   e. *Drosophila*, bacterium, single-celled eukaryote, bird

   Textbook Reference: *24.2 What Do Genomes Reveal about Evolutionary Processes?*

9. Which of the following ranks the organisms correctly in terms of the proportion of coding DNA to noncoding DNA in their genomes (from smallest proportion to largest proportion)?
   a. *E. coli*, yeast, *Drosophila*, human
   b. Human, yeast, *E. coli*, *Drosophila*
   c. Human, *Drosophila*, yeast, *E. coli*
   d. *Drosophila*, human, yeast, *E. coli*
   e. Yeast, *E. coli*, *Drosophila*, human

   Textbook Reference: *24.2 What Do Genomes Reveal about Evolutionary Processes?*

10. Which of the following would be the *least* likely result of gene duplication?
    a. The gene produces less of its product than it did before duplication.
    b. The two copies of the gene are expressed at different stages in the development of the organism.
    c. As a result of evolutionary divergence, one copy retains its original function and the other copy acquires a different function.
    d. One copy remains functional and the other copy evolves into a functionless pseudogene.
    e. All of the above are about equally likely results.

    Textbook Reference: *24.3 How Do Genomes Gain and Maintain Functions?*

11. Which of the following statements about gene families is *false*?
    a. Gene families evolve via gene duplication.

b. Pseudogenes are quickly removed from gene families by deletion.
c. Members of a gene family can include several functional genes.
d. Examples of gene families include the *engrailed* and globin gene families in vertebrates.
e. In some gene families, the members do not evolve independently of one another.

Textbook Reference: *24.3 How Do Genomes Gain and Maintain Functions?*

12. Gene duplication via the mechanism of polyploidy
    a. results in the duplication of the entire genome, apart from extranuclear DNA.
    b. has occurred in the evolutionary history of many plants.
    c. is believed not to have occurred in the evolutionary history of any animal groups.
    d. Both a and b
    e. All of the above

    Textbook Reference: *24.3 How Do Genomes Gain and Maintain Functions?*

13. Nonindependent evolution of some repeated genes within a species
    a. is called concerted evolution.
    b. can be caused by biased gene conversion.
    c. can be caused by unequal crossing over.
    d. Both a and c
    e. All of the above

    Textbook Reference: *24.3 How Do Genomes Gain and Maintain Functions?*

14. Orthologous genes are genes that can be traced back to a common _______ event.
    a. duplication
    b. substitution
    c. speciation
    d. deletion
    e. duplication and speciation

    Textbook Reference: *24.4 What Are Some Applications of Molecular Evolution?*

15. In vitro evolution
    a. can produce both nucleic acid and protein molecules unknown in living organisms.
    b. requires many rounds of production of many variant molecules and the selection of those having (or beginning to have) the desired properties.
    c. often involves techniques and molecules employed in recombinant DNA technology, such as PCR and cDNA.
    d. Both b and c
    e. All of the above

    Textbook Reference: *24.4 What Are Some Applications of Molecular Evolution?*

Answers

Key Concept Review

1.

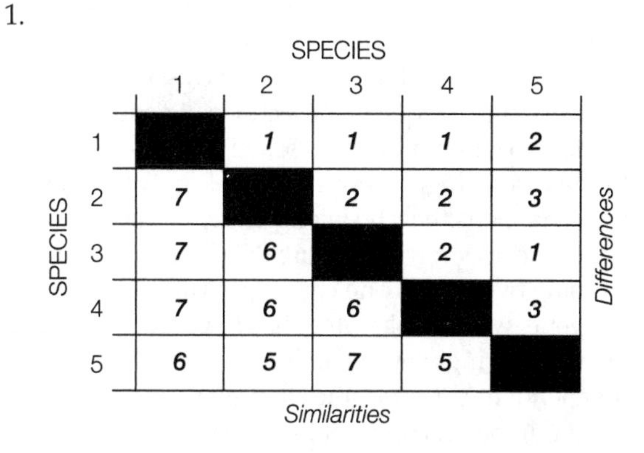

Species 2 shows an amino acid insertion (Phe) at position 3. There are no deletions.

Species 3, 4, and 5 all show amino acid substitutions.

2. Sequence alignment indicates the minimum number of nucleotide changes that must have occurred since two sequences have diverged from a common ancestral sequence. The technique is useful in determining the minimum number of changes between two DNA sequences but underestimates the actual number of changes since divergence. Changes such as multiple substitutions, coincident substitutions, parallel substitutions, or back substitutions (reversions) would not be revealed by sequence alignment

3. According to one hypothesis, the proportion of coding to noncoding DNA in the genome of a species is related to the sizes of the populations typical of the species. As discussed in Chapter 21, genetic drift has more influence on the evolution of small populations than on the evolution of large ones. If individuals of a species differ in their fitness only slightly as a function of the proportion of noncoding DNA in their genomes, and if the populations of the species are typically small, then genetic drift could lead to retention of the noncoding DNA (or even an increase in its proportion), despite its selective disadvantage. However, the effect of genetic drift is minimal in very large populations. Therefore, in species characterized by large populations, even a slight disadvantage in fitness for individuals with a greater proportion of noncoding DNA should lead to a decrease in its proportion through the action of natural selection.

4. At the molecular level, the majority of variation observed in most populations is selectively neutral. The rate of fixation of neutral mutations depends only on the neutral mutation rate and is independent of population size. Mutations that appear and are not deleterious will quickly become fixed in the population. The rate of fixation of neutral mutations is equal to the mutation rate, so as long as the underlying mutation rate

is constant, molecules evolving in separate populations should diverge in neutral changes at a constant rate.

5. As discussed in Chapter 11, during the synapsis of homologous pairs of chromosomes in meiosis I, the alignment of the homologs is normally extremely precise. This ensures that the result of crossing over will be the exchange of equivalent sections of genetic material in the two chromatids involved in the crossover event. In the case of highly repeated genes, however, it sometimes happens that the alignment of the homologs is imprecise, as shown in Figure 24.12. The result is that crossing over yields one chromatid with extra copies of the gene, whereas the other has fewer copies. Subsequently, during meiosis II, the chromatids will become independent chromosomes that now possess the altered number of copies of the repeated gene.

6. Lateral gene transfer, which allows individual genes, organelles, or fragments of genomes to move horizontally from one lineage to another, is a way in which a species can increase its genetic variation. Natural selection acts on this genetic variation, so if the transfer incurs a benefit to the species, it will persist in the population.

7.

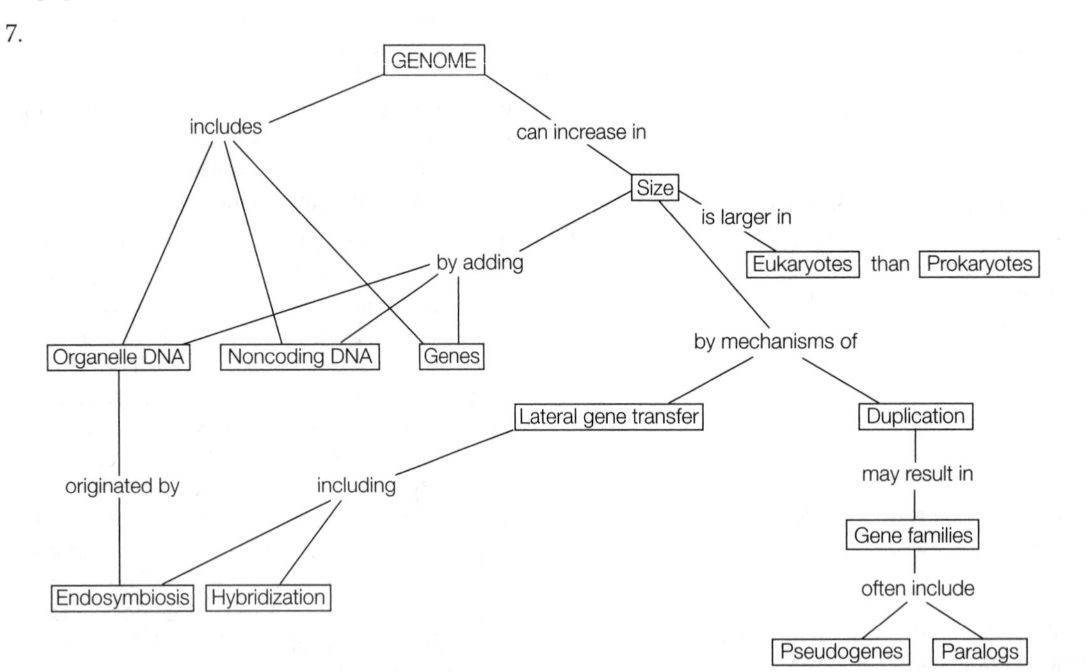

8. Variation and selection are involved in both in vitro and molecular evolution in organisms. With in vitro evolution, a huge number of variant molecules are created in the laboratory, and the experimenters select those that show any sign of having the desired property. Many repetitions of these two steps—production of variants and selection—eventually produce the targeted molecule. In molecular evolution, naturally occurring mutations in previously existing DNA sequences are the ultimate source of variation, and the environment determines through natural selection which mutations are the favorable variants. One must also remember, however, that according to the neutral theory of molecular evolution, much evolutionary change in DNA (and in the encoded proteins) is not adaptive but rather the result of genetic drift.

9. The principles of molecular evolution are critical for understanding the origin and evolutionary development of such pathogenic viruses as hantaviruses, the SARS virus, and HIV. By studying the evolutionary changes that occur in such viruses, medical researchers gain valuable insights into such questions as how some

viruses are able to switch from an animal to a human host and how vaccines can remain effective as viruses evolve. In the future, as more extensive genomic databases and evolutionary trees for viruses are developed, it will be possible to identify and treat a much wider array of human diseases.

Test Yourself

1. **c.** The genome includes not only protein-coding genes but all other DNA in the organism, including noncoding DNA and genes in organelles.
2. **d.** The sequence alignment technique can be employed to compare both amino acid sequences and nucleotide sequences. Although it enables the locations of deletions and insertions to be pinpointed, it is not capable of detecting such phenomena as back substitutions and parallel substitutions.
3. **a.** In the experiments of Rainey and Travisano, heterogeneous environments favored diversification in their bacterial cultures because of natural selection, not because of a difference in either the rates or kinds of mutations that occurred in homogeneous and heterogeneous environments. Experiments of this kind can produce observable evolutionary changes in a matter of months, at most.
4. **b.** As discussed in Chapter 14, eukaryotic genes usually contain both protein-coding regions (exons) and noncoding regions (introns). A substantial proportion of nonsynonymous substitutions occurring in an exon of a gene would most likely be deleterious and therefore eliminated by selection. Because introns and pseudogenes are not expressed, nonsynonymous substitutions occurring in them cannot be selected against.
5. **d.** A nonsynonymous substitution mutation can be selectively neutral if, as sometimes happens, it results in an amino acid change that has no significant effect on the shape (and hence the functional properties) of the protein.
6. **d.** Functionally important regions of a protein can be discovered by identifying the regions with the fewest amino acid substitutions. Because the sequences in these regions have been optimized by natural selection to accomplish the function of the molecule, random changes in these areas are not tolerated.
7. **b.** Because the animals that evolved a similar lysozyme do not share a recent common ancestor, the mechanism involved is convergent evolution. All other statements are false.
8. **a.** There is a rough relationship between the amount of coding DNA and organismal complexity.
9. **c.** As shown in Figure 24.9, multicellular eukaryotes with relatively small populations and slow rates of development have the lowest proportion of coding to noncoding DNA, whereas quickly reproducing prokaryotes such as *E. coli* have the highest proportion. It is unclear whether developmental rate or population size (or both) is chiefly responsible for this trend.
10. **a.** If there are two copies of the gene, it is more likely to produce more of its product than less of it.
11. **b.** Pseudogenes (nonfunctional DNA) may be removed via deletion, but numerous pseudogenes persist in many gene families.
12. **d.** Polyploidy is a common event in plant evolution (as discussed in Chapter 23), and it does result in the duplication of the entire nuclear genome. But it is also believed to have occurred at least twice in the lineage that gave rise to the jawed vertebrates.
13. **e.** Concerted evolution is the phenomenon in which all the copies of a highly repeated gene (such as a gene coding for ribosomal RNA) remain very similar despite experiencing nucleotide substitutions or other mutations that generally cause genes to diverge. Unequal crossing over during meiosis and biased gene conversion during DNA repair are two mechanisms that can produce concerted evolution.
14. **c.** Orthologs are homologous genes whose divergence can be traced to the speciation events that gave rise to the species in which the genes occur.
15. **e.** In vitro evolution can create novel molecules not known to occur in living organisms. The process starts with a very large pool of variant molecules (nucleic acid or protein) and involves selection on the part of the experimenters of those molecules that show some evidence of the desired property. Many rounds of production of variant molecules and selection are typically needed to produce the final product.

The History of Life on Earth

The Big Picture

- The evidence provided by nineteenth-century geologists that Earth is very ancient was crucial to Darwin's theory of evolution by natural selection.
- Radiometric dating is an excellent example of how a discovery in one field of science (nuclear physics, in this instance) can lead to advances in other fields (geology and paleontology). Prior to the development of this technique, no absolute dates could be assigned accurately to any of the important geological and evolutionary events described in this chapter.
- The concept of continental drift, first widely accepted in the late 1960s, has revolutionized our understanding of geological processes and of the history of life on Earth.
- The accumulating evidence of the sudden and catastrophic effect of the impact of large meteorites and of other rapid changes (e.g., widespread volcanic eruptions) Earth's environment has given paleontologists a new perspective on the causes of mass extinctions and on the dynamics of the evolutionary process in general.

Study Strategies

- This chapter contains many names of geological eras and periods and many dates. Instructors vary with respect to the importance they place on memorization of this information. In general, it is best to focus on broad evolutionary patterns and trends. Most instructors will also place more importance on the phylogenetic sequences (e.g., amphibians gave rise to amniotes, which gave rise to reptiles and to mammals) than on the dates at which these events occurred.
- The concept of the half-life of radioactive isotopes may seem confusing. Bear in mind that if half of a radioisotope decays in a given time period (its half-life), then at the end of the first half-life only one-half of the original quantity of the isotope is still present, one-half of which will decay in the second half-life. Thus after two half-lives, one-quarter of the original quantity of radioisotope is still present. The same line of reasoning applies to all further half-lives. Study Figure 25.1, which explains this point.
- Making a table, chart, or timeline to summarize and organize the events that occurred in each geological era and period is an excellent study strategy.
- Go to the Web addresses shown to review the following animated tutorial and activity. You can also find them in BioPortal (yourBioPortal.com), along with many additional learning resources.

  Animated Tutorial 25.1 Movement of the Continents (Life10e.com/at25.1)

  Activity 25.1 Concept Matching (Life10e.com/ac25.1)

Key Concept Review

25.1 How Do Scientists Date Ancient Events?

Radioisotopes provide a way to date fossils and rocks

Radiometric dating methods have been expanded and refined

Scientists have used several methods to construct a geological time scale

The development of the science of biology, particularly Darwin's theory of evolution by natural selection, depended on evidence supplied by geologists that Earth is very old. Short-term evolutionary changes occur rapidly enough to be studied directly. Long-term evolutionary changes involve the appearance of new species and evolutionary lineages. The fossil record provides evidence of such changes. Sedimentary rocks (formed by the accumulation of grains at the bottom of bodies of water) are deposited in strata (layers). The oldest layers are found at the bottom, and successively higher strata are progressively younger. Fossils, which are the preserved remains of ancient organisms, are useful for establishing the relative ages of the sedimentary rocks in which they occur.

The regular pattern of decay of radioactive isotopes provides a means of estimating the absolute ages of fossils and rocks. The half-life of a radioisotope is the time period during which half of the remaining radioactive material decays to become a different, stable isotope. The ratio of radioactive carbon-14 to its stable isotope, carbon-14, is used to date fossils less than 50,000 years old. The decay of potassium-40 to argon-40 is widely used to date ancient evolutionary events. Paleomagnetic dating is based on the fact that the age of sedimentary and igneous rocks can be determined because they preserve a record of Earth's magnetic field at the time they were formed. Earth's geological history is divided into eras, which are subdivided into periods. The boundaries between these divisions are marked by changes in the types of fossils found in successive layers of sedimentary rock. The first forms of life evolved early in the Precambrian era, which lasted for more than three billion years and was marked by enormous physical changes on Earth.

Question 1. ^{14}C decays to ^{12}C with a half-life of about 5,700 years. Suppose you find a fossil in which the amount of ^{14}C is only 1⁄16 of what one would find in a living organism with the same carbon content. Approximately how old is the fossil?

Question 2. Make a chart summarizing the major geological and evolutionary events of each geological period.

Question 3. In 2006, paleontologists announced that fossils of *Tiktaalik roseae*, an important link between fish and tetrapods, had been discovered in freshwater sediments in the Canadian Arctic that were 375 million years old. How were geologists able to determine the age of these sediments?

25.2 How Have Earth's Continents and Climates Changed over Time?

- The continents have not always been where they are today
- Earth's climate has shifted between hot and cold conditions
- Volcanoes have occasionally changed the history of life
- Extraterrestrial events have triggered changes on Earth
- Oxygen concentrations in Earth's atmosphere have changed over time

The theory of plate tectonics proposes that Earth's crust consists of solid lithospheric plates floating on a fluid layer of magma. Convection currents in the magma result from heat emanating from Earth's core. The currents cause continental drift, which is the gradual shift in the position of the plates and the continents they contain. The drifting of continents has had profound effects on climate, sea level, oceanic circulation, and the distributions of organisms. No free oxygen was present in the atmosphere until certain bacteria evolved the ability to use water as a source of hydrogen ions for photosynthesis. The gradual increase in oxygen concentration in the atmosphere resulted in the dominance of organisms using aerobic metabolism and in the evolution of larger eukaryotic cells and of multicellular organisms. The exceptionally high oxygen concentrations that occurred during the Carboniferous and Permian periods are associated with the evolution of giant amphibians and flying insects.

The climate of Earth has alternated between hot/humid and cold/dry conditions. Though most major climatic changes have been gradual, some have occurred within periods of 5,000 to 10,000 years or less. The rapid climate change occurring today is thought to be caused by a buildup of atmospheric CO_2, primarily from the burning of fossil fuels. The climatic shifts caused by massive volcanic eruptions associated with continental drift are implicated in several mass extinctions. The late Permian mass extinction, which occurred at the end of the Paleozoic era, was caused by volcanism triggered by the collision of continents that formed the supercontinent Pangaea. Several mass extinctions have probably been caused by collisions of Earth with meteorites or comets, such as the event 65 million years ago that is thought to have resulted in the extinction of dinosaurs.

Question 4. In what way was the evolution of eukaryotic cells linked to the increase in the oxygen concentration in the atmosphere that occurred during the Precambrian?

Question 5. Scientists believe that if there are no controls on the emission of CO_2 from the burning of fossil fuels, the concentration of this gas could double by the end of the current century, leading to a significant rise in the average temperature of Earth. What would be some of the likely evolutionary effects of this climatic change?

25.3 What Are the Major Events in Life's History?

- Several processes contribute to the paucity of fossils
- Precambrian life was small and aquatic
- Life expanded rapidly during the Cambrian period
- Many groups of organisms that arose during the Cambrian later diversified
- Geographic differentiation increased during the Mesozoic era
- Modern biotas evolved during the Cenozoic era
- The tree of life is used to reconstruct evolutionary events

All of the organisms found in a particular place or time constitute its biota. Biotas consist of flora (plants) and fauna (animals). The earliest life on Earth appeared about 3.8 billion years ago, but the fossil record of organisms that lived in the Precambrian is fragmentary. The first organisms were unicellular prokaryotes. The first eukaryotes evolved about 1.5 billion years ago. Because an oxygen-rich environment favors rapid decomposition, organisms that become fossils are likely either to have lived in a poorly oxygenated environment or to have been transported to such a site soon after death. The 300,000 known fossil species represent only a tiny fraction of the species that have ever lived. Some groups, such as hard-shelled marine animals, are much better represented in the fossil record than others. The fossil record shows that an organism of any specific type can predictably be found in rocks of a particular age. Living organisms resemble recent fossils more closely than they resemble fossils from more ancient periods.

For most of the Precambrian, which lasted for over 3 billion years, life consisted of microscopic prokaryotes. The best

Precambrian fossil deposits, dating from about 600 million years ago, contain diverse soft-bodied invertebrates, some of which may represent lineages with no living descendents.

During the Cambrian period, the oxygen concentration in the atmosphere approached its current level, and several large continents formed. The rapid increase in the diversity of multicellular life forms that occurred at this time is known as the "Cambrian explosion."

The Ordovician period was marked by a proliferation of marine filter feeders living on the sea floor. At the end of this period, massive glaciers formed over the southern continents, the sea level and ocean temperatures dropped, and the majority of animal species became extinct.

The Silurian period witnessed the evolution of swimming marine animals and the diversification of jawless fish. It also marked the appearance of the first terrestrial arthropods and vascular plants.

During the Devonian period, all major groups of fishes evolved. On land, the first insects and amphibians evolved, and forests of club mosses, horsetails, and tree ferns appeared. A mass extinction of approximately three-quarters of all marine species occurred at the end of this period, possibly caused by the collision of two large meteorites with Earth.

In the Carboniferous period, swamp forests consisting largely of giant tree ferns and horsetails were widespread. The fossilized remains of these plants formed coal. The first winged insects evolved during this period, while amphibians became better adapted to life on land and gave rise to the lineage leading to the amniotes (whose eggs can be laid in dry places).

The Permian period was marked by the formation of a single supercontinent called Pangaea. As the climate cooled drastically, amniotes split into two lineages, the reptiles and one leading to the mammals. The occurrence of the most extensive of all mass extinctions brought the Permian period to a close.

During the Triassic period, seed ferns and conifers were the dominant forms of terrestrial vegetation, and reptiles were the dominant vertebrates. A mass extinction occurred at the end of the Triassic.

The Jurassic period was marked by the complete separation of Pangaea to form Laurasia in the north and Gondwana in the south. It also witnessed the radiation of the dinosaurs, the evolution of flying reptiles, and the first appearance of mammals. The earliest fossils of flowering plants date from late in this period. In the sea, ray-finned fishes also began a great radiation.

By the early Cretaceous period, Laurasia and Gondwana had begun to break apart into the present-day continents. The flowering plants began the radiation that led to their current dominance. The dinosaurs continued as the dominant land vertebrates, despite the presence of many groups of mammals. At the end of this period, the collision of a large meteorite with Earth caused the extinction of the dinosaurs and of many other animal and plant lineages.

During the Tertiary period, the continents drifted toward their present positions. As the climate became cooler and drier, extensive grasslands appeared. Many groups of land vertebrates—especially frogs, snakes, lizards, birds, and mammals—radiated extensively.

The current geological period, the Quaternary, is divided into the Pleistocene and Holocene (Recent) epochs. Modern humans evolved during the Pleistocene, which was a time of severe climatic fluctuations, including four major periods of extensive glaciation. As humans spread geographically, many species of large birds and mammals became extinct. Phylogenetic trees help reconstruct the timing of evolutionary events and clarify relationships among modern species.

Question 6. Below is a chronologically scrambled list of important events in the history of life on Earth. Draw four lines on a sheet of paper to represent the timelines of events occurring in the Precambrian, the Paleozoic era, the Mesozoic era, and the Cenozoic era. Place the listed events on the correct line and in proper sequence.

Conifers become dominant

Cambrian "explosion"

Evolution of *Homo*

Fifth mass extinction

First eukaryotes

First flowering plant fossils

First forests ("fern" forests); first jawed fish

First fossils of multicellular animals

First mammals, dinosaurs diversify

First mass extinction

First photosynthetic eukaryotes

First vascular plants and terrestrial arthropods

Fourth mass extinction

Grasslands spread

Origin of amniotes

Origin of life

Origin of photosynthesis

Second mass extinction

Third mass extinction

Rapid radiation of mammals

Question 7. How did the mass extinctions of the Ordovician, Devonian, and Permian periods specifically impact life on Earth?

Test Yourself

1. The half-life of an isotope is the
   a. time it takes for a fixed fraction of isotope material to change from one form to another.
   b. age over which the isotope is useful for dating rocks.
   c. ratio of one isotope species to another in a sample of organic matter.
   d. Both a and b

e. None of the above
Textbook Reference: *25.1 How Do Scientists Date Ancient Events?*

2. Mountain ranges are ultimately the result of
   a. plates in Earth's crust that move against one another on top of a fluid layer of molten rock.
   b. climate changes and the movement of glacial ice sheets.
   c. leftover debris from ancient collisions with an asteroid or meteor.
   d. the breakup of Laurasia and Gondwana.
   e. Both a and b
   Textbook Reference: *25.2 How Have Earth's Continents and Climates Changed over Time?*

3. Which of the following would likely be most conducive to the occurrence of fossilization?
   a. The surf zone along a sandy beach
   b. A shallow, cool swamp with good deposition rates of mud sediments
   c. The bottom of a hot, dry cave with no running water
   d. A fast-running mountain stream
   e. It is not possible to decide based on the information provided.
   Textbook Reference: *25.3 What Are the Major Events in Life's History?*

4. The fossil record is rather detailed for
   a. soft-bodied insects.
   b. cnidarians and sponges.
   c. most terrestrial animals.
   d. hard-shelled mollusks.
   e. Both c and d
   Textbook Reference: *25.3 What Are the Major Events in Life's History?*

5. During which geological period did Pangaea became fully divided and distinctive assemblages of plants and animals begin to arise on different continents?
   a. Cambrian
   b. Tertiary
   c. Devonian
   d. Permian
   e. Jurassic
   Textbook Reference: *25.3 What Are the Major Events in Life's History?*

6. Which of the following statements about the Ordovician period is *false*?
   a. Marine filter feeders flourished.
   b. The number of classes and orders increased.
   c. Modern mammals appeared.
   d. Many groups became extinct at the end of the period.
   e. The continents were located primarily in the Southern Hemisphere.
   Textbook Reference: *25.3 What Are the Major Events in Life's History?*

7. Most of human evolution has occurred during the
   a. Paleozoic era.
   b. Devonian period.
   c. Quaternary period.
   d. Carboniferous period.
   e. Cretaceous period.
   Textbook Reference: *25.3 What Are the Major Events in Life's History?*

8. For terrestrial animals and plants, the most recent mass extinction event that occurred prior to the evolution of humans took place approximately _______ million years ago.
   a. 10
   b. 65
   c. 200
   d. 250
   e. 400
   Textbook Reference: *25.3 What Are the Major Events in Life's History?*

9. One of the main factors that distinguishes the Cambrian explosion from all others is that
   a. evolutionarily, it was the most recent explosion.
   b. many new major groups of animals appeared at this time in contrast to other explosions.
   c. it was the time when the dinosaurs became extinct.
   d. there was a dramatic drop in species diversity, especially among marine organisms.
   e. it was a time of massive volcanic eruptions.
   Textbook Reference: *25.3 What Are the Major Events in Life's History?*

10. During which of the following geological times did the greatest number of new body plans appear?
    a. Carboniferous
    b. Triassic
    c. Jurassic
    d. Devonian
    e. Cambrian
    Textbook Reference: *25.3 What Are the Major Events in Life's History?*

11. The sudden disappearance of the dinosaurs some 65 mya may have been the result of
    a. a collision between Earth and a large meteorite.
    b. slow climate changes due to planetary cooling.
    c. competition from better-adapted organisms.
    d. the rise of birds and mammals.
    e. the formation of Pangaea.
    Textbook Reference: *25.3 What Are the Major Events in Life's History?*

12. Fossil insects many times larger than any insects alive today have been dated to the Carboniferous and Permian periods. These fossils are evidence that during these periods
    a. the climate was much warmer than it is today.

b. there were fewer predators on insects than there are today.
c. the carbon dioxide concentration of the atmosphere was lower than it is today.
d. insects had fewer competitors for food than modern insects.
e. the oxygen concentration of the atmosphere was significantly higher than it is today.
Textbook Reference: *25.2 How Have Earth's Continents and Climates Changed over Time?*

13. Which of the following statements about patterns or processes in the evolution of life is *false*?
a. ^{14}C can be used to date the age of dinosaur bones.
b. The supercontinent Pangaea formed during the Permian period.
c. Mass extinctions of marine organisms have coincided with periods of low sea levels.
d. The ends of five geological periods have been marked by mass extinctions.
e. All of the above are true; none is false.
Textbook Reference: *25.1 How Do Scientists Date Ancient Events?*

14. Which of the following pairs of organisms were *not* present on Earth in their living forms at the same time?
a. Tree ferns and ray-finned fish
b. Amphibians and birds
c. Gymnosperms and insects
d. Humans and dinosaurs
e. Jawless fish and vascular plants
Textbook Reference: *25.3 What Are the Major Events in Life's History?*

15. The most severe mass extinction event, linked to the formation of Pangaea and massive volcanic eruptions, occurred at the end of the _______ period.
a. Ordovician
b. Devonian
c. Permian
d. Triassic
e. Cretaceous
Textbook Reference: *25.3 What Are the Major Events in Life's History?*

Answers

Key Concept Review

1. If only 1⁄16 of the ^{14}C remains, then four ^{14}C half-lives have passed since the fossil was formed (½ × ½ × ½ × ½ = 1⁄16). Because the half-life of ^{14}C is roughly 5,700 years, the fossil must be about 22,800 years old (4 × 5,700 = 22,800).
2. See Table 25.1.
3. Recall that radioisotope dating is used to determine the absolute age of rocks. Igneous rocks, formed from volcanic ash or lava flows, are required for this purpose. Such rocks may have been found near the discovery site in Canada, or they may have been found at one or more other sites that have sediments of the same relative age. The relative age of the sediments can be determined by the types of fossils found in them.
4. Small prokaryotic cells can obtain enough oxygen by diffusion even when oxygen concentrations are very low. Since eukaryotic cells are larger, they have a lower surface area-to-volume ratio and hence need a higher concentration of oxygen in their environment for diffusion to meet their requirement for this gas.
5. The fossil record shows that major environmental changes occurring over a short time interval sometimes lead to large-scale rapid extinctions that appear "instantaneous" in the fossil record. The possible effects of global warming on biodiversity are discussed in more detail in Chapter 59.
6. 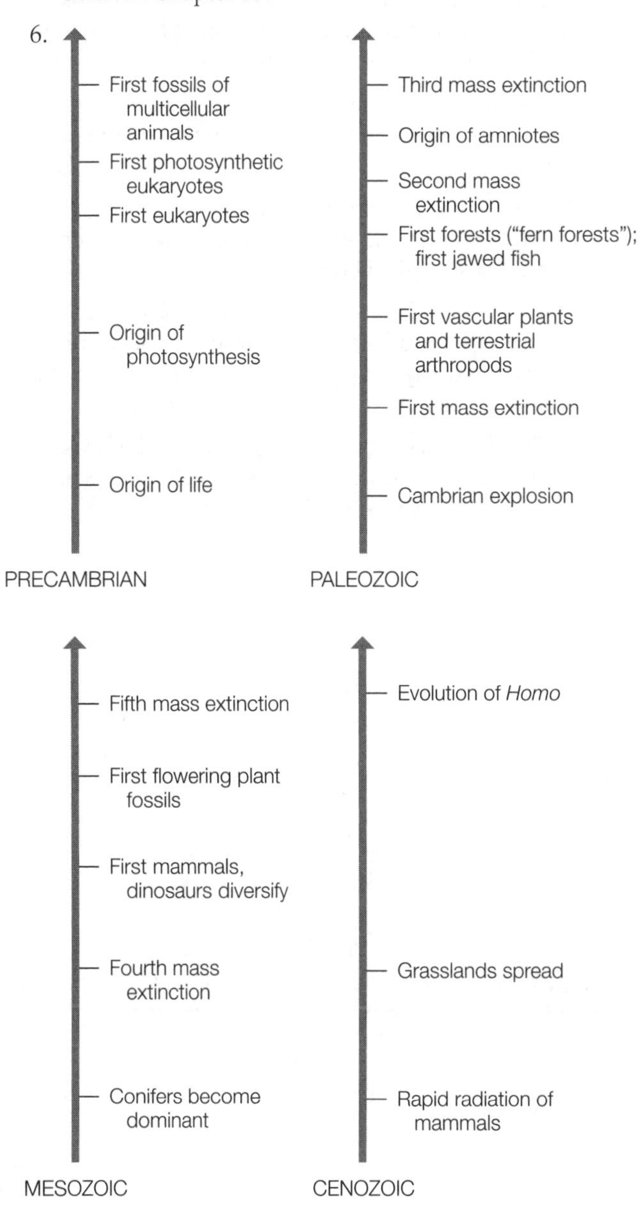

See also Figures 25.12 and 25.14 in the textbook.

7. The diversity of life was greatly reduced during all of the mass extinctions, although life rebounded as the environment changed. The mass extinction associated with the Ordovician is thought to have been caused by a drop in sea level and cooling of the oceans, both of which are associated with glaciation. The cause of the extinction during the Devonian, while uncertain, is thought to have been a meteorite collision with Earth. At the end of the Permian, major changes included volcanic eruptions (which blocked sunlight and cooled the climate), the death of forests (which depleted atmospheric oxygen), and the loss of photosynthetic organisms that would have replaced some of the oxygen. These events resulted in the disappearance of about 96 percent of multicellular species.

Test Yourself

1. **a.** Radioactive decay is measured as the time it takes for one-half the amount of a substance to spontaneously convert into another substance.
2. **a.** As the crustal plates are pushed together, one may move underneath the other, pushing up mountain ranges.
3. **b.** Fossilization occurs best in areas of low oxygen concentration and rapid sedimentation, where scavengers cannot destroy the body.
4. **d.** Hard-shelled animals such as mollusks are good candidates for fossilization because their shells can withstand decay long enough for them to be buried. Many also tend to live in quiet, shallow waters.
5. **e.** The division of Pangaea began during the Triassic period, but Laurasia and Gondwana did not become completely separated until the Jurassic.
6. **c.** The modern mammals did not appear until the Tertiary, more than 350 million years after the Ordovician.
7. **c.** The Pleistocene epoch, which is part of the Quaternary, was the time of most hominid evolution.
8. **b.** The mass extinction at the end of the Cretaceous, 65 mya, was the most recent mass extinction affecting terrestrial life before the evolution of humans.
9. **b.** The Cambrian explosion produced many novel groups of animals characterized by distinctive body plans. Later explosions caused an increase in diversity only within already existing major lineages.
10. **e.** For the reasons outlined in the previous answer, the Cambrian period is the correct answer.
11. **a.** Paleontologists believe that a collision between a large meteorite and Earth so altered conditions that the dinosaurs succumbed rapidly during what we know as the great Cretaceous mass extinction.
12. **e.** There is experimental evidence that insects can grow larger when raised in hyperoxic conditions (see Figure 25.10). The current level of atmospheric O_2 appears to limit the evolution in body size of flying insects.
13. **a.** Carbon dating is generally reliable only for fossils less than 50,000 years old.
14. **d.** Dinosaurs disappeared at the end of the Cretaceous, over 60 million years before humans evolved.
15. **c.** Though mass extinctions occurred at the end of all the periods listed, the highest percentage of species became extinct during the Permian period.

Bacteria, Archaea, and Viruses

The Big Picture

- All organisms fall within one of three domains. This chapter focuses on the domains Archaea and Bacteria. Subsequent chapters focus on Eukarya. Archaea and Bacteria are prokaryotic organisms and retain many features of the earliest forms of life on Earth. They have no membrane-enclosed organelles, circular chromosomes, or cell walls of peptidoglycans or proteins. Their structures are simple, and they exist as single cells. Some prokaryotes are stationary, while others move by means of flagella. Their metabolic processes are highly varied, as are their modes for procuring energy. Many prokaryotic organisms make valuable contributions to their environments through nitrogen fixation, photosynthesis, and other metabolic processes. The majority of prokaryotes are free-living, but some live in mutualistic relationships and others are pathogenic to other organisms.
- The evolutionary relationships among prokaryotes are still highly disputed and best understood at the gene level. The most ubiquitous extant prokaryotes are the bacteria. They are grouped according to physical and biochemical characteristics, but these groupings do not necessarily reflect their evolutionary relationships. The archaea are more closely related to the eukarya, and exist in some of the harshest known environments.
- As defined in this text, viruses do not meet the criteria for life. While there is no single definition of life used by all scientists, many do not consider viruses living organisms, even though they arose from cells. They require a host cell to carry out their basic functions, and viral genomes vary in form and structure. These include positive- and negative-stranded RNA, single-stranded and double-stranded DNA, and they can be linear or circular. Currently, the genome structure is the basic classification system for viruses, since their evolutionary history is unknown and unlikely to be determined.

Study Strategies

- This chapter is an overview of two domains of organisms and of viruses. Because it covers a large number of diverse microscopic organisms included in two of the three domains of life, the details can seem overwhelming. Many of the names of these groups are long, complex, and unfamiliar. Because many of the distinguishing features of the different groups of bacteria and archaea are biochemical, you may need to review information from Part III of the textbook.
- For the two domains that are the subject of this chapter, focus your study on what makes each one distinct and on the unique features that indicate a closer evolutionary relationship between archaea and eukaryotes than between archaea and bacteria. Then move on to learning about sample organisms from each domain and groups within each domain.
- Although the pathogenic organisms may seem particularly interesting, remember that only a very small percent of prokaryotes are pathogenic and therefore they should not be the single focus of study.
- Go to the Web addresses shown to review the following animated tutorial and activity. You can also find them in BioPortal (yourBioPortal.com), along with many additional learning resources.

  Animated Tutorial 26.1 The Evolution of the Three Domains (Life10e.com/at26.1)

  Activity 26.1 Gram Stain and Bacteria (Life10e.com/ac26.1)

Key Concept Review

26.1 Where Do Prokaryotes Fit into the Tree of Life?

The two prokaryotic domains differ in significant ways

The small size of prokaryotes has hindered our study of their evolutionary relationships

The nucleotide sequences of prokaryotes reveal their evolutionary relationships

Lateral gene transfer can lead to discordant gene trees

The great majority of prokaryote species have never been studied

All living things can be placed into one the following three domains: Archaea, Bacteria, and Eukarya. This system reflects evolutionary relationships among these monophyletic groups. All three domains arose from a common ancestor possessing a circular chromosome and machinery for transcription and translation of the DNA code to protein. Divergence from this ancestor occurred billions of years ago. The domain Archaea is more closely related to the Eukarya than to the Bacteria.

All three groups share certain characteristics, common to all living things; they possess plasma membranes and ribosomes, have a common set of metabolic pathways, use DNA as the genetic material, replicate DNA semiconservatively, and have similar genetic codes to produce proteins by transcription and translation. Domain Eukarya includes both unicellular and multicellular organisms. The two prokaryotic domains, Bacteria and Archaea, include only unicellular organisms. Some prokaryotes form colonial relationships and communities in which individual organisms can cooperate.

Prokaryotes do not have membrane-enclosed organelles; therefore, the organization and replication of DNA proceeds differently in prokaryotes than it does in eukaryotes. Prokaryotes go through an asexual process called fission after replicating their DNA. Also, the processes of cellular respiration and photosynthesis are carried out differently and rely on infoldings of the cell membrane. Finally, prokaryotes lack a cytoskeleton; therefore, true mitosis does not occur.

Our knowledge of prokaryotes has arisen relatively recently; advances in molecular biology and biochemistry have allowed scientists to discover the distinctions between Bacteria and Archaea. Archaeal cell walls (like eukaryotic cell walls) lack peptidoglycans. In bacterial prokaryotes, by contrast, the thick, rigid cell walls are composed of peptidoglycans, a polymer of amino sugars. This cell wall also is very different from that of plant and fungal cell walls. Differences in cell wall structure within the bacteria themselves, which can be identified by the Gram staining technique, also serve to separate bacteria into two separate groups. Gram-positive cells stain purple and have a thick layer of peptidoglycan. Gram-negative cells do not retain stain and appear red due to a thin layer of peptidoglycan. Some antibiotics work by interfering with the synthesis of peptidoglycan-containing cell walls.

Prokaryotes generally exist in one of three shapes: spherical (cocci), rod-shaped (bacilli), and spiral-shaped (spirilla). Each may aggregate to form chains or plates of cells. Advanced associations may result in filaments that are encased in a protective sheath.

Relationships among prokaryotes are determined with the help of various types of evidence. Ribosomal RNA is useful for this purpose because it is evolutionarily ancient, is found in all organisms, has the same function in all organisms, and is highly conserved (it evolved slowly). Scientists compare signature sequences to make evolutionary inferences. Genetic analyses of evolutionary relationships among prokaryotes are complicated because of lateral gene transfer—genetic transfer between species. Genes involved in fundamental processes are unlikely to be transferred because functional, adapted copies already exist in the target organism's genome. Genes likely to be involved in lateral gene transfer include those that confer a new adaptation and higher fitness of the receiving organism, such as those for antibiotic resistance. Another large impediment to the study of prokaryotic phylogeny is the fact that most prokaryotic species have not been isolated or studied in the lab. Environmental genomics studies individual genes present in the environment, regardless of species of origin.

Question 1. In the diagram below, label the cells as Gram-positive or Gram-negative, and identify each portion of the structure.

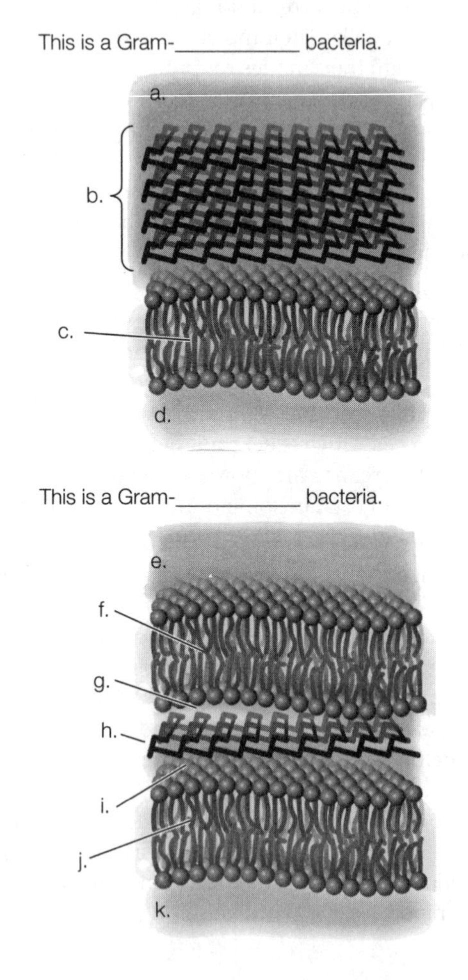

Question 2. Label the three basic shapes of bacteria shown in the micrograph below.

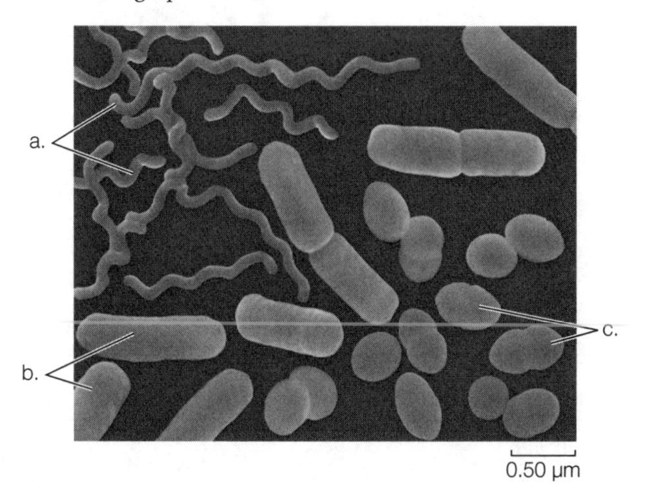

Question 3. You work for a wastewater treatment plant. An organism is getting past your treatment process, and fish are dying in a stream that the runoff enters. You are able to isolate the organism. Using a microscope, how can you tell if the organism is a prokaryote or a eukaryote?

Question 4. Describe how Gram staining depends on bacterial cell wall structure. Would Gram staining work for archaea?

26.2 Why Are Prokaryotes So Diverse and Abundant?

The low-GC Gram-positives include some of the smallest cellular organisms

Some high-GC Gram-positives are valuable sources of antibiotics

Hyperthermophilic bacteria live at very high temperatures

Hadobacteria live in extreme environments

Cyanobacteria were the first photosynthesizers

Spirochetes move by means of axial filaments

Chlamydias are extremely small parasites

The proteobacteria are a large and diverse group

Gene sequencing enabled biologists to differentiate the domain Archaea

Most crenarchaeotes live in hot or acidic places

Euryarchaeotes are found in surprising places

Korarchaeotes and nanoarchaeotes are less well known

Prokaryotes are the most numerous of all organisms. Because they are single-celled (although there are a few exceptions) and very small, they generally escape human notice. They play extremely important roles in the biosphere: for instance, in nutrient cycling. They can also metabolize many substances that eukaryotes cannot.

The diversity of low-GC Gram-positive bacteria belies the DNA sequence evidence supporting the grouping of these organisms into a monophyletic clade. These bacteria include a few that are Gram-negative or lack a cell wall entirely. Some produce endospores that are capable of surviving in extremely harsh conditions. Many medically important bacteria fall within this group, including *Clostridium botulinum* (botulism), *Staphylococcus* (the cause of many skin infections), and *Bacillis anthracis* (the causative agent of anthrax). Another subgroup, the mycoplasmas, may represent the smallest known cellular life form.

Actinobacteria, or high-GC Gram-positive bacteria, develop an elaborately branched system of filaments, some with chains of spores at the tips. This group includes the pathogen *Mycobacterium tuberculosis* (which causes tuberculosis). However, one of our most potent antibiotics, streptomycin, is produced by the actinobacteria *Streptomyces*, and most currently used antibiotics come from the actinobacteria.

Extremophiles live in what are considered extreme environments, ones that would kill most other organisms. Halophiles are one group of thermophilic extremophiles. Cyanobacteria undergo photosynthesis similar to eukaryotic photosynthesizers and contribute to the current oxygen-rich environment. These organisms frequently exist in colonies with distinct divisions of labor. Some cells are vegetative and photosynthesize. Others form resting stages called spores that can withstand harsh environmental conditions. Still others, called heterocysts, fix nitrogen.

Spirochetes are corkscrew-shaped prokaryotes characterized by the presence of axial filaments that run through their periplasmic space. Many spirochetes are parasites of humans, such as those that cause Lyme disease and syphilis. Chlamydias are the smallest bacteria and exist only as parasites. They have a two-stage life cycle and are responsible for several human diseases.

Proteobacteria represent the largest number of known species. This group is believed to be the origin of the eukaryotic mitochondria. The early ancestor of this group was probably a photoautotroph, but today the group encompasses multiple metabolic schemes. Included among the proteobacteria are *E. coli*, the nitrogen-fixing *Rhizobium*, and many human pathogens.

Domain Archaea was established based on phylogenetic relationships derived from rRNA gene sequences. Very little is known about the phylogeny of the archaea, but some of the shared characteristics distinguishing them from the bacteria are well established. They lack peptidoglycan in their cell walls, and some have cell walls composed of proteins. Archaea also have very distinctive lipids connected by ester linkages in their cell membranes. No other organisms possess this type of lipid.

Currently, Archaea has been divided into two main groups (Euryarchaeota and Crenarchaeota) and two recently discovered groups (Korarchaeota and Nanoarchaeota). Crenarchaeota live in hot, acidic environments and are thermophilic and/or acidophilic. They require temperatures above 55°C and can survive in an environment with a pH as low as

0.9 while maintaining an internal pH of 7. Some members of Euryarchaeota produce methane gas and are obligate anaerobes. These organisms are prevalent in the guts of herbivores. Others live near volcanic vents. Another group of Euryarchaeota live in extremely salty conditions and are referred to as halophiles. Korarchaeota are archaea known only from DNA isolated from hot springs. Scientists have been unable to grow any species of Korarchaeota in a pure culture. Nanoarchaeota, obtained from a deep-sea hydrothermal vent, is minute and lives attached to cells of *Ignicoccus*, a crenarchaeote.

Question 5. Complete the table below.

Domain	Group Name	Key Features	Example Genera or Associated Diseases
	Chlamydias		
	Crenarchaeota		
	Cyanobacteria		
	Euryarchaeota		
	High-GC Gram-positives		
	Korarchaeota		
	Low-GC Gram-positives		
	Nanoarchaeota		
	Proteobacteria		
	Spirochetes		

Question 6. You discover a new prokaryotic organism. What sorts of traits will you look for in order to determine whether the organism belongs in the group bacteria or archaea? (Assume you cannot sequence the genome but must use other criteria for your initial classification.)

26.3 How Do Prokaryotes Affect Their Environments?

- Prokaryotes have diverse metabolic pathways
- Prokaryotes play important roles in element cycling
- Many prokaryotes form complex communities
- Prokaryotes live on and in other organisms
- Microbiomes are critical to human health
- A small minority of bacteria are pathogens

Metabolic processes in prokaryotes are varied and diverse. Obligate anaerobes are oxygen-sensitive and can survive only in oxygen-poor conditions. Facultative anaerobes can shift between aerobic and anaerobic modes of respiration. Aerotolerant anaerobes do not respire aerobically but are not harmed by an oxygen-rich atmosphere as obligate anaerobes are. Obligate aerobes require oxygen and cannot survive in the absence of it. Photoautotrophs are photosynthetic and convert light energy to chemical energy. Some photoautotrophs use chlorophyll *a* and give off oxygen as a by-product. Others use bacteriochlorophyll and produce sulfur as a by-product. Photoheterotrophs use light for energy but also require outside carbon sources. Chemoautotrophs oxidize inorganic substances for energy. Some chemoautotrophs fix carbon dioxide into carbohydrate. Others fix nitrite or ammonia, hydrogen gas, hydrogen sulfide, and other materials. Chemoheterotrophs obtain energy and carbon by consuming complex organic molecules. Chemolithotrophs are chemotrophs that obtain energy by oxidizing inorganic substances. Some bacteria carry out respiratory electron transport using nitrogen or sulfur compounds as electron acceptors. These organisms are extremely important in nutrient cycling, and include denitrifiers, nitrogen fixers, and nitrifiers.

Prokaryotes make many contributions to their environments. The nitrogen cycle depends on nitrification by bacteria. Nitrifying bacteria use nitrogen to make ATP, which is used to fix carbon dioxide into glucose and other food molecules. Through this process, unusable nitrogen sources are converted into biologically available nitrogen sources. Cyanobacteria, using photosynthesis, have enriched the atmosphere with oxygen. As oxygen levels became too high for many existing organisms, many others evolved to fill those niches, and new organisms arose.

Although unicellular, some prokaryotes have tight associations with each other and with organisms of other species. Some prokaryotes live in mutualistic associations. For example, many human digestive processes rely on the bacteria that inhabit the intestine. A person's microbiome is increasingly understood as critical to health, since it includes prokaryotes that help the body perform regular functions such as digestion and nutrient absorption. Some prokaryotes organize into communities. One example is a biofilm. Stromatolites are made of fossilized biofilm and calcium carbonate. Prokaryotes in communities have special methods of communication, such as quorum sensing.

Although we tend to pay attention mostly to harmful bacteria, only a small number of bacteria are pathogenic. According to the criteria established by the German physician Robert Koch, in order for a bacterium to be identified as a disease-causing agent (1) it must always be found in an individual with the disease; (2) it must be taken from the host and grown in pure culture (3) a sample of the culture must produce disease if injected into a healthy individual; and (4) the newly infected individual must yield a new pure culture of the same organism.

In order to be pathogenic, an organism must be able to reach a new host, invade the host, evade the host's defenses, multiply, and infect a new host. Relatively few organisms can do this. Consequences of infection depend on the pathogen's invasiveness and toxigenicity. Bacteria sometimes cause disease because they produce toxins that are harmful to host tissues. Endotoxins are released when bacterial cells burst, but are rarely fatal. Exotoxins are secreted by bacteria and may be fatal.

Question 7. You work with the Centers for Disease Control in Atlanta. You have been called to a remote area of Uganda to study a mysterious disease that is causing respiratory ailments in a small village. You isolate a bacterium from several patients that seems like a good candidate for the pathogen. How can you determine if this bacterium is causing the illness?

Question 8. As a U.S. Department of Agriculture field representative, you counsel a young farmer to plant alfalfa in fields with soils that have low nitrogen levels. You know alfalfa roots are hosts for nitrogen-fixing bacteria. How will the alfalfa and its associated bacteria help "fertilize" this farmer's soil?

Question 9. A patient comes to your medical practice with a bacterial infection. You are terribly concerned because the bacteria responsible for the patient's infection produce an exotoxin. Why are exotoxins often more dangerous than endotoxins?

Question 10. Differentiate between the following terms:

obligate anaerobe / obligate aerobe

photoautotroph / photoheterotroph

chemolithotroph / chemoheterotroph

Question 11. A friend tells you she "hates" bacteria and declares her intention to eradicate all bacteria from her home. What do you tell her?

26.4 How Do Viruses Relate to Life's Diversity and Ecology?

Many RNA viruses probably represent escaped genomic components of cellular life

Some DNA viruses may have evolved from reduced cellular organisms

Vertebrate genomes contain endogenous retroviruses

Viruses can be used to fight bacterial infections

Viruses are found throughout the biosphere

Viruses are obligate parasites that rely on their host cell to carry out basic functions such as replication and metabolism. Viruses are capable of infecting all forms of life, from all three domains. They can replicate, mutate, and evolve. Their evolution is independent of other organisms.

Viral phylogeny is difficult to determine for many reasons. Many viral genomes are too small for many meaningful comparative analyses. Their rapid mutation and evolution can cloud evolutionary relationships. No fossil viruses have been found. Finally, viruses are very diverse, with evidence supporting a hypothesis that they have evolved repeatedly.

Viruses are classified based on their genomic structure. Currently, viral groups include negative-sense single-stranded RNA viruses, positive-sense single-stranded RNA viruses, RNA retroviruses, double-stranded DNA viruses, and double-stranded DNA mimiviruses. Some viruses are endogenous and exist within the genomes of other organisms. Bacteriophages are viruses that infect bacteria.

Viruses are the causes of many diseases of significance to humans. These include HIV, the cause of AIDS; influenza, which causes the flu; and herpes viruses, which cause chicken pox, shingles, and genital herpes. Other viruses cause diseases in important crops and can have a devastating impact on agriculture.

Question 12. Why would a physician refuse to prescribe antibiotics for a cold or the flu?

Question 13. Why are viruses classified based on structure and their type of genome? Why aren't more traditional methods for phylogenetics employed for viruses?

Test Yourself

1. Which of the following characteristics is unique to prokaryotes?
   a. Lack of membrane-enclosed organelles
   b. Presence of cell walls
   c. Presence of plasma membranes
   d. Presence of a cytoskeleton
   e. Absence of ribosomes

   Textbook Reference: *26.1 Where Do Prokaryotes Fit into the Tree of Life?*

2. Rod-shaped bacteria are referred to as
   a. bacilli.
   b. cocci.
   c. spiral.
   d. spirilla.
   e. roddi.
   Textbook Reference: *26.1 Where Do Prokaryotes Fit into the Tree of Life?*

3. Which of the following statements concerning prokaryotes is true?
   a. Because prokaryotes do not contain organelles, they cannot photosynthesize or carry out cellular respiration.
   b. Prokaryotes have no chromosomes and therefore lack DNA.
   c. Prokaryote flagella are similar in structure to eukaryote flagella.
   d. Prokaryotes undergo mitosis.
   e. None of the above
   Textbook Reference: *26.1 Where Do Prokaryotes Fit into the Tree of Life?*

4. Gram-negative bacteria stain pink because
   a. they have specialized lipids in their cell walls.
   b. their peptidoglycan layer is thin.
   c. their peptidoglycan layer is thick.
   d. they are receptive to antibiotics.
   e. their cell walls are composed largely of proteins.
   Textbook Reference: *26.1 Where Do Prokaryotes Fit into the Tree of Life?*

5. Archaea are more closely related to _______ than to any other monophyletic group.
   a. bacteria
   b. eukaryotes
   c. bacteria and eukaryotes
   d. fungi
   e. protists
   Textbook Reference: *26.2 Why Are Prokaryotes So Diverse and Abundant?*

6. The dense films laid down by many prokaryotes are
   a. endotoxins.
   b. denitrifiers.
   c. biofilms.
   d. pathogens.
   e. endospores.
   Textbook Reference: *26.3 How Do Prokaryotes Affect Their Environments?*

7. Which of the following groups of bacteria has the highest proportion of pathogens?
   a. Proteobacteria
   b. Cyanobacteria
   c. Chlamydias
   d. High-GC Gram-positives
   e. Low-GC Gram-positives
   Textbook Reference: *26.2 Why Are Prokaryotes So Diverse and Abundant?*

8. The mitochondria of eukaryotes were derived by endosymbiosis from
   a. proteobacteria.
   b. chemoheterotrophs.
   c. eukaryotes.
   d. archaea.
   e. viruses.
   Textbook Reference: *26.2 Why Are Prokaryotes So Diverse and Abundant?*

9. Autoclaves, which sterilize medical and laboratory equipment by means of pressurized heat, must pass a "spore test" in many states to demonstrate that they work correctly. The spore-producing bacteria used for this test are most likely taken from which of the following groups?
   a. Low-GC Gram-positive bacteria
   b. Proteobacteria
   c. Cyanobacteria
   d. Chlamydias
   e. High-GC Gram-positive bacteria
   Textbook Reference: *26.2 Why Are Prokaryotes So Diverse and Abundant?*

10. Archaea often live in harsh environments. Those that live in extremely salty conditions are referred to as
    a. thermophiles.
    b. halophiles.
    c. salinophiles.
    d. brinophiles.
    e. sodiophiles.
    Textbook Reference: *26.2 Why Are Prokaryotes So Diverse and Abundant?*

11. Methane gas contributes to the greenhouse effect that is raising atmospheric temperatures. A large portion of all methane emission is from grazing cattle because cows harbor methane-producing archaea from the _______ group.
    a. Crenarchaeota
    b. Euryarchaeota
    c. Anarchaeota
    d. Proteobacteria
    e. Nanoarchaeota
    Textbook Reference: *26.2 Why Are Prokaryotes So Diverse and Abundant?*

12. One of the major diagnostic characteristics that distinguishes archaea from bacteria is the absence of
    a. peptidoglycan cell walls in archaea.
    b. peptidoglycan cell walls in bacteria.
    c. ribosomes in archaea.
    d. chemoautotrophy in bacteria.

e. pseudopeptidoglycan cell walls in archaea.
Textbook Reference: *26.1 Where Do Prokaryotes Fit into the Tree of Life?*

13. A bacterium that requires a carbon source other than carbon dioxide, yet can convert light energy to chemical energy, is called a
    a. photoautotroph.
    b. photoheterotroph.
    c. chemoautotroph.
    d. chemoheterotroph.
    e. chemolithotroph.
    Textbook Reference: *26.3 How Do Prokaryotes Affect Their Environments?*

14. A bacterium that *cannot* live in the presence of oxygen is called a(n)
    a. obligate aerobe.
    b. facultative aerobe.
    c. obligate anaerobe.
    d. facultative anaerobe.
    e. aerotolerant anaerobe.
    Textbook Reference: *26.3 How Do Prokaryotes Affect Their Environments?*

15. Which of the following is *not* caused by a negative-sense single-stranded RNA virus?
    a. Measles
    b. Shingles
    c. Rabies
    d. Influenza
    e. Mumps
    Textbook Reference: *26.4. How Do Viruses Relate to Life's Diversity and Ecology?*

16. Which of the following virus types inserts a double-stranded DNA copy of its genome into the host cell's genome?
    a. Double-stranded RNA viruses
    b. Double-stranded DNA viruses
    c. Retroviruses
    d. Negative-sense single-stranded RNA viruses
    e. Positive-sense single-stranded RNA viruses
    Textbook Reference: *26.4. How Do Viruses Relate to Life's Diversity and Ecology?*

17. Which of the following is *not* one of the reasons that the study of prokaryotic phylogeny is difficult?
    a. The rapid mutation rates of rRNA genes — rRNA mutates slowly
    b. The process of lateral gene transfer
    c. The very small size of prokaryotic cells
    d. Our inability to grow many prokaryotic cells in pure culture (containing only one species)
    e. The harsh and often inaccessible habitats in which many bacteria live
    Textbook Reference: *26.1 Where Do Prokaryotes Fit into the Tree of Life?*

18. Which of the following is *not* one of Koch's postulates?
    a. The introduction of the microorganism to a new, healthy host causes the same disease that existed in the original host.
    b The microorganism is always found in the person with the disease.
    c. After the microorganism is introduced into a healthy host who gets the disease, the same microorganism can be isolated from this patient.
    d. The microorganism is susceptible to antibiotic treatment.
    e. The microorganism can be isolated from an infected individual and grown in culture.
    Textbook Reference: *26.3 How Do Prokaryotes Affect Their Environments?*

Answers

Key Concept Review

1. positive; negative
   a. Outside of cell
   b. Cell wall (peptidoglycan)
   c. Plasma membrane
   d. Inside of cell
   e. Outside of cell
   f. Outer membrane of cell wall
   g. Periplasmic space
   h. Peptidoglycan layer
   i. Periplasmic space
   j. Plasma membrane
   k. Inside of cell
2.
   a. Spirilla
   b. Bacilli
   c. Cocci
3. The most immediately apparent difference between a prokaryote and a eukaryote is the much smaller size of the prokaryote and the absence of membrane-enclosed organelles in prokaryotes. With proper staining techniques, the nuclei of any eukaryote would likely be visible with the light microscope.
4. Gram staining results depend on the arrangement and amount of peptidoglycans in the bacterial cell wall. This technique does not work with archaea because they do not have peptidoglycan in their cell walls.

5.

Domain	Group Name	Key Features	Example Genera or Associated Diseases
Bacteria	Chlamydias	Gram-negative cocci. Very small, obligate parasites. Life cycle involves two forms of cells called elementary bodies and reticulate bodies.	Cause human eye disease, the STD chlamydia, and some types of pneumonia
Archaea	Crenarchaeota	Most live in hot and/or acidic environments such as sulfur hot springs. Internal cellular environment is close to pH 7.	*Sulfolobus*
Bacteria	Cyanobacteria	Photoautotrophs. Use chlorophyll *a* for photosynthesis. May form colonies with differentiated cells (see Figure 26.9). Some also fix nitrogen. Thought to be the source of eukaryotic chloroplasts.	*Anabaena* (see Figure 26.9)
Archaea	Euryarchaeota	Many are methanogens living in the guts of ruminant herbivores or termites and cockroaches. Others are extreme halophiles. *Thermoplasma* lacks a cell wall and lives in coal deposits.	*Methanopyrus thermoplasma*
Bacteria	High-GC Gram-positives	Gram-positive, GC-rich genomes, may form spores at tips of filaments. Source of most of our antibiotics.	*Mycobacterium tuberculosis*, the cause of tuberculosis. *Streptomyces*: the source of streptomycin.
Archaea	Korarchaeota	Poorly characterized. Only DNA has been isolated directly from hot springs.	None available
Bacteria	Low-GC Gram-positives	Gram positive, although some are Gram negative and lack a cell wall. Genome is not GC-rich. May produce endospores capable of surviving extreme conditions. Mycoplasmas are the smallest cellular organisms known.	*Bacillus anthracis:* causes anthrax. *Clostridium:* a cause of food poisoning. *Staphylococcus aureus*: can colonize healthy individuals or cause infection.
Archaea	Nanoarchaeota	Poorly characterized. Lives attached to cells of *Ignicoccus*, a crenarchaeote. Discovered at a deep-sea thermal vent near Iceland's coast.	*Nanoarchaeum equitans*
Bacteria	Proteobacteria	Largest group of identified species. Very diverse group classified into five groups based on their metabolic pathways.	*Rhizobium:* important for nitrogen fixation. *Yersinia pestis:* cause of bubonic plague. *Vibrio cholerae:* cause of cholera. *Salmonella typhimurium*: cause of gastrointestinal disease. *Escherichia coli:* well characterized.
Bacteria	Spirochetes	Gram-negative, helical structure (See Figure 26.10), chemoheterotrophic. Move via axial filaments.	Some species cause syphilis and Lyme disease.

6. A Gram stain would allow you to determine if the prokaryote is a member of archaea (which will not stain at all) or bacteria (Gram-positive or Gram-negative). You could also use information about the environment in which the new organism was found. Some bacteria and archaea are known to live in extreme environments, while other bacteria can form endospores, fix nitrogen, or have unique structures like axial filaments. Using environmental clues, information from Gram staining, and the shape and motility of the cells, you would probably be able to predict which group this new organism belongs to.

7. Koch's postulates stipulate that in order for an organism to be identified as a disease-causing agent (1) it must always be found in individuals with the disease, (2) it must be taken from the host and grown in pure culture, (3) a sample of the culture must produce disease if injected into a healthy individual, and (4) the newly infected individual must yield a new pure culture of the same organism. That said, it is unethical to knowingly infect a person with a pathogen just for the purposes of identification! Therefore, a diagnosis can be made based on the first three postulates but not on the fourth one.

8. Nitrogen-fixing bacteria can convert atmospheric nitrogen into ammonia. Other bacteria can then convert the ammonia into nitrates that plants can use. (This topic is covered in greater detail in Chapter 36.)

9. Exotoxins are continuously produced by the infecting bacteria, whereas endotoxins are produced only at cell death. Only a limited number of endotoxin molecules are released, but exotoxins can be released until the host dies.

10. Obligate anaerobes die in the presence of oxygen gas, whereas obligate aerobes die without oxygen gas for cellular respiration. Photoautotrophs convert light energy to chemical energy using carbon dioxide as their carbon source; photoheterotrophs rely on an outside carbon source other than carbon dioxide but still can convert light energy to chemical energy. Chemolithotrophs can produce what they need from inorganic molecules, whereas chemoheterotrophs depend on nutrients from external sources.

11. This effort to eradicate all bacteria will not be successful, since humans are hosts to innumerable bacteria both on and in our bodies. Furthermore, most bacteria are not harmful. In fact, many are beneficial. For example, bacteria are essential in nutrient cycling in the ecosystem. In some cases the attempt to eradicate bacteria has created dangers of its own. The abuse and overuse of antibiotics, for instance, creates selective pressures that lead to an increase in antibiotic-resistant pathogenic bacteria, rendering previously effective treatments useless. Your friend would be better served by declaring a truce with the bacteria and recognizing them as part of the world we inhabit.
12. Antibiotics such as penicillin and ampicillin interfere with the synthesis of the peptidoglycan-based cell walls of bacteria. Viruses do not have a cellular structure, and they use the host cell's machinery to replicate and spread. Since the cold and flu viruses replicate in a person's cells, an antibiotic would have no effect on a cold or the flu and would indeed kill many of the "good" bacteria living in the body.
13. There are many factors that make viral phylogeny difficult to determine. Their genomes are generally too small for extensive comparisons to those of cellular organisms, they mutate quickly, there are no viral fossils, and they are extremely diverse.

Test Yourself

1. **a.** Prokaryotes are differentiated from eukaryotes by their lack of membrane-enclosed organelles. They also have a single circular piece of genomic DNA and lack a cytoskeleton.
2. **a.** A rod-shaped bacterium is known as a bacillus (plural, bacilli). Cocci are roughly spherical.
3. **e.** Bacteria do carry out both photosynthesis and multiple forms of cellular respiration. All life forms have DNA. The flagella of bacteria consist of a single protein called flagellin rather than multiple proteins as found in eukaryotes. Bacteria replicate by binary fission.
4. **b.** Gram-negative bacteria have thin peptidoglycan layers that do not retain crystal violet stain.
5. **b.** The Archaea and the Eukarya share a more recent common ancestor with each other than either does with Bacteria. One piece of evidence for this is that a signature sequence from rRNA has been found in all archaea and eukaryotes tested so far, but in none of the bacteria.
6. **c.** Biofilms are gel-like polysaccharide matrices that are laid down by prokaryotes and trap other bacteria.
7. **c.** Chlamydias are all parasitic and cause several human diseases.
8. **a.** Endosymbiosis of proteobacteria gave rise to the mitochondria of eukaryotes.
9. **a.** Among the groups listed, only low-GC Gram-positive bacteria, high-GC Gram-positive bacteria, and cyanobacteria produce spores. Low-GC Gram-positive spores can withstand extreme environmental conditions, so they would make good spores for testing an autoclave's ability to sterilize equipment.
10. **b.** The term *halophile* means "salt loving."
11. **b.** Methanogenic bacteria that reside in the guts of cows belong to the Euryarchaeota.
12. **a.** Archaea lack peptidoglycans. Instead they have a unique lipid in their cell walls. Other species have pseudopeptidoglycans in their cell walls.
13. **b.** Photoheterotrophs require an outside carbon source other than carbon dioxide, yet are able to harvest light energy.
14. **c.** Oxygen gas is toxic to obligate anaerobes.
15. **b.** Shingles is caused by a double-stranded DNA herpes virus.
16. **c.** Retroviruses, such as HIV, insert a form of their genome into the host's DNA. This provirus is replicated every time the cell divides by mitosis.
17. **a.** The mutation rates of rRNA genes are quite slow and thus rRNA has evolved slowly. This has made study of rRNA sequences quite valuable in the taxonomy of prokaryotes.
18. **d.** Koch's postulates were developed in the 1880s, when even the role of microorganisms in disease was not understood. Antibiotic therapies to target these microorganisms came later.

The Origin and Diversification of Eukaryotes

The Big Picture

- The many lineages of protists, as well as all fungi, animals, and plants, are eukaryotic. Eukaryotic cells are characterized by membrane-enclosed organelles, the presence of a cytoskeleton, and a nuclear envelope. The evolution of the eukaryotic cell was a major evolutionary milestone. The eukaryotic cell allows for compartmentalization of processes and increased adaptability. Though the evolutionary lineage is still debated, it is thought that eukaryotes evolved via multiple symbiotic events.
- Protists as a group are not monophyletic. The most recent common ancestor of protists gave rise to the plants, animals, and fungi, groups that are not included in the protists. Protists are diverse, ranging from microscopic single-celled organisms to complex multicellular organisms. Several subgroups of protists are covered in this chapter, some of which are monophyletic while others are not. The eight major eukaryotic clades are alveolates, stramenopiles, plants, amoebozoans, excavates, rhizaria, fungi, and animals.
- Certain body plans and modes of nutritional acquisition are repeated throughout the microbial eukaryote groups and are good examples of form following function. Protists as a group play a vast role in the environment and medicine and are economically valuable.

Study Strategies

- It is easy to become overwhelmed with organism names and characteristics. Focus on the trends outlined in the chapter and use lab opportunities to understand the organism groupings. For a start, see Figure 27.3.
- This chapter contains a great deal of information. It is best not to try to learn it all at once, but to break it up into smaller pieces and study it over several sessions.
- To learn the organism groups, create a chart with key characteristics. This will help you compare and contrast the groups.
- On one side of a 4 × 6 index card, write down a particular clade or subgroup; on the other side, note the defining characteristics. Do the same with sample organisms, with the genus specified on one side of the card and the clade and subgroup on the other side. Then shuffle the deck and sort them into different types of piles. For example, you can sort all of the subgroups into their proper clades. You can match sample organisms with their subgroups or clades. Or you can sort the groups according to their reproductive strategies or other criteria of your choosing.
- Go to the Web addresses shown to review the following animated tutorials and activity. You can also find them in BioPortal (yourBioPortal.com), along with many additional learning resources.

  Animated Tutorial 27.1 Family Tree of Chloroplasts (Life10e.com/at27.1)

  Animated Tutorial 27.2 Digestive Vacuoles (Life10e.com/at27.2)

  Animated Tutorial 27.3 Life Cycle of the Malarial Parasite (Life10e.com/at27.3)

  Activity 27.1 Anatomy of *Paramecium* (Life10e.com/ac27.1)

Key Concept Review

27.1 How Did the Eukaryotic Cell Arise?

The modern eukaryotic cell arose in several steps

Chloroplasts have been transferred among eukaryotes several times

The term "protist" lacks any real taxonomic meaning. It is really shorthand for all eukaryotes that are neither land plants, fungi, nor animals.

The origin of the eukaryotic cell is still being debated. The evolution of the eukaryotic cell was revolutionary and occurred at a time of great environmental change. The modern eukaryotic cell contains a flexible cell surface, a cytoskeleton, a nuclear envelope, digestive vesicles, and organelles. The first step in the path toward a eukaryotic cell may have involved loss of the rigid cell wall, which allowed infoldings of the cell surfaces to increase the surface area-to-volume ratio. This allowed cells to become larger. It also allowed vesicles to form from bits of infolded membrane that pinch off. Such infolding may have been the origin of the nuclear envelope (see Figure 27.1). Consequently, new structures evolved. Ribosomes became associated with internal membranes to form the endoplasmic reticulum, a cytoskeleton of actin and microtubules formed, and digestive vesicles developed. The presence of a cytoskeleton opened up modes of locomotion with actin/myosin-based flagella. How these structures formed is currently conjecture. It is thought that these early motile cells may have been phagocytes that engulfed prokaryotic cells. Subsequent endosymbiotic events may have led to the evolution of mitochondria and chloroplasts. See Figure 27.1 for a diagram of events in the evolution of the eukaryotic cell.

Some chloroplasts are enclosed in double membranes, and others are enclosed in triple membranes. This phenomenon can be explained by endosymbiosis. All chloroplasts can be traced to the engulfment of an ancestral cyanobacterium with a single membrane. Its engulfment in a vesicle of the phagocyte's membrane resulted in a double membrane. This initial event is called primary endosymbiosis. Primary endosymbiosis gave rise to the green and red algae. Photosynthetic euglenoids, with their triple membranes, arose from secondary endosymbiosis. In this case, a chlorophyte was ingested with its housed chloroplast, resulting in three membranes. Ultimately all of the constituents of the chlorophyte except the chloroplast were lost. Another line is thought to have arisen by means of a red algal endosymbiont. Tertiary endosymbiosis occurred when a dinoflagellate lost its chloroplast and engulfed a protist that had acquired a chloroplast through secondary endosymbiosis.

Question 1. Explain one line of thought regarding the origin of the eukaryotic cell. Why are scientists unsure of the evolutionary relationships among the protists?

Question 2. Describe the origin of double-membrane-enclosed and triple-membrane-enclosed chloroplasts.

27.2 What Features Account for Protist Diversity?

Alveolates have sacs under their plasma membranes

Stramenopiles typically have two flagella of unequal length

Rhizaria typically have long, thin pseudopods

Excavates began to diversify about 1.5 billion years ago

Amoebozoans use lobe-shaped pseudopods for locomotion

Eukaryotes began to diversify about 1.5 billion years ago, during the Precambrian. There are eight major eukaryotic clades: alveolates, stramenopiles, rhizaria, excavates, plants, amoebozoans, fungi, and animals. Five of these are covered in this chapter. (Plants, fungi, and animals, along with their close protist relatives, will be discussed in Chapters 28–33.) The five protist clades contain organisms that are very diverse in body plan and lifestyle. Some are motile, others are not; some are photosynthetic, others are heterotrophic; most are unicellular, but some are multicellular; most are microscopic, but some (such as giant kelp) are huge. The term "microbial eukaryotes" is sometimes used to refer to the microscopic, unicellular protists. Multicellularity has arisen dozens of times through eukaryotic history. Four of these gave rise to four major groups: plants, fungi, animals, and brown algae. There are also some eukaryotes that are unicellular but associate in multicellular groups. Historically, protists were classified based on their life histories and reproductive features, but the advent of genetic sequencing and electron microscopy has allowed scientists to look at other characteristics and revealed new patterns of evolutionary diversity. Lateral gene transfer also complicates attempts to classify protists.

The alveolates are a monophyletic group consisting of dinoflagellates, apicomplexans, and ciliates. The synapomorphy that defines the alveolates is the possession of cavities called alveoli just below their cell surfaces. All alveolates are unicellular, and most are photosynthetic.

Alveolates have great variation in their physical form. Dinoflagellates are mostly marine organisms with two flagella. They are yellowish in color due to photosynthetic and accessory pigments and are primary producers in marine environments. Many live in symbiosis with other organisms. They are common endosymbionts in corals, for example. Dinoflagellates such as *Pfiesteria piscicida* are responsible for "red tides" in warm marine waters and can damage fish through accumulated toxins.

Apicomplexans are parasitic protists. Their name comes from the cluster of organelles at the apex of their cells, which assists with invasion of their host organism. Apicomplexans in the genus *Plasmodium* cause malaria, a disease that claims more than a million lives annually. Apicomplexans have elaborate life cycles, with both sexual and asexual reproduction and a variety of life stages. The life cycle often alternates between two different organisms—mosquitoes and humans for *Plasmodium* species (see Figure 27.20) and cats and rats for *Toxoplasma*.

Ciliates move via cilia (which are structured like flagella, but are shorter), are characterized by the presence of two types of nuclei, and have various body plans (see Figure 27.5). Most ciliates are heterotrophic. Members of the common and well-studied ciliate genus *Paramecium* move by means of cilia. Their membranes are protected by a pellicle, and they protect themselves with trichocysts, which act as sharp darts. Paramecia contain contractile vacuoles, which are a classic example of an adaptation to living in the hypotonic environment of fresh water. Excess water obtained through osmosis is collected and expelled by the contractile vacuole. Like many other protists, paramecia engulf food and digest it in a digestive vacuole. Digested food pinches off into smaller vesicles, which increases the surface area for food absorption by the rest of the cell. Paramecia reproduce asexually by binary fission. Genetic recombination is accomplished through conjugation, which involves the exchange of equal amounts of genetic material (see Figure 27.19). The resulting recombined paramecia then go through binary fission.

Stramenopiles consist of diatoms, brown algae, oomycetes, and a few smaller groups. The defining characteristic for most is their two flagella of unequal length, with rows of hairs on the longer of the two. Those that are not flagellated have lost their flagella over the course of evolution. Diatoms are yellowish brown from their carotenoids and store carbohydrates and oils as photosynthetic products. They are most noted for their silica-containing cell walls, which occur in two halves and fit together like a petri dish. All diatoms are symmetrical and unicellular. They reproduce both asexually and sexually. Brown algae are brown because of chlorophylls *a* and *c* and the carotenoid fucoxanthin. They produce leaf-like growths or branched filaments and are almost exclusively marine. The giant kelp that grow in large oceanic "forests" are brown algae. They have specialized regions called holdfasts, which anchor them to a substrate and allow them to resist water movement. Brown algae are commercially important for the presence of alginic acid, used as a binder in many food and cosmetic products. Oomycetes are a non-photosynthetic group of stramenopiles. These are commonly known as water molds and downy mildews, but they are not fungi. Water molds are absorptive heterotrophs, aquatic, and saprobic (they feed on dead material). The downy mildews are terrestrial. Many of these are harmless, but some are infectious to plants. The Irish potato famine was caused by an oomycete. Oomycetes were once classified as fungi, but they have some key differences, such as cell wall composition (cellulose vs. chitin).

The rhizaria comprise three related groups of unicellular, mostly aquatic eukaryotes that include the cercozoans, foraminiferans, and radiolarians. Cercozoans are very diverse, with either aquatic or soil habitats. Foraminiferans live either as plankton or on the sea bottom. Their pseudopods are used to trap food. Limestone deposits are often the result of discarded foraminiferan shells made of calcium carbonate. Radiolarians have stiff, microtubule-reinforced pseudopods that help the cells float in their marine environment and provide additional surface area. Their distinctive radial symmetry and glassy endoskeletons make them instantly recognizable.

The excavates are a monophyletic group consisting of five major subgroups: the euglenids, the kinetoplastids, the heteroloboseans, the diplomonads, and the parabasalids. They are single-celled flagellates that reproduce asexually by binary fission. They have flagella, a cytoskeleton, and undulating membranes that all contribute to their locomotion. Diplomonads and parabasalids lack mitochondria. *Giardia lamblia* is a well-known diplomonad that causes the human intestinal disorder giardiasis. *Trichomonas vaginalis* is a parabasalid responsible for a sexually transmitted disease in humans. Heteroloboseans have a life cycle that alternates between an amoeboid stage and a flagellated stage. One species of *Naegleria* can cause a fatal brain infection in humans. Euglenids and kinetoplastids both are unicellular and have flagella. Their mitochondria have disc-shaped cristae, and their flagella contain a crystalline rod not seen in other species. They reproduce primarily asexually. Euglenids have anterior flagella. Many are photosynthetic and are characterized by a complex cell structure (see Figure 27.14). They are flexible in nutritional requirements and may switch between autotrophism and heterotrophism. The kinetoplastids are parasitic and are characterized by their two flagella and single large mitochondrion. The mitochondrion is unique in having a kinetoplast housing multiple circular DNA molecules and associated proteins. Many tropical human diseases are caused by kinetoplastids, including African sleeping sickness (see Table 27.1).

Amoebozoans include the loboseans, the cellular slime molds, and the plasmodial slime molds. They appear to have diverged from other eukaryotes about 1.5 billion years ago. They have distinctive lobe-shaped pseudopods that differ in form and function from rhizarian pseudopods. Loboseans live as independent, single cells and feed by phagocytosis. A few have shells (see Figure 27.16). Most are predators, parasites, or scavengers. Plasmodial slime molds form multinucleate masses (see Figure 27.17A). During the vegetative stage, an acellular slime mold is a coenocyte, a wall-less mass of cytoplasm with multiple diploid nuclei that oozes over a network of strands called a plasmodium. This phenomenon is called cytoplasmic streaming and is used to move and engulf food. If exposed to adverse environmental conditions, the slime mold will form a dormant sclerotium from which it can turn back into a plasmodium when environmental conditions again become favorable. It may also transform into a spore-bearing fruiting structure with knob-like ends called sporangia. Haploid cells form within the sporangia, which become spores that are dispersed when the fruiting structure dries. Spores germinate into swarm cells, which either divide mitotically or become gametes. Two swarm cells can fuse to make a zygote, which divides mitotically to form a new plasmodium. Cellular slime molds are made of large numbers of cells called myxamoebas with single haploid nuclei. They reproduce by mitosis and fission, and this life cycle can continue as long as conditions are favorable. Once conditions become unfavorable, myxamoebas form a motile slug (or pseudoplasmodium) that eventually produces fruiting bodies. Spores from the fruiting bodies germinate to form more myxamoebas. Sexual reproduction occurs with the fusion of two myxamoebas, which then undergo meiosis to release haploid myxamoebas (see Figure 27.18).

Question 3. Draw the phylogeny of the three domains of life and add the various lineages of protists to it. Then add the plants, animals, and fungi.

Question 4. What would be the criteria for placing a newly discovered organism within the protists?

Question 5. Most protists are motile. Describe some of the means of motility.

Question 6. Identify the structures labeled a.–m. in the *Paramecium* shown below.

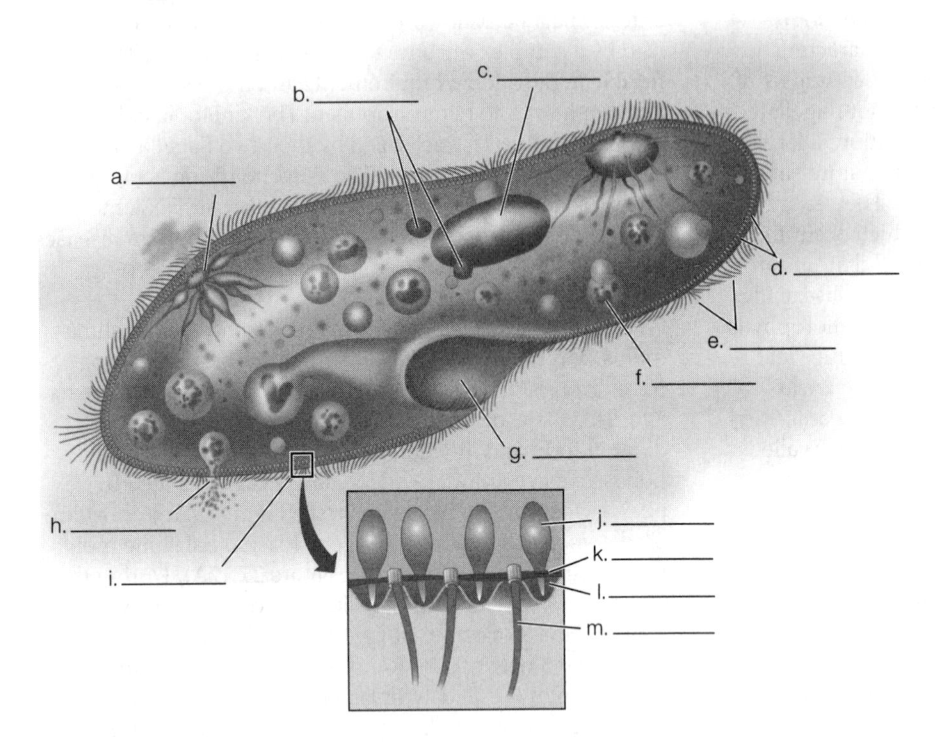

27.3 What Is the Relationship between Sex and Reproduction in Protists?

Some protists reproduce without sex and have sex without reproduction

Some protist life cycles feature alternation of generations

Most protists reproduce both sexually and asexually, but sexual reproduction has yet to be discovered in some groups. In some groups, sex and reproduction are not linked. Several asexual reproductive processes have been observed: (1) equal splitting of one cell by mitosis and cytokinesis; (2) the splitting of one cell into more than two cells; (3) the budding of a new cell from the side of another; and (4) the formation of spores that can subsequently form new individuals.

Asexual reproduction produces cells that are nearly identical to the parent cell (with differences caused only by new mutations). These groups are known as clones or clonal lineages. Sexual reproduction is quite varied among protists. Some have gametes as the only haploid cells (as in animals). Some have zygotes as the only diploid cells. In some, both haploid and diploid cells undergo mitosis and there are alternating stages of multicellular diploid and haploid states.

Paramecia are representative of the protists that reproduce without sex and have sex without reproduction. They have two kinds of nuclei, one large macronucleus and one or several micronuclei. The macronucleus has several copies of the genome arranged into groups of genes. Transcription and translation from the macronucleus regulate the life of the cell. All of the nuclei are copied before the paramecium reproduces asexually. Paramecia, and some other protists, also perform conjugation. During this sexual process, two cells meet and exchange genetic information and each individual leaves with a new assortment of genes (see Figure 27.19). Since two cells exist before and after the conjugation, it is not a reproductive process. Conjugation is essential for the health of clonal lineages of paramecia.

Many multicellular protists, all plants, and fungi undergo alternation of generations, in which a diploid sporophyte gives rise to haploid spores through meiosis to form the haploid gametophyte, which in turn produces haploid gametes. Heteromorphic alternation of generations occurs when the two generations differ morphologically, and isomorphic alternation of generations occurs when the two generations are morphologically similar. Gametes arise from sporocytes of the diploid organism, which divide meiotically to form spores. The spores may germinate and divide mitotically, forming a multicellular haploid individual that produces gametes by mitosis.

Question 7. Explain alternation of generations. Why are the protists the first group in which this process could be seen?

Question 8. Differentiate between asexual and sexual reproduction. Select one mode of microbial eukaryote sexual reproduction and explain it fully.

27.4 How Do Protists Affect Their Environments?

Phytoplankton are primary producers

Some microbial eukaryotes are deadly

Some microbial eukaryotes are endosymbionts

We rely on the remains of ancient marine protists

Microbial eukaryotes have greatly varied roles in various ecosystems. Diatoms alone are responsible for about one-fifth of the total photosynthetic output on Earth, approximately equal to all the rainforests. Diatoms are the major component of marine phytoplankton; other photosynthetic protists are also components. Phytoplankton are the primary producers of the ocean. They are eaten by heterotrophic consumers, which are eaten in turn by other consumers. Components of phytoplankton themselves can be harmful; dinoflagellates, for example, cause red tides and can contain potent toxins.

Some microbial eukaryotes are pathogenic, such as species of the apicomplexan genus *Plasmodium*, which cause malaria, a serious health concern in many parts of the world. The life cycle of *Plasmodium* alternates between mosquitoes of the genus *Anopheles* and humans (see Figure 27.20). Mosquitoes transmit the disease to humans during a feeding session; the parasites travel to the liver, where they change form and multiply, subsequently infecting red blood cells. The parasites multiply within the red blood cells, which then burst and release more parasites, which are acquired by another mosquito during another feeding session. The cells form zygotes in the mosquito that grow inside the gut and then move to the salivary glands, from where they can be injected into another human host. Plasmodium is hard to combat with medication; the best tactic is to limit mosquito breeding.

Endosymbiosis is also common among protists. Many radiolarians contain photosynthetic endosymbionts, whose carotenoids can color the radiolarians. Dinoflagellates are also often endosymbionts of other organisms, such as corals. Coral bleaching occurs when the photosynthetic endosymbionts are lost or ejected.

Ancient marine protists are important for many reasons. Layers of diatoms are a source of petroleum and natural gas. Diatomaceous earth is an important industrial product, used for insulation, metal polishing, filtration, and insecticidal activity. Foraminiferan skeletons created layers of limestone and also sand grains on beaches; the characteristic patterns of foraminiferan fossils are useful for dating and classifying layers of sedimentary rock.

Question 9. Many protists are significant human pathogens. Indentify the pathogenic protist that causes malaria, as well as the microbial eukaryote group to which it belongs, and describe its life cycle.

Question 10. In the diagram below, identify the structures labeled a.–k. using the following terms: cyst, male gametocyte, female gamete, human, male gamete, merozoites, female gametocyte, mosquito, sporozoites, and zygote. (Some terms may be used more than once.) What kind of life cycle is shown here?

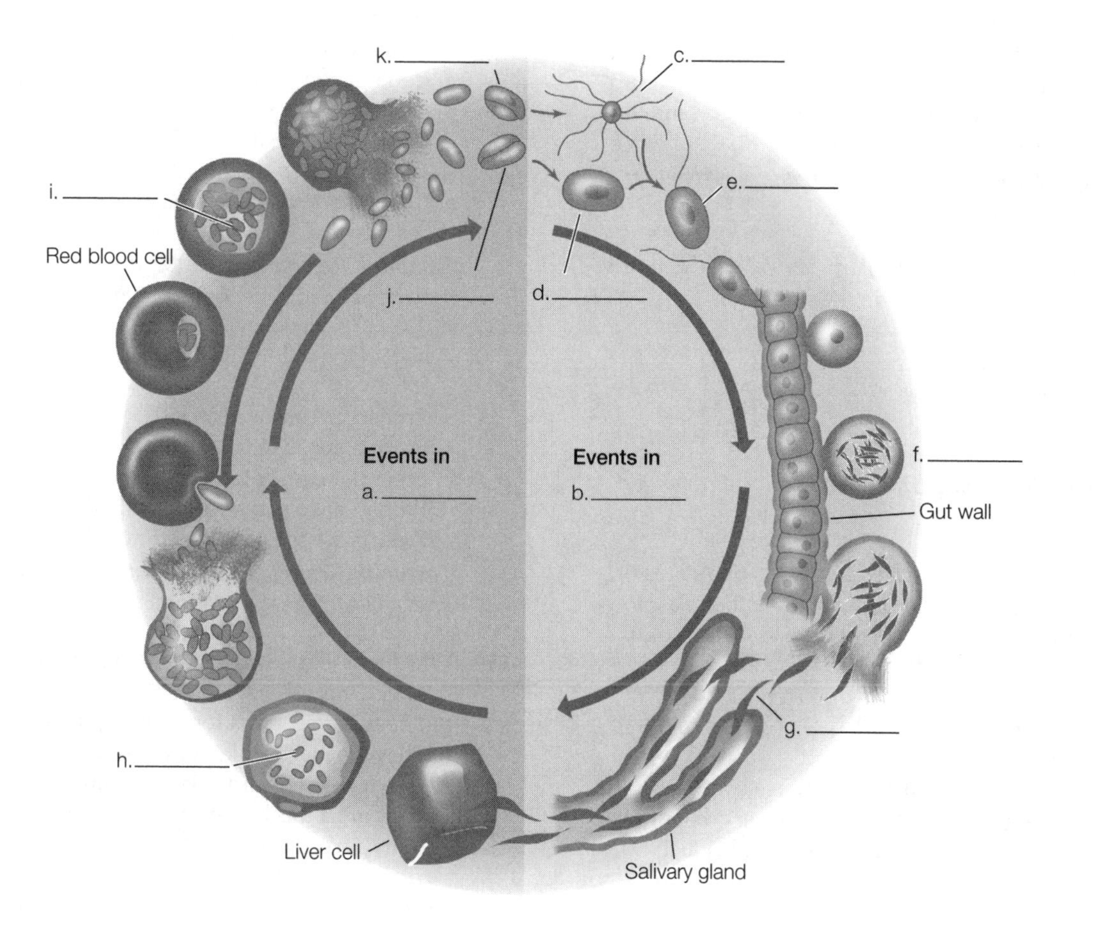

Test Yourself

1. Which of the following modes of reproduction can be found in at least some protists?
   a. Binary fission
   b. Sexual reproduction
   c. Spore formation
   d. Multiple fission
   e. All of the above
   Textbook Reference: *27.3 What Is the Relationship between Sex and Reproduction in Protists?*

2. During the evolution of eukaryotes from prokaryotes, which of the following did *not* occur?
   a. Infolding of the flexible cell membrane
   b. Loss of the cell wall
   c. A switch from aerobic to anaerobic metabolism
   d. Endosymbiosis of once free-living prokaryotes
   e. Development of a cytoskeleton
   Textbook Reference: *27.1 How Did the Eukaryotic Cell Arise?*

3. Which of the following statements about protists is *false*?
   a. Apicomplexans are the only microbial eukaryote group without parasitic representatives.
   b. Foraminiferans and radiolarians are shelled protists.
   c. Ciliates have great control over the direction in which their cilia beat.
   d. Although they appear structurally simple, amoebas are not primitive organisms.
   e. All diatoms are unicellular, although a few individual species associate in filaments.
   Textbook Reference: *27.2 What Features Account for Protist Diversity?*

4. Ciliates, as represented by *Paramecium*, have defensive organelles called _______ in their pellicles.
   a. trichonympha
   b. tridents
   c. trichomes
   d. trichocysts
   e. trochlea
   Textbook Reference: *27.2 What Features Account for Protist Diversity?*

5. A major difference between the vegetative states of cellular and plasmodial slime molds is that plasmodial slime molds _______ and cellular slime molds _______.
   a. have haploid nuclei; have diploid nuclei
   b. produce fruiting bodies; do not produce fruiting bodies
   c. undergo aggregation under adverse conditions; do not undergo aggregation under adverse conditions
   d. exist as a coenocytic mass; exist as individual myxamoebas
   e. grow almost indefinitely in favorable conditions; must aggregate into myxamoebas every seven generations
   Textbook Reference: *27.2 What Features Account for Protist Diversity?*

6. Red tides often cause massive fish kills and human illness in those eating shellfish. Which group of protists is responsible for red tides?
   a. Parabasalids
   b. Red algae
   c. Euglenozoans
   d. Dinoflagellates
   e. Radiolarians
   Textbook Reference: *27.2 What Features Account for Protist Diversity?; 27.4 How Do Protists Affect Their Environments?*

7. Holdfasts and alginic acid are characteristic of which group of protists?
   a. Parabasalids
   b. Red algae
   c. Brown algae
   d. Stramenopiles
   e. Unikonts
   Textbook Reference: *27.2 What Features Account for Protist Diversity?*

8. Which of the following statements about conjugation in *Paramecium* is *false*?
   a. Conjugation results in genetic recombination.
   b. Conjugation results in clones.
   c. Conjugation results in offspring.
   d. Conjugation is a sexual process.
   e. Conjugation results in the production of no new cells.
   Textbook Reference: *27.3 What Is the Relationship between Sex and Reproduction in Protists?*

9. Which of the following statements about protists is true?
   a. Protists are always parasitic.
   b. Protists are all single-celled.
   c. Protists are all heterotrophic.
   d. Protists are always photosynthetic.
   e. Protists are mostly aquatic, at least on a small scale.
   Textbook Reference: *27.2 What Features Account for Protist Diversity?*

10. Which of the following statements about contractile vacuoles is true?
    a. They are important for acquiring food.
    b. They are mechanism for defense against predators.
    c. They are lined with cilia to maximize surface area.
    d. They are a mechanism for removing excess water from the organism.

e. They are more important in marine protists than they are in freshwater protists.

Textbook Reference: *27.2 What Features Account for Protist Diversity?*

11. Which of the following statements about protists is *false*?
    a. They are photosynthetic and are one of the foundations of the marine food web.
    b. They are synthesizers of materials that form many sandy beaches.
    c. They neutralize the sulfuric acid in deep sea vents.
    d. They aid in ruminant herbivore digestion.
    e. They aid in the formation of crude oil.

    Textbook Reference: *27.2 What Features Account for Protist Diversity?*

12. Which of the following statements about the protist cytoskeleton is *false*?
    a. It allows for the formation of pseudopods.
    b. It is the main structural component of cilia.
    c. It is the primary component of the diatoms' outer shell.
    d. It is the main structural component of flagella.
    e. It helps in removing excess water via the contractile vacuoles.

    Textbook Reference: *27.2 What Features Account for Protist Diversity?*

Answers

Key Concept Review

1. See Figure 27.1, which details the origin of the eukaryotic cell. The evolutionary relationships are difficult to understand due to limited fossilization, lateral gene transfers, and the contributions of symbiotic prokaryotes.
2. Double-membrane-enclosed chloroplasts most likely arose from the endosymbiosis of a cyanobacteria in a process called primary endosymbiosis; triple-membrane-enclosed chloroplasts most likely arose from the endosymbiosis of a chloroplast-containing cell (see Figure 27.2).
3. Begin by drawing just the Bacteria, Archaea, and Eukarya. Now "add" the branches representing groups of protists, animals, plants, and fungi. Check your final tree against Figure 27.3.
4. The organism must be a eukaryote and not fit the criteria for land plants, animals, or fungi.
5. Protists move by flagella, cilia, and pseudopodia. Because many are aquatic, additional modifications allow for floating and movement along currents.
6.
    a. Contractile vacuole
    b. Micronuclei
    c. Macronucleus
    d. Alveoli
    e. Cilia
    f. Food vacuole
    g. Oral groove
    h. Anal pore
    i. Pellicle
    j. Trichocyst
    k. Fibrils
    l. Alveolus
    m. Cilium
7. An organism that exhibits alternation of generations exists in a haploid gamete-producing form and a diploid spore-producing form. Prokaryotes have a single chromosome, so only in eukaryotes, which have multiple copies of chromosomes, is a diploid stage of a life cycle possible.
8. Asexual reproduction results in clones of the original organism, and there is no genetic recombination or variation associated with the creation of offspring. Sexual reproduction allows for new genetic combinations. Protists undergo varied types of sexual reproduction, from fusing haploid myxamoebas in cellular slime molds, to haplontic alternation of generations, to a diplontic type of reproduction. Conjugation between paramecia is an example of sexual recombination without reproduction.
9. You may select from a variety of pathogenic protists: malaria is caused by one of the apicomplexans, African sleeping sickness caused by a kinetoplastid, or others. See Figure 27.20 for the life cycle of *Plasmodium*, the protist that causes malaria.
10. This is a protist life cycle.
    a. Human
    b. Mosquito
    c. Male gamete
    d. Female gamete
    e. Zygote
    f. Cyst
    g. Sporozoites
    h. Merozoites
    i. Merozoites
    j. Female gametocyte
    k. Male gametocyte

Test Yourself

1. **e.** Methods of reproduction are quite varied among the protists.
2. **c.** At the time the first eukaryotes evolved, the environment was becoming oxygen-rich. There was a switch from anaerobic to aerobic metabolism, not vice versa.
3. **a.** Malaria is caused by *Plasmodium*, which is an apicomplexan. In fact, all apicomplexans are parasitic.
4. **d.** Trichocysts are defensive barbs ejected from ciliates when they are disturbed.
5. **d.** The vegetative (feeding) state of an acellular slime mold is called a plasmodium; it consists of multiple diploid nuclei enclosed in a single membrane. The vegetative state of a cellular slime mold is a myxamoeba with a single haploid nucleus. Although the myxamoebas of cellular slime molds do aggregate to form

fruiting structures, individual myxamoebas never fuse to become multinucleated structures.

6. **d.** Dinoflagellates are the cause of red tides.
7. **c.** The brown algae are multicellular protists notable for their organ and tissue differentiation and the presence of alginic acid in their cell walls.
8. **d.** Conjugation does result in genetic recombination but does not result in the production of clones or offspring.
9. **e.** Protists do not have a unifying characteristic other than being eukaryotic organisms that do not fit into the kingdoms Plantae, Animalia, or Fungi. If an aquatic environment is considered to be a single droplet of water, however, protists would meet these criteria. For a single-celled organism, a droplet of water is a very large environment.
10. **d.** Contractile vacuoles remove excess water from several freshwater protists.
11. **c.** Protists are a very diverse group, but they do not release bases or neutralize environmental acids. Members of Archaea are likely to be found in an acidic environment.
12. **c.** The cytoskeleton, consisting of actin filaments, microtubules, and intermediate filaments, is a feature of eukaryotic cells. In protists, the cytoskeleton is important in anchoring organelles, movement of materials in the cell, movement of the cell in the environment, and controlling the shape of the cell. (See also Chapter 4.)

Plants without Seeds: From Water to Land

The Big Picture

- Land plants are photosynthetic eukaryotes that utilize chlorophylls *a* and *b*, undergo alternation of generations, and develop from multicellular embryos that are protected by the parent plant. The ten extant (surviving) groups can be classified as nonvascular plants (those without highly developed vascular tissue) and vascular plants (those with highly developed vascular tissue). This chapter focuses on the nonvascular plants and the nonseed vascular plants.
- In order to colonize land, plants had to evolve strategies for coping with desiccation and gravity. This involved mechanisms for extracting water from soil, means of transporting water throughout the plant, methods of ensuring fertilization, and modes of protecting developing embryos.
- This chapter introduces and describes the liverworts, hornworts, and mosses, all of which are considered nonvascular plants, and the club mosses, horsetails, and ferns, which are the nonseed vascular plants.

Study Strategies

- It is very tempting to memorize land plant types and characteristics, but it is more important to focus on evolutionary relationships among the land plants and how to differentiate one group from another. Many of the important distinctions between groups are related to adaptations for terrestrial life.
- Figure 28.7 illustrates the life cycle of a moss as an example of the nonvascular plant life cycle. You should be able to follow this life cycle and understand which stages are haploid and which are diploid. When studying life cycles, be sure to note whether the sporophyte or the gametophyte is dominant and where mitosis and meiosis take place.
- Go to the Web addresses shown to review the following animated tutorial and activities. You can also find them in BioPortal (yourBioPortal.com), along with many additional learning resources.

  Animated Tutorial 28.1 Life Cycle of a Moss (Life10e.com/at28.1)

  Activity 28.1 The Fern Life Cycle (Life10e.com/ac28.1)

  Activity 28.2 Heterospory (Life10e.com/ac28.2)

  Activity 28.3 Homospory (Life10e.com/ac28.3)

Key Concept Review

28.1 How Did Photosynthesis Arise in Plants?

Several distinct clades of algae were among the first photosynthetic eukaryotes

Two groups of green algae are the closest relatives of land plants

There are ten major groups of land plants

Primary endosymbiosis, the engulfing of any early eukaryote by a cyanobacterium, was the origin of chloroplasts. These chloroplasts became important in the evolution of land plants, and the presence of these plants provided support for the establishment of other organisms on land. Primary endosymbiosis is a synapomorphy, or key shared trait, of the group Plantae. Although we think of Plantae as including only land plants, there are clades of plants that are aquatic. We customarily refer to these clades as algae, although they are not all closely related.

There are several groups of algae that are not grouped with the protists but rather as part of Plantae. The unicellular ancestor of Plantae probably resembled the modern glaucophytes (see Figure 28.2). Glaucophytes are microscopic freshwater algae; unlike the rest of the photosynthetic eukaryotes, their chloroplasts retain some of the peptidoglycan found in the cell wall of the originating cyanobacterium. Red algae are multicellular; their characteristic color is due to the accessory pigment phycoerythrin (see Figure 28.3). Red algae are found in shallow water, deep water, and anywhere in between. Their "red" color is not always apparent, since it is due to the relative ratio of phycoerythrin to chlorophyll *a*, which is in turn due to the amount of sunlight the algae receive. The rest of the algal groups of Plantae are termed the green algae. They contain chlorophyll *a* and *b* and use starch as energy storage; in this respect they are like land plants. Collectively, the land plants and green algae are called green plants.

The largest clade of green algae is the chlorophytes. There are thousands of species of chlorophytes, and they range from microscopic unicellular species to multicellular ones that are centimeters long. *Volvox* and *Ulva* are two genera of chlorophytes. *Volvox* is a colonial unicellular species, and *Ulva* grow in thin sheets up to about 30 centimeters in width (Figure 28.4).

The remaining nonchlorophyte species of green algae, along with the land plants, are termed streptophytes. The two groups of green algae that are the closest relatives of land plants are the coleochaetophytes and the stoneworts (Figure 28.5). Some characteristics that these groups share with land plants include: (1) retention of eggs in the parent; (2) the presence of plasmodesmata; and (3) similarities in mitosis, meiosis, and cytokinesis. Stoneworts are considered to be the sister group of land plants.

All land plants (embryophytes) develop from an embryo that is protected by tissues of the parent plant. This is a key shared trait, or synapomorphy, of land plants. There are ten clades of land plants (see Table 28.1). Seven of these clades constitute the vascular plants, also called tracheophytes because they have fluid-conducting cells called tracheids. The remaining three clades—liverworts, mosses, and hornworts—are nonvascular land plants and do not together constitute a clade.

Question 1. Differentiate between glaucophytes and red algae. In what ways are these two groups similar?

Question 2. Compare and contrast the protist "algae" discussed in Chapter 27 with the Plantae "algae" discussed in this chapter. Why is the same term used for both groups?

28.2 When and How Did Plants Colonize Land?

- Adaptations to life on land distinguish land plants from green algae
- Life cycles of land plants feature alternation of generations
- Nonvascular land plants live where water is readily available
- The sporophytes of nonvascular land plants are dependent on the gametophytes
- Liverworts are the sister clade of the remaining land plants
- Water and sugar transport mechanisms emerged in the mosses
- Hornworts have distinctive chloroplasts and stalkless sporophytes

Land plants first appeared in the terrestrial environment 450–500 million years ago. Living on land provided challenges for the formerly aquatic species, most notably the threat of desiccation (drying). Other challenges included moving fluids around the inside of the plant, dispersing gametes, and support against gravity. The earliest land plants developed a number of modifications for terrestrial life: a waxy cuticle, gametangia, embryos, protective pigments, thick spore walls, mutualistic associations with fungi, and (with the exception of the liverworts) the presence of stomata. The development of a cuticle was likely one of the earliest and most important adaptations to land; waxy cuticles, though minimal in early land plants, prevented water loss from tissues as they were exposed to dry air.

The presence of land plants itself affects local soil conditions. Plants secrete acids that help break down rock, and dead plants add organic matter to the soil composition.

The living nonvascular plants (liverworts, mosses, and hornworts) grow in dense mats in moist environments. They rarely exceed a half-meter in height and hug the ground closely. The lack of a vascular system to transport water and minerals likely limits their overall height. They also lack the support of lignin. Nonvascular plants do not have leaves, stems, and roots, but they have structures that are analogous to each. They transport water via capillary action, rely on leaflike structures to trap water, and utilize diffusion to transport minerals. Maternal tissue protects embryos, and many nonvascular plants have at least a thin cuticle. Most nonvascular plants live in soil or on other plants, but some live on bare rock, fallen trees, and other minimally hospitable locations; they are able to do this because of their relationships with mycorrhizae that facilitate nutrient uptake. Most nonvascular land plants are terrestrial, and, although a few species live in fresh water, these aquatic species are descended from terrestrial ones.

Land plants undergo alternation of generations (see Figure 28.6). A diploid zygote divides via mitosis into a multicellular embryo and eventually develops into the mature sporophyte. Cells within the sporangia of the sporophyte undergo meiosis to produce unicellular, haploid spores. These haploid spores divide via mitosis and grow to become haploid gametophytes, which produce haploid gametes through mitosis. The fusion of two gametes produces another zygote and the cycle begins again. The trend in plant evolution is toward reduction of the gametophyte generation. In the nonvascular land plants, the gametophyte is larger; in those groups that appeared later in plant evolution, the sporophyte is the larger generation.

The life cycle of nonvascular land plants is dominated by the gametophyte generation. The sporophyte is very tiny and is completely dependent on the gametophyte; it may or may not be photosynthetic (see Figure 28.7). Gametes form within specialized sex organs called gametangia. The archegonium is a multicellular, flask-shaped female sex organ; it produces a single egg. The antheridium is a male sex organ in which dual-flagellated sperm are formed. In many species, antheridia and archegonia are both present and so the individual has both male and female reproductive structures. Sperm released from the antheridium have to make their way to an archegonium, on either the same plant or a neighbor. The sperm either swim or are splashed by raindrops to their destination. The archegonium also releases chemical attractants that guide the sperm. The most important aspect of this process is that it requires water. Once the sperm reach the egg inside the archegonium, fusion occurs and a zygote is formed. The zygote divides mitotically, eventually producing a diploid sporophyte. The sporophyte produces a single sporangium, which produces new spores via meiosis and thus the next generation.

There are about 9,000 species of liverworts. The liverwort gametophyte can be either leafy or a flat plate of cells (see Figure 28.8). Gametophytes produce antheridia and archegonia on their upper surfaces (sometimes on stalks) and rhizoids (rootlike structures for anchoring and water absorption) on their lower surfaces. Liverwort sporophytes are shorter than those of mosses and hornworts (often only millimeters long). In most species, the cells of the sporophyte expand, allowing the stalk to elongate. This aids in spore dispersal. In some species of liverworts, spores are not released until the surrounding sporangium rots. Other species have structures that forcefully eject spores. Liverworts can reproduce asexually by shedding small clumps of cells called gemmae. Some liverworts have gemmae cups that promote dispersal by raindrops (see Figure 28.8B).

Mosses are found in almost every terrestrial environment and are the most familiar nonvascular plants. They are often found in damp, cool conditions, and form dense mats on the ground. The mosses are a sister clade of the clade containing the hornworts and vascular plants. These clades evolved an advantage over the liverworts in the form of simple stomata: pores that help with the exchange of gas and water vapor. In some species, the gametophyte begins its development as a branched, filamentous structure called a protonema. Some of the filaments are photosynthetic and some are not. Cells at the tip of the photosynthetic filaments divide to form buds; the buds form the familiar leafy structures of mosses and are the sites of antheridia and archegonia formation.

Some mosses have specialized cells called hydroids. Hydroids are thought to be precursors to the tracheids seen in vascular plants, but they lack lignin for structural support. Like tracheids, the hydroid cell dies and leaves a tiny channel that can transport water through the plant.

As mosses of the swamp-dwelling genus *Sphagnum* die and their material accumulates, the upper layers compress the decomposing lower layers. Partially decomposed plant matter is called peat; in some parts of the world, people rely on peat for energy. Millions of years ago, continued compression of peat layers gave rise to coal.

Hornworts are so named because their sporophytes look like little horns (see Figure 28.11). Hornworts are distinguished from other nonvascular plants by two characteristics: the presence of a single platelike chloroplast in each cell, and sporophytes that do not have a stalk. Because of the absence of a growth-limiting stalk, the sporophyte is often as tall as 20 centimeters; growth is limited only by the lack of a transport system. Hornworts often have symbiotic relationships with cyanobacteria, which are able to fix atmospheric nitrogen and make it available to the hornwort.

Question 3. Label the structures in the moss life cycle in the diagram below. Indicate the sporophyte generation, gametophyte generation, and diploid or haploid status. Also indicate, in the open boxes, if the process occurring is meiosis, mitosis, or fertilization.

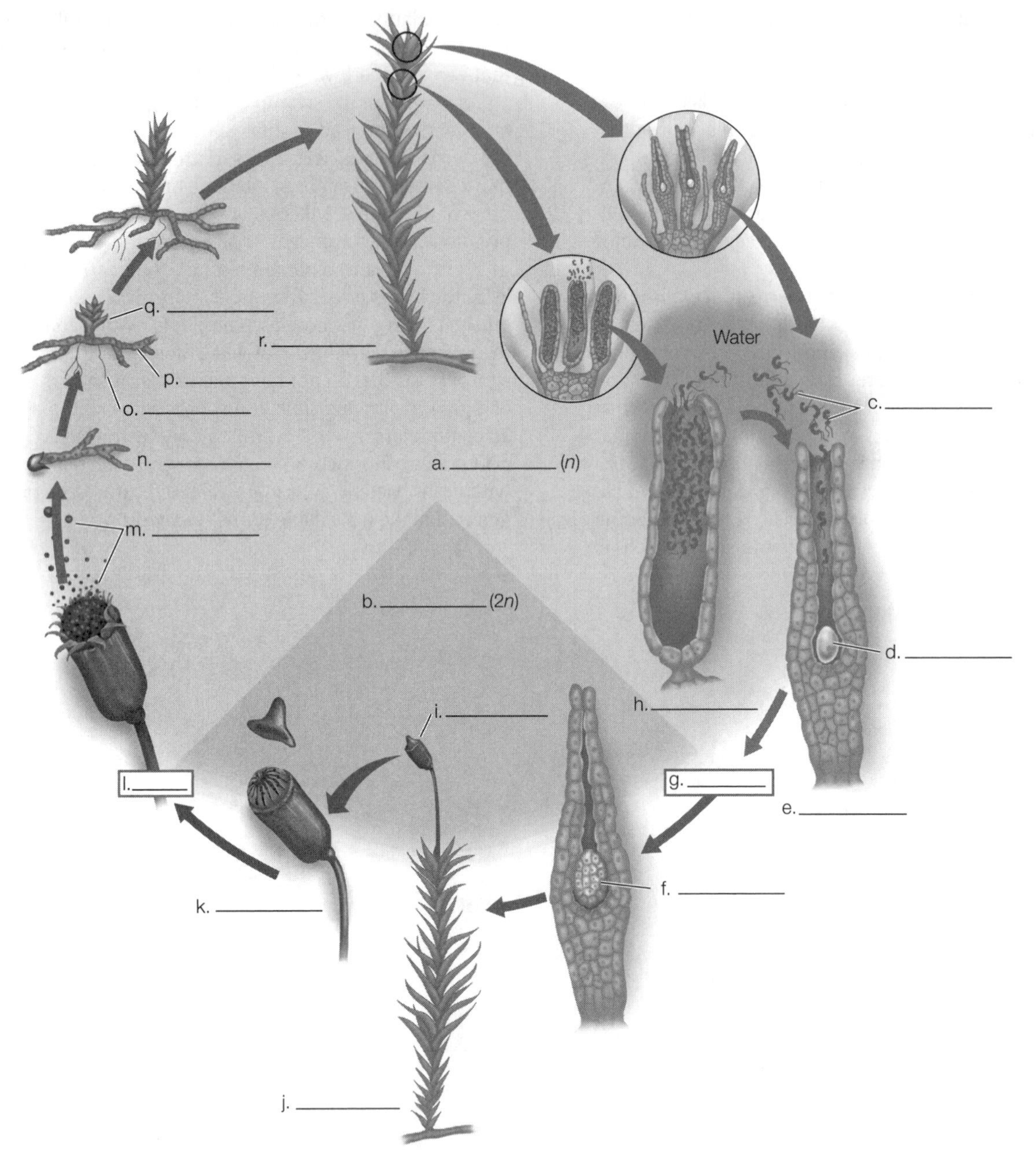

Question 4. Explain why the largest mosses are less than a meter tall.

Question 5. Discuss the challenges of terrestrial life that land plants have addressed in order to survive.

28.3 What Features Allowed Land Plants to Diversify in Form?

Vascular tissues transport water and dissolved materials

Vascular plants allowed herbivores to colonize the land

The closest relatives of vascular plants lacked roots

The lycophytes are sister to the other vascular plants

Horsetails and ferns constitute a clade

The vascular plants branched out

Heterospory appeared among the vascular plants

Vascular plants did not arise until tens of millions of years after plants first colonized land. Vascular tissues, which provide for the transport of water and nutrients throughout a plant, allowed plants to inhabit different environments and diversify. Xylem is responsible for the transport of water and minerals from the soil to the aerial parts of the plant. Phloem conducts the products of photosynthesis from the location in which they are produced to where they are stored.

In all vascular plants except the angiosperms (the flowering plants) and gnetophytes, the principal conducting element of the xylem is the tracheid, a strawlike cell that conducts water. (Angiosperms have retained tracheids as part of their vascular system, but also have a more efficient system derived from them.) The vascular system also has a rigid structural support in the form of lignin in the cell walls, which allows vascular plants to grow upward and compete for sunlight. Increased height also allows for increased dispersal of spores.

Vascular plants are also characterized by branching sporophytes. This allows for more complex development of sporophytes, including nutritional independence from the gametophyte. Among vascular plants, the sporophyte is the dominant, familiar form.

The presence of vascular plants on land allowed herbivorous species to flourish. Arthropods, vertebrates, and other animals only moved onto land once the vascular plants were established. Trees appeared in the Devonian period and flourished during the Carboniferous period. Forests of 40-meter-tall club mosses, horsetails, and tree ferns dominated the tropical swamps. Eventually the plant matter from these forests accumulated and transformed into coal. When the continents merged into Pangaea, the interior of the continent became warmer and drier and gymnosperms began to dominate.

The first vascular plants were members of a now-extinct group of land plants called rhyniophytes. Rhyniophytes had characteristics of extant vascular plants and showed the earliest features of this group, including an independent branched sporophyte and a simple vascular system of xylem and phloem, but they lacked more advanced characteristics such as true leaves and roots (see Figure 28.13). Rhizomes—horizontal portions of stem with water-absorbing filaments called rhizoids—anchored these species. The aerial branches of the rhyniophytes had a dichotomous branching pattern.

Club mosses and their close relatives (Lycophyta) diverged from the tracheid lineage relatively early and are the sister clade to the remaining vascular plants. They have simple leaflike structures called microphylls that are arranged spirally on the stem and true roots that branch dichotomously. The sporangia of many club mosses are clustered in apical strobili (see Figure 28.14A). Other club mosses have their sporangia on the upper surfaces of specialized microphylls.

The horsetails and ferns were once thought to be only distantly related, but now they form their own clade, the monilophytes. As with seed plants, there is differentiation between the main stem and side branches. Horsetails are represented by few extant species. They have true roots, their sporophytes are large and independent, and their gametophytes are highly reduced but also independent. The leaves of horsetails are simple, forming whorls around the stem (see Figure 28.14B). The ferns consist of more than 12,000 species. They are characterized by relatively large, complex leaves with branching vascular strands. The fern life cycle is dominated by the sporophyte, and the gametophyte is small and short-lived (see Figure 28.15). Like all other nonseed vascular plants and nonvascular plants, ferns today are dependent on water to carry motile sperm. In most species of ferns, the sporangia are found in clusters called sori on the underside of leaves.

Several features new to vascular plants evolved in the lycophytes and monilophytes. Roots are believed to have evolved over time from branches that grew underground. Because branches above-ground and below-ground would have been exposed to different selective pressures, underground branches might have given rise to what we know today as roots. Microphylls are present in the lycophytes. They probably evolved from sterile sporangia (see Figure 28.16A).

The monilophytes and seed plants belong to a clade called euphyllophytes ("true leaves"). An important synapomorphy of this clade is overtopping, or growth in which one branch grows beyond the others (see Figure 28.16B). Overtopping would have allowed the development of larger leaf structures, the megaphylls. These are thought to have evolved when photosynthetic tissue developed between the ends of small lateral branches. The first megaphylls were very small; it is likely that larger megaphylls could develop only after the appearance of greater numbers of stomata on the leaves to allow for the greater gas exchange necessary to support larger amounts of photosynthesis.

Most early vascular plants exhibit homospory. In homospory, spores (and the gametophytes that grow from them) are all of the same type; gametophytes produce both archegonia and antheridia. In heterospory, a system that evolved somewhat later (and appears to have evolved several times in the vascular plants), one spore type called a megaspore gives rise to the female, egg-producing megagametophyte; another spore type called a microspore develops into a male, sperm-producing microgametophyte. The sporophyte produces megaspores in small numbers in megasporangia and microspores in large numbers in microsporangia. Land plants typically produce many more microspores than megaspores.

Question 6. Label the structures in the homosporous fern life cycle in the diagram below. Indicate the sporophyte generation, gametophyte generation, and diploid or haploid status. Also indicate, in the open boxes, if the process occurring is meiosis, mitosis, or fertilization.

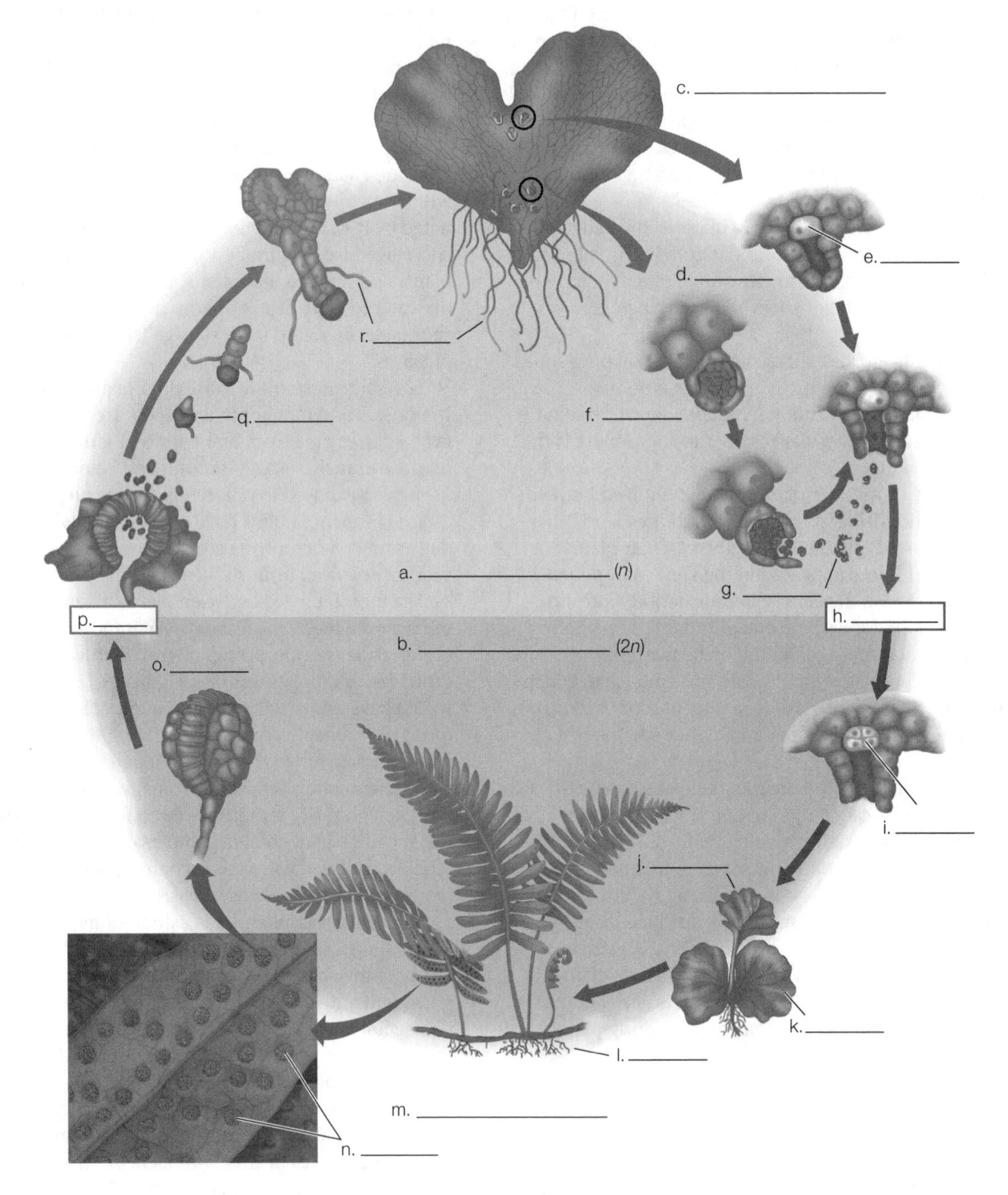

Question 7. Compare and contrast the moss life cycle and the fern life cycle.

Question 8. Draw a diagram showing the evolutionary relationships among the nonvascular plants and nonseed vascular plants. Label the major differentiating characteristics at each branch of your tree.

Question 9. Discuss how the relationship between sporophytes and gametophytes changes as one moves from the first nonvascular plants to later nonseed vascular plants.

Question 10. Compare and contrast homospory and heterospory. Which reproductive structures result from meiosis in each type of life cycle, and which structures result from mitosis?

Test Yourself

1. A land plant may be reliably distinguished from green algae by which of the following characteristics?
   a. Chlorophyll type
   b. The presence of an embryo protected by parent tissue
   c. The presence of roots
   d. Swimming sperm
   e. All of the above
   Textbook Reference: *28.1 How Did Photosynthesis Arise in Plants?; 28.2 When and How Did Plants Colonize Land?*

2. Which of the following characteristics was *not* necessary for plants to colonize land?
   a. Vascular tissue for moving water throughout the plant
   b. A waxy cuticle to reduce water loss
   c. The ability to screen ultraviolet radiation
   d. The development of thick spore walls to protect the spores from dehydration
   e. Development of embryos protected inside other tissues
   Textbook Reference: *28.2 When and How Did Plants Colonize Land?*

3. The main difference between nonvascular plants and vascular plants is that nonvascular plants
   a. lack gametophytes.
   b. produce spores.
   c. do not have tracheids.
   d. reproduce sexually.
   e. All of the above
   Textbook Reference: *28.1 How Did Photosynthesis Arise in Plants?*

4. In alternation of generations, the sporophyte generation is _______ and the gametophyte generation is _______.
   a. haploid; diploid
   b. diploid; haploid
   c. haploid; haploid
   d. diploid; diploid
   e. either haploid or diploid; either haploid or diploid
   Textbook Reference: *28.2 When and How Did Plants Colonize Land?*

5. Which of the following characteristics helps distinguish the liverworts from the mosses?
   a. The presence of hydroids
   b. Gametophyte dominance
   c. Sporophyte dominance
   d. Swimming sperm
   e. Chloroplast structure
   Textbook Reference: *28.2 When and How Did Plants Colonize Land?*

6. During a plant's life cycle, meiosis takes place in the _______ to produce _______.
   a. gametophyte; haploid gametes
   b. sporophyte; haploid gametes
   c. sporophyte; haploid spores
   d. gametophyte; diploid spores
   e. gametophyte; haploid spores
   Textbook Reference: *28.2 When and How Did Plants Colonize Land?*

7. Asexual reproduction in liverworts is accomplished by
   a. gametophytes.
   b. spores.
   c. gemmae.
   d. physical separation of gametophyte parts.
   e. Both c and d
   Textbook Reference: *28.2 When and How Did Plants Colonize Land?*

8. Which of the following limits the size of a hornwort's sporophyte?
   a. The sporophyte's ability to distribute water to all its cells
   b. The gametophyte's ability to produce enough nutrients for the sporophyte
   c. Developmental genes that prevent growth of the sporophyte beyond a certain size
   d. A tendency to collapse under its own weight once it grows beyond a certain size
   e. All of the above
   Textbook Reference: *28.2 When and How Did Plants Colonize Land?*

9. You are walking along a roadside and find a plant with the following characteristics: a very thin, waxy cuticle, stomata, simple leaves in whorls around a central stem, independent sporophytes and gametophytes, and sporangia in strobili. This plant is most likely a member of which of the following groups?
   a. Bryophyta
   b. Monilophyta
   c. Anthocerophyta
   d. Lycopodiophyta
   e. Cycadophyta
   Textbook Reference: *28.3 What Features Allowed Land Plants to Diversify in Form?*

10. Which of the following groups has a unique chloroplast?
    a. Mosses
    b. Liverworts
    c. Hornworts
    d. Club mosses
    e. Horsetails
    Textbook Reference: *28.2 When and How Did Plants Colonize Land?*

11. Which of the following groups has hydroids?
    a. Mosses
    b. Liverworts
    c. Hornworts
    d. Club mosses
    e. Horsetails
    Textbook Reference: *28.2 When and How Did Plants Colonize Land?*

12. Which of the following groups lacks stomata?
    a. Mosses
    b. Liverworts
    c. Hornworts
    d. Club mosses
    e. Horsetails
    Textbook Reference: *28.2 When and How Did Plants Colonize Land?*

13. Which of the following groups has basal growth?
    a. Ferns
    b. Mosses
    c. Liverworts
    d. Club mosses
    e. Horsetails
    Textbook Reference: *28.2 When and How Did Plants Colonize Land?*

14. In which of the following groups are sporangia arranged in strobili?
    a. Horsetails
    b. Liverworts
    c. Hornworts
    d. Club mosses
    e. Ferns
    Textbook Reference: *28.3 What Features Allowed Land Plants to Diversify in Form?*

15. Which of the following groups has large leaves with branching vascular strands?
    a. Horsetails
    b. Liverworts
    c. Hornworts
    d. Club mosses
    e. Ferns
    Textbook Reference: *28.3 What Features Allowed Land Plants to Diversify in Form?*

Answers

Key Concept Review

1. Glaucophytes are unicellular, microscopic, freshwater algae whose characteristic whitish-blue color is due to their retention of a small amount of peptidoglycan between the inner and outer membranes of their chloroplasts. Red algae are almost exclusively multicellular organisms that inhabit saltwater environments, lack peptidoglycan in their chloroplasts, and are usually red in color due to their characteristic pigment, phycoerythrin.

2. The protist algae, such as the euglenids, arose from secondary endosymbiosis. The Plantae algae discussed in this chapter are aquatic photosynthetic eukaryotes; they did not arise from terrestrial species, and they acquired their chloroplasts from primary endosymbiosis. The same term is used for both because both are small, aquatic, photosynthetic species, as distinguished from the land plants.

3.
    a. Haploid; Gametophyte generation
    b. Diploid; Sporophyte generation
    c. Sperm (*n*)
    d. Egg (*n*)
    e. Archegonium (*n*)
    f. Embryo (2*n*)
    g. Fertilization
    h. Antheridium (*n*)
    i. Sporophyte (2*n*)
    j. Gametophyte (*n*)
    k. Sporangium
    l. Meiosis
    m. Ungerminated spores
    n. Germinating spore
    o. Rhizoid
    p. Protonema
    q. Bud
    r. Gametophytes (*n*)

4. Mosses have very rudimentary water transport cells called hydroids. Hydroids lack the waterproofing and support molecule lignin. Because of this, they can carry water only short distances and cannot support tall growth.

5. Terrestrial life poses problems of support, water conduction, UV radiation, water loss, and embryo protection. The tracheid is an important evolutionary innovation that helped with both water conduction and support. Special pigments protect against UV radiation and a waxy cuticle prevents water loss from cells. Plant embryos are protected in early development by parental tissue.

6.
    a. Haploid
    b. Diploid
    c. Mature gametophyte
    d. Archegonium
    e. Egg
    f. Antheridium
    g. Sperm
    h. Fertilization
    i. Embryo
    j. Sporophyte
    k. Gametophyte
    l. Roots
    m. Mature sporophyte

n. Sori
o. Sporangium
p. Meiosis
q. Germinating spore
r. Rhizoids

7. See Figures 28.7 and 28.15. Pay particular attention to the relative dominance of the sporophyte versus the gametophyte.
8. See Figure 28.1B and Table 28.1.
9. Early nonvascular plants have reduced sporophytes that are highly dependent on the gametophyte for nutrition. Nonseed vascular plants, which evolved more recently, have independent sporophytes and gametophytes, and the gametophyte is highly reduced.
10. See Figure 28.18. In homospory, meiosis results in a single type of spore, but in heterospory, meiosis results in megaspores and microspores. Mitosis occurs throughout both types of life cycle as cells divide and plants grow, but the reproductive cells that result from mitosis in both homospory and heterospory are sperm and eggs. In homospory, the antheridium produces sperm by mitosis, and the archegonium produces eggs; in heterospory, the microgametophyte produces sperm and the megagametophyte produces eggs.

Test Yourself

1. **b.** According to the definition presented in the textbook, all land plants produce embryos that are protected by tissue of the parent plant. Green algae and land plants make use of the same types of chlorophyll. Not all land plants have roots, so a plantlike organism lacking roots will not necessarily be part of the green algae.
2. **a.** Several successful groups of terrestrial plants lack vascular tissue.
3. **c.** Tracheids are found only in the vascular plants.
4. **b.** Meiosis occurs in the diploid sporophyte to produce haploid spores that develop into the haploid gametophyte.
5. **a.** Mosses are the only group with hydroids, the precursors to true vascular tissue. Hornworts have chloroplasts that are different from those of mosses and liverworts.
6. **c.** The outcome of meiosis is four cells, each of which has half the genetic material of the parent cell. A haploid cell already has only half the normal number of chromosomes of most eukaryotes, so a cell must be diploid (or have higher ploidy) to undergo meiosis. In all plant life cycles, the sporophyte is diploid and the gametophyte is haploid; therefore, only the sporophyte can undergo meiosis. Spores are the products of sporophyte meiosis.
7. **e.** Liverworts have specialized asexual reproductive structures called gemmae; they can also reproduce by fragmentation of the gametophyte.
8. **a.** The sporophyte of a hornwort is nutritionally dependent on the gametophyte, but its growth is indeterminate—the sporophyte will continue to grow as long as water is able to reach all its cells.
9. **b.** The plant is a horsetail, which is very common along roadsides in damp ditches, particularly in the Midwest.
10. **c.** The hornworts have a single, large, platelike chloroplast.
11. **a.** Mosses have hydroids, which are cells similar to the tracheid.
12. **b.** Unlike the mosses, hornworts, and vascular plants, liverworts do not have any stomata for gas exchange.
13. **e.** This growth at the base of the stem, though uncommon in plants, is seen in both hornworts and horsetails.
14. **d.** The strobili are conelike structures of aggregated sporangia in club mosses.
15. **e.** Some ferns uncurl as their leaves grow.

The Evolution of Seed Plants

The Big Picture

- Seed plants can be divided into two main groups: the gymnosperms and the angiosperms. Seed plants have developed complex ways of protecting embryos, dispersing gametes, and moving water, nutrients, and food throughout the plant. Both gymnosperms and angiosperms are characterized by the presence of seeds for protecting the embryo until conditions favor germination. The main difference between gymnosperms and angiosperms is the presence of flowers and ovaries in angiosperms. Both groups have highly developed vascular tissue, may exhibit secondary woody growth, and show significant diversity.
- There is great variation among flower types, but the basic flower consists of the structures shown in Figure 29.4. Many plants and animals have coevolved in such a way that the two are integrally linked; the plant provides nutrition for the animal and the animal pollinates the plant.
- Both the angiosperm and gymnosperm life cycles are dominated by the sporophyte. Both groups are heterosporous. The roles of megasporangia and microsporangia are similar in both groups. The angiosperm life cycle is different from that of all other plants in that double fertilization occurs.

Study Strategies

- Remember that not all gymnosperms are pines. The examples in this chapter focus on the pine, but gymnosperms are a varied group, and cones are not the defining characteristic of this group. Cones define the conifers only.
- Remember that angiosperms are extremely varied as well. Not all flower parts are found on all plants, and some are highly modified. Also note that cultivated flowers are often sterile, have aberrant parts, and may be altered through selective breeding and treatment.
- Much of the terminology in this chapter was introduced in Chapter 28, and it is worth reviewing that chapter in light of the material presented here. The details of heterospory can be particularly confusing; reviewing Figure 28.18 can help you sort them out.
- Be sure that you understand what sets angiosperms and gymnosperms apart. Refer to Figures 29.8 and 29.16 to compare their life cycles.
- Make sure you understand how flowers can vary from the generalized structure of a perfect flower presented in Figure 29.4 (see Figures 29.10 and 29.11).
- Go to the Web addresses shown to review the following animated tutorials and activities. You can also find them in BioPortal (yourBioPortal.com), along with many additional learning resources.

  Animated Tutorial 29.1 Life Cycle of a Conifer (Life10e.com/at29.1)

  Animated Tutorial 29.2 Life Cycle of an Angiosperm (Life10e.com/at29.2)

  Activity 29.1 Flower Morphology (Life10e.com/ac29.1)

  Activity 29.2 Life Cycle of a Conifer (Life10e.com/ac29.2)

Key Concept Review

29.1 How Did Seed Plants Become Today's Dominant Vegetation?

Features of the seed plant life cycle protect gametes and embryos

The seed is a complex, well-protected package

A change in stem anatomy enabled seed plants to grow to great heights

By the Devonian period, plants shared the terrestrial environment with various arthropods and early tetrapods. Two innovations of the land plants that evolved during this time were thickened woody stems and seeds. Progymnosperms, all of which are now extinct, were among the first plants with woody stems, although they were seedless. The evolution of the seed gave plants additional protection for embryos; within this structure, the embryo can wait for favorable conditions before germinating. Now-extinct seed ferns from the Devonian were also woody. They had fernlike leaves with seeds attached. By the end of the Permian period other seed plant groups achieved domination. Today the seed plants belong to two groups, the gymnosperms (such as pines and cycads) and the angiosperms (flowering plants).

The trend of diminished gametophytes continued with the seed plants. The haploid gametophyte of seed plants develops attached to and nutritionally dependent on the sporophyte. Very few seed plants (e.g., the cycads and ginkgos) have retained swimming sperm. Most have evolved other means for dispersing male gametes.

All seed plants are heterosporous in that they produce two types of spores. One becomes the female gametophyte (megagametophyte) and the other becomes the male gametophyte (microgametophyte). Microsporangia and megasporangia are formed separately. The microsporangium produces microspores, which in turn develop into pollen grains, the male gametophytes. Pollen grains are dispersed by wind and animals. The wall of the pollen grain contains sporopollenin; sporopollenin is one of the most chemically resistant biological compounds known, and represents a major advantage for land colonization by plants. In the megasporangium, meiosis produces four megaspores, but in most seed plants only one of the four megaspores is retained. This megaspore divides by mitosis to form the multicellular (yet tiny) megagametophyte (female gametophyte), which produces the eggs. The megasporangium is surrounded by sterile sporophyte tissues (the integument). Together, the megasporangium and the integument constitute the ovule.

Because the female gametophyte is retained within sporophyte tissue, pollen grains do not have direct access to gametophytes and their eggs. The sporophyte housing a gametophyte creates tissue for receiving pollen grains; upon reaching this tissue, pollen produces pollen tubes that deliver the sperm through the sporophyte tissue to the eggs for fertilization. This process is known as pollination (see Figure 29.4). The embryo resulting from fertilization grows to a certain size and then becomes dormant within the surrounding tissues. This dormant protected embryo and its surrounding tissues constitute the seed.

Recall that alternation of generations involves a multicellular sporophyte generation that alternates with a multicellular gametophyte generation. The embryo within a seed represents the beginning of a new sporophyte generation. It is still surrounded by the female gametophyte tissue, which will provide the embryo with nutrients to begin growth (particularly if it is a gymnosperm). The tough coat, or seed coat, that surrounds the seed consists of tissue provided by the embryo's sporophyte parent. This tissue is derived from the integument of the diploid parent. Seeds protect the embryo until conditions are right for germination. Seeds can remain viable or dormant for many years, waiting for favorable conditions to germinate. Many seeds have adaptations that allow dispersal by wind or another vector.

Progymnosperms and seed ferns had thickened woody stems, which formed from the division of xylem cells. This growth in diameter is called secondary growth, which produces wood (secondary xylem). The younger portions of wood consist of vascular tissue allowing for water transport. As wood becomes older, it becomes clogged with materials and provides the plant with support. New layers of vascular tissue are produced underneath the wood. Not all seed plants have wood (e.g., grasses). During the course of evolution, some seed plants lost woody structures but gained other advantages that allow them to adapt to their environments.

Question 1. Label the flower in the diagram below.

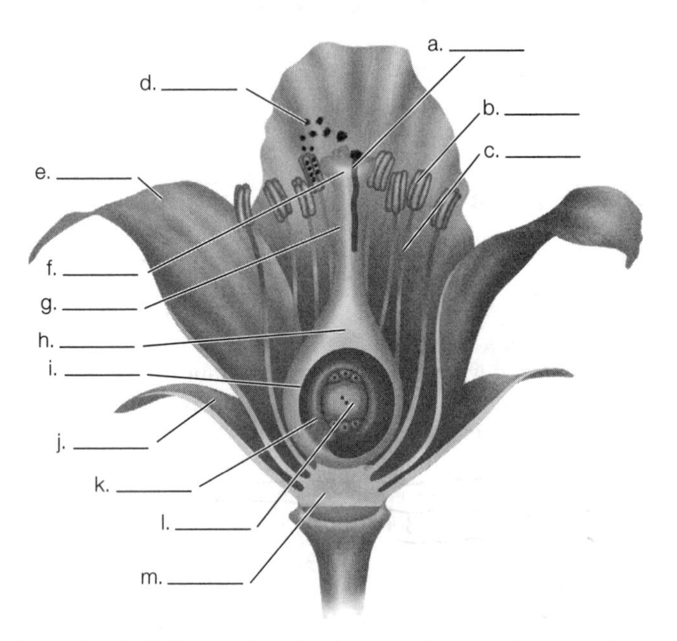

Question 2. Pollen is found only in seed-producing vascular plants. What role did pollen play in vascular plant evolution? Is pollen a sporophyte or gametophyte?

29.2 What Are the Major Groups of Gymnosperms?

There Are Four Major Groups of Living Gymnosperms

Conifers have cones and no swimming sperm

Gymnosperms are seed plants that do not form flowers or true fruits. They are diverse and live in a variety of habitats, from sparse deserts to vast forests. There are four groups within the gymnosperms: cycads, ginkgos, gnetophytes, and conifers (see Figure 29.6). Conifers are the most abundant. With the exception of the gnetophytes, gymnosperms have only tracheids as support and water-conducting cells within the xylem. Despite this simplicity, some of the largest trees known are gymnosperms. Gymnosperms dominated the Mesozoic era, and are still the dominant trees in some forests today.

Conifers differ from other gymnosperms in that they produce male and female cones for reproduction (see Figure 29.7). Female cones (megastrobili) house the megasporangia and produce megaspores, megagametophytes, and eggs; male cones (microstrobili) house the microsporangia and produce microspores, microgametophytes (pollen), and sperm. Seeds in the megastrobilus are protected by woody scales, which develop from simplified branches. The microstrobilus has scales composed of modified leaves.

The life cycle of the pine is characteristic of gymnosperms (Figure 29.8). The male gametophytes are pollen grains; pines do not have swimming sperm, and the pollen is modified for wind dispersal. Pollen lands on the megastrobilus; a pollen tube grows through the maternal sporophyte tissue to the female gametophyte, where it releases two sperm. Only one of the two sperm will fertilize an egg. The resulting zygote develops into an embryo that remains encased in the tissues of the megasporangium and sporophyte (the integument). The integument, the megasporangium, and the tissue attaching it to the sporophyte constitute the ovule. The pollen grain enters the ovule at the micropyle. Protection of the ovule comes from the scales, which are pressed together tightly. Some pines have tightly closed cones that depend on fire to open and disperse the seeds. Dispersal is aided by modifications of the seed coat. Some conifer species have soft, fleshy modifications that surround the seed. Examples include the "berries" found on yew and juniper plants. These are not true fruits, however. The fruits of angiosperms are ripened ovaries, and gymnosperms do not have ovaries.

Question 3. Illustrated below is the pine tree life cycle. Label the processes in the boxes, indicate the haploid, diploid, sporophyte and gametophyte generations, and label all of the structures indicated.

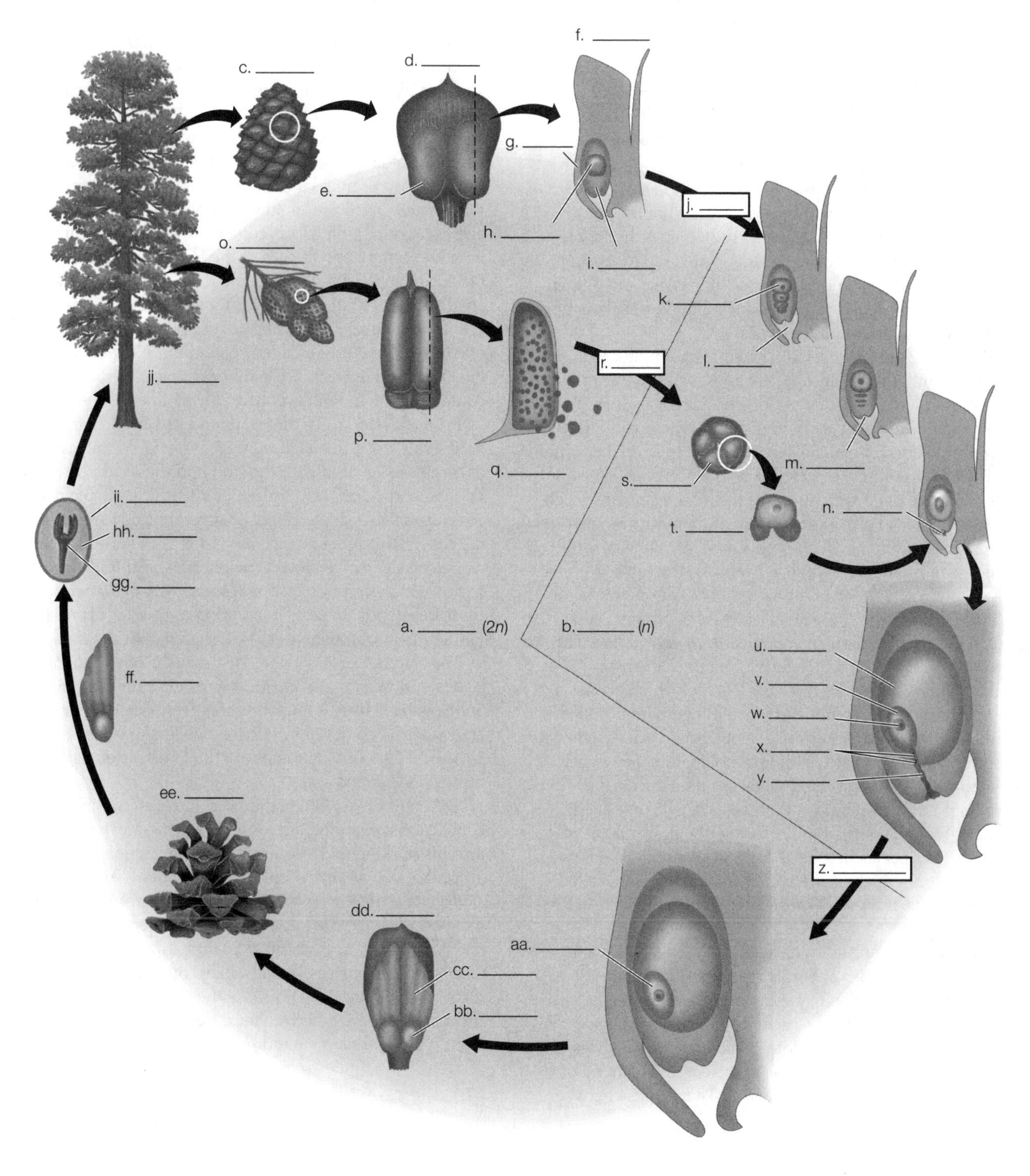

Question 4. How have gymnosperms surmounted the obstacles that all plants faced when they began to colonize terrestrial environments?

29.3 How Do Flowers and Fruits Increase the Reproductive Success of Angiosperms?

Angiosperms have many shared derived traits

The sexual structures of angiosperms are flowers

Flower structure has evolved over time

Angiosperms have coevolved with animals

The angiosperm life cycle produces diploid zygotes nourished by triploid endosperms

Fruits aid angiosperm seed dispersal

Recent analyses have revealed the phylogenetic relationships of angiosperms

Angiosperms are defined by the presence of flowers and fruits. After fertilization, the ovary of the flower develops into a fruit that protects the seeds and promotes dispersal. Angiosperms differ from all other plants in that they produce triploid endosperm (the nutritive tissue), they have double fertilization, their ovules and seeds are enclosed in a carpel, they produce flowers and fruits, and their xylem and phloem have multiple modified cell types, including vessel elements, fibers, and companion phloem cells.

Flowers are greatly diverse. They may be single or grouped in an inflorescence. There are several characteristic types of inflorescences, depending on the species (see Figure 29.10). All flower parts are modified leaf structures. Although flowers are very diverse, they are constructed from the same small set of structures (see Figure 29.4) and have evolved over time. The male structures in a flower are stamens. Each stamen consists of a filament and an anther; the anther contains microsporangia, which produce pollen. Flowers often have several stamens. Carpels are female structures that bear megasporangia. A pistil is composed of either one carpel or multiple fused carpels. The pistil consists of a stigma, a style, and an ovary. The stigma is modified to receive pollen. The style is a stalk that separates the stigma from the ovary. The ovary contains one or more ovules, each of which houses a megasporangium surrounded by two integuments. Nonreproductive floral structures include petals and sepals, which are often modified to attract animal pollinators. Sepals also serve to protect the developing flower bud. Flowers that have both megasporangia and microsporangia are called "perfect." If either structure is missing, the flower is called "imperfect." Species that have both megasporangiate flowers and microsporangiate flowers on the same plant are called "monoecious." If they are on separate plants, they are called "dioecious."

Flowers have evolved to have differentiated petals and sepals, a fixed number of floral organs, and bilateral symmetry, often with fusion of parts. Carpels and stamens probably evolved from leaflike structures with sporangia (see Figure 29.12). Natural selection has favored longer styles and filaments, probably because of increased likelihood of pollination. Perfect flowers have to strike a balance between attracting pollinators and avoiding self-pollination. The bush monkeyflower has a system that allows pollination of the stigma and subsequent exposure of anthers to pollinators.

Many plants and animals have coevolved in such a way that the nutrition of the animal and pollination of the plant are interdependent. Many flowers entice animals by providing food rewards such as nectar. Pollen grains themselves can be a food reward, and can be carried from one plant to another. Some plants have evolved methods of limiting pollination to a single species of insect, but most can be pollinated by a range of species. Bee-pollinated flowers often have nectar guides that can be seen only in the ultraviolet spectrum visible to bees (see Figure 29.16).

Like the rest of the seed plants, angiosperms are heterosporous; ovules are protected within carpels, and the male gametophytes are pollen grains. Just as in gymnosperms, pollination requires the arrival of a pollen grain on a stigma and growth of a pollen tube to the megagametophyte. The angiosperm life cycle differs from that of all other plants with the next step of pollination, fertilization. Angiosperms perform double fertilization, in which one sperm unites with the egg to produce a diploid zygote and the other sperm unites with two other haploid cells of the female gametophyte to produce a triploid cell. This triploid cell divides mitotically to create the (triploid) endosperm tissue, which provides nutrition for the developing embryo. Angiosperm embryos have one or two seed leaves called cotyledons. Cotyledon fate depends on the species. The ovule develops into the seed, containing the diploid zygote and triploid endosperm.

Fruits develop from the ovary and its supporting tissues. Fruit types (see Figure 29.17) depend on the number of carpels associated with the fruit and the extent of support tissue incorporated into the fruit structure. A simple fruit develops from a single carpel or several fused carpels (plums, peaches). An aggregate fruit develops from several separate carpels from a single flower (raspberries). Multiple fruits form from a cluster of flowers (pineapples, figs). Accessory fruits develop from parts in addition to the carpel and seeds (apples, pears, strawberries).

The number of cotyledons distinguishes the two major clades of flowering plants. Monocots have one cotyledon; eudicots have two. A few other relatively small groups of angiosperms do not fit into these two lineages (see Figures 29.18 and 29.19). Magnoliids are the likely sister clade to the monocots and eudicots.

Different phylogenetic methods have led to different conclusions regarding angiosperm phylogeny. Investigators are still working on the question of how angiosperms first arose. Molecular and morphological evidence now points to *Amborella* as the living species most similar to the first angiosperms. *Amborella* has a variable number of carpels and stamens, and it lacks vessel elements. Monocots (see Figure 29.20) include grasses, cattails, lilies, orchids, and palms. Eudicots (see Figure 29.21) include the majority of familiar seed plants such as most herbs, vines, trees, and shrubs, including oaks, willows, beans, snapdragons, roses, and sunflowers.

Question 5. Illustrated below is the angiosperm life cycle. Label the processes in the boxes, indicate the haploid and diploid generations, and label all of the structures indicated.

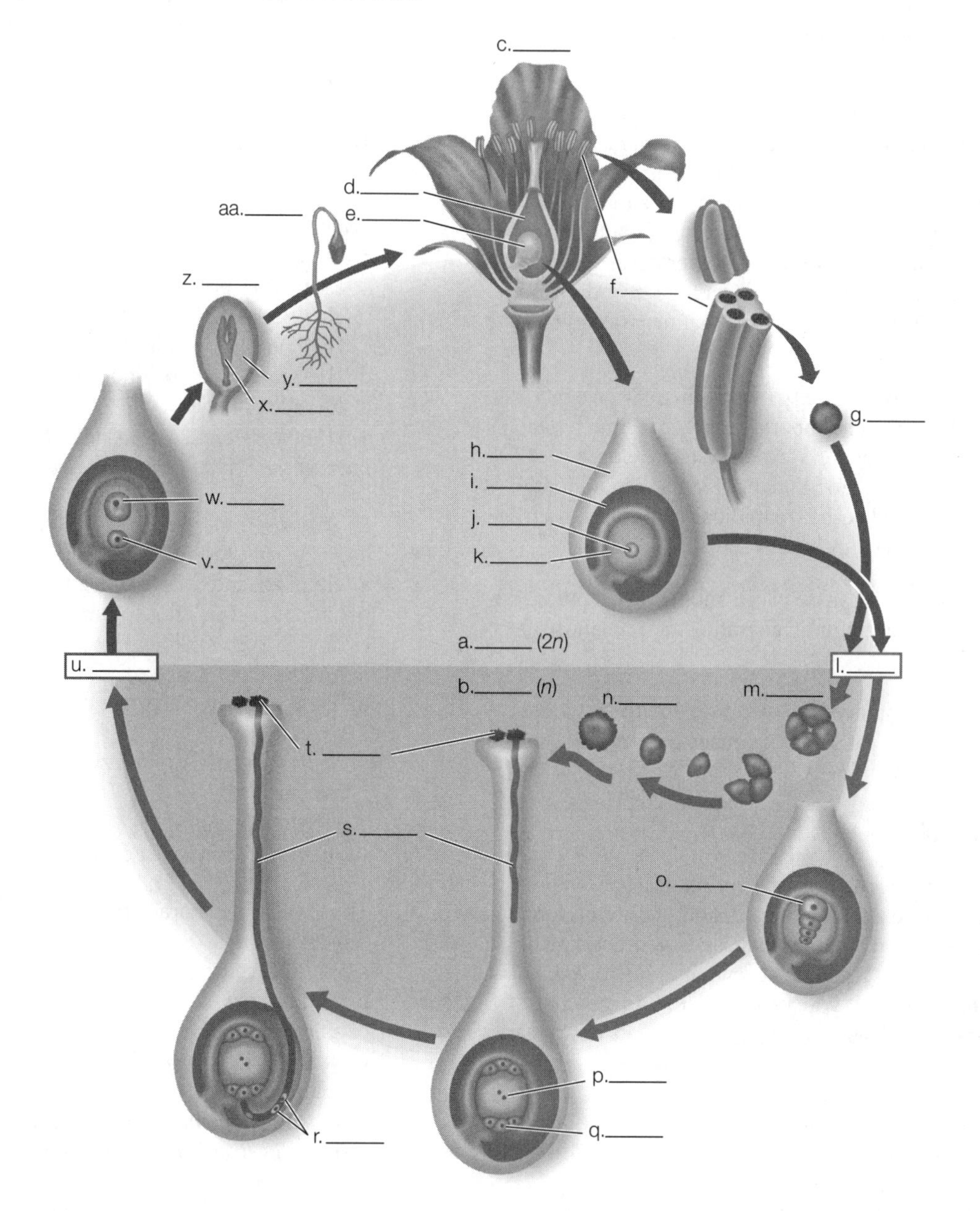

Question 6. Discuss the benefits of double fertilization in angiosperms.

Question 7. Where would you find a female gametophyte in an angiosperm? Explain your answer.

Question 8. What is a cotyledon? What role does it play in the classification of angiosperms into two monophyletic groups?

Question 9. Discuss the origin and development of the angiosperm fruit. From which types of tissues does the fruit arise?

Question 10. Match each of the following flower structures with its function.

_____ Ovule	a. Assists in attracting pollinators
_____ Anther	b. Secretes sticky material to help pollen adhere
_____ Stigma	c. Site of ovule development
_____ Style	d. Protects the immature flower bud
_____ Ovary	e. Houses the megasporangium
_____ Petal	f. Holds the stigma in position
_____ Sepal	g. Houses microsporangia

29.4 How Do Plants Benefit Human Society?

Seed plants have been sources of medicine since ancient times

Seed plants are our primary food source

Plants support environmental processes that benefit humans. They play roles in oxygen and carbon dioxide cycling, soil fertility, defense against erosion, and climate moderation. Seed plants have also been sources of medicine since ancient times (see Table 29.1). Taxol is an example of a modern medicine discovered through systematic testing of plant compounds. Ethnobotanists attempt to discover new medicines through understanding the uses of local flora by indigenous peoples. One medicine to arise in this manner was quinine as a treatment for malaria.

Seed plants are also our primary food source. Twelve different species are the most important: rice, coconut, wheat, corn, potato, sweet potato, cassava, sugarcane, sugar beet, soybean, common bean, and banana. Plants that are food sources can also provide other materials for humans, such as fertilizer, fuel, and livestock feed.

Question 11. In what ways do plants moderate climatic conditions? How might human alterations to the landscape affect climate?

Question 12. What parts of the world do you think would be highly effective places for ethnobotanists to search for putative medicinal plants and why?

Test Yourself

1. You are enjoying a stroll in the botanical gardens and you notice a plant with a beautiful flower. Upon closer inspection you find that the flower has a pistil but no stamens. You look at several more flowers on the same plant, but you are unable to find any that have stamens. This plant is _______ and its flowers are _______.
   a. monoecious; perfect
   b. monoecious; imperfect
   c. dioecious; perfect
   d. dioecious; imperfect
   e. hermaphroditic; perfect
   Textbook Reference: *29.3 How Do Flowers and Fruits Increase the Reproductive Success of Angiosperms?*

2. Gymnosperms are referred to as naked-seed plants because they
   a. lack ovules.
   b. lack ovaries.
   c. do not protect their embryos.
   d. do not have seed coats.
   e. make only very small fruits.
   Textbook Reference: *29.2 What Are the Major Groups of Gymnosperms?*

3. Which of the following statements about seeds is true?
   a. Seeds provide a mechanism for a plant's dispersal.
   b. Seeds protect the plant embryo.
   c. Seeds allow an embryo to remain dormant until optimum growth conditions are available.
   d. Seeds provide nutrients for the developing embryo.
   e. All of the above
   Textbook Reference: *29.1 How Did Seed Plants Become Today's Dominant Vegetation?*

4. Which of the following roles do animals play in the life cycle of plants?
   a. They act as pollinators.
   b. They assist in dispersal of seeds.
   c. They promote fertilization.
   d. They contribute to genetic diversity of plants.
   e. All of the above
   Textbook Reference: *29.3 How Do Flowers and Fruits Increase the Reproductive Success of Angiosperms?*

5. In gymnosperms, fertilization results in a _______. In angiosperms, fertilization results in a(n) _______.
   a. diploid zygote; haploid zygote
   b. diploid zygote; endosperm nucleus
   c. diploid zygote; diploid zygote and an endosperm nucleus
   d. haploid zygote and an endosperm nucleus; diploid zygote
   e. haploid zygote; diploid zygote and an endosperm nucleus
   Textbook Reference: *29.3 How Do Flowers and Fruits Increase the Reproductive Success of Angiosperms?*

6. The identifying characteristic of an angiosperm is the presence of
   a. multiple carpels.
   b. woody growth.
   c. a flower.
   d. secondary growth.
   e. All of the above
   Textbook Reference: *29.3 How Do Flowers and Fruits Increase the Reproductive Success of Angiosperms?*

7. Which of the following statements regarding gymnosperms is true?
   a. All gymnosperms produce cones.
   b. Gymnosperms are heterosporous.
   c. Gymnosperm seeds have no protection.
   d. Only some living gymnosperms are woody.
   e. All gymnosperms have swimming sperm.
   Textbook Reference: *29.1 How Did Seed Plants Become Today's Dominant Vegetation?*

8. Which of the following statements about the calyx is *false*?
   a. The calyx is a collection of modified leaves.
   b. The calyx functions to protect immature flower parts within the bud.

c. The calyx is a source of gametes.
d. The calyx consists of all of the sepals as a group.
e. The calyx, together with the corolla, forms the perianth.

Textbook Reference: *29.3 How Do Flowers and Fruits Increase the Reproductive Success of Angiosperms?*

9. Which of the following statements about the function of a fruit is true?
a. It aids in dispersal of seeds.
b. It protects seeds until they are mature.
c. It attracts pollinators.
d. It provides nutrients to the embryo.
e. Both a and b

Textbook Reference: *29.3 How Do Flowers and Fruits Increase the Reproductive Success of Angiosperms?*

10. Vascular tissue in angiosperms is highly developed. The purpose of this vascular tissue is to move
a. water.
b. food.
c. nutrients.
d. water and materials dissolved in the water.
e. All of the above

Textbook Reference: *29.3 How Do Flowers and Fruits Increase the Reproductive Success of Angiosperms?*

11. More than half of the world's population relies on the seeds of the _______ plant as food.
a. corn
b. rice
c. soybean
d. common bean
e. wheat

Textbook Reference: *29.4 How Do Plants Benefit Human Society?*

12. Which of the following drugs is derived from foxglove and used to strengthen contractions of the heart?
a. Atropine
b. Ephedrine
c. Morphine
d. Tubocurarine
e. Digitalin

Textbook Reference: *29.4 How Do Plants Benefit Human Society?*

13. Which of the following represents the correct path followed by the sperm from the pollen grain to the female gametophyte in an angiosperm?
a. Stigma, style, ovary, ovule, egg
b. Sepal, integument, style, ovary, ovule
c. Anther, filament, pollen tube, style, ovary
d. Ovule, ovary, style, stigma, egg
e. Receptacle, pistil, sepal, integument, egg

Textbook Reference: *29.3 How Do Flowers and Fruits Increase the Reproductive Success of Angiosperms?*

14. How many generations of material are present in a seed?
a. One: the haploid female gametophyte tissue
b. Two: the haploid female gametophyte tissue and the embryo of the diploid sporophyte
c. Two: the diploid female gametophyte tissue and the embryo of the diploid sporophyte
d. Three: integument tissue from the diploid sporophyte parent, the haploid female gametophyte tissue, and the embryo of the diploid sporophyte
e. Three: integument tissue from the diploid sporophyte parent, the diploid female gametophyte tissue, and the embryo of the diploid sporophyte

Textbook Reference: *29.1 How Did Seed Plants Become Today's Dominant Vegetation?*

15. Which of the following is *not* an ecological service performed by plants?
a. Removing CO_2 from the atmosphere
b. Reducing erosion
c. Increasing atmospheric humidity
d. Increasing soil humidity
e. Aiding in soil formation

Textbook Reference: *29.4 How Do Plants Benefit Human Society?*

Answers

Key Concept Review

1.
a. Pollen tube
b. Anther
c. Filament
d. Pollen grains
e. Petal
f. Stigma
g. Style
h. Ovary
i. Ovule
j. Sepal
k. Integument
l. Megagametophyte
m. Receptacle

2. Pollen is the male gametophyte of seed vascular plants; in other words, it is the male multicellular haploid structure of the seed-producing vascular plant's life cycle (see Figure 29.16). The most fundamental function of the pollen gametophyte is to create gametes (sperm, in this case), but it also plays a role in helping those gametes reach and fertilize female gametes. Pollen is significant in that it provides a means of dispersal that eliminates the need for water in fertilization.

3.
a. Diploid
b. Haploid
c. Immature megastrobilus

d. Scale of megastrobilus
e. Ovule
f. Section through scale
g. Integument
h. Megasporocyte
i. Megasporangium
j. Meiosis
k. Functional megaspore
l. Pollen chamber
m. Micropyle
n. Pollen grain
o. Microstrobili
p. Scale of microstrobilus
q. Section through scale
r. Meiosis
s. Microspores
t. Pollen grain
u. Female gametophyte
v. Archegonium
w. Egg
x. Sperm
y. Male gametophyte (germinating pollen grain)
z. Fertilization
aa. Zygote
bb. Seed
cc. Wing
dd. Scale of megastrobilus
ee. Mature megastrobilus
ff. Winged seed
gg. Embryo
hh. Female gametophyte (provides nutrition for developing embryo)
ii. Seed coat
jj. Sporophyte (about 10–110 m)

4. Gymnosperms have a sophisticated vascular system that allows for movement of nutrients and also provides support for large plants competing for sunlight. Wind-dispersed pollen eliminates the dependence on water for fertilization and increases the range of gymnosperms. Seeds protect embryos from desiccation and provide a mechanism for dispersal.

5.
a. Diploid
b. Haploid
c. Flower of mature sporophyte
d. Ovary
e. Ovule
f. Anther
g. Microsporocyte
h. Ovary
i. Ovule
j. Megasporocyte ($2n$)
k. Megasporangium
l. Meiosis
m. Microspores (4)
n. Pollen grain
o. Surviving megaspore (n)
p. Polar nuclei (2)
q. Egg
r. Sperm (2)
s. Pollen tube
t. Pollen grains (microgametophyte, n)
u. Double fertilization
v. Zygote ($2n$)
w. Endosperm nucleus ($3n$)
x. Embryo
y. Endosperm
z. Seed
aa. Seedling

6. Double fertilization allows for a diploid embryo and a triploid endosperm. The endosperm provides nutrition to the developing embryo at the time of seed germination.

7. The female gametophyte is highly reduced and is part of the ovule of the angiosperm.

8. A cotyledon is the first "seed leaf" produced in an angiosperm embryo. Monocots have only one seed leaf, whereas eudicots have two.

9. Angiosperm fruit arises from tissues of the sporophyte and the gametophyte. How the fruit develops depends on fruit type. Different fruits encompass more or less of the parent sporophyte tissues.

10.
e. Ovule
g. Anther
b. Stigma
f. Style
c. Ovary
a. Petal
d. Sepal

11. Plants provide shade, affect local humidity, and block wind, to name a few effects. Changes in the plant makeup of a particular area can greatly change these moderating effects and alter the normal climate conditions, such as allowing more sunlight to penetrate to the ground, allowing the sun to evaporate more moisture, and allowing wind to contribute to erosion, among other effects.

12. One potential source of medicinal plants is the tropical rain forest, mostly due to the huge diversity of species it contains.

Test Yourself

1. **d.** This is most likely a dioecious plant in which male and female flowers appear on separate specimens. The absence of stamens makes these flowers imperfect.

2. **b.** Gymnosperms lack ovaries and thus lack the ability to produce fruit. Their embryos are protected, and they have ovules and seed coats.

3. **e.** Seeds protect the embryo, provide a mechanism for dispersal, provide nutrients to the embryo, and allow an embryo to remain dormant.

4. **e.** By acting as pollination vectors, animals promote fertilization of the plant, thus increasing genetic diversity. Animals also assist with dispersal of seeds.
5. **c.** Angiosperms differ from gymnosperms in having double fertilization that results in a triploid endosperm nucleus in addition to the zygote.
6. **c.** Only angiosperms have flowers. It is true that only angiosperms can have multiple carpels, but multiple carpels are not a necessary characteristic of an angiosperm. Not all angiosperms undergo secondary (woody) growth.
7. **b.** All gymnosperms are heterosporous. Only conifers are cone producers. Only the earliest groups of angiosperms had swimming sperm.
8. **c.** The calyx is the collective term for the sepals, which are specialized leaves. The sepals can be showy and play a role in attracting pollinators, but they are not a source of gametes.
9. **e.** By the time fruit forms, a flower has already been pollinated, and there is no need to attract additional pollinators. The seed provides nutrients to the embryo, but the surrounding fruit does not participate in this process.
10. **e.** Vascular tissues in angiosperms move food, water, and nutrients dissolved in the water (such as sugars and minerals) throughout the plant.
11. **b.** Rice is the seed consumed by over one-half of the world's population. There are 12 vital crops listed in your textbook, but human consumption is not always limited to the seed tissue, as specified by the question.
12. **e.** Digitalin is derived from *Digitalis purpurea*, commonly called foxglove.
13. **a.** The pollen grain lands on the top of the pistil and two sperm move through a pollen tube from the stigma to the style, then to the ovary and into the ovule, where a sperm meets with the egg. (Chapter 38 explains how the other sperm fuses with polar nuclei to form triploid endosperm.)
14. **d.** The plant's seed coat is the first generation, the female gametophyte is the second generation, and the embryo is the third generation.
15. **d.** Plants increase atmospheric humidity by reducing soil humidity. They also reduce erosion by blocking wind and holding soil in place. Plants also consume CO_2 and use it to build organic compounds.

30 The Evolution and Diversity of Fungi

The Big Picture

- Fungi are heterotrophic organisms that absorb nutrients from the environment. Fungi can be unicellular or multicellular. The mycelium is the body of a multicellular fungus and is composed of many tubular filaments known as hyphae. Many fungi form symbiotic relationships with photosynthetic organisms, creating mycorrhizae and lichens.
- Reproduction in the fungi occurs both sexually and asexually. Asexual reproduction involves the production of spores, breakage, fission, or budding. Sexual reproduction in multicellular fungi requires the union of hyphae with two different mating types.
- Fungi are classified into six major groups: microsporidia, chytrids, Zygomycota, Glomeromycota, Ascomycota, and Basidiomycota.
  - Microsporidia are unicellular parasitic fungi. They lack true mitochondria, instead possessing structures known as mitosomes.
  - Chytrids are aquatic fungi with flagellated gametes. Many chytrids display alternation of generations.
  - Zygospore fungi (Zygomycota) have coenocytic hyphae. Sexual reproduction occurs when hyphae of two mating types join to form a zygosporangium from which a sporangium will eventually grow.
  - Arbuscular mycorrhizal fungi (Glomeromycota) are terrestrial, coenocytic, and asexual. They are the predominant arbuscular fungi and as such are essential to plant life.
  - Sac fungi (Ascomycota) produce an ascus as their reproductive structure. They reproduce by budding, fission, or the production of an ascus. Many have a dikaryotic ($n + n$) stage during reproduction.
  - Club fungi (Basidiomycota) have a reproductive structure called the basidium. The fusion of haploid hyphae forms a dikaryotic mycelium that produces the fruiting body, a basidiocarp.

Study Strategies

- Students are often confused by the numerous ways that fungi reproduce. Create flowcharts to help you understand the ploidy level at each stage of the life cycle in the different phyla of fungi. Make sure to include both the sexual and asexual stages in your charts. Comparisons of your flowcharts for the different phyla will help you learn the differences among the groups.
- Organisms are placed in particular systematic groupings because they share unique characteristics. Look for patterns that distinguish the six major fungal groups from one another. Most fungi are grouped according to the reproductive structures that they produce.
- Go to the Web addresses shown to review the following animated tutorial and activities. You can also find them in BioPortal (yourBioPortal.com), along with many additional learning resources.

  Animated Tutorial 30.1 Life Cycle of a Zygospore Fungus (Life10e.com/at30.1)

  Activity 30.1 Fungal Phylogeny (Life10e.com/ac30.1)

  Activity 30.2 Life Cycle of a Dikaryotic Fungus (Life10e.com/ac30.2)

Key Concept Review

30.1 What Is a Fungus?

Unicellular yeasts absorb nutrients directly

Multicellular fungi use hyphae to absorb nutrients

Fungi are in intimate contact with their environment

Fungi are absorptive heterotrophs, that is, they secrete digestive enzymes in order to break down food outside their bodies. Fungi can absorb the nutrition needed for their survival from dead matter (saprobes), from living hosts (parasites), and from mutually beneficial symbiotic relationships with other organisms (mutualists). Modern fungi probably evolved from a unicellular protist with a flagellum, much like the probable common ancestor of animals. The current hypothesis is that animals, choanoflagellates, and fungi share a flagellated common ancestor and form a lineage known as the opisthokonts (see Figure 30.1). A synapomorphy of opisthokonts is a posterior flagellum, whereas other eukaryotes have anterior flagella. Synapomorphies of fungi are absorptive heterotrophy and cell walls made of chitin. Unicellular fungi are commonly referred to as yeasts. Yeasts may reproduce by budding (see Figure 30.2), by fission, or sexually. The term yeast refers to a lifestyle, not a taxonomic group; some fungi have both a yeast stage and a multicellular stage. Yeasts are ideal model organisms because they are easily cultured and, as eukaryotes, are more similar to humans than bacteria.

The body of a multicellular fungus is called a mycelium. A mycelium is made up of individual tubular filaments called hyphae. Hyphae grow rapidly into a substrate; the hyphae in a single mycelium may collectively grow as much as 1 kilometer per day. Hyphal cell walls are strengthened by the polysaccharide chitin. Hyphae may be septate (divided into compartments by chitinous walls) or coenocytic (continuous and multinucleate) (see Figure 30.3). Rhizoids are hyphae that anchor fungi to their substrate. Often the mycelia of a single fungus spread over a wide area, even though this is not obvious above ground; common mushrooms are merely the fruiting structures of a large underground mycelium network. When food is abundant, fungi produce many spores, but as food supplies dwindle they often produce even more. Spores may remain dormant until conditions improve, or they may disperse to areas with new food supplies. Fungal spores are easily spread by wind and water because of their very tiny size.

Hyphae provide a large surface area-to-volume ratio, and thus a large surface area for absorption of nutrients from the substrate. Because of this, they tend to lose water quickly and are usually, but not always, found in moist environments. Fungi are tolerant of hypertonic environments and temperature extremes.

Question 1. Label the diagram below using the following terms: hypha, nuclei, cell wall, septa.

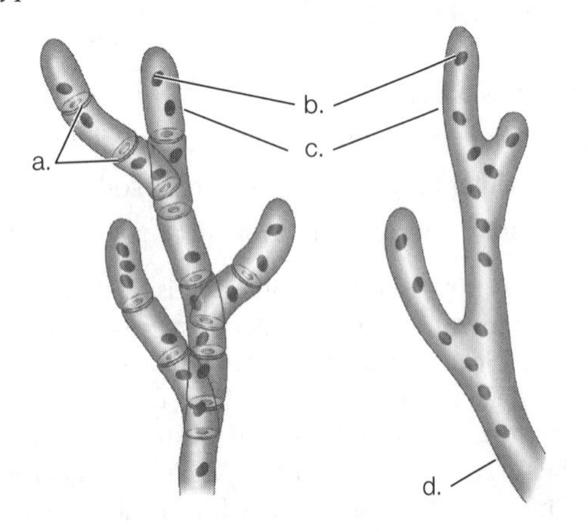

Question 2. Early taxonomists considered fungi to be members of the plant kingdom. What evidence indicates that they are in fact more closely related to animals?

30.2 How Do Fungi Interact with Other Organisms?

Saprobic fungi are critical to the planetary carbon cycle

Some fungi engage in parasitic or predatory interactions

Mutualistic fungi engage in relationships that benefit both partners

Endophytic fungi protect some plants from pathogens, herbivores, and stress

Fungi are responsible for recycling nutrients from dead organisms, and they also made possible colonization of the terrestrial environment. Saprobic fungi secrete enzymes into the environment that help in the absorption of dead matter and the decomposition and recycling of elements—especially carbon—used by living organisms. As decomposers, saprobic fungi are essential to life on Earth. There was a time in the history of Earth when the numbers of saprobic fungi declined; tropical swamps were not decomposed, instead forming peat, and over time those layers of peat became coal. Saprobic fungi use simple sugars, polysaccharide breakdown products, proteins or protein breakdown products, nitrate, or ammonium ions as nutrients. No fungus can use nitrogen gas from the environment (they cannot fix nitrogen).

Parasitic fungi are either facultative (possessing the ability to grow independently) or obligate (dependent on their living host for growth). Parasitic fungi most often infect plants and insects. Hyphae can penetrate a leaf through the stomata or through wounds. Some hyphae produce haustoria, which push through cell walls and absorb nutrients directly from the cell; they do not penetrate the cell membrane but make close contact and invaginate into it (see Figure 30.5). Some parasitic fungi not only derive nutrition from their host, but also sicken or kill the host. Such fungi are called pathogens. Pathogenic fungi cause many human diseases; people with compromised immune systems are particularly susceptible to pathogenic fungal infections. Fungi are the most important plant pathogens, causing huge crop losses every year. Some fungi are active predators that trap microscopic protists or animals in a sticky substance; others form a constricting ring around their prey (see Figure 30.6).

Fungi form two crucial types of symbiotic, mutualistic relationships: lichens and mycorrhizae. Lichens are associations of fungi with a unicellular photosynthetic alga or cyanobacteria. The fungus provides the photosynthetic partner with minerals and water, and the photosynthesizing organism provides the fungus with organic compounds. Lichens are among Earth's hardiest organisms, and can thrive in barren and extreme environments such as Antarctica. Lichens are characterized by their appearance as crustose (crusty), foliose (leafy), or fruticose (shrubby) (see Figure 30.7). Lichens can reproduce via fragmentation of the thallus (vegetative body) or through dispersal of soredia. Soredia consist of photosynthetic cells with fungal hyphae. The fungal partner of a lichen can undergo sexual reproduction, but it will disperse without the photosynthetic partner. Lichens are important colonizers of bare rock; they acidify the environment slightly, causing the breakdown of the rock and the release of minerals.

Mycorrhizae are associations between fungi and the roots of plants in which the fungus obtains the products of photosynthesis from the plant and provides minerals and water to the plant. The mycorrhizal symbiosis is essential to the survival of most plants, and the evolution of this relationship may have been the most important step in allowing plants to colonize land. Ectomycorrhizal fungi wrap their hyphae around a plant root. The hyphae of arbuscular mycorrhizal fungi penetrate the root cell walls and form treelike (arbuscular) structures outside the plasma membrane that provide the plant with nutrients (see Figure 30.9). Endophytic fungi are symbionts living within the aboveground plant parts. They help certain plants (especially grasses) resist pathogens, herbivores, and stresses such as drought and salty soil; their role in other plants is not well understood.

Question 3. Explain the different ways in which fungi are important to plants.

Question 4. How is basic research on fungi relevant to research on the prevention and cure of HIV/AIDS?

Question 5. Review the material in Chapter 7 and then list and explain three examples of cell signaling in fungi.

30.3 How Do Major Groups of Fungi Differ in Structure and Life History?

- Fungi reproduce both sexually and asexually
- Microsporidia are highly reduced, parasitic fungi
- Most chytrids are aquatic
- Some fungal life cycles feature separate fusion of cytoplasms and nuclei
- Arbuscular mycorrhizal fungi form symbioses with plants
- The dikaryotic condition is a synapomorphy of sac fungi and club fungi
- The sexual reproductive structure of sac fungi is the ascus
- The sexual reproductive structure of club fungi is the basidium

Fungi were originally distinguished by morphology and sexual reproductive processes. Most fungi belong to one of six major groups: microsporidia, chytrids, Zygomycota, Glomeromycota, Ascomycota, and Basidiomycota. Based on evidence from DNA analysis, the chytrids and zygomycota appear to be paraphyletic, while the other three groups are clades (i.e., monophyletic; see Figure 30.10 and Table 30.1). The chytrids are aquatic; the other four groups are terrestrial.

Fungi reproduce both sexually and asexually. Fungi have several means of asexual reproduction; they can produce spores enclosed within sporangia or naked spores at the tips of hyphae known as conidia. Unicellular fungi can reproduce by fission or budding. Just about any part of a mycelium is capable of living independently of the rest; therefore, simple division of a mycelium into two or more parts is a method of reproduction. Sexual reproduction is rare in some groups of fungi and common in others. Sexual reproduction involves two or more mating types. The first step in sexual reproduction is fusion of two hyphae of different mating types (called plasmogamy). Eventually, two nuclei (one from each mating type) will fuse to create a zygote nucleus (a process called karyogamy).

Microsporidia are unicellular parasitic fungi. They lack true mitochondria, possessing greatly reduced structures called mitosomes, which are derived from mitochondria but lack DNA. Some species can infect mammals (including humans), and most infections by microsporidia cause chronic disease. The spore penetrates the host cell and injects its contents; the cell is then used to make new infective spores.

Chytrids likely resemble the common ancestor of all fungi more closely than any other fungal group. Chytrids are aquatic, have chitinous cell walls, and are the only fungi with flagellated gametes. They reproduce both asexually and sexually, with the multicellular haploid stage producing independent, flagellated male and female gametes. They can be either parasitic or saprobic and are diverse in form; some are unicellular, some have rhizoids, and some have coenocytic hyphae. The genus *Allomyces* displays alternation of generations. A haploid zoospore becomes a small organism (the gametophyte) with both female and male gametangia that produce gametes (see Figure 30.14A). When the gametes fuse, they form a diploid zygote and a diploid organism (the sporophyte). The sporophyte produces diploid zoospores that grow into other diploid sporophytes. Eventually, a sporophyte produces sporangia that give rise to haploid zoospores via meiosis.

Sexual reproduction in zygomycota occurs between adjoining individuals with different mating types (see Figure 30.14B). Stimulated by chemical attractants, branches from each individual grow toward each other. These hyphae produce gametangia, specialized reproductive cells retained in the hyphae that replicate nuclei without cell division. The two multinucleate haploid gametangia fuse, forming a zygosporangium that contains many haploid nuclei of both mating types. Haploid nuclei of different mating types pair to form diploid nuclei; a thick cell wall forms around the cell, which is now called a zygospore. When environmental conditions are good, the zygospore nuclei undergo meiosis and one or more sporangiophores form, each with a sporangium. Each sporangium contains haploid nuclei incorporated into new spores. More than 1,000 species of zygospore fungi have been described, including *Rhizopus stolonifer*, common black bread mold. *Rhizopus* produces many stalked sporangiophores, each with a single sporangium containing hundreds of spores.

Glomeromycota, which form arbuscular mycorrhizae, are entirely asexual and coenocytic. They are also entirely terrestrial. The 200 species of glomeromycota form arbuscular mycorrhizae with the roots of 80 to 90 percent of all plants (see Figure 30.9B).

The last two groups of fungi, sac fungi and club fungi, form a clade known as the Dikarya. In these fungi, karyogamy occurs long after plasmogamy, and two genetically different haploid nuclei divide within the cell during this time. This stage, called a dikaryon, has a ploidy of $n + n$.

Eventually, specialized fruiting structures form, within which two different nuclei fuse; the zygote ($2n$) undergoes meiosis and then produces four haploid nuclei. Mitosis of these cells produces spores, which germinate to create the next haploid generation.

There are approximately 64,000 species of Ascomycota, which were historically divided into two groups. Based on new DNA sequence analysis, these groupings have recently been abandoned. The Ascomycota are distinguished by their ascus, a sexual reproductive structure (see Figure 30.16A). Each ascus produces haploid ascospores. The species of yeast used to make bread and alcoholic beverages, *Saccharomyces cerevisiae,* is a unicellular sac fungus. Reproduction in these yeasts is asexual and accomplished by budding. Most sac fungi are filamentous, such as the cup fungi, with cup-shaped ascomata; some ascomata are edible, such as morels and truffles. Sexual reproduction of filamentous sac fungi involves two different mating types moving into a dikaryotic stage before forming a diploid cell population. A dikaryon is produced during sexual reproduction by the fusing of two mating structures (see Figure 30.16A). From the dikaryon hyphae ($n + n$), dikaryotic asci form where the nuclei fuse. Molds are a group of filamentous sac fungi. Molds do not form ascomata but can produce asci. Many molds are parasites of flowering plants, such as chestnut blight, Dutch elm disease, and powdery mildews. Asexual reproduction in Ascomycota involves the production of conidia at the tips of hyphae (see Figure 30.18).

Club fungi (Basidiomycota) form dramatic fruiting structures called basidiomata, including mushrooms and puffballs. About 30,000 species have been described. The hyphae of club fungi are septate with small, distinctive pores. Haploid hyphae of different mating types fuse, forming dikaryotic hyphae; the dikaryon stage can persist for years. They form a sexual reproductive structure called a basidium, a swollen cell at the tip of a specialized hypha. The basidium is the site of nuclear fusion and meiosis and is analogous to the ascus and the zygosporangium. After nuclei fuse, the diploid nucleus undergoes meiosis and forms four haploid nuclei that are incorporated into basidiospores.

Question 6. Four major fungal life cycles are represented below. Label each according to the major group it represents. Indicate the haploid, diploid, and dikaryotic stages. In the boxes, label the process as mitosis, meiosis, or fertilization. Then label each structure using the appropriate terms.

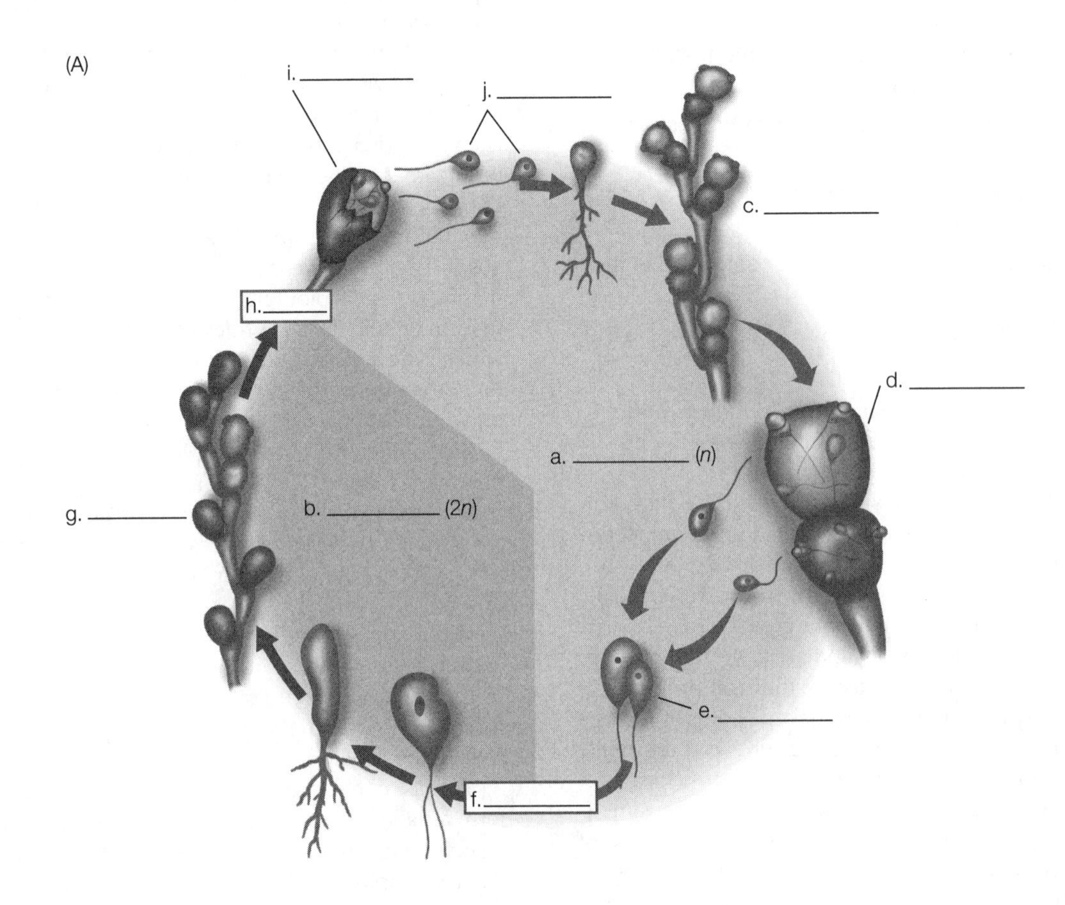

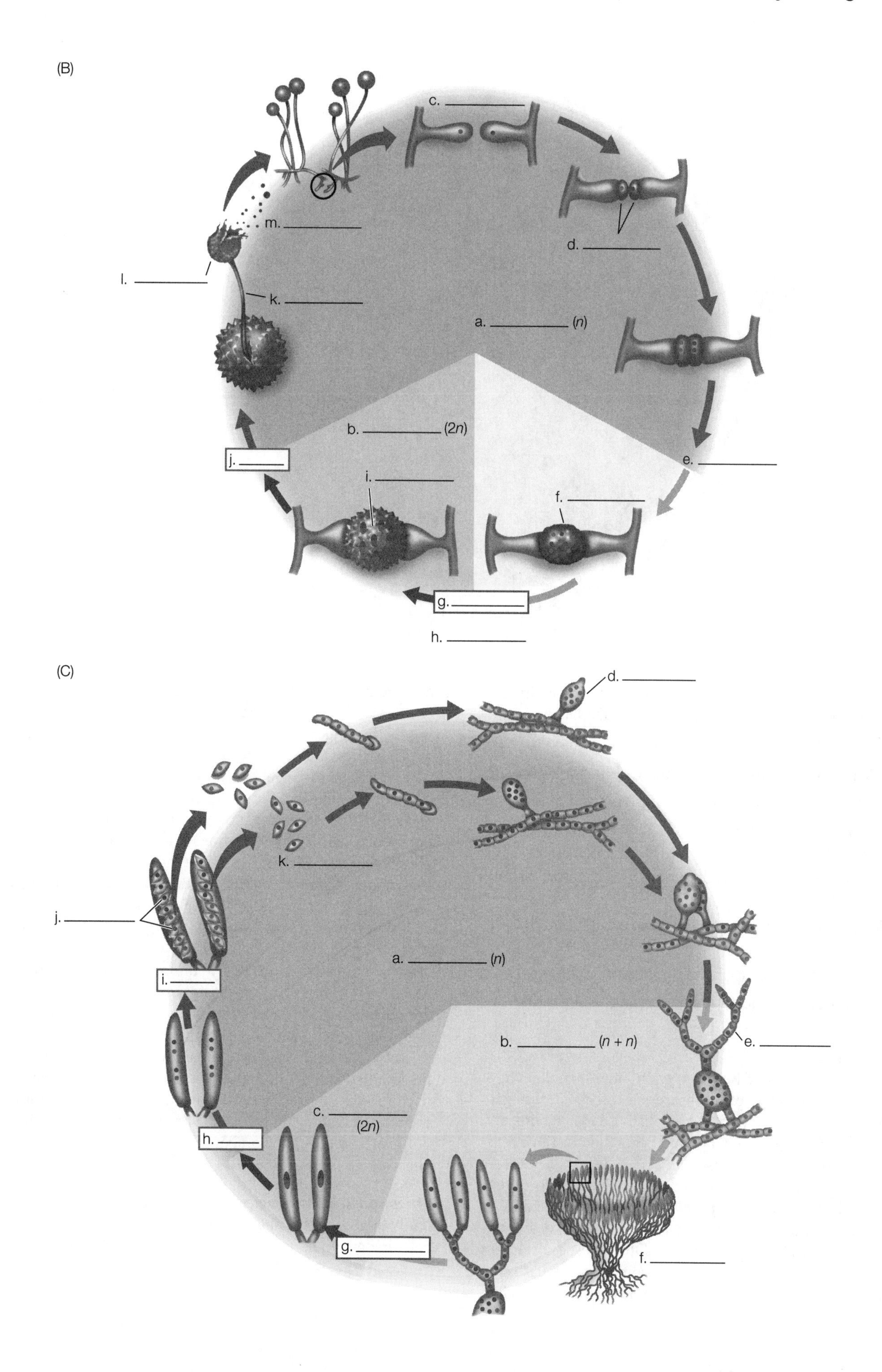
(B)
c.
d.
m.
l.
k.
a. (n)
b. (2n)
j.
i.
e.
f.
g.
h.
(C)
d.
k.
j.
a. (n)
i.
b. (n + n)
e.
c.
(2n)
h.
g.
f.

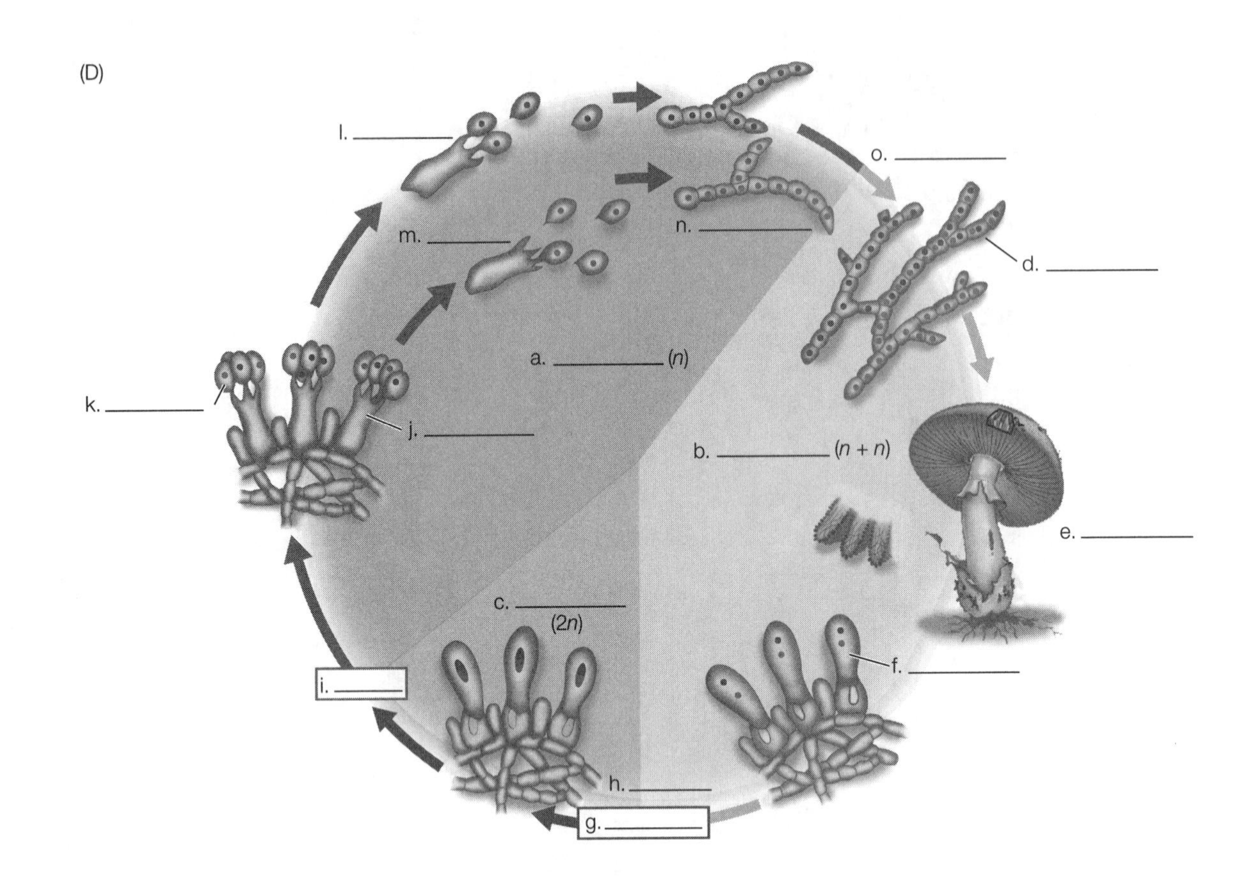

Question 7. Why have scientists abandoned the classification of fungi based on reproductive cycles? Explain and evaluate the justifications for this decision.

30.4 What Are Some Applications of Fungal Biology?

- Fungi are important in producing food and drink
- Fungi record and help remediate environmental pollution
- Lichen diversity and abundance are indicators of air quality
- Fungi are used as model organisms in laboratory studies
- Reforestation may depend on mycorrhizal fungi
- Fungi provide important weapons against diseases and pests

The baker's (or brewer's) yeast *Saccharomyces cerevisiae* is used to convert the starches from grain into ethanol in wine and beer. A by-product of this process is carbon dioxide gas, which leavens bread and gives beer its fizz. Local strains of *S. cerevisiae* give regional wines and beers their distinctive flavors. Other species of yeasts can also be used, such as *Schizosaccharomyces pombe,* named for the Swahili word for beer. Molds of the genus *Aspergillus* are also harnessed in production of food products such as soy sauce, sake, and citric acid (added to foods for tartness). Other species of *Aspergillus* produce toxic and carcinogenic compounds (aflatoxins) and are thus dangerous. *Penicillium* molds produce penicillin, and are used to make cheese (think of the blue veins in some cheeses). We also eat certain species of mushrooms, harvesting the fruiting bodies of many sac and club fungi. Lichens are also eaten in Arctic and other regions.

Collections of fungal species help scientists analyze past patterns of pollution. Fungi are also helpful in remediation and are used to clean up sites contaminated with crude oil or petroleum-derived hydrocarbons. Lichen species are highly sensitive to air pollution, and thus their relative abundance is a good indicator of pollution levels. Fungi are also used as model organisms in laboratories. Several species of sac fungi are especially important, such as *Aspergillus niger, Neurospora crassa, Saccharomyces cerevisiae,* and *Schizosaccharomyces pombe.* These species are easily cultured in the lab in small spaces and have short generation times. Their relatively small genomes make basic eukaryotic genetic studies much easier than they would be in more complex eukaryotes.

Reforestation likely depends on the mycorrhizal fungi present in the soil. Often the mycorrhizae population will decline after forests are cut down, and any efforts to restore the forest must include reestablishment of the mycorrhizae. Fungi are also used to combat crop diseases, weeds, and pests such as mosquitoes and aphids.

Question 8. Many species of fungi are used in laboratories to study biology. For what kinds of questions about eukaryotic cell biology would fungi be good study specimens?

Question 9. Imagine you are a scientist looking to improve the yield of a particular crop. What strategy involving the use of fungi might you employ to achieve this goal?

Test Yourself

1. Many species of fungi can be placed into one of six main groups based on
   a. methods of sexual reproduction.
   b. whether or not gametes have a flagella.
   c. the presence or absence of septa in the hyphae.
   d. DNA sequence analysis.
   e. All of the above
   Textbook Reference: *30.3 How Do Major Groups of Fungi Differ in Structure and Life History?*

2. Which of the following is *not* found in any of the typical fungal life cycles?
   a. Haploid nuclei
   b. Diploid nuclei
   c. Spores
   d. Chloroplasts
   e. A dikaryotic stage
   Textbook Reference: *30.3 How Do Major Groups of Fungi Differ in Structure and Life History?*

3. Fungi are absorptive heterotrophs. Which of the following is an adaptation that greatly aids this mode of nutrient procurement?
   a. Dikaryosis
   b. A large surface area-to-volume ratio
   c. Conjugation
   d. A complex life cycle
   e. A small surface area-to-volume ratio
   Textbook Reference: *30.1 What Is a Fungus?*

4. Assume that two normal hyphae of different fungal mating types meet. After a period of time, the cell walls between these hyphae will dissolve, producing a
   a. mycelium.
   b. fruiting body.
   c. zygote.
   d. spore.
   e. dikaryotic cell.
   Textbook Reference: *30.3 How Do Major Groups of Fungi Differ in Structure and Life History?*

5. Which of the following is the best way to represent the ploidy of dikaryotic hyphae?
   a. $\frac{1}{2}n$
   b. n/n
   c. n
   d. $n + n$
   e. $2n$
   Textbook Reference: *30.3 How Do Major Groups of Fungi Differ in Structure and Life History?*

6. Which of the following statements about sexual reproduction in fungi is *false*?
   a. Motile gametes are present in all fungal species.
   b. An aquatic environment is not required for fertilization to occur in most fungi.
   c. There is no true diploid tissue in the life cycle of most sexually reproducing fungi.
   d. Sexual reproduction often begins with contact between hyphae of different mating types.
   e. Alternation of generations and sexual reproduction can occur in the same species.
   Textbook Reference: *30.3 How Do Major Groups of Fungi Differ in Structure and Life History?*

7. A mycorrhiza is
   a. a specialized type of lichen.
   b. the fruiting structure of a basidiomycota.
   c. a symbiotic association between a fungus and cyanobacterium or green algae.
   d. a reproductive stage of sac fungi.
   e. a symbiotic association between a fungus and a plant.
   Textbook Reference: *30.2 How Do Fungi Interact with Other Organisms?*

8. Suppose a scientist investigating the classification of a fungus that has never been observed to reproduce sexually discovers that it has DNA sequences characteristic of basidiomycota. If this scientist could get fungi of this species to reproduce sexually, which of the following would most likely be observed?
   a. Dikaryotic hyphae segmented by septa
   b. Dikaryotic hyphae without septa
   c. Asymmetrical cell division
   d. A dikaryotic ascus
   e. Flagellated gametes
   Textbook Reference: *30.3 How Do Major Groups of Fungi Differ in Structure and Life History?*

9. Which of the following fungi have coenocytic hyphae and stalked sporangiophores?
   a. Chytrids
   b. Zygomycota
   c. Ascomycetes
   d. Basidiomycota
   e. Glomeromycota
   Textbook Reference: *30.3 How Do Major Groups of Fungi Differ in Structure and Life History?*

10. Which of the following fungi display alternation of generations?
    a. Chytrids
    b. Glomeromycota
    c. Ascomycota
    d. Basidiomycota
    e. Zygomycota
    Textbook Reference: *30.3 How Do Major Groups of Fungi Differ in Structure and Life History?*

11. A saprobe is an organism that
    a. absorbs nutrients from the sap of a host plant.
    b. reproduces in the sap of a plant.
    c. undergoes asexual reproduction.
    d. is mutualistic.
    e. absorbs nutrients from dead organic matter.
    Textbook Reference: *30.1 What Is a Fungus?*

12. Which of the following is *not* a form of asexual reproduction in fungi?
    a. Budding
    b. Formation of haploid spores in sporangia
    c. Formation of dikaryotic mycelia
    d. Fission
    e. Formation of haploid spores in conidia
    Textbook Reference: *30.1 What Is a Fungus?*

13. Which of the following statements about fungi is *false* (i.e., a common misconception)?
    a. Fungi grow only in warm, wet environments.
    b. Fungi lose water rapidly in a dry environment.
    c. Fungi can grow in environments too hypertonic to sustain bacteria.
    d Fungi are eukaryotes and have multiple mating types.
    e. Male and female fungi have no distinctive morphology.
    Textbook Reference: *30.1 What Is a Fungus?*

14. Which of the following scientific groups is correctly matched with its correct common name?
    a. Chytrids–microspore fungi
    b. Zygomycota–sac fungi
    c. Glomeromycota–mycorrhizial fungi
    d Basidiomycota–sac fungi
    e. Ascomycota–club fungi
    Textbook Reference: *30.3 How Do Major Groups of Fungi Differ in Structure and Life History?*

15. Which of the following is a human use of fungi?
    a. Food production
    b. Air quality monitoring
    c. Studying the genetics of cell biology
    d Pest control
    e. All of the above
    Textbook Reference: *30.4 What Are Some Applications of Fungal Biology?*

Answers

Key Concept Review

1.
   a. Septa
   b. Nuclei
   c. Cell wall
   d. Hypha

2. At the cellular level, fungi bear very little resemblance to plants or even to the surviving green algae that are the most likely common ancestor of all plants. They do not produce chlorophyll, nor do they have plastids for storing reserves of photosynthetic products. Fungi and plants both have cell walls, but fungal cell walls contain chitin, a polysaccharide molecule (see Chapter 3) that plants do not produce. Chitin is found in some animals (particularly among the ecdysozoans) and in choanoflagellates, the protist group most closely related to the animals. It is unlikely that this complex molecule evolved more than once, so it is likely that fungi and animals shared a chitin-producing ancestor that is more recent than any ancestor shared by fungi and plants or by animals and plants.

3. Plants rely on mycorrhizae for adequate absorption of water and nutrients from the soil. Scientists hypothesize that this is the relationship that allowed plants to colonize land in the first place. Other fungi can provide their hosts with some resistance to herbivores, increased resistance to drought, and protection from some pathogens. The mechanisms of these protective actions are not fully understood.

4. Basic research in mycology (the study of fungi) may directly improve the lives of many HIV-positive individuals who suffer from fungal infections, many of which can be fatal. Better understanding of how fungal-based pneumonia, diarrhea, and esophagitis develop in the human body will help us treat these patients and alleviate suffering. Funding for basic mycology research may also have a direct impact on the treatment of people in the later stages of AIDS.

5. Many examples of cell signaling can be found among the fungi. For example, chemical signaling is the mechanism by which plants attract fungi. Predatory fungi make use of a detection mechanism in order to contract around a nematode, and cell signaling is necessary in the coordination of many different cells to form structures such as basidiomas, ascomas, diploid chytrids, and sporangia. Even proper mating types are identified by means of cell signaling.

6.
   (A) Chytrid life cycle
   a. Haploid
   b. Diploid
   c. Multicellular haploid chytrid (n)
   d. Female gametangium
   e. Gametes
   f. Fertilization
   g. Multicellular diploid chytrid ($2n$)
   h. Meiosis
   i. Sporangium
   j. Haploid zoospores (n)

   (B) Zygospore fungi (Zygomycota) life cycle
   a. Haploid

b. Diploid
c. Hyphae
d. Gametangia (n)
e. Plasmogamy
f. Zygosporangium
g. Fertilization
h. Karyogamy
i. Multiple diploid nuclei form within the zygosporangium
j. Meiosis
k. Sporangiophore
l. Sporangium
m. Spores

(C) Sac Fungi (Ascomycota)
a. Haploid
b. Dikaryotic
c. Diploid
d. Mating structure
e. Dikaryotic mycelium ($n + n$)
f. Ascoma (fruiting structure)
g. Fertilization
h. Meoisis
i. Mitosis
j. Ascospores
k. Ascospores (n)

(D) Club fungi (Basidiomycota)
a. Haploid
b. Dikaryotic
c. Diploid
d. Dikaryotic mycelium ($n + n$)
e. Basidioma (fruiting structure)
f. Developing basidium ($n + n$)
g. Fertilization
h. Karyogamy
i. Meiosis
j. Basidium
k. Basidiospores
l. + Mating type
m. – Mating type
n. Mycelial hyphae
o. Plasmogamy

7. At a time when molecular data was not available, a classification system based on morphological features made sense, since many of these structures reflect the reproductive strategies and thus the common ancestries of different organisms. This approach is limited, however, since it does not account for changes in reproductive strategies over time. As organisms have evolved and moved into new environments, their relationships to other species may not be reflected in physical features and current reproductive strategies. DNA sequence analysis allows scientists to examine the entire genome and compare conserved sequences that may not be expressed as visible changes. This new technology thus gives us new insights into the relationships among organisms.

8. Because fungi are eukaryotes with relatively simple genomes, species such as *Saccharomyces cerevisiae* are used in many research projects exploring aspects of cell biology. The goal is to apply knowledge gained from this species to higher eukaryotes. Fungi are useful laboratory specimens in studies, for example, on gene regulation, genomic organization and modifications, cell–cell signaling, and how cells move through the cell cycle and pass through mitotic checkpoints.

9. One beginning strategy could be an examination of the mycorrhizal relationships of the particular crop. The results of this analysis might suggest beneficial supplements that could be added to seed shipments, or a soil additive containing a mycorrhizal species that could be used with the particular crop at planting. Genetic techniques might even be able to improve the relationship between the fungus and the crop, thus creating a new, higher yielding strain.

Test Yourself

1. **e.** Traditionally, a fungus's method of sexual reproduction was the primary criterion in its classification, but all the traits listed are useful in classifying fungi.

2. **d.** Fungi have haploid and diploid nuclei at different stages of their life cycle, and many have a dikaryotic stage. They also produce spores. The chloroplasts are not found in the fungi, but in plants.

3. **b.** The large surface area-to-volume ratio of the hyphae increases the ability of a fungus to absorb nutrients.

4. **e.** When two hyphae of different mating types fuse, they form a dikaryotic hyphae.

5. **d.** Dikaryotic hyphae are neither truly diploid ($2n$) nor haploid (n). Because dikaryotic hyphae include genetic material from two haploid nuclei that remain separate, the best way to represent their ploidy is $n + n$.

6. **a.** Not all fungi have motile gametes; only the gametes of chytrids are motile.

7. **e.** Mycorrhizae are associations between fungi and the roots of plants. Lichens are symbiotic relationships between fungi and cyanobacteria or green algae.

8. **a.** If this fungus is indeed a basidiomycote, the fusing of hyphae of different mating types will most likely result in dikaryotic hyphae that are segmented by septa.

9. **b.** Coenocytic hyphae are characteristic of both the chytrids and the zygomycota, but only the zygomycota regularly produce sporangiophores.

10. **a.** Among the fungi, only the chytrids have a multicellular haploid stage and a multicellular true diploid stage.

11. **e.** Saprobes are organisms that absorb nutrients from dead matter. Some bacteria are saprobes.

12. **c.** The dikaryotic mycelium is a structure formed in the sexual reproduction life cycle.

13. **a**. Fungi are plentiful in warm, wet environments, but many also survive extreme temperatures and very dry conditions.
14. **c**. The correct associations are as follows: Microsporidia/microsporidia; Chytrids / chytrids; Zygomycota / zygospore; Glomeromycota / mycorrhizae; Ascomycota / sac fungi; Basidiomycota/club fungi.
15. **e.** All of the above are uses of fungi. Fungi are used to make cheese, wine, beer, and bread; lichens are used to assess pollution levels; some fungal species are used as model organisms; fungi can be released to control pests, as well as weeds.

Animal Origins and the Evolution of Body Plans

The Big Picture

- The animals are a monophyletic group that share a number of morphological and genetic traits. Animals are motile multicellular organisms that must ingest nutrients.
- Animals are distinguished by the number of cell layers found in their embryos, their type of body cavity, and their body symmetry.
- The importance of acquiring nutrition (food) has led to the evolution of a variety of feeding strategies that maximize the available sources of nutrition. Life cycles also vary considerably among the animals, and involve a considerable number of trade-offs.
- The eumetazoans encompass all animals except the three groups of sponges. Ctenophores and cnidarians are diploblastic eumetazoans that are not bilaterally symmetrical.
- Sponges can be classified in three different groups based on molecular evidence and the morphology of their spicules.
- Both ctenophores and cnidarians are characterized by the presence of a largely inert layer of gelatinous mesoglea. Ctenophores have a complete gut with a mouth and an anus, whereas cnidarians have a blind gastrovascular cavity with a single opening that ingests food and expels wastes. Cnidarians have simple nerve nets and muscle fibers, which allow them a level of control over their movements that the ctenophores do not have.

Study Strategies

- Many of the same evolutionary "themes" can be found in widely divergent species. Venom or toxin-producing structures, for example, are found in many different animal groups, from the cnidarians to spiders to snakes. As the different animal groups are described, create a table of the various characteristics or mechanisms that consistently appear. Use this table to help you organize your thinking about evolution and the creation of diversity.
- The phylogenetic trees in the textbook are good frameworks on which you can add more information. Doing so will provide you with a point of reference when you study the different groups. Remember that animals have a set of traits shared with and inherited from their ancestors, as well as distinctive derived traits.
- It is important that you understand what features of an animal group may give it an advantage in its environment and the compromises (trade-offs) that are made in other areas.
- Don't forget that even though particular animals may be placed in a group or clade, not all species have all of the features that characterize the group. For example, think about the reasons that sponges are considered animals even though they lack many of the features of other animals.
- Go to the Web addresses shown to review the following animated tutorial and activities. You can also find them in BioPortal (yourBioPortal.com), along with many additional learning resources.

  Animated Tutorial 31.1 Life Cycle of a Cnidarian (Life10e.com/at31.1)

  Activity 31.1 Animal Body Cavities (Life10e.com/ac31.1)

  Activity 31.2 Sponge and Diploblast Classification (Life10e.com/ac31.2)

Key Concept Review

31.1 What Characteristics Distinguish the Animals?

Animal monophyly is supported by gene sequences and morphology

A few basic developmental patterns differentiate major animal groups

Although there are exceptions, the following are characteristics that are commonly associated with animals:

(1) Multicellularity: Most animal life cycles feature a complex pattern of development from a single-celled zygote.

(2) Heterotrophic metabolism: All animals are heterotrophs and must have some means of acquiring nutrients from their environment.

(3) Internal digestion: Most animals (but not all) have an internal gut where digestion takes place.

(4) Movement and nervous systems: Most animals either move to their food or have some means to bring it to them. Movement is often coordinated by a well-developed nervous system. However, some animals have stages in their life cycle in which movement does not take place.

Not all animals have all of these characteristics in all life stages. Phylogenetic analyses of animal gene sequences support the conclusion that animals are monophyletic (see Figure 31.1). A summary of the living members of the major animal groups is presented in Table 31.1.

Animals generally share the following morphological and genetic synapomorphies: extracellular matrix molecules, including collagen and proteoglycans; tight junctions, desmosomes, and gap junctions between their cells; and Hox genes that specify body pattern and axis formation. These traits were probably possessed by the common ancestor of all animals but have been lost in some groups. The common ancestor of modern animals may have been a colonial flagellated protist such as a choanoflagellate. Some cells in the colony may have become specialized for specific tasks and continued to differentiate, eventually leading to truly multicellular organisms. Evidence for evolutionary relationships can be found in fossils, patterns of embryonic development, morphology and physiology of living animals, structure of animal proteins, and gene sequences.

Differences in patterns of embryonic development, including cleavage patterns, gastrulation patterns, and the number of cell layers present, show the evolutionary relationships among animals. Early embryonic cell divisions are known as cleavage. There are different patterns of cleavage among animal embryos. Radial cleavage is the complete and even division of cells; spiral cleavage is a complicated pattern found among many lophotrochozoans. There are other unique patterns of cleavage in some animal groups, such as some ecdysozoans and reptiles. Distinct layers of cells form during early development of most animals. The cell layers differentiate into the specialized organ systems as development progresses. Diploblastic animals have two cell layers: the inner endoderm and the outer ectoderm. Triploblastic animals have a third layer in between the two, the mesoderm. Triploblastic animals form a clade (diploblastic animals do not form a clade). During early development in many animals, a hollow ball of cells indents to form a cup shape. The indentation point is called the blastopore; the fate of the blastopore is used to classify triploblasts into two groups: protostomes, in which the mouth is formed from the blastopore (the anus forms later); and deuterostomes, in which the blastopore forms the anus (the mouth is formed later). Sequencing data indicate that the protostomes and deuterostomes are two different animal clades. Together, they are known as the bilaterians and account for most of the animal species.

Question 1. Explain the fundamental difference between protostomes and deuterostomes. What evidence supports dividing organisms into these two clades?

Question 2. Imagine you are a scientist who has discovered a new organism on a scientific expedition. What specific characteristics would you investigate in order to determine whether it is an animal?

31.2 What Are the Features of Animal Body Plans?

Most animals are symmetrical

The structure of the body cavity influences movement

Segmentation improves control of movement

Appendages have many uses

Nervous systems coordinate movement and allow sensory processing

The overall organization of an animal's body is known as its body plan. The features of an animal's body plan include symmetry or asymmetry, body cavity structure, segmentation, the presence or absence of appendages, and the development of the nervous system. The overall shape of an animal is its symmetry. An animal that can be divided along at least one plane into similar halves is said to be symmetrical. Placozoans and most sponges have no plane of symmetry. The simplest symmetry is spherical symmetry, in which body parts radiate out from a central point, with an infinite number of planes of symmetry. This form is common among unicellular protists, but most animals have other types of symmetry. In radial symmetry, body parts are arranged around one axis at the body center. The sea anemone is radially symmetrical; any plane running along a sea anemone's main axis will divide it into roughly equal halves (see Figure 31.4A). The planes of symmetry all run through the main axis. Animals exhibiting bilateral symmetry can be divided into two mirror images by only one plane (see Figure 31.4B). Bilaterally symmetrical animals can be divided into left and right halves by a central midline plane running from the front (anterior) to the rear (posterior). A plane at right angles to this one divides the body into two dissimilar sides; the back is the dorsal surface and the underside is the ventral surface. Bilaterally symmetrical animals often have sense organs and nervous tissue concentrated at the anterior end. This type of organization is known as cephalization (from the Greek word for "head").

Animals have three different types of body cavities (see Figure 31.5). Acoelomates have no enclosed body cavity. Pseudocoelomates have a liquid-filled space between the mesoderm and endoderm known as the pseudocoel in which many of the internal organs are located. Coelomates have a true body cavity, a coelom, developing within the mesoderm. In coelomates, the internal organs are in pouches of the peritoneum. Fluid-filled body cavities act as hydrostatic skeletons for many animals. Other animals evolved rigid supportive skeletons that can be internal (bones or cartilage) or external (a shell or cuticle). Muscles attached to hard skeletons allow the animal to move.

Segmentation allows specialization of the different body regions and can improve control of movement. In some animals, segments are not apparent (e.g., vertebral column). In other animals, similar body segments are repeated many times, and in yet others the body segments differ (see Figure 31.6). Appendages, especially jointed limbs, enhance locomotion. Jointed limbs in the arthropods and vertebrates are a major factor in their evolutionary success. Other appendages are specialized and can be used to sense the environment (e.g., antennae) or can be used to capture prey.

Bilaterians have a well-developed nervous system to control body functions. Some other animals have nerve nets, which are more diffuse nervous systems. Still other animals (such as sponges) have no nervous system whatsoever. The central nervous system of bilaterians coordinates muscle function and thus controls appendages and body parts. It also coordinates sensory information from a variety of sensory systems, which is used for feeding, avoiding predators, and mating, among other functions.

Question 3. For each of the geometric shapes or characters shown below, determine if it has radial symmetry, bilateral symmetry, or no symmetry. Hint: draw some lines through the images to help you decide.

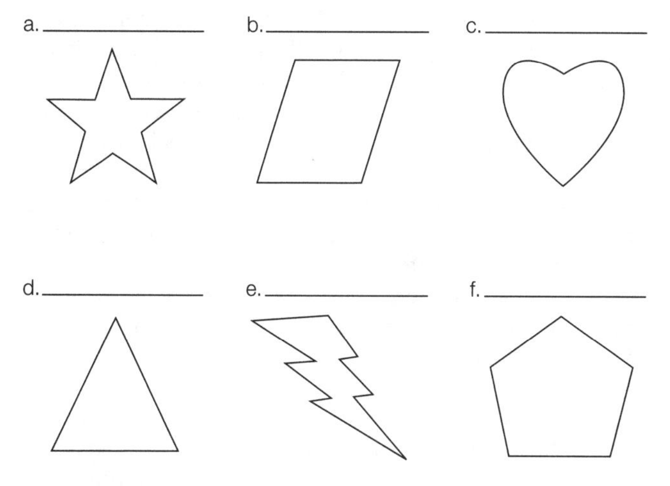

Question 4. What are some advantages to segmentation?

Question 5. What are some of the limitations imposed by a hydrostatic skeleton? Are the limitations the same for terrestrial animals?

31.3 How Do Animals Get Their Food?

Filter feeders capture small prey

Herbivores eat plants

Predators and omnivores capture and subdue prey

Parasites live in or on other organisms

Detritivores live on the remains of other organisms

All animals are ingestive heterotrophs. To obtain food, animals either move through the environment to the food or move the environment and the food to them. Most animals can move (are motile), but some are sessile (nonmoving). As heterotrophs, animals have feeding strategies that fall into a few broad categories. Animals may be filter feeders (see Figure 31.7), herbivores, predators (see Figure 31.8), parasites, or detritivores. Some animals change their feeding strategies at different developmental stages. Predators possess adaptations that allow them to capture other animals (prey). Other animals are omnivores and eat both plants and other animals. Parasites obtain nutrients from a host by living on or within them. Many parasites can consume parts of the host without killing them. Parasites can live inside a host (endoparasites) or outside a host (ectoparasites). Detritivores feed on dead bodies or wastes of other organisms (called detritus). They are also known as decomposers, and play a critical role in nutrient cycling.

Question 6. Discuss the two major strategies animals use to get food, and indicate which of these would most likely be used by sessile organisms.

Question 7. What advantage(s) might an omnivorous feeding strategy give an animal?

31.4 How Do Life Cycles Differ among Animals?

Many animal life cycles feature specialized life stages

Most animal life cycles have at least one dispersal stage

Parasite life cycles facilitate dispersal and overcome host defenses

Some animals form colonies of genetically identical, physiologically integrated individuals

No life cycle can maximize all benefits

The life cycle of an animal encompasses embryonic development, birth, growth to maturity, reproduction, and death. There are a variety of animal life cycles. In direct development, newborns look very similar to adults. In many species, however, newborns differ strikingly from adults and pass through distinct larva, pupa, and adult stages. Metamorphosis refers to the dramatic changes that can occur between the larval and adult stages, when the individual is a pupa (see Figure 31.9). In species that exhibit these stages, one stage is often specialized for feeding while the other is primarily for reproduction. If both stages feed, what is eaten can change with the stage. All life cycles have at least one dispersal stage. In general terms, dispersal refers to the movement of an organism away from the parent or from the population. As a specific example, the larva is the dispersal stage for most sessile organisms. For animals that live on the sea floor, the common larval types are the trochophore and the nauplius (see Figure 31.10). Both types feed on plankton before settling down on the ocean floor, where they develop into adults. For parasites, the dispersal stage has evolved to overcome host defenses. The life cycle of parasites can be fairly complex and involve multiple hosts and larval stages (see Figure 31.11). This complexity can facilitate parasite dispersal. In some groups of animals, asexual reproduction occurs without fission, resulting in colonies of individuals. To the eye, a colony looks like one single integrated organism. In some species colonies are composed of individuals that function alike, while in other species the individuals are specialized for different functions. The bryozoan (see Figure 31.12) and the Portuguese man-of-war (see Figure 31.19) provide examples of how the lines between an individual and a population can be blurred.

Every life cycle necessarily involves evolutionary trade-offs. Often, an adaptation that improves performance in one activity comes at a cost, which is reduced performance of another activity. Common trade-offs involve reproduction. For example, some animals produce many eggs with a small amount of nutrients or a small number of large eggs that store more nutrients (see Figure 31.13). The trade-off is the amount of offspring produced versus the amount of nutrients the offspring receive. In some bird and mammal species, offspring are altricial, meaning that the young are not fully developed and require a significant amount of parental care such as feeding. With others, the young are precocial and can care for and feed themselves soon after birth.

Question 8. Discuss two evolutionary trade-offs involving reproduction.

Question 9. Under what environmental circumstances might having precocial young be a beneficial adaptation?

31.5 What Are the Major Groups of Animals?

- Sponges are loosely organized animals
- Ctenophores are radially symmetrical and diploblastic
- Placozoans are abundant but rarely observed
- Cnidarians are specialized predators
- Some small groups of parasitic animals may be the closest relatives of bilaterians

The bilaterians make up a large, monophyletic group including all animals except sponges, ctenophores, placozoans, cnidarians, and some groups of parasites. Bilaterian synapomorphies include bilateral symmetry, three cell layers, and the presence of at least seven Hox genes. The bilaterians are divided into two major subgroups, the protostomes (see Chapter 32) and the deuterostomes (see Chapter 33). The simplest animals are the sponges, which have no body symmetry and no distinct cell layers. All animals other than the sponges make up the eumetazoans. In this chapter, ctenophores and placozoans are treated as eumetazoans; some biologists, however, do not include them as eumetazoans because they have weakly differentiated layers of tissue and no nervous system. Eumetazoans have body symmetry, a defined gut, a nervous system, and distinct organs.

Although sponges have some specialized cells, they have no distinct embryonic cell layers and no true organs. Sponges have hard skeletal elements called spicules, which can be small and simple or large and complex. The three major groups of sponges are distinguished by their spicules. Two of the groups, glass sponges and demosponges, have siliceous spicules; the third group, calcareous sponges, have calcium carbonate spicules. The sponge body plan is an aggregation of cells built around a water canal system. Choanocyte cells use flagella to divert water through the pores and filter out food particles. In addition to spicules, sponges also have an extracellular matrix composed of collagen, adhesive glycoproteins, and other molecules that hold the cells together. Most species are filter feeders, but a few sponges are carnivores and can trap prey on hook-shaped spicules that are on the outside of the body surface. Most of the 8,000 species of sponges are marine species; only about 50 freshwater species are known. The huge variety of sponge body sizes and shapes is a response to the different movement patterns of water, specifically tides and currents. Sponges that live in environments that have strong wave action tend to be firmly attached to substratum, while those that live in slowly moving water are generally flat and oriented at right angles to the current flow. Sponges reproduce sexually by producing both egg and sperm, and asexually by budding and fragmentation.

The ctenophores (comb jellies) were previously thought to be most closely related to cnidarians. However, they lack most of the Hox genes, and recent genomic studies indicate that they are among the earliest lineages to split from the remaining eumetazoans. Ctenophores are marine diploblastic animals that are radially symmetrical (see Figure 31.16). The ectoderm and endoderm of ctenophores are separated by a thick gelatinous layer called the mesoglea. Ctenophores have a complete gut (i.e., a gut with an entrance and exit, or mouth and anus). There are 250 known species, and they get their name from the rows of comblike plates of cilia, known as ctenes. The cilia are used to propel the animal through the water. Prey is caught on sticky filaments on the tentacles or body (see Figure 31.16). Most ctenophores feed on plankton, although some eat other ctenophores. They can overpopulate protected bodies of water and can cause significant damage to local ecosystems. Ctenophores have a simple life cycle and reproduce sexually. In most species, the externally fertilized egg hatches into a miniature ctenophore (i.e., direct development).

Placozoans are structurally very simple (see Figure 31.17A). Mature individuals are usually found adhered to surfaces. Placozoans do not have a mouth, gut, or true nervous system, and they consist of only a few different cell types. Based on phylogenetic analysis, their structural simplicity may have been secondarily derived (i.e., some of their common features were probably lost sometime during evolution from a common ancestor). They have upper and lower epithelial cell layers with contractile fiber cells in between. The life cycle of placozoans remains relatively unknown because their transparency makes it difficult to observe them in nature. It is known that they have a pelagic (free-swimming) stage in the ocean and that they are capable of sexual and asexual reproduction.

Cnidarians (jellyfishes, sea anemones, corals, and hydrozoans) are the largest and most diverse group of non-bilaterian animals. The cnidarian gastrovascular cavity is a blind sac, so cnidarians do not have a complete gut. There is only one opening, which serves as both mouth and anus. The gastrovascular cavity functions in food digestion, respiratory gas exchange, and circulation. It also lends support as a hydrostatic skeleton. The life cycle of most cnidarians includes a sessile polyp stage and a motile medusa stage (see Figure 31.18). The polyp stage usually reproduces asexually. The medusa stage reproduces sexually, with the fertilized egg becoming a planula larva that eventually develops into a polyp. Cnidarian tentacles have specialized cells that inject toxins into their prey with the help of special organelles called nematocysts (see Figure 31.19). Some cnidarians also host photosynthetic symbionts. Cnidarians possess muscle fibers that enable them to move and simple nerve nets that integrate their activities. They also possess the structural molecules collagen, actin, and myosin, which, along with their Hox genes, link them to the bilaterians. As in the ctenophores, a large amount of mesoglea is found between the two cell layers of cnidarians. Because of the inert nature of the mesoglea, cnidarians have low metabolic rates. All but a few of the 12,500 or so species of cnidarians are marine. The smallest individuals are almost microscopic, while individual jellyfish can be quite large. Many cnidarians are colonial. Three major clades of cnidarians contain the most species: anthozoans, scyphozoans, and hydrozoans.

Anthozoan species include sea anemones, sea pens, and corals (see Figures 31.20A, 31.20B, and 31.21). Sea anemones are solitary and occasionally motile; sea pens and corals are sessile and colonial. The coral skeleton is composed of calcium carbonate layered on top of an organic matrix. As the colony grows, old polyps die but their skeletons remain; the accumulation of these remains form islands and reefs, such as the Great Barrier Reef off the northeastern coast of Australia. Corals live in symbiosis with photosynthetic protists that provide nutrients for the coral colony through photosynthesis, contributing to the success of corals in clear, nutrient-poor tropical waters. Corals are threatened by rising CO_2 levels and the acidification of oceans.

Scyphozoans include the marine jellyfish, which have medusae with thick mesoglea. These species spend most of their lives in the medusa stage (see Figure 31.20C). Polyps are produced sexually by adult medusae. The polyps produce young medusae by budding (see Figures 31.18).

Hydrozoans have a life cycle that typically is dominated by the polyp stage, though some have only medusae and others have only polyps. The hydrozoans tend to be colonial (see Figure 31.20D), with many polyps sharing a common gastrovascular cavity (see Figure 31.22). Within the colony, some polyps have tentacles with numerous nematocysts that capture prey. Not all hydrozoans have tentacles; others have fingerlike projections that defend the colony with their nematocysts.

The orthonectids and rhombozoans are two groups of tiny parasites that may be among the closest living relatives of the bilaterians. Both groups are highly reduced parasites that lack many animal structures. As more information about their genomes is discovered, their exact phylogeny will be discerned. Two other groups, xenoturbellids and acoels, are proposed to be falling just outside the group of bilaterians. Genomic analyses suggest that these animals are actually highly specialized deuterostomes.

Question 10. In the diagram below, identify the following groups: Protostomes, Deuterostomes, Arrow worms, Calcareous sponges, Chordates, Cnidarians, Ctenophores, Demosponges, Ecdysozoans, Echinoderms, Glass sponges, Hemichordates, Lophotrochozoans, and Placozoans. Additionally, identify the distinguishing traits marked with dots on the phylogenetic tree: Bilateral symmetry along an anterior–posterior axis; three embryonic cell layers, Blastotopore develops into mouth, Blastopore develops into anus, Choanocytes; spicules, Exoskeleton molting, Notochord, Radial symmetry, Silicaceous spicules, Two embryonic cell layers; nervous system, Unique cell junctions; collagen and proteoglycans in extracellular matrix, Simplification; loss of nervous system.

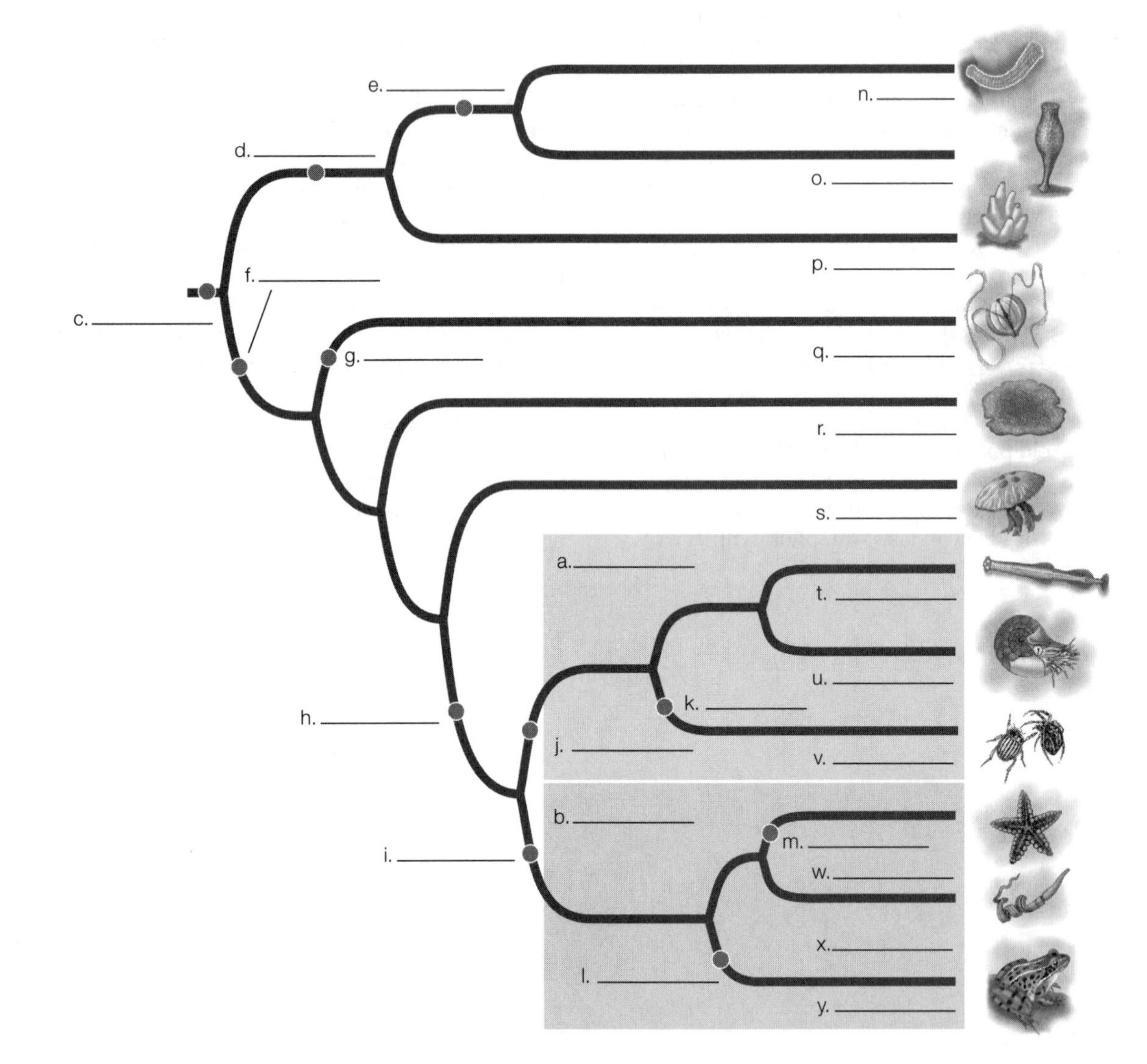

Question 11. The diagram at right shows a typical jellyfish life cycle. Identify the diploid and haploid stages, identify the process occurring in the boxes (mitosis, meiosis, fertilization, or fusion), and label the identified structures.

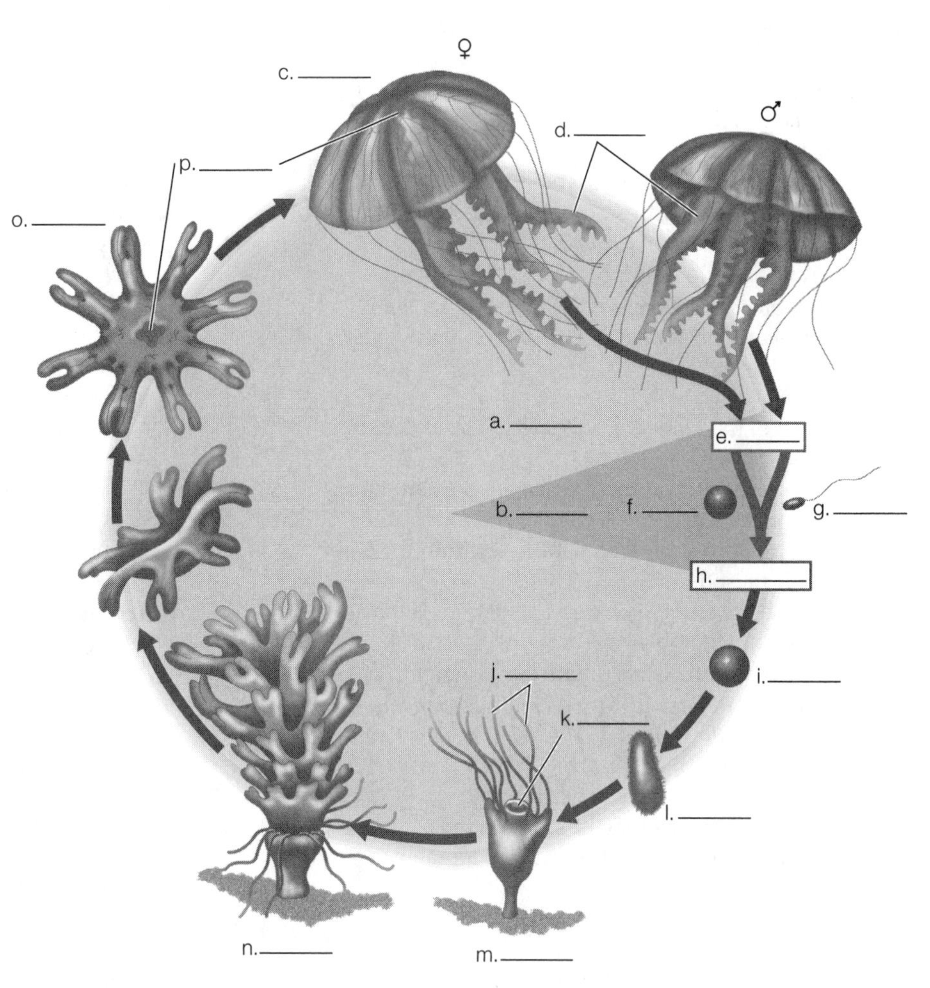

Question 12. The diagram at right shows a typical hydrozoan life cycle. Identify the diploid and haploid stages, identify the process occurring in the boxes (mitosis, meiosis, fertilization, or fusion), and label the identified structures.

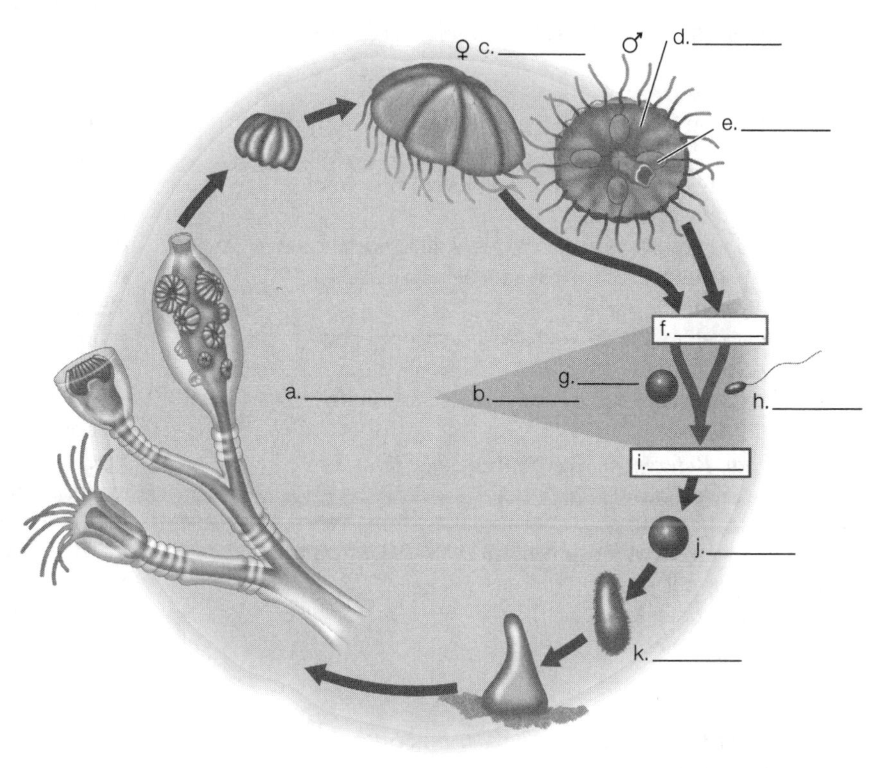

Test Yourself

1. Which of the following is *not* a derived trait that is shared by all animals?
   a. Hox genes
   b. The extracellular matrix molecule collagen
   c. Tight junctions, desmosomes, and gap junctions
   d. Bilateral symmetry
   e. Absence of a cell wall
   Textbook Reference: *31.1 What Characteristics Distinguish the Animals?; 31.2 What Are the Features of Animal Body Plans?*

2. Which of the following statements about deuterostomes is *false*?
   a. Three distinct layers of tissue are present during development.
   b. If a coelom is present, it formed within the embryonic mesoderm.
   c. The early embryonic cleavage pattern is radial.
   d. They are diploblastic.
   e. Gastrulation occurs during development.
   Textbook Reference: *31.1 What Characteristics Distinguish the Animals?; 31.2 What Are the Features of Animal Body Plans?*

3. An important factor contributing to the evolution of diversity among animals is the considerable variation in their
   a. brood sizes.
   b. cell junctions.
   c. methods of food acquisition.
   d. symmetry.
   e. segmentation patterns.
   Textbook Reference: *31.3 How Do Animals Get Their Food?*

4. Which of the following statements about the body cavity of animals is true?
   a. The body cavity of coelomates develops from the embryonic ectoderm.
   b. The body cavity of acoelomates is filled with liquid.
   c. The pseudocoel of the pseudocoelomates has a peritoneum.
   d. The acoelomates do not have an enclosed body cavity.
   e. The coelomates have a body cavity surrounded by peritoneum.
   Textbook Reference: *31.2 What Are the Features of Animal Body Plans?*

5. The mechanism of movement in a snail is by means of
   a. a hydrostatic skeleton.
   b. jointed appendages.
   c. a boney skeleton.
   d. segmentation.
   e. spicules.
   Textbook Reference: *31.2 What Are the Features of Animal Body Plans?*

6. Which of the following feeding strategies requires a life cycle in which much time and energy are devoted to dispersal?
   a. Herbivory
   b. Parasitism
   c. Predation
   d. Filter feeding
   e. Detritivory
   Textbook Reference: *31.4 How Do Life Cycles Differ among Animals?*

7. Which of the following statements about a species of bird with precocial young is true?
   a. Their eggs have a small amount of yolk.
   b. They care for their young for a long period of time.
   c. Their eggs have a long incubation period.
   d. They produce many small eggs.
   e. Their eggs have a short incubation period followed by a long period of feeding young.
   Textbook Reference: *31.4 How Do Life Cycles Differ among Animals?*

8. Which of the following statements about sponge structure or function is *false*?
   a. Choanocytes are flagellated cells that play a role in feeding.
   b. Large species are found in areas of heavy wave action where food is most abundant.
   c. Individual sponges are both male and female.
   d. Water enters a sponge through pores and exits via one or more openings.
   e. Sponges have an extensive extracellular matrix holding the cells together.
   Textbook Reference: *31.5 What Are the Major Groups of Animals?*

9. Which of the following statements about ctenophores and cnidarians is *false*?
   a. Most are marine organisms.
   b. They have radial symmetry.
   c. They have complete guts.
   d. They have feeding tentacles.
   e. Fertilization generally takes place in open water.
   Textbook Reference: *31.5 What Are the Major Groups of Animals?*

10. In what way does the mesogleal layer found in both ctenophores and cnidarians help these organisms survive, even when prey is scarce?
    a. Prey adheres to the sticky surface of the mesoglea.
    b. The mesoglea is biologically inert.
    c. The mesoglea allows the animal to capture large prey items.
    d. Mesogleal molecules provide nutrients to the host animal.
    e. The mesoglea houses photosynthetic bacteria that supplement the nutrition obtained from prey.
    Textbook Reference: *31.5 What Are the Major Groups of Animals?*

11. Which of the following traits is *not* shared by sea anemones and jellyfishes?
   a. A medusa as the dominant stage in the life cycle
   b. Possession of a gastrovascular cavity
   c. Sexual reproduction
   d. Nematocysts on the tentacles
   e. A primarily carnivorous diet
   Textbook Reference: *31.5 What Are the Major Groups of Animals?*

12. Which cnidarian group is dominated by species that live symbiotically with photosynthetic protists?
   a. Hydrozoans
   b. Ctenophores
   c. Scyphozoans
   d. Placozoans
   e. Anthozoans
   Textbook Reference: *31.5 What Are the Major Groups of Animals?*

13. Which of the following is *not* a general characteristic of animals?
   a. Multicellularity
   b. Sexual reproduction
   c. Heterotrophic metabolism
   d. Internal digestion
   e. Movement
   Textbook Reference: *31.1 What Characteristics Distinguish the Animals?*

14. Which of the following organisms is matched *incorrectly* with its feeding strategy?
   a. Rabbit: herbivore
   b. Earthworm: detritivore
   c. Tiger: predator
   d. Flamingo: carnivore
   e. Tapeworm: parasite
   Textbook Reference: *31.3 How Do Animals Get Their Food?*

15. Which of the following is the group of animals about which we know the *least*?
   a. Sponges
   b. Ctenophores
   c. Placozoans
   d. Bilateria
   e. Cnidaria
   Textbook Reference: *31.5 What Are the Major Groups of Animals?*

Answers

Key Concept Review

1. Protostomes and deuterostomes differ in the structure that becomes the anus and the structure that forms the mouth. The sequence in which these structures develop is also different. In protostomes, the mouth develops from the blastopore; in the deuterostomes, the anus develops from the blastopore. In both, the opposite structure forms later. These groups were first described according to observations of their developmental patterns, but sequence data supports this classification.

2. Some things to investigate would include the type of nutrition (autotrophy vs. heterotrophy), digestion (internal or external), movement, presence of sensory apparatus, the types junctions between cells, the presence of certain Hox genes and their organization, and the type of extracellular matrix present in the organism.

3.
   a. Radial symmetry
   b. Bilateral symmetry
   c. Bilateral symmetry
   d. Radial symmetry
   e. No symmetry
   f. Radial symmetry

4. Segmentation provides several advantages, including facilitating specialization of body regions and improved muscle coordination, as each segment can be manipulated independently.

5. Hydrostatic skeletons provide for controlled movement of body parts. In terrestrial organisms, the surrounding environment does not support the body, so most organisms with hydrostatic skeletons are very small and soft-bodied. Exoskeletons or hard interior skeletons that provide structural support and protection to soft tissues are needed for larger bodies.

6. Animals can either move through the environment to where the food is located or they can move the environment and the food to themselves. Sessile organisms would most likely have adaptations to move the environment and food to themselves.

7. Omnivory gives an animal many different opportunities to obtain a complete diet because of the variety of foods it can eat. It places the organism at various levels in the food web, which may prove advantageous if food supplies change. The omnivore could survive a dramatic change in the environment by switching its food sources.

8. One trade-off is the production of many small eggs or several large eggs. The trade-off is the amount of offspring produced versus the amount of energy resources the offspring receives. Another trade-off is the production of altricial young that hatch early but require much parental care versus the production of precocial young that require a longer incubation period but little parental care after they hatch.

9. In a situation in which it is difficult to obscure a nest in order to protect the young, having precocial young might be a beneficial adaptation.

10.
   a. Protostomes
   b. Deuterostomes
   c. Unique cell junctions; collagen and proteoglycans in extracellular matrix
   d. Choanocytes; spicules
   e. Silicaceous spicules
   f. Two embryonic cell layers; nervous system
   g. Simplification; loss of nervous system
   h. Bilateral symmetry along an anterior–posterior axis; three embryonic cell layers
   i Blastopore develops into anus
   j. Blastopore develops into mouth
   k. Exoskeleton molting
   l. Notochord
   m. Radial symmetry
   n. Glass sponges
   o. Demosponges
   p. Calcareous sponges
   q. Ctenophores
   r. Placozoans
   s. Cnidarians
   t. Arrow worms
   u. Lophotrochozoans
   v. Ecdysozoans
   w. Echinoderms
   x. Hemichordates
   y. Chordates

11.
   a. Diploid
   b. Haploid
   c. Medusa ("jellyfish")
   d. Tentacles
   e. Meiosis
   f. Egg
   g. Sperm
   h. Fertilization
   i. Fertilized egg
   j. Tentacles
   k. Mouth/anus
   l. Planula larva
   m. Polyp
   n. Mature polyp
   o. Young medusa
   p. Mouth/anus

12.
   a. Diploid
   b. Haploid
   c. Medusa
   d. Oral surface
   e. Gonad
   f. Meiosis
   g. Egg
   h. Sperm
   i. Fertilization
   j. Fertilized egg
   k. Planula larva

Test Yourself

1. **d.** Not all animals have bilateral symmetry (e.g., sponges, cnidarians, ctenophores).
2. **d.** Deuterostome embryos have three layers: the ectoderm, the mesoderm, and the endoderm. Therefore deuterostomes are triploblastic, not diploblastic.
3. **c.** Animals eat foods from all kingdoms of life. Thus, they have evolved a wide variety of methods for acquiring diverse types of food. Segmentation patterns are a product of evolution, not the cause of evolution.
4. **d.** The body cavity of coelomates develops from the mesoderm and contains a peritoneum. The acoelomates lack a body cavity.
5. **a.** In a hydrostatic skeleton, when muscles surrounding a fluid-filled body cavity contract, the fluids move to another area of the cavity. If the body tissues around the cavity are flexible, then the moving fluids can move specific body parts.
6. **b.** Because many parasites die when the host dies, parasites must have a way to disperse their progeny to new hosts.
7. **c.** Because precocial young hatch as well-developed individuals that can forage for themselves, they undergo more of their development in the egg and tend to be incubated for longer periods of time.
8. **b.** Large, upright sponges would be destroyed by heavy wave action because they are not structurally robust.
9. **c.** Ctenophores and cnidarians are mostly marine, have radial symmetry, and are composed largely of mesoglea. Fertilization occurs in open water. Ctenophores have a complete gut with a mouth and two anal pores; cnidarians have a blind gut with only one opening.
10. **b.** The biologically inert mesoglea does not require energy; thus a large portion of the body needs no nutritional support, allowing ctenophores and cnidarians to maintain a very low metabolic rate and survive for an extended period of time with minimal amounts of food.
11. **a.** Although the jellyfish have both a medusa and polyp stage in their life cycles, the sea anemones have lost the medusa stage and spend their lives as polyps.
12. **e.** Anthozoans include the corals and sea anemones, both of which contain many species that live symbiotically with photosynthetic protists. Placozoans and ctenophores are not cnidarians.
13. **b.** Not all animals perform sexual reproduction exclusively. The other characteristics, although not true of all animals, are representative of the clade of all animals.
14. **d.** Flamingos are filter feeders that strain small organisms out of the mud using their beaks.
15. **c.** Placozoans are not easily observed in nature and were unknown until 1883. They are structurally simple but very abundant in warm marine environments.

Protostome Animals

The Big Picture

- The protostomes include two major groups: the lophotrochozoans and the ecdysozoans. The arrow worms are not placed in either group; they may be sister to the protostomes as a whole, or they may be more closely related to the lophotrochozoans. Protostomes have an anterior brain and a ventral nervous system. Many species in both groups have a wormlike appearance.
- The lophotrochozoans get their name from two structures: the lophophore (a feeding and gas-exchange structure found in a number of groups in this clade) and the trochophore larvae.
- A number of lophotrochozoan groups, including the annelids (segmented worms) and mollusks, undergo spiral cleavage during early development.
- The segmentation found among the annelid worms helps improve their locomotion. The unique molluscan body plan is based on a muscular foot structure, a visceral mass containing the organs, and a mantle that covers and protects the organs. Diverse adaptations of this plan are seen in the major molluscan groups.
- The ecdysozoans include more species than all other lineages combined. They are characterized by a rigid external covering (cuticle or exoskeleton) that must be molted periodically in order to allow the animal to grow.
- The wormlike ecdysozoans, which include the priapulids and the kinorhynchs, are unsegmented and have a thin cuticle. Horsehair worms and nematodes are also unsegmented, and have a tougher cuticle.
- The arthropods are characterized by a segmented body with a hard exoskeleton and jointed appendages. They are Earth's dominant animals in both number of species and number of individuals. The trilobites are extinct arthropods. Today four arthropod groups—crustaceans, hexapods, myriapods, and chelicerates—are found in all environments. The hexapods (which include the numerous and diverse insects), chelicerates, and myriapods are found in terrestrial and aquatic environments. The crustaceans are the dominant arthropods in marine environments.

Study Strategies

- With the advent of molecular phylogenetic studies (see Chapter 24), we have had to rethink many of our previous ideas about the classification of organisms. The arrow worms, for example, used to be considered deuterostomes, but have recently been placed among the protostomes, based primarily on molecular (i.e., gene sequence) studies. Compared to our understanding of deuterostome phylogeny (which you will study in Chapter 33), our conceptions of protostome phylogeny are in a state of flux and there is a great deal of disagreement among systematists about the ancestry and monophyly—and thus, the classification—of many protostome groups.
- Try to use the phylogeny presented in Figure 32.1 as a basis for learning about different groups. Understanding that certain characteristics are shared among related groups can help you remember which characteristics define each group.
- Sometimes learning the Latin root of a group can help you organize your thinking. Although the names are unfamiliar at first, they all make sense if you understand what they mean.
- The end of the chapter provides an overview of protostome evolution. Use the overview as a guide in studying the body forms and life cycles of the different groups.
- Go to the Web addresses shown to review the following animated tutorial and activities. You can also find them in BioPortal (yourBioPortal.com), along with many additional learning resources.

  Animated Tutorial 32.1 An Overview of the Protostomes (Life10e.com/at32.1)

  Activity 32.1 Features of the Protostomes (Life10e.com/ac32.1)

  Activity 32.2 Protostome Classification (Life10e.com/ac32.2)

Key Concept Review

32.1 What Is a Protostome?

Cilia-bearing lophophores and trochophores evolved among the lophotrochozoans

Ecdysozoans must shed their cuticles

Arrow worms retain some ancestral developmental features

After the origin of diploblastic animals, a third embryological germ layer (between the ectoderm and endoderm) evolved, called the mesoderm. Triploblastic animals that possess a mesoderm fall into one of two major clades: the protostomes and the deuterostomes. Although their body patterns vary extensively from group to group, protostomes have the following general characteristics: the blastopore of the embryo develops into the mouth; they have an anterior brain that surrounds the entrance to the digestive tract; and they have a ventral nervous system consisting of paired or fused longitudinal nerve cords. The common ancestor of the protostomes had a coelom (a fluid-filled cavity within the mesoderm). However, the protostomes include some groups that are coelomate and some that are pseudocoelomate. One important group, the flatworms, is acoelomate (lacks a coelom). Two prominent protostome groups, the arthropods and the mollusks, have had secondary evolutionary modifications of the coelom. In the arthropods, the coelom has become a hemocoel; the mollusks have returned secondarily to a virtually open circulatory system. With the exception of the arrow worms, the protostomes are divided into two major clades based on DNA sequence analysis: the lophotrochozoans and the ecdysozoans (see Figure 32.1). A number of representatives in both groups have a wormlike body plan.

The lophotrochozoans share several characteristics. Several (but not all) lophotrochozoans groups are characterized by a complex U-shaped structure, the lophophore, which is used both as a feeding apparatus and in gas exchange (see Figure 32.2). These groups are distantly related, and it appears that the lophophore has evolved independently at least twice, or else it is an ancestral feature and has been lost in many groups. Nearly all animals that have a lophophore are sessile as adults. Many groups of lophotrochozoans also have a type of free-living ciliated larva known as a trochophore (hence, "lophotrochozoans"). It is likely that the trochophore form was ancestral and lost in several lineages. Several different lophotrochozoan lineages (e.g., flatworms, ribbon worms, annelids, and mollusks) exhibit a derived form of early development known as spiral cleavage. These are sometimes grouped into the spiralians, but gene sequence analysis indicates that the different spiralian groups are not monophyletic. Many lophotrochozoans have a wormlike body plan; one group, the mollusks, are an exception.

The ecdysozoans share several characteristics. Ecdysozoans have a cuticle that provides both protection and support. These animals grow by molting their cuticles (*ecdysis* is the Greek word for "shedding"). Increasing molecular evidence, including a set of Hox genes shared by all ecdysozoans, supports the monophyly of these animals and suggests that cuticle molting is a lifestyle that may have evolved only once. Before molting, the new cuticle layer is forming underneath the old one; after molting, the animal is vulnerable until the new cuticle hardens. Ecdysozoans with a wormlike body plan often have a thin and flexible cuticle that allows for gas, water, and mineral exchange. Most of these ecdysozoans are confined to moist habitats; many live in marine sediments, ingesting either sediments or capturing prey with a toothed pharynx. Others absorb nutrients through the cuticle. Still others, most notably the arthropods, have a hard cuticle known as an exoskeleton containing many layers of protein and a strong, waterproof polysaccharide called chitin. An exoskeleton limits locomotion and gas exchange. Thus, new mechanisms of locomotion and gas exchange evolved in the ecdysozoans with hard exoskeletons. Jointed appendages controlled by muscles allow for rapid locomotion as well as performing other functions, including gas exchange, food capture, copulation, and sensory perception. Over evolutionary time, different arthropod groups have experienced extensive modification of the appendages to suit life in a multitude of different environments. Arthropods are the dominant animals on Earth today, in both number of species and number of individuals.

The arrow worms are an enigmatic group. Because of their early developmental morphology, they were once classed with the deuterostomes, but molecular evidence now clearly identifies them as protostomes. It is unclear whether the arrow worms are a sister group of the entire protostome clade, or whether they are most closely related to the lophotrochozoans. Arrow worms have no circulatory system and no larval stage. They are major predators of small organisms in the open ocean (see Figure 32.5).

Question 1. The following are all characteristics of earthworms: segmented bodies, a spiral cleavage pattern, a terrestrial lifestyle, a complete gut, a blastopore that develops into a mouth, tight junctions between cells, and bilateral symmetry. In what order did these characteristics evolve?

Question 2. Scientists have not found a fossilized body of the species that is the common ancestor of all bilaterally symmetrical animals. What is the likelihood that they will do so? How have scientists been able to learn about this common ancestor?

32.2 What Features Distinguish the Major Groups of Lophotrochozoans?

Most bryozoans and entoprocts live in colonies

Flatworms, rotifers, and gastrotrichs are structurally diverse relatives

Ribbon worms have a long, protrusible feeding organ

Brachiopods and phoronids use lophophores to extract food from the water

Annelids have segmented bodies

Mollusks have undergone a dramatic evolutionary radiation

The 5,500 species of bryozoans and 170 species of entoprocts are colonial; strands of tissue connect individuals in each bryozoan colony, and in some species individuals are specialized for feeding, reproduction, defense, and support. Individual bryozoans have a great deal of control in manipulating their lophophores to increase contact with prey (see Figure 32.2). Colonies grow via asexual reproduction of the founding members. Sexual reproduction also occurs, and larvae emerge to seek suitable sites to form new colonies. Entoprocts can also reproduce either sexually or asexually. Bryozoans can cover large areas of coastal rock and can even form small reefs in shallow seas. In bryozoans, the anus is located outside the ring of tentacles of the lophophore, whereas in entoprocts the anus is located in the center of this ring. Entoprocts lack a coelom, whereas bryozoans have a three-part coelom.

Recent genomic studies indicate that the flatworms and rotifers are related, despite their structural differences. Flatworms lack respiratory organs and have only simple cells for waste removal. The flat shape of the animal helps in oxygen transport and waste removal (see Figure 32.6). The digestive tract consists of a mouth that opens into a blind sac, which has many branches that aid in nutrient absorption. Although there are some free-living flatworm species, most are parasites, most of which in turn are endoparasites. Most of the living flatworms are tapeworms and flukes, both groups of endoparasites. Parasitic flatworms feed on the nutrient-rich body tissues of their host animal and disperse their eggs in the host's feces; many endoparasitic flatworms lack digestive tracts. Schistosomiasis, a serious human disease that is common in parts of Asia, South America, and Africa, is caused by a flatworm. Most species of rotifers are very small (some smaller than ciliated protists), but they have specialized internal organs, including a complete gut and a pseudocoel that serves as a hydrostatic skeleton (see Figure 32.7A and 32.7B). Cilia are used to propel the rotifers through the water. Most species live in freshwater habitats and feed using a ciliated organ called a corona. Some species have both males and females; the bdelloid rotifers have only one sex (female) and reproduce asexually. This is the only group of animals known to have existed for millions of years without the benefits of sexual reproduction. The 800 species of gastrotrichs are tiny animals that live in marine sediments, fresh water, and in water films surrounding grains of soil. Most are simultaneous hermaphrodites, although some species have lost the male organs or have greatly reduced male organs and reproduce asexually.

Nemerteans, or ribbon worms, have a complete digestive tract with two openings. Small ribbon worms use their cilia for movement, and large ribbon worms move by means of muscle contractions. The feeding organ of the ribbon worms is a proboscis that lies within a rhynchocoel, a fluid-filled cavity (see Figure 32.8). The proboscis has a sharp stylet and can be forcefully ejected from the body to catch prey (see Figure 32.8A). Most of the 1,000 or so species are marine, although a few species are found in fresh water or on land. Most are small, but some species can be up to 20 meters long. DNA sequence analyses indicate that they are closely related to the phoronids and brachiopods.

While the lophophore of brachiopods and phoronids is similar in function to that of the bryozoans, genetic analyses suggest that these groups are not closely related. This indicates that the lophophore structure has involved more than once. Brachiopods are solitary marine animals that live attached to the substratum. Their divided shell gives them a superficial resemblance to bivalve mollusks (e.g., clams), but the two halves are dorsal and ventral instead of lateral (see Figure 32.9). The lophophore is located in the shell, and cilia help draw water and food into the shell. More than 26,000 fossil brachiopod species have been described, but only about 450 species are known to exist today. The phoronids include 10 species of tiny sessile worms that live in chitinous tubes and extract food from the water with their lophophores (see Figure 32.10).

Annelids are segmented worms. In most large annelids, the coelom in each segment is isolated from other segments. Their thin body wall serves as a surface for gas exchange, and they are restricted to marine, freshwater, and moist terrestrial environments because the thin covering causes them to lose moisture rapidly when exposed to air. A segmented body plan gives these worms extremely good control of their movement. A separate nerve center called a ganglion controls the movement of each segment, and in most cases each segment also contains an isolated coelom (see Figure 32.11). More than half of all annelids are polychaetes, meaning "many hairs"; this term is descriptive, however, rather than the name of a clade. Annelids generally are found in marine environments in the sediment. Polychaetes can have more than one pair of eyes and tentacles (see Figure 32.12A). Outgrowths called parapodia used in gas exchange extend from segments laterally over much of the body. Setae extending from the parapodia help attach the animal to the substrate and aid in movement. The pogonophorans, a relatively recently discovered polycheate group, have secondarily lost their digestive tract and secrete tubes made of chitin and other substances that come from their surroundings (see Figure 32.12B). Pogonophorans are found in the deep ocean near hydrothermal vents. They harbor a number of endosymbiont bacteria in a specialized organ known as a trophosome, and the bacteria provide much of their nutrition. There are two major clades of clitellate annelids, oligochaetes and leeches. The evolutionary relationship of the clitellates to the different polychaete groups is unclear, and the latter may be a paraphyletic group. The oligochaetes ("few hairs") include the most familiar annelids, called the earthworms. They lack parapodia, eyes, or tentacles. Oligochaetes live mainly in freshwater and terrestrial environments and are hermaphroditic, containing both male and female reproductive organs in the same individual. Sperm is exchanged between two individuals (see Figure 32.12C). The second clitellate group, the leeches, are also hermaphroditic species that live either in fresh water or on land (see Figure 32.12D). The coelom of these parasitic annelids is not segmented but is composed of undifferentiated tissue. Clusters of segments at their anterior and posterior ends are modified into suckers, which the leech attaches to the substratum for movement, or to a host mammal from which it sucks blood (its nutritional source). To keep the blood from clotting, the leech secretes an anticoagulant. Medicinal leeches are still used today, to reduce swelling and prevent blood clotting, among other purposes.

Mollusks are the most diverse group of lophotrochozoans. The mollusks have evolved into a morphologically diverse group based on a distinctive three-part body plan (see Figure 32.13) of foot, visceral mass, and mantle. All mollusk species have a large muscular foot. In some groups, such as the clams, this foot is a burrowing organ, while in squids and octopuses the foot has been modified and consists of arms and tentacles; the tentacles bear complex sensory organs. Organs such as the heart, the digestive tract, and the reproductive system are concentrated centrally in a visceral mass. A tissue fold, known as the mantle, covers a visceral mass of internal organs. In many species, the mantle secretes a hard, calcareous shell. In most species, the mantle is extended to create a mantle cavity holding the gills used in respiration. Some mollusks feed on organisms that they scrape off rocks using a razorlike body structure called the radula, while most are filter feeders. The blood vessels of the mollusks do not form a closed circulatory system. Instead, blood and other fluids empty into a hemocoel through which fluid moves around the animal to deliver oxygen to the internal organs. Mollusks have a heart that moves the blood back into the blood vessels.

Monoplacophorans were the most abundant Cambrian mollusks, but only a few species survive today. In monoplacophorans the gas exchange organs, muscles, and excretory pores are repeated over the length of the body. The molluscan body plan has resulted in a diverse array of species that fall into four major modern groups: the chitons, bivalves, gastropods, and cephalopods (see Figure 32.13). Chitons are bilaterally symmetric marine mollusks that feed on algae, bryozoans, and other organisms with a radula. A chiton's shell consists of eight overlapping plates that are surrounded by a girdle (see Figure 32.13B). Gastropods are the most species-rich and widely distributed of the mollusks, and are found in all environments. There are shelled and unshelled gastropod species, including the snails and slugs (the only terrestrial mollusks), as well as the marine nudibrachs (sea slugs), whelks, limpets, and abalones. Gastropods use their foot either to crawl or swim. Land snails and slugs are able to survive on land, as the mantle tissue is modified into a highly vascularized lung. The bivalve mollusks include the familiar clams, oysters, scallops, and mussels. They are found in both salt and fresh water, but they are all aquatic. Bivalves use an opening called an incurrent siphon to bring water into their two-part hinged shells; they are filter feeders that extract foodstuffs from these water currents. Large gills inside the shell extract the food and also function as respiratory organs. Water and gametes exit from an excurrent siphon. Among the clams, the molluscan foot has been modified into a digging device that allows the animal to burrow into the mud or sand.

Cephalopods include the octopuses, squids, and nautiluses. The excurrent siphon is modified to give cephalopods the ability to control water movement into the mantle, allowing them to use ejected water as a jet for propulsion. They capture prey with their tentacles, and their greatly enhanced mobility makes them dominant ocean predators. They are also able to control gas movement in the mantle, which helps in buoyancy control. As is typical of active, rapidly moving predators, cephalopods have a head with complex sensory organs, most notably the eyes, which are in many ways comparable to those of vertebrates.

Question 3. In the diagram below, label the following structures: lophophore tentacles, mouth, anus, outer covering (chitinous tube), and gut. Is the organism a phoronid or an ecdysozoan?

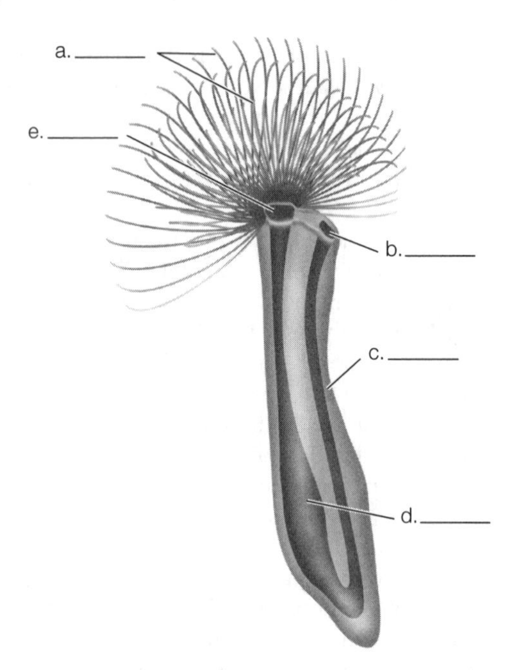

Question 4. In the diagram below of an annelid segment cross section, label the following: blood vessel, intestine, circular muscle, longitudinal muscle, setae, septum, nerve cord, coelom, and excretory organ.

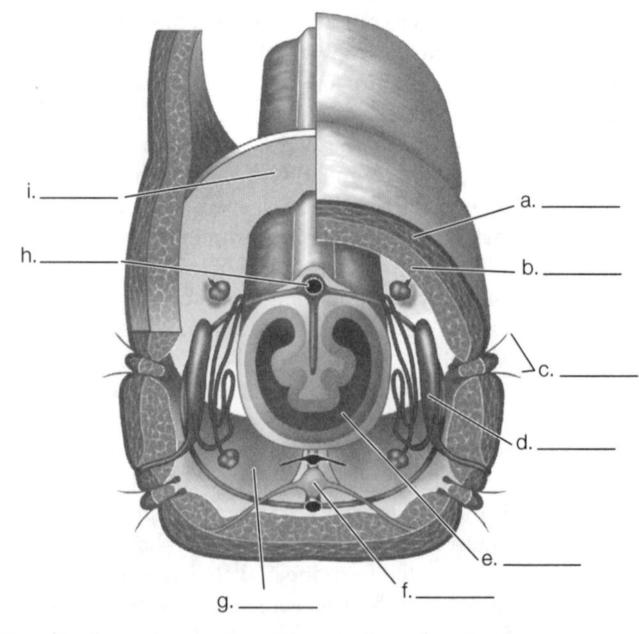

Question 5. In the diagram below, first identify the cephalopod, chiton, bivalve, and gastropod. Then identify the following structures for each organism that has them: stomach,

salivary gland, intestine, foot, anus, radula, siphon, gill(s), mantle, shell, shell plates, heart, mouth, digestive gland.

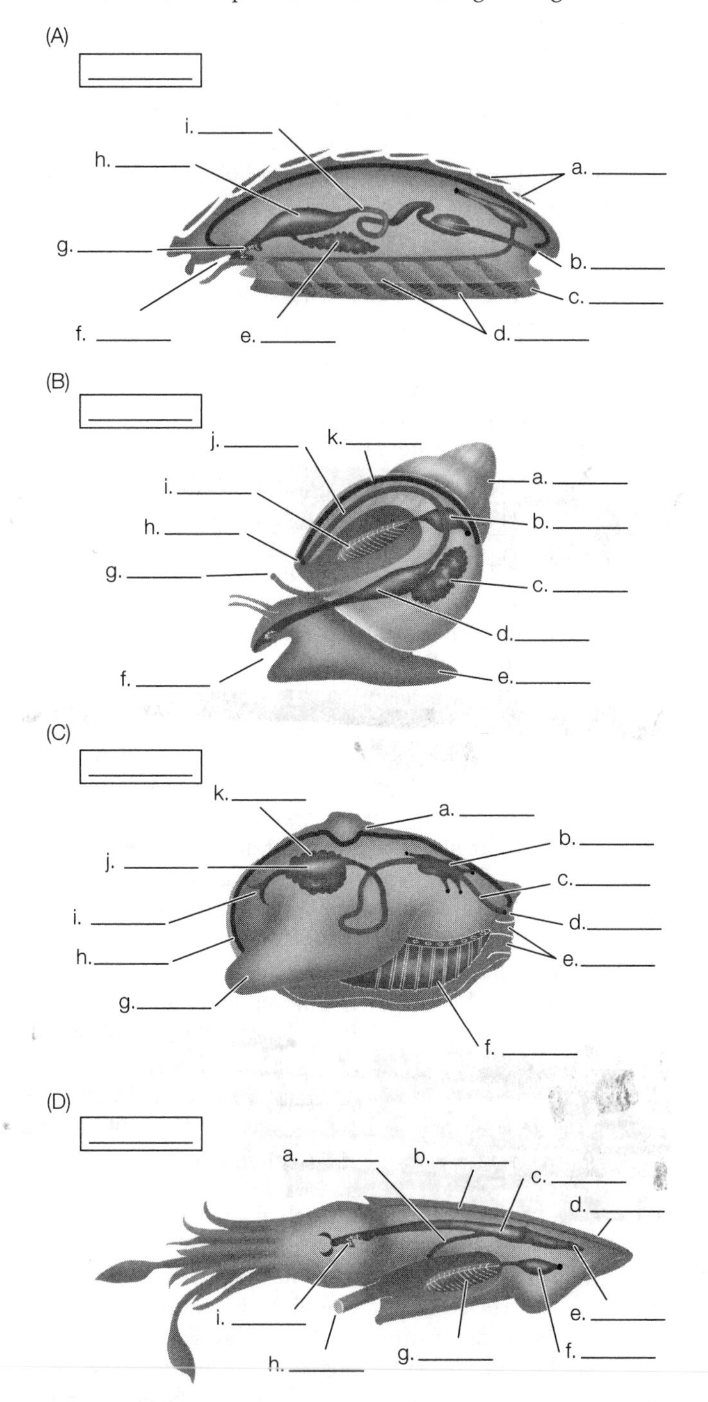

Question 6. The mollusks are a very diverse group morphologically, but they share three common morphological traits. Describe these traits and their functions.

32.3 What Features Distinguish the Major Groups of Ecdysozoans?

Several marine ecdysozoan groups have relatively few species

Nematodes and their relatives are abundant and diverse

Many ecdysozoans are wormlike, but the arthropods and related groups have limbs. The priapulids, kinorhynchs, and loriciferans have a thin exoskeleton called a cuticle that is molted periodically. The priapulids are cylindrical, unsegmented, wormlike animals with three main parts: a proboscis, trunk, and caudal appendage (tail) (see Figure 32.15A). They range from 0.5 millimeters to 20 centimeters, and live in marine sediments, preying on soft-bodied invertebrates. About 180 species of kinorhynchs have been described. They live in marine sands and are no longer than 1 millimeter. They have 13 segments, each covered with a separate cuticular plate (see Figure 32.15B). They feed by ingesting sediments and digesting the organic material. Kinorhynchs have no distinct larval stage. Loriciferans are also less than 1 millimeter in length, and were not discovered until 1983. The body is divided into a head, neck, thorax, and abdomen and is covered by six plates (see Figure 32.15C). Loriciferans live in coarse marine sediments.

Nematodes, or roundworms, have a thick, multilayered cuticle which gives the body its shape. Nematodes shed their cuticles four times during their lives. They exchange gases with the environment through the cuticle and gut wall. They range in size from microscopic to up to 9 meters in length. About 25,000 species of this diverse group have been described, but by some estimates there are more than 1 million living species. They live in soil, on the bottoms of lakes and streams, and in marine sediments. The free-living nematode *Caenorhabditis elegans* is a widely used model organism for geneticists and developmental biologists. Many species prey upon protists and other microscopic organisms, but of most significance to humans is the large number of parasitic species; several of these, including *Trichinella spiralis*, are dangerous to humans. The approximately 350 species of horsehair worms are long and very thin, as their name suggests. They live in fresh water or very damp soil near the edges of ponds and streams and feed in the larval stage as parasites of terrestrial and aquatic insects and crayfishes (see Figure 32.17). Adults have no mouth and reduced guts. In some species the adults may not feed. The adults of other species continue to grow, so it is also possible that adult worms may absorb nutrients from the environment during the time between the molting of the old cuticle and the hardening of the new one.

Question 7. What characteristics make the nematode species *Caenorhabditis elegans* an ideal model organism?

Question 8. What characteristics make priapulids, kinorhynchs, and loriciferans members of one clade, separate from the nematodes and horsehair worms?

32.4 Why Are Arthropods So Diverse?

Arthropod relatives have fleshy, unjointed appendages

Jointed appendages appeared in the trilobites

Chelicerates have pointed, nonchewing mouthparts

Mandibles and antennae characterize the remaining arthropod groups

More than half of all described species are insects

Arthropods and their relatives are ecdysozoans with paired appendages. Arthropods are the most diverse group of animals in numbers of species (more than 1.2 million and counting). Among animals, only nematodes are thought to exist in greater numbers. Arthropods have a hard exoskeleton of chitin that provides protection, waterproofing, and muscle attachment sites. Their bodies are segmented, with individual muscles attached to the exoskeleton that operate each segment. The jointed appendages that give the group its name (*arthros* = joint, *poda* = limb) allow for a great range of movement and specialization of different appendages for different purposes. Chitin is waterproof and keeps the animal from dehydrating. Four major arthropod groups exist today: the chelicerates (including the arachnids—spiders, mites, ticks, etc.); the myriapods (centipedes and millipedes); the crustaceans (crabs, lobsters, scallops, barnacles, etc.); and the hexapods (insects and their relatives). The last three groups are known together as the mandibulates.

The ancestors of the arthropods had simple, unjointed appendages, and even though not much is known about these ancestors, a few relatives with segmented bodies and unjointed appendages are still around, classified in two groups. The tardigrades (water bears) have fleshy, unjointed legs and use their fluid-filled body cavities as hydrostatic skeletons (see Figure 32.18A). Tardigrades lack a circulatory system and do not have gas exchange organs. When their environment dries out, they shrink in size drastically and can survive in a dormant state for over a decade. The onychophorans (velvet worms) were once thought to be more closely related to the annelids, but recent genetic analysis links them to the arthropods. They have unjointed legs and thin cuticles composed of chitin (see Figure 32.18B). Onychophorans have probably changed relatively little from their common ancestor with the arthropods.

The jointed appendages that characterize the arthropods appeared during the Cambrian, among a once-widespread group known as the trilobites ("three sections"). The trilobites died out during the great Permian mass extinction, but their heavy exoskeletons were readily fossilized and left an abundant record of their existence (see Figure 32.19). At least 10,000 species have been described based on the fossil record.

The chelicerates include the pycnogonids, the horseshoe crabs, and the arachnids. All chelicerates have two body parts, and most have eight legs. The pycnogonids, or sea spiders, are a group of about 1,000 exclusively marine species. Most are very small, and most are carnivorous (see Figure 32.20A). There are only four living species of horseshoe crabs. These animals have changed so little in morphology over evolutionary time that they are often referred to as "living fossils" (see Figure 32.20B). The arachnids are the most prominent chelicerates and include the spiders, scorpions, mites, and ticks (see Figure 32.21). They have a simple life cycle, with young that resemble small adults. Mites and ticks are vectors for a variety of organisms that cause diseases in plants and animals. Spiders, of which about 50,000 species have been described, are predators with hollow chelicerae which they use to inject their prey with venom. Some can chase and seize prey, and some build webs of protein threads that they use to capture prey. Spider webs are strikingly varied, often species-specific, and increase the predatory ability of spiders in many different environments.

The remaining three groups—myriapods, crustaceans, and hexapods—have mouthparts that are mandibles, not chelicerae, and they are collectively called mandibulates. Mandibles can be used to chew, bite, or hold food. Mandibulates also have sensory antennae on their heads. The myriapods include more than 3,000 described species of centipedes and 9,000 described species of millipedes. Members of both groups have similar body plans, consisting of a well-formed head and a long segmented trunk. Centipedes have one pair of legs on each trunk segment; millipedes have two pairs on each segment (see Figures 32.22A and 32.22B).

Crustaceans are the dominant marine arthropods; they are common in fresh water and some terrestrial environments. Crustaceans include many familiar animals (shrimp, lobsters, crabs, barnacles—the decapods) as well as the less-familiar isopods, copepods, branchiopods, and others. The body of most crustaceans is divided into the head, thorax, and abdomen (see Figure 32.24). The head segments are fused, and the head has five pairs of appendages. The multiple segments of the thorax and abdomen have one pair of appendages each. Crustacean appendages are specialized for walking, swimming, feeding, sensation, and gas exchange. In many species, a carapace extends dorsally from the head to protect and cover the body. Fertilized eggs of most crustacean species remain attached to the body of the female during the early stages of development. In some species the young, upon hatching, are released as larvae; in some species they are released as juveniles; still other species release the eggs into the environment.

Of the several groups of arthropods that colonized land, the most successful have been the hexapods, which include insects and their wingless relatives. Three groups of wingless hexapods—the springtails, two-pronged bristletails, and proturans—are related to the insects and probably resemble the insect ancestral form most closely (see Figure 32.25). Members of these three groups differ from insects in having internal rather than external mouthparts. Springtails are probably the most abundant hexapod. More than a million species of insects have been described so far, and many biologists believe this is only a small fraction of the species that actually exist. Like the crustaceans, the insect body has three parts: the head with a pair of antennae, the thorax with three pairs of legs, and the abdomen. In most insects the thorax also has two pairs of wings. Unlike in the crustaceans, the abdominal segments do not bear appendages (see Figure 32.26). Other distinguishing characteristics of insects include paired antennae with a sensory receptor known as Johnston's organs, and external mouthparts. Gas exchange occurs in a system composed of a series of air sacs and tubular channels called tracheae that extend from external openings called spiracles.

Table 32.2 lists the major insect groups. There are two classes of insects: the wingless apterygotes and the winged pterygotes (some species of which have secondarily become wingless). The apterygotes include the jumping bristletails and silverfish. Pterygotes typically have two pairs of wings attached to the thorax, but in some groups (e.g., parasitic lice and fleas, some beetles, and worker ants), one or both pairs of wings have been secondarily lost. Hatching insects do not look like the adults, and they undergo changes at each molt. Each stage between molts is called an instar. If the changes are gradual, the insect is said to undergo incomplete metamorphosis. If a drastic change occurs between some instars, then the insect is said to undergo complete metamorphosis. The most dramatic example is the change that occurs when a caterpillar transforms into a pupa in which the adult form develops. In insects that undergo complete metamorphosis, each stage is often specialized with regard to the environment and food source.

Adults of most flying insect species have two pairs of stiff wings attached to the thorax. True flies have one pair of wings and a pair of stabilizers (halteres). In winged beetles, one pair of wings is hardened and acts as wing covers. Two groups of pterygotes cannot fold their wings against their bodies (mayflies and dragonflies) (see Figure 32.27A); this represents the ancestral condition. Mayflies and dragonflies are not closely related; dragonflies are active predators, and mayflies lack digestive tracts in the adult form. The pterygotes that can fold their wings over their bodies are collectively known as the neopterans. Some neopterans undergo incomplete metamorphosis; these include the grasshoppers, roaches, mantids, stick insects, termites, stoneflies, earwigs, thrips, true bugs, aphids, cicadas, and leafhoppers (see Figures 32.27B and 32.27C). In these groups, the hatchlings resemble small, usually wingless adults; as they molt from one stage to the next, they gradually acquire more adult characteristics. Complete metamorphosis occurs in lacewings, caddisflies, butterflies, moths, flies, wasps, bees, and ants, among others (see Figures 32.27D, 32.27E, 32.27F, 32.27G, and 32.27H). In these groups, the larvae and adult forms are substantially different; the young pass through at least two stages (larva and pupa) before becoming adults. They form a subgroup of the neopterans called the holometabolous insects and account for about 80 percent of the known insect groups.

Insects began to diversify about 450 million years ago, around the time of the first appearance of land plants. Part of the reason for their success is their wings; pterygote insects were the first animals to achieve flight. Studies of homologous genes in insects and crustaceans suggest that insect wings are modified from a respiratory structure on the leg of an ancestral crustacean (see Figure 32.28). Flight opened up new lifestyles and food sources for insects, such as pollination of (and coevolution with) flowering plants.

Question 9. A thick cuticle provides an animal with protection, but it also can seal it off from the world, obstructing the exchange of gases. How would you design a simple, hypothetical animal with a covering that both provided protection and allowed gas exchange?

Question 10. The letters *pter* occur in the names of many insect groups. Can you discover the Greek word that is the root of this nomenclature? (Hint: the same word is the root name for a well-known dinosaur group, the pterodactyls.) Most good dictionaries give the meanings of word prefixes and roots; using such a dictionary, determine the meaning of the following names of insect groups:

Apterygota
Ephemeroptera
Orthoptera
Isoptera
Hemiptera
Homoptera
Coleoptera
Trichoptera
Lepidoptera
Diptera
Hymenoptera
Siphonaptera

Test Yourself

1. The rhynchocoel is a body plan feature found only in
   a. bryozoans.
   b. flatworms.
   c. rotifers.
   d. ribbon worms.
   e. brachiopods.
   Textbook Reference: *32.2 What Features Distinguish the Major Groups of Lophotrochozoans?*

2. Which of the following does *not* possess lophophores?
   a. Bryozoans
   b. Phoronids
   c. Annelids
   d. Brachiopods
   e. All of the above have lophophores.
   Textbook Reference: *32.1 What Is a Protostome?*

3. Which of the following statements about rotifers is *false*?
   a. They have a complete gut with an anterior mouth and posterior anus.
   b. They are coelomates.
   c. The corona is a ciliated organ used in acquiring food.
   d. They have a hydrostatic skeleton.
   e. They are very small, between 50 and 500 μm long.
   Textbook Reference: *32.2 What Features Distinguish the Major Groups of Lophotrochozoans?*

4. A restaurant appetizer of escargot (snails), clams on the half shell, and calamari (squid) would contain which types of mollusks?
   a. Chitons, bivalves, and gastropods
   b. Gastropods, bivalves, and cephalopods
   c. Chitons, gastropods, and cephalopods
   d. Bivalves and gastropods
   e. Chitons and cephalopods
   Textbook Reference: *32.2 What Features Distinguish the Major Groups of Lophotrochozoans?*

5. The combination of a true coelom and repeating body segmentation allowed the annelids to
   a. evolve complex body shapes and control movement more precisely.
   b. move through loose marine sediments.
   c. become hermaphroditic.
   d. inject paralytic poisons into their prey.
   e. become larger than their ancestors.
   Textbook Reference: *32.2 What Features Distinguish the Major Groups of Lophotrochozoans?*

6. Lobsters, millipedes, and butterflies share which of the following traits?
   a. Parapodia
   b. Setae
   c. Gas exchange across the skin
   d. Jointed appendages
   e. Spiracles
   Textbook Reference: *32.4 Why Are Arthropods So Diverse?*

7. Which of the following have a specialized internal gas exchange system?
   a. Priapulids
   b. Arrow worms
   c. Hexapods
   d. Horsehair worms
   e. Annelids
   Textbook Reference: *32.4 Why Are Arthropods So Diverse?*

8. Which of the following insect groups is made up of species that undergo complete metamorphosis and can fold their wings back over their bodies?
   a. Coleoptera (beetles)
   b. Apterygota (silverfish and springtails)
   c. Orthoptera (grasshoppers and their kin)
   d. Odonata (dragonflies and damselflies)
   e. Ephemeroptera (mayflies)
   Textbook Reference: *32.4 Why Are Arthropods So Diverse?*

9. Which of the following insect groups consists of winged species that undergo incomplete metamorphosis and can fold their wings back over their bodies?
   a. Coleoptera (beetles)
   b. Apterygota (silverfish and springtails)
   c. Orthoptera (grasshoppers and their kin)
   d. Odonata (dragonflies and damselflies)
   e. Ephemeroptera (mayflies)
   Textbook Reference: *32.4 Why Are Arthropods So Diverse?*

10. Which of the following insect groups consists of winged species that cannot fold their wings back over their bodies?
    a. Coleoptera (beetles)
    b. Apterygota (silverfish and springtails)
    c. Orthoptera (grasshoppers and their kin)
    d. Odonata (dragonflies and damselflies)
    e. None of the above
    Textbook Reference: *32.4 Why Are Arthropods So Diverse?*

11. The majority of terrestrial arthropod species are
    a. myriapods.
    b. hexapods.
    c. cirripeds.
    d. tardigrades.
    e. onychophorans.
    Textbook Reference: *32.4 Why Are Arthropods So Diverse? p. 21*

12. Which of the following does *not* represent a correct pairing of a protostome group with one of its characteristics?
    a. Bryzoans: coelomate
    b. Annelids: complete digestive tract
    c. Flatworms: complete digestive tract
    d. Rotifers: no circulatory system
    e. Nematodes: pseudocoelomate
    Textbook Reference: *32.1 What Is a Protostome?*

13. Which of the following living arthropod groups has a long fossil history showing little morphological change?
    a. Pycnogonids (sea spiders)
    b. Arachnids (mites, ticks, and spiders)
    c. Horseshoe crabs
    d. Trilobites
    e. Onychophorans
    Textbook Reference: *32.4 Why Are Arthropods So Diverse?*

14. Which of the following ecdysozoan groups is extinct?
    a. Chelicerates
    b. Crustaceans
    c. Tardigrades
    d. Trilobites
    e. Onychophorans
    Textbook Reference: *32.4 Why Are Arthropods So Diverse?*

15. Which of the following attributes is *not* seen in the arthropod body plan?
    a. Segmentation
    b. Jointed appendages
    c. A closed circulatory system
    d. A hard exoskeleton
    e. Complete digestive tract
    Textbook Reference: *32.4 Why Are Arthropods So Diverse?*

Answers

Key Concept Review

1. The first of these characteristics to evolve was the tight junction. This type of cell junction occurs in all animals and was probably present in the common ancestor of all animals. The evolution of a complete gut came next, followed by bilateral symmetry (see Figure 32.1). After the evolution of bilateral symmetry, the phylogenetic tree splits into the deuterostomes and protostomes; as protostomes, the annelids have a mouth that develops from their blastopore. The annelids belong to a group of protostomes called lophotrochozoans; some of these exhibit spiral cleavage and it is thought that the ancestral lophotrochozoan possessed this characteristic. The next characteristic to evolve was a segmented body, a characteristic that defines the annelids. Since the annelids are overwhelmingly aquatic, and even terrestrial annelids are confined to moist habitats, it seems likely that the terrestrial lifestyle was the most recent of the listed characteristics to evolve.
2. The vast majority of species that have lived on Earth have gone extinct without leaving fossils, so it is very unlikely that scientists will ever discover a fossilized body of the common ancestor of all bilaterally symmetrical animals. Some fossilized animal tracks from the Precambrian appear to have been made by a bilaterally symmetrical animal, but these offer limited information. The best source of information about the common ancestor of any group of animals is the set of characteristics that are common to all members of that group. All bilaterally symmetrical animals share a triploblastic organization and certain Hox genes, and it is probable that these were also present in the common ancestor of this group.
3. The organism shown is a phoronid.
   a. Lophophore tentacles
   b. Anus
   c. Outer covering (chitinous tube)
   d. Gut
   e. Mouth
4. a. Circular muscle
   b. Longitudinal muscle
   c. Setae (bristles)
   d. Excretory organ
   e. Intestine
   f. Nerve cord
   g. Coelom
   h. Blood vessel
   i. Septum between segments
5. (A) Chiton
   a. Shell plates
   b. Anus
   c. Foot
   d. Gills in mantle cavity
   e. Digestive gland
   f. Mouth
   g. Radula
   h. Stomach
   i. Intestine

   (B) Gastropod
   a. Shell
   b. Heart
   c. Salivary gland
   d. Stomach
   e. Foot
   f. Mouth
   g. Siphon
   h. Anus
   i. Gill
   j. Intestine
   k. Mantle

   (C) Bivalve
   a. Shell
   b. Heart
   c. Intestine
   d. Anus
   e. Siphons
   f. Gill
   g. Foot
   h. Mantle
   i. Mouth
   j. Stomach
   k. Digestive Gland

   (D) Cephalopod
   a. Intestine
   b. Shell
   c. Stomach
   d. Mantle
   e. Digestive gland
   f. Heart
   g. Gill
   h. Siphon
   i. Radula
6. The mollusk body plan has three morphological components shared by all molluscan groups. The first trait is a muscular foot used for locomotion. In cephalopods the foot has been modified into arms and tentacles. The second trait is a mantle, which covers the third trait, the visceral mass. The mantle secretes a calcareous shell in many species and often houses the gills. The visceral mass consists of the internal organs.
7. *C. elegans* is small and easy to cultivate, it has a quick generation time and a fixed number of body cells, and its genome has been completely sequenced. All of these characteristics make it a good model organism because it is easily studied in a realistic time frame, its cells can be tracked from fertilization to full-grown animal, and the entire genome can be mined for genetic information.
8. All three of these groups have relatively thin cuticles that are molted periodically as the animals grow. Each also has a retractable proboscis that is used for feeding. Most species of all three are very tiny (even microscopic), although a few priapulid species can be up to 20 centimeters long. Very few numbers of species in each group have been identified, leading to a dearth of knowledge about this clade.
9. The animal would need special mechanisms for gas exchange with the environment. The insects have overcome this problem with their system of tracheae, a network of tubes found throughout the body that is in close contact with all the cells of the animal and opens to the environment. In an aquatic environment, the animal could have an internal chamber with gills used for respiration.
10. As a root or prefix, *pter* comes from the Greek word *pteris* and refers to wings.

Apterygota: The prefix *a-* in this case means *without*. Apterygote insects have no wings.

Ephemeroptera: *Ephemeral* means *short-lived*. Adult mayflies typically live only a few hours or days.

Orthoptera: *Ortho-* means *straight*. Orthopterans tend to have very straight wings (see Figure 32.26).

Isoptera: *Iso-* means *equal* or *uniform*. All four wings on termites tend to be very similar.

Hemiptera: *Hemi-* means *half*. In many true bugs, the front half of the forewing is hardened, and they appear to have half a shell (like a beetle's) and half wings.

Homoptera: *Homo-* means *same*. Unlike their close relatives the hemipterans, homopterans have forewings and hind wings that are similar to each other.

Coleoptera: *Coleo-* means *sheath*. The forewings of beetles are modified into a protective sheath that usually covers the hind wings and abdomen.

Trichoptera: *Tricho-* means *hair*. The bodies and wings of adult caddisflies are covered in small hairs.

Lepidoptera: *Lepido-* means *scale*. Butterfly and moth wings are covered with tiny scales.

Diptera: *Di-* means *two*. Unlike most winged insects, flies have only two wings.

Hymenoptera: *Hymen-* means *membrane*. Bees, wasps, and the reproductive castes of ants have membranous wings.

Siphonaptera: *Siphon-* means *tube* or *pipe*. This is a reference to the flea's feeding method. This name, unlike the others, ends in *-aptera*, not *-optera*: Fleas do not have wings (*a-* means *without*), though as pterygotes they are descended from winged ancestors, having secondarily lost their wings over the course of evolution.

Test Yourself

1. **d.** The rhynchocoel is a fluid-filled cavity containing a proboscis, a feeding organ found in the nemerteans.
2. **c.** Annelids belong to the spiralian lineage, not the lophophorates.
3. **b.** Rotifers are pseudocoelomates and have a pseudocoel.
4. **b.** Snails are gastropods, clams are bivalves, and squid are cephalopods.
5. **a.** The segmentation of the annelids allows for more complex, coordinated movement.
6. **d.** All of these animals have jointed appendages.
7. **c.** The Hexapoda have an internal system of tracheae for gas exchange.
8. **a.** The Coleoptera, or beetles, undergo complete metamorphosis and are able to fold their wings over their bodies.
9. **c.** The Orthoptera (grasshoppers, crickets, roaches, etc.) undergo incomplete metamorphosis and are able to fold their wings over their bodies.
10. **d.** The Odonata (dragonflies and damselflies), along with the Ephemeroptera (mayflies), cannot fold their wings over their bodies. Apterygotes (silverfish and springtails) are wingless.
11. **b.** The predominant insects on Earth today are the the six-legged hexapods: the insects and their wingless relatives.
12. **c.** Flatworms have a blind gut, not a complete digestive tract with both a mouth and anus.
13. **c.** Horseshoe crabs have changed little during their fossil history; they are sometimes referred to as living fossils.
14. **d.** The chelicerates include spiders, sea spiders, and horseshoe crabs; crustaceans include decapods (crabs, shrimp, and others), barnacles, isopods, and copepods; the tardigrades are a small (but not extinct) group also known as water bears. Onychophorans (velvet worms) are not extinct. Of the groups listed, only trilobites are extinct.
15. **c.** Arthropods have a hard exoskeleton, a segmented body, jointed appendages, a complete digestive tract, and an open circulatory system (see Table 32.1).

Deuterostome Animals

The Big Picture

- The deuterostomes are not as diverse or numerous as the protostomes. Most fit into one of two groups, the echinoderms and the chordates.
- Echinoderms have an internal skeleton of calcified plates and a water vascular system used in feeding and gas exchange. They include the sea stars (starfish), sea lilies, and sea cucumbers, all of which exhibit pentaradial symmetry as adults.
- The chordates have several shared traits: pharyngeal slits, a nerve cord, a ventral heart, a tail beyond the anus, and a notochord. The chordates consist of the ascidians, lancelets, and vertebrates. Vertebrates have a dorsal vertebral column that supports a rigid endoskeleton.
- The most numerous vertebrates, both in terms of number of species and number of individuals, are the fishes. The amphibians adapted to life on land, but their life cycle requires them to return to water to reproduce.
- The evolution of the amniote egg freed reptiles from water-based reproduction. During the Carboniferous, amniotes diverged into two major clades: the reptiles and the mammals. Reptiles were the dominant vertebrates for many millennia. One reptilian group gave rise to the now-extinct dinosaurs and to the birds. Among the birds, the reptilian scales evolved into feathers, allowing for flight.
- The extinction of the dinosaurs made radiation of the mammals possible. Most mammals belong to one of 20 major groups of eutherians. The largest eutherian group is the rodents; the primate eutherians include the lemurs, monkeys, apes, and humans.
- Humans (genus *Homo*) are primates that evolved from bipedal australopithecine ancestors on the African continent. Several species of *Homo* arose and went extinct over evolutionary time. In the lineage leading to *Homo sapiens*, the only surviving species, brain size expanded greatly relative to body size, allowing humans to develop language and culture.

Study Strategies

- As you work your way through the phylogenies presented in the textbook, try to recall the representative organisms that you already know from each group and what their special traits are. If you base your learning on information with which you are already familiar, you will have an easier time learning the facts.
- You may be confused by the evolution of the chordates and vertebrates and the traits that separate the vertebrates from other chordates. Create a timeline with all the major chordate classes, and then insert the specific traits that distinguish one class from another. This will help you organize the evolution of the chordates so that memorizing it becomes a manageable task.
- Go to the Web addresses shown to review the following animated tutorials and activities. You can also find them in BioPortal (yourBioPortal.com), along with many additional learning resources.

  Animated Tutorial 33.1 An Overview of the Deuterostomes (Life10e.com/at33.1)

  Animated Tutorial 33.2 Life Cycle of a Frog (Life10e.com/at33.2)

  Activity 33.1 Deuterostome Phylogeny (Life10e.com/ac33.1)

  Activity 33.2 The Amniote Egg (Life10e.com/ac33.2)

  Activity 33.3 Deuterostome Classification (Life10e.com/ac33.3)

Key Concept Review

33.1 What Is a Deuterostome?

Deuterostomes share early developmental patterns

There are three major deuterostome clades

Fossils shed light on deuterostome ancestors

All echinoderms and vertebrates are deuterostomes. Evidence that all deuterostomes share a common ancestor not shared with the protostomes includes early developmental patterns and phylogenetic studies. Three early developmental patterns distinguish the deuterostomes: radial cleavage; development of the blastopore into an anus and formation of the mouth at the opposite end of the embryo; and development of a coelom from mesodermal pockets, rather than splitting of the mesoderm. These characteristics are not all exclusive to deuterostomes, however, and some may in fact be ancestral to all bilaterians; instead, the best evidence of the monophyly of deuterstomes comes from phylogenetic studies.

All deuterostomes are triploblastic, coelomate animals. There are many fewer species of deuterostomes than there are of protostomes, but the deuterostomes are of special interest because mammals—a group that includes the largest living animals as well as the human lineage—are deuterostomes. Skeletal support features, when present, are internal. A few species have segmented bodies, but the segments are not obvious. Deuterostomes fall into three major clades: the echinoderms, the hemichordates, and the chordates (see Figure 33.1).

Fossil beds, most of them from a 520-million-year-old site in China, have revealed insights into deuterostomate ancestry. Fossils and phylogenetic analysis of living species indicate that the earliest deuterostomes were bilaterally symmetrical, segmented animals with a slitted pharynx and external gills (see Figure 33.2). The unique pentaradial symmetry of the echinoderms likely evolved later, while other deuterostome groups retained bilateral symmetry.

Question 1. Label the different groups of organisms in the phylogenetic tree shown below and identify the key features of each branch as indicated by the dots. What are the names of the groups identified by brackets?

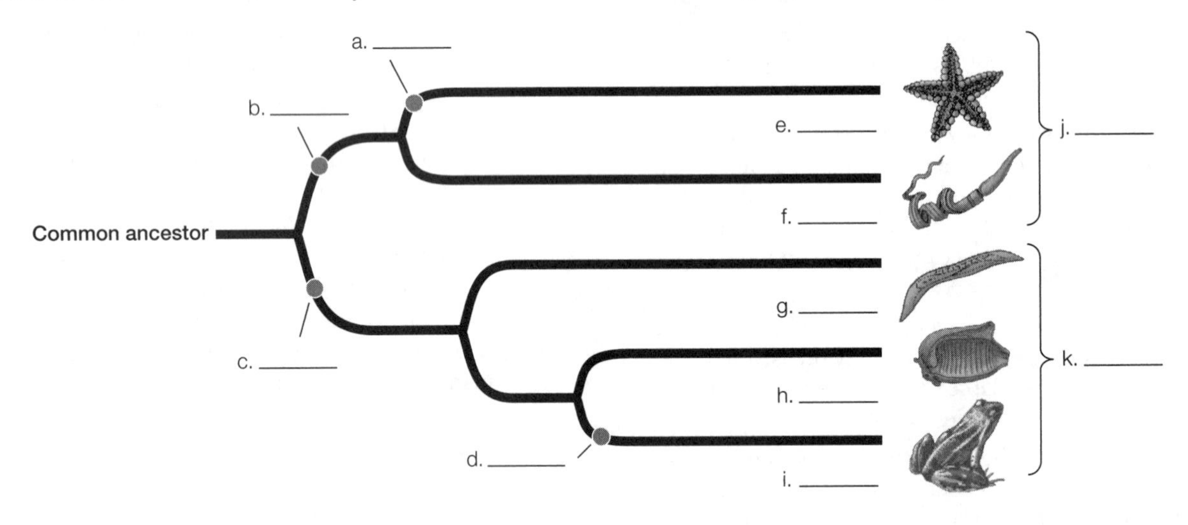

Question 2. What are the distinguishing characteristics of deuterostomes? Are these all exclusive to deuterostomes? What is the evidence for monophyly of the group we call deuterostomes?

33.2 What Features Distinguish the Echinoderms, Hemichordates, and Their Relatives?

Echinoderms have unique structural features

Hemichordates are wormlike marine deuterostome

While 23 major groups consisting of about 13,000 species of echinoderms have been described from the fossil record, only six groups containing about 7,500 species exist today. All known species live in marine environments. Only about 120 species of hemichordates are known to exist. The echinoderms and hemichordates are collectively known as ambulacrarians; they have bilateral, ciliated larvae (see Figure 33.3A). Adult hemichordates are bilaterally symmetric, but adult echinoderms develop pentaradial symmetry. Adult echinoderms have no head and move equally well in all directions, albeit slowly. They do not have anterior–posterior (head–tail) body organization, but rather an oral–aboral orientation: the bottom-facing side contains the mouth while the top-facing surface contains the anus (see Figure 33.3). Two groups of small marine organisms, the xenoturbellids and the acoels (see Figure 33.4), are thought to be the sister group to the ambulacrarians. Both groups are wormlike, with simple body plans.

In addition to their pentaradial symmetry, echinoderms have two unique features: an internal skeleton made up of thick calcified plates, and a water vascular system comprising water-filled canals that end in external tube feet. The water vascular system functions in feeding, gas exchange, and locomotion. Seawater enters through the madreporite, passes through the ring canal, and then enters the branching radial canals. One clade of echinoderms, the crinoids, includes about 80 species of sea lilies and feather stars (see Figure 33.5A). Crinoids are sessile, attached to the substrate by a stalk. About 600 living species of feather stars have been described. The remaining echinoderms are generally mobile and are divided into two main groups: the echinozoans (sea urchins and sea cucumbers; see Figures 33.5B,C) and the asterozoans (sea stars and brittle stars; see Figures 33.5D,E). Sea urchins are hemispherical and lack arms, but are covered with spines on ball-and-socket joints. Sand dollars are flattened, disc-shaped relatives of sea urchins. Sea cucumbers also lack arms, and their bodies are oriented on an anterior–posterior, as opposed to oral–aboral, axis. Sea stars, or starfish, are the most familiar echinoderms. Their gonads and digestive organs are located in the arms. Their tube feet serve as locomotion, gas exchange, and attachment. Each tube foot has an internal ampulla that is connected to an external suction cup by a muscular tube. Brittle stars have five flexible arms made of hard, jointed plates.

Echinoderms use their tube feet to catch prey in many different ways. Sea lilies orient their arms in passing currents to filter feed; the tube feet transfer the particles to grooves in the arms. Most sea urchins capture phytoplankton with their tube feet or scrape algae from rocks. Sea cucumbers capture prey with their anterior tube feet, which are large, feathery, sticky tentacles. Sea stars can capture large prey such as polychaetes, gastropod and bivalve mollusks, and small crustaceans with their many tube feet. They also can use their tube feet to pry open the shells of bivalve mollusks. Once the bivalve is open, the sea star inserts its stomach into the shell and digests the mollusk's internal organs. Most of the 2,000 species of brittle stars ingest particles from sediments and extract the organic material, although some are filter feeders and others can capture small animals.

The hemichordates include the acorn worms and the pterobranchs. Both groups have a bilaterally symmetric body plan composed of a trunk, a collar, and a sticky proboscis used for catching prey and digging (see Figure 33.6A). Acorn worms live in burrows in muddy and sandy marine sediments. In the acorn worm, cilia move food from the proboscis to the mouth. Behind the mouth are the pharynx and an intestine. The pharynx can open to the outside through pharyngeal slits that contain vascularized tissue for gas exchange. Acorn worms breathe by pumping water through the mouth and out the pharyngeal slits. Pterobranchs are sedentary marine animals that are either solitary or colonial (see Figure 33.6B). Behind the proboscis are one to nine pairs of arms that are used to capture prey and also function in gas exchange.

Question 3. Suppose that you discover the fossil of a unknown organism. It has radial symmetry, and examination of the surrounding rock and adjacent fossils suggests that it derives from a marine environment. What other features would you look for in order to determine if this fossil is an echinoderm?

Question 4. Sea cucumbers are unique among echinoderms in that they have an anterior–posterior axis rather than an oral–aboral one. What is the most likely explanation for the evolution of this bilateral-type appearance?

33.3 What New Features Evolved in the Chordates?

Adults of most lancelets and tunicates are sedentary

A dorsal supporting structure replaces the notochord in vertebrates

The phylogenetic relationships of jawless fishes are uncertain

Jaws and teeth improved feeding efficiency

Fins and swim bladders improved stability and control over locomotion

The features that reveal the evolutionary relationship between the echinoderms and chordates, as well as among the chordates, are primarily observed in the larval stage or during the early stages of development. The notochord is the main distinguishing characteristic of the chordates. There are three principal chordate clades: the lancelets, the tunicates, and the vertebrates. All three chordate clades share the following at some point during development: a dorsal hollow nerve cord; a tail that extends beyond the anus; and a dorsal supporting rod called the notochord. In tunicates the notochord is lost during the metamorphosis from larval to adult forms. In most vertebrates, the notochord is replaced by skeletal structures that provide support for the body. The pharyngeal slits that are ancestral to all deuterostomes are generally lost or greatly modified in adults. In chordates, they are separated and supported by structures called pharyngeal arches. In tunicates and lancelets the pharynx is used in filter feeding; in fishes and larval amphibians, some of the pharyngeal arches become gill arches. Some pharyngeal arches also develop into elements of the vertebrate jaw, as well as parts of the tongue, larynx, trachea, and middle ear of tetrapods; some slits care modified to become the eustachian tube and middle ear chamber.

Lancelets, also known as cephalochordates, are found in shallow marine and brackish waters worldwide and rarely exceed 5 centimeters in length. The notochord extends the entire length of the body throughout the animal's life. They are mostly sessile and covered in sand, but are able to swim. The pharynx is enlarged and becomes a pharyngeal basket, which is used to filter feed. During the reproductive season, walls of the enlarged gonads of both males and females rupture, releasing eggs and sperm into the water where fertilization occurs.

The three major tunicate groups are the ascidians (sea squirts), thaliaceans, and larvaceans. All members of these three groups are marine, and more than 90 percent of tunicate species are sea squirts. Some sea squirts form colonies via asexual budding. The body form for an adult sea squirt is baglike and surrounded by a tough tunic (hence the name "tunicate") (see Figure 33.8A). A sea squirt larva has a dorsal hollow nerve cord and notochord mostly restricted to the tail region. The larvae of most species settle on the sea floor and become sessile adults. Thaliaceans can live in tropical waters at depths up to 1,500 meters, either as a single individual or in colonies (see Figure 33.8B). Larvaceans retain a notochord and dorsal hollow nerve cord throughout adult life. They are solitary and small, but some species that live near the bottom of the ocean surround themselves with an extensive casing of mucus more than a meter wide. They capture organic particles with filters built into the mucus.

In the vertebrates, the vertebral column replaces the notochord as the primary support structure; the individual elements are called vertebrae. Four other features characterize the vertebrates: an anterior skull enclosing the brain; a rigid skeleton supported by the vertebral column; a large coelom enclosing the internal organs; and a circulatory system driven by contractions of a ventral heart. These characteristics are able to support large, active animals. An extensive muscular system supported by the skeleton can develop, driven by the oxygen supplied by the circulatory system. These features allowed many vertebrates to become large predators, driving diversification of vertebrates in general. The nonvertebrate deuterostomes are primarily marine. The lineage that led to the vertebrates is also thought to have evolved in a marine environment, possibly in estuarine waters (places where fresh and salt water mix). The first vertebrates appeared in the Cambrian; the approximately 65,000 species of vertebrates have radiated into marine, freshwater, terrestrial, and aerial environments worldwide.

The elongate, eel-like hagfishes are thought to be the sister group of the remaining vertebrates (see Figure 33.11A). Some biologists use the term "craniates" to refer collectively to the hagfishes and vertebrates. The hagfishes may be a sister group to the lampreys (collectively called the cyclostomes); if these two groups are indeed monophyletic, it is possible that during evolution the ancestor of the hagfishes lost some of the structural characteristics that define vertebrates. Hagfishes are marine animals that produce slime as a defensive mechanism. They have a weak circulatory system with three small accessory hearts and a partial cranium; they lack a stomach, and their skeleton is composed of cartilage but lacks separate vertebrae. They are virtually blind and use sensory tentacles to find food. Although they lack a jaw, they have a tonguelike structure with toothlike rasps that they use to capture prey. Hagfishes may change sex from year to year (from male to female and vice-versa).

Lampreys resemble hagfishes, but they differ biologically (see Figure 33.11B). They have a complete braincase and distinct vertebrae made of cartilage. Unlike hagfishes, which undergo direct development, lampreys undergo complete metamorphosis. The larvae, called ammocoetes, are filter feeders, while the adult forms are often parasitic. The mouth is a rasping, sucking organ and is used to attach the individual to prey. The adults of a few lamprey species do not feed as adults, surviving only long enough to breed. The known species of lampreys are either freshwater or anadromous (move into fresh water to breed).

Jawless fishes were dominant through the Devonian, but the 100 or so species of hagfishes and lampreys are the only surviving jawless vertebrates. Late in the Ordovician, some fishes evolved jaws via modifications in the skeletal arches supporting the gills (see Figure 33.12A). Those fishes and their descendants fall into the category of gnathostomes, or "jaw mouths." Jawed fishes were the major predators of the Devonian. The evolution of the jaw—and then of teeth—led to greatly enhanced feeding efficiency (see Figure 33.12B). Jaws provided access to new food sources, and teeth made for extremely effective predation. Chewing aids in chemical digestion.

Most jawed fishes have paired sets of fins, which help stabilize their position in the water and sometimes provide propulsion. Several groups of jawed fishes became abundant in the Devonian. The chondrichthyan fishes, including sharks, skates, rays, and chimaeras, have flexible, leathery skin and a skeleton composed of cartilage (see Figure 33.13). Sharks move forward by laterally undulating their body and tail (caudal) fin; skates and rays move vertically by undulating their pectoral fins. Most sharks are predators, but some are filter feeders. Most skates and rays feed on bottom-dwelling organisms. Almost all chondrichthyans are marine, but a few live in estuarine waters or migrate to lakes and rivers. One lineage of gnathostomes gave rise to the bony vertebrates, which split into two lineages: the ray-finned fishes and lobe-limbed vertebrates. Bony vertebrates have internal skeletons of bone instead of cartilage. Lunglike gas-filled sacs called swim bladders evolved in the ancestors of the ray-finned fishes. With the aid of the swim bladder, many fish can control their position in the water column with minimal energy expenditure. The outer body of a ray-finned fish is covered by thin, lightweight scales that aid in movement and provide protection. A hard flap called the operculum, which covers the gills, allows for the movement of water over the gills and thus for gas exchange. The approximately 32,000 species of ray-finned fishes encompass a remarkable variety of shapes, sizes, and lifestyles, and they exploit every kind of food source in the aquatic environment (see Figure 33.14). Some ray-finned fishes are solitary and some form aggregate groups in open waters called schools. The eggs of many fishes tend to sink, which is why the shallow coastal waters and estuary environments are important to their life cycles. Fishes such as salmon are anadromous, meaning that they live in salt water but must move to fresh water to spawn.

Question 5. Ascidians, or sea squirts, are armless and legless organisms that spend their adult lives attached to a substrate under water. They feed by pulling water into one tube, filtering out planktonic organisms, and pushing the water out another tube. Grasshoppers have legs, move around freely on land, and feed by ingesting food through a mouth. What evidence has led biologists to believe that humans are more closely related to sea squirts than to grasshoppers?

Question 6. The evolution of a hinged jaw resulted in an extensive radiation of the fishes into the many modern jawed forms. What was the main advantage and significance of hinged jaws in this radiation?

33.4 How Did Vertebrates Colonize the Land?

- Jointed limbs enhanced support and locomotion on land
- Amphibians usually require moist environments
- Amniotes colonized dry environments
- Reptiles adapted to life in many habitats
- Crocodilians and birds share their ancestry with the dinosaurs
- Feathers allowed birds to fly
- Mammals radiated after the extinction of non-avian dinosaurs

The evolution of lunglike swim bladders in some ray-finned fishes set the stage for the move to land. Some fishes may have used these sacs to supplement their oxygen supply in low-oxygen freshwater environments, and the sacs may have allowed some fishes to survive out of water. The fins of these fishes were unjointed, so they could only flop around on land. In the lobe-limbed vertebrates, the paired pelvic and pectoral fins developed into more muscular fins joined to the body by a single large bone. The modern versions of these animals include the coelacanths, lungfishes, and tetrapods. Coelacanths were thought to have died out 65 million years ago, but two species were discovered in the twentieth century. One genus, *Latimeria,* is a predator and can weigh up to 80 kilograms (see Figure 33.15A). Unlike other fishes, *Latimeria* has a skeleton composed primarily of cartilage; this is thought to be a derived feature, since the ancestors to this group had bony skeletons. Lungfishes were important predators in shallow-water habitats, but now only six species of lungfishes remain, all of them in the southern hemisphere (see Figure 33.15B). During dry periods when the water from ponds evaporates, they burrow deep in the mud and can survive in an inactive state for many months, breathing air. The change in fin structure allowed these fish to support themselves in shallow water and, eventually, to move onto land and exploit food sources there. These early land-dwelling fishes gave rise to the tetrapods—the four-legged vertebrates. In 2006, an important fossil was discovered that is believed to be an intermediate between the finned fishes and the tetrapods (see Figure 33.15C). The pectoral fins have skeletal structures found in tetrapod limbs that may have allowed for front-to-rear movement needed for walking and making brief trips out of the water. Limbs capable of movement on land evolved from the short muscular fins of aquatic ancestors (see Figure 33.16). The resulting four limbs give tetrapods their name. The tetrapod tree split relatively early into two lineages: amphibians, which mostly remain tied to moist environments, and amniotes, many of which adapted to drier conditions.

Most modern amphibians spend at least part of their life cycle in the water (see Figure 33.17). Many species return to water to lay their eggs, producing aquatic larvae. Many amphibians have evolved a variety of reproductive and parental-care modes. The approximately 7,000 species of extant amphibians fall into three groups: the wormlike, legless, tropical, burrowing or aquatic caecilians; the anurans; and the salamanders (see Figure 33.18). Anurans include the tail-less frogs and toads, and account for the vast majority of amphibian species. Some species have adapted to life in very dry, even desert, environments, while others have returned to an entirely aquatic lifestyle. The adults have a very short vertebral column and a pelvic region modified for hopping on the hind legs. Salamanders are tailed amphibians that exchange respiratory gases through both lungs and gills. One group relies on gas exchange through its skin and mouth as all amphibians do, but does not have lungs. Neoteny (retention of the juvenile form in the adult) has led to the evolution of several entirely aquatic salamander species. Most species have internal fertilization. Sperm is transferred in a jellylike capsule called a spermatophore. The social behaviors of many amphibians are quite complex; males make species-specific calls to females and can defend their own breeding territories. The amount of care given to the eggs varies. Some amphibians lay large numbers of eggs that are abandoned, while others guard a few fertilized eggs. A few species are viviparous, meaning that the adult gives birth to well-developed juveniles. Amphibian populations are declining rapidly in many parts of the world. Scientists are currently pursuing hypotheses to account for these population declines.

Amphibians are largely confined to moist environments by their need to reproduce in water. In one ancestral lineage, the amniote egg evolved (see Figure 33.19A). The egg has a protective leathery or calcium-based shell that inhibits dehydration while allowing oxygen and carbon dioxide to pass through. Within the shell, extraembryonic membranes further protect the embryo and allow gas exchange and excretion of nitrogenous wastes; large quantities of yolk provide nutrition. The amniote animals—reptiles (including birds) and mammals—were thus freed from reliance on water to reproduce. In several groups of amniotes the egg became modified to allow the embryo to grow inside the mother. In mammals, the egg has lost the shell and the embryonic membranes have been retained and modified (see Figure 33.19B). Other evolutionary adaptations to life on land appeared among the amniotes, including a tough, impermeable skin covered with scales or modified scales (i.e., hair or feathers). The excretory systems of amniotes also evolved adaptations that allowed these animals to excrete nitrogenous wastes in the form of concentrated urine with a minimal loss of valuable water. During the Carboniferous period, about 250 million years ago, the amniote animals diverged into two groups, the reptiles and the mammals. Birds, the only living members of the now-extinct dinosaurs, would evolve from one of the reptilian groups (see Figure 33.20).

The modern reptile lineage diverged from the other amniotes more than 300 million years ago. Of the more than 19,000 species of reptiles that exist today, more than half are birds. Birds are the only remaining representatives of the dinosaur lineage, the dominant terrestrial predators of the Mesozoic. The lepidosaurs are the second most species-rich clade of reptiles, and include the squamates (lizards, snakes, and amphisbaenians—another group of legless, wormlike burrowers) and the tuataras, which resemble lizards but differ from them in many anatomical characteristics. Today only two species of tuataras survive (see Figure 33.21A). The lepidosaur body is covered with scales that greatly reduce water loss but make their skin surface unavailable for gas exchange. Gases are exchanged through the lung, which is larger in surface area compared to that of the amphibians. The lepidosaurs have a three-chambered heart that partially separates oxygenated from deoxygenated blood. Most lizards are insectivores, and most walk on four limbs (see Figure 33.21B); the limbless condition of snakes is a product of secondary evolution. The major groups of limbless squamates are the snakes (see Figure 33.21C). All snakes are carnivores, and adaptations to the jaws allow them to swallow prey much larger than themselves. One relatively small reptilian group, the turtles, has a unique body plan that has changed very little over the millennia (see Figure 33.21D). In these animals, dorsal and ventral bony plates evolved from the ribs to form a protective shell. Most turtles live in aquatic environments, but groups such as the tortoises and box turtles are terrestrial. Sea turtles live at sea, except when they come ashore to lay their eggs.

The archosaurs include the extant crocodilians, the extinct pterosaurs and dinosaurs, and the "living dinosaurs" we know as birds (see Figure 33.22). The crocodilians—crocodiles, alligators, caimans, and gharials—live only in tropical and warm temperate regions. They spend most of their lives in water but build nests on land. They are all carnivores, feeding on other vertebrates, including large mammals. Dinosaurs arose among the reptiles around 215 million years ago and survived for about 150 million years. Only one group, the birds, survived the mass extinction of the dinosaurs at the end of the Cretaceous. During the Mesozoic, most terrestrial animals larger than a meter in length were dinosaurs. Both the fossil record and molecular evidence support the position of birds as a group of reptiles. Birds probably evolved from an ancestral theropod—a bipedal, predatory dinosaur that appears to have had hollow bones, a furcula (wishbone), limbs with three digits, and a backward-thrusting pelvis. Modern birds are endothermic, and some evidence suggests that the ancestral theropods were endothermic as well. Living birds diverged from a common flying ancestor and fall into one of two groups. The palaeognaths are flightless and include rheas, emus, kiwis, cassowaries, and ostriches (which are the largest birds) (see Figure 33.22B). The neognaths, most of which have retained the ability to fly, represent the remaining species of birds.

Recent fossil discoveries in China have demonstrated that some dinosaurs had scales that were highly modified to form feathers. In one species, *Microraptor gui*, the feathers were structurally similar to those of modern birds (see Figure 33.23A). *Archaeopteryx* lived 150 million years ago and represents the oldest known fossil of a bird. It was covered in feathers and had well-developed wings and a wishbone (see Figure 33.23B). It also had teeth, which were lost in later members of this lineage. Feathers are complex and strong, but also lightweight (see Figure 33.24). Their evolution was a major factor in diversification. The bones of dinosaurs and birds are hollow, and along with the development of feathers, they assisted in the evolution of flight. Along with the ability to fly came a high metabolic rate needed to fuel flight. The metabolic rate of birds means that they generate a great amount of heat, and their feathers are adapted to allow for heat loss. The lungs of birds also function differently from those of other vertebrates, another adaptation to the needs of flight. Different bird groups feed on many different types of animal and plant material, including carrion. Birds that feed on fruits and seeds are major agents of plant dispersal.

The earliest mammals lived side by side with reptiles from the first split of the two lineages early in the Mesozoic era. However, only after the large dinosaurs became extinct did mammals begin to flourish in size and number. A number of unique traits distinguish the mammals. Sweat glands in the skin produce secretions that are crucial to cooling the body. Adaptations to the sweat glands may have led to the mammary glands that provide nutrient fluid for their young and that give the group its name. Modifications of scales became the external body hair that protects and insulates most mammals. A four-chambered heart that completely separates oxygenated from deoxygenated blood evolved in the mammals. It also evolved convergently (i.e., separately) in the crocodilians and birds. In most mammals, the amniote egg became modified for growth of the embryo within the mother's uterus. In the uterus, the embryo is contained within an amniotic sac and is connected to the wall of the uterus by the placenta, which allows for nutrient and gas exchange and waste elimination via the female's circulatory system. Most mammals have a covering of hair, which has been greatly reduced in some species; cetaceans (whales and dolphins) have evolved a thick layer of blubber as insulation, and humans have learned to use clothing to keep warm.

The approximately 5,700 species of mammals are divided into two groups, the prototherians and therians. The therian clade is further divided into the marsupials and eutherians. There are five species of prototherians, found only in Australia and New Guinea. These species, the duck-billed platypus and echidnas, are egg-laying mammals with sprawling legs (see Figure 33.26). They supply milk to their young as do other mammals, but they do not have nipples on their mammary glands. The remaining mammals are classed as therians. Marsupial therians give birth to tiny underdeveloped young that they nurture externally, usually in a pouch on the mother's belly (see Figure 33.27A). Marsupials were once widespread, but today the approximately 330 marsupial species are largely found in Australia and South America, with minor representatives in North America (see Figure 33.27). The eutherian mammals develop in the mother's uterus. This group is widely known by the name placental mammals, from the placenta that provides nourishment for the growing embryo. However, some marsupials also have placentas. There are 20 major groups of living eutherians (see Table 33.1). Relationships among these groups are difficult to determine because of the explosive adaptive radiation that occurred during a relatively short time. Eutherians vary greatly because of this adaptive radiation, and today they occupy a wide range of ecological niches. The two most diverse groups of eutherians are the rodents and the bats. Rodents are defined by their tooth morphology, and bats by their evolution of flight. Grazing and browsing by many herbivorous eutherian groups have transformed the terrestrial environment. In plants, this led to the evolution of spines and tough leaves. Despite such defenses, herbivores have developed adaptations to their teeth and digestive systems in order to consume these plants. This is an example of coevolution. Eutherians exist in virtually all of Earth's environments and vary greatly in form (see Figure 33.29). Several groups, most notably the cetaceans (whales and dolphins) returned to an aquatic lifestyle. The seals, sea lions, and walruses also returned to the marine environment, and manatees and dugongs colonized estuaries and shallow seas.

Question 7. Label the different groups of organisms in the phylogenetic tree shown below and identify the key features of each branch as indicated by the dots. What are the names of the groups identified by brackets?

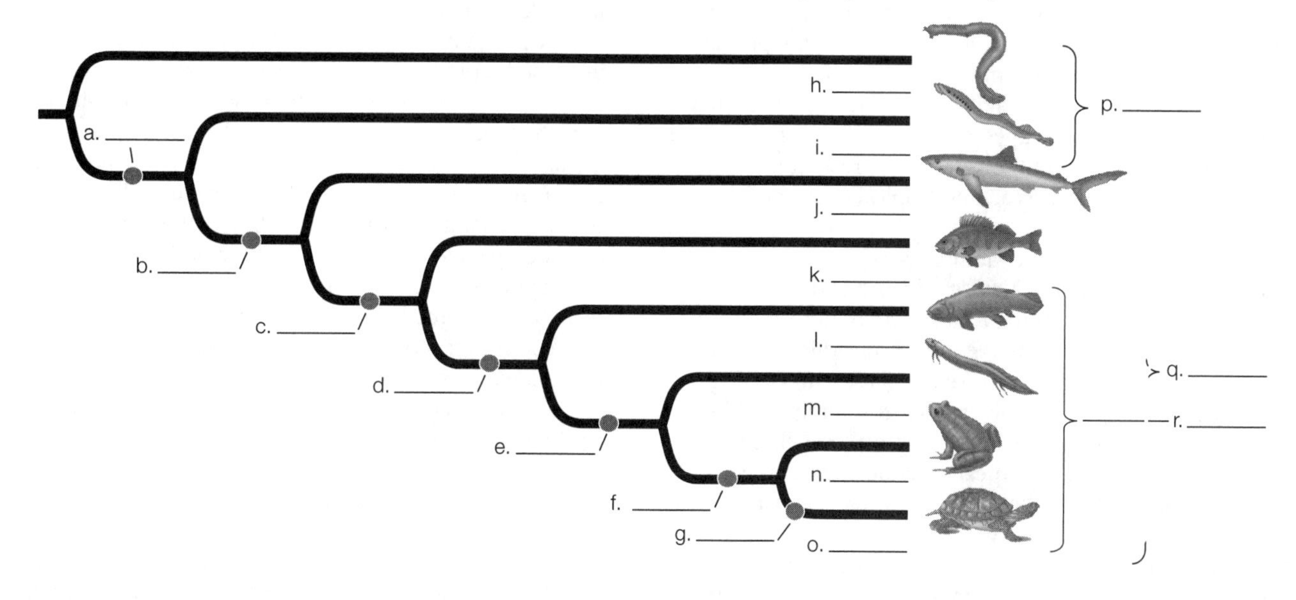

Question 8. Label the different groups of organisms in the phylogenetic tree shown below. What are the names of the groups identified by brackets? Which groups are extinct?

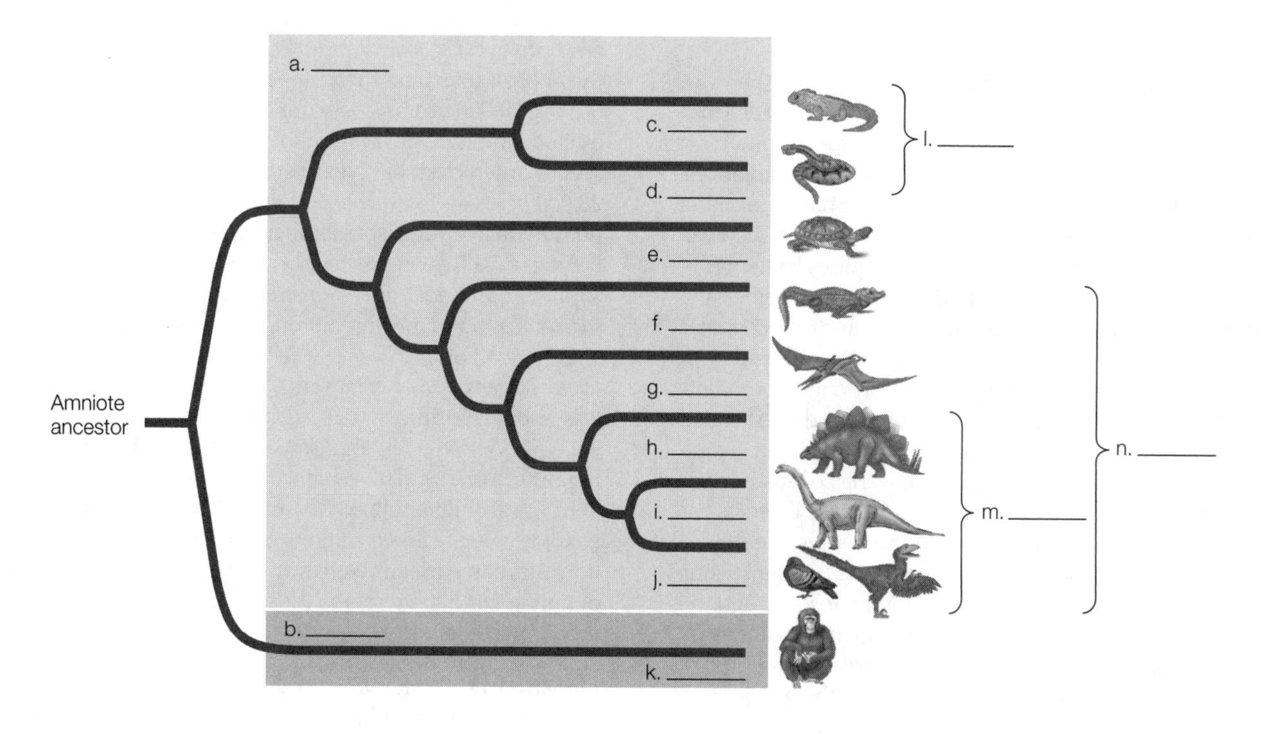

Question 9. How do mammals and reptiles address some of the challenges of a terrestrial habitat?

Question 10. Amphibians are dependent on water and moist environments for their life cycle. Which stages require an aquatic environment and why is such an environment necessary? Do any amphibians have strategies that allow them to avoid the aquatic stage? If so, describe them.

33.5 What Traits Characterize the Primates?

- Two major lineages of primates split late in the Cretaceous
- Bipedal locomotion evolved in human ancestors
- Human brains became larger as jaws became smaller
- Humans developed complex language and culture

The primate ancestor was a small, arboreal, insectivorous mammal (see Figure 33.30). Primates are identifiable by the presence of opposable digits (i.e., thumbs). Early in their evolutionary history—about 90 million years ago—the primates split into two clades, the prosimians (lemurs, lorises, and galagos) and the anthropoids (tarsiers, New World monkeys, Old World monkeys, and apes). Prosimian species were once found on all continents but today they are restricted to Africa, the island of Madagascar, and tropical Asia (see Figure 33.31).

Soon after the prosimian–anthropoid split, the anthropoids split further into the New World and Old World monkeys. The breakup of the African and South American continents meant that the two groups evolved in isolation. New World monkeys are tree dwellers (arboreal) (see Figure 33.32A). A long prehensile tail, which is unique to the New World monkeys, allows them to grasp branches. Many Old World monkeys are arboreal, but none has a prehensile tail (see Figure 33.32B). About 35 million years ago, an Old World lineage broke off that would lead to modern apes. Another split occurred between 22 and 3.5 million years ago; the Asian apes (gibbons and orangutans) descended from two groups in this lineage (see Figures 33.33A and 33.33B). Orangutans are the closest living group to the modern African apes—gorillas, chimpanzees, and humans (see Figures 33.33C and 33.33D).

A split about 6 million years ago in Africa led to the chimpanzees, on the one hand, and the hominid clade (the modern humans and their extinct relatives) on the other. Ardipithecines (the earliest protohominids) had adaptations for bipedal locomotion, which freed up their hands for other tasks and elevated the eyes which enabled them to see over tall vegetation. Bipedal movement also requires less energy. These advantages were important for the ardipithecines and their descendants, the australopithecines. *Australopithecus afarensis* is currently regarded as the ancestor of the modern humans (genus *Homo*; see Figure 33.34). The most complete australopithecine fossil skeleton was discovered in Africa in 1974—the famous "Lucy," approximately 3.5 million years old. There have been more recent discoveries of australopithecine skeletons from 4–5 million years ago. There is disagreement as to how many species are represented by australopithecine fossils. It is clear that these different groups of hominids coexisted in Africa several million years ago. The genus *Homo* arose from one of the smaller lineages of australopithecines, and the genus *Paranthropus* arose from one of the larger lineages. Early members of the genus *Homo* lived at the same time as *Paranthropus* for about a million years. The oldest known member of the genus *Homo*, *Homo habilis*, lived about 2 million years ago, and tools have been found with their fossils. *Homo erectus* appeared about 1.6 million years ago. They were as large as modern humans, used stone tools, and cooked with fire. However, their brains were smaller than those of modern humans and they had a relatively thick skull. Populations of *H. erectus* survived until at least 250,000 years ago. Some 18,000-year-old remains of a small *Homo* were discovered on the Indonesian island of Flores in 2004; since then many individuals have been found, dating from around 95,000 to 17,000 years ago. Many anthropologists believe that this species, *H. floresiensis*, was most closely related to *H. erectus*. In another hominid lineage, divergent from *H. erectus* and *H. floresiensis*, the brain increased rapidly in size and the jaw muscles decreased dramatically in size. Since these two changes were simultaneous, they may have been related in developmental terms and are another example of neoteny. Human and chimpanzee skulls are similar at birth, but chimpanzee skulls undergo dramatic changes during maturation with the jaw growing considerably in relation to the brain case (see Figure 33.35). The rapid increase in brain size in the hominin lineage may have been favored by an increasingly complex social life that thrived on ever more sophisticated communication. Any trait that allowed more effective communication would have been favored in a society based on cooperative hunting and other complex social interactions.

A number of species of *Homo* existed simultaneously over evolutionary time between 1.5 million years ago and 250,000 years ago. One species, *Homo neanderthalensis,* appeared about 500,000 years ago and became widespread in Europe and Asia. Neanderthals were short but powerfully built, and their skulls housed a brain that was somewhat larger than that of modern humans. They manufactured tools and were able to hunt large animals. By 200,000 years ago early modern humans (*H. sapiens*) were predominant in Africa, and they began to extend their range to other continents around 60,000–70,000 years ago. Around 35,000 years ago they coexisted alongside *H. neanderthalensis.* It is likely that the two species interacted, but Neanderthals abruptly became extinct 28,000 years ago, possibly exterminated by *H. sapiens*. Genetic evidence suggests that there was some interbreeding between the two species. Early modern humans used many sophisticated tools, and even created the paintings of large mammals found in some European caves. By about 20,000 years ago all other *Homo* species had been supplanted by *Homo sapiens,* the only currently existing human species. It was about this time that *H. sapiens* reached North America, via Asia. Within a few thousand years, humans had spread to South America. The expansion of language and other behavioral abilities allowed humans to develop culture and pass knowledge and traditions from generation to generation. Human societies were transformed from communities of hunters and gatherers to those of farmers and eventually urban dwellers.

Question 11. What major geologic change occurred during primate evolution and how did it alter the way primate evolution proceeded?

Question 12. What are three traits that evolved in primates and led to evolutionary success? Which ones seem to have had the greatest impact on hominid lines?

Test Yourself

1. Which of the following statements about echinoderms is *false*?
   a. They have a water vascular system.

b. They have an internal skeleton.
c. They are protostomes.
d. The larvae have bilateral symmetry.
e. The adult form lacks a head.
Textbook Reference: *33.2 What Features Distinguish the Echinoderms, Hemichordates, and Their Relatives?*

2. In which of the following echinoderm groups are the extinct members more numerous than the living members?
a. Sea stars and brittle stars
b. Sea lilies and feather stars
c. Sea urchins and sea cucumbers
d. Sea daisies
e. Asterozoans
Textbook Reference: *33.2 What Features Distinguish the Echinoderms, Hemichordates, and Their Relatives?*

3. Which of the following chordate groups evolved before the appearance of cartilaginous fishes?
a. Ray-finned fishes and sea squirts
b. Ascidians, lancelets, and hagfishes
c. Ascidians, lancelets, and ray-finned fishes
d. Lampreys and ray-finned fishes
e. Coelacanths and ray-finned fishes
Textbook Reference: *33.3 What New Features Evolved in the Chordates?*

4. Which of the following is *not* a feature of the vertebrate body plan?
a. A ventral spinal cord
b. An internal skeleton
c. A well-developed circulatory system
d. Organs suspended in the coelom
e. An anterior skull encasing a proportionally large brain.
Textbook Reference: *33.3 What New Features Evolved in the Chordates?*

5. The swim bladder of many fishes that evolved from a lunglike sac has the important function of
a. aiding in prey capture.
b. controlling swimming speed.
c. controlling buoyancy.
d. aiding in reproduction.
e. providing balance.
Textbook Reference: *33.3 What New Features Evolved in the Chordates?*

6. Which of the following traits is shared by the chondrichthyans and the ray-finned fishes?
a. Gills as the major site of gas exchange
b. A skeleton composed of cartilage
c. An outer surface covered with bony plates
d. A swim bladder
e. Propulsion by means of dorsal-ventral movements of their tails
Textbook Reference: *33.3 What New Features Evolved in the Chordates?*

7. The transition from aquatic to terrestrial lifestyles required many adaptations in the vertebrate lineage. Which of the following was *not* one of those adaptations?
a. A shift from gills to air-breathing lungs
b. Improvements in the water resistance of skin
c. An alteration in the mode of locomotion
d. Development of feathers for insulation
e. Modifications to the nitrogen elimination system
Textbook Reference: *33.4 How Did Vertebrates Colonize the Land?*

8. The amniotes evolved the ability to produce eggs that have shells. The major advantage of shelled eggs is that
a. the embryo needs only a small amount of yolk for development.
b. they do not have to be laid in a moist environment.
c. the shells increase evaporation from the egg.
d. nitrogenous wastes can be excreted across the shell.
e. gas exchange with the environment is more efficient.
Textbook Reference: *33.4 How Did Vertebrates Colonize the Land?*

9. Which of the following statements about birds and reptiles is true?
a. Birds have a lower metabolic rate than reptiles.
b. Reptiles give birth to live young, whereas birds lay eggs.
c. Birds can breathe and run at the same time, whereas reptiles cannot.
d. Birds are amniotes, reptiles are not.
e. Reptiles supplement their lung action with gas exchange through their skin, whereas birds depend entirely on their lungs for gas exchange.
Textbook Reference: *33.4 How Did Vertebrates Colonize the Land?*

10. Which of the following is *not* a trait that could be used to identify an animal as a mammal rather than an amphibian?
a. Mammary glands
b. Hair
c. Sweat glands
d. Kidneys
e. Four-chambered heart
Textbook Reference: *33.4 How Did Vertebrates Colonize the Land?*

11. Which of the following vertebrate groups is a living representative of the dinosaur lineage?
a. Crocodilians
b. Birds
c. Lobe-finned fishes
d. Snakes
e. Lepidosaurs
Textbook Reference: *33.4 How Did Vertebrates Colonize the Land?*

12. Which of the following statements about human evolution is *false?*
    a. Bipedalism was a hominid adaptation for life on land.
    b. Increases in the size of hominid brains preceded the appearance of language and culture.
    c. The extinction of the Neanderthals was caused by the emergence of *Homo habilis*.
    d. Humans are not the direct descendants of modern-day chimpanzees.
    e. *Homo sapiens* coexisted with *H. neanderthalensis* in portions of Europe and Asia.

    Textbook Reference: *33.5 What Traits Characterize the Primates?*

13. A key difference between Old World monkeys and New World monkeys is that the latter
    a. have a prehensile tail.
    b. are arboreal.
    c. have a placenta.
    d. are less closely related to tarsiers.
    e. All of the above

    Textbook Reference: *33.5 What Traits Characterize the Primates?*

14. Which of the following is *not* part of the evidence that birds are closely related to dinosaurs?
    a. The presence of feathers in representatives of both groups
    b. The presence of hollow bones in representatives of both groups
    c. DNA sequence data comparing birds to other living reptiles
    d. A bipedal stance in fossilized therapods
    e. The ability to fly

    Textbook Reference: *33.4 How Did Vertebrates Colonize the Land?*

15. Which of the following are *not* deuterostomes?
    a. Echinoderms
    b. Hemichordates
    c. Lancelets
    d. Ecdysozoans
    e. Chordates

    Textbook Reference: *33.1 What Is a Deuterostome?*

Answers

Key Concept Review

1.
    a. Radial symmetry as adults, calcified internal plates, loss of pharyngeal slits
    b. Ciliated larvae
    c. Notochord, dorsal hollow nerve cord, post-anal tail
    d. Vertebral column, anterior skull, large brain, ventral heart
    e. Echinoderms
    f. Hemichordates
    g. Lancelets
    h. Tunicates
    i. Vertebrates
    j. Ambulacrarians
    k. Chordates

2. Three early developmental patterns distinguish the deuterostomes: radial cleavage; development of blastopore into an anus and formation of the mouth at the opposite end of the embryo; development of a coelom from mesodermal pockets, rather than splitting of the mesoderm. These characteristics are not all exclusive to deuterostomes, and some may be ancestral to all bilaterians. The best evidence of monophyly of deuterostomes comes from phylogenetic studies.

3. The two traits that are defining characteristics of echinoderms are an internal skeleton and a vascular system composed of water-filled canals.

4. Echinoderms are a branch of bilaterians that have evolved pentaradial symmetry. It is likely that the sea cucumber lineage is one that secondarily evolved an elongated oral–aboral axis, which then caused their characteristic posture of lying on their side.

5. Although adult ascidians and adult humans are very different animals, their embryonic stages share certain important characteristics, including the presence of a notochord and a blastopore that develops into the anus (see Chapter 31). A look at embryonic grasshoppers, however, shows them to be fundamentally different. The embryonic grasshopper's blastopore develops into the mouth, placing grasshoppers squarely among the protostomes. Grasshoppers also have an external rather than an internal skeleton.

6. The hinged jaw of the fishes evolved during the Devonian period. The evolution of the jaw opened up a new food source for these animals. With a jaw, fish could now grasp and kill larger living prey and chew and tear body parts.

7.
    a. Vertebrae
    b. Jaws, teeth, paired fins
    c. Bony skeleton, swim bladder/lung
    d. Lobe fins
    e. Internal nares
    f. Terrestrial limbs and digits
    g. Anmiote egg
    h. Hagfishes
    i. Lampreys
    j. Chondrichthyans
    k. Ray-finned fishes
    l. Coelacanths
    m. Lungfishes
    n. Amphibians
    o. Amniotes
    p. Jawless fishes
    q. Gnathostomes
    r. Lobe-limbed vertebrates

8.
   a. Reptiles
   b. Mammals
   c. Tuataras
   d. Squamates
   e. Turtles
   f. Crocodilians
   g. Pterosaurs (extinct)
   h. Ornithischians (extinct)
   i. Sauropods (extinct)
   j. Therapods, including birds
   k. Mammals
   l. Lepidosaurs
   m. Dinosaurs
   n. Archosaurs

9. Mammals and reptiles are both amniotes. Their eggs minimize water loss and still permit gas exchange. The skin of both groups of organisms is modified to reduce water loss, and the kidneys eliminate nitrogen while minimizing water loss. Both groups of organisms have lungs designed to exchange gases with the atmosphere.

10. Amphibians generally require an aquatic environment for fertilization, larval development, and metamorphosis into a terrestrial adult. Some amphibians develop from eggs laid on land and move directly into adultlike forms, skipping the aquatic form entirely.

11. As the continental land masses separated 65 million years ago, New World monkeys were isolated from Old World monkeys and the two groups diverged.

12. Primates evolved opposable thumbs, bipedal upright locomotion, prehensile tails, and large cranium size. Opposable thumbs favored development of tool use. Bipedal locomotion increased the ability to see above vegetation and freed up hands to carry and manipulate objects. Prehensile tails increased the success of arboreal monkeys. Large cranium size permitted brain growth to accommodate language development and facilitated sophisticated social interactions.

Test Yourself

1. **c.** Species from the echinoderm clade are deuterostomes.

2. **b.** The crinoids (sea lilies and feather stars) are a relict group with many more extinct than extant species.

3. **b.** The ascidians, lancelets, and hagfishes all evolved before the cartilaginous fishes.

4. **a.** Along with an internal skeleton, well-developed circulatory system, and organs suspended in a coelom, the vertebrates have a dorsal spinal cord.

5. **c.** The swim bladder of modern-day fishes is involved in controlling buoyancy.

6. **a.** In both the ray-finned fishes and cartilaginous fishes, the major site of gas exchange is the gills. The ray-finned fishes have a skeleton of bone and a swim bladder. These traits are not shared with the cartilaginous fishes. Neither group has bony plates in its outer surface. They all move their tails laterally, in contrast to cetaceans, which move their tails vertically.

7. **d.** The move onto land did not require the development of feathers for insulation. Amphibians and reptiles do not have an insulation layer, and many mammals have hair for insulation.

8. **b.** The shelled egg of the birds and reptiles allowed them to occupy dry terrestrial habitats because the shell decreases water loss from the egg.

9. **c.** Birds and reptiles differ in the morphology of the muscles that control movement and breathing. Birds can run and breathe at the same time, whereas reptiles must stop running to take a breath.

10. **d.** Kidneys are present in reptiles, birds, fishes, and mammals. The presence of kidneys is not a trait that could be used to identify a mammal.

11. **b.** Birds are now thought to be direct descendants of a group of dinosaurs.

12. **c.** *Homo habilis* was extinct for more than a million years by the time the Neanderthals emerged.

13. **a.** Only the New World monkeys have a prehensile tail.

14. **e.** Not all birds fly and not all dinosaurs were able to fly.

15. **d.** Ecdysozoans are a type of protostome (see Figure 31.1).

The Plant Body

The Big Picture

- A plant can be thought of as having vegetative structures that carry out the major functions of day-to-day life and reproductive structures that are responsible for reproduction in the plant. You will see that there are cases in which the vegetative portions of the plant are quite adept at reproducing asexually. The vegetative plant body consists of the root system, which anchors the plant and absorbs water and nutrients, and the shoot system, which carries out photosynthesis and supports the plant against gravity. Modifications of the root and shoot systems lead to specialization of the plant.
- Plant cells are uniquely suited for support, transport, and the carrying out of cellular functions. Groups of cells form tissues that have specific roles within a plant. Patterns of development are under the control of hormones and regulatory genes. Indeterminate growth and the modular organization of plants allow for regeneration of parts lost to damage and disease.
- Plant growth occurs from meristems. Apical meristems allow for elongation of the plant, while lateral meristems allow for secondary or woody growth. Not all plants exhibit secondary growth. All tissue types arise from the meristems and go through a process of elongation, then differentiation.

Study Strategies

- This chapter covers plant anatomy. Basic plant anatomy is not difficult, but it may be new to you. Spend time looking at diagrams, pictures, and live specimens, and you will find it easier to understand anatomy. As plant anatomy is the study of structure, which is three-dimensional, it is best understood visually. As you study, think about the function of each structure. You will find that "form follows function."
- This chapter exposes you to many new vocabulary words. Make a vocabulary list to organize your study, but do not merely memorize terms. By looking at the words and understanding their roots, you will be better able to understand the terms.
- Go to the Web addresses shown to review the following animated tutorial and activities. You can also find them in BioPortal (yourBioPortal.com), along with many additional learning resources.

  Animated Tutorial 34.1 Secondary Growth: The Vascular Cambium (Life10e.com/at34.1)

  Activity 34.1 Eudicot Root (Life10e.com/ac34.1)

  Activity 34.2 Monocot Root (Life10e.com/ac34.2)

  Activity 34.3 Eudicot Stem (Life10e.com/ac34.3)

  Activity 34.4 Monocot Stem (Life10e.com/ac34.4)

  Activity 34.5 Eudicot Leaf (Life10e.com/ac34.5)

Key Concept Review

34.1 What Is the Basic Body Plan of Plants?

Most angiosperms are either monocots or eudicots

Plants develop differently than animals

Apical–basal polarity and radial symmetry are characteristics of the plant body

Plants harvest energy from sunlight and collect water and mineral nutrients from the atmosphere and the soil, sometimes over a wide area. The challenge for plants is to achieve this while not being able to move. The plant body plan also allows plants to respond to the challenge of environmental cues by growing throughout their lifetimes, for example, by extending roots toward a water supply. Plants are composed of root systems and shoot systems (see Figure 34.1). Root systems anchor the plant, absorb water and dissolved minerals, and store the products of photosynthesis. Shoot systems consist of leaves (and leaf derivatives) involved in photosynthesis and stems for support. The shoots and roots of plants are laid out in modules, or units, known as phytomers. A phytomer consists of a node and its attached leaf, the internode (a section of stem) below the node, and the axillary buds at the base of the internode. Buds are embryonic shoots. At each node where leaves meet stem, an axillary bud is produced that may generate a new branch. At each stem tip a terminal bud is responsible for elongation of that stem. In the roots, each phytomer consists of a root segment between two branches.

Most angiosperms are either monocots or eudicots, which differ in several basic characteristics. Monocots have one cotyledon and eudicots have two. The vascular bundles of eudicots are arranged in concentric circles; in monocots they are scattered. The major leaf veins are usually parallel in monocots, whereas they are reticulate (a network) in eudicots. Eudicots have taproot systems and monocots have fibrous root systems. Monocot flowers have parts that occur in threes, and eudicots have floral parts that occur in fours or fives. Monocot pollen grains have one furrow or pore, and eudicot pollen grains have three.

Four processes govern development in all organisms: determination, differentiation, morphogenesis, and growth. The same processes apply to plants but are influenced by four unique properties: apical meristems, totipotency, vacuoles, and cell walls. Plants have meristems, regions of undifferentiated cells where cell division occurs. Apical meristems are at the tips of roots and shoots and allow continuous growth throughout the plant's life. Some differentiated plant cells can dedifferentiate and become totipotent, able to differentiate into any cell type in the body. Mature plant cells usually contain a central vacuole, a watery sac containing a high concentration of solutes pumped in by transporters in the tonoplast, the vacuolar membrane. This provides the osmotic force for water uptake, and allows the expanding vacuole to exert turgor pressure on the cell wall. Each plant cell is surrounded by a cell wall, which is peppered with membrane-lined cytoplasmic channels between cells called plasmodesmata. Because plant cells are prevented from moving by the cell walls, plant morphogenesis is driven by the planes of cell division. One major way plants grow is through cell expansion, which is only possible with rearrangements of the cell wall to allow this expansion. The wall of a growing plant cell is the primary cell wall. When cell expansion stops, some types of cells deposit additional layers to strengthen the cell wall; a secondary wall. This secondary cell wall contains layers of cellulose embedded in a matrix of lignin, a very tough polymer that is the major component of wood.

The body plan of a plant is established in the embryo along a basal–apical axis and a radial axis. Two unequal daughter cells are produced as the zygote goes through a mitotic division. This results in the development of a thin suspensor and a globular embryo. The cotyledons (seed leaves) begin to form as the embryo enters the heart stage. As the cotyledons elongate, the embryo enters the torpedo stage and the internal tissues begin to differentiate. The root and shoot apical meristems develop between the cotyledons.

Question 1. Draw a typical eudicot plant. Label the root system and shoot system. Indicate on your drawing where you would find an axillary bud and where you would find a terminal bud. Label the following structures: leaf, internode, petiole, blade, node, roots, stem, branch.

Question 2. What are plasmodesmata? What is the physiological significance of the presence of plasmodesmata in plants?

34.2 What Are the Major Tissues of Plants?

- The plant body is constructed from three tissue systems
- Cells of the xylem transport water and dissolved minerals
- Cells of the phloem transport the products of photosynthesis

Tissues are composed of cells that function together. Plant tissues are grouped into the dermal, vascular, and ground tissue systems. The dermal tissue system makes up the outer covering of the plant and includes the epidermis; the stems and roots of woody plants develop a dermal tissue called periderm. Special epidermal cells include stomatal guard cells, trichomes, and root hairs. The aboveground epidermal cells secrete a protective cuticle layer made up of cutin, waxes, and polysaccharides.

The ground tissue system is found between the dermal and vascular tissue and is involved in storage, support, and photosynthesis. Thin-walled parenchyma cells have large central vacuoles and are frequently photosynthetic or used for storage. The thin cell wall consists of a primary cell wall and a shared layer between cells called the middle lamella, a layer of pectin that cements the walls together. These may continue dividing and proliferate in a wounded area. Collenchyma cells are support cells with special thickenings at the cell wall corners. They are elongated and allow for support without rigidity, which is necessary for plants in windy areas. Sclerenchyma cells have highly thickened cell walls and are either elongated fibers or variously shaped sclereids. Many undergo apoptosis and provide rigid support after they die. Fibers strengthen bark and woody stems. Densely packed sclereids are found in the shells of nuts and in some seed coats and produce the gritty texture of pears and other fruit.

The vascular tissue system is made of xylem and phloem, and is the conductive tissue of the plant. Xylem distributes water and minerals taken up by the roots; phloem performs many functions, including distribution of carbohydrates from sources (sites of photosynthesis) to sinks (sites of utilization or storage).

Xylem contains tracheary elements that have secondary cell walls and undergo apoptosis before transporting water and minerals. There are two types of tracheary elements: tracheids and vessel elements. Tracheids are evolutionarily older than vessel elements, and are the major cell type in the wood of gymnosperms. When tracheids die, the cells disintegrate and leave porous pits between the cells, which allow water and minerals to move between tracheids. Angiosperms have evolved a system of vessels formed from cells called vessel elements. These cells are laid down end-to-end, with pits of larger diameter than tracheids. Their cell walls partially break down, so that after apoptosis there exists a hollow continuous tube. The xylem of many angiosperms contains both tracheids and vessel elements.

Living phloem moves carbohydrates and nutrients. Individual phloem cells are called sieve tube elements. Plasmodesmata enlarge where sieve tube elements join, making sieve plates. Sieve tube elements may lose nuclei and other organelles to prevent clogging of the sieve plates. Adjacent companion cells regulate the function of the sieve tube elements.

Question 3. The plant body is composed of three different tissue systems. While examining a plant you find a tissue within the plant composed of dead cells that have very thick cell walls and contain lignin. What types of cells are you likely seeing and why?

Question 4. Describe the development of a xylem cell in a monocot from its origin in the apical meristem.

34.3 How Do Meristems Build a Continuously Growing Plant?

- Plants increase in size through primary and secondary growth
- A hierarchy of meristems generates the plant body
- Indeterminate primary growth originates in apical meristems
- The root apical meristem gives rise to the root cap and the root primary meristems
- The products of the root's primary meristems become root tissues
- The root system anchors the plant and takes up water and dissolved minerals
- The products of the stem's primary meristems become stem tissues
- The stem supports leaves and flowers
- Leaves are determinate organs produced by shoot apical meristems
- Many eudicot stems and roots undergo secondary growth

Plants and animals develop and function differently. Plants cannot move but can grow toward resources, both aboveground and belowground (they can grow in two directions). In most animals growth is determinate, meaning that it stops when an adult state is reached. Plants, however, have indeterminate growth and continue growing throughout their lives. Plants increase in size through primary and secondary growth. Primary growth consists of cell division followed by cell enlargement and results in the lengthening of shoots and roots. All seed plants have a primary (nonwoody) plant body, and many herbaceous plants are made entirely of primary plant body. Secondary growth increases plant thickness. Woody plants have secondary plant bodies of wood and bark.

Meristems are undifferentiated cells that are able to produce new cells indefinitely. The cells that perpetuate the meristem, the initials, are similar to human stem cells. Apical meristems contribute to primary growth. Some daughter cells of the initials become the primary meristems, which give rise to the three major tissue systems. Lateral meristems are responsible for secondary growth. Two lateral meristems, the vascular cambium and cork cambium, create the secondary plant body.

Apical meristems cause the indeterminate primary growth of the plant. Several meristems play roles in organ formation. Each branch of a plant has its own shoot apical meristem, supplying cells for new leaves and stems. These are also called vegetative meristems because they give rise to vegetative tissues. One or more shoot apical meristems are converted to inflorescence meristems when it is time to flower, and these in turn become floral meristems. Root apical meristems supply cells to extend the root; each type (taproot, lateral root, or adventitious root) has its own root apical meristem. Apical meristems give rise to primary meristems that produce the primary plant body. These primary meristems are called the protoderm (dermal tissue system), the ground meristem (ground tissue system), and the procambium (vascular tissue system). All plant parts arise from division of the apical meristems.

Root apical meristems produce root tissues. At the tip of a root, the root apical meristem forms a root cap. The zone of cell division includes the apical and primary meristems. Above this is the zone of elongation, where cells elongate and push the root through the soil. The upper layer is the zone of maturation where the cells differentiate and take on special functions. The root has three primary tissue systems. The protoderm gives rise to the epidermis and root hairs, both involved in water and mineral ion uptake. The ground meristem gives rise to the cortex and endodermis. The endodermis is specialized with a waxy coating of suberin to assist with water movement. The procambium gives rise to the stele, which houses three tissues: the pericycle, the xylem, and the phloem. In eudicots, the very center of the root is xylem, but in monocots the center is pith tissue (see Figure 34.10). Root systems develop from the radicle, an embryonic root. In most eudicots, a primary taproot develops as a single large root growing deep within the ground, extended outward via lateral roots. Together the taproot and lateral roots form a taproot system, which can also function as a nutrient storage organ. In monocots, numerous thin, adventitious roots make up a fibrous root system that absorbs water efficiently and helps the plant cling to the soil. Some plants have evolved prop roots that perform various functions, including support for the growth of the plant.

Roots and shoots are composed of repeating units called phytomers. A shoot phytomer consists of a node with attached leaf or leaves, the internode below the node, and axillary buds at the angle between the stem and the leaf. If an axillary bud develops into a branch, it produces new phytomers. The shoot primary meristem also gives rise to three primary meristems (protoderm, ground meristem, and procambium) that produce the three tissue systems. Leaves arise from leaf primordia with bud primordia forming at each leaf base. Shoot vascular tissues are arranged in vascular bundles containing both xylem and phloem. In eudicots, the vascular bundles are arranged in a cylinder, allowing for woody growth. In monocots the vascular bundles are scattered. The stem has other important storage and support tissues. The pith of eudicots lies within the ring of vascular bundles and extends between them. Outside the vascular bundles lies the cortex. Together the pith and cortex are the ground tissue system of the stem. The outermost layer is the epidermis.

Various modifications of stems also exist in nature. For instance, a potato is a tuber, an underground stem, with the eyes as axillary buds. Many desert plants have enlarged stems that retain water, such as in cacti. The runners of strawberry plants are horizontal stems with adventitious roots at a distance from the main stem.

Leaves are produced throughout most of a plant's life by vegetative meristems. Unlike stems, leaf growth is determinate; the leaf is not radial in pattern like the stem but has distinct top and bottom sides. Most eudicot leaves have two zones of photosynthetic cells called mesophyll. The upper level of cylindrical mesophyll is called palisade mesophyll, and the lower level is called spongy mesophyll. Air space around mesophyll cells is necessary for carbon dioxide to reach the photosynthesizing cells. Vascular tissue extends throughout leaves as a network of veins (see Figure 34.15B). Veins carry water to cells and transport carbohydrates to sink tissues. The entire leaf is covered by a protective epidermis and is waterproofed by a waxy cuticle. Gas exchange occurs through guarded stomata. Guard cells open and close stomata to limit water loss.

Secondary growth in eudicots involves the laying down of wood and bark by the two lateral meristems, vascular cambium and cork cambium. Vascular cambium arises from the lateral meristem and forms new secondary xylem (wood) and secondary phloem (bark). The cork cambium mainly produces waxy protective cells. The vascular cambium is initially a single layer of cells between the primary xylem and primary phloem within the vascular bundles. When these cells divide, the root or stem becomes greater in diameter, producing secondary xylem toward the inside, which becomes wood, and secondary phloem toward the outside, which contributes to bark. Stretching, breaking, and flaking of epidermis and cortex produces bark, leaving the secondary phloem at risk. Cells at the surface of the phloem become the cork cambium, which produce a protective layer of cork that is thick and reinforced with waterproof suberin. New cork is produced as secondary growth proceeds. Sometimes the cork cambium produces cells toward the inside as well as outside; these are a tissue known as phelloderm. The periderm consists of the cork cambium, cork, and phelloderm. As more secondary vascular tissue forms, these layers are lost, but the continuous formation of new cork cambia creates new corky layers. The periderm and the secondary phloem (everything external to the vascular cambium) is the bark. The areas that allow gas exchange through the bark are known as lenticels.

The annual rings seen in wood are a result of climate shifts in temperate zones, particularly water availability. As the plant changes the characteristics of the vascular tissue created, the growth can be seen as distinct rings from season to season. Tropical trees do not undergo seasonal growth and do not lay down such visible rings. Only eudicots and nonmonocot angiosperms undergo secondary growth. The few monocots that create thickened stems, such as palms, do so without true secondary growth. Dead leaf bases add diameter to the stem.

Question 5. Differentiate between apical and lateral meristems. Do all plants have apical meristems? Do all plants have lateral meristems?

Question 6. Draw a growing root. Label the primary meristems, root cap, cortex, stele, and epidermis. Discuss the function of each of these structures.

Question 7. Examine the leaf diagram in Figure 34.15 in the textbook. Which surface of that leaf faces the sun? How do you know? Why are the stomata opposite the palisade layer?

34.4 How Has Domestication Altered Plant Form?

A very simple body plan has allowed for the incredible diversity of plant species, which is not surprising due to the great genetic variability among species. Members of the same species can also show great diversity, which suggests that very simple genetic changes can underlie large differences. The corn that we eat today was domesticated from a wild grass called teosinte. The morphological differences between the two species are large: teosinte is highly branched, whereas corn has a single shoot. This difference is mostly due to the action of a single gene, *teosinte branched 1* (*tb1*), the protein product of which regulates growth of axillary buds. The allele found in modern corn represses branch formation. *Brassica oleracea*, wild mustard, is the ancestor of many crops such as kale, broccoli, Brussels sprouts, and cabbage. Humans selected and planted the seeds of variants with desirable traits and thus altered the genes of the plants until the various foods we have today were produced. For instance, Brussels sprouts have enlarged axillary buds, cabbages have short internodes and large terminal buds, and broccoli has large clusters of flower buds. The creation of these modern crops required many generations of artificial selection.

Today, we are using plant breeding programs to attempt to improve crops, or allow them to adapt to different environmental conditions. Various organizations are also organizing seed banks to store seeds of rare and extinct species, and variants within a species.

Question 8. How have humans domesticated plants? What advantages have been provided by domestication?

Question 9. List some potential targets of plant genetic engineering that would make crops more useful in certain situations.

Test Yourself

1. Which of the following is a *not* a component of a phytomer?
   a. Leaf
   b. Root hair
   c. Axillary buds
   d. Internode
   e. All of the above are components of a phytomer.
   Textbook Reference: *34.1 What Is the Basic Body Plan of Plants?*

2. Suppose you are studying tropical plants in a Costa Rican cloud forest and find a tree that is a eudicot in the forest ecosystem. This plant most likely has a(n) _______ system.

a. fibrous root
b. taproot
c. adventitious root
d. rhizoid root
e. terminal root
Textbook Reference: *34.1 What Is the Basic Body Plan of Plants?*

3. Some plants, such as sweet peas, will attach themselves to a fence by means of tendrils. Tendrils are most likely modifications of which plant structure?
a. Stems
b. Roots
c. Branches
d. Leaves
e. Seeds
Textbook Reference: *34.3 How Do Meristems Build a Continuously Growing Plant?*

4. Plant cells are easily distinguished from animal cells by their
a. rigid cell walls.
b. plastids.
c. large vacuoles.
d. chloroplasts.
e. All of the above
Textbook Reference: *34.1 What Is the Basic Body Plan of Plants?*

5. Plant cells that are photosynthetically active are _______ cells found in the _______ layer of the leaf.
a. parenchyma; mesophyll
b. parenchyma; epidermis
c. sclerenchyma; mesophyll
d. sclerenchyma; epidermis
e. mesophyll; xylem
Textbook Reference: *34.3 How Do Meristems Build a Continuously Growing Plant?*

6. Water is conducted in _______ tissue, and carbohydrates and nutrients are transported in _______ tissue.
a. xylem; phloem
b. phloem; xylem
c. parenchyma; phloem
d. parenchyma; xylem
e. mesophyll; xylem
Textbook Reference: *34.2 What Are the Major Tissues of Plants?*

7. Plants are capable of indeterminate growth because of which of the following?
a. Regions of nondividing cells
b. Meristem tissues
c. Epidermis
d. Xylem
e. All of the above
Textbook Reference: *34.3 How Do Meristems Build a Continuously Growing Plant?*

8. Which of the following best describes the origin of wood?
a. Xylem cells enlarge and deposit large amounts of lignin.
b. Primary meristems increase the amount of xylem deposited.
c. Lateral meristems contribute to continuous increases in vascular tissue.
d. Spongy mesophyll cells become cork cambium.
e. None of the above
Textbook Reference: *34.3 How Do Meristems Build a Continuously Growing Plant?*

9. Which of the following is *not* a component of the primary cell wall?
a. Cellulose
b. Hemicelluloses
c. Pectins
d. Collenchymas
e. All of the above are components of the primary cell wall.
Textbook Reference: *34.1. What Is the Basic Body Plan of Plants?*

10. Which of the following best describes the function of the cork cambium?
a. It lays down a protective cork covering over exposed phloem tissue.
b. It inhibits the sloughing off of epidermal tissue.
c. It allows for the shrinking of the diameter of stems and roots.
d. It supplies the secondary xylem.
e. All of the above
Textbook Reference: *34.3 How Do Meristems Build a Continuously Growing Plant?*

11. Sieve tube elements have sieve plates where they join other sieve tube elements. Which of the following statements about the sieve plates is true?
a. Their pores are enlargements of meristems.
b. They allow conduction between sieve tube cells through plasmodesmata.
c. They allow for the joining of cytoplasm between adjacent stomata.
d. They contain the organelles of the cell.
e. None of the above
Textbook Reference: *34.2 What Are the Major Tissues of Plants?*

12. Plants regulate gas exchange and water loss via
a. the cuticle.
b. xylem.
c. coated pits.
d. sieve plates.
e. guarded stomata.
Textbook Reference: *34.3 How Do Meristems Build a Continuously Growing Plant?*

13. The protoderm becomes the _______ tissue system.
    a. dermal
    b. ground
    c. vascular
    d. All of the above
    e. None of the above
    Textbook Reference: *34.3 How Do Meristems Build a Continuously Growing Plant?*

14. Primary growth occurs at the
    a. lateral meristems.
    b. fruit.
    c. quiescent center.
    d. apical meristems.
    e. wood.
    Textbook Reference: *34.3 How Do Meristems Build a Continuously Growing Plant?*

15. Vascular bundles are composed of _______ and _______.
    a. root hairs; xylem
    b. cork; phloem
    c. xylem; phloem
    d. wood; cork
    e. mesophyll; xylem
    Textbook Reference: *34.3 How Do Meristems Build a Continuously Growing Plant?*

Answers

Key Concept Review

1.

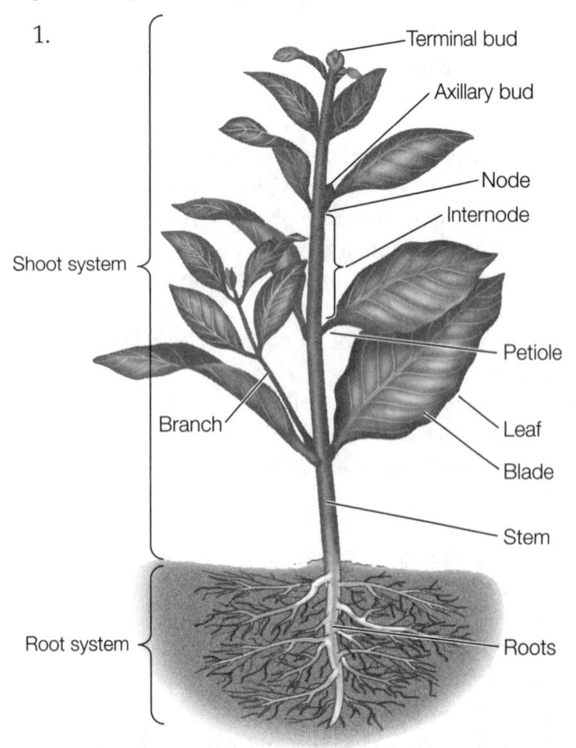

2. The plasmodesmata are membrane-lined cytoplasmic channels through the cell walls that connect the cytoplasm of multiple plant cells. Physiologically, this means that no plant cell is completely isolated from its neighbors, even though the rigid cell wall exists between them. It also means that solutes can move from cell to cell via normal molecular motion, and do not always need to be transported in and out of every cell. This has implications for how plant cells communicate with each other as compared to animal cells, which do not have cytoplasmic connections.

3. The three types of tissue are dermal, vascular, and ground. Dermal tissue acts as a covering for the plant, and its cells are typically small. It provides protection and also functions in gas exchange and nutrient and water uptake. Vascular tissue, which is composed of cells that are either dead (xylem) or alive (phloem), is involved in transport of water and nutrients within the plant. Certain types of ground tissue are involved in photosynthesis, and others provide support for the plant body. These cells are probably sclerenchyma cells, which have thick cell walls that provide support.

4. The apical meristem gives rise to the protoderm, the ground meristem, and the procambium. The procambium gives rise to the vascular tissue system, including xylem. As the vascular tissue develops, it undergoes apoptosis and becomes the tracheary elements of the xylem. Monocots do not have secondary xylem, so only primary xylem exists in monocots.

5. Apical meristems are responsible for elongation of the plant body. Lateral meristems are responsible for an increase in girth. Lateral meristems are found only in woody eudicots and are responsible for creating wood. All plants have apical meristems.

6. See Figure 34.9A for a diagram of root growth. The protoderm gives rise to the epidermis, which is adapted for protection. The ground meristem gives rise to the cortex, the cells of which serve as storage deposits. The procambium gives rise to the stele, for transport. The root cap protects the meristem as it pushes through the soil.

7. The "top" of the diagram in the textbook represents the surface that faces the sun. This is the surface where photosynthetic cells, which need maximum sun exposure, are located. The stomata are on the opposite side to reduce water loss due to evaporation during photosynthesis.

8. Humans have taken advantage of the large variation in plant shape and size within a species. They have selectively used the seeds from plants with desirable characteristics, ensuring these characteristics remain in the crop. This has allowed humans to produce more productive crops.

9. Some projects could include improving pest or disease resistance, improving drought or salt tolerance, adding genes to produce vitamins (akin to the cassava plant mentioned in the chapter introduction) or even vaccines, increasing yield (such as of wheat or corn), and

allowing use of herbicides to control weeds in fields, among other things.

Test Yourself

1. **b.** The phytomer is the repeating unit of a plant from node to node that includes the leaves, an internode, and one or more axillary buds.
2. **b.** A taproot would be necessary to anchor a plant of that size. Most eudicots have taproots.
3. **d.** Tendrils are modified leaves. Some climbing plants produce tendrils, and others produce suckers.
4. **e.** Plant cell shape is maintained by cell walls. No animal cells have cell walls. Plants have vacuoles and plastids, including chloroplasts. Though these characterisitcs are seen in some protists, no animal cells have them.
5. **a.** Palisade and spongy mesophyll cells are photosynthetically active and derived from parenchyma cells.
6. **a.** Xylem tissue transports water from the roots throughout the plant. Phloem tissue transports carbohydrates and nutrients from source tissue to sink tissue.
7. **b.** The apical and (when present) lateral meristems are regions of continuously dividing cells that can contribute to the growth of a plant throughout its life.
8. **c.** Lateral meristems are responsible for the growth of new xylem and phloem. The secondary xylem gives rise to wood.
9. **d.** The primary cell wall is composed of three polysaccharides: cellulose, hemicelluloses, and pectins.
10. **a.** The cork cambium is a meristematic region outside the secondary phloem. As girth increases and splits the epidermis, causing loss of those protective layers, the cork cambium produces new cells to cover the expanding vascular tissue. It can also produce cells known as phelloderm inside the plant.
11. **b.** The sieve plates allow for the conduction of sap from one sieve tube cell to another sieve tube cell. The pores are created by enlargements of the plasmodesmata. These cells may also lose nuclei and organelles so that carbohydrates can easily pass through the sieve tubes.
12. **e.** Guard cells at the edges of stomata respond to changes in osmotic pressure. These changes lead to the opening and closing of the stomata to regulate gas exchange and water loss.
13. **a.** The protoderm becomes the dermal tissue system of the growing plant.
14. **d.** The apical meristems are the sites of primary growth. Apical meristems are found at the tips of the roots, in stems, and in buds.
15. **c.** Vascular bundles are composed of the vascular tissue of the plant, which includes the xylem and phloem.

35 Transport in Plants

The Big Picture

- In terrestrial plants, water must be acquired from the soil, transported through the plant, and used in the leaves for photosynthesis. At the same time, the nutritional products of photosynthesis must be transported throughout the plant to nonphotosynthetic tissues. This two-way transport is achieved through specialized cells that make up the vascular tissue of the plant. Water with dissolved mineral nutrients is absorbed through the roots and transported to cells in the xylem of the plant, where it is pulled up the stem to the leaves via the transpiration–cohesion–tension mechanism. Sugars and solutes are moved out of the leaves and to the rest of the plant through cells in the phloem.
- Water uptake is regulated by osmotic and water potentials in the root cells, and the rate of transport is controlled by the rate of evaporation at the leaf surface. Guard cell activity in the leaves regulates the opening and closing of stomata to match water availability, light, and drying conditions. Sucrose movement is regulated by active transport and facilitated diffusion in the phloem tissue. The rate of sucrose transport depends on the rates of loading and unloading at source and sink tissues.

Study Strategies

- Be sure you understand the anatomy of the structures that are used in transport. Each structure is uniquely suited to its function. The process is much easier to understand if you think in terms of form coupled with function.
- Water and sucrose transport are pathways that can be understood by visually tracing a molecule of water or sucrose through a plant. Use the figures in your textbook to follow these pathways.
- In order to understand water transport, you should understand osmosis and the properties of water molecules. It may be helpful to review these concepts from earlier chapters.
- The basis of nutritional transport in plants is cell-to-cell transport. Review the sections in the textbook on diffusion, osmosis, and active transport.
- Go to the Web addresses shown to review the following animated tutorials and activity. You can also find them in BioPortal (yourBioPortal.com), along with many additional learning resources:

  Animated Tutorial 35.1 Water Uptake in Plants (Life10e.com/at35.1)

  Animated Tutorial 35.2 Xylem Transport (Life10e.com/at35.2)

  Animated Tutorial 35.3 The Pressure Flow Model (Life10e.com/at35.3)

  Activity 35.1 Apoplast and Symplast of the Root (Life10e.com/ac35.1)

Key Concept Review

35.1 How Do Plants Take Up Water and Solutes?

Water potential differences govern the direction of water movement

Water and ions move across the root cell plasma membrane

Water and ions pass to the xylem by way of the apoplast and symplast

The movement of water across a semipermeable membrane is a special type of diffusion known as osmosis (see Figure 35.2). The overall tendency of a solution to take up water across a membrane is called water potential. For osmosis to occur across a semipermeable membrane, there must be a solute potential (or difference in solute concentrations) great enough to initiate movement and a pressure potential (turgor pressure in plants) small enough to allow for movement. Water potential is the sum of the negative solute potential and the (usually) positive pressure potential. All three parameters can be measured in megapascals (MPa). Water always moves to a region of more negative water potential. If a plant cell is immersed in pure water, water will enter the cell via osmosis until the pressure potential inside the cell equals the solute potential, and the water potential will then be zero. At this point, the cell is turgid; it has a significantly positive pressure potential. The water potential of cells in a plant is dependent on the water potential of the soil. Turgid cells have no net movement of water into them. The structure of plants is maintained by osmotic phenomena. If a plant loses turgor pressure by a decrease in pressure potential, it wilts. Movement of water from cell to cell depends on the gradient of water potential. Long-distance movement depends on pressure potential and is referred to as bulk flow.

Specialized membrane channel proteins in plant cells called aquaporins can increase the rate of water movement by allowing water to cross the plasma membrane without interacting with the hydrophobic bilayer that slows water flow. Though aquaporins can increase the rate of osmosis, they cannot influence the direction of flow. Mineral ion uptake from the soil solution requires active transport via proteins. When mineral concentrations are greater in the soil solution than in the plant, they are taken up by facilitated diffusion. If minerals are in smaller concentrations outside the plant than inside the plant, or if they must be moved against an electrochemical gradient, then the plant must rely on active transport. Plants rely on a proton pump for active transport of minerals into cells. Plants actively pump protons out of cells, causing the area outside the cell to become more positive. This assists facilitated diffusion of positive ions through protein channels. It also drives the movement of negatively charged ions into the cell by active transport (see Figure 35.4). The result of this pumping action is that the internal environment of the plant cell becomes highly negative compared to its environment. The proton gradient that develops across the membrane can also facilitate secondary active transport of ions such as Cl^-.

Water and ions follow two paths for reaching the vascular tissue: the rapid apoplast or the slower symplast. The apoplast is formed by cell walls and intercellular spaces and is exclusively outside the plasma membrane. Water and minerals may move unregulated through this space without ever having to cross a membrane. The symplast is the living portion of the plant and is enclosed in plasma membranes. Movement of water and minerals in the symplast is highly regulated. Water and minerals can travel through the apoplast as far as the endodermis. At the endodermis, water and minerals are stopped by the Casparian strip, which is a waxy region surrounding the endodermal cells. Because of this, water can reach the stele only via the symplast. The transport proteins in the endodermal cells determine which minerals enter the stele. Once past the endodermal barrier, water and minerals can again leave the symplast and move back to the apoplast with the aid of parenchyma cells. As the concentration of ions in the apoplast increases, water moves out of the cells into the apoplast via osmosis. Ultimately, water and minerals from the soil solution end up in xylem cells and are referred to as xylem sap.

Question 1. Create a flow chart of the path taken by a water molecule as it moves from the soil solution to the stele of a plant. Identify where the molecule is traveling through the apoplast and where it is traveling through the symplast.

Question 2. Describe the role of the proton pump in moving minerals into the root.

35.2 How Are Water and Minerals Transported in the Xylem?

The transpiration–cohesion–tension mechanism accounts for xylem transport

Water and nutrients are moved through the plant in the xylem and must move up the plant to reach the other structures. The tracheids and vessel elements that make up the xylem are dead cells; when fused end-to-end, they form xylem vessels. It was originally thought that a pumping mechanism for the movement of fluids might be active in plants. This was shown to be false by experiments in 1893. Tree trunks immersed in poison showed progressive death of all living cells as the poison progressed through the plant. This experiment led to three important conclusions: (1) Because movement continued even as cells were killed, no "pumping" cells were active; (2) Leaves are critical to transport because transport continued until the leaves died; (3) Transport does not depend on the roots because it occurred in the absence of roots. Water moves up a column via capillary action, but this was ruled out as a major transport mechanism because the vessels are too wide to bring water up a 15-meter tree.

The current model of xylem transport has three components: transpiration (evaporative water loss from leaves), tension in the xylem sap (resulting from the transpiration), and cohesion of water molecules in the xylem sap. Pulling forces known as tension, which are due to transpiration at the leaf surface, are responsible for the movement of water through the xylem. Water evaporates from mesophyll cells during transpiration, creating tension on the water associated with the mesophyll cell wall. Transpiration generates tension on the water molecules in the xylem water column, which pulls them up from the roots and through the apoplast of the leaves. Water molecules are cohesive, meaning that they stick together enough to resist the tension, and they pull other water molecules along because of their hydrogen bonding. Some evidence supporting this theory includes the difference in water potential between the soil solution and the air (on the order of 100 MPa), the continuity of the water column in the xylem, and the measurements of xylem pressures as negative pressure potentials, indicating tension. The entire mechanism that pulls water up from the roots through the plant is known as the transpiration–cohesion–tension mechanism. This is a passive process requiring no energy input by the plant. Minerals are drawn passively along with the water column. Transpiration also assists with temperature regulation through evaporative cooling of the leaves.

Question 3. Under what conditions does transpiration occur most rapidly? What effect does increased transpiration have on water flow in a plant? What happens if adequate water for the plant is not available?

Question 4. Explain how transpiration, cohesion, and tension work together to move water in a large plant.

35.3 How Do Stomata Control the Loss of Water and the Uptake of CO_2?

The guard cells control the size of the stomatal opening

Plants can control their total numbers of stomata

Leaf surfaces are covered with a waxy cuticle to prevent excessive water loss. However, the leaf must take up CO_2 for photosynthesis. Whenever a plant surface is open enough to allow gas exchange, water is lost to the environment. Stomata with guard cells are pores that regulate gas exchange and water loss from a leaf. Most plants open their stomata only when the light is sufficient for photosynthesis. Stomata are ancient structures seen in fossils more than 400 million years old.

Opening and closing of stomata are regulated by several factors, including light, CO_2 levels, temperature, and water availability. Guard cells open when light is sufficient to maintain photosynthesis or when carbon dioxide levels are low. Changes in turgor pressure in the guard cells, which arise due to changes in K^+ concentration, are responsible for the speed with which the stomata can open. Blue light stimulates a proton pump that helps regulate guard cell activity. Guard cells open when potassium ions diffuse into the cell as a result of the electrical gradient set up by the proton pump. High potassium levels cause water to move in by osmosis. Pressure potential builds in the guard cells, and they are pulled apart to reveal the stoma. When the proton pumps becomes less active, potassium and chloride ions move out of the guard cells. Water follows, the cells go limp, and the stomata close (see Figure 35.9). Stomata respond to water availability as well. The water potential of the mesophyll cells is the cue that causes closure of the stomata, even in sunlight. Although this prevents photosynthesis, it also prevents the plant from wilting. CAM plants in the desert open their stomata at night to accumulate CO_2. Plants can lose large amounts of water through the stomata because of the huge numbers of them on leaves: up to 250,000 per square inch. Plants can regulate water loss by closing stomata, but they can also regulate the number of stomata on the leaves. This process takes days or weeks and can involve shedding leaves or modified numbers of stomata on new leaves. CO_2 levels can also affect stomatal density.

Question 5. Why might it be advantageous for a plant to reduce the number of stomata on new leaves under high CO_2 conditions?

Question 6. Explain the roles played by an electrochemical gradient and osmosis in the opening and closing of stomata.

35.4 How Are Substances Translocated in the Phloem?

Sucrose and other solutes are carried in the phloem

The pressure flow model appears to account for translocation in the phloem

The carbohydrate products of photosynthesis, primarily created in the leaf, move to the nearest small vein via diffusion, where they are actively transported into sieve tube elements of the phloem. The movement of solutes through the phloem is called translocation. The products of photosynthesis are called photosynthates, and the phloem contents are called phloem sap. Sources are organs that produce more sugars than are used by metabolism, storage, and growth. Sinks are organs that do not make enough sugar for their own growth or storage needs. Sugars, amino acids, minerals, and other substances are translocated between sources and sinks in the phloem. An organ can be a source or a sink at different times; in sugar maples, sugars are stored in the roots for the winter and then returned to the branches to support new growth in the spring. Malpighi proved that phloem carries sucrose and other solutes by "girdling" a tree, removing a ring of bark. The bark contained the phloem, and the solution above the cut was trapped, causing swelling. The bark below the girdle died, presumably because it lacked nutrients. Eventually the roots and then the entire tree died.

Sieve tube elements are the cells of the phloem. They are connected end-to-end by sieve plates containing plasmodesmata, allowing for movement between cells. Because the sieve tube elements do not contain organelles, they rely on companion cells to provide them with all they need to survive. Scientists can sample the sieve tube sap of a single sieve tube member using aphids. An aphid drills into a single cell, and sap is forced out. Once the aphid begins eating, it is frozen and its feeding organ is used as a tap to collect the sap from a single sieve tube. Sucrose makes up 90 percent of the solutes in phloem sap. Other contents include hormones, small molecules such as amino acids, mineral nutrients, and viruses. The flow rate can be up to 100 cm per hour. Different sieve tube elements flow in different directions, and the overall movement is bidirectional. Unlike xylem sap, the movement of phloem sap requires living cells.

The pressure flow model explains how materials move through the phloem. Energy is required for the loading of the sieve tubes at the sources and the unloading of solutes at the sink. Sucrose is actively transported into sieve tubes at the sources, decreasing the solute potential of the phloem. This causes water to move into sieve tubes by osmosis from xylem, thereby increasing the pressure potential at the source end and pushing the contents toward the sink end. The result is bulk flow, which is passive. Sugars and other solutes move from mesophyll cells to the phloem by either the apoplastic pathway or the symplastic pathway. In the apoplastic pathway, solutes move out of the mesophyll cells and diffuse through the apoplast to the sieve tubes. Active transport into the sieve tube elements allows regulation of the composition of the phloem sap. In the symplastic pathway solutes remain in the symplast and pass through plasmodesmata to the sieve tube cells. The solutes are loaded into the phloem sap by mechanisms other than active transport.

At the sink, solutes are unloaded passively and through active transport, maintaining the solute and pressure potentials, in turn maintaining phloem sap flow. This also allows for delivery of nutrients to developing organs, and for storage of proteins and carbohydrates in storage organs.

An important experiment showed that sink strength, the relative ability to attract photosynthates, was shown to be very important in potato tubers. Invertase catalyzes the hydrolysis of sucrose to glucose and fructose, and fructose is in turn converted to glucose. An important study showed that accumulation of invertase in the apoplast caused large amounts of sucrose to be hydrolyzed, and the resulting glucose was taken up by the tuber cells and converted to starch. The invertase lowered the sucrose concentration and increased phloem unloading; this in turn caused the tubers to grow larger than those of the control plants. Thus the sucrose concentration gradient between the phloem and the apoplast influences the rate of phloem unloading.

Question 7. Differentiate between source and sink tissues. What happens relative to phloem in each?

Question 8. Describe the pressure flow model of phloem transport.

Test Yourself

1. The function of the Casparian strips is to
   a. divert water and minerals through the membranes of endodermal cells.
   b. prevent water and minerals from entering the stele through the apoplast.
   c. provide regulation for water and mineral movement in the plant.
   d. All of the above
   e. None of the above
   Textbook Reference: *35.1 How Do Plants Take Up Water and Solutes?*

2. The primary difference between the apoplast and the symplast is that
   a. the apoplast consists of nonliving spaces and cell walls, whereas the symplast consists of living cells.
   b. apoplast movement is tightly regulated and symplast movement is not.
   c. the symplast consists of nonliving spaces and cell walls, whereas the apoplast consists of living cells.
   d. apoplast movement is slow and symplast movement is fast.
   e. the apoplast transports only ions and the symplast transports only water.
   Textbook Reference: *35.1 How Do Plants Take Up Water and Solutes?*

3. Which of the following statements about water transport is true?
   a. Root pressure is sufficient to drive xylem sap movement.
   b. Bulk flow is not a mechanism by which water and minerals are transported.
   c. The cohesive nature of water is central to water movement in a plant.
   d. Water transport is an active process.
   e. None of the above
   Textbook Reference: *35.2 How Are Water and Minerals Transported in the Xylem?*

4. Tension in the xylem is a result of
   a. transpiration at the leaf surface.
   b. the cohesive nature of water.
   c. the narrowness of the xylem tube.
   d. the surface area of the phloem.
   e. All of the above
   Textbook Reference: *35.2 How Are Water and Minerals Transported in the Xylem?*

5. The fact that water transport continues as long as leaves are alive and active indicates that
   a. leaves pump water.
   b. leaves are necessary for transport of water.
   c. roots are active.
   d. water is not needed for leaves to remain alive.
   e. sieve tube elements are inactive.
   Textbook Reference: *35.2 How Are Water and Minerals Transported in the Xylem?*

6. Which of the following statements regarding transport in phloem is true?
   a. It always moves in the direction of leaves to roots.
   b. It proceeds from source tissue to sink tissue.
   c. It requires no energy inputs from the plant.
   d. It is the same as the transport process in xylem.
   e. None of the above
   Textbook Reference: *35.4 How Are Substances Translocated in the Phloem?*

7. If the pressure potential of a plant's cells is 0.16 megapascals (MPa) and the solute potential is –0.24 MPa, then the water potential would be
   a. 0.04 MPa.
   b. 0.08 MPa.
   c. –0.08 MPa.
   d. –0.24 MPa.
   e. –0.04 MPa.
   Textbook Reference: *35.1 How Do Plants Take Up Water and Solutes?*

8. Which of the following represents the correct ordering of the water potential of these root cells or regions, from least negative to most negative?
   a. Xylem, cortex apoplast, stele apoplast, soil next to root
   b. Soil next to root, xylem, stele apoplast, cortex apoplast
   c. Xylem, stele apoplast, cortex apoplast, soil next to root
   d. Stele apoplast, cortex apoplast, xylem, soil next to root
   e. Soil next to root, cortex apoplast, stele apoplast, xylem
   Textbook Reference: *35.1 How Do Plants Take Up Water and Solutes?*

9. The movement of water up the stems of tall plants is *least* dependent on which of the following factors?

a. The exudation of xylem sap on leaves in humid conditions
b. Transpiration
c. Cohesion of water molecules
d. Tension within water columns
e. All of the above are of equal importance.
Textbook Reference: *35.2 How Are Water and Minerals Transported in the Xylem?*

10. Which of the following statements about xylem transport and phloem transport is true?
a. Both are passive processes that do not require energy from the plant.
b. Both rely on only living cells.
c. Both rely on a water potential gradient.
d. The direction of flow can reverse in both.
e. The driving force for both is in the leaves.
Textbook Reference: *35.2 How Are Water and Minerals Transported in the Xylem?*

11. Stomatal opening and closing are regulated by
a. abscisic acid levels.
b. light levels.
c. carbon dioxide concentrations.
d. All of the above
e. None of the above
Textbook Reference: *35.3 How Do Stomata Control the Loss of Water and the Uptake of CO_2?*

12. The opening and closing of the stomata are accomplished by the
a. sieve tube.
b. guard cells.
c. process of translocation.
d. aquaporins.
e. xylem.
Textbook Reference: *35.3 How Do Stomata Control the Loss of Water and the Uptake of CO_2?*

13. Regulators of stomatal opening and closing work by activating the
a. proton pump in guard cells.
b. proton pump in stomata.
c. sodium–potassium pump in guard cells.
d. sodium–potassium pump in stomata.
e. All of the above
Textbook Reference: *35.3 How Do Stomata Control the Loss of Water and the Uptake of CO_2?*

14. Mineral ions enter the cell due to the force of an electrochemical gradient set up by the pumping of _______ out of the cells.
a. K^+
b. Ca^{2+}
c. Na^+
d. H^+
e. Cl^-
Textbook Reference: *35.1 How Do Plants Take Up Water and Solutes?*

15. In the pressure flow model of translocation, the movement of water by osmosis occurs
a. from the xylem to the phloem at the sink.
b. from the phloem to the xylem at the source.
c. from the xylem to the phloem at the source.
d. in both directions at the sink.
e. None of the above
Textbook Reference: *35.4 How Are Substances Translocated in the Phloem?*

Answers

Key Concept Review

1.

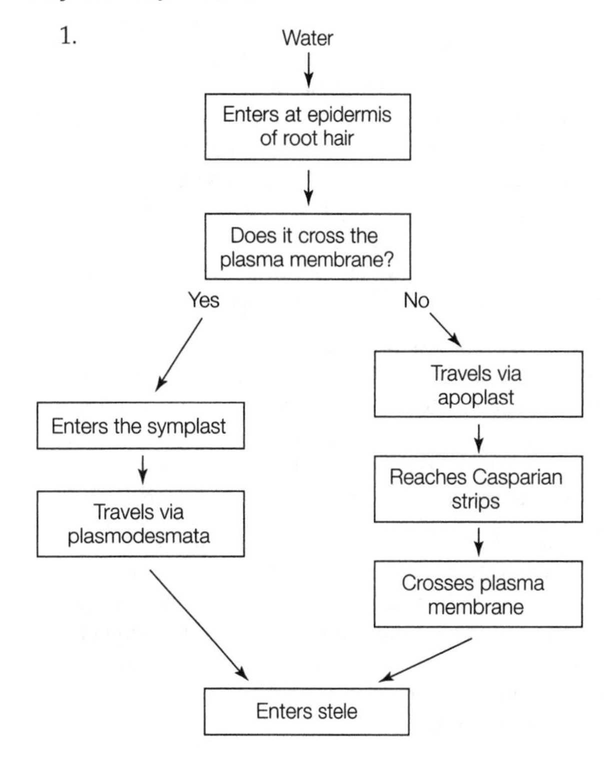

2. Plants rely on a proton pump for active transport of minerals into cells. Plants pump protons out of cells, which causes the area outside the cell to be more positive. This assists facilitated diffusion of positive ions through protein channels. It also drives the movement of negatively charged ions into the cell by active transport (see Figure 35.4).

3. Transpiration occurs most rapidly in high light conditions when stomata are open, along with high wind conditions and low humidity when evaporation is greatest. This results in faster bulk flow through the xylem and increased water demands by the plant. If water is not available, plant cells lose turgor and the plant wilts.

4. The transpiration–cohesion–tension mechanism pulls water from the roots up through the plant. Water evaporates from mesophyll cells during transpiration. This puts tension on the film of water associated with the mesophyll cell wall. The tension at the mesophyll

cell draws water from the xylem of the nearest vein. This creates tension in the entire xylem column, and the column is drawn upward from the roots.

5. Under high CO_2 conditions, the plant may take in proportionally more CO_2 per volume of air moved through the leaf. Therefore, it can reduce the number of stomata on the leaves and expose itself to less water loss through the leaves.
6. When the guard cells are stimulated by blue light, they activate a proton pump which pumps H^+ ions outside the guard cells. This creates an electrochemical gradient that drives K^+ ions into the guard cells to offset the charge difference. This makes the water potential of the guard cells more negative. Other ions, such as Cl^-, flow with the K^+ to offset the charge difference. This causes the solute potential of the guard cells to change, and osmosis allows water to enter the guard cells. With this increase in pressure potential, the shape of the guard cells changes and a gap appears between them, creating the stoma. When the blue light signal is gone, the proton pump is less active, K^+ and Cl^- ions diffuse out, water follows by osmosis, and the cells return to their original shape, closing the gap.
7. Source tissues produce sugars in excess of what can be used and stored. Phloem loading occurs in source tissues and creates a pressure potential that results in bulk flow of sieve tube sap toward sink tissues. Sink tissues produce fewer sugars than can be stored or used and unload phloem through active transport.
8. The difference in solute concentration between sources and sinks creates a pressure potential along sieve tubes, resulting in bulk flow. For this to occur, sugars must be loaded at the source tissue and unloaded at the sink tissue through active transport, and the sieve plates must remain open and unclogged along the phloem column.

Test Yourself

1. **d.** Not all minerals that enter the apoplast of a plant's root are beneficial to the plant. The Casparian strips prevent water and minerals from reaching the stele through the apoplast, diverting them instead through the plasma membranes of the endodermal cells. Channel proteins in these plasma membranes determine which minerals can enter the symplast, and from there, the stele. The Casparian strips thus contribute to the regulation of water and mineral movement in the plant.
2. **a.** The intercellular spaces and cell walls of the plant constitute the apoplast.
3. **c.** Water movement depends on the cohesive nature of water and its capacity to withstand the tension placed on the water column by transpiration.
4. **a.** Transpiration causes tension.
5. **b.** Leaves are necessary for transpiration to take place.
6. **b.** Transport in phloem does not always go from leaf to root, but it always proceeds from source tissue to sink tissue. The plant must contribute energy to create the water pressure gradient by pumping solutes into the phloem at the source and out of the phloem at the sink.
7. **c.** Water potential is equal to pressure potential plus solute potential.
8. **c.** The xylem would be the most negative, followed by the stele, then the cortex, then the area outside the root.
9. **a.** Exudation of xylem sap occurs under extremely humid conditions when water is plentiful.
10. **c.** Both xylem transport and phloem transport depend on water potential.
11. **d.** Abscisic acid, light, and carbon dioxide levels all regulate stomatal opening and closing.
12. **b.** Guard cells are specialized epidermal cells that regulate the opening and closing of the stomata by covering the stomata opening.
13. **a.** Stomatal regulators work by activating and deactivating the proton pump in guard cells.
14. **d.** Cells pump H^+ ions out into the soil with the help of proton pumps.
15. **c.** In the pressure flow model, water moves by osmosis from the xylem into the phloem at the source and from the phloem to the xylem at the sink. This movement is all driven by the solute concentrations at each location.

Plant Nutrition

The Big Picture

- Plants require specific macro- and micronutrients. Deficiencies in any of these challenge the health of the plant. Essential nutrients must be available, and no substitutions will sustain the plant. Nutrients are procured from the soil solution that bathes the roots of a plant. The availability of nutrients depends on the quantity, solubility, and structure of soil. Many agricultural practices deplete soils of nutrients, and these must be replenished through fertilization.
- Plants interact with fungi and bacteria that help them obtain needed nutrients. Nitrogen availability is essential to plant growth. Bacteria and plants are intrinsically linked in the nitrogen cycle (see Figure. 36.10). Agriculture makes use of biological nitrogen fixation in the practice of crop rotation, in which farmers plant nonharvested crops that harbor nitrogen-fixing bacteria in their root nodules. Commercial nitrogen fixation by means of chemical fertilizers is extremely energy demanding.
- A small number of plants do not photosynthesize. These heterotrophic plants are often parasites of other plants and acquire nutrients solely through their hosts.

Study Strategies

- The interactions between nitrogen-fixing bacteria and plants can get confusing. Remember that it is a mutualistic relationship and that the bacteria have the enzymes necessary to fix atmospheric nitrogen.
- The most difficult concept of this chapter is the nitrogen cycle. The figures in the textbook should help you visualize this process and understand how organisms interact with the environment.
- Be sure you understand the consequences of nutrient shortages in plants.
- Go to the Web addresses shown to review the following animated tutorial and activity. You can also find them in BioPortal (yourBioPortal.com), along with many additional learning resources.

  Animated Tutorial 36.1 Nitrogen and Iron Deficiencies (Life10e.com/at36.1)

  Activity 36.1 The Nitrogen Cycle (Life10e.com/ac36.1)

Key Concept Review

36.1 What Nutrients Do Plants Require?

All plants require specific macronutrients and micronutrients

Deficiency symptoms reveal inadequate nutrition

Hydroponic experiments identified essential elements

The basic nutrient requirements for all living things are carbon, hydrogen, oxygen, and nitrogen. These elements are the fundamental building blocks of all macromolecules. Plants are autotrophs and incorporate carbon through photosynthesis. Oxygen and hydrogen enter plants as water. Nitrogen's entry into plants is dependent on nitrogen-fixing bacteria in the soil. Mineral nutrients are essential to life. Sulfur, phosphorus, magnesium, iron, and calcium are all essential components of many macromolecules. Mineral nutrients enter biological organisms through soil solutions, which plants take up through their roots. A plant nutrient is an essential element if it is absolutely required for the life cycle or normal growth. A deficiency in any essential nutrient leads to an unhealthy plant. Macronutrients are essential elements required in concentrations of 1 g per 1 kg of dry plant matter, and micronutrients are essential elements required in concentrations of 100 mg per 1 kg of dry plant matter. See Table 36.1 for a list of macro- and micronutrients, their functions, and symptoms of their deficiencies. Though many nutrient deficiencies ultimately lead to plant death, specific deficiency symptoms are evident in a plant before it dies. Deficiencies can be corrected by the addition of fertilizers to supplement nutrients in soils. Scientists determined which elements are essential by growing plants hydroponically, with roots suspended in nutrient solutions instead of soil, allowing for manipulation of the nutrients that are available to the plant.

Question 1. Explain how scientists determined which plant nutrients are the essential nutrients.

Question 2. Differentiate between micro- and macronutrients. How are most of these nutrients acquired?

36.2 How Do Plants Acquire Nutrients?

Plants rely on growth to find nutrients

Nutrient uptake and assimilation are regulated

Plants are sessile, meaning that they cannot move around, and therefore they cannot "search" for nutrients. They overcome this problem by growing toward new resources. A plant grows taller to procure more sunlight and to outcompete nearby plants, and its roots spread to acquire mineral nutrients in the soil. As it grows, a plant needs to deal with a variable environment. These microenvironments encourage or discourage root growth and thus guide the plant.

Nutrients must cross the plasma membrane of cells in order to be used by the plant. Polar molecules cross via specialized transport systems. In most cases ions are actively transported into the symplast because their concentrations in the soil are generally lower than the concentrations in the cells. Plants have specialized systems for the uptake of specific ions, including nitrate, ammonium, and phosphate ions. This uptake is highly regulated at the transcription level for the transport protein genes. Assimilation of nutrients is also regulated according to the plant's needs; for instance nitrate and ammonium are incorporated into amino acids more frequently when nitrogen is abundant. Nitrogen uptake and assimilation are also stimulated by photosynthesis.

Question 3. Explain how plants "grow into their nutrients."

Question 4. Do nutrients enter the plant from the soil via diffusion? Why or why not?

36.3 How Does Soil Structure Affect Plants?

Soils are complex in structure

Soils form through the weathering of rock

Soils are the source of plant nutrition

Fertilizers can be used to add nutrients to soil

Most terrestrial plants grow in soil, which provides mechanical support, mineral nutrients and water, and O_2 for root respiration. Soils are composed of living and nonliving matter. The living portion of soil contains roots, protists, bacteria, fungi, and many small animals (such as insects and worms). The nonliving portion consists of rock fragments, clay, water, air spaces, and dead organic matter. The air spaces provide the O_2 needed for a plant to survive. All soils have a soil profile consisting of two or more horizons (layers). Water-soluble nutrients are leached to deeper horizons through rainfall. Three major horizons (also called zones), termed A, B, and C, can be identified in soils. The A horizon is topsoil. It is organically rich, very biologically active, and the most agriculturally important layer. The B horizon is the subsoil, which holds many leached nutrients. The C horizon is parent rock, which roots cannot penetrate.

Soil fertility is a soil's ability to support plant growth. A loam is a topsoil with an optimal mixture of sand, silt, and clay; it has high nutrient content, plentiful water, and adequate air spaces. Soils with too much sand typically do not hold nutrients or water well, and clays are too dense for the trapping of air. Humus is the breakdown product of organic matter in the soil; it is used as a food source by microbes that break it down further and release simple molecules into the soil solution. Humus also provides air spaces that increase O_2 availability.

Rocks are weathered into soils by mechanical and chemical weathering. Mechanical weathering is caused by freeze and thaw cycles, rain, and drying. Chemical weathering is caused by oxidation, hydrolysis, or breakdown by acids, and it can change the composition of rock. Chemical weathering is very important in clay formation. Mineral nutrients are tied to clay particles in the soil. Because many nutrients are positively charged cations, clays with a negative charge can hold these nutrients and make them available to plants. Roots release protons into the soil; these bind to clay particles and cause the release of cation minerals by a process called cation exchange. CO_2 released from the roots can also form bicarbonate and free protons, which bind with clay. Some soil particles are positively charged and can undergo anion exchange, but often soil pH is not low enough to allow this to happen. Nutrients that are negatively charged and therefore do not participate in anion exchange are rapidly leached from soil. Thus, nitrate and sulfate are often not available to plants.

Fertilizers can restore soil fertility by replenishing nutrients lost through use and leaching. Shifting agriculture is an old method of ensuring that crops are planted in fertile soil. A field that lies fallow will replenish the nutrients naturally through the addition of organic material from naturally growing plants and the weathering of parent rock. This process usually takes many years, and so it is not efficient for the large yields needed by modern farmers. Microorganisms in the soil break down organic matter into nutrients that become available to plants. One such process is the conversion of proteins from dead plant leaves into ammonium and then nitrate by different bacterial species. Farmers can increase the nutrient content of soil by adding compost or manure, which are organic fertilizers. Organic fertilizers slowly release ions as the materials decompose, and may act too slowly if a soil is to be used every year. Inorganic fertilizers can be used if the soil must be used every year. Inorganic fertilizers supply mineral nutrients in available form, such as ammonia, nitrate, phosphate, and sulfate. The use of inorganic fertilizers has contributed to the huge agricultural yields in recent decades, but it has environmental drawbacks.

Question 5. Draw the different horizons of the soil and indicate what each is comoposed of.

Question 6. What kinds of environmental problems can result from the use of inorganic fertilizers?

36.4 How Do Fungi and Bacteria Increase Nutrient Uptake by Plant Roots?

Plants send signals for colonization

Mycorrhizae expand the root system

Soil bacteria are essential in getting nitrogen from air to plant cells

Nitrogenase catalyzes nitrogen fixation

Biological nitrogen fixation does not always meet agricultural needs

Plants and bacteria participate in the global nitrogen cycle

Soil can contain tens of thousands of bacterial species and hundreds of meters of fungal hyphae. Plants maintain mutualistic associations with some of these species. Formation of mycorrhizae is stimulated by the release of strigolactones by the root. A prepenetration apparatus helps guide the fungi as they grow into the root. Arbuscules in the cortical cells of the roots are the site of nutrient exchange. The cytoplasms never mix but are separated by the fungal and plant plasma membranes.

Legumes can form symbiotic associations with several genera of bacteria collectively known as rhizobia. Flavonoids released by the plant attract the rhizobia. In response to the flavonoids, the bacteria turn on the production of Nod factors, which cause formation of the nodule by the plant. Cells in the root cortex divide, forming a primary nodule meristem, which gives rise to the root nodule. Bacteria enter the root through an infection thread and are eventually released into the cytoplasm of the nodule cells. They are enclosed in membrane vesicles, inside of which they differentiate into bacteroids, which can fix nitrogen. There is increasing evidence that the establishment of both root nodules and arbuscular mycorrhizae depend on similar mechanisms of invagination and entry of symbiotic organisms.

The fungi in mycorrhizae help expand the surface area of the root to increase nutrient uptake. The fungi help plants take up phosphorus, while the plants provide the fungi with sugars for energy. Nitrogen gas is readily available in the atmosphere, but plants are unable to break the triple bonds between the two nitrogen atoms. Only a few bacteria species can fix nitrogen gas into biologically usable ammonia through nitrogen fixation. Nitrogen fixers fix approximately 170 million metric tons of nitrogen per year. The fixation of nitrogen gas requires the enzyme nitrogenase to catalyze the reaction, lots of energy (ATP), and a strong reducing agent. Nitrogenase is inhibited by oxygen and therefore is active only under anaerobic conditions. The nitrogen-fixing bacteria are typically found in root nodules that maintain very low oxygen levels. The protein leghemoglobin binds with oxygen to keep oxygen levels low. Bacterial nitrogen fixation is not adequate to supply all the nitrogen needed for agriculture. One method to enrich soil nitrogen includes crop rotation (including legumes in the rotation), but this often does not meet soil nitrogen needs for agriculture. Currently, nitrogen fertilizers are produced via industrial fixation through the Haber process, an energy-expensive process.

The global nitrogen cycle includes four key steps: fixation of nitrogen into NH_3 and NH_4^+, nitrification to nitrate, nitrate reduction by plants, and denitrification of nitrate back into N_2 by bacteria. Nitrifiers oxidize ammonia to nitrate ions (NO_3^-) via nitrification. Nitrate ions are taken up under basic conditions, and ammonium is taken up under more acidic ones. To use nitrate, a plant must reduce it to ammonium by nitrate reduction. Animals cannot reduce nitrogen and so depend on plants for reduced nitrogenous compounds. The nitrogen cycle is essential for life on Earth. (See Figure 36.10 for a review of the nitrogen cycle.)

Question 7. Diagram the nitrogen cycle.

Question 8. Why do many farmers plant crops such as alfalfa and soybeans without harvesting them?

36.5 How Do Carnivorous and Parasitic Plants Obtain a Balanced Diet?

Carnivorous plants supplement their mineral nutrition

Parasitic plants take advantage of other plants

The plant–parasite relationship is similar to plant–fungus and plant–bacteria associations

Examples of carnivorous plants are Venus flytraps, pitcher plants, and sundews. Carnivorous plants acquire nitrogen from the proteins of trapped flies and other small animals that get trapped in them. Carnivorous plants are typically found in boggy habitats, which are typically acidic and nutrient-deficient. Pitcher plants have deep leaves that trap animals looking for rainwater collected in them. Sundews have sticky leaves that trap insects. Venus flytraps have pressure-sensitive leaves that fold shut when an insect lands on them. All of these plants secrete enzymes to digest their prey after they are trapped. Carnivorous plants can perform photosynthesis and absorb soil nutrients like other plants. Although these plants can survive and grow without consuming insects, they thrive when they have a continuous supply of them.

About 1 percent of flowering plants are parasitic and derive some or all of their water and minerals and sometimes photosynthates from other plants. They have evolved absorptive organs, called haustoria, to tap into the host tissues. There are two broad classes of parasitic plants: hemiparasites and holoparasites. Hemiparasites can still perform photosynthesis but take water and mineral nutrients from other plants; one such group is the mistletoes. Holoparasites cannot perform photosynthesis and thus also take photosynthates from the host plant. Holoparasites are a very diverse group; some do not have leaves or stems because they live underground and break through the soil only to reproduce.

Parasitic plants detect signals from nearby plants and grow toward them to take their nutrients. This is similar to the plant–bacteria and plant–fungi relationships, except that it is not mutualistic. The holoparasite *Striga* is an example of opportunistic evolution: the repurposing of an already-established process rather than the creation of a new one, to have a new effect. A strigolactone compound secreted by plant roots causes seed germination in this species. Scientists hypothesize that *Striga* evolved to exploit the secretion of these compounds for their own purposes, to find suitable host plants to attack.

Question 9. Most carnivorous plants are found in boggy, wet, acidic environments. What is the effect of such an environment on nutrient availability?

Question 10. Differentiate between hemiparasitic and holoparasitic plants. How would you expect them to differ morphologically?

Test Yourself

1. Which of the following nutrients is *not* considered essential for plant growth?
   a. Cadmium
   b. Nitrogen
   c. Manganese
   d. Potassium
   e. All are essential.
   Textbook Reference: *36.1 What Nutrients Do Plants Require?*

2. Compared to micronutrients, macronutrients are
   a. larger.
   b. needed in greater quantities.
   c. more essential.
   d. of equal importance.
   e. less essential.
   Textbook Reference: *36.1 What Nutrients Do Plants Require?*

3. Nitrogen and potassium are acquired from
   a. the soil solution.
   b. heterotrophs.
   c. air.
   d. micronutrients.
   e. All of the above
   Textbook Reference: *36.2 How Do Plants Acquire Nutrients?*

4. A tomato plant whose young leaves are very yellow, but whose older leaves are still green, most likely has a(n) _______ deficiency.
   a. nitrogen
   b. carbon
   c. water
   d. iron
   e. phosphorus
   Textbook Reference: *36.1 What Nutrients Do Plants Require?*

5. Years of cotton farming in the South have stripped away much of the A horizon of the soils. Subsequent agriculture
   a. has been problematic, because the A horizon contains the most available nutrients.
   b. has not been affected, because the B horizon contains significantly more available nutrients.
   c. has been affected slightly, because the C horizon is most conducive to root growth.
   d. has been problematic, because stripping of the A horizon leaches the C horizon of its nutrients.
   e. All of the above
   Textbook Reference: *36.3 How Does Soil Structure Affect Plants?*

6. Clay particles in soils are important for
   a. holding the soil together.
   b. ion exchange.
   c. holding water.
   d. All of the above
   e. None of the above
   Textbook Reference: *36.3 How Does Soil Structure Affect Plants?*

7. Most clays form from the _______ of rock.
   a. mechanical weathering
   b. chemical weathering
   c. heaving
   d. grinding
   e. All of the above
   Textbook Reference: *36.3 How Does Soil Structure Affect Plants?*

8. Which of the following descriptions of soil components is *false*?
   a. The A horizon is the topsoil.
   b. The B horizon is the subsoil.
   c. The C horizon is the parent rock.
   d. Loam consists mostly of clay.
   e. Humus is produced from the breakdown product of organic matter.
   Textbook Reference: *36.3 How Does Soil Structure Affect Plants?*

9. The relationship between rhizobium bacteria and the roots of legumes can best be described as
   a. parasitic.
   b. one-sided.
   c. mutualistic.
   d. carnivorous.
   e. detrimental.
   Textbook Reference: *36.4 How Do Fungi and Bacteria Increase Nutrient Uptake by Plant Roots?*

10. Nitrogen gas is reduced to ammonia by which of the following enzymes or processes?
    a. Rhizobium
    b. Nitrogenase
    c. Nitrification
    d. Denitrification
    e. Rhizobase
    Textbook Reference: *36.4 How Do Fungi and Bacteria Increase Nutrient Uptake by Plant Roots?*

11. Plants are able to take up and use nitrogen in the form of _______ and _______.
    a. ammonia; nitrate
    b. ammonia; nitrogen gas
    c. nitrogen gas; nitrate
    d. nitrogen gas; nitrous oxide
    e. All of the above
    Textbook Reference: *36.4 How Do Fungi and Bacteria Increase Nutrient Uptake by Plant Roots?*

12. Carnivorous plants are often found in acidic and nutrient-poor environments. The main selective pressure for carnivory is

a. lack of nitrogen and phosphorus sources.
b. lack of iron and calcium sources.
c. incomplete ion exchange.
d. lack of water sources.
e. All of the above

Textbook Reference: *36.5 How Do Carnivorous and Parasitic Plants Obtain a Balanced Diet?*

13. Nitrate and sulfate tend to leach from the soil because
a. they bind with ions such as K^+ and Mg^{2+}.
b. the H^+ ions released by the roots push them out.
c. they are unable to bind with the negatively charged clay particles.
d. they bind with the positively charged clay particles.
e. All of the above

Textbook Reference: *36.3 How Does Soil Structure Affect Plants?*

14. A plant lowers the pH of soil by means of
a. ion exchange.
b. leaching.
c. Na^+ pumping.
d. proton pumping.
e. All of the above

Textbook Reference: *36.3 How Does Soil Structure Affect Plants?*

15. The role of leghemoglobin is to maintain _______ levels in the root nodule.
a. high O_2
b. high CO_2
c. low O_2
d. low CO_2
e. high N_2

Textbook Reference: *36.4 How Do Fungi and Bacteria Increase Nutrient Uptake by Plant Roots?*

Answers

Key Concept Review

1. Plants were grown in cultures that lacked specific nutrients. If a plant could not complete its life cycle, the missing nutrient was said to be essential. These experiments were well-controlled hydroponic studies because nutrient content can easily be manipulated in water.

2. Macronutrients are needed at concentrations of 1 g/ 1 kg of dry plant tissue. Micronutrients are needed at concentrations of 100 mg/1 kg of dry plant tissue. Most of these nutrients are in soil solution and are taken up as water is drawn into roots.

3. Plants cannot move to areas with greater nutrients. They are limited to extending their roots into soils that contain a larger nutrient reserve. When nutrients are limited, plant growth matches the availability of resources.

4. Plants take up minerals via active transport. This is mostly because minerals exist in the cell at a higher concentration than they exist in the soil, and thus transport has to fight the concentration gradients.

5.

A horizon
Topsoil

B horizon
Subsoil

C horizon
Weathering parent rock

6. Inorganic fertilizers tend to leach from the soil when introduced in excess, which can create runoff with high nutrient density. Excess nitrogen fertilizer runoff contributes to algal blooms that disrupt the balance of aquatic ecosystems.

7.

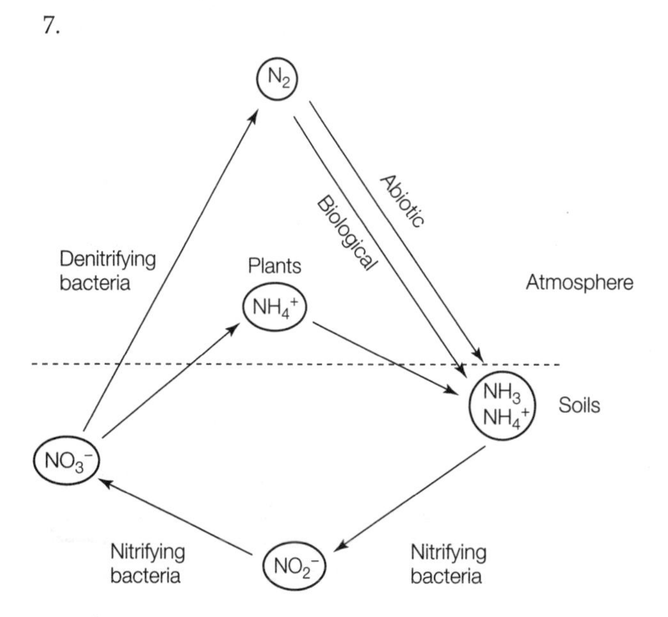

8. By rotating crops (and especially by rotating and plowing under legume crops) farmers can organically add nitrogen to depleted soils. This results in significantly larger yields of the harvested crops.

9. Acidic environments limit decomposition and thus nitrogen sources. They also limit ion exchange. Both conditions result in reduced nutrient availability that can be offset by carnivory.
10. Although both types of plants are parasitic, hemi-parasites can perform photosynthesis while holopara-sites cannot. For this reason, one would expect most hemiparasites to be green from the chlorophyll that performs photosynthesis, while holoparasites would be colorless or some color other than green. Holoparasites might not be visible above the soil surface and might be greatly reduced in complexity compared to a photo-synthetic plant.

Test Yourself

1. **a.** Cadmium is not one of the 14 essential micro- and macronutrients.
2. **b.** The main difference between micronutrients and macronutrients is in the quantity of each needed by a plant for survival.
3. **a.** Nitrogen and mineral nutrients are acquired in soil solution (dissolved in water).
4. **d.** If the older leaves look normal but the younger leaves are yellow, an iron deficiency should be suspected.
5. **a.** Topsoil, or the A horizon, is most conducive to root growth. Ample nutrients are available, as are air spaces and water for ease of root growth.
6. **d.** Clay particles are critical for ion exchange. They are also important for retaining water and for the integrity of the soil.
7. **b.** Chemical weathering leads to clay formation.
8. **d.** A loam is an optimal mix of sand, silt, and clay that has sufficient levels of air, water, and nutrients for plants.
9. **c.** The relationship is mutualistic, in that both the plant and the bacteria benefit from the association.
10. **b.** Nitrogenase catalyzes the reduction of nitrogen gas to ammonia. This is an energy-expensive process.
11. **a.** Plants can take up and use nitrogen that is in the form of ammonia or nitrate.
12. **a.** Carnivory supplements insufficient nitrogen and phosphorus availability.
13. **c.** Nitrate and sulfate are negatively charged anions. Without any positively charged molecules in the soil with which they can interact, they leach from the soil.
14. **d.** Plants decrease the pH in the surrounding soil by pumping protons out of the roots.
15. **c.** Leghemoglobin is a protein produced by the root nodule of plants to help maintain low levels of O_2 and thus provide nitrogen-fixing bacteria with an anaerobic environment.

Regulation of Plant Growth

The Big Picture

- Plant growth is controlled by interactions among a plant's environment, hormones, and genetic makeup. Changes in plant growth are dependent on the information transmitted by hormones to hormone receptors, the subsequent activation of signal transduction pathways, and the eventual alteration of gene expression. Receptors are often highly specific and respond to select hormones. Hormones are produced in a specific region of the plant body and are translocated throughout the plant; therefore, their effects are often concentration dependent. All growth in a plant is a result of changes in cell division, cell expansion, and cell differentiation.
- Seed germination, growth of the vegetative structures, reproduction, and senescence all proceed in defined patterns. Seed dormancy is broken, and development begins when the seed coat is abraded, the seed imbibes water, and inhibitory chemicals are diluted. This begins a series of events that mobilize nutrients and induce growth. Once a seedling emerges from the soil, light begins to influence subsequent development under the control of hormones.
- Hormones exist in several classes, each with its own effects on growth and development. Some of these effects are antagonistic, and therefore control is maintained via relative concentration of several hormones. Hormones influence every step of development, from the breaking of seed dormancy to senescence.
- Light regulates plant processes through photoreceptors. Photoreceptors respond to very specific wavelengths and induce cascades that lead to changes in a plant.

Study Strategies

- Understanding how phytochrome shifts from red form to far-red form frequently gives students difficulty. This is understandable, because phytochromes have puzzled researchers for many years.
- There is a tendency when studying hormones simply to memorize functions. You need to understand the effects in a plant of different ratios of hormones, not what an individual hormone does.
- The amount of information in this chapter may seem overwhelming. Take your time in learning how growth is regulated and the effects of the various hormones.
- Avoid focusing too much on details. Try to assimilate the big picture of how environment, receptors, hormones, and genome interact. From there, begin to work toward the details. A big mistake is to jump in and memorize functions of hormones or sequences of development without understanding the broader picture.
- Go to the Web addresses shown to review the following animated tutorials and activities. You can also find them in BioPortal (yourBioPortal.com), along with many additional learning resources.

  Animated Tutorial 37.1 Tropisms (Life10e.com/at37.1)

  Animated Tutorial 37.2 Went's Experiment (Life10e.com/at37.2)

  Animated Tutorial 37.3 Auxin Affects Cell Walls (Life10e.com/at37.3)

  Activity 37.1 Monocot Shoot Development (Life10e.com/ac37.1)

  Activity 37.2 Eudicot Shoot Development (Life10e.com/ac37.2)

  Activity 37.3 Events of Seed Germination (Life10e.com/ac37.3)

Key Concept Review

37.1 How Does Plant Development Proceed?

In early development, the seed germinates and forms a growing seedling

Several hormones and photoreceptors help regulate plant growth

Genetic screens have increased our understanding of plant signal transduction

Plants have unique features that allow them to obtain the resources they need; these features are meristems, post-embryonic organ formation, and differential growth. The factors involved in regulating plant growth are environmental cues, receptors, hormones, and regulatory proteins and enzymes (encoded by the genome). Plant seeds are dormant and remain so until seed germination. Dormancy, which lasts for different periods of time depending on the plant, involves exclusion of water or oxygen from the embryo by the seed coat, mechanical restraint of the embryo by the seed coat, and chemical inhibition of the embryo. Dormancy ensures survival through adverse conditions. Some seeds rely on other environmental cues before germination, such as cold temperatures, light, or the passage of a specific amount of time. These cues help ensure that the seed will germinate in the correct location and at the correct time. Quiescence occurs when a seed does not germinate because conditions are unfavorable for growth. Germination is triggered by one or more mechanical or environmental cues. To germinate, a plant must imbibe water and draw polysaccharides, fats, and protein nutrients from the endosperm or cotyledons. Dormancy can be broken by mechanical abrasion or fire. Prolonged exposure to water may leach chemical inhibitors away from the seed and induce germination. A plant is considered a seedling when its radicle, the embryonic root, breaks through the seed coat to end germination.

Hormones are chemical signals that act at a distance from the location where they are produced. The effects of the hormones are determined by their relative concentrations, and they play multiple regulatory roles in plants. Interactions among hormones can be complex. Several hormones control the growth and development of the plant; others play roles in defense against herbivores and microorganisms. Photoreceptors detect changes in quality and direction of light, which are important cues for developmental processes.

Genetic screens can be used to analyze signal transduction pathways. Disruption of a pathway in a mutant indicates that the affected gene plays a role in that signaling pathway. *Arabidopsis thaliana* is a major plant genetic model organism. A screen involves creation of a library of random mutations across the genome using mutagens or transposons. The treated plants are examined for a phenotype of interest, and their genotypes are compared with those of wild-type plants.

Question 1. Trace the steps that occur between the planting of a pea seed and the emergence of the pea plant.

Question 2. What is the difference between seed dormancy and seed quiescence?

37.2 What Do Gibberellins and Auxin Do?

Gibberellins have many effects on plant growth and development

Auxin plays a role in differential plant growth

Auxin affects plant growth in several ways

At the molecular level, auxin and gibberellins act similarly

Two important plant hormones were discovered from observations of natural phenomena. Chemicals were isolated that could cause the phenomena, and then mutations eliminating production of the hormones were found with the expected phenotypes. The hormone families of gibberellins and auxins both have similar effects: lack of either hormone causes dwarfism. Gibberellins were discovered by observations of rice plants with a fungal disease that causes them to be overly tall. Gibberellic acid was isolated from the fungus and also discovered in plants themselves. Tomato plant mutants that do not make gibberellic acid are dwarf, and the phenotype can be reversed by application of gibberellic acid. It was observed that seedlings bend toward a light source. When applied to the tips of seedlings, indole-3-acetic acid, the major auxin, mimics the bending effect. *Arabidopsis* plants that do not make auxin are also short, but adding auxin reverses the phenotype.

Gibberellins are involved in stimulating elongation of the plant stem, and inhibitors of gibberellin synthesis cause reduced stem elongation. Controlling these hormones makes it possible to grow plants in specific ways (for instance, limiting the height of chrysanthemums and creating wheat plants that do not fall over from the weight of the grains). Gibberellins also regulate the growth of fruits; they are sprayed on grape vines to increase fruit size. In the developing seed, gibberellins stimulate the aleurone layer, a tissue layer under the seed coat, to secrete enzymes that break down the seed coat. Abscisic acid acts as an antagonistic hormone to gibberellins and maintains seed dormancy.

The discovery of auxin (indole-3-acetic acid) was the result of work by Charles and Francis Darwin. They were interested in how plants grow toward light by phototropism and which part of an emerging plant coleoptile is responsible for sensing light. They found that the tip is the light-receptive portion, but that the growing region (responsible for the bending of the plant to or away from light) is some distance below the receptor region. From this observation they reasoned that some chemical must be transmitted from the tip to the growing region. Additional experiments in which tips were removed and placed on gelatin blocks indicated that a chemical does indeed move from the tip to the growing region. Subsequent experiments by other researchers showed that gelatin exposed to the tips is sufficient to cause altered growth. The chemical was later isolated and determined to be auxin.

Auxin movement in plant tissues is unidirectional and polar, from apex to base. Auxin enters the cell in its nonpolar acid form by passive diffusion, and proton pumps transport H^+ out of the cell, causing the nonpolar auxin to become an anion. Auxin anion efflux carriers are carrier proteins at the basal end of the cell responsible for export of auxin anions from cells, and they contribute to the unidirectional movement of auxin. Redistribution of auxin laterally is responsible for directional plant growth. Asymmetric placement of a coleoptile results in excess growth on that side, even in the absence of light. Auxin carrier proteins are responsible for lateral redistribution of auxin. Auxin at the tip of a coleoptile moves laterally toward the shaded side. This asymmetry is maintained through the apical–basal polar transport of auxin, and thus cell elongation is sped up on the shaded side, bending the plant toward the light. Gravity also affects the distribution of auxin. Growth in a direction determined by gravity is called gravitropism.

Phototropin in the membranes of plants is responsible for the phototropic response by stimulating the transport of auxin to the cells on the shaded side of the plant. The gravitational settling of starch-rich plastids may be the trigger for the release of auxin from the bottom of the root or shoot. Under both conditions, higher auxin concentrations on one side of the plant cause increased rates of growth along that side and lead to bending.

Auxin affects vegetative growth by initiating root growth in cuttings and promoting and maintaining apical dominance (apical buds inhibit the growth of axillary buds). Auxin also inhibits abscisson, the dropping of leaves. Auxin can stimulate unfertilized fruit to form (parthenocarpy). Auxin is involved in cell expansion by acting on cell walls. Cells grow by taking up water. The amount of water that can be taken up is restricted by a rigid cell wall. Cell walls must loosen, stretch, and add polysaccharides and cellulose to maintain structure as the cell expands. According to the acid growth hypothesis of cell expansion, auxin stimulates the production and insertion of proton pumps into the plasma membrane. The subsequent decrease in pH stimulates proteins called expansins to alter polysaccharide bonding so that they slide past one another during expansion.

Scientists have studied mutant plants to understand the mechanisms by which gibberellins and auxin work. The two main mutant types are excessively tall plants and dwarf plants. In the tall plants, the pathway by which the hormones act is always turned on. In the dwarf plants, the pathway by which the hormones act is always off. These differences are due to mutations in a repressor gene of a transcription factor for a growth-promoting gene. In normal plants, the hormone removes this repressor, allowing the growth-promoting gene to be transcribed. In the tall mutants the repressor is never functional, and in the dwarf mutants the repressor is always functional.

Question 3. Both gibberellins and auxin act in a similar fashion at the molecular level. Create a flow chart describing the signal transduction pathways for gibberellins and auxin. Note their similarities and their differences. (Remember that the pathways work by the same mechanisms; the differences are in the actual receptors, repressors, and transcription factors involved in each one.)

Question 4. Discuss how gibberellins were discovered. How did researchers determine that they are chemical in nature?

Question 5. Explain how auxin distribution regulates phototropism and gravitropism.

37.3 What Are the Effects of Cytokinins, Ethylene, and Brassinosteroids?

Cytokinins are active from seed to senescence

Ethylene is a gaseous hormone that hastens leaf senescence and fruit ripening

Brassinosteroids are plant steroid hormones

Plant cells, unlike animal cells, retain the ability to divide after differentiation. Hormones are responsible for stimulating these cells to divide. Early plant cell culture experiments determined that there is a molecule in coconut milk that stimulates plant cell division. It was eventually discovered that a derivative of adenine (kinetin) would produce the effect, and it was termed a cytokinin. Kinetin is not present in plant cells, but a related adenine derivative called zeatin was isolated from corn endosperm. More than 150 different cytokinins have been isolated. Cytokinins are powerful stimulators of cell division, bud formation, and shoot formation; they also aid in seed germination, inhibit stem elongation, and delay leaf senescence. They are synthesized primarily in the roots and are translocated throughout the plant. Cytokinins act through a two-component system similar to that found in bacteria. A receptor (AHK) phosphorylates proteins, and a target transcription factor (ARR) acts as an effector. An intermediate protein (AHP) transfers the phosphate from the receptor to the transcription factor. More than 20 genes are expressed in response to cytokinin signaling.

Ethylene is a gaseous hormone that promotes leaf senescence and fruit ripening. In many instances it is given off by rotting fruit. The use of ethylene spray to promote fruit ripening is a common commercial practice. Ethylene scrubbers are used in fruit storage to prevent ethylene from accumulating and causing fruit to spoil. Silver thiosulfate and 1-methylcyclopropene are used in the flower industry to inhibit ethylene's effects on flower senescence. Ethylene plays a role in maintaining the apical hook on emerging eudicots by inhibiting cells on the inner portion of the hook. Ethylene inhibits stem elongation, promotes lateral swelling of stems, and inhibits sensitivity to gravitropic stimulation. These three responses are known as the triple response. The signal transduction pathway by which ethylene exerts its effects is initiated by the binding of ethylene to a receptor on the endoplasmic reticulum (see Figure 37.14). This binding initiates a cascade, starting with the activation of endoplasmic reticulum channels. Ultimately the activation of a transcription factor promotes the expression of genes, resulting in physiological changes.

Brassinosteroids were originally isolated from a member of the Brassicaceae. They have been shown to stimulate cell elongation, pollen tube elongation, and vascular tissue differentiation, and to inhibit root elongation, all of which are similar to the effects of auxin. The brassinosteroid receptor is found on the cell wall and initiates a transduction pathway that influences gene expression.

Question 6. Describe the hormonal influences affecting a pea plant from the moment it is planted until its emergence.

Question 7. Auxins and cytokinins appear to cancel out the effects of each other. Explain why a hormone that affects bud growth is produced in the roots, while the one that affects root growth is produced in the shoots. How does this relate to the polar distribution of hormones?

Question 8. Genetic engineering has produced fruits that are deficient in the ability to produce ethylene. How is this deficiency useful in the storage and marketing of fruits?

Question 9. Suppose that a scientist is experimenting with tissue culture methods. Pith tissue that was isolated has grown an undifferentiated mass of cells (called a callus), and that tissue has been divided up and placed in the following culture media: (1) necessary nutrients plus indole-3-acetic acid; (2) necessary nutrients plus zeatin (a cytokinin); (3) necessary nutrients plus equivalent concentrations of indole-3-acetic acid and zeatin; (4) necessary nutrients plus an excess of zeatin and minimal indole-3-acetic acid. Explain what will happen to the mass of tissue under each condition.

37.4 How Do Photoreceptors Participate in Plant Growth Regulation?

Phototropins, cryptochromes, and zeaxanthin are blue-light receptors

Phytochromes mediate the effects of red and far-red light

Phytochrome stimulates gene transcription

Circadian rhythms are entrained by light reception

Photoreceptors in plants interpret intensity, duration, and wavelength of light. Light regulates a wide variety of plant processes, including germination, flower production, and shoot elongation. Three or more blue-light receptors mediate the effects of high-intensity blue light, and phytochrome mediates the effect of red light. Phototropin, a blue-light receptor, is a protein kinase that stimulates cell elongation by auxin. Phototropin is activated when a flavin mononucleotide associated with the protein absorbs blue-light energy and causes a shape change in the protein. The activated protein kinase initiates a signal transduction pathway that results in the stimulation of stem elongation by auxin. Zeaxanthin and phototropin act together to regulate light-stimulated opening of the stomata. Cryptochromes absorb blue and ultraviolet light and influence seedling development and flowering.

Red light stimulates photomorphogenesis in plants. These developmental and physiological events can include germination, flowering, and production of chlorophyll in seedlings. Exposure to far-red light reverses the effects of exposure to red light, and vice-versa. This "switching" occurs because phytochrome can be shifted from the P_r form to the P_{fr} form upon absorption of the correct wavelength of light. Phytochromes are composed of a protein chain that interacts with transcription factors and a pigment chromophore. Red light changes the conformation of the protein from the P_r to P_{fr} form and exposes a nuclear localization sequence. The P_{fr} form then moves into the nucleus, where it stimulates gene expression through a transcription factor. It can also act as a kinase and phosphorylate other proteins. When exposed to far-red light, P_{fr} returns to the P_r conformation. The ratio of P_r to P_{fr} form present determines whether a phytochrome-mediated response will occur.

Biological organisms exhibit a daily cycle in their functioning known as a circadian rhythm. Two characteristics define a circadian rhythm: the period (length of one cycle) and the amplitude (magnitude of the change through a cycle). The phytochromes are probably involved in the circadian rhythms of plants.

Question 10. You have produced a plant with a mutation that has only one isoform for the photoreceptor phytochrome. This plant flowers continuously and develops more shoots than your wild-type control plants. What would happen if you were to expose this plant to either red light or far-red light? What isoform is this plant producing and how do you know this?

Question 11. Design an experiment to test how long circadian rhythms can persist in bean plants.

Test Yourself

1. Plant growth is regulated by
   a. environmental cues.
   b. hormones.
   c. signal transduction pathways.
   d. the expression of the plant's genome.
   e. All of the above
   Textbook Reference: *37.1 How Does Plant Development Proceed?*

2. The mechanism by which gibberellins act involves
   a. the adding of proton pumps to the plasma membrane.
   b. the phosphorylation of proteins.
   c. binding with a transcription factor.
   d. the removal of a repressor from a transcription factor.
   e. None of the above
   Textbook Reference: *37.2 What Do Gibberellins and Auxin Do?*

3. Which of the following triggers germination of a seed?
   a. The imbibing of water
   b. Its release from the fruit
   c. Chemical changes
   d. The exclusion of water
   e. All of the above
   Textbook Reference: *37.1 How Does Plant Development Proceed?*

4. Which of the following may have the effect of breaking dormancy in seeds?
   a. Penetration of the seed coat
   b. Leaching of inhibitory compounds by water
   c. Exposure to fire
   d. Passing through an animal's digestive tract
   e. All of the above
   Textbook Reference: *37.1 How Does Plant Development Proceed?*

5. Which of the following hormones is responsible for breaking winter dormancy of buds in the spring in deciduous trees?
   a. Auxins
   b. Cytokinins
   c. Gibberellins
   d. Ethylene

e. Brassinosteroids
Textbook Reference: *37.1 How Does Plant Development Proceed?*

6. Which of the following is *not* involved in the acid growth hypothesis for the regulation of cell expansion by auxin?
a. The pumping of protons into the cell wall
b. The pumping of protons into the cytosol
c. Increased gene expression of the proton pump gene
d. Increased insertion of proton pumps into the plasma membrane
e. All of the above are involved.
Textbook Reference: *37.2 What Do Gibberellins and Auxin Do?*

7. Which of the following is *not* involved in the polar transport of auxin?
a. Diffusion across a plasma membrane
b. Membrane protein asymmetry of auxin transport carriers
c. Proton pumping from the cytosol
d. Ionization of auxin as a weak acid
e. All of the above are involved.
Textbook Reference: *37.2 What Do Gibberellins and Auxin Do?*

8. A homeowner has installed an outdoor gas-burning grill on her back patio next to her favorite camellia bush. After the first few nights of using the grill, she notices that the camellia is beginning to lose its leaves. Which of the following is the best explanation for what is happening?
a. The bush is getting too warm next to the grill.
b. Ethylene is a by-product of the burning gas and is causing senescence in the plant.
c. Abscisic acid is a by-product of the burning gas and is causing senescence in the plant.
d. The plant is a biennial and is bolting.
e. Auxin production is being inhibited.
Textbook Reference: *37.3 What Are the Effects of Cytokinins, Ethylene, and Brassinosteroids?*

9. Cytokinins interact with which other hormone?
a. Ethylene
b. Abscisic acid
c. Gibberellins
d. Auxins
e. Brassinosteroids
Textbook Reference: *37.3 What Are the Effects of Cytokinins, Ethylene, and Brassinosteroids?*

10. Which of the following light receptors is responsible for absorbing both blue and ultraviolet light?
a. Phytochrome P_r
b. Phytochrome P_{fr}
c. Cryptochrome
d. Phototropin
e. Zeaxanthin
Textbook Reference: *37.4 How Do Photoreceptors Participate in Plant Growth Regulation?*

11. Etiolated seedlings are produced by germinating seeds that are kept in total darkness. Plants that are kept in the dark will begin to synthesize chlorophyll after they are given a pulse of
a. blue light.
b. red light.
c. red light followed by a pulse of far-red light.
d. far-red light followed by a pulse of red light.
e. ultraviolet light.
Textbook Reference: *37.4 How Do Photoreceptors Participate in Plant Growth Regulation?*

12. Ethylene is produced by which part of a plant?
a. The seedling
b. The leaves
c. The fruit
d. All of the above
e. None of the above
Textbook Reference: *37.3 What Are the Effects of Cytokinins, Ethylene, and Brassinosteroids?*

13. Auxin transport within a plant is said to be _______ and is dependent on the action of _______ pumps.
a. polar; proton
b. nonpolar; potassium
c. polar; potassium
d. bidirectional; proton
e. polar; sodium
Textbook Reference: *37.2 What Do Gibberellins and Auxin Do?*

14. Which of the following is *not* initiated by auxin?
a. Stimulation of root initiation
b. Inhibition of leaf abscission
c. Stimulation of leaf abscission
d. Maintenance of apical dominance
e. Auxin is involved in all of the above.
Textbook Reference: *37.2 What Do Gibberellins and Auxin Do?*

15. Red light activation of the phytochrome into the P_{fr} state leads to which of the following events?
a. Inhibition of chlorophyll
b. Leaf expansion
c. Hook folding
d. Hook unfolding
e. Both b and d
Textbook Reference: *37.4 How Do Photoreceptors Participate in Plant Growth Regulation?*

Answers

Key Concept Review

1. After the pea seed is planted, watering of the seed promotes the leaching of germination inhibitors. The

seed also begins to imbibe water, resulting in metabolic changes. DNA synthesis is halted until the radicle emerges from the seed coat. Breakdown of starch and protein reserves in the cotyledons and the endosperm begins to provide nutrients for the developing embryo. The apical hook begins to push through the soil. Upon its emergence and exposure to light, chlorophyll synthesis begins.

2. Seed dormancy is the state in which a seed is prevented from germinating even when conditions are favorable. Seed quiescence is the state in which a seed cannot germinate, even when dormancy is broken, due to unfavorable environmental conditions.

3.

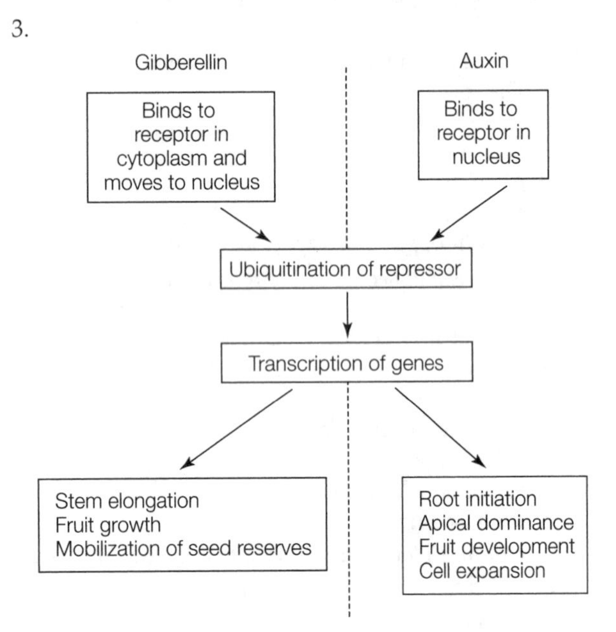

4. The first gibberellin was isolated from a fungus that infects rice plants and causes them to grow tall and spindly. Initial experiments indicated that the medium in which the fungus was grown was sufficient to cause this effect; therefore, the agent that caused the phenomenon had to be a chemical produced by the fungus.

5. Lateral distribution of auxin controls phototropism and gravitropism. Auxin accumulates in the shaded portions of a stem and stimulates cell growth. This uneven cell growth results in a bending of the stem toward the light. The same mechanism works in response to gravity. An accumulation of auxin occurs where the gravitational pull is the strongest. Cells grow in response to auxin and stems bend upward, away from the gravitational force.

6. Dormancy of the seed is maintained by abscisic acid until it is leached from the seed by water. Once water is imbibed, cytokinins begin to influence germination. Gibberellins assist with the mobilization of storage products to the growing embryo. Ratios of auxins and cytokinins balance root and bud formation. The apical hook is maintained by ethylene.

7. Roots need shoots, and vice-versa. Increased root development requires increased shoot development for photosynthesis. Regulation of the development of one by the other keeps growth in tandem. Because the distribution of the molecules is polar, a concentration gradient can be established. It is this relative gradient that controls development.

8. Fruits can be kept from ripening until they reach their destination. Once at market, they can be sprayed with ethylene to stimulate ripening. The result is fruit that can be shipped more easily yet can be ripe at any time in the market.

9. The results of the different treatments are the following: (1) roots will develop from the callus; (2) buds will develop from the callus; (3) both roots and buds will develop from the callus; (4) buds will develop from the callus.

10. In a plant there are two isoforms of phytochrome that respond to red light and far-red light. Because the mutant plant only has one isoform for the phytochrome, exposure either to one type of light or the other will have no effect. The isoform that responds to far-red light is the active isoform that stimulates flowering and shoot development. Because the plant flowers continuously and has more shoots than the control, its phytochrome must be the active isoform.

11. One manifestation of circadian rhythm in bean plants is the raising and lowering of its leaves. Therefore, you could test how long this rhythm is maintained by entraining bean plants to a particular light/dark cycle, and then placing the plants in darkness. Observation of the movements of the leaves would indicate how long that rhythm persists without light cues.

Test Yourself

1. **e.** The growth of a plant is regulated by all of these factors.

2. **d.** Gibberellin brings about a response in the plant cell by removing the repressor from a transcription factor, thus allowing transcription to occur.

3. **a.** The uptake of water by a seed begins the processes that lead to seed germination.

4. **e.** Mechanical abrasion, leaching of inhibitors by water, exposure to fire, and passing through a digestive tract may all trigger germination. Actual germination cannot begin until a seed imbibes water.

5. **c.** Gibberellins are responsible for breaking winter dormancy of buds in deciduous trees.

6. **b.** The acid growth hypothesis involves an increase in the number of proton pumps in the plasma membrane, increased gene expression of the proton pump gene, and the pumping of protons into the cell wall.

7. **e.** All of the characteristics ensure that the transport of auxin is polar.

8. **b.** Ethylene gas promotes senescence and is one of the by-products of burning gas.
9. **d.** Cytokinins interact with auxins. The auxin-to-cytokinin ratio controls the bushiness of plants.
10. **c.** Cryptochromes respond to blue and ultraviolet light wavelengths.
11. **b.** The pulse of red light converts P_r to P_{fr}. Subsequent pulses of far-red light reverse the effects of red light pulses.
12. **d.** Ethylene can be produced by the fruit, seedling, or leaves in most plants.
13. **a.** Auxin transport is a polar process, meaning that it moves in only one direction with the help of proton pumps and auxin anion efflux carriers (see Figure 37.7).
14. **c.** Auxin inhibits leaf abscission rather than stimulating it. Auxin is involved in all of the other processes.
15. **e.** The P_{fr} phytochrome stimulates chlorophyll synthesis, hook unfolding, and leaf expansion.

38 Reproduction in Flowering Plants

The Big Picture

- Though plants can reproduce both asexually and sexually, maintenance of genetic variability depends on sexual reproduction. The flower is the basis of sexual reproduction in plants. The flower not only produces the necessary gametes, but it is also integrally involved in ensuring pollination. Eggs are produced through megagametogenesis, and pollen is produced through microgametogenesis. Angiosperms exhibit double fertilization, which results in a diploid embryo and a triploid endosperm. The seed protects the embryo, and in many cases a fruit is formed for protection until maturation and to aid in dispersal of the mature seed.
- Hormones and signaling cascades are involved in a plant's transition from the vegetative state to the reproductive state. Meristem identity genes and floral organ identity genes encode genes necessary for flowering. The photoperiod sets the shoot apical meristem on the path to flowering. Some plants are short-day plants and flower when the day is shorter than a critical maximum while others are long-day plants and flower once the day length reaches a critical minimum. The actual key for photoperiod is the length of the night (continuous darkness), and not the day length. The gene FT codes for florigen and is involved in photoperiod signaling. Plants may flower, set seed, and die in one growing season or two growing seasons, or they may continue to do so for even longer periods of time.

Study Strategies

- Megagametogenesis is probably the most difficult concept in this chapter, but it is relatively simple to understand if you look at the source of each cell.
- Recall that flowering, like all other plant processes, is a result of environmental signals, receptors, enzyme cascades mediated by hormones, and alterations in gene expression.
- Refer to the figures in the textbook to understand flower structure, megagametogenesis, microgametogenesis, and fertilization.
- Spend some time thinking about the advantages and disadvantages of sexual and asexual reproduction, as related questions are often given in exams.
- Be sure you understand the particular features of double fertilization.
- Go to the Web addresses shown to review the following animated tutorials and activity. You can also find them in BioPortal (yourBioPortal.com), along with many additional learning resources.

  Animated Tutorial 38.1 Double Fertilization (Life10e.com/at38.1)

  Animated Tutorial 38.2 The Effect of Interrupted Days and Nights (Life10e.com/at38.2)

  Activity 38.1 Sexual Reproduction in Angiosperms (Life10e.com/ac38.1)

Key Concept Review

38.1 How Do Angiosperms Reproduce Sexually?

The flower is an angiosperm's structure for sexual reproduction

Flowering plants have microscopic gametophytes

Pollination in the absence of water is an evolutionary adaptation

A pollen tube delivers sperm cells to the embryo sac

Many flowering plants control pollination or pollen tube growth to prevent inbreeding

Angiosperms perform double fertilization

Embryos develop within seeds contained in fruits

Seed development is under hormonal control

Sexual reproduction confers an evolutionary advantage because it encourages genetic diversity within a population. Sexual reproduction in plants is significantly different from that of animals. Plant sexual reproduction involves the alternation of multicellular diploid and haploid generations, whereas there is no multicellular haploid stage in animals. Meiosis in plants produces spores, in which mitosis produces gametes; in animals, meiosis usually produces gametes directly. The cells that will form gametes in a plant are determined in the adult, whereas in an animal the germline cells are determined before birth. In angiosperms, the plant that we see is a sporophyte, and the gametophytes are contained within the flowers. A complete flower includes four groups of organs—carpels, stamens, petals, and sepals—all of which are modified leaves. The stamen and carpel are the male and female parts of the flower, respectively. Flowers with both a carpel and stamen are known as "perfect." "Imperfect" flowers contain either the carpel or stamen, but not both, and thus are either female or male. Monoecious plants have both male and female flowers on each individual plant. Dioecious plants have either male or female flowers on a given individual, but not both.

The flower produces haploid spores that develop into gametophytes. The female gametophytes, called embryo sacs, develop in the ovule. One or more ovules are contained in the ovary, the lower part of the carpel. Male gametophytes, called pollen grains, develop in the anther, which is part of the stamen. Within the ovule, a megasporocyte produces four haploid megaspores through meiosis. Only one of these megaspores survives, and it divides mitotically to produce eight nuclei within a single large cell. The nuclei migrate to either end of the cell, but two remain in the middle. Cell walls form, isolating the three nuclei at either end into individual cells; the two nuclei in the middle remain together in one cell. At one end, the three cells become two synergid cells and one egg cell; at the other are three antipodal cells, which eventually degenerate. In the large central cell are the two polar nuclei. This seven-celled embryo sac is the megagametophyte (see Figure 38.2). Pollen grains develop when a microsporocyte undergoes meiosis. All products of meiosis are retained and undergo mitosis to form a two-celled pollen grain composed of the tube cell and the generative cell. Further development is halted until after the pollen is transferred from the anther to the stigma during pollination. After pollination, the generative cell divides to form two sperm cells and the tube cell forms the pollen tube that delivers the sperm to the embryo sac.

Pollination provides a means for fertilization to occur without water, allowing for the evolution of plants on land. Pollen grains are carried to female flowers via wind, animals, or other vectors. When a pollen grain lands on a stigma, it germinates. Germination begins when the pollen grain begins to take up water from the stigma. Chemical signals, in the form of small proteins from synergids within the ovule, direct the growth of the pollen tube. Some plants are able to self-fertilize within the same flower. However, most plants prevent self-fertilization by separation in either space or time of male and female gametophytes. Plants can also prevent self-fertilization by means of genetic self-incompatibility. A single gene, the *S* gene, regulates self-incompatibility and prevents self-fertilization in many plants. If the pollen expresses an *S* allele that matches the stigma, then the pollen is rejected. Rejected pollen either fails to germinate or fails to produce a pollen tube.

The pollen grain consists of two cells at the time of pollination—the tube cell and the generative cell. The tube cell controls the growth of the pollen tube. As the tube is growing, the generative cell undergoes meiosis and produces two haploid sperm cells. Once the pollen tube enters the embryo sac, the two sperm cells are released into the remains of a degenerated synergid. One nucleus fuses with the egg cell, the other with the polar nuclei of the central cell. This results in the zygote ($2n$) and a triploid ($3n$) cell that becomes the endosperm. All other cells disintegrate as the embryo begins to develop. Double fertilization refers to these two fusion events. The second fertilization, of the sperm with the central cell to form a triploid cell, is a defining characteristic of angiosperms (see Figure 38.6).

Fertilization begins the growth and development of the embryo, endosperm, integuments, and carpel. Ultimately, a seed coat develops from the integuments and protects the dormant embryo. As the embryo and seed are developing, the ovary begins to form the fruit. Other parts of the flower and plant may be included in the fruit, but to be considered a fruit, only the ovary wall and the seed need be involved. The fruit disperses by various means, including by traveling on the coats of animals or by being consumed and later deposited. The fruit also protects the seed from animals and microbial diseases. Abscisic acid regulates the development of the seed and the production of proteins that will protect against desiccation. Abscisic acid also inhibits germination of the seed on the plant. The condition of premature germination, in which a seed germinates while still on the plant, is called vivipary. The general effect of abscisic acid in seeds is to extend dormancy.

Question 1. Outline egg formation in angiosperms, beginning with the sporophyte.

Question 2. It takes a great deal of effort to detassel corn (removing male flowers). Why can't corn plants simply be sprayed with a meiosis inhibitor to halt pollen production? (Hint: Male and female flowers occur on the same corn plant.)

Question 3. Describe double fertilization. What is the ploidy level of the products of double fertilization?

Question 4. Define a fruit. Which of the following are fruits: tomato, pear, potato, banana, cucumber, snow pea, peanut, sunflower seed?

Question 5. Identify the following structures as occurring in the sporophyte or gametophyte generation.

a. Embryo sac __________
b. Antipodal cells __________
c. Polar nuclei __________
d. Integument __________
e. Receptacle __________
f. Ovary __________
g. Anther __________
h. Pollen grain __________

38.2 What Determines the Transition from the Vegetative to the Flowering State?

Shoot apical meristems can become inflorescence meristems

A cascade of gene expression leads to flowering

Photoperiodic cues can initiate flowering

Plants vary in their responses to photoperiodic cues

Night length is a key photoperiodic cue that determines flowering

The flowering stimulus originates in a leaf

Florigen is a small protein

Flowering can be induced by temperature or gibberellin

Some plants do not require an environmental cue to flower

The beginning of flowering involves reallocation of the energy in the plant away from leaves and stem and toward flowers and gametes. Plants fall within three different life cycle patterns. Annual plants go from seed to seed set and die within one growing season. Biennial plants require one vegetative growing season before reproducing. Perennial plants flower repeatedly and live for many years.

Flowers may appear on a plant individually or in a cluster called an inflorescence. The first transition from vegetative growth to floral production is the transition of the apical meristem into an inflorescence meristem. The inflorescence meristem can produce floral meristems. The floral meristem differs from the apical meristem in that growth is determinate rather than indeterminate. The floral meristem is programmed to produce four consecutive whorls of flower organs. A gene cascade leads to flower formation, which begins with the activation of a set of meristem identity genes. Two genes important in switching from vegetative growth to reproductive growth are *LEAFY* and *APETALA1*. The expression of floral organ identity genes, such as *AGAMOUS*, specify successive whorls of floral organs.

The gene cascade leading to flower development is controlled by environmental cues. Seasonal flowering is a result of changes in photoperiod, the relative lengths of light and dark. Control of a plant's responses by length of day or night is called photoperiodism. Each plant type has a critical day length corresponding to the light availability that induces flowering. Short-day plants flower when the amount of available light is less than the critical maximum. Long-day plants flower only when the day is longer (more light available) than the critical minimum. Experiments have shown that plants actually respond to the length of the night, rather than the amount of daylight. In experiments in which daylight was interrupted, there was little effect on flowering, but in experiments in which dark was interrupted, there were significant effects on flowering. Some plants require complex combinations of day lengths. Plants, like all other organisms, have an internal mechanism for measuring the length of continuous dark periods. The duration of dark periods appears to be detected by special phytochromes that detect red light. *CONSTANS* (*CO*) is a gene whose expression follows a circadian rhythm; *CO* encodes a transcriptional regulator of flowering genes.

The stimulus for flowering comes from the leaf of the plant. Florigen (FT), a small protein that can travel through plasmodesmata, is the hormone responsible for flowering. The gene *FLOWERING LOCUS T* codes for the FT protein. High levels of FT induce the plant to flower. The gene *CONSTANS* codes for a transcription factor (CO) that stimulates FT synthesis in the phloem cells. *FLOWERING LOCUS D* is a gene that codes for a transcription factor (FD) in the apical meristem that increases transcription of *APETALA1*, a meristem identity gene. Transgenic plant studies have provided strong evidence that FT controls flowering. Vernalization is the induction of flowering by low temperatures. A transcription factor (FLC) encoded by *FLOWERING LOCUS C* inhibits the FT pathway described above. Cold temperatures decrease the production of the FLC protein, leading to a functional FT pathway. Gibberellin can also stimulate a plant to flower through the actions of the meristem identity gene *LEAFY*. Those plants that do not require an environmental cue to flower rely on an "internal clock" to trigger flowering. This most likely works through relative changes in FLC concentrations in the plant, stimulating the FT–FD pathway described above.

Question 6. Florigen is a plant hormone actively involved in the initiation of flower formation. Draw a flow chart showing its mechanism of action, including the interaction of the three genes involved in initiating flower formation.

Question 7. Flowering is stimulated when light sets off a gene cascade. Explain how this may be hormonally controlled.

38.3 How Do Angiosperms Reproduce Asexually?

Many forms of asexual reproduction exist

Vegetative reproduction has a disadvantage

Vegetative reproduction is important in agriculture

Asexual reproduction accounts for many of the plants on Earth. Asexual reproduction does not involve genetic recombination. During the process of vegetative reproduction, asexual reproduction occurs through the modification of a vegetative organ. Stolons, tubers, rhizomes, bulbs, corms, and suckers are all modifications of stems or roots that allow vegetative reproduction. Some plants, such as dandelions, reproduce asexually through seeds in a process called apomixis. Apomictic plants skip over meiosis and fertilization and produce diploid seeds with a genetic makeup that is identical to that of the maternal plant. The major disadvantage of vegetative reproduction is that if the environment changes significantly, the plant population may not be able to adapt to the new environment and thus disappear from an area.

Vegetative reproduction is used in agriculture. Cuttings are commonly used in horticulture. Grafting is widely used in fruit crops in which the root-containing stock is grafted to a scion (shoot system). This allows for a hardy root stock to be combined with a good but often fragile fruit producer. Meristem culture (production of entire new plants from pieces of shoot apical meristem) allows for production of virus-free plants and/or a uniform population of plants.

Question 8. Compare and contrast asexual and sexual reproduction in plants. In which category does self-fertilization belong?

Question 9. When is asexual reproduction beneficial to plants? Under what conditions is sexual reproduction beneficial?

Test Yourself

1. You manage a greenhouse that produces roses for Valentine's Day. Roses normally bloom in June. Which of the following will most likely be the best lighting schedule for your roses?
   a. 16 hours of light, followed by 8 hours of interrupted dark
   b. 16 hours of light, followed by 8 hours of uninterrupted dark
   c. 10 hours of light, followed by 14 hours of uninterrupted dark
   d. 10 hours of light, followed by 14 hours of interrupted dark
   e. None of the above
   Textbook Reference: *38.2 What Determines the Transition from the Vegetative to the Flowering State?*

2. After setting the correct photoperiod in your greenhouse, you still do not have blooming roses. Which of the following might have contributed to the problem?
   a. The heating system allowed for fluctuations in temperature between 20°C and 25°C.
   b. The furnace mechanic accidentally turned off the lights for an hour two days in a row.
   c. The cleaning crew turned the lights on for an hour three nights in a row.
   d. All of the above
   e. None of the above
   Textbook Reference: *38.2 What Determines the Transition from the Vegetative to the Flowering State?*

3. You have moved into a new house. During the first summer you notice that many of the plants in the yard do not bloom. During the second summer the yard is a sea of blooms. It is now spring of the third year, and there are no plants. Which of the following is the most likely explanation?
   a. The plants are annuals.
   b. The plants are biennials.
   c. The plants are perennials.
   d. The plants are being affected by drought.
   e. The nights are too long for the plants.
   Textbook Reference: *38.2 What Determines the Transition from the Vegetative to the Flowering State?*

4. You notice that a new houseplant sends out long stems with what look like "little plants" attached. You allow one of these to rest in a cup of water and note that roots form. You have observed an example of
   a. asexual reproduction.
   b. apomixis.
   c. heterospory.
   d. parthenogenesis.
   e. vivipary.
   Textbook Reference: *38.3 How Do Angiosperms Reproduce Asexually?*

5. Self-pollination in plants that produce both pollen and eggs is prevented by
   a. self-incompatibility genes.
   b. physical barriers.
   c. production of pollen and eggs at different times.
   d. All of the above
   e. None of the above
   Textbook Reference: *38.1 How Do Angiosperms Reproduce Sexually?*

6. Which of the following about the horticultural practice of grafting is false?
   a. The piece grafted on is called the "scion."
   b. The host plant is called the "stock."
   c. It involves attaching a piece of a plant to the root or root-bearing stem of another plant.
   d. Another term for grafting is meristem culture.
   e. Once two plants are grafted, a continuous cambium is formed.
   Textbook Reference: *38.3 How Do Angiosperms Reproduce Asexually?*

7. The induction of flowering by means of exposure to low temperature is called
   a. vernalization.
   b. frigidation.
   c. apomixis.
   d. vivipary.
   e. None of the above
   Textbook Reference: *38.2 What Determines the Transition from the Vegetative to the Flowering State?*

8. The *LEAFY* and *APETALA1* genes are examples of _______ genes.
   a. floral organ identity
   b. meristem identity
   c. viviparity
   d. photoperiod
   e. inflorescence
   Textbook Reference: *38.2 What Determines the Transition from the Vegetative to the Flowering State?*

9. The production of seeds without fertilization is called
   a. apomixis.
   b. parthenogenesis.
   c. conception.
   d. circadian rhythm.
   e. vernalization.
   Textbook Reference: *38.3 How Do Angiosperms Reproduce Asexually?*

10. In the transition from vegetative growth to floral growth, the _______ must be transformed into the _______. This involves a shift from _______ growth to _______ growth.
    a. apical meristem; floral meristem; indeterminate; determinate

b. lateral meristem; floral meristem; indeterminate; determinate
c. apical meristem; floral meristem; determinate; indeterminate
d. apical cambium; floral cambium; determinate; indeterminate
e. floral meristem; apical cambium; determinate; indeterminate

Textbook Reference: *38.2 What Determines the Transition from the Vegetative to the Flowering State?*

11. Which of the following best describes the fate of the generative cell of the pollen grain?
    a. It coordinates growth of the pollen tube.
    b. It divides by meiosis to produce two sperm nuclei.
    c. It divides by mitosis to produce two sperm nuclei.
    d. It forms the pollen tube.
    e. None of the above

    Textbook Reference: *38.1 How Do Angiosperms Reproduce Sexually?*

12. Which of the following is *not* part of a megagametophyte?
    a. Pollen grain
    b. Synergids
    c. Antipodal cells
    d. Polar nuclei
    e. All of the above are part of the megagametophyte.

    Textbook Reference: *38.1 How Do Angiosperms Reproduce Sexually?*

13. During double fertilization two sperm cells fuse (one each) with
    a. an egg cell.
    b. the two polar nuclei.
    c. a pollen grain.
    d. Both a and b
    e. Both a and c

    Textbook Reference: *38.1 How Do Angiosperms Reproduce Sexually?*

14. Which of the following statements about plants and photoperiodic cues is true?
    a. Short-day plants flower when the day is longer than a critical maximum.
    b. Short-day plants only reproduce asexually.
    c. Long-day plants flower when the dark night is longer than a critical maximum.
    d. Long-day plants flower when the day is longer than a critical maximum.
    e. Both a and c

    Textbook Reference: *38.2 What Determines the Transition from the Vegetative to the Flowering State?*

15. Which of the following is *not* a part of the flower?
    a. Carpels
    b. Petals
    c. Sieve tube
    d. Stamens
    e. Sepals

    Textbook Reference: *38.1 How Do Angiosperms Reproduce Sexually?*

Answers

Key Concept Review

1. See Figure 38.2 in the textbook.
2. Meiosis occurs in the megasporocyte as well as the microsporocyte. Inhibition of pollen formation via a meiosis inhibitor would not be useful because it would also inhibit egg formation.
3. Double fertilization occurs when the two sperm nuclei produced from the generative cell of the pollen grain unite with the egg nucleus and two polar nuclei, respectively. The resulting zygote is diploid and the endosperm is triploid.
4. A fruit is the seed and the ovary wall, and it may include other structures of a flowering plant. With the exception of the potato, all of the listed structures are fruits.
5.

a. Embryo sac:	gametophyte
b. Antipodal cells:	gametophyte
c. Polar nuclei:	gametophyte
d. Integument:	sporophyte
e. Receptacle:	sporophyte
f. Ovary:	sporophyte
g. Anther:	sporophyte
h. Pollen grain:	gametophyte

6.

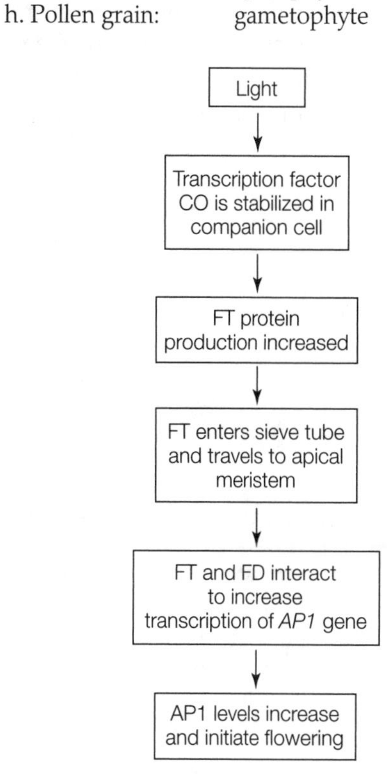

7. Light induction begins with the leaves. A single isolated leaf may stimulate floral production throughout the plant. In experiments in which leaves from one plant were grafted onto other plants, an "induced" leaf caused a plant to flower even though the plant had not been exposed to the required amount of darkness. Therefore, the leaf must send a signal (the protein florigen) that begins the gene cascade leading to flower formation.
8. Asexual reproduction does not involve meiosis, fertilization, or genetic recombination. The offspring from asexual reproduction are genetically identical to the parent plant. Sexual reproduction requires a meiotic event and fertilization. Self-fertilization is sexual reproduction, even though genetic recombination is limited. This is because a meiotic event occurs, and fertilization is necessary for reproduction to take place.
9. Asexual reproduction is beneficial in stable environments in which many genetically identical plants are sustainable. This process can help colonize a habitat or help a population of plants to spread. The disadvantage is lack of genetic diversity, which can be detrimental if the environment changes rapidly and adaptability of the population is required.

Test Yourself

1. **b.** Flowering is regulated by darkness. Interruptions in darkness can prevent flowering.
2. **c.** The interruptions in the dark cycle by the cleaning crew could have affected the signals that tell the plant to begin flowering.
3. **b.** The plants are most likely biennials that spend one season growing vegetatively and one season producing flowers before dying.
4. **a.** The runner (stolon) is propagating this plant asexually.
5. **d.** Some plants have self-incompatibility genes that prevent the growth of the pollen tube through the style. Others have mechanical barriers to their own pollen. Still others produce eggs and pollen at different times.
6. **d.** Meristem culture involves culturing pieces of shoot meristem on growth media to produce platelets, which are then planted in the field.
7. **a.** Vernalization, which is noted in winter wheat, is used agriculturally to grow high-yielding wheat.
8. **b.** These two genes are meristem identity genes that code for proteins involved in the initiation of flower formation.
9. **a.** Dandelions and other plants produce seeds by apomixis, without meiosis and fertilization. These seeds are genetically identical to the parent plant.
10. **a.** Apical meristems and floral meristems differ in that growth from the apical meristem is indeterminate and growth from the floral meristem is determinate, leading to four whorls of floral structures.
11. **c.** The two sperm nuclei involved in double fertilization are derived from the haploid generative cell of the pollen grain via mitosis.
12. **a.** The megagametophyte is the female gametophyte. It is initially called the embryo sac and contains three antipodal cells, two synergid cells, and two polar nuclei at the seven-cell stage.
13. **d.** Double fertilization involves the fertilization of the egg cell and the two polar nuclei by two sperm cells.
14. **d.** The cue for flowering is the length of the night, so long-day plants flower when the day is longer than a critical maximum and when the night is shorter than a critical minimum.
15. **c.** The carpels, petals, stamens and sepals are all part of the flower.

39 Plant Responses to Environmental Challenges

The Big Picture

- Plants must respond to invasion by pathogens, physical damage by natural events and herbivory, variable water supplies, temperature fluctuations, and variable soil conditions. Natural selection has made plants uniquely suited to the areas they occupy. Some thrive in very harsh environments, but others cannot.
- Interactions between plants and pathogens stimulate a series of chemical changes that ward off further infection. Plant strategies to isolate pathogens include sealing plasmodesmata, releasing proteins that interact with pathogens, and signaling other parts of the plant.
- Herbivory can be detrimental to plants. In some cases, a plant produces chemicals to prevent herbivory and mounts defenses around the damaged tissues.
- Specific adaptations allow plants to withstand adverse conditions. Saline environments, water loss, and high temperatures can be survived if water uptake is maximized and water loss is minimized. Sunken stomata, reduction of surface area, and increased water potential of cells are adaptations to these conditions. Extremes in temperature are survived by means of heat shock proteins that stabilize metabolic proteins during temperature extremes. Heavy metal tolerance includes the ability to store the metals and avoid their toxic effects.

Study Strategies

- Gene-for-gene resistance can be difficult to understand, but remember that in any pathogen situation, the pathogen and the host are continually interacting; therefore, over time, natural selection has allowed interplay among the genomes.
- There are many ways plants deal with pathogens and herbivory. This chapter provides many examples, but these are not exhaustive.
- The best strategy is to study the various components of this chapter one at a time so that all of the defenses/modifications do not seem to run together. Focus first on pathogen interaction, then herbivory resistance, and so on.
- Think about how form follows function as you look at how plants are adapted to specific harsh environments.
- Go to the Web addresses shown to review the following animated tutorial and activity. You can also find them in BioPortal (yourBioPortal.com), along with many additional learning resources.

  Animated Tutorial 39.1 Signaling Between Plants and Pathogens (Life10e.com/at39.1)

  Activity 39.1 Concept Matching (Life10e.com/ac39.1)

Key Concept Review

39.1 How Do Plants Deal with Pathogens?

Physical barriers form constitutive defenses

Plants can seal off infected parts to limit damage

General and specific immunity both involve multiple responses

Specific immunity involves gene-for-gene resistance

Specific immunity usually leads to the hypersensitive response

Systemic acquired resistance is a form of long-term immunity

There are thousands of plant diseases that can be caused by pathogens of many different types, including bacteria, fungi, protists, nematodes, and viruses. Plants have many different responses to fight off attacking pathogens. Plants and pathogens evolve together, with pathogens evolving new ways to attack plants and plants evolving new defenses for these attacks. Plants have mechanical and chemical defenses, which can be constitutive (always present) or induced (produced in reaction to an attack). The first line of defense is the prevention of entry by pathogens. Cutin, suberin, and waxes cover the epidermis of the plant to exclude pathogens. Plants cannot repair damaged and invaded tissues, so they seal them off to prevent systemic infection. Upon recognition of a pathogen, additional polymers are deposited in the cell wall and seal off the plasmodesmata to form a barrier between cells. Subsequently, lignin is deposited, which reinforces the barrier. As a bonus, lignin precursors are toxic to some pathogens.

Induced responses are controlled by receptors on plant cells. Various molecules have been identified as elicitors—molecules that induce a response by a plant. The plant responses to these elicitors are the "plant immune system." There are two forms of immunity, general and specific. General immunity is an overall response triggered by general elicitors called pathogen associated molecular patterns (PAMPs). PAMPs are recognized by pattern recognition receptors on plant cells, which activate signal transduction pathways (PTI, PAMP-triggered immunity). Specific elicitors called effectors trigger specific immunity (ETI, effector-triggered immunity). Effectors include many different pathogen-produced molecules; when these molecules enter the cell, they bind to cytoplasmic receptors called R proteins that trigger specific immunity. The signaling pathways lead to various responses, including formation of reactive oxygen and nitric oxide, callose deposition, hormone signaling, and changes in gene expression. Phytoalexins are nonspecific antifungal and antibacterial compounds produced by plants in response to pathogens. One such compound is camalexin, which appears to function by disrupting the cell membranes of invaders; it is induced by signaling pathways of either general or specific immunity that upregulate genes that convert tryptophan to camalexin. Pathogenesis-related proteins (PR proteins) include enzymes that function to break down the walls of pathogens, antifungal peptides, and molecules that signal a warning to other cells in the plant.

Gene-for-gene resistance is a highly specific type of resistance that is a response to evolution of effectors that disable plant immune defenses. Intracellular receptors (R proteins) recognize specific effectors and activate signal transduction pathways of the specific immune response. R proteins are encoded by resistance (*R*) genes. Pathogens have associated *Avirulence* (*Avr*) genes that encode effectors. If a plant possesses an *R* gene (receptor) that matches the product of a particular *Avr* gene (effector), it is then resistant to that particular pathogen. This is known as the gene-for-gene concept. Invaded plants begin a programmed cell death response to the pathogen called the hypersensitive response. In the hypersensitive response, pathogen-containing tissue and surrounding tissue undergo apoptosis upon infection and become necrotic lesions. This serves to contain and isolate the pathogen. Systemic acquired resistance protects a plant against further infection. Salicylic acid produced at the site of an infection stimulates PR protein production. Salicylic acid is also transported to other parts of the plant to signal the presence of an invader. Plants can use interference RNA (RNAi, also known as posttranscriptional gene silencing) to respond to attack by RNA viruses. A plant produces small pieces of small interfering RNA (siRNA) from interactions with the viral RNA. These siRNAs degrade the viral mRNA.

Question 1. It has long been a practice in rural areas to nail fencing to trees. Diagram how the act of nailing a fence to a tree trunk can introduce pathogens into the tree. Be sure to show what measures the tree takes to ward off disease.

Question 2. Explain the concept of gene-for-gene resistance.

39.2 How Do Plants Deal with Herbivores?

The foraging activities of herbivores do damage to plants and sometimes spread plant diseases. A majority of herbivores are insects, but each major class of vertebrates contains at least a few herbivorous species. Plants have many ways of defending themselves against this damage. Some plants use mechanical defenses to protect themselves from herbivores, including morphological features such as thorns, spines, and hairs. Some plants produce latex in response to injury from an herbivore; insects are trapped by the latex and starve to death. Plants also produce secondary metabolites that provide defense from grazers. The effects of these chemicals are diverse, ranging from neurotoxins to hormone mimics (see Table 39.1). Canavanine is a defensive secondary metabolite that plays multiple roles in a plant. Because it is similar to arginine, it is incorporated into the proteins made by the herbivore after it is consumed. Canavanine alters the tertiary structure of proteins and can be lethal to the organism consuming the plant.

Plants respond to herbivory with induced defenses as well. There are several classes of elicitors that function in defense against herbivory. Some are produced by the herbivore, and some are the products of digestion of plant tissues. The signal transduction pathways that are activated by herbivory have several key components. Changes in the electrical potential of the plasma membrane occur in the damaged area and can be transmitted throughout the plant; generation of reactive oxygen species (such as superoxide and hydrogen peroxide) that act as signaling molecules; and production of hormones (such as jasmonic acid, or jasmonate). Jasmonate triggers systemic defenses against herbivores. When the plant senses an herbivore-produced elicitor, jasmonate and jasmonate derivatives are produced through breakdown of certain membrane lipids. Jasmonate binds a transcriptional inhibitor and targets it for degradation, thus allowing expression of previously inhibited genes (see Figure 39.7). Plants also produce protease inhibitors, which interfere with the digestion of proteins by herbivores and stunt their growth.

Plants protect themselves from their own defenses by isolating toxic secondary metabolites in compartments. Water-soluble toxins are stored in vacuoles; hydrophobic poisons are stored in laticifers along with latex; other toxins are dissolved in the waxes covering the epidermis. Some plants store precursors of toxic compounds separately from the enzymes that convert them to active poison; the two meet only after the plant is damaged enough to cause mixing. Some plants have modified versions of proteins that lower the toxicity of the defensive agent. In plants that use the canavanine amino acid in defense, the plant's enzymes are altered to be able to distinguish between canavanine and arginine and thus prevent incorporation of canavanine into proteins.

Sometimes the herbivore can overcome the plant's defenses. Milkweeds store defensive chemicals in latex in specialized tubes called laticifers that are adjacent to veins in the leaves. When the leaf is bitten by an herbivore (usually an insect), the latex is released, protecting the plant from the insect. One beetle that feeds on milkweeds has evolved a method of disarming the laticifers; they cut some of the leaf veins and drain the latex before beginning to feed. Because plant defenses involve a continuous arms race, it is possible that the plant would evolve even more sophisticated defenses against such sophisticated attacks.

Question 3. Describe strategies used by plants to prevent herbivory.

Question 4. Explain what is meant by coevolution of plants and herbivores. Create a hypothesis for how grazing might increase the productivity of some plants.

Question 5. Many plants produce toxins that damage eukaryotic predators. Considering that the cell structure of both the predator and the plant is similar, why is the plant not affected by its own toxins?

39.3 How Do Plants Deal with Environmental Stresses?

Some plants have special adaptations to live in very dry conditions

Some plants grow in saturated soils

Plants can respond to drought stress

Plants can cope with temperature extremes

Plants cope with stresses from the environment in two basic ways: (1) adaptation, acquiring genetically encoded properties that resist the stress, or (2) acclimation, increasing their tolerance for the stress due to prior exposure. Plants have evolved different methods to cope with very dry environments. Some desert plants are drought avoiders. They survive in these areas by completing their entire life cycle in the short period that water is available, or they remain dormant until conditions are favorable. Xerophytes are specifically adapted for growth in dry areas. Three basic structural adaptations are found in their leaves: specialized anatomy that reduces water loss, a thick cuticle and hairs, and trichomes that diffuse sunlight to resist damage to the photosynthetic apparatus. Some xerophytes have stomata located in cavities below the leaf surface known as stomatal crypts. Succulents have water-storing leaves. Roots can also be adapted for drought conditions. Long taproots can reach water supplies far underground. Shallow, fibrous root systems that grow only during rainy seasons are also adapted to desert growth. Xerophytic plants are able to extract more water when it is available by changing the osmotic and water potential of their root cells. They do this by storing proline or secondary metabolites in their vacuoles. Too much water can be as dangerous to plant survival as too little. Roots submerged in water do not get adequate oxygen to sustain respiration. Some plants that grow in standing water overcome this by producing pneumatophores, which are root extensions that grow above water and provide oxygen to the entire root system. Others have leaf modifications called aerenchyma for buoyancy and oxygen storage.

When conditions become so dry that even xerophytes are stressed, plants activate inducible responses to drought stress. The same responses are found in plants that are not adapted to dry conditions. Inadequate water supply results in changes in membrane integrity and in the three-dimensional structure of proteins. Drought conditions stimulate the roots to release abscisic acid, which travels to the leaves. Abscisic acid closes the stomata and initiates gene expression that leads to physiological changes that conserve water. One such set of genes encodes the late embryogenesis abundant (LEA) proteins. The LEA proteins stabilize membranes and proteins.

Extremes in temperature can damage and even kill plants. High temperatures denature proteins and destabilize membranes. Cold temperatures decrease membrane fluidity and permeability, and freezing causes rupturing of membranes if ice crystals form. Plants have evolved a number of adaptations to deal with heat, including hairs and spines to dissipate heat, changes in leaf profile to catch less sun, and CAM photosynthesis. Plants also have inducible responses, including the production of heat shock proteins that act as chaperonins to stabilize protein structure against denaturation. Plants can adjust to cold through a process of cold-hardening, which involves repeated exposure to cool, nondamaging temperatures. This alters the saturated and unsaturated fatty acid composition of membranes, allowing them to remain fluid at cooler temperatures and making them more resistant to rupture. Protein chaperonins are also produced in response to cold temperatures. Some plants have antifreeze compounds that prevent ice crystal formation.

Question 6. Pansies are planted in the winter in the southeastern United States. To ensure that the plants are ready to enter cool ground, nurseries cold-harden the plants for several days. Describe this process and explain how this helps the plants survive cooler temperatures.

Question 7. A somewhat neglectful houseplant owner has African violets that are rarely watered. The plants often go weeks or months without watering. When they are eventually watered, they often flower soon after. Why do the plants have this response to finally receiving water?

39.4 How Do Plants Deal with Salt and Heavy Metals?

Most halophytes accumulate salt

Some plants can tolerate heavy metals

Saline environments restrict the growth of angiosperms. Plants living in saline environments are exposed to an osmotic challenge. In addition, sodium has some toxic properties of its own, inhibiting enzymes and protein synthesis. Halophytes are adapted to saline environments. These plants accumulate sodium and chloride ions, which they store in leaf vacuoles. The increased salt concentration of the halophyte means it has a more negative water potential, allowing it to take up water from the saline environment. Some plants are able to excrete salt so that it does not reach toxic proportions. Salt glands move salt to the leaf surface, where it can be lost to wind or rain. Salt glands on the leaf can also assist with water procurement from the roots and with reduction of water loss to evaporation. Many modifications such as succulence, thick cuticles, and crassulacean acid metabolism allow for survival in both drought and saline conditions.

Heavy metals such as chromium, mercury, lead, and cadmium are poisonous to most plants. Some geographic areas are rich in these metals as a result of normal geological processes, and others have been contaminated by human activity. Plants living in these areas have adapted to allow them to accumulate large quantities of heavy metals. These plants, known as hyperaccumulators, increase ion transport into the roots, increase translocation of ions to the leaves, accumulate ions in shoot vacuoles, and resist the toxin. These plants are important in bioremediation cleanup efforts.

Question 8. At a T-intersection in a small town, a local business owner plants several ornamental shrubs. During the winter, snowplows push snow from the intersection around the shrubs. In an attempt to save the shrubs, the business covers the shrubs with burlap. Will this save the shrubs? What conditions in the intersection are likely to cause the greatest damage to the shrubs?

Question 9. Bioremediation of mine and ore refinery sites is an area of intense study. How can plants help with the cleanup of sites contaminated with heavy metals?

Test Yourself

1. Defensive strategies in plants that are always turned on are called _______ defenses.
   a. induced
   b. heat shock protein
   c. alternating
   d. constitutive
   e. All of the above
   Textbook Reference: *39.1 How Do Plants Deal with Pathogens?*

2. Plants acquire systemic resistance in much the same way that people acquire resistance to pathogens; however, the mechanism of systemic acquired resistance is quite different in plants. Which of the following does *not* have a role in acquired resistance in plants?
   a. Salicylic acid
   b. *R* genes
   c. Methyl salicylate
   d. PR proteins
   e. All of the above have a role in acquired resistance in plants.
   Textbook Reference: *39.1 How Do Plants Deal with Pathogens?*

3. Gene-for-gene resistances depend on which of the following?
   a. Compatible alleles in plant and pathogen
   b. Incompatible alleles in plant and pathogen
   c. Recessive *Avr* genes
   d. Recessive *R* genes
   e. PR proteins
   Textbook Reference: *39.1 How Do Plants Deal with Pathogens?*

4. Upon infection by a pathogen, plant cells increase synthesis of cell wall polymers. The function of these polymers is to
   a. destabilize the cell walls.
   b. synthesize antibodies to the pathogen.
   c. isolate the pathogen in the invaded tissue.
   d. break down the wax barrier.
   e. signal the existence of the pathogen to the leaves.
   Textbook Reference: *39.1 How Do Plants Deal with Pathogens?*

5. Canavanine is toxic to many herbivores but not to plants. Which of the following statements regarding this differential toxicity is true?
   a. Canavanine is confused with arginine in plants but not in animals.
   b. Canavanine is incorporated into plant cells and causes them to fold properly.
   c. Plants are able to differentiate between arginine and canavanine, whereas animals cannot.
   d. Plants store canavanine in vacuoles, whereas animals metabolize it.
   e. None of the above
   Textbook Reference: *39.2 How Do Plants Deal with Herbivores?*

6. Which of the following best describes how plants produce their own insecticide?
   a. Wounded cells release an elicitor that directly induces synthesis of protease inhibitors, and protease inhibitors act as insecticides.
   b. Wounded cells release jasmonates, jasmonates stimulate elicitor synthesis, elicitor causes the production of protease inhibitors, and protease inhibitors act as insecticides.
   c. Wounded cells release an elicitor that causes membrane breakdown, membrane breakdown releases jasmonates, and jasmonates act as insecticides.
   d. Wounded cells release an elicitor that causes membrane breakdown, membrane breakdown releases jasmonates, jasmonates induce synthesis of protease inhibitors, and protease inhibitors act as insecticides.
   e. None of the above
   Textbook Reference: *39.2 How Do Plants Deal with Herbivores?*

7. Which of the following is *not* an adaptation to drought conditions?

a. Water-storing tissues
b. Leaf loss
c. Sunken stomata
d. Increased stomata number
e. CAM photosynthesis
Textbook Reference: *39.3 How Do Plants Deal with Environmental Stresses?*

8. Halophytes are different from all other types of plants in that they
a. can accumulate sodium and chloride ions.
b. have a positive water potential.
c. contain no sodium or chloride ions.
d. contain no stomata.
e. All of the above
Textbook Reference: *39.4 How Do Plants Deal with Salt and Heavy Metals?*

9. Which of the following conditions stimulates the production of heat shock proteins?
a. Abnormally high temperatures only
b. Abnormally low temperatures only
c. Both abnormally high and abnormally low temperatures
d. Heat shock proteins are continually available in a plant, no matter what the temperature.
e. None of the above
Textbook Reference: *39.3 How Do Plants Deal with Environmental Stresses?*

10. The main function of heat shock proteins is to
a. stabilize proteins necessary to a cell's survival.
b. reinforce membranes that lose fluidity.
c. cause the plant to enter dormancy.
d. act as an antifreeze compound.
e. stimulate bolting.
Textbook Reference: *39.3 How Do Plants Deal with Environmental Stresses?*

11. Cold-hardening does *not* involve
a. production of antifreeze proteins.
b. increased production of saturated fatty acids.
c. production of phytoalexins.
d. production of heat shock proteins.
e. All of the above are involved in cold-hardening.
Textbook Reference: *39.3 How Do Plants Deal with Environmental Stresses?*

12. Plants that produce toxins for defense avoid poisoning themselves by
a. storing the toxic substances throughout all plant tissues.
b. storing the toxic substances in laticifers.
c. producing the toxic substances continuously by every cell.
d. storing the toxic substances in the same place as enzymes that convert them to the active form.
e. None of the above; plants are slowly poisoned by the toxic substances.
Textbook Reference: *39.2 How Do Plants Deal with Herbivores?*

13. Which of the following is *not* a secondary plant metabolite?
a. Phenolics
b. Alkaloids
c. Terpenes
d. Glucosinolates
e. PR proteins
Textbook Reference: *39.2 How Do Plants Deal with Herbivores?*

14. Which of the following play a role in the hypersensitive response?
a. PR proteins
b. Phytoalexins
c. Secondary metabolites
d. Both a and b
e. All of the above
Textbook Reference: *39.1 How Do Plants Deal with Pathogens?*

15. A plant's immune response to RNA viruses involves
a. heat shock proteins.
b. secondary metabolites.
c. small interfering RNA.
d. PR proteins.
e. All of the above
Textbook Reference: *39.1 How Do Plants Deal with Pathogens?*

Answers

Key Concept Review

1.

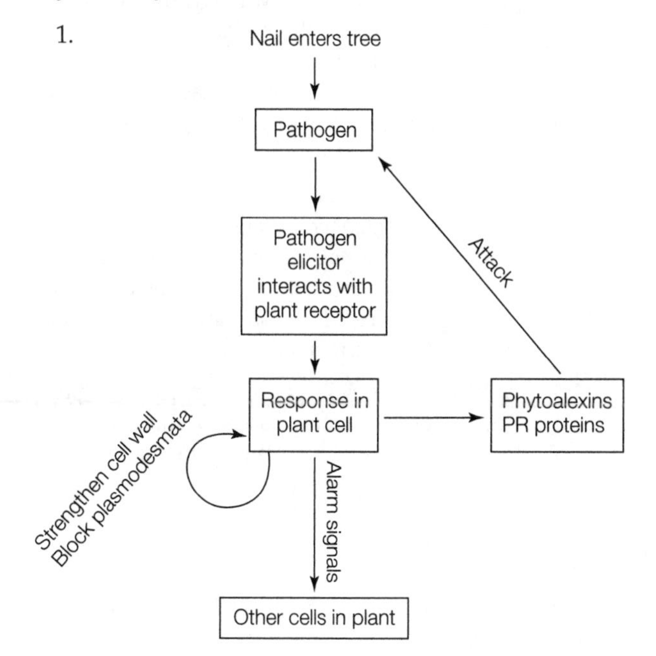

2. Gene-for-gene resistance is based on recognition of effector molecules by specialized receptors on the plant cells. The pathogenic *Avirulence* (*Avr*) genes encode the effectors. The plant resistance (*R*) genes encode the receptors. If the plant receptor recognizes the effector, then the plant is resistant to that particular pathogenic individual. Thus a particular *Avr* gene makes the pathogen "avirulent" on the plant with a corresponding *R* gene.
3. Plants prevent herbivory by producing secondary compounds that affect potential herbivores. These compounds may act as neurotoxins, inhibit digestion, or have a number of other consequences (see Table 39.1).
4. Coevolution of herbivores and plants occurs when the herbivore has resistance to toxins that the plant may produce, and the plant develops strategies to cope with herbivory. In some cases, regular grazing increases photosynthetic productivity of the plant by reducing shading and allowing better partitioning of resources.
5. Plants have a number of strategies to protect themselves from their own toxins. They may isolate the toxins in vacuoles within their cells so that the toxins do not come into contact with the active machinery of the cells. They may restrict toxin production to already-damaged tissue. They may develop enzymes that differentiate toxins and render them harmless. They can also separate toxin precursors and the enzymes that convert them to active form, so that these mix only when there is damage to the cell.
6. Cold-hardening is gradual exposure to adverse temperatures. This allows for structural changes in cell membranes to help plants cope with extreme temperatures. It also allows for the production of heat shock proteins and other compounds necessary for the plant's survival at temperature extremes.
7. The neglectful plant owner is unwittingly mimicking the natural conditions in which the African violets evolved. It is likely that the plants respond to the watering as if it were the rainy season in their natural habitat, and they are employing drought avoidance adaptations to reproduce at times when there is sufficient water present.
8. The burlap will most likely do little for the shrubs. The roots are in the most danger. Because streets are heavily salted in the winter, the snow around them creates a saline environment. If the shrubs are not adapted to a saline environment, there is little that can be done to help them survive the salt accumulation around their roots.
9. Only plants that have evolved a strategy for taking up and coping with heavy metals can grow in these contaminated sites. Deliberately planting them at these sites initiates the slow process of removing the metals from the soil.

Test Yourself

1. **d.** If something is always turned on, it is said to be constitutive.
2. **b.** Plants that have been exposed to pathogens wall off invaded tissue. This tissue releases salicylic acid to the rest of the plant, which stimulates PR proteins. The tissue also releases airborne methyl salicylate, which causes release of PR proteins in distant regions of the plant and even in neighboring plants.
3. **a.** The plant must have an *R* allele, and the pathogen must have an *Avr* allele. Compatibility between these alleles leads to resistance.
4. **c.** Polysaccharides seal plasmodesmata and reinforce cell walls. The purpose of this is to contain the pathogen in a minimal number of cells.
5. **c.** Canavanine's mechanism of action is that it is taken up by the herbivore and incorporated into its protein structures. Because the structure of canavanine is different from that of the arginine it replaces, the herbivore's proteins then have a structure that is lethal to the herbivore. The plant itself and some herbivores have enzyme mechanisms to distinguish canavanine from arginine and prevent its incorporation into proteins.
6. **d.** See Figure 39.7 in the textbook.
7. **d.** Water storage tissues allow the plant to take advantage of any water that is available and hold on to it for lean times. When photosynthesis cannot be sustained, leaf loss prevents loss of resources. Sunken stomata prevent excessive water loss to evaporation. CAM photosynthesis conserves water.
8. **a.** They can accumulate sodium and chloride ions. This makes their water potential more negative, and they are able to extract more water from their surroundings than plants that do not store the ions.
9. **c.** Heat shock proteins are produced in response to extreme high and low temperatures.
10. **a.** Heat shock proteins are chaperonin proteins. They function to stabilize and prevent denaturation (unfolding) of essential proteins.
11. **c.** Cold-hardening involves production of antifreeze proteins, heat shock proteins, and an increase in saturated fatty acids in the membrane. The production of phytoalexins is involved in chemical defenses against pathogens.
12. **b.** Plants that produce toxins as a defense typically store them in special locations such as laticifers, where they will not damage tissue.
13. **e.** Alkaloids, phenolics, terpenes, and glucosinolates are all examples of secondary plant metabolites. PR proteins are proteins that function in defense.
14. **d.** The hypersensitive response involves the release of antibiotic phytoalexins and PR proteins. Secondary metabolites are a response to herbivory and are not involved in the hypersensitive response.
15. **c.** Plants that are invaded by RNA viruses use the RNA of the invading virus to interfere with and block viral replication.

Physiology, Homeostasis, and Temperature Regulation

The Big Picture

- Tissues (epithelial, muscle, connective, and nervous), organs, and organ systems all work to maintain homeostasis within the body's cells. Negative feedback helps regulate these physiological systems to keep intracellular conditions relatively constant. Temperature affects the rate of metabolic reactions, so most animals regulate their temperature and/or acclimatize to adjust to different temperatures.
- Endotherms generate their own body heat whereas ectotherms have body temperatures that are environmentally determined. The metabolic rate of endotherms increases when environmental temperatures fall below or rise above a narrow thermoneutral zone, whereas the metabolic rate of ectotherms is temperature-dependent and falls with decreases in environmental temperature. Endotherms have higher metabolic rates than ectotherms and many adaptations for body temperature regulation. Larger mammals have greater total basal metabolic rates, but their per gram metabolic rates are lower than those of smaller mammals.
- Control of body temperature occurs by both behavioral and physiological mechanisms. In mammals, the brain's hypothalamus is the thermostat for temperature regulation.

Study Strategies

- Understanding the difference between ectotherms and endotherms can be difficult. These terms refer to the *source* of heat that determines an animal's temperature. Whereas body temperatures of ectotherms are determined primarily by external sources of heat, those of endotherms are determined by heat generated metabolically (i.e., within their bodies).
- Refer to Figure 40.2 to help you understand how homeostasis is maintained through regulatory systems. Think through what would happen if the temperature dropped or increased. Physiologically, what represents the sensor? The thermostat? The controlled systems?
- Review the organization of physiological systems, from cells to tissues to organs to organ systems. This will help you when you study each of the organ systems in later chapters.
- Go to the Web addresses shown to review the following animated tutorial and activity. You can also find them in BioPortal (yourBioPortal.com), along with many additional learning resources.

  Animated Tutorial 40.1 The Hypothalamus (Life10e.com/at40.1)

  Activity 40.1 Thermoregulation in an Endotherm (Life10e.com/ac40.1)

Key Concept Review

40.1 How Do Multicellular Animals Supply the Needs of Their Cells?

An internal environment makes complex multicellular animals possible

Physiological systems are regulated to maintain homeostasis

Single-celled organisms and small multicellular organisms, in which all the cells are only a few layers from the environment, can exchange nutrients and waste with the environment without the help of specialized cells. In larger multicellular organisms, specialized cells and groups of cells help control the makeup of the internal environment and maintain the differences between the internal and external environments. Most of the water within an animal is intracellular fluid (water within cells); the rest is extracellular fluid, which includes plasma (the liquid portion of blood) and interstitial fluid (fluid between cells). This fluid environment within and around cells is maintained within certain physical and biochemical parameters. This regulated constancy of the internal environment is called homeostasis.

Physiological regulatory systems require both a set point at which the parameter is to be held, and feedback, which provides information about the state of the system (see Figure 40.2). Differences between the set point and the feedback information indicate to the regulatory mechanism which corrective measures are required to restore a parameter to its set point. Physiological control systems must take information from the regulatory systems and effect changes. Negative feedback results in a slowdown or reversal of a process, returning the variable controlled by that process back toward its set point or regulated level. In contrast, positive feedback—which is less common than negative feedback—results in amplification of a response. Feedforward information can change the set point in preparation for a change in conditions.

Question 1. Use your knowledge of cells and cell functions to explain why it is important for an organism to maintain homeostasis.

Question 2. During lactation (breast-feeding), increased sucking by the baby causes the release of hormones that cause the mother's breasts to increase milk production. What type of feedback—negative or positive—is involved in this situation?

Question 3. In response to the smell of food, the stomach produces hydrochloric acid in anticipation of receiving the food. Is this an example of negative feedback, positive feedback, or feedforward information? Explain your answer.

40.2 What Are the Relationships between Cells, Tissues, and Organs?

Epithelial tissues are sheets of densely packed, tightly connected cells

Muscle tissues generate force and movement

Connective tissues include bone, blood, and fat

Neural tissues include neurons and glial cells

Organs consist of multiple tissues

A tissue consists of a group of similar cells with the same form and function. There are four types of tissue: epithelial, muscle, connective, and nervous. Epithelial tissues line the inner and outer body surfaces, such as the lining of the various organs and the skin. Muscle tissues can contract to exert force. Skeletal muscle attaches to bone and is responsible for body movement; smooth muscle generates force in blood vessels and internal organs, such as the stomach and bladder; and cardiac muscle generates the heartbeat and pumps blood. Connective tissue (cartilage, bone, adipose tissue and blood) consists of an extracellular matrix that holds together a dispersed group of cells. This extracellular matrix typically includes protein fibers such as collagen and elastin. Nervous tissue includes neurons, which generate and conduct electrochemical signals, and glial cells, which support and protect neurons.

Each organ is made of two or more tissues that work together to carry out a specific body function. A number of organs that are integrated to carry out a specific function form an organ system (e.g., the digestive system).

Question 4. Label the organs and four main tissue types in the diagram below.

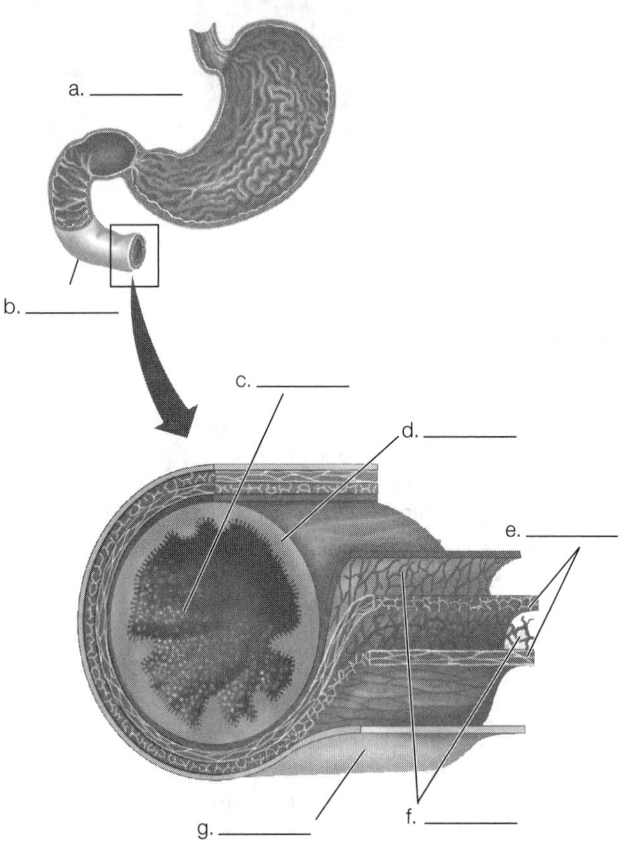

Question 5. Your tongue includes all four tissue types. Use your knowledge of tissue types and their functions to deduce how each tissue type might help form your tongue.

40.3 How Does Temperature Affect Living Systems?

Q_{10} is a measure of temperature sensitivity

Animals acclimatize to seasonal temperatures

Cellular functions are typically limited to a temperature range between 0°C and 40°C since ice crystals form in cells below this range and proteins start to denature above this range. In fact, many animals have an even smaller range of body temperatures that they can survive. Physiological processes are sensitive to temperature such that increases in temperature usually increase the rate of a given process. Q_{10} describes how a physiological process changes as temperature increases or decreases by 10°C (see Figure 40.8). Q_{10} is calculated by dividing the rate of a process at one temperature (R_T) by the rate of the process at a temperature that is 10°C lower (R_{T-10}). A Q_{10} equal to 2 means that the reaction rate doubles when temperature increases by 10°C. Most biological Q_{10} values range between 2 and 3. Biochemical reactions connect in linked networks, so organisms either prevent large temperature changes or compensate for them in order to maintain balance between processes with different Q_{10} values that respond differently to temperature change.

Acclimatization is the change in a physiological process that occurs over time in response to a change in the environment. For example, a fish experimentally exposed in summer to decreasing water temperatures will show slowed physiological functions. However, the physiological functions of the same fish in winter are not as slow as they were during the summer experiment because the fish has adjusted its physiology to maintain its function during the winter season's lower temperature. For example, fish can acclimatize by producing and using different enzyme versions (isozymes) during different seasons.

Question 6. The metabolic rate of an organism is the total energy used by the animal and is usually measured as oxygen consumption. You measure the metabolism of an animal and find that its metabolic rate is 5.5 ml of oxygen per hour at 15°C and 11.5 ml of oxygen per hour at 25°C. Calculate the Q_{10} for the metabolic rate of this animal.

Question 7. How does seasonal acclimatization affect an animal's long-term versus short-term sensitivity to environmental temperature changes?

40.4 How Do Animals Alter Their Heat Exchange with the Environment?

Endotherms produce substantial amounts of metabolic heat

Ectotherms and endotherms respond differently to changes in environmental temperature

Energy budgets reflect adaptations for regulating body temperature

Both ectotherms and endotherms control blood flow to the skin

Some fish conserve metabolic heat

Some ectotherms regulate metabolic heat production

Endotherms regulate their body temperature with internal heat production by means of a high metabolic rate and effective insulation, whereas ectotherm body temperature is determined largely by the temperature of the surrounding environment. Heterotherms switch between endotherm-like and ectotherm-like body temperature regulation. All biological energy transformations release some energy as heat, but the cells of endotherms release more heat than those of ectotherms because they are less efficient in their energy transformations. For example, endotherm membranes leak more ions so that membrane ion pumps have to work harder to maintain ion gradients. Endothermy may have arisen when increased metabolic heat production due to leaky membranes allowed an animal to stay active longer as temperatures dropped in the evening.

Ectotherms have body temperatures that track environmental temperatures. In contrast, endotherms maintain a constant body temperature even as the environmental temperature changes. When exposed to cold temperatures, metabolic heat production increases in endotherms and decreases in ectotherms (see Figure 40.10). Both endotherms and ectotherms seek out favorable thermal environments using behaviors like basking, burrowing, nest construction, and huddling.

Heat can be gained or lost to the environment through several routes (see Figure 40.12). Evaporation from the skin or respiratory tract cools the body. Solar radiation from the sun heats the body. Thermal radiation, convection (heat transfer from movement of surrounding medium), and conduction (direct heat transfer) can either warm or cool the body, depending on the thermal environment. The total balance of heat production and heat exchange for an animal is its energy budget. To maintain a constant body temperature, the heat entering an animal must equal the heat leaving. Generally, $heat_{in}$ is generated by metabolism and absorbed radiation (R_{abs}), and $heat_{out}$ occurs through radiation emitted (R_{out}), convection, conduction, and evaporation.

An energy budget is greatly influenced by an animal's surface temperature. In both endotherms and ectotherms, blood flow to the skin increases when body temperature rises in order to vent heat to the environment. Blood vessels close when body temperature is low or the environment is cold in order to reduce heat loss. For example, iguanas decrease their heart rate and skin blood flow when diving, but increase these when warming on land. Fur insulates well, so mammals have hairless skin surfaces where surface blood flow can more easily lose heat.

Most fish have limited control of their body temperature because heat picked up from metabolically active muscles is lost when blood is pumped from the heart to the gills, where it comes in contact with cold water. However, some large fishes can maintain a body temperature that is higher than the temperature of the surrounding water. These fish possess special vascular countercurrent heat exchangers that retain metabolically produced heat in the muscle mass (see Figure 40.14). In countercurrent heat exchangers, heat is exchanged between blood vessels carrying blood in opposite directions.

Some ectotherms even generate heat to raise body temperature. For example, flying insects warm themselves by contracting their flight muscles. Other insects, such as honey bees, form clusters in winter and regulate temperature by adjusting their own metabolic heat production and the density of the cluster.

Question 8. In the figure below, graph the lizard's predicted body temperature based on your knowledge of how ectotherms regulate their temperature with behavior.

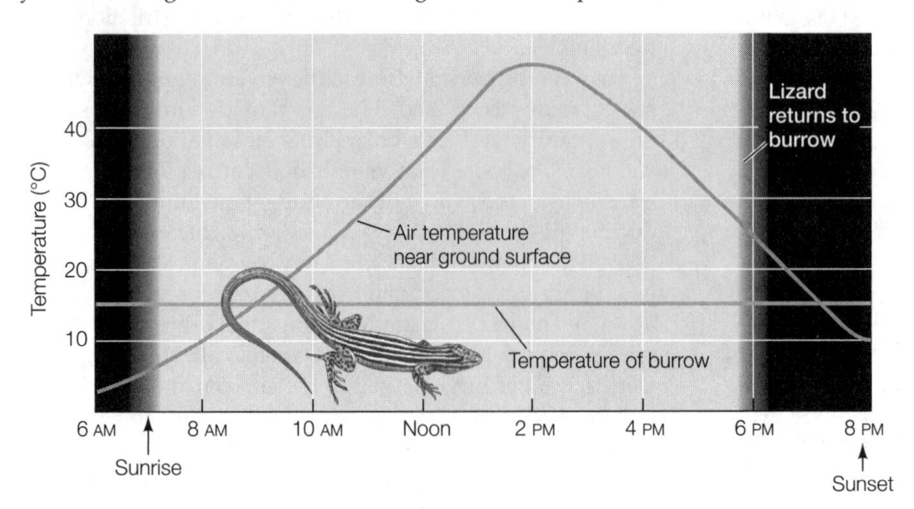

Question 9. Provide an example of one way in which both endotherms and ectotherms influence their energy budget.

40.5. How Do Endotherms Regulate Their Body Temperatures?

- Basal metabolic rates correlate with body size
- Endotherms respond to cold by producing heat and adapt to cold by reducing heat loss
- Evaporation of water can dissipate heat, but at a cost
- The mammalian thermostat uses feedback information
- Fever helps the body fight infections
- Some animals conserve energy by turning down the thermostat

There is a narrow range of environmental temperatures over which endotherm metabolic rate does not change. Within this thermoneutral zone, metabolic rate is at the minimum level needed to maintain physiological functions essential to homeostasis in the resting animal; this level is called the basal metabolic rate (BMR). Large mammals have greater total metabolic rates, but their per gram metabolic rates are lower than those of smaller mammals (see Figure 40.16). When a mammal experiences temperatures within its thermoneutral zone, it is able to regulate body temperature using mainly passive strategies. If the environmental temperature drops below the lower critical temperature or rises above the upper critical temperature, the mammal must add active thermoregulatory responses that use energy and thus increase the metabolic rate above the basal level (see Figure 40.17).

When the environmental temperature falls below the lower critical temperature, metabolic heat can be produced by shivering thermogenesis (skeletal muscles contract) and nonshivering thermogenesis (heat produced by uncoupling oxidative phosphorylation from ATP production in the mitochondria of brown fat). Birds do not have brown fat, but large pads of brown fat are found in mammalian babies, hibernators, and small mammals living in cold climates. Other adaptations found in endotherms living in cold climates are reduced surface-to-volume ratios, increased thermal insulation, and countercurrent heat exchange and/or reduced blood flow in extremities. Oil secretions that protect against skin wetting also help maintain effective insulation.

Evaporative cooling by sweating or panting is an essential heat loss strategy when temperatures rise above the upper critical temperature. Since sweating and panting require loss of body water and expenditure of metabolic energy, animals also use passive strategies like increasing skin blood flow and seeking shade and cool breezes. The major integrative center for thermoregulation in mammals is the hypothalamus, a structure at the bottom of the brain. The hypothalamus has a set point temperature; changes in temperature of the hypothalamus result in thermoregulatory responses to offset any temperature change (see Figures 40.19 and 40.20). The hypothalamus also integrates thermal information received from the skin. Other factors such as the sleep–wake cycle may influence set points. For example, during an infection, molecules called pyrogens initiate an increase in the hypothalamic set point, causing fever. Low-level fevers actually help your body fight infections.

Hypothermia is a general state of below-normal body temperature typically resulting from starvation, cold exposure, illness, or anesthesia. Unregulated hypothermia is potentially dangerous, but some animals engage in regulated hypothermia in order to conserve energy. Regulated hypothermia on a daily schedule is called daily torpor; regulated hypothermia that lasts longer is called hibernation. Arousal from hibernation occurs when the hypothalamic set point returns to normal.

Question 10. Why would an endothermic animal's metabolic rate be higher at 40°C than at 30°C? What physiological phenomenon would explain why its metabolic rate rises when the environmental temperature is above the upper critical limit?

Question 11. Design a study to determine whether a lizard has a thermostat located in the hypothalamus.

Test Yourself

1. Which of the following statements about interstitial fluid is true?
   a. It is found within cells.
   b. It represents most of the water in an animal's body.
   c. It makes up blood plasma.
   d. It does not exchange molecules with intracellular fluid.
   e. It is found between the cells of the body.
   Textbook Reference: *40.1 How Do Multicellular Animals Supply the Needs of Their Cells?*

2. In an environment with an ambient temperature lower than an animal's core body temperature, which of the following would be an *inappropriate* physiological or behavioral response of an animal that needed to eliminate excess body heat?
   a. Wallowing in a pool of water
   b. Inhibiting blood flow to peripheral vessels
   c. Sweating
   d. Decreasing physical activity
   e. All of the above would be inappropriate responses.
   Textbook Reference: *40.4 How Do Animals Alter Their Heat Exchange with the Environment?*

3. A lizard lives in a temperate desert environment where the temperature is cold at night and warm during the day. In order for the lizard to maintain the most stable body temperature, it could
   a. stay in a burrow during the night and shuttle between the sun and shade on the surface during the day.
   b. stay on the surface during the night and move to a burrow during the day.
   c. increase its metabolism and heat production during the night and decrease it during the day.
   d. decrease its metabolism and heat production during the night and increase it during the day.
   e. move into a state of hypothermia during the night and a state of torpor during the day.
   Textbook Reference: *40.4 How Do Animals Alter Their Heat Exchange with the Environment?*

4. On a hot day, an old dog pants heavily and visits his water bowl frequently in between naps. Which of the following statements about the dog's condition or behavior is true?
   a. The environmental temperature is within his thermoneutral zone.
   b. He is operating at a basal metabolic rate.
   c. He is lowering his core body temperature by means of evaporative water loss.
   d. His hypothalamic set point must have been raised.
   e. His behavior has nothing to do with thermoregulation.
   Textbook Reference: *40.5 How Do Endotherms Regulate Their Body Temperatures?*

5. Evaporative cooling is an effective way to increase heat loss, but it carries the physiological drawback of
   a. an increased use of ATP and substantial water loss.
   b. the lowering of the hypothalamic thermal set point.
   c. the decreased use of ATP and substantial water gain.
   d. the exhaustion of the supply of brown fat.
   e. a lowered basal metabolic rate.
   Textbook Reference: *40.5 How Do Endotherms Regulate Their Body Temperatures?*

6. Which of the following statements about fever is true?
   a. It is a higher-than-normal body temperature that is always dangerous.
   b. It decreases the metabolic rate of the body to conserve energy.
   c. It is a regulated resetting of the body's thermostat to a higher set point.
   d. It causes the production of pyrogens.
   e. It should always be controlled with fever-reducing drugs.
   Textbook Reference: *40.5 How Do Endotherms Regulate Their Body Temperatures?*

7. Which of the following is a characteristic unique to connective tissue?
   a. Highly modified cells that show the special property of contractibility
   b. Diverse anatomy with distinct specializations for information transfer
   c. Densely packed cells that cover inner and outer body surfaces
   d. A loose array of cells embedded in an extracellular matrix
   e. Possession of cilia that move substances over surfaces
   Textbook Reference: *40.2 What Are the Relationships between Cells, Tissues, and Organs?*

8. Elephants use their ears to release heat to the environment. What mechanisms might they employ to increase heat loss from the ears?
   a. Increased conduction from their ears
   b. Countercurrent exchange in their ears
   c. Nonshivering thermogenesis in their ears
   d. Covering their ears with dust
   e. Increased blood flow to their ears
   Textbook Reference: *40.4 How Do Animals Alter Their Heat Exchange with the Environment?*

9. Mechanical systems, like physiological systems, can be regulated to maintain homeostasis. Imagine an automated system that allows water to flow from a storage tank into a water bowl for your pet. In this system, the valve that opens when the water level drops below the desired water level is the
   a. sensor.
   b. effector.
   c. error signal.

d. feedforward information.
e. set point.
Textbook Reference: *40.1 How Do Multicellular Animals Supply the Needs of Their Cells?*

10. What is the difference between negative feedback mechanisms and positive feedback mechanisms?
a. Negative feedback mechanisms exist only in the circulatory system.
b. Negative feedback mechanisms return a system to a set point, whereas positive feedback mechanisms amplify a response.
c. Negative feedback mechanisms move a system away from a set point, whereas positive feedback mechanisms stabilize a system toward a set point.
d. Negative feedback mechanisms stabilize a system toward a set point, whereas positive feedback mechanisms reset the set point.
e. There is essentially no difference between the two feedback systems.
Textbook Reference: *40.2 What Are the Relationships between Cells, Tissues, and Organs?*

11. A number of physiological processes can undergo acclimatization. Which of the following statements about acclimatization is *false*?
a. It occurs in response to seasonal temperature changes.
b. Acclimatization of metabolic rate occurs because enzyme expression changes.
c. It involves the changing of a set point.
d. Acclimatization and adaptation are the same thing.
e. Acclimatization is primarily seen in organisms with body temperatures tightly coupled to the environmental temperatures.
Textbook Reference: *40.3 How Does Temperature Affect Living Systems?*

12. In fast-swimming cool-water fishes such as sharks and tunas, which of the following contribute(s) to the generation and maintenance of core body temperatures in excess of ambient water temperatures?
a. Specializations in the size and arrangement of blood vessels
b. Low rates of activity in the swimming muscles
c. An ability to acclimatize rapidly to the surrounding water
d. Metabolic rates that are insensitive to temperature change
e. An insulating layer of fat located beneath the scaled skin
Textbook Reference: *40.4 How Do Animals Alter Their Heat Exchange with the Environment?*

13. Which of the following is most likely to occur if a baby bat's hypothalamus is experimentally cooled using an implanted probe?
a. A drop in the bat's core body temperature
b. Stimulation of nonshivering thermogenesis in brown fat
c. Sweating or panting to increase evaporative water loss
d. Decreased metabolic rate
e. None of the above
Textbook Reference: *40.5 How Do Endotherms Regulate Their Body Temperatures?*

14. You measure metabolic rates of several four-toed salamanders at 20°C and at 30°C and calculate a Q_{10} equal to 2 for the species. Based on this information, which of the following statements is true?
a. Metabolic rate is not temperature sensitive in this species.
b. 30°C must be above the thermoneutral zone for this species.
c. This lizard shows acclimatization over this temperature range.
d. Metabolic rate at 20°C is twice that at 30°C for this species.
e. Metabolic rate at 20°C is half that at 30°C for this species.
Textbook Reference: *40.3 How Does Temperature Affect Living Systems?*

15. Which of the following statements about nervous tissue is true?
a. It cushions internal organs and insulates the body.
b. Its glial cells fire nerve impulses.
c. It is contractile.
d. Its glial cells support and protect neurons.
e. Its glial cells are a type of connective tissue holding neurons in place.
Textbook Reference: *40.2 What Are the Relationships between Cells, Tissues, and Organs?*

16. Which tissue type would form the inner layer of your tear duct?
a. Adipose tissue
b. Connective tissue
c. Epithelial tissue
d. Muscle tissue
e. Nervous tissue
Textbook Reference: *40.2 What Are the Relationships between Cells, Tissues, and Organs?*

Answers

Key Concept Review

1. The body functions with the help of proteins and enzymes. Changes in, for example, pH, temperature, glucose level, and oxygen and carbon dioxide levels of the internal environment can affect cellular function. Loss of homeostasis can lead to improper function of proteins and cell membranes, and ultimately to cell death.

2. Increased lactation in response to increased sucking is an example of a positive feedback.
3. This is an example of feedforward information. Feedforward information is predictive of a future change in the internal environment and functions to change the set point.
4.
   a. Stomach
   b. Small intestine
   c. Epithelial cells
   d. Mucosa (mostly connective tissue)
   e. Smooth muscle
   f. Nervous tissue
   g. Epithelial cells and connective tissue
5. Epithelial cells cover the surface of the tongue and include the taste receptors. Connective tissue can be found underneath the epithelial layer and also includes the blood that you might see if you bite your tongue. You can move your tongue because much of the bulk of the tongue is skeletal muscle. Finally, neurons connect to taste receptors and allow the tongue to sense touch.
6.
   $Q_{10} = (R_T/R_{T-10})$

Metabolic rate at 15°C	Metabolic rate at 25°C	Q_{10}
5.5 ml O_2/hour	11.5 ml O_2/hour	2.1 (= 11.5/5.5)

7. Animals undergo acclimatization in response to changes in seasonal conditions such as temperature. Often these changes are brought about by the production of enzymes that function better at the new temperature. Because of acclimatization, metabolic functions are less sensitive to long-term changes in temperature than they are to short-term changes.
8. During the day the lizard can regulate its body temperature behaviorally.

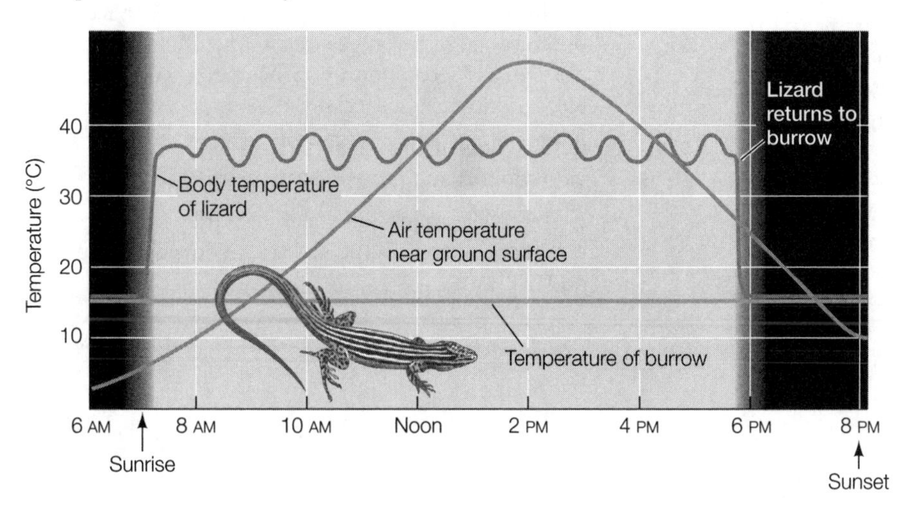

9. Both endotherms and ectotherms alter their blood flow patterns to increase or decrease the amount of blood that flows near the skin surface. Skin blood flow can influence heat loss via radiation, convection, conduction, and evaporation.
10. An endotherm in a hot environment will attempt to keep core body temperature from increasing. The heat-loss mechanisms available to an endotherm—sweating and panting—require the use of ATP and result in an increased metabolic rate. Endotherms will expend energy to protect body temperature and lose heat to the environment.
11. Two possible approaches would be to conduct lesion experiments and thermal stimulation experiments. Lesion experiments would involve destroying the region of the hypothalamus thought to function as the thermostat to see what happens to the lizard's ability to regulate body temperature behaviorally. Thermal stimulation experiments (similar to the one shown in Figure 40.19) could involve cooling or warming the hypothalamus and then monitoring body temperature and behavioral changes associated with thermoregulation. One experiment suggesting lizard set points can change showed that lizards with a bacterial infection had higher body temperatures than uninfected lizards, when able to choose from a variety of temperature microenvironments.

Test Yourself

1. **e.** Interstitial fluid is found between the cells of the body.
2. **b.** Since the environment is cooler than the animal, the appropriate response would be to lose heat to the environment across the skin. An increase in peripheral blood flow moves heat from the core of the animal to the surface, where it is lost to the cooler environment. Therefore, an inhibition of this process would be an *inappropriate* response.
3. **a.** For an ectothermic lizard in a temperate desert to maintain the most stable body temperature, it should remain in a burrow during the night, where temperatures do not drop very much, and then shuttle between the sun and shade during the day. If the lizard were to stay on the surface overnight, its activity would likely be too low for it to forage effectively. Ectotherms do not regulate temperature using metabolic heat production. (See Figure 40.11.)
4. **c.** The environmental temperature must be above the dog's thermoneutral zone because it is panting even though it is not particularly active. Panting is a type of active thermoregulation, so the dog's metabolic rate must be higher than its basal metabolic rate. There is no indication that the hypothalamic set point has been raised, but the napping and water-bowl visiting behaviors appear to be related to thermoregulation.
5. **a.** Evaporative cooling is a costly means to lower body temperature. Sweating and panting both use ATP. Water loss can also be high during evaporative cooling.
6. **c.** Fevers are brought about by a resetting of the thermoregulatory set point in the hypothalamus. Pyrogens cause the change in the set-point rather than being an effect of this change.
7. **d.** Connective tissues are blood, cartilage, and bone, all of which contain a loose population of cells embedded in an extracellular matrix.
8. **e.** Elephants use both physiological mechanisms (increased blood flow to the ears) and behavioral mechanisms (flapping of ears or wetting of ears to increase evaporation) to release heat into the environment. Ear flapping increases heat loss by convection, whereas conduction requires direct heat transfer between two objects, which is difficult to accomplish with ears. Countercurrent exchange helps conserve heat rather than release it, and nonshivering thermogenesis produces metabolic heat.
9. **b.** The effector is the part that causes the change in the environment, so the valve is the effector—when it opens, the water level increases. The sensor would detect the difference, the error signal would transmit the information, and feedforward information would be a dial that lets you change the desired water level or set point.
10. **b.** Negative feedback loops use information about the state of a system to bring the internal environment back toward the set point values. Positive feedback loops amplify the response of a system, moving the system away from the set point.
11. **d.** Physiological processes acclimate in response to a seasonal change in the environment. Acclimatization may involve changes in enzyme expression and a resetting of a physiological set point. Acclimatization occurs in individuals as their bodies adjust to environmental conditions, whereas adaptation occurs within a population over evolutionary time.
12. **a.** Large fishes such as tuna are able to maintain core body temperature above ambient temperature by means of a number of mechanisms. They have a special countercurrent heat exchanger built into their blood vessels and high metabolic rates; thus they produce heat in their swimming muscles.
13. **b.** Experimental cooling of the hypothalamus will stimulate thermoregulatory mechanisms that would help raise body temperature. Thermogenesis in brown fat is the only listed strategy/effect that would help raise body temperature. Most baby mammals have brown fat, and nonshivering thermogenesis in this organ is one common way for baby mammals to produce metabolic heat.
14. **e.** A Q_{10} of 2 means that the rate of a process doubles for each 10°C rise in temperature. Since 20°C is 10°C colder than 30°C, the rate at this temperature would be half that measured at 30°C. The lizard is an ectotherm and thus does not have a thermoneutral zone. While the lizard might show acclimatization over this temperature range, Q_{10} measures temperature sensitivity, not metabolic compensation. Metabolic compensation will tend to make processes less temperature sensitive.
15. **d.** Glial cells are a type of nervous tissue cell. They provide support and protection to neurons.
16. **c.** Epithelial tissues line ducts since they form the boundaries between the inside and outside of the body.

41 Animal Hormones

The Big Picture

- Hormones regulate functions ranging from growth to sexual maturity. In numerous ways, the endocrine system complements the nervous system, both of which are the major regulators of body function and homeostasis.
- Circulating hormones travel via the bloodstream to distant sites within the body to affect target tissues. Other endocrine signals act more locally. Target cells can respond to an endocrine signal because they have receptors. Receptors for water-soluble proteins, peptides, and amines are found on the cell surface, whereas receptors for lipid-soluble steroids and amines are found inside a cell.
- Hormone action is controlled by several different mechanisms. In many instances there is a "hormone cascade," in which a hormone controls the release of another hormone, which controls the release of another hormone, and so on. In a situation of negative feedback, a released hormone inhibits (negatively feeds back on) the tissue that may have stimulated its release in the first place. Tropic hormones influence other endocrine glands.

Study Strategies

- The many hormones may seem to constitute a long list of random compounds. Make flashcards with the hormone name on one side and the function and target tissue on the other. As you study, sort the flashcards into groups by the organ in which they are produced, then by the target organ.
- Some hormones have extremely specific actions (e.g., follicle-stimulating hormone), whereas others have broad effects on many target tissues (e.g., epinephrine and thyroxine). Learning which hormones are "specialists" and which are "generalists" will help you better understand the endocrine system.
- The pituitary gland produces numerous regulatory hormones. The functions of the anterior and posterior pituitary are quite distinct, and keeping them separate is important for understanding the endocrine system. See if you and your classmates can devise a mnemonic to help you remember which hormones go with which part of the pituitary gland.
- Remember that male sex steroids (androgens) and female sex steroids (estrogens and progesterone) are made and used by both sexes, but the relative levels vary in the two sexes.
- Go to the Web addresses shown to review the following animated tutorials and activities. You can also find them in BioPortal (yourBioPortal.com), along with many additional learning resources.

  Animated Tutorial 41.1 Complete Metamorphosis (Life10e.com/at41.1)

  Animated Tutorial 41.2 The Hypothalamic–Pituitary–Endocrine Axis (Life10e.com/at41.2)

  Animated Tutorial 41.3 Hormonal Regulation of Calcium (Life10e.com/at41.3)

  Activity 41.1 The Human Endocrine Glands (Life10e.com/ac41.1)

  Activity 41.2 Concept Matching: Vertebrate Hormones (Life10e.com/ac41.2)

Key Concept Review

41.1 What Are Hormones and How Do They Work?

Endocrine signaling can act locally or at a distance

Hormones can be divided into three chemical groups

Hormone action is mediated by receptors on or within their target cells

Hormone action depends on the nature of the target cell and its receptors

Individual cells within a multicellular organism must be able to communicate with one another. Typically, one cell releases a chemical signal that travels to target cells with appropriate receptors to trigger a response. The endocrine system consists of cells that make and release chemical signals into the extracellular fluid (ECF). Hormones are chemical signals that enter the blood and affect cells distant from their release site. Paracrine signals affect nearby cells and autocrine signals affect the cell that releases them (see Figure 41.1). Endocrine cells may exist as single cells scattered within an organ or as large aggregates known as endocrine glands. Secretions from exocrine glands are released into ducts that empty onto the surface of the skin or into a body cavity instead of into the ECF. Most neurotransmitters have a very local effect, but if they diffuse into the blood or act on distant target cells they are called neurohormones.

Three classes of hormones are the peptides or protein hormones, the steroid hormones, and the amine hormones. Most hormones are peptides, which are usually released from vesicles by exocytosis. Peptide hormones are water-soluble and easily transported in the blood. Their receptors are on the exterior of the target cell. Steroid hormones are derived from cholesterol and are lipid-soluble. Steroid hormones easily pass through cell membranes, but they require carrier proteins to transport them in the blood. Receptors for steroid hormones are often inside target cells. Amine hormones are small molecules made from single amino acids. Some are water-soluble and some are lipid-soluble.

Target cells respond to endocrine signals because they have receptors that recognize and bind to a signal, eventually triggering a cell response. Receptors for water-soluble hormones have a binding domain on the outside of the plasma membrane, a transmembrane domain, and a cytoplasmic domain. The cytoplasmic domain starts the cell response, often via second messenger action on protein kinases and phosphatases that activate or inactivate enzymes. Lipid-soluble hormones typically act by altering the target cell's gene expression. In all cases, the type of response depends on the type of receptor and the type of cell, such that the same hormone can cause different responses in different cell types. For example, the amine hormone epinephrine, released as part of the fight-or-flight response, increases heartbeat rate, decreases blood flow to digestive organs, and stimulates glycogen breakdown in the liver and fat breakdown by fat cells.

Question 1. Describe the modes of action of lipid-soluble and water-soluble hormones. Which type of hormone would likely produce more rapid effects?

Question 2. Why do different organs have such different responses to epinephrine?

41.2 What Have Experiments Revealed about Hormones and Their Action?

The first hormone discovered was the gut hormone secretin

Early experiments on insects illuminated hormonal signaling systems

Three hormones regulate molting and maturation in arthropods

Chemical signaling between cells was critical for the evolution of multicellularity. Even sponges, the most primitive of multicellular animals, have forms of intercellular chemical communication. Over evolutionary time, there have been relatively minor changes in hormone structure but many major changes in hormone function. For example, prolactin's structure is similar in all vertebrates, but this hormone stimulates milk production in female mammals, migration in salmon, and nest building in birds.

Secretin was the first hormone described. It stimulates digestive fluid secretion by the pancreas. Bayliss and Starling demonstrated the existence of a diffusible chemical signal by removing all nerves from the pancreas and showing that the pancreas secretion could be stimulated by injection of (1) stomach acid into the gut or (2) ground-up gut tissue into the bloodstream. Similarly, addition of culture fluid from "exercised" muscle cells stimulated metabolic changes in fat cells, whereas addition of culture fluid from control muscle cells did not. The substance of interest produced by the "exercised" muscle cells is a hormone called irisin.

Experiments performed by Sir Vincent Wigglesworth shed light on the hormonal control of molting in insects. Insects have hard exoskeletons that must be shed (molted) and regrown as the insect grows. The growth stage between two molts is called an instar. Sir Wigglesworth decapitated the blood-sucking bug *Rhodnius* at various times after feeding (see Figure 41.6). He found that a substance secreted in the heads of the bugs stimulated molting after a blood meal. If the head was removed immediately after a blood meal, molting did not occur, but if the head was removed a week after the meal, molting did occur. A substance produced in the insect's head shortly after feeding diffused slowly, eventually triggering molting. In further experiments, unfed decapitated fifth-instar bugs were connected to blood-fed, partly decapitated fourth-instar bugs by a glass tube. Both bugs molted into juvenile forms, showing that a chemical from the rear part of the fourth-instar bug could diffuse through the tube and cause the fifth-instar bug to molt into another juvenile instead of becoming an adult. The brains of *Rhodnius* and other insects produce prothoracicotropic hormone (PTTH), which is stored in the corpora cardiaca attached to the brain. In *Rhodnius* the blood meal causes PTTH release. PTTH diffuses to the prothoracic gland, which releases a steroid hormone called ecdysone. Ecdysone then diffuses to target tissues where it affects gene expression to cause the production of enzymes involved in molting.

Juvenile hormone, which is secreted by the corpora allata, prevents premature maturation. In insects such as *Rhodnius* with incomplete metamorphosis, high levels of juvenile hormone result in the molting bug's becoming a slightly larger juvenile. When levels of juvenile hormone drop, the bug becomes an adult. In insects such as butterflies or moths that undergo complete metamorphosis, the levels of juvenile hormone decrease as the larva molts a fixed number of times. When juvenile hormone falls below a certain level, the larva spins into a cocoon and molts into a pupa. No juvenile hormone is secreted during the pupal stage, so the pupa metamorphoses into the adult (see Figure 41.7).

Question 3. What would happen to a fourth-instar *Rhodnius* if it were either partially or fully decapitated one week after a blood meal?

Question 4. In vertebrates, the hormone testosterone is produced and released mainly by the testes. Use your knowledge of experimental approaches to demonstrating hormonal action to design a study to determine the effects of testosterone on the behavior of male rats.

41.3 How Do the Nervous and Endocrine Systems Interact?

The pituitary is an interface between the nervous and endocrine systems

The anterior pituitary is controlled by hypothalamic neurohormones

Negative feedback loops regulate hormone secretion

The pituitary gland is a key meeting point for nervous and endocrine systems. Located just below the brain and in constant communication with the hypothalamus, the pituitary gland produces hormones that control many other endocrine glands. The anterior pituitary is derived from gut epithelium (an outpocketing of the embryonic mouth) and the posterior pituitary is derived from neural tissue. These two parts function somewhat differently: endocrine cells in the anterior pituitary are controlled by neurohormones from the hypothalamus, whereas posterior pituitary hormones are made by hypothalamic neurons and then transported down axons that extend into the posterior pituitary where the hormones are stored until they are released (see Figure 41.10).

The two neurohormones made in the hypothalamus and released from the posterior pituitary are oxytocin and antidiuretic hormone (ADH, also known as vasopressin). Antidiuretic hormone acts on the kidneys to stimulate reabsorption of water when blood pressure drops or salt increases in the blood. Oxytocin stimulates uterine contractions during childbirth and ejection of milk from mammary glands. It can be secreted simply in response to the sight and sound of a baby—an example of how an external stimulus received by the nervous system can control a hormonal process. Oxytocin can also promote pair bonding and maternal bonding.

The anterior pituitary releases four tropic hormones: thyroid-stimulating hormone (TSH), which controls thyroid function; luteinizing hormone (LH), which controls the function of the testes; follicle-stimulating hormone (FSH), which controls ovarian function; and adrenocorticotropic hormone (ACTH), which controls the adrenal cortex. A different cell type produces each of the tropic hormones. The anterior pituitary also produces several other peptide hormones, including growth hormone and prolactin. Growth hormone (GH) stimulates the liver to release chemical signals that stimulate bone and cartilage growth. GH also stimulates amino acid uptake by other cells. Over- or underproduction of GH can cause gigantism or pituitary dwarfism, respectively. The anterior pituitary also produces small amounts of endorphins and enkephalins.

The endocrine cells of the anterior pituitary make and release their hormones in response to neurohormones secreted by the hypothalamus and transported to the anterior pituitary by portal blood vessels (see Figure 41.10). For example, thyrotropin-releasing hormone (TRH) causes the release of thyrotropin by the anterior pituitary. Thyrotropin, in turn, stimulates the thyroid gland to release thyroxine. Similarly, gonadotropin-releasing hormone (GnRH) causes the release of tropic hormones that control gonadal activity. Other hypothalamic neurohormones that control anterior pituitary secretion include prolactin-releasing and release-inhibiting hormones, growth hormone-releasing and release-inhibiting hormones, and corticotropin-releasing hormone.

The anterior pituitary is under negative feedback control by hormones of the target glands it stimulates (see Figure 41.11). For example, cortisol is released by the adrenal glands in response to corticotropin from the anterior pituitary. As cortisol levels rise, release of corticotropin is inhibited. Cortisol also inhibits release of corticotropin-releasing hormone (CRH) by the hypothalamus.

Question 5. Alcohol inhibits release of antidiuretic hormone (ADH). What effect would alcohol consumption have on urination?

Question 6. Make a flow chart depicting how breast-feeding an infant soon after birth helps a mother's uterus return to its prepregnancy size.

41.4 What Are the Major Endocrine Glands and Hormones?

The thyroid gland secretes thyroxine

Three hormones regulate blood calcium concentrations

PTH lowers blood phosphate levels

Insulin and glucagon regulate blood glucose concentrations

The adrenal gland is two glands in one

Sex steroids are produced by the gonads

Melatonin is involved in biological rhythms and photoperiodicity

Many chemicals may act as hormones

The following examples of hormonal influence on physiology focus on humans, but they are similar in all mammals. The thyroid gland wraps around the front of the trachea, expanding into a lobe on each side. It helps regulate metabolic rate through the actions of thyroxine (T_4) and triiodothyronine (T_3), both of which increase metabolic rate in mammals. The precursor to T_4, thyroglobulin, is made by epithelial cells that surround a round hollow structure known as a follicle, where the thyroglobulin is stored and has iodine added to it. When the thyroid gland is stimulated by thyrotropin (also called thyroid-stimulating hormone or TSH) produced by the anterior pituitary gland, the cells surrounding the follicle take up thyroglobulin. Thyroglobulin is then cleaved into a dipeptide consisting of two tyrosine residues (see Figure 41.12). In T_4 all four possible iodination sites have an ion atom, while in T_3 only three out of the four possible sites are bound to an ion atom. T_4 is more abundant, but T_3 is the more active form of thyroid hormone. Target cells have diodinase enzymes that can convert T_4 to T_3, thereby setting their own sensitivity to the hormone.

Thyroxine is lipid-soluble, so it enters cells on its own and binds to receptors in the nucleus. Changes in gene transcription caused by thyroxine cause changes in protein synthesis that provide the metabolic machinery to increase metabolic rates. Thyroxin levels increase following days of cold exposure. Thyroxin is also very important during development and growth because it promotes uptake of amino acids and synthesis of proteins. Insufficient thyroxine during prenatal or early postnatal development can cause cretinism, a condition characterized by delayed physical and mental development.

Hyperthyroidism (thyroxine excess) or hypothyroidism (thyroxine deficiency) can result in an enlarged thyroid gland, a condition known as goiter. Hyperthyroidism is often the result of an autoimmune response that activates TSH receptors, causing uncontrolled production and release of thyroxine. In hypothyroidism there is insufficient circulating thyroxine to turn off TSH production via negative feedback. This most often occurs when lack of dietary iodine causes more and more thyroglobulin production without enough iodination to produce enough active thyroxine.

Cells found in the spaces between thyroid follicles produce calcitonin, one of three hormones involved in regulating blood calcium levels. Careful regulation of calcium levels is essential because slight decreases in plasma calcium can cause muscle spasms and seizures and slight increases can cause weak muscles and a depressed nervous system. Almost 99 percent of the body's calcium is in bone tissue, about 1 percent is found in cells, and only 0.1 percent is in extracellular fluid. The body has three strategies for changing blood calcium levels: deposition or resorption of bone, excretion or retention by the kidneys, and absorption from the digestive system. Calcitonin lowers blood calcium levels by inhibiting osteoclasts (which break down bone) and thus allowing more bone deposition by osteoblasts. In adults, bone turnover is not very high, so the other mechanisms of calcium regulation have greater importance.

The second hormone that affects blood calcium is parathyroid hormone (PTH) produced by four small glands found on the posterior surface of the thyroid gland. PTH increases blood calcium by stimulating the bone-turnover activities of osteoclasts and osteoblasts and by increasing the reabsorption of calcium by the kidneys. Circulating calcium binds to a surface receptor on the parathyroid cells, which inhibits PTH synthesis and release (negative feedback). PTH also acts to lower blood phosphate levels by stimulating the kidneys to eliminate phosphate in the urine. This helps prevent the precipitation of calcium phosphate salts within the body. The third hormone affecting blood calcium is calcitriol, which stimulates cells of the digestive tract to absorb calcium from food. The liver and kidneys convert vitamin D (also called calciferol) into calcitriol. Strictly speaking, vitamin D is not a vitamin because skin cells can convert cholesterol into vitamin D in the presence of sufficient ultraviolet wavelengths from sunlight.

Endocrine cells in the pancreas called the islets of Langerhans produce the hormones insulin and glucagon, which regulate blood glucose levels. Glucose enters cells by diffusion, but it does not pass the lipid membrane easily. Proteins in the cells called glucose transporters assist in the entry of glucose into the cell. When insulin binds to receptors on cells, the glucose transporters move from the cytoplasm to the cell membrane, allowing glucose to enter the cell from the bloodstream. When insulin is absent or the receptors become insensitive to it, glucose accumulates in the blood, resulting in the disease diabetes mellitus. With type I diabetes there is a lack of insulin, whereas in type II diabetes the insulin receptors become unresponsive to insulin. In either case, the increased concentration of glucose in the blood causes water to move out of the cells. The rise in blood volume increases urine production. Glucose in the tubules of the kidneys also pulls more water into the urine. Individuals with diabetes can become dehydrated, and they do not have adequate metabolic fuel.

Insulin is produced by beta cells within the pancreatic islets of Langerhans. When blood glucose rises after a meal, these cells release insulin. Increased plasma membrane glucose transporters cause cells to use more glucose as fuel and to convert it into storage products like glycogen and fat. When blood glucose falls, insulin is no longer released and cells start using more glycogen and fat as fuel. If blood glucose falls further, other alpha islet cells release glucagon—a hormone produced by these cells that has effects opposite to those of insulin. Glucagon stimulates the liver to break down glycogen and release glucose into the blood. Delta islet cells produce the hormone somatostatin, which is released following rapid increases in blood glucose and amino acids. Somatostatin inhibits insulin and glucagon release and also slows digestion in the gut. Somatostatin is also made in the hypothalamus and released into portal blood vessels, where it acts to inhibit the release of growth hormone and thyrotropin from the anterior pituitary.

The adrenal glands have two regions: the adrenal medulla makes up the core of the glands, and the adrenal cortex surrounds the medulla. Epinephrine and norepinephrine are produced by the adrenal medulla and are involved in the fight-or-flight response. The medulla develops from and is controlled by the nervous system. Epinephrine and norepinephrine are water-soluble and bind to cell surface adrenergic receptors. The α-adrenergic receptors respond more strongly to norepinephrine, but β-adrenergic receptors respond equally to both hormones. "Beta blocker" drugs inhibit the β-adrenergic receptors that largely mediate the fight-or-flight response without affecting the physiological regulatory processes controlled by norepinephrine's interactions with the alpha-adrenergic receptors (see Figure 41.16).

The adrenal cortex produces mineralocorticoids, glucocorticoids, and relatively small amounts of sex steroids. The main mineralocorticoid is aldosterone, which stimulates the kidneys to reabsorb sodium and excrete potassium. The main glucocorticoid in humans is cortisol, which mediates our response to stress by directing cells that are not necessary for the fight-or-flight response to decrease their use of glucose and to use fats and proteins as energy sources instead. Cortisol also inhibits the immune system and has long-term effects due to its gradual release from blood carrier proteins and its action on target cell gene expression. Corticotropin (also called adrenocorticotropic hormone or ACTH) from the anterior pituitary controls cortisol synthesis and release. Corticotropin's release, in turn, is controlled by the hypothalamus's corticotropin-releasing hormone.

Gonads produce hormones as well as sperm (in testes) or ova (in ovaries). The male steroid hormones are androgens (primarily testosterone) and the female steroids are progesterone and estrogens (primarily estradiol). Both sexes make and use androgens and estrogens, but the relative levels differ between the sexes because females have high levels of an enzyme, aromatase, that converts testosterone to estradiol. Early in development, the sex organs of human embryos are similar. At about week 7, expression of genes on the Y chromosome in male embryos causes the undifferentiated gonads to begin producing androgens. In response to androgens, the reproductive system develops into a male system. In the absence of androgens, the reproductive system develops into a female system (see Figure 41.17).

Luteinizing hormone (LH) and follicle-stimulating hormone (FSH) from the anterior pituitary gland (collectively called gonadotropins) control the production of sex steroids. The production of these tropic hormones by the anterior pituitary is controlled by the hypothalamic gonadotropin-releasing hormones (GnRH). Puberty begins when decreased hypothalamic sensitivity to negative feedback causes GnRH levels to increase. The GnRH stimulates LH and FSH production and release, which stimulates the gonads to produce more sex hormones. These sex steroids promote development of secondary sexual characteristics.

Finally, the pineal gland produces the amine hormone melatonin. The release of melatonin by the pineal gland occurs in the dark; light inhibits melatonin release. Melatonin is involved in biological rhythms, including photoperiodicity, the phenomenon whereby changing day length across the seasons prompts changes in physiology.

Question 7. Calcitonin is considered to be most important during childhood, and possibly important at certain times in adulthood, such as during the late stages of a pregnancy. Why might calcitonin be important at these particular times in life?

Question 8. Would the oversecretion or undersecretion of aldosterone produce high blood pressure? Explain.

Question 9. Would castration (removal of the testes) lead to an immediate and complete absence of testosterone in the person's bloodstream? Why or why not?

Question 10. In Diagram A below, identify and match the gland with its location in the body. In the case of gonads, you may use the same location as long as you identify the sex of the individual. In Diagram B, label the structures indicated. In Diagram C, list the hormones secreted by each gland pictured.

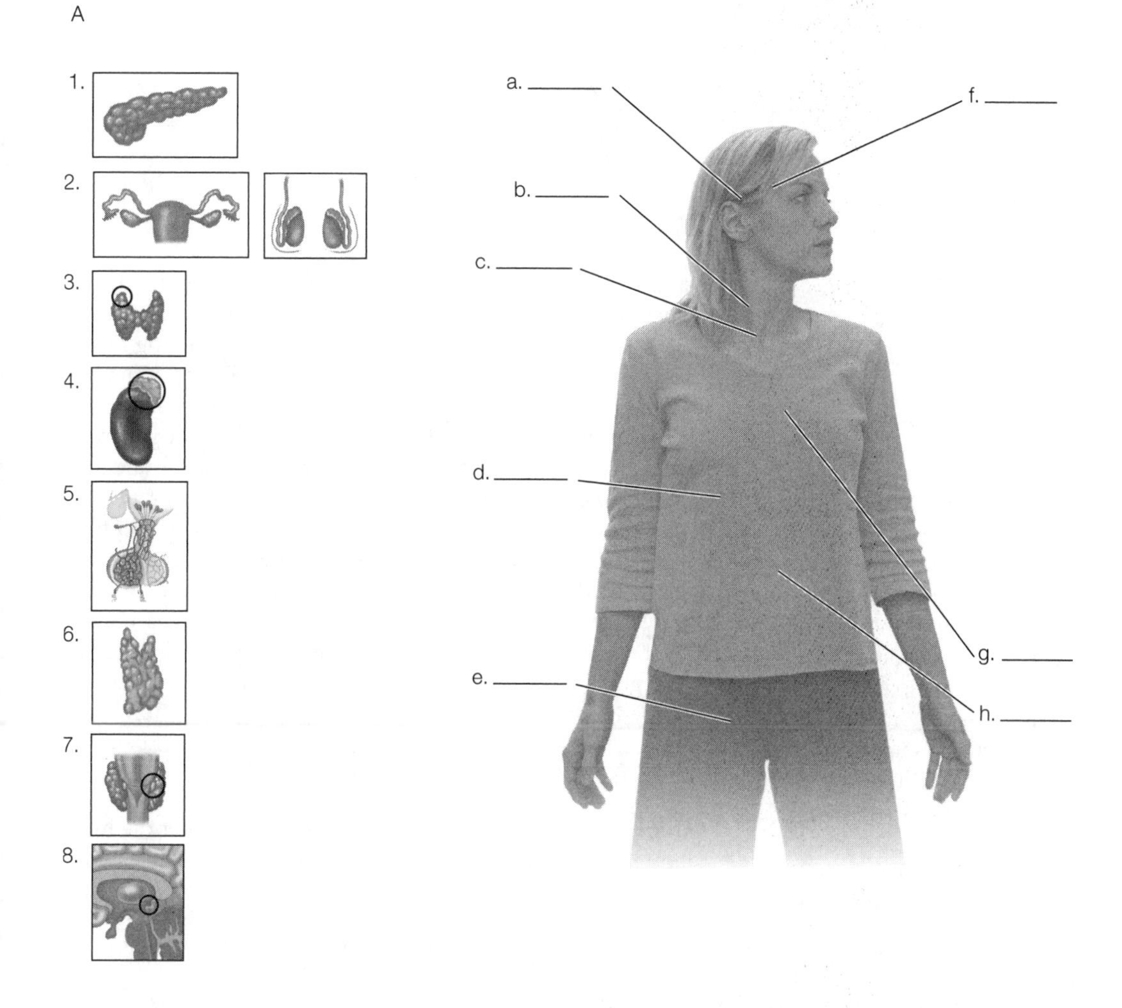

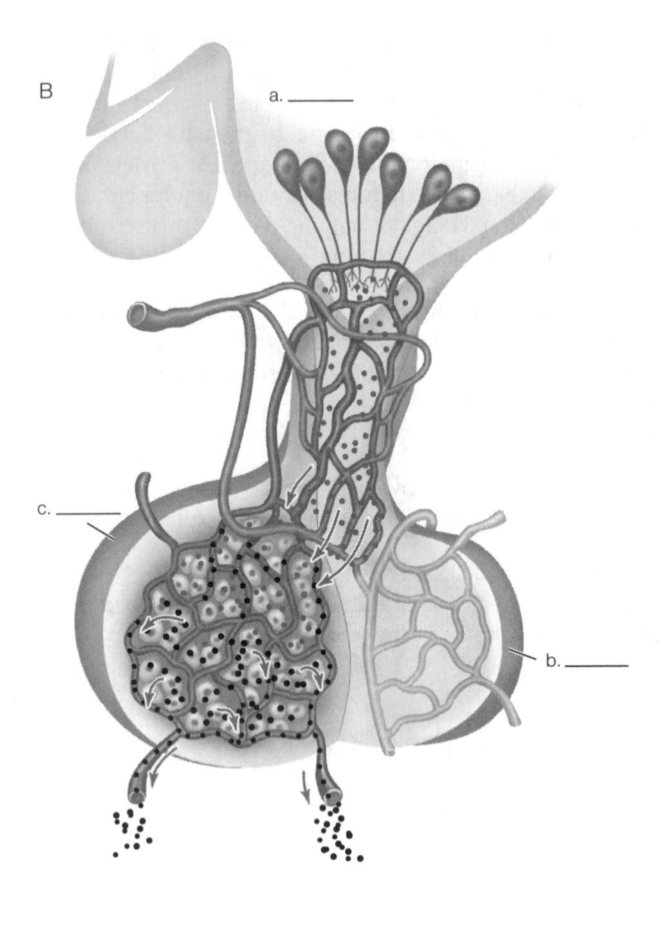

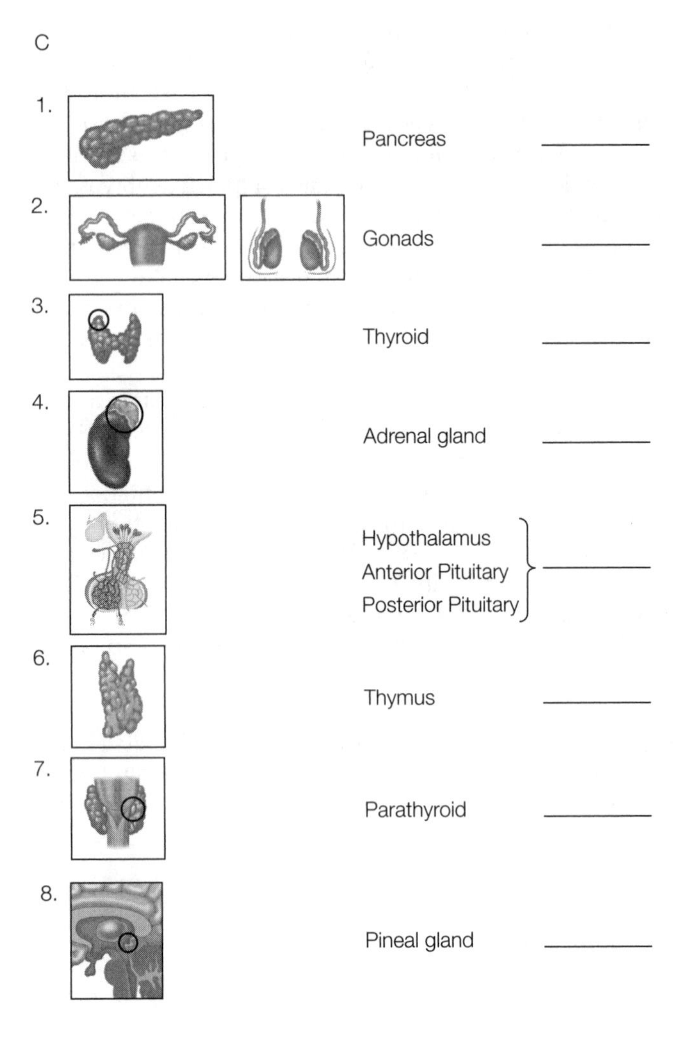

41.5 How Do We Study Mechanisms of Hormone Action?

Hormones can be detected and measured with immunoassays

A hormone can act through many receptors

Since many hormones are produced in small quantities, scientists have developed specialized tests to measure their effects and their interactions with their receptors. Hormone concentrations are measured by immunoassays (see Figure 41.19). This technique analyzes the amount of labeled hormone that binds to a given amount of antibody-binding sites when different amounts of unlabeled hormone are present. Results from an unknown sample are compared with a standard curve made with known hormone amounts to determine the hormone concentration in the sample. Immunoassays allow scientists to determine the time course or half-life of hormone action and to measure dose–response relationships (see Figure 41.20).

A single hormone can have many different receptors, and efforts to characterize receptors often rely on biochemical separation techniques such as affinity chromatography. In this technique, the hormone is attached to beads in a column and selected cells are poured through the column. If they have receptors for that hormone, the receptors will bind to the hormone and can then be isolated. Understanding how hormone receptors differ from cell type to cell type permits development of drugs that selectively block or stimulate very specific responses. Many receptors can be grouped in molecular families that share structural features. Researchers use knowledge about shared DNA coding sequences within receptor families to search for previously unknown receptors.

The abundance of hormone receptors is sometimes regulated through negative feedback mechanisms. Downregulation involves a decrease in the number of receptors in response to high levels of a hormone (e.g., type II diabetes). Upregulation can occur when levels of a hormone are suppressed; in response, a target cell may increase the number of receptors for that hormone (e.g., long-term use of beta blockers).

Question 11. Why is a standard curve necessary for the interpretation of immunoassay results on an unknown sample?

Question 12. Why is it advantageous for an animal to have multiple different types of receptors for the same hormone?

Test Yourself

1. A paracrine signal
   a. circulates in the bloodstream and affects distant cells.
   b. always acts on a wide variety of target tissues.
   c. acts on nearby cells.
   d. acts on neuronal cells only.
   e. acts on glands only.
   Textbook Reference: *41.1 What Are Hormones and How Do They Work?*

2. What result would you expect from an experiment in which a fed fifth-instar bug is completely decapitated one week after a blood meal and connected by glass tubing to an unfed, partially decapitated, fourth-instar bug?
   a. Neither bug would molt.
   b. Only the fifth-instar bug would molt and it would become an adult.
   c. Only the fifth-instar bug would molt and it would remain a juvenile.
   d. Both bugs would molt and only the fifth-instar bug would become an adult.
   e. Both bugs would molt and both bugs would remain juveniles.
   Textbook Reference: *41.2 What Have Experiments Revealed about Hormones and Their Action?*

3. Which of the following is the definition of "upregulation"?
   a. An increase in hormone receptors in response to low hormone levels
   b. An increase in hormone receptors in response to high neurotransmitter levels
   c. An increase in hormone levels produced by an increase in hormone receptors
   d. A decrease in hormone levels produced by a decrease in hormone receptors
   e. A change in hormone levels in response to a change in receptor levels
   Textbook Reference: *41.5 How Do We Study Mechanisms of Hormone Action?*

4. Like many hormones, growth hormone (GH) participates in a negative feedback loop that regulates its own secretion. Which of the following processes is most likely part of this negative feedback loop?
   a. Growth hormone stimulates release of growth hormone-releasing hormone.
   b. Growth hormone stimulates release of growth hormone release-inhibiting hormone.
   c. Growth hormone stimulates release of corticotropin-releasing hormone.
   d. Growth hormone inhibits release of all tropic hormones.
   e. Growth hormone inhibits blood flow through the portal blood vessels.
   Textbook Reference: *41.3 How Do the Nervous and Endocrine Systems Interact?*

5. In insects, juvenile hormone
   a. is produced by the corpora cardiaca.
   b. is also known as brain hormone.
   c. prevents maturation.
   d. is produced only by those with complete metamorphosis.
   e. All of the above
   Textbook Reference: *41.2 What Have Experiments Revealed about Hormones and Their Action?*

6. You are performing affinity chromatography on cultured liver cells with the peptide hormone gastrin attached to the column beads, but when you analyze the beads, you do not detect anything bound to them. Which of the following is the best conclusion?
   a. Your standard curve is not accurate.
   b. There was not enough labeled gastrin in the sample.
   c. Gastrin hormone acts directly on tissues instead of via a receptor.
   d. Liver cells must express an unknown gastrin receptor.
   e. Liver cells must not have any gastrin receptors.
   Textbook Reference: *41.5 How Do we Study Mechanisms of Hormone Action?*

7. Which of the following glands has one part derived from gut epithelium and a second part derived from neural tissue?
   a. Adrenal gland
   b. Thyroid gland
   c. Hypothalamus
   d. Pituitary gland
   e. Parathyroid gland
   Textbook Reference: *41.3 How Do the Nervous and the Endocrine Systems Interact?*

8. Which of the following glands regulate(s) daily rhythms?
   a. Pineal gland
   b. Thyroid gland
   c. Adrenal glands
   d. Ovaries
   e. Pituitary gland
   Textbook Reference: *41.4 What Are the Major Endocrine Glands and Hormones?*

9. Which of the following represents the correct sequence of steps leading to the release of thyroxine by the thyroid gland?
   a. Thyroid-stimulating hormone, thyrotropin-releasing hormone, thyroxine
   b. Adrenocorticotropic hormone, thyrotropin-releasing hormone, thyroxine
   c. Thyrotropin-releasing hormone, thyroid-stimulating hormone, thyroxine

d. Thyroid-stimulating hormone, aldosterone, thyroxine
e. Prolactin, thyroid-stimulating hormone, thyroxine
Textbook Reference: *41.4 What Are the Major Endocrine Glands and Hormones?*

10. The target tissues of hormones are those tissues that
a. can be penetrated by the particular hormones.
b. have specific enzymes with which the hormones interact directly.
c. have high concentrations of the "second messenger."
d. have receptors for the particular hormones.
e. have the particular genes that the hormones can express.
Textbook Reference: *41.1 What Are Hormones and How Do They Work?*

11. Which of the following vertebrate hormones are produced in the anterior pituitary gland?
a. Somatostatin, antidiuretic hormone, and insulin
b. Prolactin, growth hormone, and enkephalins
c. Oxytocin, prolactin, and adrenocorticotropin
d. Estrogen, progesterone, and testosterone
e. Growth hormone, gonadotropin-releasing hormone, and thyroid-releasing hormone
Textbook Reference: *41.3 How Do the Nervous and Endocrine Systems Interact?*

12. Hormones that are secreted by one endocrine gland and control the activities of another endocrine gland are called _______ hormones.
a. growth
b. obstructive
c. tropic
d. selective
e. paracrine
Textbook Reference: *41.3 How Do the Nervous and Endocrine Systems Interact*

13. Which of the following statements about parathyroid hormone (PTH) is true?
a. It increases calcium in the blood.
b. It decreases calcium in the blood.
c. It alters metabolic rate.
d. It increases blood sodium.
e. It initiates the fight-or-flight response.
Textbook Reference: *41.4 What Are the Major Endocrine Glands and Hormones?*

14. Which of the following tumors would cause weakened bones?
a. A tumor of the parathyroid glands
b. A tumor of the thyroid glands
c. A tumor of the adrenal cortex
d. A tumor of the adrenal medulla
e. A tumor of the pancreas
Textbook Reference: *41.4 What Are the Major Endocrine Glands and Hormones?*

15. Which of the following statements about steroid hormones is true?
a. They are water-soluble.
b. They are produced by the thyroid gland.
c. They are lipid-soluble.
d. They are derived from the amino acid tyrosine.
e. They are associated with second, messenger cascades.
Textbook Reference: *41.1 What Are Hormones and How Do They Work?*

Answers

Key Concept Review

1. The plasma membrane of a cell is hydrophobic. Lipid-soluble hormones, which (as their name suggests) dissolve in lipids (fats), can move through the plasma membrane and act on receptors inside the target cells. Water-soluble hormones cannot cross the lipid-based plasma membrane and can only act on receptors on the outside of target cells. Receptor-binding by water-soluble hormones initiates changes in the cell by activating enzymes. Because lipid-soluble hormones cross the plasma membrane and enter the cell, they tend to remain active for a longer time period than do water-soluble hormones. Lipid-soluble hormones often cause changes in gene expression, and it takes longer to make new proteins than simply to activate or deactivate enzymes that are already present. Thus, response to water-soluble hormones is usually more rapid than response to lipid-soluble hormones.
2. Different organs may have different receptors that cause different cell responses. Alternatively, even the same receptor with the same second messenger could cause a different cell response, since different cell types have different enzymes. Activation or deactivation of tissue and organ-specific enzymes will cause tissue and organ-specific responses.
3. The insect *Rhodnius* can live for long periods of time after it has been decapitated. If the insect is partially decapitated, leaving the corpora allata, it will molt into a fifth-instar juvenile, because the corpora allata produces juvenile hormone, which prevents maturation into an adult. An insect that is fully decapitated will molt into an adult rather than another juvenile instar. The fully decapitated animal has enough prothoracicotropic hormone (brain hormone) circulating one week after a meal to allow for molting; however, in the absence of juvenile hormone, it molts into an adult.
4. One method would be to remove the testes, the main glands that produce testosterone, and record the behavior of the castrated rats once they have recovered from surgery. Later, testosterone could be replaced through injections or implants and the behavior of the rats could be examined to see if it returned to presurgery levels. Another method would be to monitor

levels of testosterone over the course of one year while simultaneously monitoring the behavior of the rats to see if there is a correlation between high levels of testosterone and certain behaviors, such as aggression. A third method would be to administer a drug that temporarily and reversibly blocks the effects of testosterone to see if it changed the behavior of the rats.

5. Because alcohol temporarily inhibits secretion of antidiuretic hormone (ADH) by the posterior pituitary, it will increase the amount of urine produced. The main function of ADH is to conserve body water by decreasing urine output.

6.

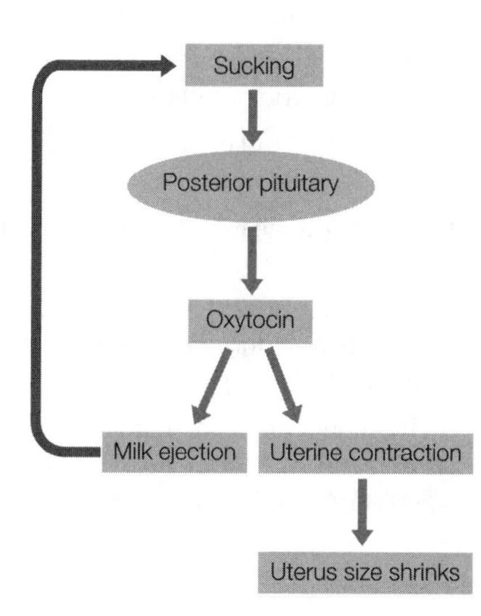

7. When blood calcium is high, calcitonin stimulates the absorption of calcium by bone and inhibits the breakdown of bone. These activities would be especially important during periods of intense growth, such as childhood and during pregnancy when the fetus is growing rapidly.

8. Aldosterone from the adrenal cortex stimulates the kidneys to conserve sodium and return it to the blood. Because water follows sodium, this will result in increased blood volume and blood pressure.

9. Castration would not lead to an immediate absence of testosterone in the bloodstream. Like many hormones, testosterone has a half-life of days to weeks. In addition, small amounts of the sex steroids are produced by the adrenal glands in both males and females.

10. Diagram A
    - a. 8. Pineal gland
    - b. 3. Thyroid gland
    - c. 7. Parathyroid gland
    - d. 4. Adrenal gland
    - e. 2. Gonads (ovaries; testes)
    - f. 5. Hypothalamus/Posterior pituitary/ Anterior pituitary
    - g. 6. Thymus
    - h. 1. Pancreas

    Diagram B
    - a. Hypothalamus
    - b. Posterior pituitary
    - c. Anterior pituitary

    Diagram C

    Pancreas: Insulin, glucagon, somatostatin
    Gonads:
      Ovaries: Estrogens, progesterone
      Testes: Testosterone
    Thyroid: Thyroxine (T_3 and T_4), calcitonin
    Adrenal gland:
      Cortex: Cortisol, aldosterone, testosterone (in both sexes), estrogen (in both sexes)
      Medulla: Epinephrine, norepinephrine
    Hypothalamus: Antidiuretic hormone (ADH), oxytocin
    Anterior pituitary: Thyrotropin (TSH), follicle stimulating hormone (FSH; produced in both sexes), luteinizing hormone (LH, produced in both sexes), adrenocorticotropic hormone (ACTH), growth hormone (GH), prolactin, melanocyte-stimulating hormone (MSH), endorphins, enkephalins.
    Posterior pituitary: Releases (but does not make) ADH and oxytocin.
    Thymus: Thymosin
    Parathyroid: Parathyroid hormone (PTH)
    Pineal gland: Melatonin

11. The unknown sample will have unlabeled hormone in it, but the immunoassay detects the amount of labeled hormone that binds to antibodies. In order to estimate the amount of hormone in the unknown sample, one must use a standard curve that describes how the labeled hormone binds in the presence of different amounts of unlabeled hormone.

12. Multiple types of receptors for the same hormone allows for increased flexibility in cellular response. As shown in Figure 41.16B, different receptors cause different cell responses (e.g., one receptor inhibits adenylyl cyclase and another receptor activates phospholipase C). Thus, organs with different receptors for the same hormone can respond to that hormone in different ways.

Test Yourself

1. **c.** Paracrine signals act on nearby cells. Autocrine signals act on the same cells that secrete them. Circulating hormones enter the bloodstream and affect distant cells.

2. **e.** Both bugs would molt because the fifth instar was not decapitated until a week after the blood meal. By this time the PTTH released following the blood meal has already stimulated the prothoracic gland to secrete ecdysone, the hormone that causes molting. Both bugs would remain juvenile because the unfed fourth-instar bug is continuously releasing juvenile hormone from the corpora allata (which are not removed in a partial decapitation).

3. **a.** Upregulation of hormone receptors on a cell is the production of more receptors when a hormone is at low levels over time in the blood or in other fluids surrounding the cell.
4. **b.** By increasing the amount of growth hormone release-inhibiting hormone that is secreted by the hypothalamus, growth hormone (GH) output decreases (negative feedback). If GH were to stimulate growth hormone-releasing hormone, it would be an example of positive feedback. Corticotropin-releasing hormone does not affect GH, and it would not make sense for GH to have a global effect on all tropic hormones either directly or by effects on the portal blood vessels.
5. **c.** Juvenile hormone is secreted continuously by the corpora allata of insects and prevents maturation. Prothoracicotropic hormone (PTTH) is secreted by the corpora cardiaca and is involved with ecdysone in molting.
6. **e.** Affinity chromatography is used to isolate hormone receptors. If nothing binds to the beads, the liver cells must not have any gastrin receptors. Remember that not all cells respond to a given hormone. Options **a** and **b** refer to things that could go wrong with immunoassays, and option **c** is not a possibility, since peptide hormones cannot even enter cells and must act via a membrane receptor.
7. **d.** The pituitary gland has an anterior lobe that develops from gut epithelium and a posterior lobe that develops from neural tissue.
8. **a.** The pineal gland secretes melatonin, a hormone that controls daily rhythms.
9. **c.** The hypothalamus releases thyrotropin-releasing hormone (TRH), which stimulates the anterior pituitary to release thyroid-stimulating hormone (TSH), which stimulates the thyroid gland to produce and release thyroxine.
10. **d.** For a hormone to act on a target cell, the target cell must have the receptors for the specific hormone to bind and trigger the hormonal action.
11. **b.** Prolactin, growth hormone, and enkephalins are produced in the anterior pituitary.
12. **c.** Hormones that control endocrine gland function are known as tropic hormones.
13. **a.** Parathyroid hormone increases the concentration of calcium in the blood.
14. **a.** A tumor on a parathyroid gland could cause oversecretion of parathyroid hormone (PTH). Because PTH pulls calcium from bone, weakened bones could result.
15. **c.** Steroid hormones are lipid-soluble and are produced by the gonads and adrenal glands. They do not act through second-messenger systems.

Immunology: Animal Defense Systems

The Big Picture

- Animals have two main ways to defend against pathogens: innate immune responses and adaptive immune responses. Innate immune responses are nonspecific and include barriers, phagocytic cells, specialized proteins, and inflammation. Adaptive immune responses are specific, slow to develop, and long-lasting. The two forms of adaptive responses are humoral (B cells and antibodies) and cellular (T cells).

Study Strategies

- The adaptive immune response may seem complex initially, but there are actually only two major components: B cells, which secrete antibodies, and T cells, which regulate the activities of other white blood cells and thereby help direct the immune response.
- MHC proteins may also seem complicated at first, but they have only a few essential functions in the immune system. These proteins are specific for each individual and are essential for presenting antigens to the two different types of T lymphocytes: class I MHC proteins (found on all nucleated cells in mammals) present antigens to T_C cells, while class II MHC proteins (found on macrophages, B cells, and dendritic cells) present antigens to T_H cells.
- The specificity of the immune response can be understood by reviewing cellular communication. Antibodies of B cells bind with antigen on the surface of pathogens; T cell receptors bind antigen presented on MHC I and MHC II proteins on the surface of cells. Once these membrane proteins interact, cytokines are secreted, and cells begin to divide and proliferate. B cells secrete antibodies.
- The rearrangement of DNA in the nucleus of the B cell seems strange because we typically think that the integrity of genetic information should be preserved. But this change in the genetic makeup of activated B and T cells is essential to providing the organism with diverse and specific antibodies and cell receptors.
- Make a list of the innate (nonspecific) responses that a human body can make to an invading pathogen, and then list cells that are part of the adaptive (specific) responses. Include proteins that are crucial to generating that specificity. Review how genetic rearrangements generate a diverse set of antibodies for B cells.
- Diagram an antibody and label the important components. Highlight those areas where mutations and alterations occur that change the amino acid sequence of the heavy and light chains.
- Review the expression and function of MHC I and MHC II proteins.
- Make a list of the functions of T_H cells and T_C cells and note ways in which these cells are similar and ways in which they are different.
- List some disorders of the immune system and their causes.
- Go to the Web addresses shown to review the following animated tutorials and activities. You can also find them in BioPortal (yourBioPortal.com), along with many additional learning resources.

  Animated Tutorial 42.1 Cells of the Immune System (Life10e.com/at42.1)

  Animated Tutorial 42.2 Pregnancy Test (Life10e.com/at42.2)

  Animated Tutorial 42.3 Humoral Immune Response (Life10e.com/at42.3)

  Animated Tutorial 42.4 A B Cell Builds an Antibody (Life10e.com/at42.4)

  Animated Tutorial 42.5 Cellular Immune Response (Life10e.com/at42.5)

  Activity 42.1 The Human Defense System (Life10e.com/ac42.1)

  Activity 42.2 Inflammatory Response (Life10e.com/ac42.2)

  Activity 42.3 Immunoglobulin Structure (Life10e.com/ac42.3)

Key Concept Review

42.1 What Are the Major Defense Systems of Animals?

Blood and lymph tissues play important roles in defense

White blood cells play many defensive roles

Immune system proteins bind pathogens or signal other cells

Pathogens are harmful organisms and viruses that cause disease. Animals defend themselves against these pathogens by (1) identifying the pathogen as nonself in the recognition phase, (2) mobilizing defenses in the activation phase, and (3) eliminating the invading pathogen in the effector phase.

Innate and adaptive defenses act together to protect mammals from pathogens. Innate, nonspecific defenses are inherited mechanisms that act rapidly. Innate defenses include physical barriers and cellular and chemical defenses. Most animals have innate responses. Adaptive, specific defenses are mechanisms aimed at particular targets. The specific defenses include humoral responses (B cells and antibodies) and cellular responses (T cells) (see Table 42.1). These responses develop slowly, last a long time, and are found only in vertebrates.

Blood and lymph tissues play important roles in defense. White blood cells and platelets are found in both blood plasma and lymph. There are two main types of white blood cells (leukocytes): phagocytes and lymphocytes (see Figure 42.2). Phagocytes, such as macrophages, engulf pathogens and are part of innate and adaptive responses. Lymphocytes, such as B cells and T cells, are involved in the adaptive response. When lymph passes through lymph nodes, lymphocytes detect any invading pathogens or toxins and initiate an immune response.

Cells and other components of the adaptive immune system interact with one another to fight off pathogens. There are four main players in the immune system response. Antibodies, which are produced by B cells, are proteins that bind to substances identified as nonself (antigens). Binding can inactivate pathogens and toxins and mark them for attack by immune system cells. Major histocompatibility complex (MHC) proteins present antigens to other immune system cells. MHC I proteins are found on the surface of most mammalian cells, and MHC II proteins are found on the surface of immune system cells. T cell receptors are proteins on the surface of T cells that recognize and bind to antigens displayed by the MHC proteins of other cells. Cytokines are soluble signal proteins released by many cell types. They bind to cell surface receptors and can activate or inactivate B cells, macrophages, and T cells.

Question 1. Deer ticks transmit the bacterium that causes Lyme disease. The test for Lyme disease involves drawing blood and looking for antibodies to the bacterium. Why would a doctor refuse to test for Lyme disease in a patient who found a deer tick attached to her body and asked to be tested immediately?

Question 2. Does urine play a role in innate or adaptive defense? Explain.

42.2 What Are the Characteristics of the Innate Defenses?

Barriers and local agents defend the body against invaders

Cell signaling pathways stimulate the body's defenses

Specialized proteins and cells participate in innate immunity

Inflammation is a coordinated response to infection or injury

Inflammation can cause medical problems

The innate defenses encountered first by a pathogen on the skin include the physical barrier of the skin, the saltiness of the skin, and the presence of normal flora (bacteria and fungi) with which the pathogen must compete.

Innate defenses encountered by pathogens that enter the nose or other internal organs include mucus (a secretion that traps microorganisms), lysozyme (an enzyme that causes bacteria to burst), harsh conditions (for example, acidic gastric juice in the stomach), and defensins (peptides that are toxic to many bacteria, enveloped viruses, and microbial eukaryotes).

Once a pathogen is inside the body, it is recognized as an invader (nonself) by the pattern recognition receptors (PRRs) of cells responsible for internal innate defenses (macrophages, dendritic cells, natural killer cells). PRRs recognize molecules called pathogen associated molecular patterns (PAMPs), which are commonly found in many types of microbes. The PAMPs are like a signal, and cell response to PRR activation involves a signal transduction pathway. One group of PRRs, the toll-like receptors, activates the transcription factor NF-κB, which initiates transcription of genes for defensive proteins.

About 20 complement proteins act as antimicrobial proteins in vertebrate blood. They function as follows: one protein binds to an invading cell so that phagocytes can recognize and destroy it; other proteins activate the inflammatory response; still other proteins lyse the invading cell.

A group of cytokines called interferons are small signaling proteins produced primarily in response to viral infection. They bind to uninfected cells and stimulate signaling pathways that inhibit viral reproduction if the cells later become infected. They also stimulate cells to digest bacterial and viral proteins into smaller peptides; this is an important first step in adaptive immunity.

Phagocytes ingest cells and viruses, which are then killed by either hydrolysis or defensins inside the phagocyte. Natural killer cells are white blood cells that recognize virus-infected or cancerous cells and cause them to die. Natural killer cells also interact with the adaptive defense mechanisms by lysing antibody-labeled target cells. Dendritic cells assist the innate response and release signals that activate adaptive immune response pathways. After dendritic cells endocytose and digest pathogens, they use their MHC II proteins to display selected digested PAMP fragments to cells involved in the adaptive response.

Mast cells help defend the body against infectious agents by releasing tumor necrosis factor, prostaglandins, and histamine as part of an inflammation response (see Figure 42.5). These chemicals help activate other cells involved in the immune response and make blood vessels wider and leakier so that fluid and immune system cells more easily enter the affected tissue. Although sometimes painful, inflammation can help promote healing. For example, certain cytokines released by phagocytes in the inflamed area signal the brain to increase body temperature (fever). This increased temperature speeds the immune response and can slow pathogen growth.

Sometimes an inappropriately strong inflammation response causes allergies or autoimmune diseases (see Section 42.6). An overactive inflammatory response can also cause sepsis, a condition in which blood vessels throughout the body dilate, causing a severe drop in blood pressure. Excessive pain or other problems associated with inflammation can often be treated with drugs that act on chemicals and signal transduction pathways activated as part of the inflammatory response.

Question 3. Pathogenic bacteria in the digestive tract often cause diarrhea. Why might diarrhea be considered a protective response by the body? Also, what other characteristic of the large intestines might help protect against invaders?

Question 4. Table 42.1 lists defensins as part of the early innate immune response and presents interferons as acting slightly later. Why do defensins act earlier than interferons?

42.3 How Does Adaptive Immunity Develop?

Adaptive immunity has four key features

Two types of adaptive immune responses interact: an overview

Adaptive immunity develops as a result of clonal selection

Clonal deletion helps the immune system distinguish self from nonself

Immunological memory results in a secondary immune response

Vaccines are an application of immunological memory

Adaptive immunity develops in response to a pathogen and is specific to that pathogen. For example, serum from guinea pigs exposed to diphtheria toxin protects against lethal doses of the same strain, but not against lethal doses of different strains. Adaptive immunity relies on antibody production by the humoral immune response and the destruction of infected cells by the cellular immune response. The four key features of the adaptive immune response are: (1) specificity, (2) distinction between self and nonself, (3) ability to respond to many different nonself molecules, and (4) immunological memory.

The adaptive immune response specifically targets pathogens present in the body. Each pathogen has several proteins or polysaccharides that can serve as antigenic determinants. These antigenic determinants, or epitopes, are small parts of a pathogen's molecule. Each type of T cell receptor and/or antibody produced by B cells binds to one specific antigenic determinant (or "antigen"). This specific binding initiates and targets the adaptive immune response.

One key feature of adaptive immunity is its capacity to distinguish self from nonself. Animals are able to tolerate their own antigens due to clonal deletion, negative selection, and Treg cell actions. Another feature of adaptive immunity is its diversity. Pathogens come in many shapes and sizes, so the immune system must be able to respond to millions of different antigens. A final feature of adaptive immunity is its immunological memory (i.e., it "remembers" a pathogen), which allows it to respond even more effectively in subsequent encounters with the same pathogen.

There are two interactive immune responses against invaders. The humoral immune response relies on B cells that make antibodies. The cellular immune response relies on cytotoxic (killer) T cells that bind to and destroy self cells that are mutated or infected by pathogens. These systems act in concert and share some of the same mechanisms (see Figure 42.7). A key event in both the humoral and cellular responses is presentation of the antigen to the immune system. Usually the "antigen-presenting cells" are dendritic cells and T-helper (T_H) cells which bind to the presented antigen and stimulate further humoral and cellular immune responses.

Antibodies bind to pathogen antigens in blood, lymph, and tissue fluids at the start of the humoral response. The first time an antigen is detected by T cell receptors, its B cell counterpart multiplies so that all daughter cells make and secrete the antibody. In the cellular immune response, T cell receptors stimulate an immune response after they bind to antigens from virus-infected or mutated cells.

The body's huge range of antigen-binding lymphocytes comes from DNA changes that occur when B and T cells are made in bone marrow. Millions of B and T cells develop, each producing just one type of antibody. These cells are ready and waiting before antigens appear. When an antigen is presented on a dendritic cell, it stimulates division of only those lymphocytes that can respond to that antigen. This proliferation of one specific type of cell makes a clone—a group of genetically identical cells. This process is called clonal selection. During early differentiation of T and B cells, those that respond to any self antigens are programmed to die in a process known as clonal deletion.

In the primary immune response, activated B and T cells produce effector cells and memory cells. Effector cells attack a pathogen by producing specific antibodies (called plasma cells in the case of B cells) or cytokines (in the case of T cells). Memory B and T cells live much longer than effector cells and are able to start dividing again if necessary. During a primary immune response the lymphocytes that bind the antigen multiply. This allows the body to mount a faster and stronger secondary immune response.

Vaccination is the intentional stimulation of the immune system with an antigen form that does not cause disease. This stimulates the primary immune response, produces memory cells, and ensures that any further exposure stimulates maximal immune response. Strategies used to produce harmless antigen sources include inactivation (killing the pathogen with heat or chemicals), attenuation (selecting for nonpathogenic bacterial strains), and recombinant DNA technology. Vaccination has eliminated many serious diseases worldwide.

Question 5. Imagine that a particular bacterium has surface molecules that resemble membrane proteins found on heart valves. What would happen to an individual infected with this bacterium if the infection were not treated?

Question 6. In the diagrams below, identify the cellular and humoral immune responses. Label each cell type and MHC protein with its correct class. Describe what is occurring in each step of the process for both diagrams.

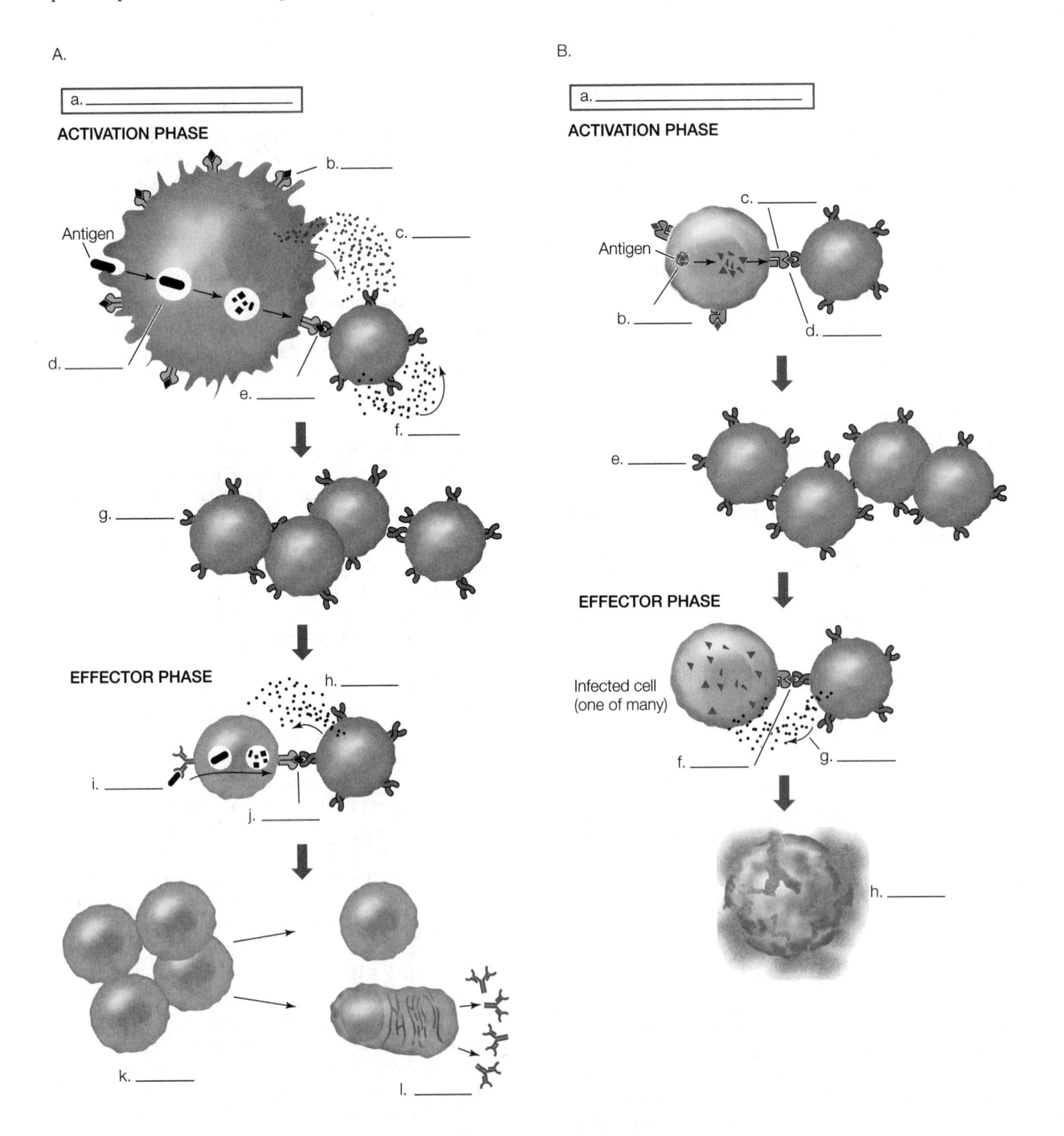

42.4 What Is the Humoral Immune Response?

Some B cells develop into plasma cells

Different antibodies share a common structure

There are five classes of immunoglobulins

Immunoglobulin diversity results from DNA rearrangements and other mutations

The constant region is involved in immunoglobulin class switching

Monoclonal antibodies have many uses

Antigen binding to a specific B cell receptor facilitates proliferation of plasma and memory cells derived from that B cell. T-helper (T_H) cell signals are also usually necessary for B cell division. These T_H signals are released when the same antigen binds to a T-helper cell whose specificity matches that of the activated B cell. Following B cell division and differentiation, the resulting plasma cells secrete antibodies that bind to the original activating antigen. Plasma cells have many ribosomes and a highly developed endoplasmic reticulum, which allow the cells to produce large amounts of antibody proteins.

Antibodies, like other immunoglobulin proteins, consist of two identical heavy polypeptide chains and two identical light polypeptide chains held together by disulfide bonds. Each polypeptide has a constant region (which is similar from one immunoglobulin to another within a class) and a variable region that forms the antigen-binding site (see Figure 42.10). The variable region produces the specificity of the millions of antibodies. The two antigen binding sites of each immunoglobulin are identical, so antibodies are bivalent and can bind two antigens at once. Because most antigens also have multiple epitopes, antibodies can form large complexes with antigens. These complexes are easy targets for phagocytes.

The constant region determines the class of the immunoglobulin: IgG, IgM, IgD, IgA, or IgE. IgG, secreted by B cells, is the most abundant immunoglobulin. During a secondary immune response, IgG molecules bind to antigens, signaling macrophages to phagocytose and destroy the antigens.

Differentiating B cells have a limited number of alleles for each antibody region, but the many different combinations of these alleles generate millions of unique antibodies. DNA fragments are rearranged and joined during B cell development to make multiple-allele supergenes (one for the heavy chain and another for the light chain) that code for unique antibodies. The variable region of the light chain originates from two families of genes, and the variable region of the heavy chain originates from three families of genes (*V*, *D*, and *J*). In mice there are 100 *V*, 30 *D*, and 6 *J* genes for the heavy chain and therefore $100 \times 30 \times 6 = 18{,}000$ different possible unique *VDJ* heavy chain combinations alone. Various types of mutation and recombination events achieve additional diversity. Similar processes lead to similar diversity among T cell receptors.

B cells typically make only one type of immunoglobulin at a time, but certain genetic deletion events can cause class switching. For example, a deletion that repositions the variable region genes next to a new constant region gene will cause the cell to make a new immunoglobulin class (but with the same antigen specificity). Cytokines secreted by T_H cells can initiate this type of class switching by B cells.

Immune responses are normally polyclonal because most antigens have many epitopes, each of which stimulates a different B cell clone. For research purposes, monoclonal antibodies that bind to only one antigenic determinant are more useful because the researcher can pick the one that binds to only the specific molecule of interest. Monoclonal antibodies are used in immunoassays to detect tiny amounts of molecules in fluids and tissues and in immunotherapy to specifically target cancer cells with radiation, a drug, or an immune response.

Question 7. What might happen to a person receiving a blood transfusion if the donor blood contained foreign antigens?

Question 8. What might be the consequence of a mutation that disables or impairs the function of an enzyme responsible for *VDJ* supergene assembly?

42.5 What Is the Cellular Immune Response?

T cell receptors bind to antigens on cell surfaces

MHC proteins present antigen to T cells

T-helper cells and MHC II proteins contribute to the humoral immune response

Cytotoxic T cells and MHC I proteins contribute to the cellular immune response

Regulatory T cells suppress the humoral and cellular immune responses

MHC proteins are important in tissue transplants

T cell membranes have glycoprotein receptors that are made of two different polypeptide chains. These receptors also contain a variable and a constant region (see Figure 42.14). The variable region provides the T cell receptor's antigen-binding specificity. T cell receptors only bind antigen fragments that are displayed by MHC proteins. Antigen binding causes T cells to divide and form clones of cytotoxic T cells (T_C cells) and T-helper cells (T_H cells). The T_C cells bind to and kill infected or mutated cells, whereas T_H cells contribute to the cellular and humoral immune responses.

MHC proteins are plasma membrane glycoproteins that present antigens to the two different types of T lymphocytes. Class I MHC proteins are present on the surface of every nucleated cell in the vertebrate body. They present antigens (fragments of virus proteins in virus-infected cells or abnormal proteins made by cancer cells) to T_C cells. Class II MHC proteins are found on the surfaces of B cells, macrophages, and dendritic cells. They present antigens to T_H cells after ingesting and breaking down pathogens within phagosomes (see Figure 42.15). T_C cells have a CD8 cell surface protein that binds them to MHC I proteins, whereas T_H cells have a CD4 cell surface protein that binds them to MHC II proteins. In the thymus gland, only T cells that can correctly bind to an MHC protein are retained during development (positive selection). This ensures that T cells bind only to antigen–MHC complexes. Next, negative selection eliminates any T cells that bind to self antigens. This clonal deletion keeps the adaptive immune system from responding to self molecules.

In the humoral immune response, T-helper cells that survive selection leave the thymus and enter lymphoid tissue, where they become activated if their specific receptor binds to an antigen-presenting cell. Activation causes the T_H cell to release cytokines which, in turn, cause the T_H cell to divide and produce a clone of T_H cells. T_H cells then activate only those naïve B cells that bind to the same antigen. Recall that B cells can, themselves, present antigens using their class II MHC proteins. When this happens, the cytokines released by the T_H cell directly stimulate the B cell to make a clone of plasma cells. Both pathways end with a clone of plasma (effector B) cells secreting antibodies that bind to the original antigen (see Figure 42.16A).

Activation of a T_H cell stimulates B cells to divide and produce antibodies against the antigen, but activation of a T_C cell (by binding to an MHC I–antigen complex) results in the production of clones with that same T cell receptor. These T_C cells bind to the mutated or virus-infected cells carrying the antigen on their class I MHC proteins. Once bound, the T_C cell kills the antigen-carrying cell; it accomplishes this by producing perforins that lyse the cell and/or by stimulating the cell to undergo apoptosis (see Figure 42.16B).

Regulatory T cells (Tregs), also made in the thymus, express a T cell receptor and are activated if they bind to MHC proteins complexed with self antigens. When Tregs are activated, they secrete a cytokine called interleukin-10 that blocks T_C and T_H cell activation and causes the T cells to undergo apoptosis (see Figure 42.17). Therefore, Tregs contribute to self-recognition. Treg destruction (as seen in experiments on mice or in human gene mutations) leads to an overactive immune system that responds to self protein.

Organ transplantation introduces new cells that present antigens from donor self proteins that act as nonself antigens in the transplant recipient. Only if the transplant is performed immediately after birth (or comes from an identical twin) will the donor antigens be learned/identified as self. Organ rejection can be combated by treating the patient with immune-suppressing drugs that prevent T cell development.

Question 9. Why are organ transplants more successful when the donor is a relative of the recipient?

Question 10. Draw a flow chart depicting the body's three lines of defense: the first line of innate mechanisms, the second line of innate mechanisms, and the third line of adaptive mechanisms. Within each of the three lines of defense, include the component mechanisms.

42.6 What Happens When the Immune System Malfunctions?

Allergic reactions result from hypersensitivity

Autoimmune diseases are caused by reactions against self antigens

AIDS is an immune deficiency disorder

Sometimes the immune system overreacts to an antigen. Immediate hypersensitivity in allergic reactions is caused, first, by initial exposure to an environmental antigen (or allergen) that stimulates production and release of large amounts of IgE. When this happens, basophils and mast cells bind the constant end of the IgE. If there is further exposure, binding of the allergen to the IgE causes the basophils and mast cells to release histamine (see Figure 42.18). Pollen allergies can be treated by desensitization, in which a small amount of allergen is injected to stimulate IgG production. If IgG reacts before IgE can interact with the allergen, the allergic response will be minimized. This approach generally does not work for food allergens. Delayed hypersensitivity occurs when antigen-presenting cells process an antigen and initiate a T cell response hours after exposure.

Autoimmunity occurs when T cells direct their response against self antigens. Potential origins of autoimmunity include the failure to delete a clone that makes antibodies against a self antigen and molecular mimicry, in which a T cell that recognizes a nonself antigen from a virus also recognizes a self antigen that has a similar structure. Examples of autoimmune diseases include systemic lupus erythematosis, rheumatoid arthritis, Hashimoto's thyroiditis, and insulin-dependent diabetes mellitus.

The human immunodeficiency virus (HIV) destroys T_H cells and causes acquired immune deficiency syndrome (AIDS). This retrovirus is transmitted through body fluids, including blood and semen. HIV initially infects macrophages, T_H cells, and antigen-presenting dendritic cells. Initially there is an immune response to the viral infection, and some T_H cells are activated. However, T_H cells eventually decline: HIV itself kills them and T_C cells lyse infected T_H cells. Extensive production of HIV by infected cells activates the humoral response, antibodies bind HIV, and the level of HIV in the blood decreases (see Figure 42.19). Nevertheless, the infection remains at a low level because of the depletion of T_H cells. A person may remain at this "set point" for 8 to 10 years. Eventually, the T_H cells are destroyed, and the patient becomes susceptible to opportunistic infections. Drug treatments generally focus on inhibiting HIV proteins involved in viral assembly and replication.

Question 11. If cytotoxic T cells were eliminated from a person's array of immune defenses, what kinds of disease would the person be susceptible to?

Question 12. The polio vaccine comes in two forms: an injected dose of inactivated (dead) virus and an oral vaccine of attenuated virus. Which would be the safer vaccine for a child with an immunodeficiency disorder? Explain your reasoning.

Test Yourself

1. Which of the following statements about phagocytes is true?
   a. They are derived from T and B cells.
   b. They present antigen on MHC II complexes.
   c. They digest nonself materials.

d. They are a type of erythrocyte.
e. They secrete antibodies.
Textbook Reference: *42.1 What Are the Major Defense Systems of Animals?*

2. In 2012 there was a serious meningitis outbreak due to contamination of injectable steroids. Interestingly, the fungus involved is commonly found in the environment. Why was this fungus much more likely to cause meningitis when found in the steroid medication than when contacted in the general environment?
a. The adaptive immune system cannot respond to an injected substance.
b. The fungus was able to avoid most innate defenses when injected.
c. Defensins cannot act on fungi.
d. Environmental fungi are recognized as nonself by PRRs, but are recognized as "self" when injected.
e. The injected fungus activated natural killer cells.
Textbook Reference: *42.2 What Are the Characteristics of the Innate Defenses?*

3. Which of the following is an innate, nonspecific defense of the immune system?
a. Macrophages
b. Natural killer cells
c. Complement proteins
d. Mucus
e. All of the above
Textbook Reference: *42.2 What Are the Characteristics of the Innate Defenses?*

4. When the receptor of a T_H cell binds to a pathogen presented on a macrophage, it
a. inactivates itself.
b. secretes cytokines.
c. inactivates B cells.
d. inactivates the macrophage.
e. becomes a T_C cell.
Textbook Reference: *42.5 What Is the Cellular Immune Response?*

5. Part of the normal immune response includes
a. the production of B memory cells.
b. the production of memory macrophages.
c. antibody secretion by eosinophils.
d. the production of B cells that attack the individual's own cells.
e. the production of complement proteins as a specified immune response.
Textbook Reference: *42.3 How Does Adaptive Immunity Develop?*

6. If there were only two versions each for the *V*, *D*, and *J* variable region antibody genes, how many different unique *VDJ* combinations could be made?
a. 2
b. 3
c. 6
d. 8
e. 16
Textbook Reference: *42.4 What Is the Humoral Immune Response?*

7. Which of the following statements about cytotoxic T (T_C) cells is true?
a. They release cytokines that activate B cells.
b. They attack pathogens by binding to cell surface antigens on those pathogens.
c. They destroy pathogens by engulfing them.
d. They destroy host cells that are infected with a virus.
e. They stimulate the classical complement pathway.
Textbook Reference: *42.5 What Is the Cellular Immune Response?*

8. When would a B cell present an antigen on an MHC I protein?
a. When the B cell is infected with a virus
b. As part of the effector phase of any immune response
c. After activation by a T_H cell
d. After ingesting a pathogen in a phagosome
e. After stimulation by cytokines
Textbook Reference: *42.5 What Is the Cellular Immune Response?*

9. An autoimmune disease is
a. active when organ transplantation is successful.
b. caused by viruses.
c. a response in which the immune cells attack the body's own tissues.
d. a result of the destruction of the immune system.
e. inflammation that is always targeted to a specific location in the body.
Textbook Reference: *42.6 What Happens When the Immune System Malfunctions?*

10. Patients with HIV are susceptible to a variety of infections because
a. the virus produces cell surface receptors that bind to pathogens, making it easier for those pathogens to be infective.
b. the synthesis of a DNA copy of the viral genome makes a person susceptible to infection.
c. HIV attacks and destroys the T_H cells, which are central to mounting an effective immune response.
d. HIV destroys B cells, so antibodies cannot be made in response to invading pathogens.
e. HIV mutates the B cells, so they cannot make the huge array of antibodies needed for an effective immune response.
Textbook Reference: *42.6 What Happens When the Immune System Malfunctions?*

11. DNA rearrangements in the B cell
a. are responsible for generating single B cells that can express many different antibodies.

b. lead to mutations in T cells, resulting in the elimination of essential T cell genes.
c. occur only in B memory cells.
d. are responsible for generating many different antibodies, with each B cell expressing only one set of identical antibodies.
e. occur only in fully mature B cells.

Textbook Reference: *42.4 What Is the Humoral Immune Response?*

12. Major histocompatibility proteins function in the immune system by
a. engulfing pathogens.
b. presenting antigens to T_C and T_H cells.
c. generating antibodies to different pathogens.
d. presenting antigen fragments to B cells.
e. presenting macrophages to T_H cells.

Textbook Reference: *42.5 What Is the Cellular Immune Response?*

13. The process by which immature B or T cells with the potential to respond strongly to self antigens undergo apoptosis is called
a. the primary immune response.
b. the secondary immune response.
c. immunological memory.
d. clonal selection.
e. clonal deletion.

Textbook Reference: *42.3 How Does Adaptive Immunity Develop?*

14. Inflammation occurs when _______ release _______.
a. mast cells; histamine
b. neutrophils; toxins
c. B cells; histamine
d. mast cells; toxins
e. neutrophils; antihistamine

Textbook Reference: *42.2 What Are the Characteristics of the Innate Defenses?*

15. When the sea urchin's genome was sequenced, researchers concluded that this non-vertebrate deuterostome had an exceptionally strong immune system. Which of the following observations most likely led researchers to this conclusion?
a. Sea urchins make more antibodies than humans do.
b. Sea urchins have many more toll-like receptors than humans do.
c. Sea urchins do not have clonal deletion of B cells, so they can respond to more different types of pathogens.
d. Sea urchins make a compound with antimicrobial properties that lyses bacteria.
e. Sea urchins have more types of MHC proteins than humans do.

Textbook Reference: *42.1 What Are the Major Defense Systems of Animals?; 42.2 What Are the Characteristics of the Innate Defenses?*

Answers

Key Concept Review

1. Antibodies are part of the adaptive, specific immune response, and such responses typically take at least 96 hours to develop (see Table 42.1). Thus, it would not make sense to test immediately for antibodies to the bacterium that causes Lyme disease. Ideally, the test would be performed at least 96 hours after the deer tick was found.
2. Urine plays a role in innate defense because it flushes pathogens from the urinary tract. The chemical composition of urine (e.g., its acidity) also makes the urinary tract inhospitable to some pathogens.
3. Diarrhea helps move the pathogen out of the digestive tract quickly. The large intestine is home to many bacteria that keep invaders in check.
4. Defensins are present all the time in mucus and within phagocytes, but interferons must be made in response to the presence of a virus. Defensins also act directly on pathogens, whereas interferons act via their effects on signaling pathways.
5. If the bacterial infection was not treated, the individual's immune response to this pathogen would include the secretion of antibodies. Such antibodies could cross-react with the individual's own membrane proteins on the cells of the heart valve, leading to an autoimmune response and possibly heart disease. This is an example of autoimmunity induced by molecular mimicry.
6. Diagram A
a. Humoral Immune Response
b. MHC class II receptor presents fragments of antigen.
c. Cytokines produced by the macrophage in response to phagocytosis of an antigen help activate T_H cells.
d. An antigen is taken into the macrophage by phagocytosis and broken down.
e. Antigen is presented to the CD4 receptor on the T_H cell
f. Cytokines released by the T_H cell stimulate it to proliferate.
g. The T_H cell divides to form a clone.
h. Cytokines are produced by the T_H cell. This will stimulate the B cell to proliferate and generate both memory cells and plasma cells.
i. An antigen binds to a B cell receptor (IgM). This triggers endocytosis of the antigen, followed by degradation and presentation of fragments on the MHC class II receptor.
j. The B cell MHC class II receptor presents the antigen fragment to the T_H cell's CD4 receptor.
k. The proliferating B cells give rise to both memory cells and plasma cells.
l. The plasma cell produces antibodies.

Diagram B
a. Cellular Immune Response

b. The viral proteins produced in the infected cell are degraded.
c. The viral proteins are displayed on the class I MHC protein.
d. The T_C cell's CD8 receptor binds to the MHC class I receptor that is presenting the viral antigen.
e. The T_C cell proliferates after binding to an antigen presented on the infected cell's MHC class I receptor.
f. The activated T_C cell recognizes an infected cell by the antigen presented on the infected cell's MHC class I receptor. This MHC receptor binds to the T_C cell's CD8 receptor.
g. The T_C cell releases perforin to destroy the infected cell.
h. The infected cell lyses, halting the viral replication process in the infected cell.

7. If a person receives a blood transfusion with donor blood that contains foreign antigens, antibodies in the recipient's blood will cause the donor cells to clump together. This can be damaging and possibly fatal.

8. Problems with *VDJ* supergene assembly would severely impair the immune system by greatly limiting the number and variety of heavy chain combinations.

9. Organ transplants are more successful if the MHC proteins and other cell surface markers are similar in the donor and recipient, making the immune system of the recipient more tolerant of the foreign tissue. Since close relatives have more genes in common than unrelated people do, they are more likely to have surface antigens on their cells that are similar. Organ transplants from unrelated individuals require the administration of immunosuppressive drugs to the recipient so that the patient's own immune system does not reject the foreign tissue. The drug cyclosporin, which inhibits T cell development, is often used.

10.

First line of defense (innate)
Physical barriers (e.g., skin, mucus membranes)
Chemical barriers (e.g., acid conditions, lysozyme)

↓

Second line of defense (innate)
Defensive cells (e.g., phagocytic cells, natural killer cells)
Defensive proteins (e.g., complement system, interferons)
Inflammation
Fever

↓

Third line of defense (adaptive)
Humoral immune response (B cells make antibodies)
Cellular immune response (T_C cells)

11. Cytotoxic T cells target virally infected cells and some cancer cells. If T_C cells were eliminated, the individual would be much more susceptible to viral infections and cancer.

12. The injected dose of inactivated poliovirus would be safer for a patient with a compromised immune system, because an inactivated pathogen cannot cause disease. In contrast, the attenuated vaccine contains a mutant form of the virus that does reproduce itself in the patient, although at a very slow rate. However, it is possible for the virus to mutate back to a virulent form (which does happen occasionally) and cause disease.

Test Yourself

1. **c.** Phagocytes are nonspecific cells (not B and T cells) that digest nonself materials and present protein fragments of those nonself materials on their surface via MHC class II complexes. They do not have antibodies.

2. **b.** An injected fungus will bypass many of the body's innate responses. This fungus is not normally pathogenic because it does not normally get past the skin barrier. Other options do not explain the increased risk; the adaptive immune response would still be able to respond, some defensins can act on fungi (a type of microbial eukaryote), injection would not affect self versus nonself recognition, and natural killer cells respond to cancers and virus-infected cells, not to fungi.

3. **e.** Macrophages, natural killer cells, cilia on mucus membranes, and complement proteins are all part of the nonspecific response of the immune system. B and T cells are part of the specific response of the immune system.

4. **b.** When the T_H cell binds antigen being presented on a macrophage, it secretes cytokines, which stimulate other immune cells to divide.

5. **a.** In a normal immune response, B memory cells are produced, allowing the organism to mount a faster and more effective response to any subsequent encounter with the pathogen. Memory macrophages do not exist. Eosinophils do not secrete antibodies. In an abnormal immune response, B cells that attack the individual's own cells can be activated, as seen in autoimmune diseases. Complement proteins are part of the innate, nonspecific immune response.

6. **d.** The total number of possible combinations is determined by multiplying the gene numbers of each region times each other. If there were only two gene possibilities for each gene family, the number of combinations would be $2 \times 2 \times 2 = 8$. In reality, there are many more genes in each gene family.

7. **d.** Cytotoxic cells bind to virus antigen presented on MHC I protein and destroy those cells. T_H cells release cytokines to activate B cells. The antibodies of B cells bind cell surface antigens on pathogens. T_C cells do not engulf pathogens. T cells have no involvement in the classical complement pathway.

8. **a.** Class I MHC proteins are present on all nucleated cells and present antigens when a cell is infected with a virus or has become cancerous. Thus, B cells have both

MHC I and MHC II proteins and would present antigens on MHC I proteins if they were infected with a virus. B cells present antigens on MHC II proteins as part of the effector phase of the humoral immune response, but this precedes T_H cell activation. B cells endocytose receptor-bound antigens, whereas macrophages ingest whole pathogens. Cytokines stimulate B cell proliferation (not antigen presentation).

9. **c.** Autoimmune diseases occur when the immune cells attack the body's own cells. Transplant rejection can occur if the body identifies the transplanted tissue as nonself, so this is not an autoimmune response. The immune system is damaged, not overactive, in an immune deficiency disorder.

10. **c.** An HIV-infected individual is more susceptible to a variety of infections because the virus destroys T_H cells, which are essential for mounting an effective immune response. HIV does not bind to pathogens and does not destroy B cells or cause mutations in their DNA that alter antibody production.

11. **d.** DNA rearrangements occur in B cell precursors and result in the expression of a unique kind of antibody by each mature B cell. DNA rearrangements also occur in T cells to generate T cell receptors. This rearrangement does not destroy essential T cell genes. Memory B cells are cells in which DNA rearrangement has already occurred.

12. **b.** MHC proteins present antigens to T_C and T_H cells. The cytokines of T cells (and not MHC proteins) activate B cells. MHC proteins do not generate antibodies. Macrophages are antigen presenting cells, but they are not themselves presented as antigens to T_H cells.

13. **e.** Clonal deletion is the process by which immature B or T cells that show the potential to mount a strong immune response to self antigens undergo apoptosis.

14. **a.** Inflammation occurs in response to an infection or tissue injury. Once an infection or injury occurs, the local mast cells release histamine. The histamine in turn makes the capillaries leak, and fluid accumulating in the area produces the inflammation.

15. **b.** Sea urchins are invertebrates, so they do not have an adaptive immune system. Thus, they do not make antibodies, have MHC proteins, or have clonal deletion of B cells (which would impair self-recognition rather than increasing pathogen response). However, sea urchins do have many more toll-like receptors than humans do (more than 20 times more!). Recall that toll-like receptors are part of the innate immune response and are found in both vertebrates and invertebrates. Sea urchins can make compounds with antimicrobial properties, but so do humans and many other organisms (e.g. defensins have antimicrobial properties), so this is not evidence of an exceptionally strong immune system.

Animal Reproduction

The Big Picture

- Animals reproduce both asexually and sexually. Asexual reproduction includes budding, regeneration, and parthenogenesis. Sexual reproduction has an advantage over asexual reproduction in that it increases genetic diversity. Sexual reproduction has three main stages: gametogenesis, mating (or spawning), and fertilization.
- Gametogenesis is the process by which the sex cells are produced. In males, haploid sperm develop from diploid spermatogonia by the process of spermatogenesis. In females, the haploid egg develops from diploid oogonia by the process of oogenesis.
- Fertilization is the union of one egg and one sperm to form a diploid zygote. External fertilization is typical of many aquatic animals. Internal fertilization, often with the help of accessory sex organs to ensure fertilization, occurs in terrestrial animals and some aquatic animals. The eggs and sperm of a given species recognize one another.
- The reproductive system of the human male produces semen, which contains sperm (produced in the testes) and substances secreted from seminal vesicles, the prostate gland, and bulbourethral glands. Hormones such as luteinizing hormone, follicle-stimulating hormone, inhibin, and testosterone control the male reproductive system. The human female has two interrelated cycles: the ovarian cycle, during which an egg is produced and released from the ovary, and the uterine cycle, during which the uterus is prepared for implantation and pregnancy. These two cycles occur in parallel and are coordinated by changes in hormone levels. Luteinizing hormone, follicle-stimulating hormone, estrogen, and progesterone all play a role in these cycles.
- Several methods are available for preventing pregnancy and for overcoming infertility.

Study Strategies

- Understanding the ploidy level of the developing egg and sperm can be very confusing, but it is essential in determining how genetic inheritance works.
- Recognize that although sperm and egg formation are completely separate processes, each with its own characteristics, the same general events are happening in both. This will help you recognize and remember the stages of spermatogenesis and oogenesis. Be sure though, you also understand the differences between spermatogenesis and oogenesis with respect to timing and number of gametes produced.
- Following the hormonal changes and cues associated with human ovarian and menstrual cycles can be difficult. Create a chronological sequence of hormones and the events they trigger during both cycles.
- Go to the Web addresses shown to review the following animated tutorials and activities. You can also find them in BioPortal (yourBioPortal.com), along with many additional learning resources.

  Animated Tutorial 43.1 Fertilization in a Sea Urchin (Life10e.com/at43.1)

  Animated Tutorial 43.2 The Ovarian and Uterine Cycles (Life10e.com/at43.2)

  Activity 43.1 The Human Male Reproductive Tract (Life10e.com/ac43.1)

  Activity 43.2 Spermatogenesis (Life10e.com/ac43.2)

  Activity 43.3 The Human Female Reproductive Tract (Life10e.com/ac43.3)

Key Concept Review

43.1 How Do Animals Reproduce without Sex?

Budding and regeneration produce new individuals by mitosis

Parthenogenesis is the development of unfertilized eggs

A variety of animals, mostly invertebrates, employ asexual reproduction, which produces offspring that are genetically identical to one another and to the parent. Individuals that reproduce asexually often are either sessile or members of a sparse population; both of these conditions make searching for and finding a mate challenging. Asexual reproduction has two advantages: (1) time and energy are not wasted on mating; and (2) every member of the population can produce offspring. The main disadvantage of asexual reproduction is that it does not generate genetic diversity, which can be essential if the environment changes.

New offspring can be produced either by budding off outgrowths to produce a new individual or by regenerating a new individual from pieces of an animal (see Figure 43.1). These methods rely on mitosis. Another way that animals can produce new individuals asexually is through parthenogenesis—the development of unfertilized eggs into new offspring. Sometimes sexual behavior is required for asexual reproduction; this is the case, for example, in a species of parthenogenetic whiptail lizard (see Figure 43.2).

Question 1. Some animals can reproduce both sexually and asexually. When and why might these animals switch between sexual and asexual reproduction?

Question 2. The ovaries produce estrogen and progesterone. How might removal of the ovaries influence sexual behavior of female parthenogenetic whiptail lizards? What would you predict about their behavior if you then gave these females injections of either estrogen or progesterone?

43.2 How Do Animals Reproduce Sexually?

- Gametogenesis produces eggs and sperm
- Fertilization is the union of sperm and egg
- Getting eggs and sperm together
- Some individuals can function as both male and female
- The evolution of vertebrate reproductive systems parallels the move to land
- Animals with internal fertilization are distinguished by where the embryo develops

Sexual reproduction has three stages: gametogenesis (producing gametes), mating or spawning (bringing gametes together), and fertilization (fusing gametes). Gametogenesis occurs in an animal's gonads. Male gametes—sperm—are produced in the testes, and female gametes—eggs or ova—are produced in the ovaries. Gametogenesis produces haploid gametes by meiotic cell division. During meiosis, crossing over between homologous chromosomes and independent assortment of chromosomes contribute to genetic diversity. Sexual reproduction also produces genetic diversity when two haploid gametes join at fertilization to form a diploid cell.

Males produce sperm by spermatogenesis, which begins with a male germ cell ($2n$) that proliferates through mitosis to produce spermatogonia (see Figure 43.3A). Spermatogonia mature into primary spermatocytes ($2n$) that enter meiosis. The first meiotic division produces secondary spermatocytes (n), and a second meiotic division produces four spermatids (n). The spermatids differentiate and mature into sperm cells in a process called spermiogenesis.

Females produce eggs by oogenesis, which begins with a female germ cell ($2n$) that proliferates through mitosis to produce oogonia (see Figure 43.3B). Oogonia mature into primary oocytes ($2n$). The primary oocytes enter prophase of the first meiotic division and in human females remain arrested in prophase for at least 10 years (until puberty). Upon exiting prophase, the primary oocyte completes the first meiotic division to produce a secondary oocyte (n) and the first polar body. The second meiotic division produces an ootid (n) and the second polar body. This second meiotic division may not be completed until fertilization in some species. The polar bodies degenerate.

Fertilization is the joining of a haploid sperm with a haploid egg to form a diploid zygote. For successful fertilization to occur, the sperm and egg must recognize one another. Specific recognition molecules on the gametes mediate recognition. During the acrosomal reaction, the plasma membrane covering the head of the sperm breaks down, as does the underlying acrosomal membrane, releasing enzymes that digest a path for the sperm through the protective layers surrounding the egg. Fusion of the plasma membranes of the sperm and egg triggers blocks to polyspermy that prevent more than one sperm from entering the egg. Soon after entry by a sperm, the egg is metabolically activated and stimulated to begin development. The diploid nucleus of the zygote is created when the haploid nucleus of the egg fuses with the haploid nucleus of the sperm.

Many aquatic animals have external fertilization of eggs. The eggs and sperm are released into water (in a process called spawning), where fertilization may occur. Terrestrial animals employ internal fertilization, in which sperm and egg fuse within the female reproductive tract rather than in the external environment. Some aquatic animals (e.g., sharks) also have internal fertilization.

Gonads (testes and ovaries) are primary sex organs. Organs other than gonads that are part of an animal's reproductive system are considered accessory sex organs. Copulation is the physical joining of male and female accessory sex organs. Dioecious species have separate male and female individuals. In monoecious species, however, a single individual may produce both eggs and sperm; such individuals are called hermaphrodites. Hermaphroditism is associated with having a low probability of encountering a potential mate.

The move to land required a reproductive system that functions in a dry environment. Although amphibians were the first vertebrates to live on land, they still require water to reproduce. Many reptiles and all birds have solved the problem of a dry environment by being oviparous; they lay protective shelled eggs (termed amniote eggs) that contain food (yolk) and a watery environment within which the embryo develops. Internal fertilization must occur before shell formation. Among mammals, monotremes are oviparous and all other mammals are viviparous, retaining the embryo within the uterus where it is nourished via the placenta. Most non-mammalian viviparous animals are ovoviviparous, with embryos developing in shelled eggs that are retained in females until hatching.

Question 3. Create flowcharts for both spermatogenesis and oogenesis that show how ploidy changes from germ cell to mature gamete. Be sure to label each stage of gamete formation (e.g., spermatogonium, primary spermatocyte, etc.) and include its ploidy.

Question 4. How are spermatogenesis and oogenesis similar? How are they different?

Question 5. Why might species-specific recognition molecules of sperm and ova be somewhat more important in species with external fertilization than in those with internal fertilization?

43.3 How Do the Human Male and Female Reproductive Systems Work?

- Male sex organs produce and deliver semen
- Male sexual function is controlled by hormones
- Female sex organs produce eggs, receive sperm, and nurture the embryo
- The ovarian cycle produces a mature egg
- The uterine cycle prepares an environment to receive a fertilized egg
- Hormones control and coordinate the ovarian and uterine cycles
- FSH receptors determine which follicle ovulates
- In pregnancy, hormones from the extraembryonic membranes take over
- Childbirth is triggered by hormonal and mechanical stimuli

Males produce semen containing sperm (n) and fluids that support the sperm and facilitate fertilization. In the testis, sperm cells are produced by spermatogenesis in the seminiferous tubules. Sertoli cells surround developing sperm cells and provide protection and nourishment (see Figure 43.9C). Immature sperm move from the seminiferous tubules to the epididymis, where they mature, become motile, and are stored. The epididymis connects to the vas deferens and then to the urethra, which opens to the outside of the body at the tip of the penis. In addition to containing sperm, semen contains fluids from accessory glands (bulbourethral glands, seminal vesicles, and prostate gland).

Testosterone, produced in the Leydig cells of the testes, controls sexual function and sperm production (see Figure 43.10). Testosterone also prompts development and maintenance of secondary sexual characteristics. Luteinizing hormone (LH) from the anterior pituitary stimulates the Leydig cells to increase production of testosterone. Follicle-stimulating hormone (FSH) from the anterior pituitary works with testosterone to influence Sertoli cells and stimulate spermatogenesis.

The ovaries release eggs into the abdominal cavity, where they enter the oviduct and move toward the uterus (see Figure 43.11A). Sperm swim up the vagina, through the opening to the uterus (cervix), through the uterus, and into the oviduct. Fertilization typically occurs in the oviduct; embryos develop in the uterus. The initial division of the zygote produces a blastocyst that moves into the uterus, where it burrows into the uterine wall (the endometrium); this process is called implantation. The blastocyst interacts with the endometrium to form the placenta.

Mature eggs are produced during a 28-day ovarian cycle (see Figure 43.12). Prior to ovulation, the primary oocyte undergoes a meiotic division to become a secondary oocyte and is expelled from the ovary. The remaining follicle cells become the corpus luteum, a glandular structure that produces progesterone and estrogen. The uterine cycle, in concert with the ovarian cycle, prepares the endometrium to receive the blastocyst. If a blastocyst does not implant by around day 14 of the uterine cycle, the endometrium is broken down and expelled in the process of menstruation. Luteinizing hormone (LH) and follicle-stimulating hormone (FSH) control the timing of ovarian and uterine cycles (see Figure 43.13).

If the egg is fertilized and the blastocyst implants in the uterus, cells surrounding the blastocyst produce human chorionic gonadotropin, which keeps the corpus luteum functional. Continued production of estrogen and progesterone by the corpus luteum supports development and maintenance of the endometrium, thereby preparing the uterus for pregnancy. The placenta eventually replaces the corpus luteum.

The uterine contractions of childbirth are triggered by pressure from the fetal head on the maternal cervix. This mechanical stimulus triggers release of oxytocin by the mother and fetus. Oxytocin stimulates increased strength and frequency of uterine contractions in a positive feedback cycle (see Figure 43.15B). Stages of childbirth include dilation of the cervix, delivery of the baby, and delivery of the placenta.

Question 6. If a man fails to produce sufficient gonadotropin-releasing hormone (GnRH), what general health issues will be a likely result?

Question 7. Two different types of follicle cells, each arranged in a layer, surround a developing egg. Thecal cells make up the outer layer and produce testosterone when stimulated by luteinizing hormone (LH). Granulosa cells make up the layer closest to the egg and they grow and mature in response to follicle-stimulating hormone. Thecal cells are similar to which type of cell in males? Granulosa cells are similar to which type of cell in males?

43.4 How Can Fertility Be Controlled?

- Humans use a variety of methods to control fertility
- Reproductive technologies help solve problems of infertility

Methods of contraception differ in their mode of action and failure rate (see Table 43.1). Whereas some are nontechnological methods, others form barriers or interfere with ovulation or implantation. Additionally, both males and females can be sterilized.

Several factors may prevent couples from conceiving. A male may not produce adequate numbers of sperm, or the sperm may lack motility. In the female, the environment of the uterus may be poor for hosting sperm or the fertilized egg, or the oviducts may be blocked, preventing passage of gametes. The options available to infertile couples include artificial insemination, in vitro fertilization, and intracytoplasmic sperm injection. In conjunction with in vitro fertilization, genetic testing may be performed on cells taken from early blastocysts in a procedure known as preimplantation genetic diagnosis (PGD).

Question 8. In the diagram of the reproductive tract of the human female below, label the following structures: vagina, cervix, endometrium of uterus, ovary, and oviduct. Then choose, from the following list, the particular birth control method that acts (or is applied) at each of the five anatomical structures: diaphragm, tubal ligation, Mifepristone (RU-486), spermicidal jelly, the pill.

Question 9. You are a physician, and one of your male patients is considering a vasectomy. He is concerned, however, with the possibility of decreased testosterone production and its effects on sexual function. What should you tell your patient?

Question 10. A carrier is a person who carries the gene for a particular disease but does not have the disease. Suppose that both members of a couple know they are carriers for a disease, and any children they produce could be free of the disease, be a carrier, or die from the disease. What reproductive technologies might their physician recommend?

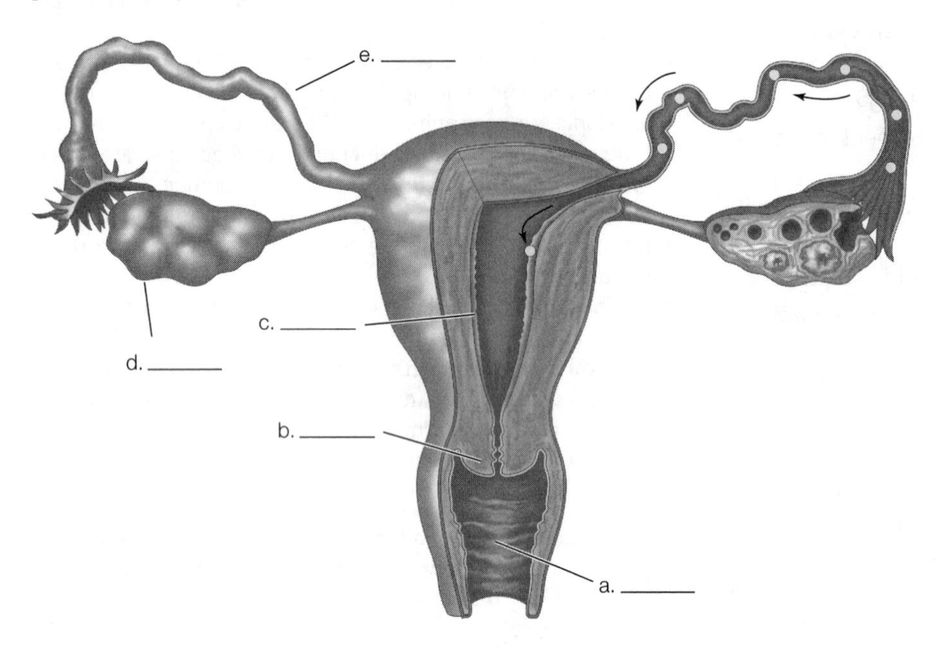

Test Yourself

1. Asexual reproduction is an effective strategy in stable environments because
   a. gametogenesis is most efficient under these conditions.
   b. the resulting offspring, which are genetically identical to their parents, are preadapted to their environment.
   c. asexual parthenogenesis produces a large amount of genetic diversity.
   d. animal cells tend to be more totipotent under stable conditions.
   e. sessile animals and sparse populations are more common under stable conditions.

   Textbook Reference: *43.1 How Do Animals Reproduce without Sex?*

2. If you compared the genetic makeup of a female animal produced by parthenogenesis with that of its mother, which of the following would you expect?
   a. About 100 percent genetic similarity
   b. About 50 percent genetic similarity
   c. No genetic similarity
   d. About 25 percent genetic similarity
   e. None of the above; parthenogenetic animals do not have mothers.

   Textbook Reference: *43.1 How Do Animals Reproduce without Sex?*

3. An important difference between a sperm and an egg concerns
   a. their size.
   b. the amount of cytoplasm they contain.
   c. whether or not they are motile.
   d. whether or not protective layers exist outside the plasma membrane.
   e. All of the above

   Textbook Reference: *43.2 How Do Animals Reproduce Sexually?*

4. External fertilization
   a. is typical of terrestrial animals.
   b. requires a penis.
   c. can occur when a spermatophore is transferred.
   d. occurs in some, but not all, aquatic animals.
   e. is confined to invertebrates.

   Textbook Reference: *43.2 How Do Animals Reproduce Sexually?*

5. Which of the following represents an *incorrect* match between mode of reproduction and vertebrate group?
   a. Oviparity; birds
   b. Viviparity; therian mammals
   c. Ovoviviparity; crocodilians
   d. Oviparity; monotremes mammals
   e. Ovoviviparity; some snakes

   Textbook Reference: *43.2 How Do Animals Reproduce Sexually?*

6. Which of the following is *not* an accessory sex organ?
   a. Penis
   b. Prostate gland
   c. Ovary
   d. Vagina
   e. Uterus

   Textbook Reference: *43.2 How Do Animals Reproduce Sexually?*

7. Spermatogenesis
   a. results in two sperm cells from each primary spermatocyte.
   b. involves a period of arrest during the first meiotic division.
   c. results in diploid gametes.
   d. apportions cytoplasm equally among resulting cells.
   e. apportions cytoplasm unequally among resulting cells.

   Textbook Reference: *43.2 How Do Animals Reproduce Sexually?*

8. Which of the following statements about fertilization is *false*?
   a. The egg permits several sperm to enter it.
   b. The plasma membranes of sperm and egg fuse.
   c. The egg is activated and stimulated to begin development.
   d. Species-specific recognition occurs between egg and sperm.
   e. During the acrosomal reaction, enzymes spill from a cap on the head of the sperm and digest a path through the protective layers surrounding the egg.

   Textbook Reference: *43.2 How Do Animals Reproduce Sexually?*

9. Which of the following statements about oogenesis is *false*?
   a. The polar bodies degenerate after the second meiotic division.
   b. The ovum produced is haploid.
   c. The major growth phase of the primary oocyte occurs in prophase I.
   d. The primary oocyte is haploid.
   e. The secondary oocyte is haploid.

   Textbook Reference: *43.2 How Do Animals Reproduce Sexually?*

10. For approximately how long during the human female's menstrual cycle are progesterone concentrations high enough to maintain the uterus in a proper condition for pregnancy?

a. The entire duration of the cycle
b. No portion of the cycle
c. During the first half of the cycle
d. During the second half of the cycle
e. For a 5-day window mid-cycle
Textbook Reference: *43.3 How Do the Human Male and Female Reproductive Systems Work?*

11. Which of the following best represents the normal path of a sperm cell as it makes its way from the point of entry into a female's reproductive tract to the location where fertilization typically occurs?
a. Cervix, vagina, ovary, oviduct
b. Vagina, cervix, uterus, oviduct
c. Uterus, cervix, vagina, oviduct
d. Vagina, uterus, cervix, oviduct
e. Cervix, vagina, oviduct, ovary
Textbook Reference: *43.3 How Do the Human Male and Female Reproductive Systems Work?*

12. The function of the seminal vesicle is to
a. produce a solution of fructose to provide energy for the sperm.
b. secrete alkaline fluids that neutralize the acidity of a female's reproductive tract.
c. initiate the muscular contractions that lead to emission.
d. produce lubrication for the tip of the penis.
e. serve as the location for spermatogenesis.
Textbook Reference: *43.3 How Do the Human Male and Female Reproductive Systems Work?*

13. Which of the following is an example of positive feedback control in the reproductive cycle of males or females?
a. The increased response of the hypothalamus and anterior pituitary gland in response to estrogen
b. The decreased response of the hypothalamus and anterior pituitary gland in response to estrogen
c. The inhibition of luteinizing hormone by high levels of testosterone
d. The stimulation of luteinizing hormone by low levels of testosterone
e. The inhibition of oxytocin release by increased uterine contractions during childbirth
Textbook Reference: *43.3 How Do the Human Male and Female Reproductive Systems Work?*

14. Which of the following statements about the birth control pill is *false*?
a. It works by preventing ovulation.
b. It works by preventing implantation.
c. It suspends the ovarian cycle.
d. It contains low doses of estrogen and progesterone.
e. It allows the uterine cycle to continue because hormones are not taken for one week every 21–28 days.
Textbook Reference: *43.4 How Can Fertility Be Controlled?*

15. Which of the following methods of contraception has the lowest failure rate?
a. Vaginal ring
b. Rhythm method
c. Diaphragm
d. Coitus interruptus
e. Condom
Textbook Reference: *43.4 How Can Fertility Be Controlled?*

Answers

Key Concept Review

1. Asexual reproduction does not waste time and energy on mating, and every member of the population can produce offspring. However, asexual reproduction does not yield genetic diversity, which can be beneficial when environmental conditions change. Sexual reproduction generates genetic diversity. Thus, we would expect animals that can reproduce by either method to reproduce asexually when environmental conditions are stable and favorable, and to switch to sexual reproduction when environmental conditions are unstable and unfavorable.
2. Removal of the ovaries would result in female whiptail lizards that show no sexual behavior. If these females were given injections of estrogen, they would likely show female sexual behavior. If these females were given injections of progesterone, they would likely show male sexual behavior.

3. **Spermatogenesis:**

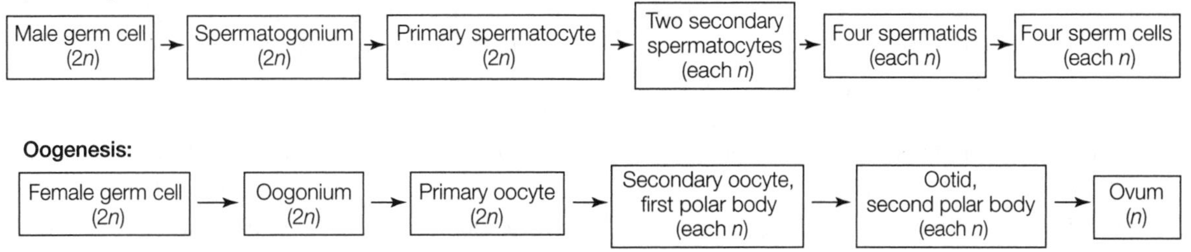

4. Both spermatogenesis and oogenesis are processes that occur in gonads, and they include the same general events. For example, once in the gonads, embryonic germ cells proliferate by mitosis to produce cells that eventually will enter meiosis, and the end result of both processes is the production of haploid cells. Two main differences between spermatogenesis and oogenesis are: (1) spermatogenesis continues to completion once the primary spermatocyte has differentiated, whereas oogenesis is characterized by a period of arrest during the first meiotic division; and (2) cytoplasm is apportioned equally to resulting cells in spermatogenesis and unequally to resulting cells in oogenesis (polar bodies receive much less cytoplasm than do secondary oocytes or ootids).
5. Species with external fertilization release their gametes into the external environment where gametes from other species may be present, so recognition molecules are critical. In contrast, species with internal fertilization release gametes into the female reproductive tract where gametes of other species usually are not present because of species-specific courtship and mating behaviors.
6. Because GnRH prompts the anterior pituitary to release luteinizing hormone (LH) and follicle-stimulating hormone (FSH), a man with low GnRH will produce little testosterone (LH stimulates the Leydig cells to produce testosterone) and have problems with sperm production (testosterone and FSH act on Sertoli cells to maintain spermatogenesis).
7. Because thecal cells of the follicle are stimulated by luteinizing hormone (LH) to produce testosterone, they are most similar to Leydig cells in the male. Because granulosa cells of the follicle are stimulated by follicle-stimulating hormone (FSH) to grow and mature, they are most similar to Sertoli cells in the male.
8. a. Vagina; spermicidal jelly
   b. Cervix; diaphragm
   c. Endometrium of uterus; Mifepristone (RU-486)
   d. Ovary; the pill
   e. Oviduct; tubal ligation
9. A vasectomy involves cutting and tying off the vasa deferentia so that sperm cannot reach the urethra. Sperm production continues, but the sperm are reabsorbed. The surgery has no impact on the testes, so production of testosterone by the Leydig cells continues as it did before surgery and sexual function should be unaffected.
10. Their physician might recommend in vitro fertilization (IVF) coupled with preimplantation genetic diagnosis (PGD). In vitro fertilization combines eggs and sperm outside the body, so if fertilization takes place, then any resulting embryo can be screened to see whether it carries a harmful gene. During PGD, a cell can be removed from a very young embryo (at the 4- or 8-cell stage) and screened without harming the developing embryo. If the harmful gene is not detected, the embryo can be injected into the female's reproductive tract.

Test Yourself

1. **b.** The parents that have survived to reproduce asexually are able to survive in the current stable environment. Therefore, the offspring should be preadapted for this stable environment.
2. **a.** Parthenogenesis is a form of asexual reproduction in which offspring develop from unfertilized eggs; there is no sexual recombination of genes, so a female animal born parthenogenetically will be nearly identical to its mother genetically. In many hymenopterans (bees, wasps, and ants), only males are born from unfertilized eggs. Such males are haploid, so they are not genetically identical to their diploid mothers.
3. **e.** There are many differences between eggs and sperm. Eggs are larger and contain more cytoplasm and organelles. Eggs are immotile, whereas sperm have a flagellum and are motile. Eggs, but not sperm, have protective layers outside their plasma membranes.
4. **d.** External fertilization occurs in some, but not all, aquatic animals. It occurs in invertebrates (e.g., oysters and sea urchins) and vertebrates (e.g., lampreys and salmon), and does not require a penis. Transfer of a spermatophore (sperm packet) results in internal fertilization without copulation.
5. **c.** Crocodilians are oviparous (egg-laying). Ovoviviparous animals retain the fertilized egg in the female reproductive tract, which is not characteristic of crocodilians.

6. **c.** The ovary is a gonad and therefore is not an accessory sex organ. Gonads are the primary sex organs producing the sperm and the egg.
7. **d.** During spermatogenesis cytoplasm is apportioned equally among resulting cells. This process differs from oogenesis, in which cytoplasm is apportioned unequally among daughter cells.
8. **a.** During fertilization, mechanisms in the egg that operate as blocks to polyspermy prevent more than one sperm from entering it.
9. **d.** During oogenesis, the primary oocyte is diploid; after the first meiotic division into the secondary oocyte the cell becomes haploid.
10. **d.** High levels of progesterone are needed to maintain the uterus in the proper condition for pregnancy. The levels of progesterone are high only during the second half of the uterine cycle.
11. **b.** A sperm is ejected by the male into the vagina. From the vagina the sperm move through the cervix into the uterus and finally into the oviduct, where fertilization typically occurs.
12. **a.** The seminal vesicles contribute substances to semen. One of the components of seminal fluid is fructose, which serves as an energy source for the sperm.
13. **a.** During days 12–14 of the ovarian cycle, estrogen exerts positive feedback control on the anterior pituitary, prompting it to release LH and FSH.
14. **b.** The birth control pill interferes with the maturation of the follicles and the ova, inhibiting the release of an egg. Thus, it blocks ovulation, not implantation.
15. **a.** Among those methods listed, the vaginal ring has the lowest failure rate (<1%).

Animal Development

The Big Picture

- Animal development involves several steps: fertilization, cleavage, gastrulation, and organogenesis (for example, neurulation).
- Fertilization is the joining of the egg and sperm to form a zygote. Following fertilization, the cells of the embryo begin to divide in a process known as cleavage. Eggs undergo either complete cleavage or incomplete cleavage, depending on the amount of yolk present. Those that undergo complete cleavage form a blastula, and some of those that undergo incomplete cleavage form a flat blastodisc.
- Cleavage in mammals is different from cleavage in other groups: it occurs more slowly, and the products of embryonic genes play a role in directing its course. In other animals, cleavage is directed entirely by molecules present in the egg prior to fertilization.
- During gastrulation, the germ layers—endoderm, ectoderm, and mesoderm—form and move into specific positions. In the sea urchin, an archenteron forms; this will become the gut, with specific cells forming the three germ layers. In frogs, gastrulation begins at the gray crescent and a dorsal lip forms. The dorsal lip is the primary embryonic organizer in amphibians because it establishes body axes. The blastodisc of reptiles, birds, and mammals goes through a much different pattern of gastrulation.
- Neurulation occurs early in organogenesis. Ectoderm overlying the notochord forms a neural plate. The neural plate forms a neural tube, which will become the brain and spinal cord. Somites produce the repeating segments of the vertebrate body plan. Several genes play a large role in the differentiation of tissues along the body axes. For example, Hox genes control differentiation along the anterior–posterior axis.
- Reptiles and birds have four extraembryonic membranes: yolk sac, amnion, chorion, and allantois. In mammals other than monotremes, some extraembryonic membranes contribute to the placenta, which is the organ where nutrients, gases, and wastes are exchanged between the mother and the developing fetus.

Study Strategies

- All fertilized cells must divide to grow into adult animals. Sorting out the features of division that are common to all animals from those specific to certain taxonomic groups can be difficult and confusing until you begin to recognize the basic patterns.
- Many students find gastrulation difficult to picture and understand, especially given the different ways in which gastrulation occurs in different organisms. Look carefully at the figures depicting gastrulation in various animals in the chapter and then view Tutorial 44.1.
- The structure and role of the placenta can be confusing. Is it embryonic? Is it maternal? And how do nutrients, gases, and wastes travel between the mother and embryo or fetus? Look carefully at Figure 44.18, which depicts the structure of the placenta.
- This chapter contains a great deal of terminology. Many terms are common to many animals, and some apply only to specific animals. Create a list of new terms from this chapter. Determine which are the more general terms applicable to numerous animals (e.g., "somites," "gastrulation") and learn these first. Then tackle those referring to development in specific animals (e.g., "dorsal lip of the blastopore," "primitive streak").
- A key point in understanding animal development is that all specialized tissues arise from the three germ layers: endoderm, ectoderm, and mesoderm. More important than simply committing these three layers to memory is learning the specialized tissues that derive from them. This will allow you to make predictions about tissue functions and to compare body plans among animals.
- Go to the Web addresses shown to review the following animated tutorials and activity. You can also find them in BioPortal (yourBioPortal.com), along with many additional learning resources.

  Animated Tutorial 44.1 Gastrulation (Life10e.com/at44.1)

  Animated Tutorial 44.2 Tissue Transplants Reveal the Process of Determination (Life10e.com/at44.2)

  Activity 44.1 Extraembryonic Membranes (Life10e.com/ac44.1)

Key Concept Review

44.1 How Does Fertilization Activate Development?

The sperm and the egg make different contributions to the zygote

Rearrangements of egg cytoplasm set the stage for determination

Development in sexually reproducing animals begins when a haploid sperm joins with a haploid egg. However, fertilization does more than produce a diploid zygote; it also activates development. The egg contributes most of the cytoplasm and organelles to the zygote. The sperm contributes its haploid nucleus and a centriole. The centriole becomes the centrosome, which forms the mitotic spindles necessary for cell division.

In an unfertilized frog egg, nutrients are concentrated in the lower half of the egg, called the vegetal hemisphere, and the haploid nucleus is found in the upper half of the egg, known as the animal hemisphere. Upon fertilization, the cytoplasm of the frog egg undergoes rearrangement (see Figure 44.1). A sperm enters in the animal hemisphere at what will become the ventral side of the frog and causes rotation of the cortical cytoplasm to create a gray crescent, which will become the dorsal region of the frog. This helps establish body axes of the developing animal. During movement of the cytoplasm, the transcription factor β-catenin is degraded in parts of the frog egg so that it becomes concentrated in the dorsal side of the embryo (see Figure 44.2). This happens because the protein GSK-3 targets β-catenin for degradation, but another protein inhibits the action of GSK-3 on the dorsal side. As a result, the concentration of β-catenin is higher on the dorsal side.

Question 1. What two organelles does the sperm contribute at fertilization and why is each one important?

Question 2. A researcher working with frog zygotes experimentally disables the protein that inhibits GSK-3. What will be the immediate result?

44.2 How Does Mitosis Divide Up the Early Embryo?

Cleavage repackages the cytoplasm

Early cell divisions in mammals are unique

Specific blastomeres generate specific tissues and organs

Germ cells are a unique lineage even in species with regulative development

During cleavage, the first cell divisions occur rapidly with little growth or differentiation. The embryo divides into smaller and smaller cells, producing first a solid ball of cells and then a blastula with a central fluid-filled cavity called the blastocoel. Individual cells in the blastula are called blastomeres. The amount of yolk in the egg determines how cleavage proceeds (see Figure 44.3). Yolk impedes cleavage furrows. Frog eggs have small amounts of yolk and complete cleavage in which early cleavage furrows completely divide the egg. The eggs of fishes, reptiles, and birds have large amounts of yolk and incomplete cleavage. In these organisms, the embryo develops from a disc of cells (known as the blastodisc) on top of the yolk mass. Some insects undergo superficial cleavage, a variation of incomplete cleavage in which a syncytium (single cell with many nuclei) is produced early in development when cycles of mitosis occur without cell division.

The early cell divisions of placental mammals occur more slowly than in other animals, and are asynchronous. Additionally, genes of placental mammals are expressed during cleavage and influence the process; this differs from other animals in which molecules present in the egg direct cleavage. During cleavage from the 16-cell to the 32-cell stage, the cells separate into an inner cell mass and the trophoblast. The inner cell mass will become the embryo proper, and the trophoblast will become part of the placenta and attach to the uterine lining during implantation. Cells of the inner cell mass are pluripotent embryonic stem cells. At the 32-cell stage, the mass of mammalian cells is called a blastocyst (see Figure 44.4).

During cleavage, cytoplasm is distributed to the cells of the blastula in such a way that cells in different regions have different levels of nutrients and informational molecules. Specific blastomeres will become specific tissues and organs, and this can be mapped out in the developing cells as a fate map of the blastula (see Figure 44.6). The blastomeres become determined (committed to specific fates) at different times for different species. Animals with mosaic development have blastomeres that are set very early to contribute to specific parts of the embryo. In animals with regulative development, the blastomeres can be removed and other cells will compensate for the loss. The lineage of cells that eventually will give rise to either spermatogonia or oogonia is distinct early in development.

Question 3. If you were developing an embryology lab and wanted to include animal species that display diverse cleavage patterns (complete and incomplete, with different orientations of mitotic spindles), which species might you include? If possible, select model species and list why you chose each.

Question 4. What medical condition can result if a human blastocyst hatches too early from the zona pellucida?

Question 5. Explain how identical and non-identical twins form in humans. Are conjoined twins identical or non-identical twins? Do humans display mosaic or regulative development?

44.3 How Does Gastrulation Generate Multiple Tissue Layers?

Invagination of the vegetal pole characterizes gastrulation in the sea urchin

Gastrulation in the frog begins at the gray crescent

The dorsal lip of the blastopore organizes embryo formation

Transcription factors and growth factors underlie the organizer's actions

The organizer changes its activity as it migrates from the dorsal lip

Reptilian and avian gastrulation is an adaptation to yolky eggs

The embryos of placental mammals lack yolk

During gastrulation, three germ layers form and position themselves in the embryo. The inner layer, or endoderm, will become the lining of the digestive tract, the lining of the respiratory tract, and some organs, such as the pancreas and liver. The outer layer, or ectoderm, will become the epidermis (outer layer of the skin), hair, nails, glands (sweat and oil), and the nervous system. The middle layer, or mesoderm, will become the heart, blood vessels, muscle, and bone. Gastrulation makes possible the inductive interactions between cells, which trigger differentiation and organ formation.

Gastrulation in sea urchins involves invagination of the vegetal pole, which flattens and then moves inward to form the endoderm and primitive gut (archenteron). Some cells from the vegetal pole break free to become the mesenchyme cells that make up the mesoderm. The opening of the archenteron is the blastopore, and it will become the anus. The place where the tip of the archenteron meets the ectoderm will become the mouth (see Figure 44.7).

The frog embryo has more yolk than the sea urchin, and gastrulation is more complex (see Figure 44.8). Initially, cells near the gray crescent begin to bulge into the blastocoel. These initial cells (bottle cells) move along the interior of the blastula and pull the outer surface of cells along with them, creating a dorsal lip; this process is called involution. The first cells to move in are prospective endoderm, and they form the archenteron. Another group of cells moves between the endoderm and outermost cells to form mesoderm. By the end of gastrulation, the cells have become fate-determined and are layered (from the outside in) as the ectoderm, mesoderm, and endoderm.

Hans Spemann examined the timing and fate of cells during gastrulation in salamanders to determine if they were totipotent or able to direct development (see Figure 44.9). Spemann and Hilde Mangold found that when the dorsal lip was transplanted to another early gastrula, the result was a second site of gastrulation and eventually two embryos attached at the belly. They concluded that the dorsal lip acts as the primary embryonic organizer (see Figure 44.10). Presence of β-catenin and a complex series of interactions between growth and transcription factors create the organizer and lead to induction of the body plan. As the organizer migrates from the dorsal lip, it inhibits various growth factors along the way to achieve different patterns of differentiation along the anterior–posterior axis.

In birds and reptiles, cleavage produces a blastodisc on top of the large amount of yolk in the egg. The blastula is a circular layer of cells composed of an outer epiblast layer and an inner hypoblast layer. The epiblast gives rise to the embryo proper, and the hypoblast contributes to extraembryonic membranes. The fluid-filled space between the two layers is the blastocoel. The primitive groove (also known as the chick blastopore) forms, and cells move through it to become endoderm and mesoderm (see Figure 44.13). There is no archenteron in the developing chick; instead, endoderm and mesoderm move forward and form gut structures. A group of cells at the anterior end of the primitive groove, known as Hensen's node, acts in the same manner as the dorsal lip of the frog blastopore.

In placental mammals, just as in birds and reptiles, the cells of the embryo segregate into an epiblast and a hypoblast with a blastocoel in between. Extraembryonic membranes, including that which contributes to the placenta, develop from the hypoblast. The embryo and some extraembryonic membranes develop from the epiblast. Gastrulation in mammals occurs just as it does in birds, with the formation of a primitive groove through which epiblast cells migrate to form endoderm and mesoderm.

Question 6. In the diagram below of a sea urchin gastrula, label the ectoderm, endoderm, primary mesenchyme, secondary mesenchyme, blastopore, and archenteron.

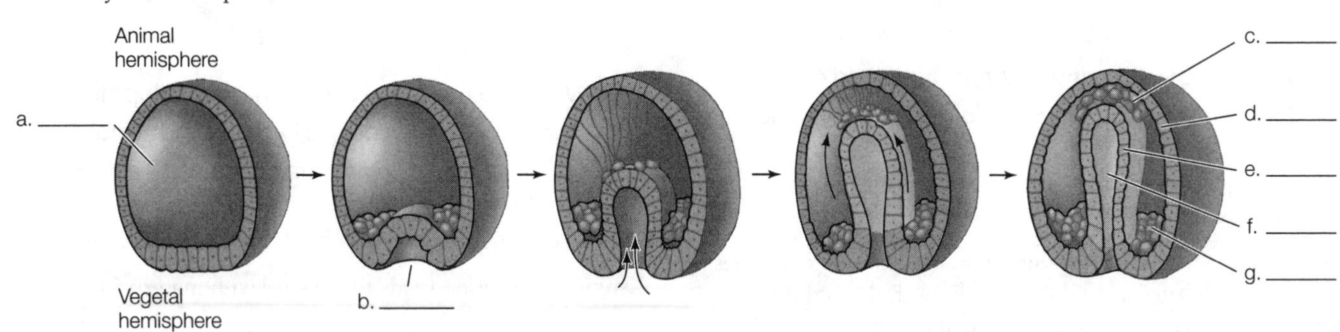

Question 7. What will happen if a researcher, working with developing amphibians, transplants a piece of ectoderm destined to become the epidermis (outer layer of skin) from one early gastrula to a location destined to be neural tissue in another early gastrula? What will happen if the transplant is made between late gastrulas?

Question 8. Although details of gastrulation differ among sea urchins, frogs, and chickens, what is the common result in all of these animals?

44.4 How Do Organs and Organ Systems Develop?

The stage is set by the dorsal lip of the blastopore

Body segmentation develops during neurulation

Hox genes control development along the anterior-posterior axis

After gastrulation, the organs begin to develop through the process of organogenesis. Chordamesoderm forms a rod of mesoderm known as the notochord. The notochord provides structural support to the embryo, and ultimately is replaced by the vertebral column in most vertebrates. Neurulation—the initial development of the nervous system—begins early in organogenesis. Ectoderm overlying the notochord flattens and thickens, forming the neural plate. The edges of the neural plate form neural folds, which eventually fuse to form the neural tube (see Figure 44.14). The anterior end of the neural tube will become the brain, with the rest of the neural tube forming the spinal cord. Neural crest cells break away from the neural tube and migrate inward to become peripheral nerves and other structures.

The repeating pattern of the vertebrate body plan forms from blocks of mesoderm known as somites that develop on both sides of the neural tube (see Figure 44.15). The somites will become ribs, vertebrae, and trunk muscles. Body segments differentiate as the embryo develops, and Hox genes control this differentiation. Hox genes are found on different chromosomes and are expressed along the anterior–posterior axis in their order on the chromosomes (see Figure 44.16). Dorsal–ventral differentiation is controlled by a separate set of genes. For example, the *Sonic hedgehog* gene, expressed in the notochord, prompts the most ventral cells in the overlying neural tube to become motor neurons. After segmentation occurs, organs and organ systems develop rapidly.

Question 9. Spina bifida is a birth defect in humans in which part of the spinal cord, as well as the adjacent area of the spine, develops abnormally. What process has been disturbed during prenatal development in individuals with spina bifida?

Question 10. Design an experiment to determine which substance is responsible for the development of spinal cord circuits.

44.5 How Is the Growing Embryo Sustained?

Extraembryonic membranes form with contributions from all germ layers

Extraembryonic membranes in mammals form the placenta

Extraembryonic membranes surround the embryos of reptiles, birds, and mammals. In birds, the yolk sac is derived from the hypoblast and nearby mesoderm. It surrounds the entire yolk to help retrieve nutrients from the yolk and deliver them to the embryo via blood vessels (see Figure 44.17). Cells from the ectoderm and mesoderm form the amnion and chorion. The amnion helps provide an aqueous environment, and the chorion, which develops just under the shell, regulates water, oxygen, and carbon dioxide exchanges across the shell. The allantoic membrane is derived from endoderm and mesoderm, and forms a sac to store metabolic wastes.

In placental mammals, the first extraembryonic membrane to form, the trophoblast, interacts with the endometrium to attach to the uterine wall and begin implantation of the blastocyst. The hypoblast cells interact with the trophoblast to form the chorion. The chorion of the embryo and endometrium of the mother form the placenta (see Figure 44.18). The amnion of the developing mammal surrounds the embryo to produce a closed, fluid-filled environment. The allantois of mammals has the function of removing nitrogenous wastes, but its function is relatively minor in some mammals, including humans. The tissues of the allantois help to form the umbilical cord, which carries major blood vessels that provide a route for exchanges of nutrients, wastes, carbon dioxide, and oxygen between the mother and fetus.

Question 11. Chorionic villus sampling, one method of prenatal genetic testing, involves taking a small sample of chorionic villi from the placenta. Why would cells from the placenta reveal the genetic makeup of the fetus?

Question 12. Consider the following statement: "The placenta is the organ where fetal and maternal blood supplies mix." Is this statement correct or incorrect, and why?

Question 13. In the diagram below of a 9-day chick embryo, label the embryo, yolk sac, allantoic membrane, allantois, amnion, and the chorion.

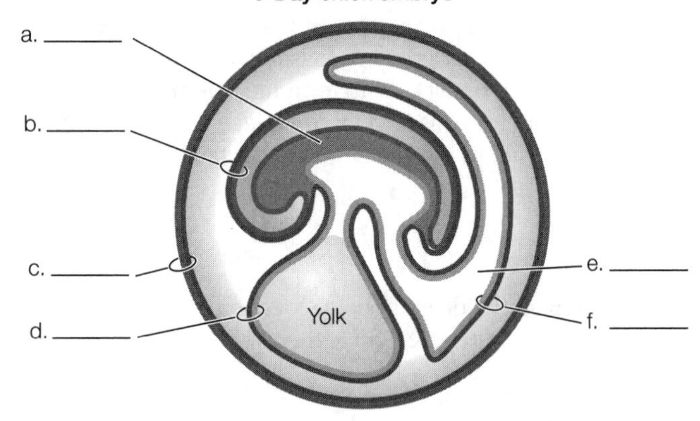

44.6 What Are the Stages of Human Development?

Organ development begins in the first trimester

Organ systems grow and mature during the second and third trimesters

Developmental changes continue throughout life

Length of gestation tends to increase with maternal body size. For example, pregnancy lasts about 21 days in mice, 266 days in humans, and 600 days in elephants.

Human pregnancy can be divided into trimesters. During the first trimester, cell division and tissue differentiation are rapid, and organ development begins. At this stage, the developing human is most sensitive to environmental disrupters. By the end of the first trimester, the embryo is considered a fetus. During the second trimester, organ systems continue to grow and mature, and the mother begins to feel fetal movements. The third trimester is marked by extremely rapid growth of the fetus and further maturation of the internal organs. Development continues after birth.

Question 14. Why would exposure to a harmful chemical during the first trimester be more likely to produce a major birth defect than exposure to the same chemical during the last trimester?

Question 15. Obstetric ultrasounds can be used to evaluate fetal structure, function, and growth. When during pregnancy would a physician perform an ultrasound in order to evaluate each of the following in a human fetus: limb development; kidney function; overall growth? Explain your answer.

Test Yourself

1. Which of the following statements about the respective contributions of the sperm and egg to the zygote is *false*?
   a. The sperm contributes most of the organelles.
   b. The egg contributes most of the cytoplasm.
   c. Both the sperm and the egg contribute a haploid nucleus.
   d. The sperm contributes a centriole, which becomes the centrosome of the zygote.
   e. All of the above are false, none is true.
   Textbook Reference: *44.1 How Does Fertilization Activate Development?*

2. In which of the following animals is the cleavage pattern rotational?
   a. Frog
   b. Mammal
   c. Sea urchin
   d. Mollusk
   e. Bird
   Textbook Reference: *44.2 How Does Mitosis Divide Up the Early Embryo?*

3. Which of the following statements about complete cleavage versus incomplete cleavage is true?
   a. Incomplete cleavage occurs in species with small volumes of cytoplasm.
   b. Complete cleavage is found in mammals and is the more evolved characteristic.
   c. Incomplete cleavage occurs in species with large amounts of yolk.
   d. Complete cleavage occurs only in eggs that have been fertilized by two sperm.
   e. Incomplete cleavage occurs in species with small amounts of yolk.
   Textbook Reference: *44.2 How Does Mitosis Divide Up the Early Embryo?*

4. A fate map can be used to map out the tissues and organs that eventually will develop from specific germ layers. On the fate map, ectoderm will become
   a. the lining of the gut.
   b. the nervous system.
   c. muscle.
   d. the heart.
   e. the lining of the respiratory tract.
   Textbook Reference: *44.2 How Does Mitosis Divide Up the Early Embryo?*

5. The location of the _______ determines the anterior–posterior axis of the sea urchin embryo.
   a. primitive streak
   b. blastopore
   c. vegetal hemisphere
   d. hypoblast
   e. tertiary mesenchyme
   Textbook Reference: *44.3 How Does Gastrulation Generate Multiple Tissue Layers?*

6. Which of the following represents the correct order of the germ layers, from the inside to the outside?
   a. Mesoderm, ectoderm, endoderm
   b. Endoderm, ectoderm, mesoderm
   c. Ectoderm, mesoderm, endoderm
   d. Endoderm, mesoderm, ectoderm
   e. Mesoderm, endoderm, ectoderm
   Textbook Reference: *44.3 How Does Gastrulation Generate Multiple Tissue Layers?*

7. Hans Spemann called the dorsal lip of the blastopore the embryonic organizer for amphibians because it
   a. is the point where gastrulation begins.
   b. becomes part of the nervous system.
   c. becomes part of the notochord.
   d. leads to the establishment of the embryonic axes.
   e. is the location at which the sperm enters the egg.
   Textbook Reference: *44.3 How Does Gastrulation Generate Multiple Tissue Layers?*

8. The primary embryonic organizer of amphibians is most likely initiated by
   a. the yolk.
   b. Tcf-3 protein.
   c. β-catenin.
   d. cAMP.
   e. None of the above
   Textbook Reference: *44.3 How Does Gastrulation Generate Multiple Tissue Layers?*

9. The dorsal lip of the blastopore organizes embryo formation in frogs. The equivalent structure in chickens is
   a. the epiblast.
   b. the hypoblast.
   c. Hensen's node.
   d. the bottle cell.
   e. the gray crescent.
   Textbook Reference: *44.3 How Does Gastrulation Generate Multiple Tissue Layers?*

10. Which of the following statements about avian extra-embryonic membranes is *false*?
    a. The yolk sac surrounds the yolk and provides nutrients.
    b. The amnion and chorion are derived from ectoderm and mesoderm.
    c. The allantoic membrane forms a sac that stores nutrients.
    d. The chorion exchanges gases and water between the embryo and the environment.
    e. The allantois is derived from endoderm and mesoderm.

    Textbook Reference: *44.5 How Is The Growing Embryo Sustained?*

11. Which of the following statements about the mammalian blastocyst is *false*?
    a. The trophoblast gives rise to the embryo proper.
    b. Maternal genes are expressed during cleavage.
    c. The blastocyst typically implants in the mother's uterus.
    d. Early mammalian cleavage is relatively slow.
    e. The trophoblast contributes to formation of the chorion.

    Textbook Reference: *44.2 How Does Mitosis Divide Up the Early Embryo?; 44.5 How Is the Growing Embryo Sustained?*

12. The _______ eventually develop into vertebrae, ribs, and trunk muscles, and are found along the sides of the _______.
    a. somites; neural tube
    b. neural tube cells; notochord
    c. blastopore cells; dorsal lip
    d. neural crest cells; dorsal lip
    e. neural plate cells; archenteron

    Textbook Reference: *44.4 How Do Organs and Organ Systems Develop?*

13. During its development, the human embryo is contained within a fluid-filled chamber enclosed by the extraembryonic membrane called the
    a. yolk sac.
    b. amnion.
    c. chorion.
    d. allantois.
    e. trophoblast.

    Textbook Reference: *44.5 How Is the Growing Embryo Sustained?*

14. The third trimester of human prenatal development is characterized by
    a. extremely rapid growth.
    b. the formation of major organs.
    c. the greatest sensitivity to damage from drugs and radiation.
    d. gastrulation.
    e. the formation of the placenta.

    Textbook Reference: *44.6 What Are the Stages of Human Development?*

15. Which of the following statements about human development is *false*?
    a. At about 3 months gestation, the developing human is called a fetus.
    b. Mothers first feel fetal movements during the second trimester.
    c. Substantial developmental changes occur in the brain between birth and adolescence.
    d. Humans have the longest gestation period among mammals because of the developmental requirements of their relatively large brains.
    e. Development continues after postnatal growth has stopped.

    Textbook Reference: *44.6 What Are the Stages of Human Development?*

Answers

Key Concept Review

1. The sperm contributes a haploid nucleus, which, when combined with the egg's haploid nucleus, produces a diploid zygote. The sperm also contributes a centriole, which becomes the centrosome of the zygote. The centrosome functions as a microtubule-organizing center.
2. GSK-3 degrades β-catenin. Normally, the protein that inhibits GSK-3 moves to the gray crescent area, where it prevents the degradation of β-catenin. If this protein is experimentally disabled, then β-catenin will be degraded throughout the cytoplasm.
3. To develop an embryology lab demonstrating diversity in animal cleavage patterns, the following species could be included:
   (1) Fruit fly (incomplete cleavage; superficial)
   (2) Chick or zebrafish (incomplete cleavage discoidal)
   (3) Frog (complete cleavage; radial)
   (4) Snail (complete cleavage; spiral)
   (5) Mouse (complete cleavage; rotational)
4. Normally the blastocyst moves down the oviduct and into the uterus, where it hatches from the zona pellucida and implants in the uterine lining (endometrium). If a blastocyst hatches early, it can embed itself in the wall of the oviduct. This is a dangerous medical condition called an ectopic pregnancy.
5. Identical (monozygotic) twins are formed during an early stage of cleavage when the mass of cells splits. Non-identical twins form when two secondary oocytes are released from the ovaries and are fertilized by two different sperm. Conjoined twins are identical twins formed when splitting of the mass of cells is

incomplete. The occurrence of identical twins provides evidence that humans display regulative development.

6.
a. Blastocoel
b. Blastopore
c. Secondary mesenchyme
d. Ectoderm
e. Endoderm
f. Archenteron
g. Primary mesenchyme

7. When transplants occur between early gastrulas, the transplanted pieces will develop into tissues appropriate for the location in which they are placed. In this case, therefore, the transplanted tissue will develop into neural tissue. When transplants occur between late gastrulas, however, the fates of the transplanted cells have already been determined. So, in this case, the transplanted tissue will develop into epidermis.

8. Although the details of gastrulation differ among sea urchins, frogs, and chickens, the common result is the formation of germ layers from which all tissues and organs will form.

9. Spina bifida is a neural tube defect characterized by failure of the neural tube to develop and close properly during prenatal development. It represents a disturbance in the process of neurulation.

10. The experiment should be directed at blocking Sonic hedgehog, the transcription factor released by the notochord. Sonic hedgehog diffuses into the ventral region of the neural tube, where it directs development of spinal cord circuits.

11. The placenta is formed from the chorion of the embryo and the endometrium (uterine lining) of the mother. The chorion is formed when the yolk sac contributes mesodermal tissues that interact with the trophoblast. Thus, the chorionic villi are the embryo's contribution to the placenta and cells taken from them would have the same genetic makeup as cells of the fetus.

12. Under normal circumstances there is no direct mixing of fetal and maternal blood supplies at the placenta. Instead, all exchanges of materials (nutrients, gases, wastes) occur across fetal capillaries within chorionic villi.

13.
a. Embryo
b. Amnion
c. Chorion
d. Yolk sac
e. Allantois
f. Allantoic membrane

14. Exposure to a harmful chemical during the first trimester would be more likely to produce a major birth defect than exposure to the same chemical during the last trimester because tissues and organs are forming during the first trimester. The last trimester involves mostly rapid growth.

15. Limb development can be assessed by week 8. Kidney function and overall growth can best be assessed toward the end of the third trimester.

Test Yourself

1. **a.** The egg donates most of the organelles to the zygote. In most species, the sperm contributes a centriole.

2. **b.** Cleavage in mammals occurs in a rotational pattern.

3. **c.** Incomplete cleavage occurs because the cleavage furrows cannot completely penetrate the yolk. Incomplete cleavage occurs in animals such as birds and reptiles, whose eggs have large amounts of yolk.

4. **b.** Ectoderm will eventually become the nervous system.

5. **b.** The blastopore, which will eventually become the anus, marks the posterior of the sea urchin embryo.

6. **d.** The germ layers that are formed during gastrulation are the inner layer of endoderm, the middle layer of mesoderm, and the outer layer of ectoderm.

7. **d.** The dorsal lip leads to the establishment of the embryonic axes.

8. **c.** β-catenin is thought to be the initiator of organizer activity during early amphibian development.

9. **c.** In chickens, Hensen's node is the equivalent of the dorsal lip of the frog blastopore.

10. **c.** During development, an avian embryo produces wastes, but because birds develop in a shell, the wastes must be stored in the egg. The allantoic membrane forms a sac that is used for the storage of metabolic wastes.

11. **a.** In the mammalian blastocyst, the trophoblast forms the fetal part of the placenta (chorion). A disc-shaped portion of the inner cell mass becomes the embryo.

12. **a.** Somites are located along the neural tube in the developing vertebrate. These cells will develop into the vertebrae, ribs, and trunk muscles.

13. **b.** The amnion is the membrane that surrounds the developing mammalian fetus in its fluid-filled amniotic cavity.

14. **a.** Extremely rapid growth characterizes the third trimester. Gastrulation, formation of major organs, and formation of the placenta occur during the first trimester, which is also the period of greatest sensitivity to drugs and radiation.

15. **d.** The human gestation period of nine months is not the longest among mammals. Length of gestation is related to overall body size, not relative brain size.

Neurons, Glia, and Nervous Systems

The Big Picture

- The neuron, with support from surrounding glia, is the functional unit of the nervous system. The neuron is composed of a cell body, dendrites, and an axon. The membrane of a neuron has a difference in voltage across it. Nerve impulses are passed down the axon of a neuron as action potentials. An action potential is a temporary disruption of the "battery-like" state of the resting neuron membrane due to the opening and closing of sodium and potassium voltage-gated channels. The action potential is an all-or-none response that occurs when the depolarization of an axon reaches a threshold level.
- At the end of the axon synapse, information flow is controlled through excitation or inhibition of synapses. There are many different neurotransmitters that transmit a nerve impulse from the presynaptic cell to the postsynaptic cell. The action of a specific neurotransmitter depends on the receptor to which it binds.
- The nervous systems of animals range in complexity from the relatively simple nerve nets of cnidarians, to ganglia of earthworms, to the central nervous system (brain and spinal cord) and peripheral nervous system (sensory and effector neurons and supporting cells) of vertebrates.

Study Strategies

- In studying the nervous system, one of the most difficult challenges is understanding how the resting membrane potential and action potential are produced. Remember that the neuron at rest is like a battery, and the charge of the membrane is dependent on the ions that are on either side and moving across the membrane. Learn what each important anion and cation does at rest and during the action potential. Then, piece by piece, put together an understanding of how the neuron membrane is charged at rest and what happens during an action potential.
- You may have difficulty understanding the function of a synapse and how the action potential is transmitted across it. Remember that in many cases the signal goes from an electrical signal, to a chemical signal, and back to an electrical signal.
- You should understand the concept of "presynaptic" and "postsynaptic" neurons. Try to remember the sequence of events and relate it to the function of the neuron transmitting a stimulus from one cell to another.
- Go to the Web addresses shown to review the following animated tutorials and activity. You can also find them in BioPortal (yourBioPortal.com), along with many additional learning resources.

  Animated Tutorial 45.1 The Resting Membrane Potential (Life10e.com/at45.1)

  Animated Tutorial 45.2 The Action Potential (Life10e.com/at45.2)

  Animated Tutorial 45.3 Synaptic Transmission (Life10e.com/at45.3)

  Animated Tutorial 45.4 Neurons and Synapses (Life10e.com/at45.4)

  Animated Tutorial 45.5 Information Processing in the Spinal Cord (Life10e.com/at45.5)

  Activity 45.1 Neurotransmitters (Life10e.com/ac45.1)

Key Concept Review

45.1 What Cells Are Unique to the Nervous System?

The structure of neurons reflects their functions

Glia are the "silent partners" of neurons

The nervous system is made up of two major types of cells: neurons, which generate and transmit electric signals, and glia, which support neurons and modulate their activity. A neuron is composed of four parts: the cell body, dendrites, axon, and axon terminals (see Figure 45.1). The cell body contains the nucleus, with dendrites projecting from it and receiving information from other neurons. The axon conducts action potentials away from the cell body. Axon terminals interact with the neuron's target cells to form a synapse. When information is passed from one neuron to another neuron at a synapse, the first neuron is the presynaptic neuron and the second is the postsynaptic neuron. Synapses can be either chemical or electrical.

Glial cells far outnumber neurons in the brain, have diverse structures, and serve many functions. Some supply nutrients to neurons and remove wastes, while others contribute to the blood–brain barrier, which protects the brain from some toxins in the blood. Glial cells that insulate axons in the central nervous system are oligodendrocytes, and those that insulate axons in the peripheral nervous system are Schwann cells (see Figure 45.3). The covering produced by Schwann cells and oligodendrocytes is called myelin; myelinated axons conduct action potentials more rapidly than do axons lacking myelin. Star-shaped glial cells, called astrocytes, release neurotransmitters and thereby alter the activity of neurons. Astrocytes also contact blood vessels and communicate changes in blood composition to neurons, and modulate synaptic activity (see Figure 45.4).

Question 1. Label the following structures on the neuron shown below: axon, axon hillock, axon terminals, cell body, and dendrites. What is the direction and manner in which an action potential is conducted along the neuron?

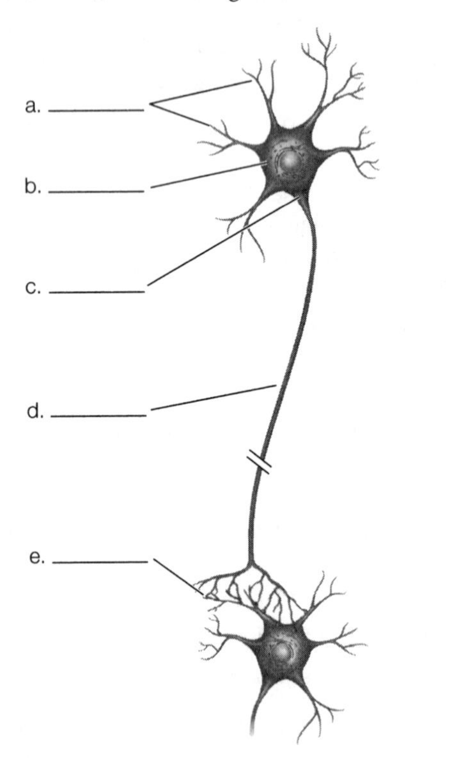

Question 2. If you worked for a drug company and were asked to develop a drug that could reach the brain, would you want the drug to be fat soluble or water soluble? Explain your answer.

Question 3. Two patients have been diagnosed with demyelinating diseases, one of the brain and spinal cord and the other in parts of the nervous system outside the brain and spinal cord. What are demyelinating diseases and which cells are responsible in each of the two cases?

45.2 How Do Neurons Generate and Transmit Electric Signals?

Simple electrical concepts underlie neural function

Membrane potentials can be measured with electrodes

Ion transporters and channels generate membrane potentials

Ion channels and their properties can now be studied directly

Gated ion channels alter membrane potential

Graded changes in membrane potential can integrate information

Sudden changes in Na^+ and K^+ channels generate action potentials

Action potentials are conducted along axons without loss of signal

Action potentials jump along myelinated axons

Resting neurons have a negative charge inside and a positive charge outside, resulting in a difference in electrical charge across the membrane known as the membrane potential. In an unstimulated neuron, this voltage difference is called a resting potential, and is typically between –60 and –70 mV (see Figure 45.5). The electrical charge across the membrane at rest is due to differences in concentrations of the charged ions sodium (Na^+), chloride (Cl^-), potassium (K^+), and calcium (Ca^{2+}). Ions move across the plasma membrane through channels or by ion pumps, both of which are formed by proteins. The sodium–potassium pump transports Na^+ out of the cell and K^+ into it, thereby maintaining higher concentrations of Na^+ ions outside the cell and higher concentrations of K^+ ions inside it (see Figure 45.6A). At rest, neurons have a specific charge due to K^+ movement to the outside of the cell, resulting in a resting potential. K^+ channels are the most common open channels (sometimes called leak channels), allowing K^+ to diffuse out of the cell down the concentration gradient that has been set up by the Na^+–K^+ pump (see Figure 45.6B).

Ion channels are selective pores in the plasma membrane that allow specific ions to diffuse across the membrane. Whereas some are always open (such as K^+ channels), others are gated (open under certain conditions and closed under other conditions). Ion channels can be voltage-gated (responding to changes in the voltage across the membrane), chemically gated (responding to the presence of a specific chemical), or mechanically gated (responding to mechanical force applied to the membrane). Patch clamping allows the recording of voltage differences due to the movements of ions through channels in an isolated patch of plasma membrane (see Figure 45.8).

Membranes can be depolarized or hyperpolarized (see Figure 45.9). Depolarization occurs when the inside of a neuron becomes less negative compared to the resting potential. Hyperpolarization occurs when the inside of a neuron becomes more negative compared to the resting potential. Changes in polarity of the plasma membrane due to opening and closing of ion channels are passed down along an axon to transmit a signal as an action potential.

Action potentials are very short-lived, but large, changes in membrane potential (see Figure 45.10). They are generated when the membrane reaches a threshold potential,

Na^+ voltage-gated channels open, and Na^+ enters the cell to make the inside of the axon positive. Voltage-gated K^+ channels then open, allowing positively charged K^+ to leave the axons to help return the membrane potential back to the resting level. As the K^+ channels open, the Na^+ channels close and cannot be opened for about 1 to 2 milliseconds, which is the refractory period. The Na^+–K^+ pump helps return the concentration of ions back to the resting levels. The Na^+ ions that enter during an action potential flow to adjoining regions of the axon, stimulating depolarization and the movement of the action potential along the axon (see Figure 45.11). The refractory period keeps an action potential moving in one direction, away from the cell body. An action potential is an all-or-none response; the depolarization must reach a threshold level for an action potential to occur. An action potential is also a self-regenerating response; once an action potential occurs at one location on an axon, it stimulates the adjacent area to generate an action potential.

In the nervous systems of invertebrates, the conduction velocity of axons increases with increasing diameters of axons. In the nervous systems of vertebrates, conduction velocity of axons increases primarily by myelination. Myelination is produced by glia that wrap themselves around some axons; nodes of Ranvier are gaps in the myelin wrapping at which depolarization can occur. Depolarization jumps from node to node (saltatory conduction), increasing the speed of transmission of an action potential along the axon (see Figure 45.12).

Question 4. In order to determine the role of the potassium channels in a neuron, a researcher has knocked out all of the functional potassium channels and depolarized the membrane potential. What will happen to the membrane potential after depolarization?

Question 5. Explain how an action potential travels more quickly down an axon wrapped in myelin than it does down an unmyelinated axon.

Question 6. In an invertebrate such as a squid, the largest-diameter neurons would most likely innervate which one of the following structures: eyes; eye muscles; digestive tract; muscles used in locomotion? Explain your answer.

45.3 How Do Neurons Communicate with Other Cells?

The neuromuscular junction is a model chemical synapse

The arrival of an action potential causes the release of neurotransmitter

Synaptic functions involve many proteins

The postsynaptic membrane responds to neurotransmitter

Synapses can be excitatory or inhibitory

The postsynaptic cell sums excitatory and inhibitory input

Synapses can be fast or slow

Electrical synapses are fast but do not integrate information well

The action of a neurotransmitter depends on the receptor to which it binds

To turn off responses, synapses must be cleared of neurotransmitter

The diversity of receptors makes drug specificity possible

In electrical synapses, the action potential spreads directly from presynaptic to postsynaptic cell. A chemical synapse uses a chemical messenger, or neurotransmitter, to communicate between the presynaptic and postsynaptic cells. The neurotransmitter is released into the synaptic cleft, crosses it, and binds with receptors on the surface of the postsynaptic cell. The neurotransmitter acetylcholine (ACh) is the chemical messenger carrying information between motor neurons and muscle cells at neuromuscular junctions (see Figure 45.13). Acetylcholinesterase breaks down acetylcholine in the synapse to halt the action of the released acetylcholine (see Figure 45.14).

Synapses in vertebrates can be excitatory or inhibitory. Excitatory synapses depolarize the postsynaptic membrane and inhibitory synapses hyperpolarize it. Neurons may receive synaptic inputs from many neurons. Excitatory and inhibitory postsynaptic potentials are summed over space and time. Spatial summation adds up simultaneous potentials. Temporal summation adds up the rapid firing of postsynaptic potentials at a particular sight (see Figure 45.15).

The axon hillock is the "decision-making" area of a neuron. If the axon hillock is depolarized, an axon will fire an action potential.

The two general categories of neurotransmitter receptors are ionotropic and metabotropic. Ionotropic receptors are ion channels on the postsynaptic membrane that are activated by the binding of the neurotransmitter. They allow fast, short-lived responses. Metabotropic receptors are not ion channels, but they act by initiating signaling cascades, which eventually cause changes in ion channels. When mediated by metabotropic receptors, postsynaptic cell responses are usually slower and longer-lived than those generated by ionotropic receptors.

Electrical synapses are formed by direct contact between adjacent neurons; these synapses contain numerous gap junctions. Two neurons forming an electrical synapse are joined by connexins, which are tunnels (pores) between the two neurons that allow ions to pass between the two cells. Electrical synapses can transmit an action potential in either direction, and are good for rapid communication. They are less common than chemical synapses in vertebrate nervous systems, require large areas of contact, cannot be inhibitory, and do not allow temporal summation of synaptic inputs.

There are more than 50 neurotransmitters, including amino acids, peptides, gases (e.g., nitric oxide), purines, and monoamines. One neurotransmitter can act on several different receptors, and its particular action depends on the receptor to which it binds. The actions of neurotransmitters can be stopped in several ways. First, enzymes may destroy a neurotransmitter. Second, a neurotransmitter may simply diffuse away from the synaptic cleft. Third, nearby cell membranes may take up a neurotransmitter using active transport.

Question 7. The active ingredients in many nerve gases belong to a class of chemicals called anticholinesterases (chemicals that block acetylcholinesterase). Suggest a possible synaptic mechanism to explain how these chemicals can damage an animal's nervous system.

Question 8. Clinical depression is thought to be due, in part, to insufficient levels of the neurotransmitter serotonin. Drugs known as selective serotonin reuptake inhibitors (SSRIs) can be used to treat depression. Taking the name of this class of drugs as a clue, propose a mechanism by which they might act.

45.4 How Are Neurons and Glia Organized into Information-Processing Systems?

Nervous systems range in complexity

The knee-jerk reflex is controlled by a simple neural network

The vertebrate brain is the seat of behavioral complexity

Neurons are organized into neural networks. Afferent neurons carry sensory information into the nervous system. Efferent neurons carry information from the nervous system to effectors, such as muscles or glands. Interneurons facilitate communication between afferent and efferent neurons.

Nervous systems differ in complexity among animals (see Figure 45.17). Simple animals, such as the sea anemone, possess a nerve net. The nervous systems of more complex animals, such as earthworms and squid, contain clusters of neurons (ganglia) distributed throughout the body. Vertebrates have a central nervous system—the brain and spinal cord—and a peripheral nervous system—the neurons and supporting cells in the rest of the body.

Cell bodies are found in gray matter and axons are found in white matter of the nervous system. The white color is due to myelin. The spinal cord converts some afferent information from the peripheral nervous system into efferent information sent back to the peripheral nervous system in a process known as a spinal reflex (see Figure 45.18). A monosynaptic reflex, such as the knee-jerk reflex, involves only an afferent neuron, an efferent neuron, and one synapse. More complicated reflexes involve interneurons and additional synapses. The brains of vertebrates vary in size (even among species with similar body masses) and complexity. The size of the cerebrum, the part of the brain responsible for complex behaviors, is especially variable across taxa (see Figure 45.19).

Question 9. In Lou Gehrig's disease, motor neurons die and muscles no longer receive neural messages. Affected individuals gradually lose control over their limbs and body, and the eventual cause of death is respiratory failure. The disease does not affect sensory neurons and interneurons and individuals do not experience loss of cognitive function. Explain in terms of the neurological processes why cognition is unimpaired.

Question 10. You are asked to develop a laboratory display demonstrating diversity in vertebrate brains, and how size of particular brain areas relates to physiological specializations (particularly sensory systems) and behavioral complexity. Which groups would you have represented in your display? What factors would you consider when selecting the groups? Explain your selections.

Test Yourself

1. The extensions of postsynaptic neurons that provide the main receptive surface for presynaptic neurons are the
   a. nuclei.
   b. somas.
   c. axons.
   d. dendrites.
   e. glia.
   Textbook Reference: *45.1 What Cells Are Unique to the Nervous System?*

2. The substance that wraps around the axon of many neurons and provides for increased conduction speed is
   a. a bipolar cell.
   b. histamine.
   c. acetylcholine.
   d. myelin.
   e. microglia.
   Textbook Reference: *45.1 What Cells Are Unique to the Nervous System?*

3. The long extension from the cell body of a neuron that provides the pathway for action potentials to the synapse is the
   a. dendrite.
   b. Schwann cell.
   c. axon.
   d. presynaptic membrane.
   e. nerve net.
   Textbook Reference: *45.1 What Cells Are Unique to the Nervous System?*

4. Which of the following statements pairing types of glial cells with their function is *false*?
   a. Oligodendrocytes wrap around axons of neurons in the brain and spinal cord.
   b. Astrocytes generate action potentials.
   c. Microglia function in immune defense.
   d. Schwann cells wrap around axons of neurons outside the brain and spinal cord.
   e. All of the above are true, none is false.
   Textbook Reference: *45.1 What Cells Are Unique to the Nervous System?*

5. The threshold of a neuron is the
   a. amount of inhibitory neurotransmitter required to inhibit an action potential.
   b. membrane voltage at which an axon potential will be suppressed.

c. amount of excitatory neurotransmitter required to elicit an action potential.
d. membrane voltage at which the membrane potential develops into an action potential.
e. closing of numerous sodium channels.

Textbook Reference: *45.2 How Do Neurons Generate and Transmit Electrical Signals?*

6. When a membrane is at the resting potential, the concentration of
a. sodium and potassium ions is higher on the inside of the membrane than on the outside.
b. sodium and potassium ions is higher on the outside of the membrane than on the inside.
c. sodium ions is higher on the inside of the membrane and the concentration of potassium ions is higher on the outside.
d. sodium ions is higher on the outside of the membrane and the concentration of potassium ions is higher on the inside.
e. sodium equals the concentration of potassium inside the cell.

Textbook Reference: *45.2 How Do Neurons Generate and Transmit Electrical Signals?*

7. The rapid depolarization of a neuron during the first half of an action potential is due to the
a. exit of K^+ ions from the cell through gated potassium channels.
b. rapid reversal of ion concentration caused by the action of the Na^+–K^+ pump.
c. entry of Na^+ ions into the cell through gated sodium channels.
d. movement of both Na^+ and K^+ ions through appropriate open channels.
e. closing of sodium channels.

Textbook Reference: *45.2 How Do Neurons Generate and Transmit Electrical Signals?*

8. The refractory period of a neuron
a. is the period when the Na^+–K^+ pump is nonfunctional.
b. results from activation of voltage-gated chloride channels.
c. results from the closing of inactivated voltage-gated sodium channels.
d. occurs when the action potential reaches the synapse.
e. lasts about a minute.

Textbook Reference: *45.2 How Do Neurons Generate and Transmit Electrical Signals?*

9. A particular disease of the nervous system specifically involves the Ca^{2+} channels at the chemical synapses of motor neurons where neurotransmitter is stored and released. This disease therefore affects the
a. axon terminals of the presynaptic cell and the release of acetylcholine.
b. axon terminals of the postsynaptic cell and the release of K^+.
c. movement of Na^+ out of the postsynaptic cell.
d. axon terminals of the presynaptic cell and the release of K^+.
e. axon terminals of the postsynaptic cell and the release of Cl^-.

Textbook Reference: *45.3 How Do Neurons Communicate with Other Cells?*

10. Which of the following statements about electrical synapses is *false*?
a. Connexins form molecular tunnels between two cells.
b. Electrical synapses cannot be inhibitory.
c. Electrical synapses do not allow for temporal summation.
d. The transmission capacity of electrical synapses is very slow.
e. Electrical synapses allow for transmission either toward or away from the cell.

Textbook Reference: *45.3 How Do Neurons Communicate with Other Cells?*

11. Which of the following statements about neurotransmitter receptors is *false*?
a. Ionotropic receptors are ion channels.
b. The acetylcholine receptor of the motor end plate is a metabotropic receptor.
c. Metabotropic receptors are not ion channels.
d. Metabotropic receptors induce signaling cascades in the postsynaptic cell.
e. Responses in the postsynaptic cell mediated by metabotropic receptors are usually slower than those mediated by ionotropic receptors.

Textbook Reference: *45.3 How Do Neurons Communicate with Other Cells?*

12. Which of the following statements about the process of summation in a neuron is *false*?
a. Slight perturbations of the membrane potential spread across the postsynaptic cell body.
b. Axons that terminate closer to the axon hillock have more influence on the summation process than those that do not.
c. The process is essentially a comparison of all the excitatory and inhibitory postsynaptic inputs.
d. The concentration of voltage-gated sodium channels is highest in the dendrites of the postsynaptic cell.
e. Spatial summation adds up the simultaneous influences of synapses at different locations on the postsynaptic cell.

Textbook Reference: *45.3 How Do Neurons Communicate with Other Cells?*

13. Which of the following statements about neurotransmitters is *false*?
a. Gases, such as nitric oxide, can act as neurotransmitters.
b. Each neurotransmitter has a single type of receptor.
c. Amino acids and their derivatives, monoamines, function as neurotransmitters.

d. Neurotransmitters have different effects in different tissues.
e. Some neurotransmitters are cleared from synapses by enzymes that destroy them.

Textbook Reference: *45.3 How Do Neurons Communicate with Other Cells?*

14. The electrical events called excitatory postsynaptic potentials (EPSPs) are the result of _______ of the _______ membrane.
a. hyperpolarization; postsynaptic
b. depolarization; postsynaptic
c. hyperpolarization; presynaptic
d. depolarization; presynaptic
e. repolarization; presynaptic

Textbook Reference: *45.3 How Do Neurons Communicate with Other Cells?*

15. Which of the following statements about the knee-jerk reflex is *false*?
a. It is a monosynaptic reflex.
b. It causes the leg extensor muscle to contract.
c. Chemoreceptors sense a physician's hammer tap.
d. The afferent nerve travels from the receptor to the spinal cord.
e. The motor neuron leaves via a ventral root of the spinal cord.

Textbook Reference: *45.4 How Are Neurons and Glia Organized into Information-Processing Systems?*

Answers

Key Concept Review

1.
a. Dendrites
b. Cell body
c. Axon hillock
d. Axon
e. Axon terminals

The action potential travels down the axon, away from the axon hillock.

2. Astrocytes are glial cells that help form the blood–brain barrier by surrounding tiny, very permeable blood vessels in the brain. However, because the barrier is made of plasma membranes, fat-soluble substances such as anesthetics and alcohol can pass through it. So, any drug designed to reach the brain should be fat-soluble.

3. Demyelinating diseases affect myelin. Multiple sclerosis is an example; this autoimmune disease is caused by the production of antibodies to proteins in the myelin of the brain and spinal cord. The particular glial cells involved are oligodendrocytes. Guillain-Barre syndrome is another example of a demyelinating disease; this disease, usually caused by an infection, affects myelin outside the brain and spinal cord. The particular glial cells involved are Schwann cells.

4. The potassium voltage-gated channels are responsible for setting up the resting potential of a membrane. Potassium ions have a tendency to diffuse out of the cell, leaving a negative charge inside. Knocking out the function of the potassium voltage-gated channels would result in the cell's being unable to maintain resting potential. If the cell was depolarized by the opening of sodium voltage-gated channels, it might not repolarize because the potassium channels that help repolarize the membrane would not be functioning.

5. The conduction of an action potential down a myelinated axon is called saltatory conduction. The myelin acts to insulate areas of the axon, preventing depolarization. The areas of the axon between the myelin sheaths are known as nodes of Ranvier. Depolarization can occur only at these nodes. As the action potential moves down a myelinated axon, the influx of sodium ions at one node diffuses down the axon. This results in the depolarization of the next node of Ranvier. Depolarization can occur only in the downstream nodes because the upstream nodes are in a refractory period. As a result, the action potential moves quickly down the axon to the synapse.

6. The largest-diameter neurons would most likely innervate the muscles used in locomotion. Increased diameter translates into increased conduction velocity, which would be necessary for muscles used in locomotion during escape responses.

7. Acetylcholine is the neurotransmitter used by all neuromuscular synapses in vertebrates. It transmits the action potential from a presynaptic cell to a postsynaptic cell. The enzyme acetylcholinesterase is found in the synaptic cleft, and it cleaves acetylcholine to help remove it from the synaptic cleft after an action potential. A nerve gas with components that block the action of acetylcholinesterase would cause acetylcholine to build up in the synaptic cleft. This buildup would mean that the receptors on the postsynaptic cell would remain bound with acetylcholine, resulting in prolonged muscle contraction.

8. Selective serotonin reuptake inhibitors such as Paxil, Zoloft, and Prozac increase the level of serotonin at the synapse by reducing its rate of removal.

9. A person with Lou Gehrig's disease would have no cognitive deficits because only the motor neurons (concerned with output) are affected by the disease; sensory neurons (concerned with input) and interneurons (concerned with integration) are unaffected.

10. Ideally, you would want a representative from each of the major vertebrate groups (jawless fishes, cartilaginous fishes, bony fishes, amphibians, reptiles, birds, and mammals). To the extent possible, you would select representatives of similar body mass. Additionally, for each major sensory system (olfaction, vision, etc.) you would want to compare the sizes of the particular brain area involved in a vertebrate in which the sensory

system is highly developed and a vertebrate in which the sensory system is poorly developed. For example, you might compare the size of the olfactory lobe in a mammal (most mammals have an excellent sense of smell) with that of a bird (most birds have a poor sense of smell). Because complex behavior and learning are more typical of birds and mammals than other vertebrates, you could compare the size of the cerebrum in these two groups with those of other groups of vertebrates. Across all of your representative vertebrates, the brainstem should be much more similar than the cerebrum.

Test Yourself

1. **d.** The neuron is composed of a cell body, an axon, and dendrites. The dendrites form synapses with presynaptic cells to create the junction where information from one neuron is transferred to another neuron.
2. **d.** The glial cells that coat the axon of some neurons form myelin.
3. **c.** The neuron is composed of the cell body, the dendrite, and the axon. The axon carries action potentials away from the cell body to the synapses.
4. **b.** Astrocytes perform many functions, which include contributing to the blood–brain barrier, removing neurotransmitter from a synapse, providing nutrients to neurons, and releasing neurotransmitters. They do not, however, generate action potentials.
5. **d.** For an action potential to occur in an axon, the membrane must be depolarized above a certain level. This level is known as the threshold.
6. **d.** The resting potential of a neuron membrane occurs when the sodium ion concentration is higher on the outside and the potassium ion concentration is higher on the inside.
7. **c.** The first step in an action potential is the influx of Na^+ leading to a depolarization of the axon membrane. Na^+ rushes into the cell due to the higher concentration outside of the cell and the negative membrane potential.
8. **c.** After the spike of the depolarization, the sodium voltage-gated channels close. One of the properties of these channels is that they will open again only after a short delay. This short delay (about 1–2 milliseconds) is known as the refractory period, during which the voltage-gated Na^+channels are inactive.
9. **a.** If the disease acts on a chemical synapse where the neurotransmitter is stored and released, it is affecting the axon terminals of the presynaptic cell. Ca^{2+} channels are involved in regulating the release of acetylcholine by allowing Ca^{2+} to enter the presynaptic cell and promoting the fusing of acetylcholine-containing vesicles to the membrane.
10. **d.** Electrical synapses join two cells with protein tunnels known as connexins. These synapses provide for very fast transmission between cells.
11. **b.** The acetylcholine receptor of the motor end plate is an ionotropic receptor.
12. **d.** Dendrites, and most of the cell body, have few gated sodium channels. These channels mediate the action potentials that travel down the axon, where their levels are high.
13. **b.** Each neurotransmitter has multiple types of receptors.
14. **b.** Excitatory postsynaptic potentials (EPSPs) make it easier for an action potential to occur, so they depolarize the postsynaptic membrane.
15. **c.** The knee-jerk reflex is an example of a monosynaptic reflex. Stretch receptors (not chemoreceptors) sense the hammer tap on the tendon.

Sensory Systems

The Big Picture

- Sensory structures work by converting some form of stimulus—mechanical, chemical, light—into action potentials in the nervous system, which are then interpreted by the central nervous system.
- Receptors are named on the basis of their sensitivity. For example, chemoreceptors respond to chemical stimulation, mechanoreceptors respond to mechanical stimulation, and photoreceptors respond to light.
- Different animals have different types of senses and different sensitivities of senses.

Study Strategies

- The senses have what appear to be very different mechanisms for the transmission of information to the brain. To help sort this out, remember that there are only a few types of receptors that respond to stimuli and that they all generate action potentials. Recall also that the action potentials produced in the neurons of the ear, eye, and other sense organs, as well as in the knee or stomach, are identical. The action potentials coming from the eye, for example, are interpreted as light because of the region of the brain that receives and analyzes them.
- The route by which sound travels in the ear can be very confusing. View the cochlea in the uncoiled form in Figure 46.10. This will help you visualize how pressure waves of different wavelengths produce different sounds.
- Go to the Web addresses shown to review the following animated tutorials and activities. You can also find them in BioPortal (yourBioPortal.com), along with many additional learning resources.

  Animated Tutorial 46.1 Sound Transduction in the Human Ear (Life10e.com/at46.1)

  Animated Tutorial 46.2 Mechanoreceptors (Life10e.com/at46.2)

  Animated Tutorial 46.3 Photosensitivity (Life10e.com/at46.3)

  Activity 46.1 Structures of the Human Ear (Life10e.com/ac46.1)

  Activity 46.2 Structure of the Human Eye (Life10e.com/ac46.2)

  Activity 46.3 Structure of the Human Retina (Life10e.com/ac46.3)

Key Concept Review

46.1 How Do Sensory Receptor Cells Convert Stimuli into Action Potentials?

Sensory transduction involves changes in membrane potentials

Sensory receptor proteins act on ion channels

Sensation depends on which neurons receive action potentials from sensory cells

Many receptors adapt to repeated stimulation

Sensory receptor cells (also called sensors or receptors) convert physical and chemical stimuli into neural signals, which are transmitted to different sites in the central nervous system (CNS). Sensory transduction begins when receptor proteins of the sensory receptor cell open or close ion channels in response to a particular stimulus. Change in the flow of ions produces a change in membrane potential, called the receptor potential. Receptor potentials produce action potentials either by prompting release of a neurotransmitter that induces an associated neuron to generate action potentials, or by generating action potentials within the sensory receptor cell itself.

Sensory organs, such as eyes and ears, are groups of sensory cells that, along with other cells, collect, filter, and amplify stimuli. Sensory cells, their associated structures, and the networks of neurons that process the information form a sensory system. Even though all sensory systems process information as action potentials, we perceive different sensations (e.g., pain, light, sound) because messages from the different sensory systems go to different areas of the CNS. The frequency of action potentials encodes the intensity of sensation. Some sensory cells exhibit adaptation, whereby their responses gradually diminish with repeated stimulation.

Question 1. Although photoreceptors respond best to light, they can also respond to pressure. Explain why, when you press gently on your closed eyelids, you see spots of light.

Question 2. The term "perception" describes the conscious awareness of sensations; it occurs when the cerebral cortex of the brain integrates sensory information. Are humans consciously aware of all information transmitted by sensory cells?

Question 3. When you put on a pair of shoes, you feel the shoes at first, but the sensation diminishes. However, if you had a blister on your heel, you might think about the shoes throughout the day. What phenomenon is occurring in the first situation (wearing shoes without a blister) and not in the second (wearing shoes with a blister)? Why is there a difference?

46.2 How Do Sensory Systems Detect Chemical Stimuli?

- Olfaction is the sense of smell
- Some chemoreceptors detect pheromones
- The vomeronasal organ contains chemoreceptors
- Gustation is the sense of taste

Olfaction is the sense of smell. The olfactory sensors of vertebrates are olfactory receptor neurons (ORNs) with axons extending to the olfactory bulb of the brain; the dendrites of these neurons end in olfactory cilia that are exposed to the environment within the epithelium of the nasal cavity (see Figure 46.3). Olfactory receptor proteins are found on the cilia, and each receptor binds with specific odorants. Binding of the odorant generates action potentials, which are transmitted to glomeruli in the olfactory bulb. The ability to discriminate many different odors is due to the large number of specific receptors. The strength of a smell relates to the number of odorant molecules that bind to receptors.

Pheromones are chemicals used during within-species communication. Insects use pheromones to attract mates by remotely stimulating their target's chemoreceptors (see Figure 46.4). The concentration of the pheromone released by a female creates a gradient that provides information about her specific location. Amphibians, reptiles, and some mammals have a vomeronasal organ, a paired structure located in the nasal epithelium. In mammals, the vomeronasal organ senses pheromones and conveys information to the accessory olfactory bulb in the brain. In snakes, the forked tongue presents odorant molecules from the environment to the chemoreceptors of the vomeronasal organ on the roof of the mouth; thus, in snakes, the tongue is used in smell and not in taste.

Gustation, the sense of taste, relies on clusters of chemoreceptor cells called taste buds (see Figure 46.5). Binding of the stimulus to receptor proteins on the microvilli of sensory cells causes a change in membrane potential and the release of neurotransmitters that stimulate sensory neurons at the base of the taste bud. Humans can perceive five general tastes: sweet, sour, salty, bitter, and umami (meaty).

Question 4. One of the usual symptoms of the common cold is a diminished sense of smell. What is the cause of this loss of smell?

Question 5. Would removal of the vomeronasal organ of a rodent disrupt its ability to find food, recognize a potential mate, or both? Explain your answer.

46.3 How Do Sensory Systems Detect Mechanical Forces?

- Many different cells respond to touch and pressure
- Mechanoreceptors are found in muscles, tendons, and ligaments
- Hair cells are mechanoreceptors of the auditory and vestibular systems
- Auditory systems use hair cells to sense sound waves
- Flexion of the basilar membrane is perceived as sound
- Various types of damage can result in hearing loss
- The vestibular system uses hair cells to detect forces of gravity and momentum

Mechanical force causes distortion of the membranes of mechanoreceptors, causing ion channels to open and generation of an action potential. The skin has several different types of mechanoreceptors (see Figure 46.6), including Meissner's corpuscles, which are very sensitive but adapt rapidly, and Merkel's discs, which adapt slowly and provide information about objects touching the skin. Pacinian corpuscles and Ruffini endings are deeper in the skin and respond to vibrations. Muscle spindles are mechanoreceptors (specifically, stretch receptors) in skeletal muscles that perceive muscle stretch. Golgi tendon organs are mechanoreceptors in the tendons and ligaments that provide information about forces generated during muscle contraction. Collectively, these mechanoreceptors provide information on limb position as well as stresses on muscles and joints (see Figure 46.7).

Hair cells are mechanoreceptors of the auditory system. Bending of the stereocilia projecting from their surface causes changes in ion channels of the hair cell plasma membrane (see Figure 46.8). Bending in one direction opens the ion channels, causing depolarization and the release of neurotransmitters. Bending in the other direction closes ion channels. In the mammalian auditory system, the pinna collects sound waves and directs them into the auditory canal. The tympanic membrane at the end of the auditory canal vibrates and transmits sound waves to tiny bones (ossicles) in the middle ear, which transmit sound waves to the oval window. Sound travels through the oval window into the fluid-filled cochlea (in the inner ear), where pressure waves are turned into action potentials. Movement of the oval window generates pressure waves in the cochlear fluid, which cause the basilar membrane to vibrate. The organ of Corti, which sits on the basilar membrane, contains hair cells with stereocilia that are in contact with the overhanging tectorial membrane. When the basilar membrane vibrates, the hair cell stereocilia of the organ of Corti are pushed against the tectorial membrane. Movements of the stereocilia are transduced into action potentials that are carried to the brain by the vestibulocochlear nerve (see Figure 46.9). The round window relieves pressure created by movements of the oval window.

In the inner ear of mammals, hair cells also are present in the organs of the vestibular system. The vestibular system consists of three semicircular canals and two chambers called the saccule and utricle; the entire system is filled with the fluid endolymph. Changes in position of the head cause shifts in the fluid within semicircular canals, which push on the gelatinous cupulae of hair cells, causing their stereocilia to bend (see Figure 46.11A). In the saccule and utricle, the stereocilia are bent by gravitational forces on otoliths, which are granules of calcium carbonate that sit on top of the gelatinous mass overlying the hair cells (see Figure 46.11B).

Question 6. In the diagram of the human ear below, label each of the following structures: tympanic membrane, malleus, incus, stapes, oval window, round window, cochlea, semicircular canal of the vestibular system, and vestibulocochlear nerve.

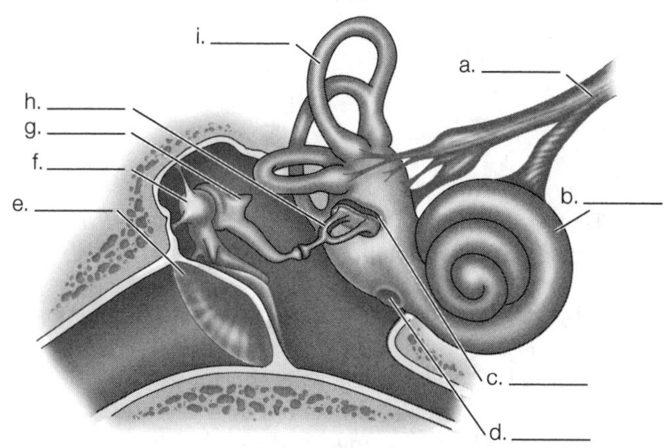

Question 7. Design a study to determine the lowest threshold of hearing for a mammal other than a human.

Question 8. An infection that causes vertigo (dizziness) would be located in which sensory organ and in which particular part of the organ?

46.4 How Do Sensory Systems Detect Light?

Rhodopsin is a vertebrate visual pigment

Invertebrates have a variety of visual systems

Image-forming eyes evolved independently in vertebrates and cephalopods

The vertebrate retina receives and processes visual information

Rod and cone cells are the photoreceptors of the vertebrate retina

Information flows through layers of neurons in the retina

Photoreceptors contain light-sensitive pigments. Rhodopsin is a family of pigments made up of the protein opsin and the light-absorbing nonprotein group 11-*cis*-retinal (see Figure 46.12). The 11-*cis*-retinal absorbs photons of light and changes conformation to all-*trans*-retinal, causing opsin to change conformation and become photoexcited rhodopsin. Photoexcited rhodopsin triggers a G protein cascade, which ultimately leads to changes in membrane potential and the photoreceptor's response to light.

Animals display diverse visual systems. Flatworms have photoreceptor cells organized into eye cups, which help the animal orient away from light sources. Arthropods have compound eyes with many ommatidia (optical units) that contain photoreceptors called retinula cells (see Figure 46.13). The compound eye communicates a low-resolution image to the CNS. Cephalopod mollusks and vertebrates independently evolved image-forming eyes (see Figure 46.14). The vertebrate eye is surrounded by the sclera, formed from connective tissue. The cornea is the transparent sclera through which light passes. The iris controls the amount of light entering the eye through the pupil; it also gives the eye its color. Mammals and birds focus on near and far objects by changing the shape of the lens (see Figure 46.15). Fishes, amphibians, and reptiles focus by moving their lenses closer to or farther from their retinas.

In vertebrate eyes, rod cells are found in the retina, along with a layer that transduces visual information into action potentials. Rod cells become hyperpolarized in response to light and respond by decreasing the levels of neurotransmitter released (see Figure 46.17). The retina also contains cone cells, which absorb light of various wavelengths, allowing for color vision. The human retina contains three types of cone cells that differ in the wavelengths of light that they absorb best (see Figure 46.19).

There are five layers of neurons in the retina, with the photoreceptive rods and cones in the last layer, farthest from the lens (see Figure 46.20). The first layer of cells consists of ganglion cells (which create the action potential), the axons of which form the optic nerve. Bipolar cells are stimulated by neurotransmitters from the photoreceptors to transmit the signal from the photoreceptor to the ganglion cells. Thus, information flow in the retina is from photoreceptor cells at the back to bipolar cells to ganglion cells, which send the information to the brain. The two other layers in the retina are the horizontal cells (which connect adjoining groups of photoreceptors and bipolar cells) and amacrine cells (which connect adjoining groups of bipolar cells and ganglion cells); these cells are interneurons responsible for lateral communication across the retina.

Question 9. In the diagram of the vertebrate eye below, label each of the following structures: sclera, cornea, iris, pupil, lens, retina, fovea, vitreous humor, and optic nerve.

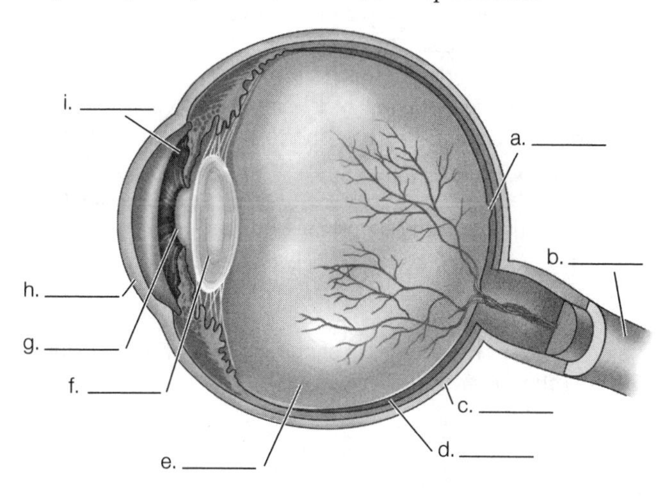

Question 10. You have just given a presentation in your biology class that had many elaborate red- and green-colored slides. Afterward, a male friend tells you that he could not see any of the differences you were reporting. Why could your friend not see the differences?

Question 11. Gray squirrels are diurnal (active during the day) and southern flying squirrels are nocturnal (active at night). How would you expect their retinas to differ?

Test Yourself

1. An electrode is inserted into a chemosensory nerve leading away from a taste bud in the mouth of a dog. A mild acid solution is then flushed continuously over the sensors associated with this nerve. Initially, the nerve responds to this stimulation, but over time it ceases to carry action potentials. Which of the following processes would best explain this observation?
   a. Translocation
   b. Adaptation of the sensory cells
   c. Depletion of neurotransmitter in the sensory nerve
   d. Second messenger influences that increase cell membrane potentials
   e. Action potentials arriving at the wrong area in the CNS

   Textbook Reference: *46.1 How Do Sensory Receptor Cells Convert Stimuli into Action Potentials?*

2. Which of the following statements about sensory cells is *false*?
   a. Mechanoreceptors respond to physical force.
   b. Chemoreceptors monitor aspects of the internal environment.
   c. Mechanoreceptors are metabotropic.
   d. Chemoreceptors are metabotropic.
   e. Sensory cells involved in touch adapt to repeated stimulation.

   Textbook Reference: *46.1 How Do Sensory Receptor Cells Convert Stimuli into Action Potentials?*

3. Which of the following statements about sensory receptor proteins is *false*?
   a. Ionotropic receptor proteins are either ion channels themselves or they directly influence the opening of ion channels.
   b. Photoreceptors are ionotropic.
   c. Electrosensors may lack receptor proteins.
   d. Thermoreceptors respond to temperature and are ionotropic.
   e. Metabotropic receptors influence ion channels indirectly through second messengers.

   Textbook Reference: *46.1 How Do Sensory Receptor Cells Convert Stimuli into Action Potentials?*

4. Which of the following statements about receptor potentials is *false*?
   a. They are changes in the resting membrane potential of a sensory cell in response to a stimulus.
   b. The receptor potential spreads from the cell body of a sensory cell to the axon hillock, where action potentials are generated.
   c. They must be converted to action potentials to travel long distances.
   d. A receptor potential always prompts the release of a neurotransmitter that induces an associated neuron to generate an action potential.
   e. They are graded membrane potentials.

   Textbook Reference: *46.1 How Do Sensory Receptor Cells Convert Stimuli into Action Potentials?*

5. Silkworm moths use chemosensory signals known as _______ for mate attraction.
   a. general odorants
   b. hormones
   c. pheromones
   d. G proteins
   e. locally acting chemical messengers

   Textbook Reference: *46.2 How Do Sensory Systems Detect Chemical Stimuli?*

6. Which of the following statements about human gustation is *false*?
   a. Taste bud cells are relatively short lived because of the high degree of abrasion they encounter.
   b. Taste buds are confined to the oral cavity.
   c. Changes in the membrane potential of the taste bud sensory cells cause them to release neurotransmitter onto the dendrites of sensory neurons.
   d. Humans perceive only three categories of taste: sweet, sour, and bitter.
   e. Most taste buds are found on the papillae of the tongue.

   Textbook Reference: *46.2 How Do Sensory Systems Detect Chemical Stimuli?*

7. Which of the following statements about the detection of chemical stimuli is *false*?
   a. Snakes use their tongues to smell.
   b. Many mammals have a vomeronasal organ to detect pheromones.
   c. A greater frequency of action potentials is associated with perception of a more intense smell.
   d. Taste buds are confined to the oral cavity in aquatic animals.
   e. Chemoreceptors monitor aspects of the internal environment.

   Textbook Reference: *46.2 How Do Sensory Systems Detect Chemical Stimuli?*

8. Stretch receptors in the aorta and carotid artery sense changes in arterial pressure. These receptors are therefore considered
   a. chemoreceptors.

b. thermoreceptors.
c. electroreceptors.
d. mechanoreceptors.
e. muscle spindles.
Textbook Reference: *46.3 How Do Sensory Systems Detect Mechanical Forces?*

9. Which of the following does *not* employ hair cells as its transducer?
a. Meissner's corpuscle
b. Utricle
c. Organ of Corti
d. Semicircular canal
e. Saccule
Textbook Reference: *46.3 How Do Sensory Systems Detect Mechanical Forces?*

10. Which of the following structures is *not* found in the inner ear?
a. Vestibular membrane
b. Tectorial membrane
c. Tympanic membrane
d. Basilar membrane
e. Semicircular canal
Textbook Reference: *46.3 How Do Sensory Systems Detect Mechanical Forces?*

11. Which of the following statements about the photosensitive molecule rhodopsin is *false*?
a. Opsin is converted from the 11-*cis* to the all-*trans* form upon absorbing a photon of light.
b. The retinal is the light-absorbing group.
c. Photoexcited rhodopsin triggers a cascade of reactions that ultimately alters the membrane potential of a photoreceptor cell.
d. Opsin is a protein; retinal is not a protein.
e. 11-*cis*-retinal is covalently bonded to opsin.
Textbook Reference: *46.4 How Do Sensory Systems Detect Light?*

12. In the human visual system, _______ send information directly to the brain.
a. amacrine cells
b. bipolar cells
c. ganglion cells
d. rods and cones
e. horizontal cells
Textbook Reference: *46.4 How Do Sensory Systems Detect Light?*

13. Through which of the following cell layers must a photon of light pass before striking a cone cell in the eye of a human?
a. Amacrine
b. Bipolar
c. Ganglion
d. Horizontal
e. All of the above
Textbook Reference: *46.4 How Do Sensory Systems Detect Light?*

14. Which of the following animals change the shape of their lenses in order to focus?
a. Fishes
b. Reptiles
c. Birds
d. Mammals
e. Both c and d
Textbook Reference: *46.4 How Do Sensory Systems Detect Light?*

15. Which of the following statements about visual systems is *false*?
a. The human retina contains five kinds of cone cells, each with slightly different opsin molecules.
b. Cephalopod mollusks and vertebrates independently evolved image-forming eyes.
c. Nocturnal animals have a white reflective layer behind their retinas that reflects photons back onto photoreceptors.
d. Arthropods have compound eyes with many optical units called ommatidia.
e. Birds have two foveae in each eye.
Textbook Reference: *46.4 How Do Sensory Systems Detect Light?*

Answers

Key Concept Review

1. When you press gently on your closed eyelids, you see spots of light because the pressure stimulates photoreceptors, which send action potentials to the visual cortex. This is one illustration of the way in which sensation depends on the particular part of the brain that receives the nerve impulses.
2. Humans are not consciously aware of all information transmitted by sensory cells. Some sensory cells transmit information about internal conditions in the body (e.g., about blood pressure and carbon dioxide concentration in the blood, or limb position) that is outside of conscious awareness.
3. The phenomenon called sensory adaptation is occurring in the first situation. A person will feel the shoes at first, but touch receptors adapt quickly and the sensation of wearing shoes diminishes. In the second situation, pain receptors are stimulated and these typically adapt very little.
4. The sense of smell depends on olfactory cilia that line the surface of the nasal epithelium. The cilia's receptors bind with odorant molecules, triggering an action potential that is sent to the olfactory bulb of the brain. Usually this epithelium is covered with a thin layer of

protective mucus. However, when you have a cold, the production of mucus increases and mucus covers the epithelium and the olfactory cilia, making it more difficult for odorant molecules to reach the cilia. Thus the sense of smell is decreased.

5. Removal of the vomeronasal organ (VNO) in a rodent would likely impair its ability to recognize a potential mate, because the VNO of mammals typically detects pheromones, which are chemical signals involved in communication between individuals of the same species. Also, experiments indicate that the VNO plays a role in gender identification and sexual behavior. The ability to find food would probably not be impaired because the odorants associated with food are likely sensed by receptors in the olfactory epithelium of the main olfactory system and not by the VNO.

6.
   a. Vestibulocochlear nerve
   b. Cochlea
   c. Oval window (under stapes)
   d. Round window
   e. Tympanic membrane
   f. Malleus
   g. Incus
   h. Stapes
   i. Semicircular canal of the vestibular system

7. The lowest threshold of human hearing can be determined by presenting auditory stimuli to subjects and asking them to respond with a yes or no answer as to whether a particular stimulus can be heard. In other mammals, the electrical activity of the cochlear nerve and auditory regions of the brain can be monitored directly. The lowest threshold of hearing can be determined by presenting sounds that stimulate the ear, cochlear nerve, and parts of the brain involved with hearing and then recording, by means of electrodes placed strategically on particular locations of the head, the differences in electrical potentials elicited by the different sounds. Subjects are typically anesthetized during the procedure.

8. Vertigo is caused by infection in the ear, specifically the inner ear, which contains the organs of equilibrium (in addition to the cochlea).

9.
   a. Fovea
   b. Optic nerve
   c. Sclera
   d. Retina
   e. Vitreous humor
   f. Lens
   g. Pupil
   h. Cornea
   i. Iris

10. Your friend has red–green color blindness. The cones in our eyes allow us to see color. We have cones for red, green, and blue. A lack of one of type of cone, or a reduced number, can cause color blindness, as can problems in the functioning of one type of cone. Red–green color blindness is a sex-linked trait that is more common in men than in women.

11. Cones are responsible for color vision. Rods are responsible for highly sensitive black-and-white vision. We would expect diurnal species (such as the gray squirrel) to have mostly cones in their retinas, and nocturnal species (such as the flying squirrel) to have mostly rods.

Test Yourself

1. **b.** When a sensor cell is stimulated by an unchanging, steady-state stimulus, it will adapt to that stimulus. This allows the sensory system to ignore the unchanging stimulus while still being able to respond to new information.

2. **c.** Mechanoreceptors are ionotropic, not metabotropic.

3. **b.** Photoreceptors are metabotropic because they influence ion channels indirectly, through G proteins and second messengers.

4. **d.** The receptor potential does not always prompt the release of a neurotransmitter to induce an associated neuron to generate an action potential. Sometimes the receptor potential generates action potentials within the sensory cell itself.

5. **c.** Pheromones are chemical signals used in communication within a species. The female silkworm moth releases a pheromone (bombykol) into the environment. The male uses chemoreceptors to follow the pheromone to the source.

6. **d.** Humans can perceive five tastes: sweet, salty, sour, bitter, and umami (a meaty taste). The combination of taste and smell provides the complex subtle flavors of the food we eat.

7. **d.** Taste buds are confined to the oral cavity in terrestrial animals. However, some aquatic animals, such as fish, have taste buds in their skin.

8. **d.** The stretch receptors of the aorta, which detect changes in blood pressure, are examples of mechanoreceptors.

9. **a.** Meissner's corpuscle of the skin does not use hair cells to sense a stimulus. The cell membranes of the Meissner's corpuscle deform in response to light touching of the skin.

10. **c.** Although the tympanic membrane is found in the human ear, it is not found in the inner ear. The tympanic membrane is the membrane that transmits sounds from the auditory canal to the middle ear.

11. **a.** Rhodopsin contains two groups: the protein opsin and the light-sensitive group retinal. Retinal, not opsin, is converted from the 11-*cis* to the all-*trans* form upon

absorbing a photon of light. Opsin does change conformation in response to a change in the rhodopsin to signal the detection of light.

12. **c.** The ganglion cells transmit information from the bipolar cells to the brain. The axons of the ganglion cells form the optic nerve.
13. **e.** The photoreceptive cells are located at the back of the retina. Light must pass through a layer of ganglion cells, a layer of amacrine and bipolar cells, and a horizontal cell layer.
14. **e.** Mammals and birds change the shape of their lenses in order to focus. Fishes, amphibians, and reptiles move their lens closer to or farther from their retinas to focus.
15. **a.** The human retina contains *three* kinds of cone cells, each with slightly different opsin molecules.

The Mammalian Nervous System: Structure and Higher Functions

The Big Picture

- The organization of the mammalian nervous system can be described in anatomical or functional terms. The anatomical divisions are the central nervous system (CNS) and peripheral nervous system (PNS). The functional divisions relate to the direction in which information is flowing and the type of information. Afferent nerves of the PNS carry information from sensory receptor cells toward the CNS, and efferent nerves of the PNS carry information from the CNS to muscles and glands.
- The brain can be divided anatomically and functionally into many different regions, each responsible for its many vital actions. The "higher" brain centers of the cerebrum are responsible for conscious thought and deliberate (voluntary) movements. The "lower" brain centers such as the cerebellum, pons, and medulla regulate involuntary movements and help maintain homeostasis throughout the body.

Study Strategies

- Several key terms occur repeatedly in this chapter: parasympathetic, sympathetic, afferent, efferent, preganglionic, and postganglionic. Until you master this vocabulary, it will be difficult to put together a comprehensive picture of the nervous system. Create your own list of the terms you see repeatedly, and make sure that you understand their meanings. It is especially easy to confuse "afferent" and "efferent," and it is important to be able to differentiate them. Think of afferent as arriving, and efferent as exiting the reference point—here, the CNS.
- Brain anatomy is complex. Start by organizing the structures into forebrain, midbrain, and hindbrain regions. Your understanding will be more complete if you learn the general functions of each brain section and structure as you go along. That is, rather than learning the anatomy of the brain and then starting over to learn the functions, learn the two at the same time.
- It is often difficult for students to appreciate that, when dealing with sensory input to the brain, specific regions of the body "map onto" specific regions of the cerebrum. Similarly, with regard to control of movements, specific regions of the brain "map onto" specific regions of the body. Moreover, the amount of brain matter devoted to a particular body region depends on the amount of muscle control and sensors contained in that body region. Thus, a relatively small area of the cerebral hemispheres is devoted to the upper leg (which has a relatively limited range of movement and sensation), whereas the tongue, with its many sensory receptors and high degree of mobility, commands more of the tissue of the cerebrum.
- It is important to recognize that although the neurotransmitters acetylcholine and norepinephrine always have antagonistic effects on each other, there is no universal pattern as to which stimulates tissue and which inhibits tissue. For example, acetylcholine causes the smooth muscle in blood vessels in many regions of the body to relax, but it causes the smooth muscle in the stomach and intestines to contract.
- Go to the Web addresses shown to review the following animated tutorials and activities. You can also find them in BioPortal (yourBioPortal.com), along with many additional learning resources:

  Animated Tutorial 47.1 Visual Receptive Fields (Life10e.com/at47.1)

  Animated Tutorial 47.2 Information Processing in the Retina (Life10e.com/at47.2)

  Activity 47.1 The Human Cerebrum (Life10e.com/ac47.1)

  Activity 47.2 Language Areas of the Cortex (Life10e.com/ac47.2)

  Activity 47.3 Structures of the Human Brain (Life10e.com/ac47.3)

Key Concept Review

47.1 How Is the Mammalian Nervous System Organized?

Functional organization is based on flow and type of information

- The anatomical organization of the CNS emerges during development
- The spinal cord transmits and processes information
- The brainstem carries out many autonomic functions
- The core of the forebrain controls physiological drives, instincts, and emotions
- Regions of the telencephalon interact to control behavior and produce consciousness
- The size of the human brain is off the curve

The mammalian nervous system consists of the central nervous system (CNS; the brain and spinal cord) and the peripheral nervous system (PNS; the part of the nervous system outside the brain and spinal cord, including cranial and spinal nerves). The peripheral nervous system is composed of afferent nerves that carry information to the CNS and efferent nerves that carry information from the CNS to muscles and glands (see Figure 47.1). We are conscious of some information carried by afferent pathways (e.g., sounds, light), but not all of it (e.g., blood pressure). Efferent pathways are either voluntary (executing our conscious movements) or involuntary (autonomic, controlling physiological functions). A nerve is a bundle of axons that carries information in both directions between the organs of the body and the CNS.

The CNS develops from a tube of neural tissue running the length of a vertebrate embryo in its early developmental stages (see Figure 47.2). The anterior end of the neural tube develops into the brain and the rest of the neural tube develops into the spinal cord. The brain consists of three parts: (1) the forebrain (the diencephalon, which includes the thalamus and hypothalamus, and the telencephalon, which is the cerebrum); (2) the midbrain (structures that process visual and auditory information); and (3) the hindbrain (the medulla, pons, and cerebellum). The spinal cord is a bidirectional neural pathway for information flow between the peripheral nervous system and the brain. The spinal cord participates in reflexes ranging from simple (the knee-jerk reflex) to more complex (the withdrawal reflex), and contains complex motor programs, such as central pattern generation.

A nucleus is an anatomically distinct group of neurons in the CNS. The nuclei of the reticular activating system are located in the brainstem and regulate sleep and wakefulness. The limbic system regulates instincts and emotions (see Figure 47.3). The cerebrum is the largest portion of the mammalian brain. The cerebral hemispheres are covered by a convoluted cerebral cortex, a sheet of gray matter that processes sensory information and higher-order information in the association areas (see Figure 47.4). The convolutions are ridges known as gyri and the valleys are sulci. Underneath the gray matter is white matter containing axons that connect cell bodies in the cortex and other areas of the brain. Each cerebral hemisphere has four main regions: the temporal lobe, the frontal lobe, the occipital lobe, and the parietal lobe. Each lobe has association areas that integrate information. Among vertebrates, humans (and dolphins) stand out as having larger brains than would be predicted by their body sizes (see Figure 47.8). Degree of convolution of the cerebral cortex (a measure of the area of cortex) is greatest in humans, as is the percentage of cortex that is association cortex (a measure of the area of cortex devoted to the integration of information).

Question 1. Imagine that you have been eyeing candy in a dish and finally decide to unwrap a piece and eat it. As you begin to suck on the candy, your salivary glands begin to secrete saliva. What parts (divisions or branches) of the nervous system are involved in this sequence of events?

Question 2. During a boxing match, a sharp punch to the jaw of a boxer may cause loss of consciousness. Which area of the brain is likely to have been affected by such a knockout punch?

Question 3. Label the following structures in the diagram of the adult brain below (midsagittal section, right cerebral hemisphere): medulla, pons, midbrain, pituitary, hypothalamus, thalamus, cerebrum, cerebellum, and spinal cord.

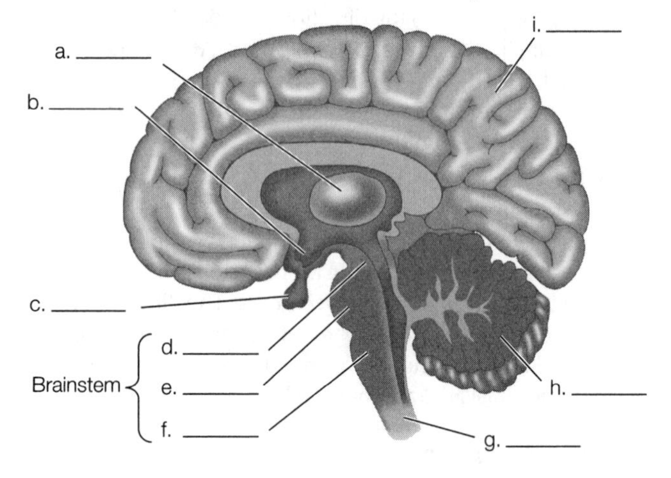

47.2 How Is Information Processed by Neural Networks?

- Pathways of the autonomic nervous system control involuntary physiological functions
- The visual system is an example of information integration by the cerebral cortex
- Three-dimensional vision results from cortical cells receiving input from both eyes

The sympathetic and parasympathetic divisions of the autonomic nervous system have antagonistic effects on the organs they innervate and differ in their anatomy and neurotransmitters. The sympathetic division is involved in the fight-or-flight response of increased heart rate, blood pressure, and cardiac output. The parasympathetic division slows down heart rate and decreases blood pressure and cardiac output; however, it accelerates digestive activities. The parasympathetic system has preganglionic neurons that come from the brain stem and the sacral region of the spinal cord. In the sympathetic system, preganglionic neurons come from the lumbar and thoracic regions of the spinal cord (see Figure 47.9). Whether sympathetic or parasympathetic, preganglionic neurons of the autonomic efferent pathway use the neurotransmitter acetylcholine. Norepinephrine is the neurotransmitter in postganglionic neurons of the sympathetic system, whereas acetylcholine is the neurotransmitter in postganglionic neurons of the parasympathetic system.

Our sense of vision involves complex interactions of neurons. In the retina, one ganglion cell receives and integrates the information from many groups of photoreceptors that make up a circular receptive field with a center and a surround (see Figure 47.10). Light falling on the center of an on-center receptive field excites ganglion cells; light falling on the surround of an on-center receptive field inhibits ganglion cells. The opposite pattern characterizes off-center receptive fields. Information from the retina is transferred by the optic nerve to the thalamus and then to the visual cortex in the occipital lobe. The visual cortex is composed of cells with specific receptive fields associated with areas of the retina that respond to specific light patterns. Mental images of the world are determined by analyzing the edges of the patterns of light falling on the retina. The two optic nerves (one from each eye) meet at the optic chiasm, where half of the axons from one eye cross over to the opposite side of the brain (see Figure 47.11). Binocular vision occurs because the overlap in the field of view for each eye is transmitted to the same location in the visual cortex, the binocular cells. Binocular cells receive overlapping visual information from both eyes and interpret the disparity between the overlapping information from the two eyes to produce a three-dimensional image.

Question 4. Humans have the ability to see things in three dimensions. What allows us to have binocular vision?

Question 5. What would happen if a surgeon accidentally cut the right optic nerve of a patient? What would a patient experience if she had a tumor in the right visual cortex?

Question 6. Create a table summarizing the differences between the parasympathetic division and sympathetic division of the autonomic nervous system. Be sure to include information on the following characteristics: location of origin of preganglionic neurons (= origin); location of most ganglia; actions; and neurotransmitters released by preganglionic and postganglionic neurons.

47.3 Can Higher Functions Be Understood in Cellular Terms?

- Sleep and dreaming are reflected in electrical patterns in the cerebral cortex
- Language abilities are localized in the left cerebral hemisphere
- Some learning and memory can be localized to specific brain areas
- We still cannot answer the question "What is consciousness?"

Several tools are used to study sleep. An electroencephalogram (EEG) measures the electrical activity of neurons in the cerebral cortex. An electromyogram (EMG) records electrical activity of muscles, and an electroocculogram (EOG) records eye movements. There are two main states of sleep in humans: rapid-eye movement (REM) sleep and non-REM sleep (see Figure 47.12). Non-REM sleep has four states, progressing from stage 1 to the restorative stages 3 and 4. During REM sleep, dreams and nightmares occur along with near complete paralysis of skeletal muscles. Neurons that were hyperpolarized during non-REM sleep return to waking levels, allowing information to be processed. Afferent and efferent pathways are inhibited during REM sleep. During a typical night of sleep, the brain cycles between REM sleep and non-REM sleep, with 80 percent of sleep being non-REM sleep.

In most people, the ability to produce and interpret language occurs in the left cerebral hemisphere. The two cerebral hemispheres are connected by the corpus callosum, which allows communication between the two hemispheres. Severing an individual's corpus callosum results in the person's inability to express in language the knowledge that is in the right hemisphere. Several areas have been located that are important for language (see Figure 47.13). The frontal lobe contains Broca's area, which is involved in motor aspects of speech. Wernicke's area, located in the temporal lobe, influences sensory aspects of language. The angular gyrus integrates spoken and written language. Language involves the flow of information among these areas of the left cerebral cortex, so damage to any one can result in aphasia, a deficit in using or understanding words.

Learning occurs when behavior is modified as a result of experience. Long-lasting synaptic changes characterize learning. Long-term potentiation (LTP) occurs when high-frequency electrical stimulation makes certain circuits more sensitive to subsequent stimulation. Long-term depression (LTD) occurs when certain circuits become less sensitive as a result of repetitive, low-level stimulation. Associative learning involves linking two unrelated stimuli, as in the conditioned reflex of Pavlov's dogs. The dogs were conditioned to associate eating with the ringing of a bell, so that even in the absence of food the mere ringing of a bell stimulated salivation. Memory, which can be classified as immediate, short-term, or long-term, is the phenomenon whereby the nervous system retains what has been experienced. Repetition or reinforcement enhances the transfer of short-term memory to long-term memory. Declarative memory is memory of people, places, events, and things. Procedural memory is memory of how to perform motor tasks, such as riding a bicycle. The insular cortex (insula) of the forebrain is expanded in humans and the great apes, and may be related to self-awareness and conscious experience.

Question 7. Describe the neurological basis of the phenomenon of sleepwalking. A person who is sleepwalking is in which state of sleep?

Question 8. What are the differences between the brain functions involved in reading a written sentence out loud and those involved in repeating a sentence one has just heard?

Question 9. Narcolepsy is a condition in which a person who is awake falls suddenly into REM sleep, which may last about 15 minutes. How would the 15-minute "nap" during the day differ in structure from a typical night of sleep?

Question 10. A pianist experiences a traumatic brain injury and loses the ability to play the piano and tie her shoes. Which form of learning and memory has been impaired by the brain injury?

Test Yourself

1. A man has damage to his brain that affects his ability to recognize the faces of people he knows. The damage must have occurred in the
   a. hypothalamus.
   b. temporal lobe.
   c. parietal lobe.
   d. frontal lobe.
   e. occipital lobe.
   Textbook Reference: *47.1 How Is the Mammalian Nervous System Organized?*

2. Which of the following is *not* part of the central nervous system?
   a. Brain stem
   b. Spinal gray matter
   c. Cerebellum
   d. Neuronal cell body of a sensory afferent
   e. Pons
   Textbook Reference: *47.1 How Is the Mammalian Nervous System Organized?*

3. The primary motor cortex of the cerebrum
   a. is mapped from the head region on the lower side of the cortex to the lower part of the body on the upper side of the cortex.
   b. receives touch and pressure information from the body.
   c. is located in the parietal lobe.
   d. occurs behind the central sulcus.
   e. is located in the temporal lobe.
   Textbook Reference: *47.1 How Is the Mammalian Nervous System Organized?*

4. Translating visual experiences into language occurs in
   a. Broca's area.
   b. the spinal cord.
   c. an association area of the occipital cortex.
   d. the reticular system.
   e. the amygdala.
   Textbook Reference: *47.1 How Is the Mammalian Nervous System Organized?*

5. Which of the following statements about the peripheral nervous system is *false*?
   a. Afferent portions carry information to the CNS.
   b. Efferent portions carry information from the CNS.
   c. It communicates only with the circulatory and digestive systems.
   d. Efferent portions contain voluntary and involuntary divisions.
   e. It includes cranial and spinal nerves.
   Textbook Reference: *47.1 How Is the Mammalian Nervous System Organized?*

6. Which of the following statements about the developing CNS is *false*?
   a. The CNS develops from a solid neural cylinder.
   b. The midbrain becomes part of the brain stem.
   c. The forebrain develops into both the diencephalon and the telencephalon.
   d. The hindbrain develops into the medulla, pons, and cerebellum.
   e. In humans, the telencephalon develops into the largest part of the brain.
   Textbook Reference: *47.1 How Is the Mammalian Nervous System Organized?*

7. Which cortical lobes contain association areas?
   a. Temporal
   b. Parietal
   c. Occipital
   d. Frontal
   e. All of the above
   Textbook Reference: *47.1 How Is the Mammalian Nervous System Organized?*

8. Which of the following groups of vertebrates would likely have the smallest ratio of telencephalon size to body size?
   a. Fishes
   b. Amphibians
   c. Mammals
   d. Reptiles
   e. Birds
   Textbook Reference: *47.1 How Is the Mammalian Nervous System Organized?*

9. When compared with brains of other vertebrates, the human brain
   a. is larger than body size might lead one to predict.
   b. has proportionately more of the cortex devoted to information integration.
   c. has proportionately more cerebral cortex.
   d. has more convolutions in the cortex.
   e. All of the above
   Textbook Reference: *47.1 How Is the Mammalian Nervous System Organized?*

10. The secretion of hormones from an endocrine gland is most directly under the control of which of the following components of the nervous system?
    a. Autonomic
    b. Voluntary
    c. Afferent portion of the PNS that carries information of which we are conscious

d. Limbic
e. Afferent portion of the PNS that carries information of which we are unconscious

Textbook Reference: *47.2 How Is Information Processed by Neural Networks?*

11. Which of the following statements about the sympathetic division of the autonomic nervous system is *false*?
a. It increases heart rate.
b. It relaxes the urinary bladder.
c. It stimulates digestion.
d. It increases blood pressure.
e. It relaxes airways.

Textbook Reference: *47.2 How Is Information Processed by Neural Networks?*

12. The processing of visual information by the retina involves
a. a convergence of information.
b. the telencephalization.
c. long-term depression.
d. long-term potentiation.
e. a divergence of information.

Textbook Reference: *47.2 How Is Information Processed by Neural Networks?*

13. Observations of people with aphasia indicate that
a. only Broca's area is essential for normal language skills.
b. language skills depend on proper flow of neural information between the temporal lobes and the motor cortex.
c. in humans, the right hemisphere is dominant in the production and use of language.
d. Broca's area influences the sensory aspects of language and Wernicke's area influences the motor aspects.
e. lateralization does not pertain to language ability.

Textbook Reference: *47.3 Can Higher Functions Be Understood in Cellular Terms?*

14. A friend wakes you from sleep, and you have the sensation of having just experienced a vivid dream. Which of the following statements about the state of sleep from which you were awakened is *false*?
a. Your hands and feet were twitching slightly.
b. Your eyes were twitching.
c. Most of your voluntary body muscles were inactive.
d. Your cerebral cortex was not as active as it is when you are awake.
e. You were in the state of sleep that accounts for about 20 percent of your sleep.

Textbook Reference: *47.3 Can Higher Functions Be Understood in Cellular Terms?*

15. The insular cortex is
a. located in the hindbrain.
b. greatly expanded in fishes.
c. most active in humans during times of mild emotion.
d. unrelated to perception of self.
e. greatly expanded in humans and great apes.

Textbook Reference: *47.3 Can Higher Functions Be Understood in Cellular Terms?*

Answers

Key Concept Review

1. The peripheral system contributed to both seeing the candy and the movements of the arms and legs that you used to pick it up. The parasympathetic branch of the autonomic nervous system stimulated salivation. The central nervous system was involved in the recognition and decision to unwrap and eat the candy.

2. A sharp punch to the jaw will cause the head to turn sharply, and this is likely to twist the medulla and reticular activating system. The reticular system is a network of neurons in the brain stem that, unless inhibited by other regions of the brain, activates the cerebral cortex and causes consciousness. A sharp blow can affect the reticular system and cause temporary loss of consciousness.

3.
a. Thalamus
b. Hypothalamus
c. Pituitary
d. Midbrain
e. Pons
f. Medulla
g. Spinal cord
h. Cerebellum
i. Cerebrum

4. The right side of your brain receives visual information from the left visual field, and the left side receives information from the right visual field. In the visual cortex, the cells are organized in columns that alternate between receiving information from the right and left eyes. At the borders of the columns, the inputs from the right and left eyes overlap. The cells that receive the overlap are called binocular cells, and they interpret the disparity between what the two eyes sense. This provides a three-dimensional image.

5. If a surgeon accidentally cut the right optic nerve of a patient, the patient would lose vision in the right eye (i.e., lose right and left visual fields from the right eye only). If a patient had a tumor in her right visual cortex, this would affect the left visual field.

6.

Characteristic	Parasympathetic Division	Sympathetic Division
Origin	Cranial nerves of brainstem and sacral region of spinal cord	Thoracic and lumbar regions of spinal cord
Location of ganglia	Close to target organs	In two chains alongside the spinal cord
Action	Rest and digest	Fight or flight
Neurotransmitters	All preganglionic and postganglionic neurons release acetylcholine	All preganglionic neurons release acetylcholine and most postganglionic neurons release norepinephrine

7. Sleepwalking takes place during non-REM sleep. During REM sleep, the skeletal muscles of the body become paralyzed and the sleeper is unable to move. A sleepwalker will not exhibit the eye movements typical of REM sleep.
8. Both speaking written language and repeating heard language involve similar pathways in the brain. The main difference has to do with the initial region of the brain perceiving the word. In reading a word, the area at the back of the cerebrum is used to visualize it. The spoken word uses an area of the cerebrum just behind the area used for speech. Once the word has been processed by the initial centers, the path used for speaking the word is the same. Wernicke's area is stimulated, followed by Broca's area, and then the motor area.
9. In a typical sleep pattern, a person first enters stage 1 non-REM sleep, gradually enters stages 2, 3, and 4 of non-REM sleep, and then eventually experiences the first episode of REM sleep. Four or five cycles of non-REM and REM sleep typically occur throughout a night. In contrast, a person with narcolepsy immediately enters REM sleep.
10. The brain injury impaired procedural memory, which is the memory of how to perform a motor task. In contrast, declarative memory involves consciously recalling people, places, and events.

Test Yourself

1. **b.** The temporal lobe is involved in recognition of people and objects. A person who has had damage to the temporal lobe will not be able to identify someone by face and must use other cues.
2. **d.** The cell bodies of the sensory neurons are located in the periphery and send their axons to the CNS.
3. **a.** The motor neurons in the head region control specific parts of the body. Parts of the body can be mapped on the primary motor cortex, from the head region on the lower side to the lower part of the body at the top.
4. **c.** Translating visual experiences into language occurs in an association area of the occipital cortex.
5. **c.** The peripheral nervous system is in contact with every tissue in the body sending and receiving information to the CNS.
6. **a.** The CNS develops from a hollow tube composed of neural tissues.
7. **e.** Association areas are present in all of these regions of the cerebral cortex.
8. **a.** Fishes have the most undeveloped telencephalon of the vertebrates listed, resulting in the smallest ratio.
9. **e.** Relative to the brains of other vertebrates, the human brain is larger than body size might lead one to predict, and it exhibits a greater degree of convolution of the cerebral cortex. The human brain also has proportionately more cerebral cortex and association areas within the cortex.
10. **a.** Secretion of hormones by endocrine glands is involuntary. The autonomic nervous system consists of efferent pathways that link the CNS with many physiological functions.
11. **c.** The sympathetic division of the autonomic nervous system inhibits digestion rather than stimulating it.
12. **a.** In the retina, the information from over 100 million photoreceptors is integrated by about 1 million ganglion cells; this type of processing is called convergence of information.
13. **b.** Speaking a written or heard word requires neural flow from Wernicke's area in the temporal lobe to Broca's area in the frontal lobe. Therefore, language skills depend on proper flow of neural information between the temporal lobes and the motor cortex.
14. **d.** During REM sleep the body is paralyzed, except for the twitching of muscles. During this stage of sleep the brain is as active as when you are awake.
15. **e.** The insular cortex is greatly expanded in humans and great apes. This area of the forebrain may be involved with self-recognition and conscious experience.

Musculoskeletal Systems

The Big Picture

- Actin and myosin are the "universal" proteins for motion. Whether they are located in a unicellular animal or the leg muscle of a human, their molecular interactions produce movement in the structures in which they reside.
- The sliding filament theory provides a model for how actin and myosin filaments slide past each other, resulting in the shortening (contraction) or lengthening (relaxation) of muscle cells. The process of actin and myosin interaction is tightly regulated by the movement of calcium ions into and out of the intracellular spaces of muscle cells, which in turn is activated by the arrival of action potentials in motor neurons. Muscle contraction requires the use of energy in the form of ATP.
- In vertebrates, muscles act in concert with an internal skeleton made of bone. Bone is living tissue that is constantly remodeled. Bones are articulated, forming joints that provide for specialized directional movements of the tissues supported by the bones. Muscles controlling joint movement often occur in pairs that act antagonistically, with one set of muscles causing bending, or flexion, of the joint and the other causing straightening, or extension, of the joint. Some invertebrates have a hydrostatic skeleton (a fluid-filled body cavity surrounded by muscles), while others have an exoskeleton (a rigid outer covering).

Study Strategies

- The sarcomere is the functional unit of the muscle cell, and until you understand its fine structure—Z lines, H lines, etc.—it will be difficult to appreciate how the sarcomere shortens through the actions of actin and myosin.
- The interactions of myosin, actin, troponin, tropomyosin, and calcium and the role of action potentials in stimulating muscle contraction make up a complex, multistep process. First, break the process down into its constituents and learn their locations and general structures. Second, determine how actin and myosin move relative to each other through a series of power strokes. Finally, examine how calcium ions released from the sarcomeres by action potentials initiate and maintain the process of muscle contraction.
- You may think of bone as tissue that is not living. However, bone is a living tissue that is constantly remodeled. Become familiar with the three types of cells in bone: osteocytes, osteoblasts, and osteoclasts.
- The way that movements of multiple muscle groups cause both flexion and extension of a joint can be confusing. Thinking of joints in terms of levers may help you understand their actions.
- Go to the Web addresses shown to review the following animated tutorials and activities. You can also find them in BioPortal (yourBioPortal.com), along with many additional learning resources.

  Animated Tutorial 48.1 Molecular Mechanisms of Muscle Contraction (Life10e.com/at48.1)

  Animated Tutorial 48.2 Smooth Muscle Action (Life10e.com/at48.2)

  Activity 48.1 The Structure of a Sarcomere (Life10e.com/ac48.1)

  Activity 48.2 The Neuromuscular Junction (Life10e.com/ac48.2)

  Activity 48.3 Joints (Life10e.com/ac48.3)

Key Concept Review

48.1 How Do Muscles Contract?

Sliding filaments cause skeletal muscle to contract

Actin–myosin interactions cause filaments to slide

Actin–myosin interactions are controlled by calcium ions

Cardiac muscle is similar to and different from skeletal muscle

Smooth muscle causes slow contractions of many internal organs

Vertebrates have three types of muscle: skeletal, cardiac, and smooth. In all three types, contraction is due to the interaction between the contractile proteins actin and myosin. Skeletal muscles are voluntary muscles made of large striated muscle fibers with many nuclei (see Figure 48.1). Within the muscle fibers, bundles of actin and myosin filaments are arranged into myofibrils. The contracting unit of myofibrils is the sarcomere. In a sarcomere, the actin filaments are anchored by the Z lines, and myosin filaments are found at the center in the A band (see Figure 48.1). In relaxed muscle, the H zone and I band are the regions in which there is no overlap of actin and myosin. During muscle contraction, the Z lines move toward one another, and the H zone and I band shrink in size due to the sliding of actin filaments along the myosin filaments. These are the components of the sliding filament theory of muscle contraction (see Figure 48.2).

Myosin molecules are made of two polypeptide chains wrapped around each other, each with a globular head at one end. A myosin filament is composed of many myosin molecules (see Figure 48.3). Actin filaments are composed of two monomer chains in a helical arrangement. The proteins tropomyosin and troponin are associated with actin (see Figure 48.3). Myosin heads change conformation when they bind to actin filaments, forming a cross-bridge connection. The conformational change in the myosin pulls the actin in toward the middle of the sarcomere. ATP then binds to myosin, resulting in actin's release from myosin and myosin's return to the original conformation.

All the fibers activated by a single motor neuron constitute a motor unit. Action potentials spread deep into the sarcoplasm (i.e., the cytoplasm) of the muscle fiber through transverse tubules (T tubules) that are in contact with the sarcoplasmic reticulum (i.e., what is called the ER, or endoplasmic reticulum, in nonmuscle cells; see Figure 48.5). The sarcoplasmic reticulum takes up and releases Ca^{2+} ions into the sarcoplasm, thereby controlling relaxation and contraction of the myofibrils. In relaxed muscle, tropomyosin and troponin cover the myosin binding sites on actin filaments, preventing muscle contraction. Calcium regulates contraction by binding with troponin, causing the tropomyosin to change conformation and expose the myosin binding sites on the actin filaments (see Figure 48.6)

Cardiac muscle is found in the heart and is composed of branched muscle cells that form a strong meshwork (see Figure 48.7). Cardiac muscle cells are smaller than skeletal muscle cells, and each has a single nucleus. Intercalated discs add additional strength by holding the cells together. Gap junctions within the intercalated discs allow cardiac muscle cells to be electrically coupled. Heartbeats originate at the pacemaker cardiac muscle cells and spread rapidly through gap junctions. The mechanism of excitation–contraction coupling in cardiac muscle cells is called Ca^{2+}-induced Ca^{2+} release.

Smooth muscles are involuntary muscles composed of long spindle-shaped cells, each with a single nucleus. The stretching of smooth muscle cells depolarizes the membranes, and the resulting action potentials initiate the contractile mechanism. The strength of the contraction is proportional to the stretch of the muscle. Smooth muscle cells are arranged in sheets. Gap junctions allow electrical contact between the cells and promote coordinated contraction of cells in a sheet. Contraction is controlled by a calmodulin–Ca^{2+} complex. This complex activates myosin kinase, which phosphorylates the myosin head to cause contraction. Myosin phosphatase works in the opposite direction by dephosphorylating myosin and stopping interactions between actin and myosin (see Figure 48.9).

Question 1. Label the following structures in the figure below: I band, muscle, single muscle fiber, myosin filament, A band, single myofibril, single sarcomere, M band, sarcomere, actin filament, Z line, tendons, H zone. Also label any or all of the structures (actin filament, myosin filament, Z line, I band, M band) on the myofibril and on the enlargement of the sarcomere, wherever they exist.

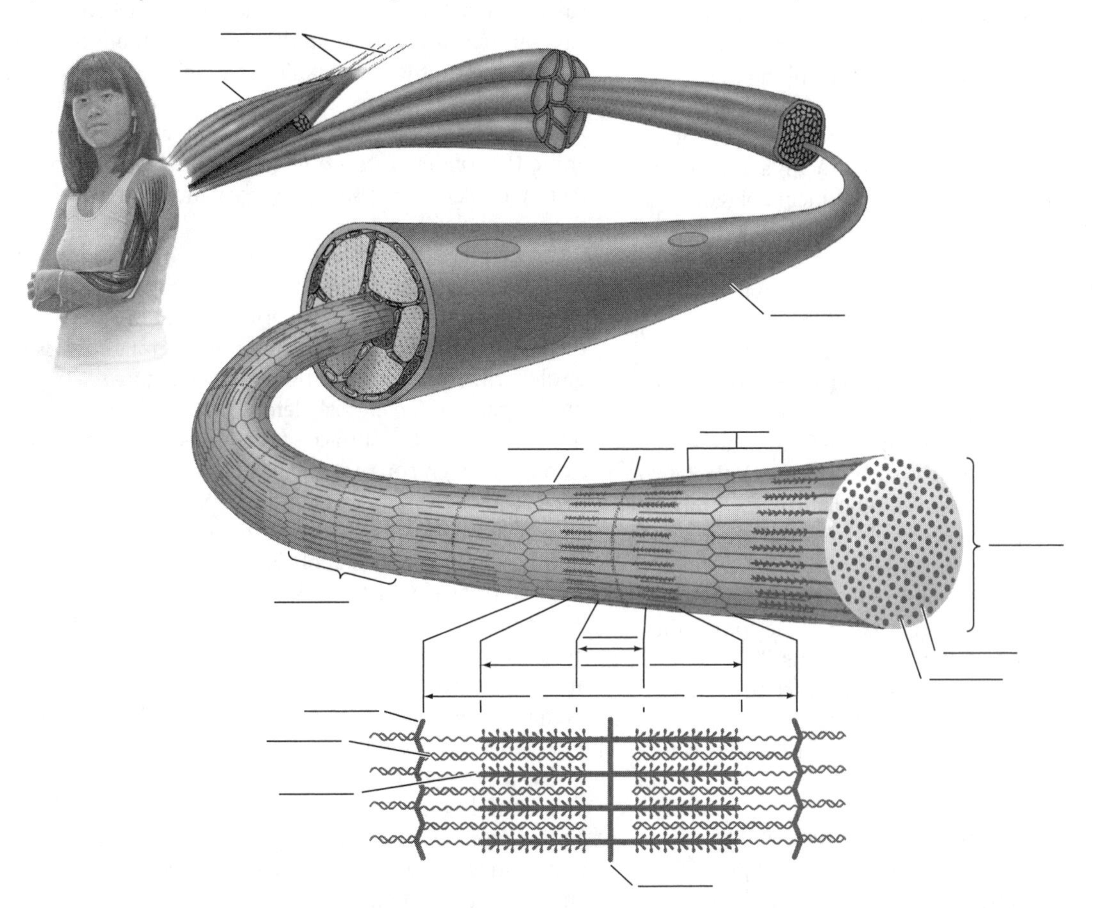

Question 2. Smooth muscle contracts involuntarily, whereas skeletal muscle contraction is under voluntary control. How do the mechanisms that control smooth muscle and skeletal muscle contraction differ?

Question 3. Smooth muscle contracts involuntarily in the digestive tract, blood vessels, and urinary bladder. Describe the two main ways in which smooth muscle contraction and the membrane potential of smooth muscle are controlled.

Question 4. What causes rigor mortis (i.e., the stiffening of muscles after death)?

Question 5. Curare is a poison from South America that is applied to the tips of poison arrow darts. Mammals hit by the darts die by asphyxiation because their respiratory muscles cannot contract. Suggest a mechanism by which curare might work.

48.2 What Determines Skeletal Muscle Performance?

The strength of a muscle contraction depends on how many fibers are contracting and at what rate

Muscle fiber types determine endurance and strength

A muscle has an optimal length for generating maximum tension

Exercise increases muscle strength and endurance

Muscle ATP supply limits performance

Insect muscle has the greatest rate of cycling

An action potential in a skeletal muscle fiber causes a minimal contraction called a twitch. Twitches can occur as discrete contractions, or if they occur frequently enough, they can be summed (see Figure 48.10).

The level of tension generated by a muscle depends on the number of motor units activated and the frequency with which the motor units fire. Maximum muscle tension, or tetanus, occurs when there is a high rate of stimulation by action potentials. Muscle tone reflects the small but changing number of motor units active in a muscle at any given time.

The strength and endurance of muscles depend on fiber type. Slow-twitch muscle fibers are highly resistant to fatigue because they are well supplied with myoglobin (an oxygen-binding protein similar to hemoglobin), mitochondria, and blood vessels. They are also called "red" or "oxidative" muscle (see Figure 48.11). Fast-twitch muscle fibers rapidly develop maximum tension but fatigue quickly. They have fewer mitochondria and blood vessels than slow-twitch fibers, little or no myoglobin, and are called "white," or "glycolytic," muscle (see Figure 48.11).

The amount of force a sarcomere can generate depends on its resting length (see Figure 48.12). The stretching of a muscle causes the sarcomeres to lengthen, resulting in less overlap between actin and myosin filaments and less force. Anaerobic exercise, such as weight lifting, increases strength. Such exercise induces the formation of new actin and myosin filaments in existing muscle fibers, thus producing bigger fibers and bigger muscles. Aerobic exercise, such as jogging, increases endurance. This form of exercise increases myoglobin, the number of mitochondria, the density of capillaries, and the enzymes involved in energy utilization.

Muscles use the immediate, glycolytic, and oxidative systems to obtain ATP needed for contraction (see Figure 48.13). The immediate system utilizes preformed ATP and creatine phosphate. The glycolytic system follows the immediate system within seconds and metabolizes carbohydrates to lactate and pyruvate. The oxidative system is fully activated within about one minute and completely metabolizes carbohydrates or fats to water and carbon dioxide.

The striated muscle of vertebrates and many invertebrates is described as synchronous because cycling of the contractile mechanism is tied to the firing of motor neurons. The flight muscles of insects are asynchronous because cycling of the contractile mechanism is not tied to the firing rate of the flight motor neurons; this allows for very high rates of cycling and wingbeat frequencies.

Question 6. White muscle and red muscle are found in different parts of the body and are used for different types of movement. What are the physiological and morphological characteristics that distinguish the two types of muscle?

Question 7. In turkeys, the breast (flight) muscles are light in color, while in mallard ducks they are dark. What does this tell you about the flight capabilities of these two birds?

Question 8. Some athletes hoping to improve their performance take creatine as a dietary supplement. There are some reports that creatine supplements boost performance in activities that require short bursts of energy, such as sprinting, but not in those that require endurance. Why might this be the case?

Question 9. You are an athletic trainer with two clients, one of whom wants advice on strength training and the other on endurance training. Design a training program for each client.

48.3 How Do Skeletal Systems and Muscles Work Together?

- A hydrostatic skeleton consists of fluid in a muscular cavity
- Exoskeletons are rigid outer structures
- Vertebrate endoskeletons consist of cartilage and bone
- Bones develop from connective tissues
- Bones that have a common joint can work as a lever

Many soft-bodied invertebrates have a fluid-filled body cavity that acts as a hydrostatic skeleton. Earthworms have circular muscles and longitudinal muscles that oppose each other to act on the hydrostatic skeleton and control elongation and shortening of body segments (see Figure 48.14). Arthropods have a rigid exoskeleton, or cuticle, composed of chitin that offers protection and provides sites for muscle attachment. As arthropods grow, they must shed and replace their exoskeleton.

The vertebrate endoskeleton provides a frame for support and movement. The human skeleton is composed of a central axial skeleton (skull, vertebral column, sternum, and ribs) and an appendicular skeleton (pectoral girdle, pelvic girdle, arms, hands, legs, and feet) (see Figure 48.15). The pliable parts of the endoskeleton, such as the framework of the nose and the surface of the joints, are composed of cartilage, which contains the protein collagen. The bone in the endoskeleton is strong and composed mainly of collagen and calcium phosphate. Bone is living tissue that is constantly remodeled by osteoblasts, which lay down new bone, and osteoclasts, which break down bone (see Figure 48.16). When an osteoblast becomes enclosed by the matrix it is laying down, it stops forming matrix and exists within a lacuna; at this stage, the cell is called an osteocyte. Osteocytes communicate with one another and influence the activities of osteoblasts and osteoclasts.

Developing bone is created as membranous bone growing on a scaffolding of connective tissue or as cartilage bone that ossifies from an initial cartilage model. The long bones of the arms and legs are cartilage bones that ossify first at the center and then at each end; elongation occurs at epiphyseal plates (see Figure 48.17). Compact bone is solid, whereas cancellous bone is lightweight, with many cavities. In mammals, most compact bone is composed of concentric rings with blood vessels and nerves running through a central canal. These structural units are known as Haversian systems (see Figure 48.18).

There are six types of joints where bones meet: ball-and-socket, pivot, saddle, ellipsoid, hinge, and plane (see Figure 48.19). The muscles attached to bones at joints work antagonistically (see Figure 48.20). The flexor muscles bend the joints and the extensor muscles straighten them. Two types of connective tissues hold joints and bones together: ligaments, which hold bone to bone, and tendons, which hold muscle to bone. The bones that make up a joint can be viewed as lever systems.

Question 10. Describe how an earthworm's hydrostatic skeleton is used to move the earthworm through the soil.

Question 11. A patient with a family history of osteoporosis is advised by her physician to engage in weight-bearing exercise. How might such exercise help to prevent osteoporosis?

Question 12. Hydrostatic skeletons are found in annelids, such as earthworms, and cnidarians, such as hydras. Could a terrestrial animal that moves by walking function with a hydrostatic skeleton? Why or why not?

Question 13. Armadillos dig powerfully with their forelimbs and horses run swiftly. Given these differences in function, what differences would you expect in the sites of attachment of forelimb muscles?

Test Yourself

1. A motor unit is best described as
   a. all the nerve fibers and muscle fibers in a single muscle bundle.
   b. one muscle fiber and its single nerve fiber.
   c. a single motor neuron and all the muscle fibers that it innervates.
   d. the neuron that provides the central nervous system with information about muscle contraction.
   e. a neuron that communicates information from sensory to motor neurons.
   Textbook Reference: *48.1 How Do Muscles Contract?*

2. Which of the following statements about muscle tissue is *false*?
   a. Cardiac muscle is striated.
   b. Smooth muscle does not contain actin.
   c. Skeletal muscle is considered voluntary muscle.
   d. Smooth muscle is found in the digestive tract and the walls of the bladder.
   e. A single skeletal muscle cell has many nuclei.
   Textbook Reference: *48.1 How Do Muscles Contract?*

3. The action potential that triggers a muscle contraction travels deep within the muscle cell by means of
   a. sarcoplasmic reticulum.
   b. transverse (or T) tubules.
   c. synapses.
   d. motor end plates.
   e. neuromuscular junctions.
   Textbook Reference: *48.1 How Do Muscles Contract?*

4. A sarcomere is a
   a. moveable structural unit within a myofibril bounded by H zones.
   b. fixed structural unit within a myofibril bounded by Z lines.
   c. fixed structural unit within a myofibril bounded by A bands.
   d. moveable structural unit within a myofibril bounded by Z lines.
   e. collection of myofibrils.
   Textbook Reference: *48.1 How Do Muscles Contract?*

5. ATP provides the energy for muscle contraction by allowing for the
   a. formation of an action potential in the muscle cell.
   b. breaking of actin–myosin bonds.
   c. formation of actin–myosin bonds.
   d. release of calcium by the sarcoplasmic reticulum.
   e. formation of T tubules.
   Textbook Reference: *48.1 How Do Muscles Contract?*

6. Ca^{2+} binds to _______ in skeletal muscle and leads to exposure of the binding site for _______ on the _______ filament.
   a. troponin; myosin; actin
   b. troponin; actin; myosin
   c. actin; myosin; troponin
   d. tropomyosin; myosin; actin
   e. myosin; actin; troponin
   Textbook Reference: *48.1 How Do Muscles Contract?*

7. Which of the following proteins moves tropomyosin?
   a. Calmodulin
   b. Acetylcholine
   c. Actin
   d. Troponin
   e. Titin
   Textbook Reference: *48.1 How Do Muscles Contract?*

8. Which of the following statements about exercise and muscle strength and endurance is *false*?
   a. Strength training induces the formation of new actin and myosin filaments in existing muscle fibers.
   b. Satellite cells are muscle stem cells.
   c. Aerobic exercise increases the number of mitochondria in muscle cells and the density of capillaries in muscle.
   d. A common effect of strength training is the production of more muscle fibers.
   e. Aerobic exercise increases myoglobin in skeletal muscle cells.
   Textbook Reference: *48.2 What Determines Skeletal Muscle Performance?*

9. The oxygen-binding molecule in skeletal muscle is
   a. myoglobin.
   b. hemoglobin.
   c. ATP.
   d. glycogen.
   e. creatine phosphate.
   Textbook Reference: *48.2 What Determines Skeletal Muscle Performance?*

10. Summation of frequent muscle twitches to give maximum contraction is called
    a. motor unit summation.
    b. supercompensation.
    c. facilitation.
    d. tetanus.
    e. muscle tone.
    Textbook Reference: *48.2 What Determines Skeletal Muscle Performance?*

11. Which of the following statements about systems that supply ATP for muscle contraction is *false*?
    a. The glycolytic system produces relatively more ATP than the oxidative system.

b. Oxidative metabolism makes ATP available at a slower rate than glycolysis does.
c. Together, the immediate system and the glycolytic system provide energy for less than one minute.
d. The glycolytic system leads to the accumulation of lactic acid.
e. Preformed ATP and creatine phosphate make up the immediate system.

Textbook Reference: *48.2 What Determines Skeletal Muscle Performance?*

12. Whereas _______ lay down new bone matrix, _______ resorb bone.
a. satellite cells; Haversian systems
b. osteoblasts; osteocytes
c. osteoclasts; osteocytes
d. osteoblasts; osteoclasts
e. myoblasts; osteoclasts

Textbook Reference: *48.3 How Do Skeletal Systems and Muscles Work Together?*

13. Which of the following statements about exoskeletons is *false?*
a. They are found in mollusks and arthropods.
b. They are located on the outside of the body.
c. They can have joints.
d. They require molting as the animal grows.
e. They consist of cartilage and bone.

Textbook Reference: *48.3 How Do Skeletal Systems and Muscles Work Together?*

14. A soccer player who has suffered a knee injury that damages the tissue holding his upper and lower leg bones together has most likely damaged _______ tissue.
a. muscle
b. tendon
c. ligament
d. cartilage
e. membrane

Textbook Reference: *48.3 How Do Skeletal Systems and Muscles Work Together?*

15. Sites of elongation between the ossified regions of long bones are called
a. glue lines.
b. fulcrums.
c. hinge joints.
d. epiphyseal plates.
e. Haversian systems.

Textbook Reference: *48.3 How Do Skeletal Systems and Muscles Work Together?*

16. Which of the following statements about endoskeletons is *false?*
a. In humans, cartilage is the principal component of the embryonic skeleton.
b. Some vertebrates retain a cartilaginous endoskeleton into adulthood.
c. Cancellous bone has numerous cavities.
d. Physical stress on bones causes them to become thinner.
e. Calcitonin and parathyroid hormone regulate the deposition of calcium in bone.

Textbook Reference: *48.3 How Do Skeletal Systems and Muscles Work Together?*

Answers

Key Concept Review

1.

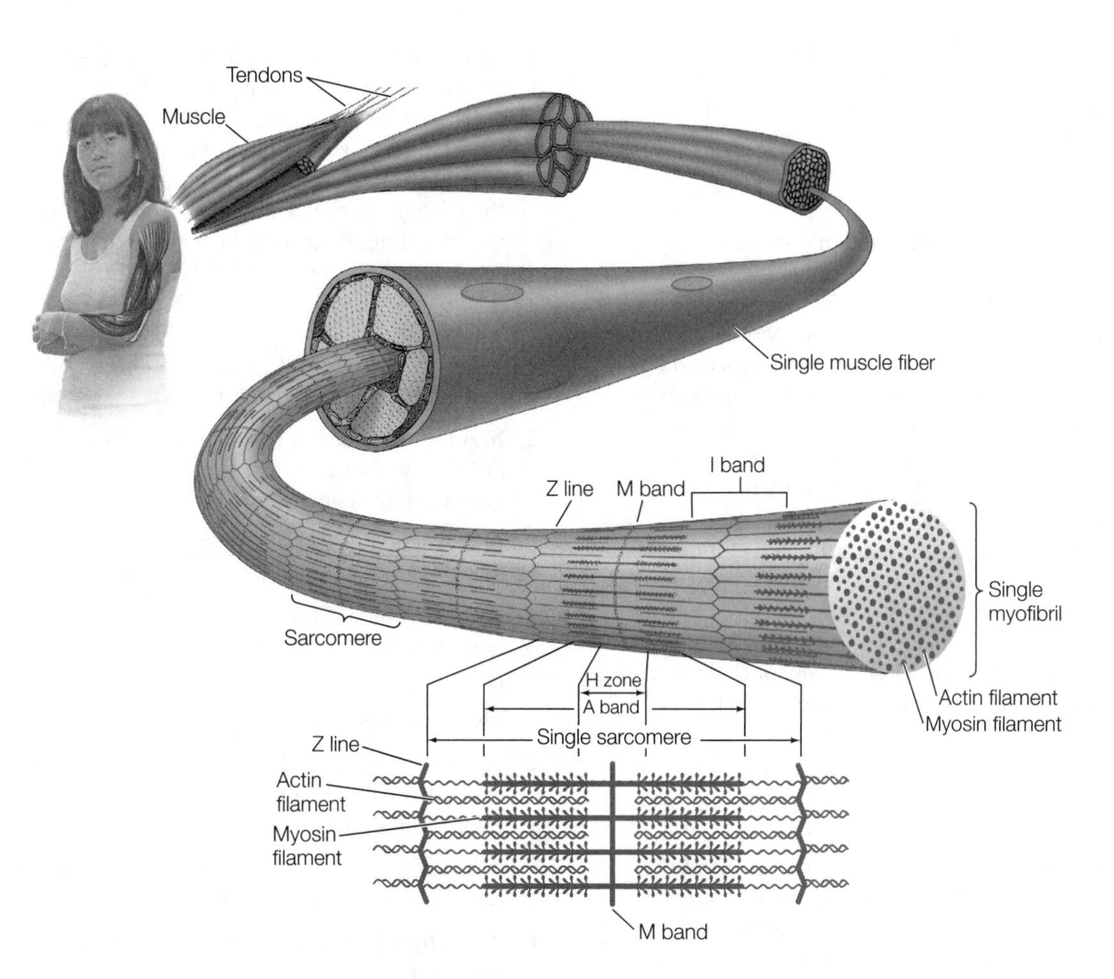

2. Smooth muscle contraction and skeletal muscle contraction are regulated by the presence of Ca^{2+}. In smooth muscle, Ca^{2+} joins with the protein calmodulin to activate myosin kinase in the sarcoplasm. Myosin kinase then phosphorylates the myosin head, allowing the myosin head to bind with actin. In skeletal muscle, Ca^{2+} binds with troponin on the actin filaments. This binding of Ca^{2+} causes a conformational change in tropomyosin and uncovers the myosin binding sites on the actin, allowing myosin to bind with actin.

3. In response to stretching, smooth muscle will depolarize and contract, with the strength of the contraction proportional to the amount of stretch in the muscle. Parasympathetic inputs of acetylcholine also cause depolarization of the muscle membrane resulting in contraction. Norepinephrine, by contrast, hyperpolarizes the muscle membranes and decreases the force of contraction.

4. ATP is needed to break the actin–myosin bonds. ATP production stops at death, so the actin–myosin bonds cannot be broken and muscles stiffen, causing rigor mortis. Eventually, the proteins deteriorate and the muscles soften.

5. Skeletal muscles are used in breathing. Acetylcholine released by motor neurons at the neuromuscular junction initiates action potentials in skeletal muscle. Curare acts by preventing acetylcholine from binding to the postsynaptic membrane, where it would normally cause ion channels in the motor end plate to open.

6. Fast-twitch fibers are known as white muscle and have few mitochondria, small amounts of myoglobin, and few blood vessels. Muscles with many fast-twitch fibers are good for short-term work that requires maximum strength. Slow-twitch fibers are known as red muscle. Red muscle has many mitochondria, large amounts of myoglobin, and many blood vessels. Muscles with many slow-twitch fibers function well in endurance activities.

7. The breast muscles of turkeys are fast-twitch, or white, muscle ("white meat"), and their leg muscles are slow-twitch, or red, muscle ("dark meat"). This tells us that turkeys fly only in short bursts and typically walk or run. In contrast, the breast muscles of mallard ducks are slow-twitch, or red, muscle, indicating that they are capable of prolonged flight.

8. Taking creatine supplements may increase stores of creatine phosphate in muscle. Creatine phosphate stores energy in a phosphate bond, which it can transfer to ADP to form ATP. The energy is available immediately, but the supply is quickly exhausted (creatine phosphate is part of the immediate system for supplying ATP to muscle). Thus, we would predict that creatine supplements would be most useful for activities such as sprinting, in which fast-twitch fibers generate a lot of force quickly.
9. The training program for the client interested in strength training should emphasize anaerobic exercises, such as weight lifting or exercises such as pull-ups. These activities repeatedly contract particular muscles under heavy loads; typically, 8–12 repetitions are performed in a set until the muscle is completely fatigued. Several sets are usually performed. This type of activity induces formation of new actin and myosin filaments in existing muscle fibers, thus increasing overall muscle size. The training program for the client interested in endurance training should emphasize aerobic exercises such as jogging, walking, cross-country skiing, cycling, and swimming. These activities increase the oxidative capacity of muscles by increasing the number of mitochondria, density of capillaries, amount of myoglobin, and enzymes associated with energy use.
10. The hydrostatic skeleton of the earthworm is an incompressible fluid-filled cavity surrounded by longitudinal and circular muscles. Contraction of the longitudinal muscles causes the segments to contract and the body to shorten. Contraction of the circular muscles causes the body segments to elongate and the body to lengthen. Alternating contractions between the longitudinal and the circular muscles move the animal in a push and pull manner. Bristles on the body help hold the animal in place after elongation, and the body is pulled forward during shortening.
11. Osteoporosis is a decrease in bone density that occurs when the destruction of bone by osteoclasts outpaces the formation of new bone by osteoblasts, leading to thin, brittle bones. Weight-bearing exercises place stress on bone, ultimately altering the interplay of osteoblast and osteoclast activity to induce thickening of bone.
12. Hydrostatic skeletons work well for aquatic animals, such as hydras, and for terrestrial animals such as earthworms that move by crawling through a substrate. However, a hydrostatic skeleton would not work for most terrestrial animals. Such skeletons provide little or no protection against drying out (dehydration is always a danger on land), and they do not provide sufficient support for a large animal that walks by holding its body off the ground.
13. You would expect armadillos to have lever systems in their forelimbs to maximize force. Thus, the forelimb muscles used in digging should be attached in such a way that there is a long effort arm relative to the load arm. Horses, in contrast, should have lever systems in their forelimbs to maximize speed. Thus, their forelimb muscles should attach in such a way that there is a short effort arm relative to the load arm.

Test Yourself

1. **c.** A motor unit is a single motor neuron and all the muscle fibers that it innervates.
2. **b.** Although contraction of smooth muscle is controlled differently from that of skeletal muscle, smooth muscle does contain actin and myosin.
3. **b.** The action potential arriving to the muscle travels into the muscle through the transverse (or T) tubules.
4. **d.** A sarcomere is a structural unit within a myofibril bounded by Z lines; it contains actin and myosin.
5. **b.** ATP provides energy that is used to break actin–myosin bonds.
6. **a.** Calcium is released from the sarcoplasmic reticulum and binds with troponin, resulting in exposure of the myosin-binding site on actin.
7. **d.** The binding of calcium with troponin causes a conformational change in tropomyosin.
8. **d.** Strength training typically produces bigger, rather than more, muscle fibers.
9. **a.** Myoglobin is the main oxygen-carrying molecule in skeletal muscle.
10. **d.** Tetanus is the maximum level of muscle contraction.
11. **a.** The oxidative system produces relatively more ATP than the glycolytic system.
12. **d.** Osteoblasts lay done new bone matrix and osteoclasts resorb bone.
13. **e.** Endoskeletons (such as those of mammals), not exoskeletons, consist of cartilage and bone.
14. **c.** Ligaments hold bones together.
15. **d.** Epiphyseal plates are the sites of elongation in long bones.
16. **d.** Physical stress on bones causes them to thicken.

Gas Exchange

The Big Picture

- Cellular metabolism requires O_2 and produces CO_2 as a waste product that must be eliminated. Animals have evolved diverse structures and mechanisms for exchanging these gases with the environment. Tracheal systems in insects, gills in fishes, and lungs in terrestrial vertebrates are three examples of gas exchange organs. Very short diffusion distances and very large surface areas—adaptations designed to maximize gas exchange, characterize all of these systems.
- Oxygen is transported by the protein hemoglobin in vertebrates. The P_{O_2} levels surrounding hemoglobin determine O_2 binding. If P_{O_2} is high, hemoglobin accepts O_2, but it readily gives up its O_2 when the P_{O_2} falls. These properties allow hemoglobin to bind O_2 in the gas exchange organ, transport it to the metabolizing tissues, and then release it for consumption in cellular metabolism.

Study Strategies

- Refer to Fick's law of diffusion as a guideline for understanding gas exchange. The various components of the law—distance for diffusion, partial pressure gradient, surface area over which gas exchange occurs—provide a good framework for understanding why gas exchange organs have evolved with certain common characteristics. For example, gas exchange organs tend to have large surface areas and very short diffusion distances, and experience large partial gradients for O_2 across their surfaces. Fick's law of diffusion also governs the movement of O_2 and CO_2 in the body, and can help you to remember where and why these gases are picked up and released in the body.
- The structure and pattern of air flow through the avian respiratory system is complex and very different from the mammalian pattern. Study Figures 49.7 and 49.8, and be sure to remember that avian air sacs are not sites of gas exchange.
- A common error is the conception of O_2 and CO_2 exchange as a "two-way street," with O_2 taking the immediate place of CO_2 in the lungs and CO_2 then taking the place of O_2 in the tissues. This is incorrect. In fact, the mechanisms for exchange of these two gases are completely different.
- Understanding how O_2 is picked up, transported by, and released from hemoglobin can be confusing. Think of the O_2 dissociation curve as a "tool." Think first of the situation at the top of the curve, where P_{O_2} is high (in the lungs or gills, for example). If the P_{O_2} is high, hemoglobin will maximally bind O_2. Think, then, of transporting that blood to a region with a given P_{O_2}, and picture on the curve what must happen to O_2 saturation under the new P_{O_2} level. The O_2 from hemoglobin is now available for tissue respiration.
- Go to the Web addresses shown to review the following animated tutorials and activities. You can also find them in BioPortal (yourBioPortal.com), along with many additional learning resources.

  Animated Tutorial 49.1 Airflow in Birds (Life10e.com/at49.1)

  Animated Tutorial 49.2 Airflow in Mammals (Life10e.com/at49.2)

  Animated Tutorial 49.3 Hemoglobin: Loading and Unloading (Life10e.com/at49.3)

  Activity 49.1 The Human Respiratory System (Life10e.com/ac49.1)

  Activity 49.2 Oxygen-Binding Curves (Life10e.com/ac49.2)

  Activity 49.3 Concept Matching (Life10e.com/ac49.3)

Key Concept Review

49.1 What Physical Factors Govern Respiratory Gas Exchange?

Diffusion of gases is driven by partial pressure differences

Fick's law applies to all systems of gas exchange

Air is a better respiratory medium than water

High temperatures create respiratory problems for aquatic animals

O_2 availability decreases with altitude

CO_2 is lost by diffusion

The respiratory gas oxygen (O_2) is required by cells to produce energy in the form of ATP. The respiratory gas carbon dioxide (CO_2) is an end product of ATP production that must be eliminated from an animal's body. In the respiratory systems of animals, transfer of these gases occurs by simple diffusion. The concentrations of gases in a mixture are known as the partial pressures of those gases. The partial pressure gradient is the difference between the partial pressure of a gas at two locations. The partial pressure gradients of O_2 and CO_2 drive the movement of O_2 into the body and CO_2 out of the body. Rate of diffusion, described by Fick's law of diffusion, depends in part on a diffusion coefficient that varies in relation to temperature, the medium, and the diffusing molecules. Rate of diffusion also depends on the cross-sectional area and path length over which a gas is diffusing and the partial pressure gradient.

As compared to air, water has several disadvantages as a respiratory medium. Because water is denser than air, it is more energetically expensive to move over a respiratory surface. Water also contains far less O_2 than air, and O_2 diffuses much more slowly in water than in air. In addition, water-breathing animals face respiratory problems when water temperature rises. This is because there is less O_2 in warm water than in cold water, and an animal's need for O_2 increases with temperature (see Figure 49.2). Getting rid of CO_2 typically is not a problem for water-breathing animals.

The tendency for a gas to move by diffusion across gills, skin, or lungs depends on its partial pressure. The sum of all the gases' partial pressures in a gas mixture is the total partial pressure, which in the atmosphere equals the atmospheric pressure. O_2 makes up 20.9 percent of the atmospheric pressure. As elevation increases and atmospheric pressure decreases, the total amount of O_2 in air decreases. In air, diffusion of CO_2 is rapid because of the large partial pressure gradient between the blood and the atmosphere. The partial pressure gradient for CO_2, unlike that of O_2, changes little with altitude.

Question 1. How can an aquatic animal with no specialized respiratory surfaces get enough O_2 to all of its cells to survive?

Question 2. Why would a tropical fish in a fish tank face severe respiratory problems (and possibly death) if the heater in the tank malfunctioned and caused extremely high water temperatures?

Question 3. In humans, the effective thickness of the respiratory membrane increases dramatically when fluid accumulates in the lungs, such as during pneumonia. Using Fick's law of diffusion to frame your answer, what general health concern would arise under such conditions?

49.2 What Adaptations Maximize Respiratory Gas Exchange?

Respiratory organs have large surface areas

Ventilation and perfusion of gas exchange surfaces maximize partial pressure gradients

Insects have airways throughout their bodies

Fish gills use countercurrent flow to maximize gas exchange

Birds use unidirectional ventilation to maximize gas exchange

Tidal ventilation produces dead space that limits gas exchange efficiency

Although diverse types of respiratory organs have evolved in animals, respiratory membranes are typically thin (to minimize diffusion path length), with large surface areas for gas exchange (see Figure 49.3). Animals actively move the respiratory medium over gas exchange surfaces (ventilation), and move the internal medium (e.g., blood), over the internal side of the exchange surface (perfusion). Insects have tracheae—air tubes that end at the cells as air capillaries and open to the environment through spiracles. Gases diffuse through the tracheae into the air capillaries, aided by movement of the animal's body (see Figure 49.4). Fish have internal gills that they ventilate with unidirectional flowing water. Each gill has hundreds of gill filaments with folds, or lamellae, that act as the respiratory gas exchange surfaces. Countercurrent flow (blood in the lamellae flowing in the direction opposite to that of water flowing over the lamellae) maximizes the P_{O_2} gradient between the water and the blood (see Figures 49.5 and 49.6). As blood flows through the lamellae it is always in contact with water that has a higher O_2 level, resulting in a continuous O_2 gradient that maximizes the uptake of O_2 into the blood. Birds have a unidirectional flow of air through their lungs. Gas exchange occurs in the air capillaries of the parabronchi of the lungs; it does not occur in the associated air sacs (see Figure 49.7). Air requires two cycles of inhalation and exhalation to move from the posterior air sacs, to the lungs, to the anterior air sacs, and out of the respiratory tract (see Figure 49.8).

Mammalian lungs have alveoli, which are dead-end sacs. Ventilation is tidal, meaning that air flows in and exhaled gases flow out by the same route. The tidal breathing of mammals can be described in terms of specific lung volumes measured with a spirometer (see Figure 49.9). The normal volume of air moved during one cycle is known as the tidal volume. Vital capacity consists of the resting tidal volume, plus the additional volume of air that can be taken in with a large breath (inspiratory reserve volume) and the additional volume of air that can be forced out with a large exhale (expiratory reserve volume). Not all of the air can be forced out of the lungs. Some air remains in the bronchi and trachea, making up the dead space; this air is called the residual volume. Total lung capacity is the sum of the vital capacity and the residual volume. The residual volume cannot be measured directly with a spirometer, but it can be measured indirectly by means of the helium dilution method. Because fresh air entering the lungs mixes with stale air, tidal breathing limits the partial pressure gradient needed to drive the diffusion of O_2 from air into blood. Nevertheless, mammalian lungs have a very large surface area and a very short path length for diffusion.

Question 4. The gills of fish employ a countercurrent exchange mechanism to ensure that O_2 is efficiently extracted from the water. Describe how the countercurrent exchange works.

Question 5. Why would you be more likely to find birds than mammals at high altitudes?

Question 6. Why would the internal gills of fishes fail to work as respiratory organs on land?

Question 7. Smoking can lead to emphysema, a condition in which the walls of the alveoli break down, leading to fewer and larger alveoli. One consequence of these changes is increased residual volume. What would be another consequence?

49.3 How Do Human Lungs Work?

Respiratory tract secretions aid ventilation

Lungs are ventilated by pressure changes in the thoracic cavity

Air enters the human respiratory system at the oral cavity or nasal passage, travels down the pharynx through the larynx, and then through the trachea, which branches into two bronchi, one leading to each lung. In the lung, more branching occurs to produce bronchioles and finally the site of gas exchange—small air sacs, the alveoli (see Figure 49.10). O_2 diffuses across the thin-walled alveoli into many surrounding capillaries. The alveoli have surface tension due to an aqueous layer that makes inflation difficult. This is overcome by a surfactant produced by the cells forming the alveolar walls. The aqueous layer is needed to ensure the diffusion of gases across the membrane. Mucus, produced along the lung's larger airways, catches and removes inhaled dust and microorganisms. Cilia along the airways move the mucus up the airway and into the throat, acting as a mucus escalator.

During inhalation, lungs are inflated by negative pressure created by contraction of the diaphragm muscle at the bottom of the thoracic cavity. Closed pleural membranes line the thoracic cavity and allow for the development of negative pressure; as the thoracic cavity expands, air rushes in. During exhalation, the diaphragm relaxes and the thoracic cavity contracts (see Figure 49.11). Inhalation is an active process involving contraction of the diaphragm. At times of rest, exhalation is a passive process, occurring when the diaphragm relaxes. At times of strenuous exercise, however, intercostal muscles also help change the volume of the thoracic cavity and both inhalation and exhalation are active processes.

Question 8. An asthma attack occurs when smooth muscles lining the bronchioles spasm, obstructing the flow of air and making breathing difficult. Why is this problem limited to the bronchioles?

Question 9. The trachea is held open by C-shaped rings of cartilage. As mentioned in the textbook, you can feel these rings by running your fingers down your throat, just below your larynx. What do the open ends of the rings face and why is this important?

49.4 How Does Blood Transport Respiratory Gases?

Hemoglobin combines reversibly with O_2

Myoglobin holds an O_2 reserve

Hemoglobin's affinity for O_2 is variable

CO_2 is transported as bicarbonate ions in the blood

Gases diffuse between the respiratory organ and the circulatory system, where they are transported in blood. The liquid blood plasma transports only a small amount of dissolved O_2, with most O_2 transported by the O_2-binding pigment hemoglobin found in the red blood cells of vertebrates. Hemoglobin contains four polypeptide subunits, each with a heme group that reversibly binds O_2. The amount of O_2 bound to hemoglobin depends on the partial pressure of O_2. At the high blood P_{O_2} that is characteristic of blood leaving the lungs, hemoglobin is almost 100 percent saturated with O_2. At the tissues, the P_{O_2} is lower and hemoglobin releases approximately 25 percent of its bound O_2. The relationship between the P_{O_2} of the environment and the percentage of hemoglobin bound with O_2 is known as the S-shaped O_2-binding curve (see Figure 49.12). At low levels of P_{O_2}, only one of the hemoglobin subunits can bind O_2. Once it binds a single O_2 molecule, the conformation of the hemoglobin changes, making it easier for the other three sites on the hemoglobin to bind O_2. This process is known as positive cooperativity. In muscle, myoglobin is the O_2-carrying molecule and functions primarily to store O_2 intracellularly. The O_2 affinity is higher in myoglobin than in hemoglobin, but myoglobin has only one unit to bind one O_2 molecule.

Several factors influence hemoglobin's ability to bind and release oxygen, including type of hemoglobin, pH, and presence of certain metabolites. Fetal hemoglobin differs structurally from adult hemoglobin and has a higher affinity for O_2. This allows the efficient movement of O_2 from maternal to fetal blood. The Bohr effect is the shifting of the hemoglobin-binding curve to the right in response to a decrease in pH of the blood. This results in the release of more O_2 at a given environmental P_{O_2} at the lower pH. Hemoglobin's affinity for O_2 is lowered by 2,3-bisphosphoglyceric acid (BPG), a metabolite of glycolysis. BPG binds with hemoglobin to cause a conformational change in the hemoglobin, resulting in a rightward shift in the O_2-binding curve.

Tissues produce CO_2 as a by-product of metabolism. CO_2 must be moved from the tissues and excreted into the environment. Significant amounts of CO_2 dissolve in the plasma and are carried in this form to the lungs. Most of the CO_2 is transported as bicarbonate ions. CO_2 reacts with water to make bicarbonate ions and a proton. Carbonic anhydrase speeds up the conversion of CO_2 to carbonic acid, which then dissociates into bicarbonate ions.

Question 10. Why is carbon monoxide (CO) such a dangerous gas?

Question 11. In the diagram below, indicate the lines that represent each of the following: fetal hemoglobin, adult hemoglobin at pH 7.4, adult hemoglobin at pH 7.2, and myoglobin. Which line also represents increased BPG, a metabolite of glycolysis?

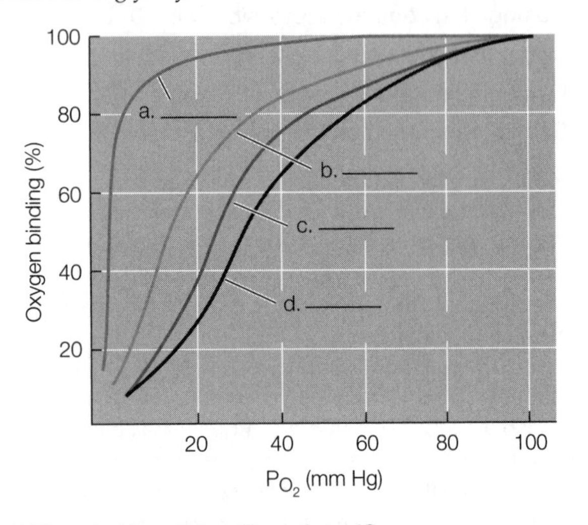

49.5 How Is Breathing Regulated?

Breathing is controlled in the brainstem

Regulating breathing requires feedback

Neurons in the medulla generate the basic breathing rhythm. Brain areas above the medulla modify breathing in response to speaking, eating, and emotional states.

In air-breathing animals, chemoreceptors on the medulla are sensitive to CO_2 levels and the pH of cerebrospinal fluid. Ventilation rate increases when CO_2 levels increase and pH decreases. Control of breathing is relatively insensitive to blood O_2 levels, but chemosensors on large arteries leaving the heart (aorta and carotid) can signal the medulla in response to low levels of O_2. In water-breathing animals, O_2 is the primary feedback stimulus for gill ventilation.

Question 12. During an episode of hyperventilation, a person breathes very deeply and rapidly, and large amounts of CO_2 are eliminated from the blood. What happens next and why?

Question 13. Design a study to investigate whether brain areas above the medulla modify breathing patterns in rats. What three variables would you measure?

Test Yourself

1. Which of the following statements is *false*?
   a. Compared to given volume of water, the same volume of air is moved more easily across a respiratory surface.
   b. Compared to water, air holds more O_2 per unit volume.
   c. Water breathers have a difficult time ridding themselves of CO_2 because CO_2 does not dissolve well in water.
   d. Temperature increases affect the O_2 content of water more than they do that of air.
   e. O_2 diffuses more rapidly in air than in water.

   Textbook Reference: *49.1 What Physical Factors Govern Respiratory Gas Exchange?*

2. According to Fick's law of diffusion, which of the following does *not* play a role in the diffusion of O_2 across a membrane?
   a. Surface area
   b. Volume
   c. Difference in concentration or partial pressure
   d. Diffusion distance
   e. Diffusion coefficient

   Textbook Reference: *49.1 What Physical Factors Govern Respiratory Gas Exchange?*

3. The barometric pressure at sea level is 760 mm Hg. Because of the relatively high altitude of Antonito, Colorado, the town has a normal barometric pressure of about 600 mm Hg. The partial pressure of O_2 in Antonito's air is therefore approximately ______ mm Hg.
   a. 75
   b. 126
   c. 160
   d. 76
   e. 21

   Textbook Reference: *49.1 What Physical Factors Govern Respiratory Gas Exchange?*

4. The movement of O_2 and CO_2 between the blood in the tissue capillaries and the cells in tissues depends most directly upon
   a. active transport of O_2 and CO_2.
   b. total atmospheric (barometric) pressure differences across the cell membranes.
   c. diffusion of O_2 and CO_2 down a concentration gradient.
   d. diffusion of O_2 and CO_2 down a partial pressure gradient.
   e. osmosis across cell membranes.

   Textbook Reference: *49.1 What Physical Factors Govern Respiratory Gas Exchange?*

5. External gills, tracheae, and lungs share which of the following sets of characteristics?
   a. They are parts of gas-exchange systems, they exchange both CO_2 and O_2, and they increase surface area for diffusion.
   b. They are used by water breathers, their functioning is based on countercurrent exchange, and they make use of negative pressure for breathing.
   c. They exchange only O_2, they are associated with a circulatory system, and they are found in vertebrates.
   d. They are found in insects, they employ positive-pressure pumping, and their functioning is based on crosscurrent flow.

e. They are parts of gas-exchange systems, they exchange both CO_2 and O_2, and they decrease surface area for diffusion.

Textbook Reference: *49.2 What Adaptations Maximize Respiratory Gas Exchange?*

6. Which of the following represents a larger volume of air than is normally found in the resting tidal volume of a human lung?
   a. Residual volume
   b. Inspiratory reserve volume
   c. Expiratory reserve volume
   d. Vital capacity
   e. All of the above

   Textbook Reference: *49.2 What Adaptations Maximize Respiratory Gas Exchange?*

7. Both bird and mammal lungs
   a. require two cycles of inhalation and exhalation in order for air to move.
   b. contain alveoli at the terminal ends.
   c. have an anatomical dead space.
   d. exchange O_2 and CO_2 with blood in capillaries.
   e. have a unidirectional flow of air moving through them.

   Textbook Reference: *49.2 What Adaptations Maximize Respiratory Gas Exchange?*

8. Air flows into the lungs of mammals during inhalation because the
   a. pressure in the lungs falls below atmospheric pressure.
   b. volume of the lungs decreases.
   c. pressure in the lungs rises above atmospheric pressure.
   d. diaphragm moves upward toward the lungs.
   e. internal intercostal muscles contract.

   Textbook Reference: *49.3 How Do Human Lungs Work?*

9. The alveoli of the lungs do not contain air with 20.9 percent O_2 (i.e., the percentage in dry air) because
   a. we normally do not ventilate our lungs at a high enough rate.
   b. the lungs have too many alveoli to ventilate.
   c. there is dead space in the trachea and bronchi.
   d. the trachea and bronchi are too small in volume to contain this amount of O_2.
   e. some O_2 has been exchanged before reaching the alveoli.

   Textbook Reference: *49.3 How Do Human Lungs Work?*

10. Which of the following statements about hemoglobin is *false*?
    a. Hemoglobin allows the blood to carry a large amount of O_2.
    b. Hemoglobin contains a single polypeptide chain with very high affinity for O_2.
    c. Hemoglobin is packaged inside red blood cells.
    d. Fetal hemoglobin is structurally different from adult hemoglobin.
    e. Compared with adult hemoglobin, fetal hemoglobin has a higher affinity for O_2.

    Textbook Reference: *49.4 How Does Blood Transport Respiratory Gases?*

11. The Bohr shift describes the
    a. outward movement of Cl^- from a blood cell in exchange for HCO_3^- moving into the cell.
    b. leftward shift of the entire O_2 equilibrium curve when temperature rises.
    c. rightward shift of the entire O_2 equilibrium curve when pH rises.
    d. rightward shift of the entire O_2 equilibrium curve when pH falls.
    e. leftward shift of the entire O_2 equilibrium curve when BPG increases.

    Textbook Reference: *49.4 How Does Blood Transport Respiratory Gases?*

12. The presence of CO_2 in blood will lower blood pH because CO_2 combines with _______ to form _______. The rate of the reaction is increased by _______.
    a. H_2O; H^+ and HCO_3^-; carbonic anhydrase
    b. H_2O; only HCO_3^-; carbonic anhydrase
    c. H_2O; only H^+; carbonic ions
    d. H^+; HCO_3^-; oxyhemoglobin
    e. H_2O; H^+ and HCO_3^-; fetal hemoglobin

    Textbook Reference: *49.4 How Does Blood Transport Respiratory Gases?*

13. The largest proportion of CO_2 carried by the blood is in the form of
    a. bicarbonate ions (HCO_3^-) carried in the plasma.
    b. molecular CO_2 dissolved in the plasma.
    c. bicarbonate ions (HCO_3^-) carried within the red blood cells.
    d. molecular CO_2 chemically bound to hemoglobin.
    e. molecular CO_2 chemically bound to myoglobin.

    Textbook Reference: *49.4 How Does Blood Transport Respiratory Gases?*

14. Which of the following is involved in the control of breathing?
    a. Neurons in the medulla
    b. Chemoreceptors on the surface of the medulla
    c. Chemosensors on the aorta
    d. Chemosensors on the carotid arteries
    e. All of the above

    Textbook Reference: *49.5 How Is Breathing Regulated?*

15. Which of the following statements is *false*?
    a. Air-breathing animals are remarkably insensitive to falling levels of O_2.
    b. In water-breathing animals, O_2 is the primary feedback stimulus for breathing.

c. In humans, chemoreceptors for CO_2 are located in the medulla.
d. Areas of the brain above the medulla have no impact on breathing rhythm.
e. The basic breathing rhythm is an involuntary function.

Textbook Reference: *49.5 How Is Breathing Regulated?*

Answers

Key Concept Review

1. An aquatic animal without a specialized respiratory system must rely on simple diffusion of O_2 from the environment into its cells. For simple diffusion to be an effective means of O_2 transport, the animal must be very small, with all of its cells only a few cell layers from the environment. Some animals without respiratory systems are thin and flat to increase their external surface area for gas exchange, and still others have a central cavity through which water can circulate to bring oxygen close to their cells. The metabolic rates of these animals are likely very low in order to accommodate the lack of any special respiratory system.

2. The tropical fish would be facing the double bind of needing more O_2 as the water temperature increased (due to its increasing metabolic rate with increasing water temperature) along with lower levels of O_2 in the warmer water (since warm water holds less dissolved O_2 than does cold water).

3. An increase in the effective thickness of the respiratory membrane translates into an increase in the diffusion path length (*L*) in Fick's law. Such an increase would result in slower rates of diffusion (*Q*). From a health perspective, this might result in inadequate gas exchange and oxygen deprivation of body tissues.

4. In the countercurrent exchange in the gills of fish, water and blood move in opposite directions. This results in blood with low O_2 content entering the gills and flowing past water with a higher O_2 content. O_2 diffuses down the partial pressure gradient into the blood. As blood moves along the lamella, it picks up O_2, but it always has less O_2 than the water moving in the opposite direction. In this way the blood is able to optimize O_2 uptake during the entire trip through the lamella.

5. Birds have an extremely efficient respiratory system characterized by the continuous unidirectional flow of air through the lungs. In contrast, mammals use tidal ventilation, a less efficient method in which air flows in and exhaled gases flow out the same route. Tidal ventilation results in the mixing of fresh air with stale air. O_2 availability decreases with altitude, and thus we would expect to see birds rather than mammals at very high elevations.

6. The internal gills of fishes consist of thin, delicate tissue. On land, such gills would collapse and clump together, significantly reducing the surface area for gas exchange.

7. Another consequence of the breakdown and merging of the walls of alveoli is decreased surface area for gas exchange. The shortness of breath characteristic of emphysema indicates both the decreased surface area and the increased residual volume (dead space).

8. The problem occurs in the bronchioles because the bronchioles, unlike the trachea and bronchi, do not have cartilage supports to keep them open.

9. The trachea lies in front of the esophagus. The open ends of the C-shaped rings of cartilage in the trachea face the side of the trachea next to the esophagus. This allows the esophagus to expand when food is swallowed.

10. CO binds to hemoglobin much more readily than O_2 does, so it prevents hemoglobin from binding and transporting oxygen. This can be fatal.

11.
    a. Myoglobin
    b. Fetal hemoglobin
    c. Adult hemoglobin (pH 7.4)
    d. Adult hemoglobin (pH 7.2)

    The curve for adult hemoglobin at pH 7.2 could also represent adult hemoglobin with increased BPG.

12. Following hyperventilation, breathing temporarily stops because so much CO_2 has been removed from the blood that the respiratory control center temporarily stops sending signals to the diaphragm and intercostal muscles. Breathing begins again when the CO_2 builds to sufficient levels in the blood to prompt the respiratory center to send its signals.

13. Brain areas above the medulla can modify the basic breathing pattern to allow for activities such as eating and drinking (swallowing temporarily closes the opening to the respiratory system). This could be studied in rats by simultaneously monitoring the following: (1) the activity of neurons from higher brain centers that bring such information into the pons, which then communicates with the respiratory center in the medulla; (2) the behavior of the rats to keep track of when they are eating or drinking; and (3) the respiratory rates of the rats.

Test Yourself

1. **c.** Water breathing is more difficult than air breathing because the higher density of water makes it more expensive to move across the respiratory surfaces than air, it has less O_2 than air, and the O_2 content is dependent on the temperature of the water. CO_2, however, dissolves easily in water and it is easily expelled, even in stagnant water.

2. **b.** The volume of the gas is not included in Fick's law of diffusion. All the other parameters are important in

determining the rate of diffusion of gases in the respiratory system of animals.

3. **b.** Air at sea level with an atmospheric pressure of 760 mm Hg has a partial O_2 pressure of 160 mm Hg (21% of 760). Therefore, the partial pressure of O_2 at Antonito, Colorado, is about 126 mm Hg (21% of 600).

4. **d.** Movement of O_2 and CO_2 from the blood to the tissues always occurs by diffusion of O_2 and CO_2 down their partial pressure gradients.

5. **a.** The external gill, tracheae, and lungs are all examples of gas-exchange systems used by various animals to exchange both CO_2 and O_2. One of the important properties of all respiratory systems is that they increase surface area for diffusion.

6. **e.** All have a larger volume than the resting tidal volume.

7. **d.** The lungs of birds and mammals are both sites for the exchange of O_2 and CO_2 with blood in capillaries. The lungs of birds require two cycles in order for air to move through them, whereas the mammalian lung requires only one cycle. Only the mammalian lung ends in alveoli and has an anatomical dead space. Flow of air is unidirectional in birds and tidal in mammals.

8. **a.** Inhalation in the mammalian lung occurs by means of negative pressure produced by contraction of the diaphragm. Thus the pressure in the lungs falls below atmospheric pressure.

9. **c.** The alveoli do not contain "air" with 20.9 percent O_2 because incoming air is mixed with air left in the dead space of the trachea and bronchi, which has had some of the O_2 removed by the lungs.

10. **b.** Hemoglobin is composed of four subunits, which act with positive cooperativity to bind O_2. O_2-binding myoglobin found in muscle has only one unit.

11. **d.** The Bohr effect describes the action of pH on the O_2-binding curve. Decreases in pH result in a net rightward shift of the entire O_2 equilibrium curve.

12. **a.** Carbon dioxide combines with H_2O in the plasma to form H^+ and HCO_3^-. The enzyme carbonic anhydrase catalyzes the reaction.

13. **a.** Most of the CO_2 is carried as bicarbonate ions (HCO_3^-) in the plasma.

14. **e.** All are involved in the control of breathing.

15. **d.** Areas of the brain above the medulla modify the breathing rhythm with respect to speech, eating, coughing, and emotional state.

Circulatory Systems

The Big Picture

- Circulatory systems are needed when animals grow too large to acquire nutrients and eliminate wastes by diffusion alone. Open circulatory systems have relatively low pressures and large sinuses that bathe tissues and organs in circulating fluid. Closed circulatory systems have an interconnected series of vessels that keep blood separated from interstitial fluid but allow gases, nutrients, and wastes to diffuse through vessel walls.
- During the evolution of vertebrates, the heart changed from the simple, two-chambered hearts of fishes, which rely on gills, to the complex, four-chambered hearts of crocodilians, birds, and mammals. Reptiles (turtles, lizards, snakes, and crocodilians) often breathe intermittently; their hearts allow shunting of blood away from the lungs (pulmonary circuit) and out to the body (systemic circuit). Amphibians pick up oxygen at their lungs and skin.
- The human heart is really two pumps in one. The left pump provides oxygenated blood under high pressure to the systemic tissues, and the right pump sends deoxygenated blood under low pressure to the lungs. A cardiac pacemaker sets the cardiac rhythm.
- The flow of blood through capillaries is highly regulated at the local level by changes in local conditions; blood flow through arteries and arterioles is regulated by the combined action of neural activity and circulating hormones.

Study Strategies

- The hearts of amphibians and reptiles are often thought of as primitive structures awaiting "repair through evolution" into the more highly evolved hearts of birds and mammals. In fact, the hearts of amphibians and reptiles are highly adapted to suit their particular lifestyles, specifically, their intermittent breathing and relatively low metabolic rates. By being able to shunt blood from pulmonary to systemic circuits, these animals can save the energy that would otherwise be wasted in sending blood to the temporarily nonventilated lungs. Because we are unaccustomed to the idea of breathing intermittently, the topic of shunting blood past the lungs can be challenging. It is best to learn the conditions under which these vertebrates perform this function (e.g., when under water) and then learn the routes by which they do so.
- It can be difficult to appreciate that the human heart functions as two entirely distinct pumps—the right heart pumping blood to the lungs and the left heart pumping blood to the body. The two pumps just happen to be packaged in the same structure—the heart. In fact, cardiologists refer to the "left heart" and "right heart," emphasizing the separate nature of these two pumps. Learning the route by which blood flows through the human heart becomes easier when it is considered from this perspective. In order to understand the pattern of blood flow through the human heart, it is helpful to envision blood circulation as a great circle with two pumps at opposite sides and a vascular bed between the pumps. Once you have mastered this concept, you will see that the only way to place the two pumps in proximity in the same structure (the heart) is to twist the circle into a folded figure eight—the typical diagram of the circulation in the textbook. Study the various chambers and valves as sequences of structures in the right and left pumps of the heart, rather than trying to learn them as a list of terms printed on a complex, anatomically accurate diagram of the heart.
- Go to the Web addresses shown to review the following animated tutorials and activities. You can also find them in BioPortal (yourBioPortal.com), along with many additional learning resources.

Animated Tutorial 50.1 The Cardiac Cycle (Life10e.com/at50.1)

Animated Tutorial 50.2 Blood Pressure and Heart Rate Regulation (Life10e.com/at50.2)

Activity 50.1 Vertebrate Circulatory Systems (Life10e.com/ac50.1)

Activity 50.2 The Human Heart (Life10e.com/ac50.2)

Activity 50.3 Structure of a Blood Vessel (Life10e.com/ac50.3)

Key Concept Review

50.1 Why Do Animals Need a Circulatory System?

Some animals do not have a circulatory system

Circulatory systems can be open or closed

Open circulatory systems move extracellular fluid

Closed circulatory systems circulate blood through a system of blood vessels

Circulatory systems transport nutrients to cells and wastes away from cells. In very small or flat multicellular animals, every cell within the body is close to the environment. These typically aquatic animals do not require a circulatory system because materials can efficiently move into and out of cells by diffusion alone. Some slightly larger aquatic invertebrates, such as sponges, have a gastrovascular system, which consists of a highly branched central cavity through which water circulates and where exchange of gases and nutrients occurs.

Large, active animals need a circulatory system. Circulatory systems can be classified as open or closed. In an open circulatory system, blood and other tissue fluids are not separated; the fluid is called hemolymph (see Figures 50.1A and B). A pump may be present to help move hemolymph through the various large compartments, sometimes assisted by movements of the animal. In a closed circulatory system, blood is pumped through a series of interconnected closed vessels; it is kept separate from interstitial fluid (see Figure 50.1C). There are numerous advantages to closed systems compared with open ones: more rapid transport of fluid in the closed vessels, the ability to divert the flow of blood between different tissues, and direct hormone and nutrient transport to the tissues.

Question 1. Many animals with open circulatory systems are relatively inactive. Explain how insects, with their open circulatory systems, maintain their high levels of activity.

Question 2. What is the difference between interstitial fluid and extracellular fluid in a closed circulatory system?

Question 3. What are three advantages of a closed circulatory system over an open system?

50.2 How Have Vertebrate Circulatory Systems Evolved?

Circulation in fish is a single circuit

Lungfish evolved a gas-breathing organ

Amphibians have partial separation of systemic and pulmonary circulation

Reptiles have exquisite control of pulmonary and systemic circulation

Birds and mammals have fully separated pulmonary and systemic circuits

Mammals and birds have two circuits: a systemic circuit in which blood flows from the heart to the body tissues, and a pulmonary circuit in which blood flows from the heart to the lungs. In fishes that rely on gills as respiratory organs, there is a single circuit: blood is pumped from the heart to the gills to the tissues and then back to the heart. The fish heart, in this case, has one atrium and one ventricle. African lungfish have reduced gills and a lung for gas exchange. A heart with a partially divided atrium helps separate oxygenated blood returning from the lung and deoxygenated blood returning from the tissues as they enter the single ventricle. The lungfish heart and vascular system exhibit adaptations that partially separate systemic and pulmonary circuits.

Amphibians have two atria and one ventricle. One atrium receives deoxygenated blood from the tissues and the other receives oxygenated blood from the lungs. Anatomical features of the ventricle help direct deoxygenated blood preferentially to the lungs and skin (which also acts as a gas exchange organ in amphibians) and oxygenated blood to the systemic body tissues. Even though the two circuits are not fully divided, mixing of high oxygen and low oxygen blood is minimal.

Turtles, snakes, and lizards have two atria and a ventricle that is internally divided into subchambers that allow for control of the distribution of blood to the lungs and body. When the animal is breathing, blood from the right side of the ventricle is preferentially pumped to the lungs because there is a higher resistance in the systemic circuit than in the pulmonary circuit. Also, there is an asynchronous ejection from the heart because pumping occurs first from the right and then the left side of the ventricle. At the stage in the cardiac cycle in which blood from the left side leaves the heart, the resistance is higher in the pulmonary circuit. Thus, the last blood leaving the heart tends to be deoxygenated blood and is preferentially directed to the pulmonary circulation. When a reptile stops breathing, vessel constriction in the lungs causes pulmonary circuit resistance to increase, causing blood from the right side to largely bypass the pulmonary circulation and flow to the systemic circuit. This process is called shunting.

Birds and mammals have a four-chambered heart that allows for complete separation and no mixing of systemic and pulmonary blood. The pulmonary circulation operates at low pressure, protecting the delicate lung membranes, whereas the systemic circuit operates at high pressures, allowing the perfusion of tissues and organs distant from the heart. Alligators and crocodiles have a four-chambered heart, resembling that of a bird or mammal. However, unlike birds and mammals, alligators and crocodiles have two vessels leaving the right ventricle: one vessel, the pulmonary artery, leads to the lungs and the other, an aorta, serves the body. There is a small passageway between this aorta and the aorta leaving the left ventricle, which also serves the body. When breathing, crocodiles and alligators can operate their heart like that of a bird or mammal, distributing blood to both the pulmonary and systemic circulations. When not breathing, however, they can largely bypass the lungs by directing most blood into the aorta that leaves the right ventricle rather than into the pulmonary artery.

Question 4. What are the important differences between the circulatory system of a fish and that of a turtle?

Question 5. In crocodilians, which circuit—pulmonary or systemic—would receive the most blood during a long dive? During such a dive, in which circuit is resistance high, and why? Finally, what is the key anatomical difference in the circulatory systems of crocodilians and birds that allows shunting in crocodilians but not in birds?

Question 6. Design a method to temporarily disable the shunt in a crocodilian heart.

50.3 How Does the Mammalian Heart Function?

- Blood flows from right heart to lungs to left heart to body
- The heartbeat originates in the cardiac muscle
- A conduction system coordinates the contraction of heart muscle
- Electrical properties of ventricular muscles sustain heart contraction
- The ECG records the electrical activity of the heart

The human heart has two atria and two ventricles (see Figure 50.2). Between each atrium and its adjacent ventricle is an atrioventricular valve, which stops backflow of blood from the ventricle into the atrium. The pulmonary valve resides between the right ventricle and the pulmonary artery, and the aortic valve resides between the left ventricle and the aorta; these two valves prevent backflow of blood into the ventricles when they relax. Deoxygenated blood flows from the body through the superior vena cava and inferior vena cava into the right atrium and into the right ventricle by passive filling during heart relaxation. During heart contraction, blood flows from the right ventricle to the pulmonary artery and the lungs, where it becomes oxygenated. Oxygenated blood from the lungs flows back to the heart via pulmonary veins into the left atrium, then into the left ventricle, and finally out through the aorta to the systemic tissues. Resistance in the blood vessels of the body's systemic tissues is higher than in the lungs, so the left ventricle must be larger and more muscular than the right ventricle to generate more force while pumping the same volume of blood.

Systole is contraction of the ventricles, and diastole is relaxation of the ventricles. At the very end of diastole, the atria contract. These phases make up the cardiac cycle. Pressure is highest during systole and lowest during diastole (see Figure 50.3). Blood pressure can be measured using a sphygmomanometer (see Figure 50.4). A typical reading for a healthy young adult is 120 (systolic pressure) over 80 (diastolic pressure).

The human heartbeat is generated in the heart's pacemaker, or sinoatrial node, which is a small section of highly modified muscle tissue located at the junction of the superior vena cava and right atrium. The pacemaker rhythmically produces action potentials that pass to the rest of the heart and stimulate a heartbeat (see Figure 50.7). Gap junctions connect cardiac muscle cells and thus allow the action potentials to spread rapidly from cell to cell. The action potential moves from the atria to the ventricles through the atrioventricular node (AV node), passes rapidly down the bundle of His, and to the Purkinje fibers. From the Purkinje fibers the action potential spreads through both ventricles, causing contraction to occur slightly after the contraction of the atria.

The autonomic nervous system also influences the heartbeat. Sympathetic activity increases heart rate, and parasympathetic activity decreases heart rate (see Figure 50.6).

The electrocardiogram (ECG or EKG) measures the electrical activity of the heart as a complex wave pattern (see Figure 50.10). The wave has five parts: P (depolarization and contraction of the atria); Q, R, and S (depolarization and contraction of the ventricles); and T (relaxation and repolarization of the ventricles).

Question 7. The diagrams below trace the course of the heartbeat. Label the following structures on the left diagram: sinoatrial node, atrioventricular node, bundle of His, Purkinje fibers, atria, and ventricles. On the middle and right diagrams, indicate with arrows the path of the action potentials at the particular stage indicated at the top of each diagram.

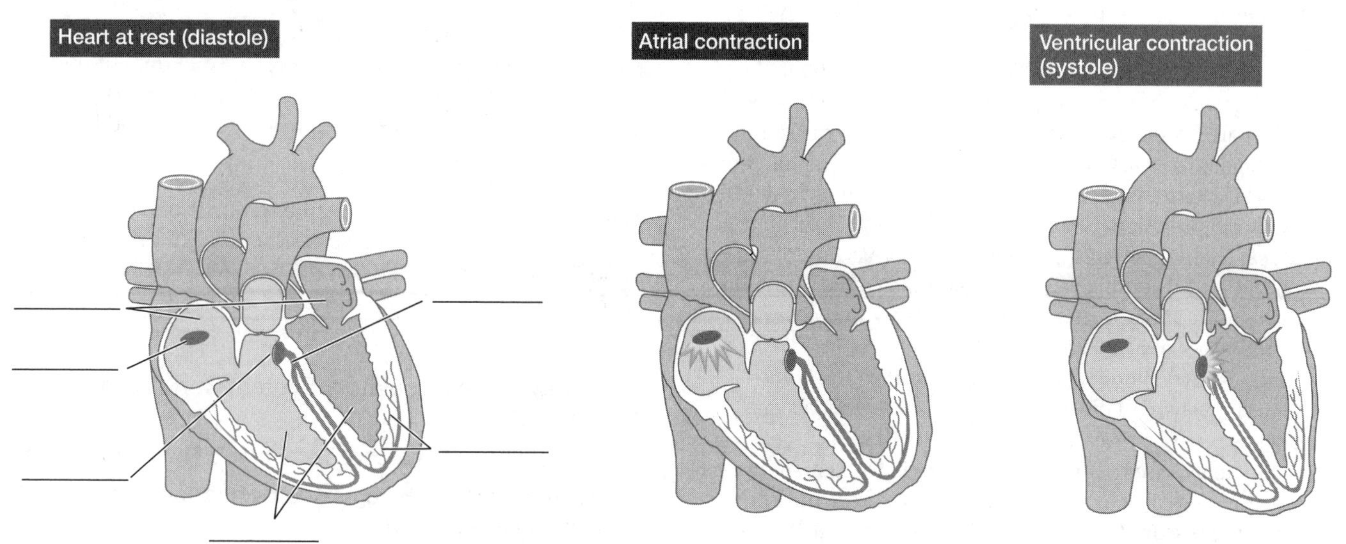

Question 8. Given the role of the heart's conduction system in coordinating the contraction of heart muscle, what problems might arise if the conduction system were impaired?

Question 9. Heart block is a very serious condition in which action potentials fail to move from the atria to the ventricles. As a result, the ventricles beat at their own intrinsic rate, which is insufficient. Heart block results from damage to which structure?

Question 10. A physician tells her patient that the "lub" sound of his heart sounds somewhat abnormal. What structure(s) might be impaired? Are the ventricles in systole or diastole at the time the structure is malfunctioning?

50.4 What Are the Properties of Blood and Blood Vessels?

- Red blood cells transport respiratory gases
- Platelets are essential for blood clotting
- Arteries withstand high pressure, arterioles control blood flow
- Materials are exchanged in capillary beds by filtration, osmosis, and diffusion
- Blood flows back to the heart through veins
- Lymphatic vessels return interstitial fluid to the blood
- Vascular disease is a killer

The fluid matrix of blood is known as plasma. It contains water, salts, proteins, and hormones that are carried by the blood to target cells. The composition of plasma is very similar to that of interstitial fluid, except that proteins are present in higher concentrations in plasma. The hematocrit (also called packed-cell volume) is the percentage of blood volume composed of cells. A normal hematocrit is about 42 percent for women and 46 percent for men. Several different cell types occur in blood, where they carry out diverse functions (see Figure 50.11). Erythrocytes, or red blood cells, transport respiratory gases. In mammals, but not other vertebrates, mature red blood cells lack a nucleus and other organelles. Red blood cells are produced by stem cells in bone marrow. Erythropoietin, a hormone produced by the kidneys, stimulates red blood cell production. Once released into the bloodstream, red blood cells survive for about 120 days. The spleen removes old or damaged red blood cells. Platelets are produced by the breaking off of cell fragments from megakaryocytes in the bone marrow. Platelets function in blood clotting by binding to the edges of a break in a vessel wall and providing a mechanical plug, and by starting the chemical chain reaction that leads to the conversion of prothrombin into thrombin. Thrombin stimulates the formation of sticky fibrin threads that form a meshwork and cover the hole in the vessel (see Figure 50.12).

Arteries carry blood away from the heart to the body under relatively high pressure. Arteries and arterioles are known as resistance vessels because smooth muscle cells in their walls cause them to constrict or dilate, changing resistance. Veins carry blood toward the heart from the body under low pressure (see Figure 50.13). Veins that act against gravity have valves, and blood flowing toward the heart is helped by the movements of surrounding skeletal muscles (see Figure 50.16). Veins are called capacitance vessels because of their ability to stretch and store blood. Capillaries have a large cross-sectional area and extremely thin walls to allow the diffusion of gases, nutrients, and wastes (see Figure 50.14). Blood pressure from the arterial side of a capillary bed forces fluids, ions, and small molecules out of the capillaries through small holes called fenestrations. Near the venule end of the capillary bed, the difference in osmotic potential between the plasma inside the capillaries and the surrounding fluids creates an osmotic pressure that results in fluid recovery back into the capillaries. Thus, the net flow of water between the plasma and tissue fluid is determined by the difference in blood pressure and osmotic pressure (see Figure 50.15). These two opposing forces are known as Starling's forces.

The lymphatic system returns tissue fluid not reclaimed by the capillaries to the blood. Once interstitial fluid enters the lymphatic system, it is called lymph. In humans, lymph is moved through the lymphatic vessels up the body by surrounding skeletal muscle contractions. Lymph is returned to the blood through the thoracic ducts that empty into large veins at the base of the neck. Lymph nodes exist in mammals and birds and contribute to body defenses.

Hardening of the arteries is known as atherosclerosis. Plaque deposits begin to form at sites of endothelial damage, lipids accumulate on the deposits, and then connective tissue invades, making the wall of the artery less elastic. Blood platelets stick to the plaque, and this can lead to the formation of a stationary blood clot, or thrombus. The coronary arteries that supply the heart with blood are highly susceptible to blockage, which can result in a heart attack (myocardial infarction). An embolus is a piece of a thrombus that breaks off and lodges in a vessel. When this happens in the brain, it is known as a stroke.

Question 11. Blood moves through the arteries and veins at a faster velocity than the blood moving through the capillaries. What is the cause and effect of the slowing down of blood as it moves through the capillaries?

Question 12. The rate of production of red blood cells in the body usually matches the rate of destruction, but under certain circumstances the body speeds up the rate of production. Under what circumstances would it do so, and how is this accomplished?

Question 13. A chronically fast heart rate is one of the symptoms that would lead a physician to test for anemia, a condition characterized by a reduction in the blood's ability to carry oxygen. What is the link between heart rate and anemia?

50.5 How Is the Circulatory System Controlled and Regulated?

- Autoregulation matches local blood flow to local need
- Arterial pressure is regulated by hormonal and neural mechanisms

The circulatory system is under hormonal, neural, and local control. Blood flow in the capillary beds of many tissues is controlled locally by autoregulatory mechanisms. Precapillary sphincters at the arterial end of the capillaries control the flow of blood, with contraction of these sphincters limiting or stopping blood flow through the capillary bed (see Figure 50.18). Low levels of O_2 and high levels of CO_2 can open the precapillary sphincters. The increased supply of blood brings in more O_2 and removes more CO_2; this response is called hyperemia (excess blood).

Norepinephrine from sympathetic neurons and epinephrine released from the adrenal medulla cause arteries and arterioles to constrict. In response to a decline in blood pressure, the kidneys release renin, which is converted to angiotensin. Angiotensin stimulates thirst and also constricts arterioles. All of these hormones reduce blood flow to less essential organs (such as those of the gut) and increase central blood pressure, thereby ensuring that essential organs (such as the brain, heart, and kidneys) have sufficient blood.

Autonomic control of heart rate and constriction of blood vessels occurs from the medulla. Stretch receptors (known as baroreceptors) in the aorta and carotid arteries help regulate blood pressure. With rising blood pressure they inhibit sympathetic output and cause the heart to slow and arterioles to dilate. When pressure is low, sympathetic output is stimulated, increasing heart rate and constriction of arterioles. Chemoreceptors in the aorta and carotid arteries sense changes in blood composition. In response to a fall in arterial pressure (signaled by decreased activity of baroreceptors), the posterior pituitary releases antidiuretic hormone (ADH, also known as vasopressin). ADH stimulates the kidneys to reabsorb more water, and this results in increased blood volume and pressure (see Figure 50.19).

Question 14. You are about to take a final exam in your biology course and are very nervous. How will your circulatory system respond to this emotional state?

Question 15. How do precapillary sphincters and arterioles react to low oxygen, high carbon dioxide, and high lactate in tissue? How are total peripheral resistance and mean arterial pressure affected by the actions of the precapillary sphincters and arterioles?

Test Yourself

1. Which of the following is *not* one of the reasons that closed circulatory systems are more efficient than open circulatory systems?
   a. Open systems rely exclusively on fluid percolating through tissues, whereas closed systems rely on muscular pumps (e.g., hearts) to move fluid.
   b. Transport within closed systems is more rapid than in open systems.
   c. Blood can be directed to specific areas in closed systems, but not in open systems.
   d. Compared to open systems, closed systems operate better under higher pressure.
   e. In closed systems molecules and cells that transport hormones and nutrients can be kept in vessels until they unload their goods at specific tissues.

   Textbook Reference: *50.1 Why Do Animals Need a Circulatory System?*

2. When turtles stop breathing (e.g., on extended dives),
   a. blood is shunted from the systemic circuit to the pulmonary circuit.
   b. blood is shunted specifically to the brain.
   c. blood is shunted from the pulmonary circuit to the systemic circuit.
   d. blood is shunted to the skin for gas exchange.
   e. there is no change in pattern of circulation.

   Textbook Reference: *50.2 How Have Vertebrate Circulatory Systems Evolved?*

3. In which of the following would the highest blood pressure be recorded?
   a. In the ventricle supplying blood to the gills of a fish
   b. In the vessels leaving the gills of a fish
   c. In the pulmonary vein of a frog
   d. In the ventricle supplying blood to the systemic circuit of a bird
   e. In the ventricle supplying blood to the pulmonary circuit of a bird

   Textbook Reference: *50.2 How Have Vertebrate Circulatory Systems Evolved?*

4. Which of the following statements about vertebrate circulatory systems is *false*?
   a. Crocodilians have a four-chambered heart.
   b. Lungfishes have a partially divided atrium.
   c. The ventricle of turtles, snakes, and lizards is partly divided by a septum.
   d. Amphibians have two atria and one ventricle.
   e. In fishes, blood passes from the heart to the gills, returns to the heart, and then is pumped to the body.

   Textbook Reference: *50.2 How Have Vertebrate Circulatory Systems Evolved?*

5. Which of the following represents the correct sequence of structures through which cardiac action potentials pass in the mammalian heart?
   a. Purkinje fibers, AV node, SA node, bundle of His, atrial fibers
   b. AV node, atrial fibers, SA node, bundle of His, Purkinje fibers
   c. SA node, bundle of His, atrial fibers, AV node, Purkinje fibers
   d. Purkinje fibers, bundle of His, AV node, atrial fibers, SA node
   e. SA node, atrial fibers, AV node, bundle of His, Purkinje fibers

   Textbook Reference: *50.3 How Does the Mammalian Heart Function?*

6. In the mammalian heart, the AV node _______, and the Purkinje fibers _______.

a. creates simultaneous atrial and ventricular depolarization; speed up transmission of the cardiac impulse into the ventricle
b. delays ventricular depolarization relative to atrial depolarization; insulate the cardiac impulse from the general ventricular fibers
c. delays ventricular depolarization relative to atrial depolarization; ensure that the cardiac impulse spreads rapidly and evenly throughout the ventricles
d. delays atrial depolarization relative to ventricular depolarization; transmit the cardiac impulse to very small, localized groups of ventricular fibers
e. initiates each heartbeat; transmit impulses to the atria.

Textbook Reference: *50.3 How Does the Mammalian Heart Function?*

7. In the mammalian heart, the atrial walls are _______ the ventricular walls, and pressure generated in the atrial chambers is _______ the pressure in the ventricles.
a. thinner than; higher than
b. thinner than; lower than
c. thicker than; higher than
d. thicker than; lower than
e. the same thickness as; the same as

Textbook Reference: *50.3 How Does the Mammalian Heart Function?*

8. The left ventricle of the mammalian heart exceeds the right ventricle in the
a. amount of blood that enters during each heart contraction.
b. volume expelled during each heart contraction.
c. pressure developed during each heart contraction.
d. speed with which it contracts.
e. All of the above

Textbook Reference: *50.3 How Does the Mammalian Heart Function?*

9. The fluid fraction of the blood consists of _______ and the solid fraction consists of _______.
a. plasma; water, erythrocytes, and platelets
b. erythrocytes; leukocytes, macrophages, and platelets
c. plasma; erythrocytes, platelets, and leukocytes
d. leukocytes; erythrocytes and platelets
e. interstitial fluid; stem cells

Textbook Reference: *50.4 What Are the Properties of Blood and Blood Vessels?*

10. Which of the following statements about red blood cells is true?
a. They are biconcave cells containing hemoglobin.
b. They are spherical cells containing hemoglobin.
c. They are spherical cells capable of amoeboid motion and containing hemoglobin.
d. They are biconcave cells whose primary function is the formation of blood clots.
e. They are biconcave cells that function in the immune response.

Textbook Reference: *50.4 What Are the Properties of Blood and Blood Vessels?*

11. Blood clotting pathways cause the conversion of _______ to _______.
a. vitamin K; prothrombin
b. fibrin; fibrinogen
c. thrombin; prothrombin
d. prothrombin; thrombin
e. fibrinogen; thrombin

Textbook Reference: *50.4 What Are the Properties of Blood and Blood Vessels?*

12. Which region of the vascular bed is the precise site of gas exchange with surrounding tissue?
a. Arteries
b. Capillaries
c. Lymphatic vessels
d. Veins
e. Venules

Textbook Reference: *50.4 What Are the Properties of Blood and Blood Vessels?*

13. In the human lymphatic system, the _______ act(s) primarily as a filter for detecting and destroying microorganisms in lymph traveling through major lymph vessels.
a. lymph nodes
b. thymus
c. lymph capillaries
d. tonsils, but not the appendix
e. thoracic ducts

Textbook Reference: *50.4 What Are the Properties of Blood and Blood Vessels?*

14. The net loss of fluid from blood capillaries increases if
a. plasma filtration decreases.
b. the osmotic pressure of plasma increases.
c. blood pressure increases in the capillaries.
d. the osmotic pressure of interstitial fluid decreases.
e. blood pressure drops below osmotic pressure.

Textbook Reference: *50.4 What Are the Properties of Blood and Blood Vessels?*

15. Which of the following statements about the control of circulation in humans is *false*?
a. Sympathetic nerve input to skeletal muscle causes the blood vessels in the muscle to dilate.
b. Blood flow can be regulated by autonomic nerve signals emanating from cardiovascular control centers in the medulla of the brain.
c. Carotid artery chemoreceptors promote increased blood pressure when they detect low O_2 levels in the blood.
d. Hormones such as angiotensin and vasopressin cause venules to constrict.

e. Autoregulatory mechanisms can cause constriction or dilation of arterioles.

Textbook Reference: *50.5 How Is the Circulatory System Controlled and Regulated?*

Answers

Key Concept Review

1. Despite having an open circulatory system, insects are able to maintain high levels of activity because they do not rely on a circulatory system for the exchange of respiratory gases. Instead, they rely on a system of air-filled tubes, called tracheae, which serve as sites for gas exchange.
2. Interstitial fluid is the fluid around the cells. Extracellular fluid includes the interstitial fluid plus the blood plasma.
3. Three advantages of closed circulatory systems over open systems are: (1) the circulatory fluid can flow more rapidly; (2) the diameters of vessels can be changed to control the flow of blood to particular tissues; and (3) specialized cells and molecules that transport oxygen, hormones, and nutrients are kept in the vessels and can be directed to particular locations.
4. Fishes have a two-chambered heart (one atrium and one ventricle), with blood flowing from the heart to the gills and then directly to the body. It does not return to the heart for additional pumping. Turtles have a more complex three-chambered heart (two atria and one ventricle). The ventricle, however, is partially divided, so outflow can be directed to the pulmonary circuit and to the systemic circuit. Thus the circulatory system of a fish delivers blood to the respiratory organ and the body tissues in succession, whereas the circulatory system of a turtle delivers blood to the respiratory organ and the body tissues in parallel.
5. During a dive, crocodilians are not breathing, so they shunt most blood away from their lungs (the pulmonary circuit) and to the body (the systemic circuit). When crocodilians stop breathing, vessels in the pulmonary circuit constrict, resistance increases in that circuit, and blood flows out of the aorta leaving the right ventricle rather than out of the pulmonary artery to the lungs. The key anatomical difference is that crocodilians have two vessels leaving their right ventricle: the pulmonary artery (to the lungs) and an aorta (to the body; there is a second aorta leaving the left ventricle). Birds have a single vessel leaving their right ventricle (the pulmonary artery to the lungs) and cannot shunt blood away from the lungs to the body.
6. The shunt in a crocodilian heart could be experimentally disabled by surgically sealing the aorta leaving the right ventricle. This procedure would force blood in the right ventricle to go out the pulmonary artery to the lungs. The surgical procedure could then be reversed by removing the sutures and allowing blood to flow through the aorta once again.
7. 

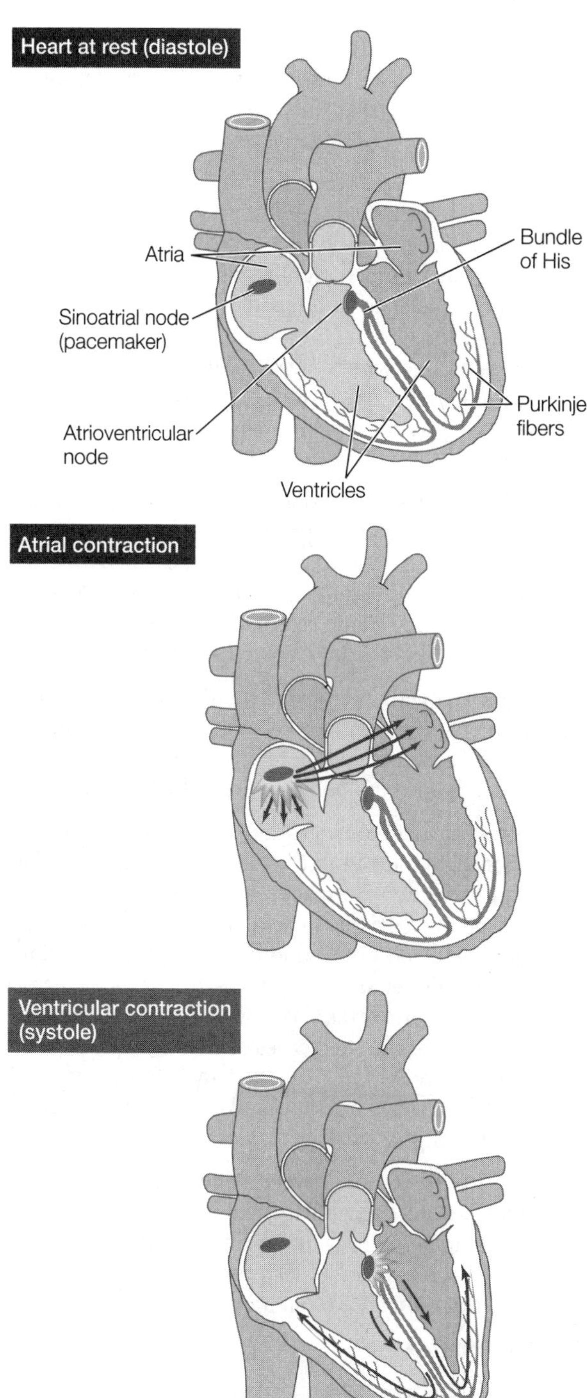

8. If the heart's conduction system is faulty, then cells can begin to contract independently. This cellular independence can cause rapid, irregular contractions of the ventricles, which in turn, could make the ventricles useless as pumps and circulation would stop. A defibrillator can be used to electrically shock the heart and prompt the SA node to function normally.

9. The atrioventricular (AV) node, stimulated by atrial depolarization, generates action potentials that pass to the ventricles. Heart block results from damage to the AV node.
10. The "lub" sound represents closing of the atrioventricular valves at the beginning of ventricular systole. Thus, an abnormal "lub" sound suggests problems with an atrioventricular valve.
11. Blood is pumped from the heart into the arteries, where it has a relatively fast velocity. Upon entering the capillaries, much of the velocity is lost. The capillaries have a much higher total cross-sectional area than the arteries, and blood flow velocity is inversely proportional to the cross-sectional area; thus, the more area through which the blood flows, the more slowly it will go. The slow velocity of blood through the capillaries allows for greater exchange of materials with the tissues.
12. During times of blood loss, the body might speed its rate of red blood cell production. In response to a decreased supply of oxygen to the cells, the kidneys release erythropoietin, a hormone that extends the lives of mature red blood cells and stimulates production of new red blood cells in the bone marrow.
13. A person with anemia has an insufficient number of oxygen-transporting red blood cells. An anemic person's heart rate will therefore become faster to compensate for the blood's reduced ability to carry oxygen.
14. Stress and nervousness elicit the fight-or-flight response. Signals from the medullary cardiovascular control center stimulate sympathetic inputs from the autonomic nervous system. In response to sympathetic inputs, the adrenal gland releases epinephrine into the bloodstream. Epinephrine stimulates an increased heart rate and arterial pressure. The blood flow to the smooth muscles decreases due to constriction of blood vessels, and blood is diverted away from areas such as the digestive tract to the skeletal muscle needed for fight or flight.
15. In response to low oxygen, high carbon dioxide, and high lactate, precapillary sphincters would open, increasing blood flow to the area. Arterioles would dilate. These actions would decrease total peripheral resistance and mean arterial pressure.

Test Yourself

1. **a.** Both closed systems and open systems rely on pumping mechanisms to distribute fluid throughout the body.
2. **c.** During nonbreathing periods, the blood is shunted from the pulmonary circuit to the systemic circuit. The shunt results from an increase in resistance in the pulmonary circuit.
3. **d.** Of the vertebrate groups, mammals and birds tend to have the highest blood pressures, and the ventricle supplying blood to the systemic circuit has higher pressures than the ventricle supplying the pulmonary circuit. Therefore, the ventricle supplying blood to the systemic circuit of a bird has the highest pressure of those listed.
4. **e.** In fishes, blood passes from the heart to the gills and directly out to the body; after gas exchange at the gills, blood does not return to the heart for additional pumping.
5. **e.** The cardiac action potential passes through the SA node, atrial fibers, AV node, bundle of His, and finally the Purkinje fibers.
6. **c.** The AV node delays the ventricular depolarization relative to atrial depolarization, so atrial contraction occurs before ventricular contraction. The Purkinje fibers ensure that the action potential spreads rapidly and evenly throughout the ventricles.
7. **b.** The atrium has thinner walls and generates lower pressures than the ventricles.
8. **c.** The pressure generated by the left ventricle in the blood flowing to the systemic circuit is greater than the pressure generated by the right ventricle in blood flowing to the pulmonary circuit.
9. **c.** The fluid portion of blood is the plasma; three components of the solid fraction are erythrocytes, platelets, and leukocytes.
10. **a.** Red blood cells are biconcave cells that contain the oxygen-binding hemoglobin.
11. **d.** Blood clotting involves the conversion of prothrombin to thrombin.
12. **b.** Gas exchange at the tissues occurs across the capillaries.
13. **a.** The lymph nodes filter and destroy microorganisms that are traveling through the lymphatic system.
14. **c.** Changes in blood pressure will change the amount of filtration that occurs at the capillaries. Thus, increases in blood pressure will result in greater rates of filtration.
15. **d.** Angiotensin and vasopressin control constriction of the arterioles, not the venules.

Nutrition, Digestion, and Absorption

The Big Picture

- Animals have evolved diverse mechanisms for acquiring energy and raw materials for metabolism. All animals are heterotrophs, acquiring nutrients and energy by eating plants, animals, or both. An animal's digestive system is specialized for the type of food it must break down. The relatively short guts of carnivores have acid and protein-reducing enzymes. The longer guts of herbivores have numerous holding chambers for the laborious process of breaking down plant material. The teeth of carnivores and herbivores also reflect dietary differences.
- Food passing down the length of the gut is first broken down mechanically into small pieces and then broken down chemically into small molecules that can be easily absorbed across the gut. As nutrients and water are absorbed from ingested plant and animal material, the remaining material forms feces, which are eventually eliminated from the body.
- The process of digestion is under both neural and hormonal control. The brain and hormones regulate food intake, and the digestive tract has an intrinsic nervous system to regulate digestive functions. Additional hormones control the peristaltic movements of the gut and the secretion of enzymes and other chemicals needed in digestion.

Study Strategies

- Although the guts of higher animals are structurally complex, they can be divided functionally into a foregut, midgut, and hindgut. In your study of digestive structures and their functions, it can be helpful to break down the gut into these sections and determine which physiological process occurs in each section.
- Because different types of food are digested and absorbed by different mechanisms, you should learn how a particular type of food is digested and be able to match the appropriate enzymes to that process. For example, it will be much easier to remember how proteins are digested, rather than to recount a long list of digestive enzymes and then try to pick the one that might break down protein. In short, learn the process, not the list!
- Go to the Web addresses shown to review the following animated tutorials and activities. You can also find them in BioPortal (yourBioPortal.com), along with many additional learning resources.

  Animated Tutorial 51.1 The Digestion and Absorption of Fats (Life10e.com/at51.1)

  Animated Tutorial 51.2 Insulin and Glucose Regulation (Life10e.com/at51.2)

  Animated Tutorial 51.3 Parabiotic Mice (Life10e.com/at51.3)

  Activity 51.1 Mineral Elements Required by Animals (Life10e.com/ac51.1)

  Activity 51.2 Vitamins in the Human Diet (Life10e.com/ac51.2)

  Activity 51.3 Mammalian Teeth (Life10e.com/ac51.3)

  Activity 51.4 The Human Digestive System (Life10e.com/ac51.4)

Key Concept Review

51.1 What Do Animals Require from Food?

Energy needs and expenditures can be measured

Sources of energy can be stored in the body

Food provides carbon skeletons for biosynthesis

Animals need mineral elements for a variety of functions

Animals must obtain vitamins from food

Nutrient deficiencies result in diseases

Animals are heterotrophs, meaning that they must get their energy and nutrients from the fats, carbohydrates, and proteins of plants and other animals. The energy that animals acquire from food is used to make ATP. The conversion of ATP during cellular work creates heat as a by-product. The calorie and kilocalorie (Calorie) are measures of heat energy. Metabolic rate is a measure of the total energy used by an animal. Basal metabolic rate is the resting metabolic rate or the energy consumption needed for all essential physiological functions. Energy needed for metabolism can be obtained from food taken in or from stored food. Fats, carbohydrates, and proteins are the components of food that provide energy. Fat has the highest energy content and is the most important form of energy stored in an animal's body. Carbohydrates are stored as glycogen in the liver and muscles, and structural proteins are used for energy as a last resort (see Figure 51.3). An animal is undernourished when it takes in too little food to meet its energy requirements and overnourished when it consistently takes in more food than it needs.

Animals must take in certain organic molecules. The acetyl group supplies the carbon skeleton for larger organic molecules (see Figure 51.4). Amino acids needed to make proteins are obtained by breaking down proteins in food. Some amino acids can be synthesized, but the essential amino acids can only be obtained from outside sources. Essential fatty acids, such as linoleic acid, must also be acquired in food and are necessary for producing membrane phospholipids. The nutrients needed for survival are categorized as either macronutrients or micronutrients, depending on the quantity needed (see Table 51.1). Vitamins are carbon compounds that are required in small amounts and often function as coenzymes. Humans require 13 vitamins. Malnutrition is caused by the lack of any essential nutrient in the diet, and when chronic, leads to deficiency diseases (see Table 51.2).

Question 1. Ingested food typically contains many useful proteins. However, these proteins are not used in the form in which they are ingested but are broken down into smaller peptides and amino acids. Why does the digestive system break down these ingested proteins?

Question 2. From a nutritional standpoint, what is the best way to cook a vegetable containing folic acid—boiling it in water or steaming it?

Question 3. What are some of the likely reasons that basal metabolic rate is higher in human males than in females?

Question 4. Why would taking megadoses of vitamin A be potentially more risky for your health than taking megadoses of niacin?

51.2 How Do Animals Ingest and Digest Food?

- The food of herbivores is often low in energy and hard to digest
- Carnivores must find, capture, and kill prey
- Vertebrate species have distinctive teeth
- Digestion usually begins in a body cavity
- Tubular guts have an opening at each end
- Digestive enzymes break down complex food molecules

Animals acquire nutrition in diverse ways. Two types of organisms feed on dead matter: saprobes (also called saprotrophs) absorb nutrients from decaying organic matter, whereas detritivores (also called decomposers) actively feed on dead matter. Predators feed on other organisms, and there are three main types: herbivores feed on plants, carnivores feed on animals, and omnivores feed on plants and animals. Vertebrates use their teeth to acquire and process food. Teeth are composed of a hard outer surface (enamel), a bony layer (dentine), and a pulp cavity that contains blood vessels and nerves (see Figure 51.6A). Mammals have four types of teeth: incisors, canines, premolars, and molars. The shapes, numbers, and organization of mammalian teeth reflect different diets (see Figure 51.6B).

Digestion typically occurs in either a cavity, such as a gastrovascular cavity, or a tube. In simple animals with a gastrovascular cavity, a single opening functions as both a mouth and an anus, connecting the cavity to the environment. More complex animals have tubular guts with a mouth that takes in food and an anus through which digestive wastes are eliminated (see Figure 51.7). Food enters the mouth and, in some animals, is broken up by grinding mechanisms. After grinding, food moves to a storage chamber, such as the stomach or crop, where digestion may or may not occur. It then enters the midgut or intestine, where most nutrients are absorbed. The absorptive surfaces have increased surface area in the form of fingerlike projections known as villi, which in turn have microscopic projections called microvilli (see Figure 51.8). The hindgut absorbs water and ions and stores waste until it is expelled from the anus during defecation. Enzymes break down the macromolecules of food. Most enzymes are produced in an inactive form called a zymogen and are converted to the active form at the site of use. Enzymes do not digest the gut because of a protective layer of mucus that lines it.

Question 5. Some mammals have very high-crowned cheek teeth (premolars and molars). Predict whether these animals are carnivores, herbivores, or omnivores. What is the basis for your prediction?

Question 6. Distinguish between the stomach and small intestine with respect to location in the gut and function in digestion.

51.3 How Does the Vertebrate Gastrointestinal System Function?

- The vertebrate gut consists of concentric tissue layers
- Mechanical activity moves food through the gut and aids digestion
- Chemical digestion begins in the mouth and the stomach
- The stomach gradually releases its contents to the small intestine
- Most chemical digestion occurs in the small intestine
- Nutrients are absorbed in the small intestine
- Absorbed nutrients go to the liver
- Water and ions are absorbed in the large intestine
- Herbivores rely on microorganisms to digest cellulose

Four layers of different cell types form the wall of the gut (see Figure 51.10). The innermost layer is the mucosa. Cells of the mucosal epithelium have secretory functions; some secrete mucus whereas others secrete enzymes, hormones, or hydrochloric acid. In some regions of the gut, cells of this layer have absorptive functions. The next layer is the submucosa, which contains blood and lymph vessels for nutrient transport. This layer also contains a network of nerves that regulates gut activities. Smooth muscle surrounds the submucosa in circular and longitudinal layers. These layers help move the contents of the gut by peristalsis. The peritoneum lines the abdominal cavity and covers the gut and other abdominal organs.

Food enters the mouth where the enzyme amylase, secreted by the salivary glands, begins the breakdown of starch. Food then passes through the pharynx into the esophagus and then into the stomach (see Figure 51.11). In the stomach, the major enzyme is pepsin, which breaks proteins into shorter peptides. Pepsin is initially secreted by gastric glands of the stomach in the zymogen form, pepsinogen (see Figure 51.12). Low pH activates the conversion of pepsinogen to pepsin through autocatalysis, a positive feedback process. The secretion of hydrochloric acid by the gastric glands in the wall of the stomach maintains a low pH. Small amounts of a few substances are absorbed across the walls of the stomach. The acidic mixture of digesting food and gastric juices, called chyme, slowly passes into the small intestine, where carbohydrate and protein digestion continue. Fat digestion begins in the small intestine with the help of secretions from the liver (bile) and pancreas (several enzymes, initially secreted as zymogens). The pancreas also helps maintain a slightly alkaline pH in the small intestine by secreting bicarbonate ions. This helps neutralize the acidic chyme entering from the stomach. Most nutrients are absorbed in the small intestine. The hepatic portal vein carries blood from the digestive organs to the liver, where nutrients are either stored or converted to needed molecules. Material entering the large intestine has had most of the nutrients removed, but still contains important ions and water. The water and ions are absorbed across the large intestine, leaving behind semisolid feces.

Although cellulose is the main component of plants, most herbivores cannot produce cellulase enzymes to break down cellulose. Instead, herbivores rely on microorganisms in their digestive tracts to digest it. Ruminants have a large four-chambered stomach in which fermentation occurs (see Figure 51.15). The rumen and reticulum, the first two chambers, contain microorganisms that break down cellulose. The contents of the rumen are periodically regurgitated into the mouth for rechewing; this process is often described as "chewing the cud." Food eventually passes through the omasum, where water absorption occurs, and into the abomasum, or true stomach. In the abomasum, hydrochloric acid and proteases kill the microorganisms and digest them. Some herbivores, such as rabbits, have a microbial fermentation chamber, called the cecum, extending from the large intestine. Many of these species produce two types of feces, one that is pure waste and the other that still contains some nutrients. Coprophagy is the reingestion of this second type of feces to digest and absorb additional nutrients.

Question 7. Suppose you have just eaten pizza with a lot of cheese. Because this food is high in lipids, your digestive system will have to perform specific tasks to digest and absorb the fats. What are the steps that your digestive system will take to digest and absorb the fats in this meal?

Question 8. In ruminants such as cows and bison, microorganisms that break down cellulose are found in two chambers of the greatly enlarged stomach. In other herbivores, including rabbits, the microorganisms that break down cellulose are found in the cecum, a chamber off the large intestine. In which type of herbivore would the absorption of nutrients from cellulose be more efficient?

Question 9. Create a flow chart of the following organs, indicating the order in which they occur from the start to the end of the human gut: small intestine, anus, esophagus, large intestine, mouth, and stomach. Indicate in which of these organs chemical digestion occurs, and the particular nutrient (e.g., fat, protein) being broken down. Also indicate in which of the organs absorption occurs, and the particular nutrient or substance absorbed.

51.4 How Is the Flow of Nutrients Controlled and Regulated?

Hormones control many digestive functions

The liver directs the traffic of the molecules that fuel metabolism

The brain plays a major role in regulating food intake

Salivation and swallowing are unconscious reflexes. The digestive tract has an intrinsic nervous system that coordinates motility and digestive activities. Hormones control many digestive functions (see Figure 51.16). For example, the duodenum of the small intestine releases secretin, which stimulates the pancreas to release bicarbonate ions into the small intestine to control pH. Fats and proteins stimulate the mucosa of the small intestine to secrete cholecystokinin, which stimulates the release of bile and pancreatic digestive enzymes. The stomach releases gastrin into the blood, where it is transported to the upper regions of the stomach to stimulate stomach secretions and movement.

When fuel molecules are abundant in the blood, the liver stores them as glycogen or fat. When fuel molecules are scarce in the blood, the liver delivers them back into the blood. The liver can convert monosaccharides into glycogen or fat, and vice versa. The liver also can convert amino acids and certain other molecules into glucose; this process is called gluconeogenesis. The liver controls fat metabolism through its production of lipoproteins. Very low-density lipoproteins (VLDLs) contain mostly triglyceride fats, which they transport to cells of adipose tissue; these lipoproteins are only 3 percent cholesterol. Low-density lipoproteins (LDLs) contain 50 percent cholesterol, which they carry around the body for storage or use. High-density lipoproteins (HDLs) remove cholesterol from tissues and carry it to the liver for bile synthesis; they are 15 percent cholesterol. LDLs and VLDLs are associated with cardiovascular disease.

The absorptive period is the period after a meal when digestion occurs. The postabsorptive period is the period when the gut is empty and the body uses energy reserves. During the absorptive period, the liver takes up glucose from the blood and converts it to glycogen and fat, fat cells take up glucose, and the body preferentially uses glucose as the energy source. During the postabsorptive period, the liver breaks down glycogen into glucose, fatty acids are supplied to the blood by adipose tissues, and the body preferentially uses fatty acids as an energy source. The pancreatic hormones insulin and glucagon control metabolic fuel use (see Figure 51.17). Regulation of food intake involves the hypothalamus and signals from hormones (such as leptin and ghrelin) and enzymes (such as AMP-activated protein kinase).

Question 10. During the absorptive period, the body digests and absorbs nutrients, whereas during the postabsorptive period, the body uses stored fuel. What are the main differences in fuel use and control mechanisms during these two periods?

Question 11. You have just discovered a new hormone that may control appetite. Design at least two methods to experimentally determine its effects in rats.

Question 12. Analyze why salivation is controlled by nervous stimulation rather than by hormonal stimulation.

Test Yourself

1. Certain amino acids are essential to the diet of animals because they
   a. prevent overnourishment.
   b. are cofactors and coenzymes that are required for normal physiological function.
   c. cannot be directly synthesized by animals through the transfer of an amino group to an appropriate carbon skeleton.
   d. are needed to make stored fats that are used during hibernation and migration.
   e. are the most important source of stored energy.
   Textbook Reference: *51.1 What Do Animals Require from Food?*

2. Which of the following statements about sheep digestion is true?
   a. Sheep are saprobes; they engulf food and perform intracellular digestion.
   b. Sheep are autotrophs; they synthesize organic nutrients and perform extracellular digestion.
   c. Sheep are herbivores; they ingest food and perform extracellular digestion.
   d. Sheep are detritivores; they ingest food and perform intracellular digestion.
   e. Sheep are omnivores; they ingest plants and animals and exhibit intracellular digestion.
   Textbook Reference: *51.2 How Do Animals Ingest and Digest Food?*

3. Which of the following statements about mammalian teeth is true?
   a. Enamel covers the crown and the root of a tooth.
   b. Teeth lack nerves and blood vessels.
   c. The teeth of omnivores are more specialized than the teeth of carnivores and herbivores.
   d. Carnivores have large canines, whereas herbivores have large premolars and molars.
   e. Dentine is restricted to the crown of a tooth.
   Textbook Reference: *51.2 How Do Animals Ingest and Digest Food?*

4. Which of the following does *not* contribute to the large surface area available for nutrient absorption in the small intestines of humans?
   a. Villi
   b. Intestinal length
   c. Microvilli
   d. Typhlosole
   e. Surface folds
   Textbook Reference: *51.2 How Do Animals Ingest and Digest Food?*

5. Which of the following protects the walls of the stomach against the action of its own digestive juices?
   a. An antienzyme chemical formed by the gastric glands
   b. The nervous reactions of the lining of the stomach
   c. Control by a center in the medulla of the brain
   d. Mucus coating on the stomach's inner surface
   e. Alkaline secretions of the gastric glands that neutralize acidic secretions
   Textbook Reference: *51.3 How Does the Vertebrate Gastrointestinal System Function?*

6. Chylomicrons are produced in the
   a. mouth.
   b. stomach.
   c. lumen of the small intestine.
   d. epithelial cells of the small intestine.
   e. liver.
   Textbook Reference: *51.3 How Does the Vertebrate Gastrointestinal System Function?*

7. The gallbladder
   a. produces bile.
   b. is part of the liver.
   c. stores bile produced by the liver.
   d. produces cholecystokinin.
   e. is vestigial in humans.
   Textbook Reference: *51.3 How Does the Vertebrate Gastrointestinal System Function?*

8. The pancreas
   a. is exclusively an endocrine gland responsible for the production and release of insulin and glucagon.
   b. produces salivary amylase.
   c. contains villi to increase surface area.
   d. secretes cholecystokinin.

e. produces exocrine products involved in chyme digestion.

Textbook Reference: *51.3 How Does the Vertebrate Gastrointestinal System Function?*

9. Hydrochloric acid
   a. is secreted by the gastric glands of the liver.
   b. is secreted by parietal cells of the stomach.
   c. produces a low pH in the small intestine.
   d. promotes the growth of microorganisms in the stomach.
   e. is secreted by chief cells of the stomach.

   Textbook Reference: *51.3 How Does the Vertebrate Gastrointestinal System Function?*

10. Bile produced in the liver is associated with which of the following?
    a. Emulsification of fats into tiny globules in the small intestine
    b. Digestive action of pancreatic amylase
    c. Emulsification of fats into tiny globules in the stomach
    d. Digestion of proteins into amino acids
    e. Emulsification of fats into tiny globules in the large intestine

    Textbook Reference: *51.3 How Does the Vertebrate Gastrointestinal System Function?*

11. Most of the chemical digestion of food in humans is completed in the
    a. mouth.
    b. appendix.
    c. ascending colon.
    d. stomach.
    e. small intestine.

    Textbook Reference: *51.3 How Does the Vertebrate Gastrointestinal System Function?*

12. Waves of muscle contractions that move the intestinal contents are
    a. caused by contraction of skeletal muscle.
    b. regulated by liver secretions.
    c. called peristalsis.
    d. voluntary.
    e. caused by the relaxation of smooth muscles of the gut in response to their being stretched.

    Textbook Reference: *51.3 How Does the Vertebrate Gastrointestinal System Function?*

13. What is the function of enterokinase?
    a. It converts pepsinogen to pepsin.
    b. It converts trypsinogen to trypsin.
    c. It digests proteins.
    d. It activates HCl secretion.
    e. It prompts opening of the lower esophageal sphincter.

    Textbook Reference: *51.3 How Does the Vertebrate Gastrointestinal System Function?*

14. Digestive enzymes responsible for breaking down disaccharides include
    a. pepsin, trypsin, and trypsinogen.
    b. amylase, pepsin, and lipase.
    c. sucrase, lactase, and maltase.
    d. pepsin, trypsin, and chymotrypsin.
    e. nuclease and aminopeptidase.

    Textbook Reference: *51.3 How Does the Vertebrate Gastrointestinal System Function?*

15. Which of the following statements about the large intestine is true?
    a. It has almost no bacterial populations.
    b. It contains chyme.
    c. It absorbs much of the water remaining in waste materials.
    d. It is the site of most of digestion.
    e. It receives blood from the digestive tract via the hepatic portal vein.

    Textbook Reference: *51.3 How Does the Vertebrate Gastrointestinal System Function?*

16. The innermost layer of the digestive tract is the
    a. peritoneum.
    b. mucosa membrane.
    c. submucosa membrane.
    d. lumen.
    e. circular muscle.

    Textbook Reference: *51.3 How Does the Vertebrate Gastrointestinal System Function?*

17. Which of the following hormones stimulates gluconeogenesis?
    a. Glucagon
    b. Insulin
    c. Estrogen
    d. Secretin
    e. Gastrin

    Textbook Reference: *51.4 How Is the Flow of Nutrients Controlled and Regulated?*

18. Which of the following statements about lipoproteins is *false?*
    a. Lipoproteins move fats from storage sites to sites where they are used.
    b. Lipoproteins have a hydrophobic core of cholesterol and fat.
    c. High-density lipoproteins are "good" lipoproteins.
    d. Low-density lipoproteins are "bad" lipoproteins.
    e. The liver is solely responsible for synthesis of lipoproteins.

    Textbook Reference: *51.4 How Is the Flow of Nutrients Controlled and Regulated?*

Answers

Key Concept Review

1. Proteins in digested food are not absorbed and used in their original form for several reasons. First, the proteins are too large to be easily absorbed by the cells of the digestive tract, whereas amino acids are readily absorbed. Second, proteins are species-specific, and thus a protein that is optimal for a prey species may not be optimal for the predator species. Finally, the proteins of another species would be considered foreign and attacked by the immune system.
2. Folic acid is a water-soluble vitamin found in vegetables. When vegetables are boiled, the folic acid is lost. Therefore, the best cooking method is steaming.
3. Human males have higher basal metabolic rates than females do because muscle uses more energy than fat does and a male's body typically has more muscle and less fat than a female's. The difference cannot be explained by any potential differences in activity, because the basal metabolic rate is a measure of the minimum energy required at rest.
4. Niacin is a water-soluble vitamin, so excesses are eliminated in the urine. Vitamin A, however, is fat-soluble and when taken in excess, it can accumulate in body fat and reach toxic levels in the liver.
5. Mammals, such as horses, have very high-crowned cheek teeth. Even though enamel is very hard, the teeth of horses wear down over time because of their abrasive grass diet. Animals with high-crowned cheek teeth are herbivores. Carnivores and omnivores have low-crowned cheek teeth.
6. The stomach is in the foregut and functions primarily in the physical breakdown and storage of food. The small intestine is in the midgut and is the main site for enzymatic digestion of food and nutrient absorption.
7. The digestion of fats begins in the small intestine. Lipases are water soluble, but lipids tend to aggregate into large droplets in an aqueous environment. Bile helps to increase the exposed surface area of the lipids by preventing the formation of large lipid droplets and promoting small micelles. This allows lipases to break down the fats into free fatty acids and monoglycerides. The fatty acids dissolve into the plasma membrane of the intestinal epithelial cells and are absorbed across this surface. In mucosal cells, the free fatty acids and monoglycerides are turned into triglycerides, which are incorporated into chylomicrons. The chylomicrons can then pass out of the mucosal cells and into the lymphatic system (see Figure 51.14).
8. In vertebrates, most nutrients are absorbed in the small intestine. Because ruminants break down cellulose in chambers of their stomach (which comes before the small intestine in position along the gut), the nutrients from cellulose are available for absorption in the small intestine and absorption is relatively efficient. In non-ruminant herbivores, however, cellulose is broken down in the cecum after it has passed through the small intestine, so nutrient absorption is less efficient. Some non-ruminant species produce two types of feces and eat the one containing cecal material to gain greater access to nutrients.
9.

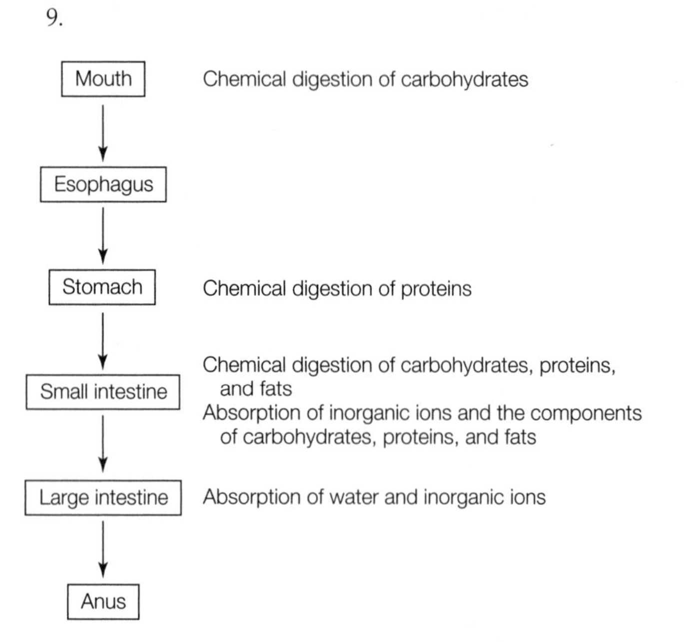

10. When an animal's body is in the absorptive state, it attempts to store nutrients and fuel. The preferred fuel source at this time is glucose, and insulin is released to help facilitate glucose uptake by cells. When the body is in the postabsorptive state, cells tend to metabolize fats, keeping the blood glucose reserves for the nervous system. Insulin levels are low during this period.
11. There are several methods for determining the function of the newly discovered hormone. One approach is to simultaneously monitor levels of the hormone and feeding behavior of the rats. The goal here is to look for changes in feeding behavior that parallel fluctuations in the level of the hormone. A second approach is to inject rats with the hormone and then monitor their feeding behavior. A third approach is to block the effects of the hormone, possibly with a drug that temporarily and reversibly suppresses its action, and then to monitor feeding behavior. In all three approaches, body mass would likely be tracked as well.
12. Food does not spend much time in the mouth, so salivation must occur quickly to be effective. Salivation is under neural control because nervous stimulation is generally faster than hormonal stimulation. Hormones come into play in areas of the gut where food stays for longer periods of time.

Test Yourself

1. **c.** Essential amino acids must be acquired through diet because an animal cannot directly synthesize all of the amino acids needed for protein production.
2. **c.** Sheep are herbivores that ingest plant material for nutrition. Digestion occurs extracellularly in a sheep's digestive tract.
3. **d.** Carnivores have large canines for holding and killing prey animals. Herbivores have large premolars and molars for grinding plant material.
4. **d.** The length of the intestines, surface folds, and the presence of microvilli and villi increase the surface area of the small intestine. The typhlosole does not; it is found in earthworms.
5. **d.** The stomach is protected from digestive enzymes and low pH by the mucus secreted over its inner surface.
6. **d.** The epithelial cells of the small intestine produce chylomicrons by combining triglycerides with cholesterol and phospholipids, allowing lipids to pass into the lymphatic system.
7. **c.** Bile is produced by the liver, stored in the gallbladder, and released into the small intestine to aid in lipid digestion.
8. **e.** The pancreas produces exocrine products such as lipases, nucleases, amylases, and trypsin, all of which are involved in chyme digestion. Note that it also produces hormones, so it has both endocrine and exocrine functions.
9. **b.** Hydrochloric acid is a strong acid secreted by parietal cells of the stomach. It lowers the pH of the stomach fluid.
10. **a.** Bile aids in the digestion of lipids in the small intestine by changing fat droplets into small fat particles.
11. **e.** The small intestine is the main site for chemical digestion in humans.
12. **c.** Food is moved through the digestive tract by wavelike contractions of smooth muscle called peristalsis.
13. **b.** Enterokinase converts the zymogen trypsinogen to trypsin.
14. **c.** The digestive enzymes sucrase, lactase, and maltase are involved in the breakdown of disaccharides into the monosaccharides glucose, galactose, and fructose.
15. **c.** The large intestine is the site of water and ion absorption. The large populations of bacteria found in the large intestine contribute useful vitamins to their hosts.
16. **b.** The membranes of the digestive tract are, from the inside to the outside: mucosa, submucosa, and circular and longitudinal muscles. The peritoneum surrounds the digestive tract.
17. **a.** Glucagon increases the rate of gluconeogenesis, and insulin reduces the rate of gluconeogenesis.
18. **e.** Chylomicrons are lipoproteins synthesized in the cells of the small intestine. Thus, the liver is not the sole producer of lipoproteins.

Salt and Water Balance and Nitrogen Excretion

The Big Picture

- Different animals experience different types of stresses related to water and salt balance. Animals living in marine environments have to overcome problems with water loss, whereas animals living in freshwater environments have to overcome problems with water gain. Some invertebrates circumvent these problems by allowing their body fluid ion concentration to reflect that of their surroundings. Other invertebrates, and almost all vertebrates, retain their body fluids at more constant levels, expending energy to eliminate excess water and/or salts.
- All animals produce metabolic wastes (notably nitrogenous waste) as a by-product of metabolism. Invertebrates have evolved a variety of structures to eliminate wastes, whereas the vertebrates have evolved variations on the kidney. The nephron is the "functional unit" of the vertebrate kidney. By eliminating wastes without losing valuable nutrients and salts, the kidneys cleanse the blood of metabolic waste products.
- Kidneys work by differentially processing nitrogenous wastes, ions, and water. In the mammalian kidney, large volumes of plasma are filtered from the blood into the interior of nephrons. As the fluid passes through the various specialized regions of the nephron, the desirable components to be retained (specific ions, nutrients, and water) are pulled back into the blood, leaving behind the waste products in an ever more concentrated form of urine. As urine leaves the nephron, water content is adjusted through hormonal alteration of the water permeability of collecting ducts. If the collecting ducts are highly water permeable, water will be drawn by osmosis out of the collecting ducts into the region of high salt concentration created by the action of the loop of Henle.

Study Strategies

- Depending on whether they are living in fresh water or salt water, fishes may excrete copious amounts of dilute urine or small amounts of more concentrated urine. Given that living in either of these environments creates osmoregulatory challenges, think about whether the tissues have higher or lower solute concentration relative to the surrounding water, and then predict water and ion movements on that basis.
- Understanding how the nephron functions, and in particular how the loop of Henle works as a "countercurrent multiplier," is one of the greater challenges in this chapter. Nephron function can be understood more easily if you place yourself in the position of a single Na^+ ion in the blood entering the nephron. Follow your pathway—and your potential pathways—as you traverse the nephron, eventually ending up in the urine. Note especially that it may take you some time to pass through the loop of Henle as you leave the ascending limb, only to recycle back into the descending limb—a process that could be repeated many times before finally escaping the countercurrent multiplier. Repeat this imaginary journey again as a water molecule.
- Go to the Web addresses shown to review the following animated tutorials and activities. You can also find them in BioPortal (yourBioPortal.com), along with many additional learning resources.

  Animated Tutorial 52.1 The Mammalian Kidney (Life10e.com/at52.1)

  Animated Tutorial 52.2 Kidney Regulation (Life10e.com/at52.2)

  Activity 52.1 Annelid Metanephridia (Life10e.com/ac52.1)

  Activity 52.2 The Vertebrate Nephron (Life10e.com/ac52.2)

  Activity 52.3 The Human Excretory System (Life10e.com/ac52.3)

  Activity 52.4 The Major Organ Systems (Life10e.com/ac52.4)

Key Concept Review

52.1 How Do Excretory Systems Maintain Homeostasis?

Water enters or leaves cells by osmosis

Excretory systems control extracellular fluid osmolarity and composition

Aquatic invertebrates can conform to or regulate their osmotic and ionic environments

Vertebrates are osmoregulators and ionic regulators

Excretory organs control the volume, composition, and concentration of extracellular fluids. Water movement in the body occurs either by pressure differences or differences in solute potential that produce water movement by osmosis. There is no active transport of water in excretory systems. Excretory systems control water and solute balance by means of filtration across membranes, reabsorption, and secretion.

Osmolarity is the number of moles of osmotically active solutes per liter of solvent. Osmoconformers do not actively regulate the osmolarity of their tissues; rather, they allow them to come to equilibration with their environment. Marine invertebrates tend to be osmoconformers unless the concentrations in the environment become extreme. Osmoregulators actively regulate the osmolarity of their tissues. The brine shrimp maintains its tissue fluid osmolarity below that of the environment in high osmolarity waters and above that of the environment in low osmolarity waters (see Figure 52.1). Ionic conformers allow the ionic composition of their extracellular fluid to match that of the environment. Ionic regulators employ active transport mechanisms to keep ions at optimal concentrations in their extracellular fluid. Almost all vertebrates are osmoregulators and ionic regulators. Exceptions include hagfishes, which are osmosconformers and ionic conformers, and chondrichthyans, which are osmoconformers but not ionic conformers.

Question 1. Why are terrestrial animals always osmoregulators?

Question 2. Why do osmoconformers inhabit marine environments but not freshwater environments?

Question 3. A scientist wishes to study whether and how animals adjust to the osmotic concentration of their environment. Suggest invertebrate species for the scientist to study; include osmoconformers and osmoregulators. Then do the same for vertebrate species.

52.2 How Do Animals Excrete Nitrogen?

Animals excrete nitrogen in a number of forms

Most species produce more than one nitrogenous waste

The end products of the metabolism of proteins and nucleic acids contain nitrogen and are called nitrogenous wastes. Ammonia (NH_3) is a highly toxic nitrogenous waste that either must be excreted or converted to the less toxic urea or uric acid (see Figure 52.3). Ammonotelic animals excrete ammonia to the aquatic environment, usually across gill membranes. Examples include most bony fishes and aquatic invertebrates. Ureotelic animals excrete nitrogenous waste as water-soluble urea. Examples include mammals, cartilaginous fishes, and most amphibians. To conserve water, uricotelic birds and reptiles excrete nitrogenous waste as insoluble uric acid. Insects are also uricotelic. Most animals produce more than one nitrogenous waste.

Question 4. Why do water-breathing animals typically excrete ammonia as their primary nitrogenous waste?

Question 5. Discuss two treatment strategies that could be pursued by a pharmacology lab working on the development of a new medication for gout.

Question 6. Ammonia, urea, and uric acid are three forms of nitrogenous waste. First, rank these forms from least to most nitrogen in a molecule. Second, rank these forms from least to most soluble in water. Finally, which of the three forms is most toxic?

52.3 How Do Invertebrate Excretory Systems Work?

The protonephridia of flatworms excrete water and conserve salts

The metanephridia of annelids process coelomic fluid

Malpighian tubules of insects use active transport to excrete wastes

Reflecting their diversity, invertebrates employ a variety of excretory mechanisms. Freshwater flatworms have protonephridia consisting of tubules and flame cells that conserve ions and excrete water to produce dilute urine (see Figure 52.4). Annelid worms filter blood across the capillaries into the coelom, where the fluid enters a metanephridium (a pair of metanephridia occurs in each segment) through a funnel-like opening called a nephrostome. Tubules of the metanephridia actively secrete and absorb various ions and end in nephridiopores, which open to the environment to excrete dilute urine containing nitrogenous wastes and other solutes (see Figure 52.5).

Malpighian tubules are the excretory organs of insects; they are located at the junction of the midgut and hindgut. Malpighian tubules actively transport uric acid and sodium and potassium ions from the extracellular fluid into the tubules, with water passively following (see Figure 52.6). In the hindgut and rectum, sodium and potassium ions are actively transported back to the extracellular fluid, with water following. Insects eliminate semisolid matter containing uric acid and other wastes.

Question 7. The excretory systems of flatworms and annelids share one major similarity. What is the similarity, and why are the excretory systems of insects different in this regard?

Question 8. Explain where filtration, reabsorption, and secretion occur in the excretory system of annelids.

52.4 How Do Vertebrates Maintain Salt and Water Balance?

Marine fishes must conserve water

Terrestrial amphibians and reptiles must avoid desiccation

Mammals can produce highly concentrated urine

The nephron is the functional unit of the vertebrate kidney

Blood is filtered into Bowman's capsule

The renal tubules convert glomerular filtrate to urine

Kidneys are the main excretory organs of vertebrates, and nephrons are the functional units of kidneys. Vertebrates exhibit diverse excretory adaptations. Freshwater fishes live in an environment hypotonic to their own body fluids, with water tending to move into the body by osmosis. To counter this, they excrete large amounts of dilute urine. Marine bony fishes maintain their extracellular fluids at one-third to one-half the osmolarity of seawater. They live in an environment hypertonic to their body fluids, so water tends to be drawn out of the body by osmosis. To cope with this environment, marine bony fishes produce a small amount of concentrated urine and actively secrete salts across the gills. Cartilaginous fishes, almost all of which are marine, allow urea and trimethylamine oxide concentrations to increase, which increases the osmolarity of their tissues. Cartilaginous fishes are not ionic conformers; they have a rectal gland that excretes salts. Amphibians living in or near fresh water produce large amounts of dilute urine, whereas those living in terrestrial environments have skin with reduced water permeability and may estivate, relying on dilute urine in their bladder. Reptiles lay shelled eggs, which allows them to live in a fully terrestrial environment (i.e., they do not have to return to water to reproduce). They have scaly, dry skin to decrease evaporative water loss, and they excrete nitrogenous wastes as uric acid with little water. Mammals display several adaptations for water conservation, including the ability to produce urine that is more concentrated than their extracellular fluids.

The nephron of vertebrate kidneys is composed of the renal tubule and two capillary beds, which are the glomerulus and the peritubular capillaries (see Figure 52.7). Three main processes lead to the formation of urine: filtration, tubular reabsorption, and tubular secretion. Filtration occurs at the glomerulus and tubular reabsorption and secretion occur along the renal tubule. Blood enters the glomerulus at the afferent arteriole and exits at the efferent arteriole. From the efferent arteriole the peritubular capillaries of the second capillary bed emerge and surround the tubule component of the nephron (see Figure 52.7). The renal tubule begins with Bowman's capsule, which encloses the glomerulus. Podocyte cells of Bowman's capsule wrap around the capillaries of the glomerulus. The endothelial walls of the capillaries in the glomerulus have pores that allow water and small molecules to leave the blood, enter Bowman's capsule, and form filtrate. Arterial blood pressure is the driving force for filtration at the glomerulus. The filtrate entering the renal tubule is similar to blood plasma, but as it moves through the tubules, ions and molecules are actively reabsorbed and secreted, thus altering the composition of the filtrate.

Question 9. In the diagram below, label the following structures: afferent arteriole, renal tubule, Bowman's space, Bowman's capsule, collecting duct, efferent arteriole, peritubular capillaries, and glomerulus. Also label each of the three boxes to indicate where the following processes occur in the nephron: filtration; reabsorption and secretion; excretion.

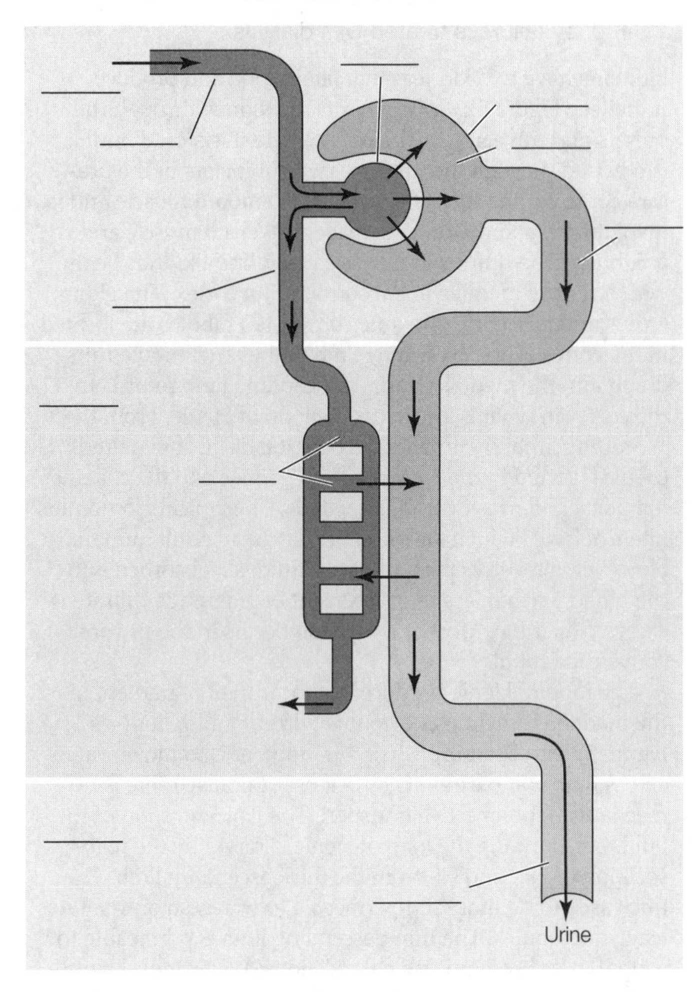

Question 10. Blood pressure drives glomerular filtration. Given what you know about the arrangement of blood vessels at the glomerulus, suggest two changes in these vessels that would increase glomerular filtration rate.

Question 11. Chondrichthyans and bony fishes that live in the ocean ingest salt with their food. How does each group deal with the excess salt taken in?

52.5 How Does the Mammalian Kidney Produce Concentrated Urine?

Kidneys produce urine and the bladder stores it

Nephrons have a regular arrangement in the kidney

Most of the glomerular filtrate is reabsorbed by the proximal convoluted tubule

The loop of Henle creates a concentration gradient in the renal medulla

Water permeability of kidney tubules depends on water channels

The distal convoluted tubule fine-tunes the composition of the urine

Urine is concentrated in the collecting duct
The kidneys help regulate acid–base balance
Kidney failure is treated with dialysis

Humans have two kidneys that filter blood and produce urine (see Figure 52.9). Urine exits the kidney through the ureters and travels to the bladder, where it is stored until it is released through the urethra. Two sphincters in the urethra control urination. One is made of smooth muscle and is involuntary, and the other is made of skeletal muscle and is voluntary. The human kidney is shaped like a kidney bean, with a central medulla and a surrounding cortex. The glomerulus and adjoining proximal convoluted tubules are located in the cortex. The descending limb of the renal tubule runs down into the medulla, and the ascending limb returns to the cortex in what is known as the loop of Henle. From the ascending limb, the tubules become the distal convoluted tubule (located in the cortex), which connects to the collecting ducts and runs down the medulla. The typical glomerulus filters about 180 liters of blood per day in an adult human. However, almost 99 percent of this fluid is reabsorbed into the blood, so that less than 2 percent of glomerular filtrate is excreted as urine. Most reabsorption occurs in the proximal convoluted tubules.

The loop of Henle produces a concentration gradient in the medulla by acting as a countercurrent multiplier (see Figure 52.10). The purpose of this gradient is to move water across fluid compartments by osmosis, because there is no mechanism for its active transport. The concentration of the fluids surrounding the loop of Henle is raised by the reabsorption of Na^+ and Cl^- from the thick ascending limb. The thick ascending limb is impermeable to water, so only solutes leave the tubules. The thin descending limb is permeable to water but not sodium and chloride ions. Water moves out of the tubules into the surrounding tissues due to the higher concentration of Na^+ and Cl^- in them. In the thin ascending limb, the fluid is more concentrated than the fluid in the surrounding tissues, so Na^+ and Cl^- move out of the tubule. Water cannot move out because of the low permeability of the thin ascending limb. The fluid reaching the distal convoluted tubule is less concentrated than blood plasma. Aquaporins, which are membrane proteins that form water channels, account for the high water permeability of some regions of the renal tubule. The collecting duct passes through the medulla, which has a high solute concentration set up by the loop of Henle. Water moves out of the collecting duct and into the tissue, resulting in concentrated urine. The permeability of the collecting ducts can be adjusted by hormones to regulate the amount of water excreted.

In addition to filtering blood, the kidneys help regulate blood pH. Bicarbonate ions are the major buffer in the blood. The kidneys remove H^+ from the blood and return bicarbonate ions to the blood (see Figure 52.11). Kidney (renal) failure severely disrupts homeostasis, causing high blood pressure (due to the retention of salts and water), uremic poisoning (due to the retention of urea), and acidosis (due to the retention of metabolic acids). A dialysis machine can replace kidney function until a kidney transplant can be arranged (see Figure 52.13).

Question 12. What would happen to the composition of excreted urine if all active transport processes in the nephron came to a halt?

Question 13. Where in the renal tubule does most reabsorption of water and solutes occur? What characteristics of the cells of the renal tubule make it ideal for reabsorption?

52.6 How Are Kidney Functions Regulated?

Glomerular filtration rate is regulated
Regulation of GFR uses feedback information from the distal tubule
Blood osmolarity and blood pressure are regulated by ADH
The heart produces a hormone that helps lower blood pressure

Glomerular filtration rate is a function of the blood flow and pressure to the kidneys. The kidneys have autoregulatory mechanisms that help maintain their high filtration rate. In response to decreasing blood pressure, the afferent renal arterioles open, or dilate, to increase flow to the glomerulus. If this does not increase glomerular filtration rate sufficiently, the kidneys release the enzyme renin into the blood. Renin converts an angiotensin precursor into active angiotensin. Angiotensin constricts the efferent renal arterioles to increase pressure through the glomerulus. Angiotensin also constricts peripheral blood vessels, stimulates the release of aldosterone from the kidney, and stimulates thirst (see Figure 52.14).

Several hormones influence kidney function. Aldosterone stimulates the reabsorption of sodium by the kidney, which increases water reabsorption and results in increased blood volume and pressure. Antidiuretic hormone (ADH) is produced in the hypothalamus and released from the posterior pituitary gland (see Figure 52.15). ADH controls the permeability of the collecting ducts to water by stimulating production of aquaporins over the long term, and insertion of aquaporins into the membrane in the short term. Stretch receptors in the aorta and carotid arteries inhibit the release of ADH in response to increased blood pressure, causing water to leave the body and blood volume to decrease. Osmosensors in the hypothalamus stimulate the release of ADH in response to increased osmolarity, resulting in increased water reabsorption and dilution of the blood. ADH also causes constriction of peripheral blood vessels. The heart releases atrial natriuretic peptide (ANP) in response to high blood volume and pressure. At the kidneys, this hormone decreases sodium reabsorption, resulting in decreased water reabsorption and a lowered blood volume and pressure.

Question 14. A certain drug inhibits antidiuretic hormone secretion from the pituitary. How does this change glomerular filtration rate, urine flow, and urine concentration in patients taking this medication?

Question 15. You have just eaten a large number of very salty potato chips and the osmolarity of your blood has increased. What response will your body have in order to bring your blood osmolarity back to homeostasis?

Question 16. You are interested in studying the autoregulatory mechanisms that keep glomerular filtration rate constant when blood pressure drops in rats. Create a list of variables that you might measure or monitor to determine whether autoregulatory mechanisms are functional.

Test Yourself

1. The sole mechanism for water reabsorption by the renal tubules is
   a. active transport.
   b. osmosis.
   c. cotransport with sodium ions.
   d. cotransport with bicarbonate ions.
   e. None of the above
   Textbook Reference: *52.1 How Do Excretory Systems Maintain Homeostasis?*

2. Which of the following vertebrates groups is an osmoconformer?
   a. Birds
   b. Freshwater bony fishes
   c. Hagfishes
   d. Mammals
   e. Marine bony fishes
   Textbook Reference: *52.1 How Do Excretory Systems Maintain Homeostasis?*

3. Which of the following statements about nitrogenous wastes is *false*?
   a. Chondrichthyans are ammonotelic.
   b. Most bony fishes are ammonotelic.
   c. Mammals are ureotelic.
   d. Birds are uricotelic.
   e. Although urea is the principal nitrogenous waste of humans, ammonia and uric acid also are produced.
   Textbook Reference: *52.2 How Do Animals Excrete Nitrogen?*

4. Which of the following statements about the excretory system of insects is *false*?
   a. Active transport moves uric acid and ions from the extracellular fluid into the Malpighian tubules.
   b. Water moves into the rectum for excretion.
   c. Reabsorption of salts takes place in the hindgut and rectum.
   d. Water reabsorption takes place by osmotic movement only.
   e. Insects excrete uric acid.
   Textbook Reference: *52.3 How Do Invertebrate Excretory Systems Work?*

5. In order to maintain homeostasis, a marine bony fish
   a. excretes small amounts of water.
   b. excretes large amounts of water.
   c. converts nitrogenous wastes to urea.
   d. excretes uric acid.
   e. produces urine that is more concentrated than its extracellular fluids.
   Textbook Reference: *52.4 How Do Vertebrates Maintain Salt and Water Balance?*

6. Which of the following represents the correct pathway of water and solutes traveling through a nephron?
   a. Glomerulus, Bowman's capsule, proximal tubule, loop of Henle, distal tubule, collecting ducts
   b. Bowman's capsule, glomerulus, distal tubule, loop of Henle, proximal tubule, collecting ducts
   c. Glomerulus, Bowman's capsule, distal tubule, loop of Henle, proximal tubule, collecting ducts
   d. Glomerulus, Bowman's capsule, proximal tubule, collecting ducts, distal tubule, loop of Henle
   e. Glomerulus, Bowman's capsule, proximal tubule, distal tubule, collecting ducts, loop of Henle
   Textbook Reference: *52.4 How Do Vertebrates Maintain Salt and Water Balance?*

7. Which of the following is *not* a normal constituent of the glomerular filtrate?
   a. Amino acids
   b. Urea
   c. Sodium ion
   d. Glucose
   e. Red blood cells
   Textbook Reference: *52.4 How Do Vertebrates Maintain Salt and Water Balance?*

8. Which of the following animals does *not* have an excretory system that conserves water?
   a. Mammal
   b. Bird
   c. Freshwater bony fish
   d. Marine bony fish
   e. Lizard
   Textbook Reference: *52.4 How Do Vertebrates Maintain Salt and Water Balance?*

9. As the glomerular filtrate passes through the renal tubule, about _______ percent is returned to the blood.
   a. 2
   b. 10
   c. 50
   d. 75
   e. 99
   Textbook Reference: *52.5 How Does the Mammalian Kidney Produce Concentrated Urine?*

10. NaCl is actively transported out of the renal tubules from _______ and _______.
    a. Bowman's capsule; the descending limb of the loop of Henle
    b. the descending limb of the loop of Henle; the thin ascending limb of the loop of Henle

c. the proximal convoluted tubule; the thick ascending limb of the loop of Henle
d. the descending limb of the loop of Henle; the collecting duct
e. the proximal convoluted tubule; the thin ascending limb of the loop of Henle

Textbook Reference: *52.5 How Does the Mammalian Kidney Produce Concentrated Urine?*

11.–15. Refer to the diagram below to complete the statements about the mammalian nephron that follow.

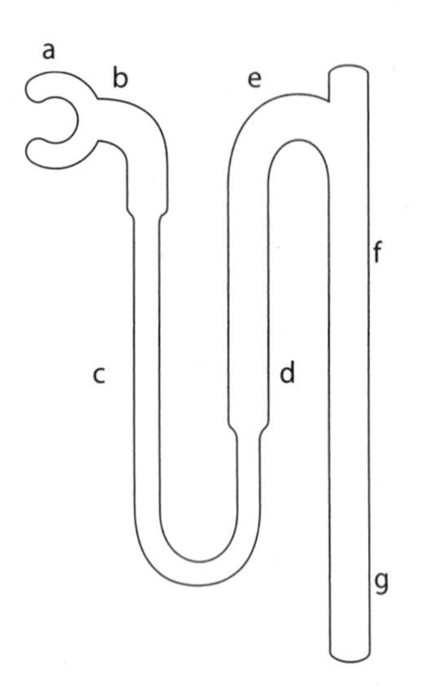

11. The composition of the filtrate would be most like plasma in the tubule next to letter _______.
12. The NaCl concentration in the extracellular fluid would be greatest in the area of letter _______.
13. The osmolarity of the filtrate next to letters _______ is similar to the osmolarity of blood plasma.
14. The urine would be most concentrated in the collecting duct next to letter _______.
15. Most of the glomerular filtrate is reabsorbed into the blood in peritubular capillaries next to letter _______.

Textbook Reference: *52.5 How Does the Mammalian Kidney Produce Concentrated Urine?*

16. Several hormones help regulate water and solute uptake and release in the nephron. Antidiuretic hormone (ADH) promotes _______ in response to _______.
    a. active transport of Cl^-; increased solute concentration
    b. active transport of Na^+; increased blood pressure
    c. increased permeability of the collecting duct to water; decreased blood pressure
    d. decreased permeability of the collecting duct to water; increased solute concentration
    e. decreased permeability of the collecting duct to water; decreased blood pressure

    Textbook Reference: *52.6 How Are Kidney Functions Regulated?*

17. Which of the following statements about atrial natriuretic peptide is *false*?
    a. It is released by muscle fibers of the atria.
    b. It decreases reabsorption of sodium in the kidneys.
    c. It causes the production of large amounts of dilute urine.
    d. It decreases reabsorption of water in the kidneys.
    e. It is released when blood volume and pressure are low.

    Textbook Reference: *52.6 How Are Kidney Functions Regulated?*

Answers

Key Concept Review

1. On land, water and salts are usually in short supply, so terrestrial animals are osmoregulators, actively regulating the osmolarity of their extracellular fluid.
2. Osmoconformers allow their extracellular fluid to equilibrate with their surroundings. This is possible in marine environments because there are many solutes in saltwater, some of which are necessary to support life. In addition, many marine osmoconformers have the ability to regulate the concentrations of certain ions in their extracellular fluids. In fresh water, however, there are few solutes, so the body fluids of an osmoconformer would be too dilute to support life.
3. Invertebrate study species might include mussels (osmoconformers) and brine shrimp (osmoregulators). Vertebrate study species might include hagfishes and chondrichthyans (osmoconformers) and any other aquatic or terrestrial species (osmoregulators).
4. Water-breathing animals typically excrete ammonia as their primary nitrogenous waste because ammonia is highly soluble in water and diffuses rapidly. Ammonia excretion occurs continuously across the gills.
5. The symptoms of gout are caused by the precipitation of uric acid crystals in joints due to high levels of uric acid in the extracellular fluid. Two potential approaches for new medications would involve blocking the production of uric acid by the body or improving the removal of uric acid from the body.
6. Ranking from least to most nitrogen in a molecule: (1) ammonia, (2) urea, (3) uric acid. Ranking from least to most soluble in water: (1) uric acid, (2) urea, (3) ammonia. The most toxic nitrogenous waste is ammonia.
7. In the excretory systems of flatworms and annelids, pressure differences drive filtration. Because insects have an open circulatory system, pressure differences cannot be used to filter extracellular fluid into the Malpighian tubules. Instead, the cells of the Malpighian tubules use active transport to move substances (sodium and potassium ions, and uric acid) from the extracellular fluid into the tubules.

8. In annelids, blood is filtered as it passes through capillary walls into the coelom. Reabsorption and secretion occur in the metanephridium.

9.

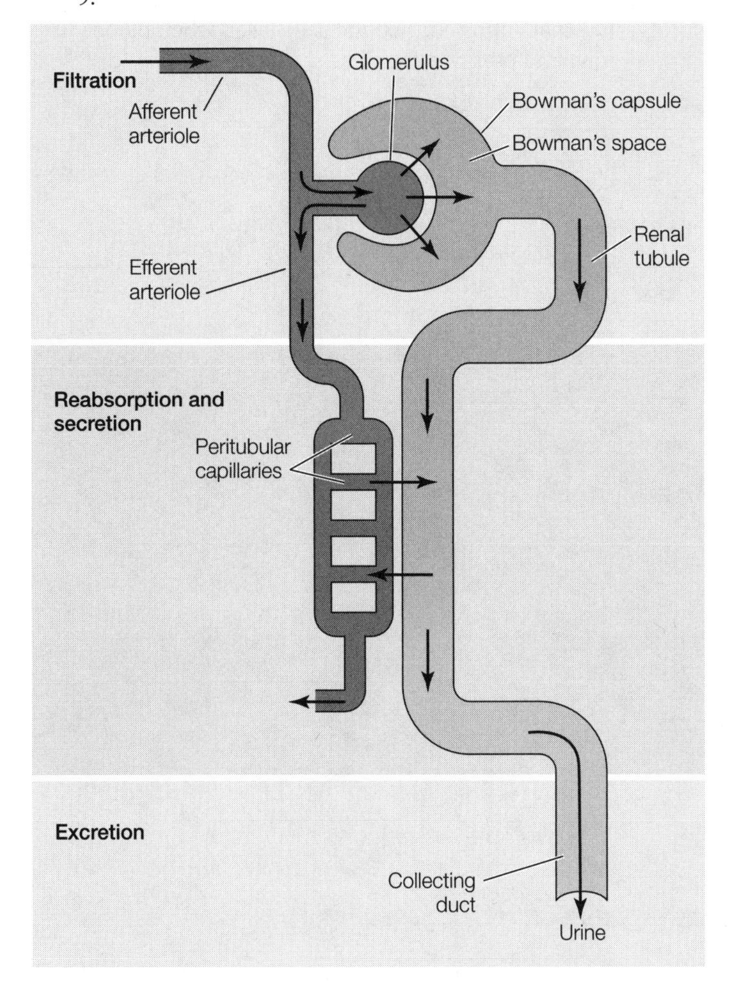

10. Dilation of the afferent arteriole and constriction of the efferent arteriole would increase glomerular filtration rate.

11. Chondrichthyans have a rectal gland that actively secretes NaCl. Marine bony fishes do not absorb certain ions from their guts, and actively excrete NaCl across their gill membranes.

12. Active transport of Na^+ and Cl^- in the nephron provides the ions that set up the countercurrent multiplier, allowing for the production of concentrated urine in mammals. If active transport in the nephrons were to stop, the urine produced would eventually be isotonic with blood plasma because the countercurrent multiplier would disappear.

13. Most reabsorption of water and solutes occurs in the proximal convoluted tubule. Cells in this location have numerous microvilli that increase the surface area for reabsorption. The cells also have numerous mitochondria, which indicates they are metabolically active.

14. Inhibition of antidiuretic hormone by this drug affects the permeability of the collecting ducts to water. This results in increased urine flow because water will not be reabsorbed across the collecting ducts. Blockage of ADH has no effect on the glomerular filtration rate.

15. In response to increased blood osmolarity, osmosensors in the hypothalamus stimulate the release of ADH from the pituitary into the blood. ADH acts to increase the permeability of the collecting ducts to water so that increased amounts of water can be reabsorbed to bring down the blood osmolarity. The osmosensors will also stimulate thirst, causing you to increase your water intake.

16. Some of the variables that might be measured include: diameter of the afferent renal arteriole; level of renin in the blood; level of angiotensin in the blood; diameter of the efferent renal arteriole; diameters of some peripheral blood vessels; blood pressure at one or more peripheral locations; level of aldosterone in the blood; and amount of water consumed.

Test Yourself

1. **b.** The sole mechanism for water reabsorption in the renal tubules is by osmosis.

2. **c.** Hagfishes are osmoconformers.

3. **a.** Chondrichthyans are ureotelic.

4. **b.** Most water moves out of the rectum and is conserved, not excreted.

5. **a.** Marine bony fishes live in an environment in which water tends to be drawn out of the body. To counter these effects, these fishes excrete small amounts of water.

6. **a.** The route of water and solutes through the nephron is from the glomerulus, to Bowman's capsule, to the proximal tubule, to the loop of Henle, to the distal tubule, to the collecting ducts.

7. **e.** Red blood cells are too large to be filtered out of the blood at the glomerulus and thus will not be found in the filtrate.

8. **c.** Water tends to move into the bodies of freshwater bony fish, so they produce large quantities of dilute urine.

9. **e.** About 99 percent of the filtrate is returned to the blood.

10. **c.** Two sites at which NaCl is actively transported out of the renal tubule are the proximal convoluted tubule and the thick ascending limb of the loop of Henle.

11. **a.** At Bowman's capsule the filtrate is most similar to plasma.

12. **g.** The sodium concentration is the highest in the extracellular fluid near the middle of the medulla.

13. **a**, **b**, and **e.** The osmolarity of the filtrate is similar to that of plasma in the cortex of the kidney (including the glomerulus and Bowman's capsule), the proximal convoluted tubule, and the distal convoluted tubule.
14. **g.** The highest concentration of the filtrate in the collecting ducts will be near their ends, deep in the medulla.
15. **b.** The bulk of the water and solute are reabsorbed at the proximal convoluted tubule.
16. **c.** Antidiuretic hormone acts on the collecting ducts by increasing their permeability to water. Antidiuretic hormone secretion is stimulated by a decrease in blood pressure.
17. **e.** Atrial natriuretic peptide is released when blood volume and pressure are high.

53 Animal Behavior

The Big Picture

- Tinbergen outlined the challenges animal behaviorists face as concerning causation, development, function, and evolution. Causation and development refer to proximate causes of behavior. Function and evolution refer to ultimate causes of behavior. In their research, behaviorists have typically performed laboratory experiments, using a small number of model species, and studied learning and memory. In contrast, ethologists have conducted comparative studies on diverse species, often under field conditions, and focused on instinctive behaviors.
- Animals exhibit diverse species-specific behaviors that can be either genetically based or environmentally determined. Genetically based behaviors, such as fixed action patterns, require triggers or releasers. Gene cascades control most behaviors. Patterns of behavior typically have both genetic and environmental influences.
- Hormones are important in controlling the behavior of animals. Sex steroids organize and activate sexually dimorphic behaviors, such as the copulatory behavior of male and female rats and singing by male songbirds. For some behaviors to develop, an animal must be exposed to certain stimuli during a brief window of time called the critical (or sensitive) period. Parents often come to recognize their offspring during a critical period of exposure.
- Behaviors have benefits and costs. Behavioral ecologists study how animals make behavioral choices, such as when foraging, defending a territory, or selecting a habitat in which to settle.
- Animals find their way around their environment by several mechanisms, including piloting, distance-and-direction navigation, and bicoordinate navigation.
- Communication is an integral part of animal behavior. Communication signals can be chemical, visual, acoustic, or mechanosensory. Many animal displays contain signals from two or more sensory modalities.
- Because standard Darwinian theory holds that only traits that contribute to individual fitness are favored by natural selection, explaining the evolution of altruistic behavior has presented a major challenge to evolutionary biologists. The concepts of inclusive fitness and kin selection have proved helpful to understanding altruism in animals and have also been applied to many aspects of human social behavior (though not without controversy).

Study Strategies

- It can sometimes be difficult to distinguish between instinctive behaviors and those that are learned during the course of an animal's lifetime. Further muddying the waters, the ability to learn new behavior is, of course, a heritable trait! As you learn about various patterns of behavior, take time to consider whether a behavior is primarily genetically determined or primarily environmentally determined.
- The categories of animal orientation—piloting, distance-and-direction navigation, and bicoordinate navigation—can be confusing. Refer to Figure 53.17 in the textbook to consolidate your understanding of how the sun is used as a time-compensated compass.
- Learn to distinguish the various behavior cycles associated with internal controls via "biological clocks" as opposed to behaviors that are directly stimulated by an immediate event in an animal's environment.
- To study the properties of the various sensory modes of communication (visual, chemical, etc.), make a table that compares each one with respect to characteristics such as cost of production, effective signaling distance (and how signaling distance is affected by environmental conditions), durability, and information content.
- Go to the Web addresses shown to review the following animated tutorials and activities. You can also find them in BioPortal (yourBioPortal.com), along with many additional learning resources.

  Animated Tutorial 53.1 The Costs of Defending a Territory (Life10e.com/at53.1)

  Animated Tutorial 53.2 Foraging Behavior (Life10e.com/at53.2)

  Animated Tutorial 53.3 Circadian Rhythms (Life10e.com/at53.3)

Animated Tutorial 53.4 Homing Simulation (Life10e.com/at53.4)

Animated Tutorial 53.5 Time-Compensated Solar Compass (Life10e.com/at53.5)

Activity 53.1 Honey Bee Dance Communication (Life10e.com/ac53.1)

Activity 53.2 Concept Matching (Life10e.com/ac53.2)

Key Concept Review

53.1 What Are the Origins of Behavioral Biology?

Conditioned reflexes are a simple behavioral mechanism

Ethologists focused on the behavior of animals in their natural environment

Ethologists probed the causes of behavior

One early approach to the study of animal behavior was behaviorism. This approach, exemplified by the work of Pavlov and Skinner, emphasized the study of learned behavior in a few model species (e.g., the albino rat) under laboratory conditions. Pavlov discovered the conditioned reflex through his experiments on the salivation response of dogs (see Figure 53.1). Before conditioning, food is an unconditioned stimulus that elicits salivation, which is an unconditioned response. If the unconditioned stimulus is presented immediately after a neutral stimulus, such as a particular sound, the dog eventually salivates solely in response to the sound. The sound has become a conditioned stimulus and the salivation response to sound is a conditioned response. Skinner showed that any random action of an animal could become a conditioned response to a stimulus if a reward was associated with the action and the stimulus. This type of learning is called operant conditioning.

Ethology, another approach to the study of animal behavior, arose around the same time as behaviorism. Ethologists such as Lorenz, von Frisch, and Tinbergen focused on instinctive behaviors (fixed action patterns) in diverse species and often studied animals in their natural habitats. Fixed action patterns are performed without learning and cannot be modified by learning. Fixed action patterns are also stereotypic, meaning that they are performed the same way each time. Releasers trigger fixed action patterns. In general, releasers are very simple subsets of the information available in the environment (see Figure 53.2). Tinbergen suggested organizing animal behavior questions into four categories: causation, development, function, and evolution. Questions about causation and development are proximate questions, and questions about function and evolution are ultimate questions.

Question 1. Three-spined sticklebacks are bony fish that occupy coastal waters. During the breeding season, the ventral surface of the male turns bright red and his eyes turn blue. During the breeding season, the aggressive reaction of a male three-spined stickleback to the arrival of a male intruder in his territory is a fixed action pattern. Design an experiment to determine the releaser for this fixed action pattern.

Question 2. The phenomenon by which a young animal leaves its place of birth is called natal dispersal. Develop four research questions about natal dispersal in gray squirrels using Tinbergen's four questions.

53.2 How Do Genes Influence Behavior?

Breeding experiments can produce behavioral phenotypes

Knockout experiments can reveal the roles of specific genes

Behaviors are controlled by gene cascades

Behavioral geneticists conduct breeding experiments to analyze if particular behaviors are inherited. Breeding experiments alone cannot identify the particular genes influencing a behavior or how they influence the behavior, so molecular genetic approaches are needed. Most behaviors are complex traits involving many genes. Quantitative trait analysis makes use of genetic markers, which are unique DNA sequences, to identify multiple genes influencing a particular trait in species with well-studied and mapped genomes.

Once a gene has been identified, a knockout experiment can be conducted in which biologists inactivate the particular gene and investigate the behavioral effects of its loss. A knockout mouse is one in which a particular gene is targeted and inactivated to eliminate the gene product, possibly a receptor. For example, inactivating particular receptors in the vomeronasal organ of male mice disrupts their ability to discriminate female from male conspecifics (see Figure 53.3). Molecular genetic approaches also have been used to study the *Drosophila* mutant *fruitless,* in which males are unable to tell the sexes apart and therefore court both males and females. The *fruitless* gene has been cloned and sequenced, and experimentally altered. These studies have shown that the gene product is a transcription factor that controls the expression of many genes. This example illustrates that gene cascades control patterns of behavior and offer many points at which a change in a single gene can influence behavior (see Figure 53.4).

Question 3. How can a change in a single gene have a major impact on behavior?

Question 4. The hormone progesterone is thought to mediate the aggression shown by male mice toward infant mice. Design a genetic knockout experiment to test this hypothesis.

53.3 How Does Behavior Develop?

Hormones can determine behavioral potential and timing

Some behaviors can be acquired only at certain times

Birdsong learning involves genetics, imprinting, and hormonal timing

The timing and expression of birdsong are under hormonal control

Hormones can determine the development of a behavioral potential at an early age and the expression of that behavior at a later age. For example in rats, the sex steroids present during early development determine the pattern of sexual behavior in adulthood, but sex steroids present in adults control the expression of those patterns (see Figure 53.5). In imprinting, an animal learns a specific set of stimuli during a critical period. Imprinting of offspring on parents or of parents on offspring is a learned response that helps in recognition. The critical period for imprinting is often determined by the developmental or hormonal state of the animal.

Birds use songs in territorial displays and in courtship. In some species, imprinting of the species-specific song is required in the male nestling in order for it to sing the song as an adult, even though the juvenile bird never sings the song. Imprinting forms a memory of the song that is recalled as the bird approaches adulthood. In males, both the initial imprinting of the song as a juvenile and the ability to match the song with auditory feedback as an adult are necessary to sing the correct song (see Figure 53.7). Social experience and the reaction of females to song also influence song development.

Hormones also control singing behavior. In male birds, testosterone induces certain regions of the brain to grow during the breeding season. During the nonbreeding season, those regions of the brain that are associated with singing are reduced in size. Female birds that are treated with testosterone during the spring will also develop the species-specific song, as this stimulates growth in areas of the brain associated with singing.

Question 5. Much like patterns of sexual behavior in rats (females display lordosis and males display mounting), urinary posture in dogs is sexually dimorphic (females typically squat and males lift a leg). In both sexual behavior in rats and urinary posture in dogs, testosterone early in development organizes the potential for particular behaviors in adulthood. Would testosterone have the same effect on the timing of expression of the behaviors in adulthood? (Hint: Testosterone is necessary for the expression of sexual behavior in adult male rats. Is testosterone necessary for the expression of the male urinary posture in adult male dogs?)

Question 6. Explain why the experimental deafening of a young songbird and an adult songbird would have very different effects in terms of the quality of their song.

53.4 How Does Behavior Evolve?

Animals are faced with many choices

Behaviors have costs and benefits

Territorial behavior carries significant costs

Cost–benefit analysis can be applied to foraging behavior

Animals make a series of choices over their lifetimes. One choice is where to live. The habitat of an animal is the environment in which it normally lives. In choosing habitat, animals use cues that are good predictors of general conditions suitable for future survival and reproduction. The success of already-settled individuals of the same species (conspecifics) may be used as an indicator of habitat quality.

Natural selection molds behavior in accordance with costs and benefits. There are three aspects to the cost of a behavior: (1) energetic cost—the amount of energy the animal expends during the behavior; (2) risk cost—the amount of risk the behavior entails for the animal; and (3) opportunity cost—the benefits the animal forfeits by not engaging in other behaviors instead. The territory of an animal is an area from which it excludes other individuals of its own species (and sometimes individuals of other species). By establishing a territory, an animal (usually a male) may improve its fitness by gaining exclusive use of the resources of part of its habitat. Cost–benefit analysis explains the variety of territorial behaviors characteristic of different species (see Figure 53.9). Some animals defend all-purpose territories in which nesting, mating, and foraging take place. Other animals cannot establish feeding territories but defend nest sites or areas that provide access to females. Still other animals defend territories (leks) that are used only for mating (see Figure 53.10).

Cost–benefit analysis can also be applied to foraging behavior (i.e., what food an animal selects and when and where it searches for it). Optimal foraging theory predicts that the foraging choices of many animals result in their maximizing their rate of energy intake (see Figure 53.11). Because minerals and foods with medicinal value are important in the diets of many animals, they may sometimes forage in a way that deviates from the energy-maximization model (see Figure 53.12).

Question 7. Your textbook describes how some species use presence of conspecifics as a cue during habitat selection. A similar phenomenon exists whereby some species use presence of individuals of another species as a cue when selecting a place to settle. One example of this concerns the attraction of migrant birds of one species to resident birds of other species. Why might a migrant species use presence of a resident species as a cue during habitiat selection?

Question 8. Describe the experiments performed with bluegill sunfish preying on water fleas and the implications of these studies for foraging theory.

53.5 What Physiological Mechanisms Underlie Behavior?

Biological rhythms coordinate behavior with environmental cycles

Animals must find their way around their environment

Animals use multiple modalities to communicate

The nervous and endocrine systems control behavior. Execution of behavior involves effector mechanisms. Behavioral responses to the environment must be timed appropriately. Circadian rhythms are daily cycles in activity, sleep, foraging, and other physiological processes and behaviors that are controlled by an endogenous clock (see Figure 53.13). The length of one cycle in a rhythm is defined as one period. Any point in the cycle is a phase. Two cycles can be in phase if the rhythms match, or they can be phase-advanced or phase-delayed if they do not match. Circadian rhythms can be reset by environmental cues, such as the light–dark cycle, during entrainment. In constant conditions an animal's circadian clock is free-running and will have a natural period that is different from the 24-hour period of the day. In mammals, the master circadian "clock" is located in the suprachiasmatic nuclei (SCN) of the brain. Day length can trigger seasonal rhythms, which are important in controlling the timing of breeding and migrating. Hibernators and equatorial migrants cannot rely on changes in day length; they rely instead on circannual rhythms—built-in neural calendars that keep track of the time of the year.

Animals find their way around their environment by diverse mechanisms, including piloting, distance-and-direction navigation, and bicoordinate navigation. Piloting uses landmarks. It is the mechanism used by some species that migrate or that are capable of homing (the ability to return to a specific location). Homing and migrating species that are able to take direct routes to their destinations through environments they have never experienced must use mechanisms other than piloting. Many animals appear to have a compass sense, which allows them to use environmental cues to determine direction, and some appear to have a map sense, which allows them to determine their position. Distance-and-direction navigation involves knowledge of direction and distance to a destination. The position of the sun and stars can be a source of directional information. Pigeons, for example, have the ability to determine direction by means of a time-compensated solar compass (see Figure 53.17). The stars offer two sources of information about direction: moving constellations and a fixed point (the point directly over the axis on which Earth turns). Bicoordinate navigation (also known as true navigation) requires knowledge of the map coordinates of both the current position and the destination.

Communication signals benefit both the sender and the receiver. Pheromones are chemical signals used to communicate among individuals of a species. Because of their diverse molecular structures, pheromones can communicate very specific, information-rich messages, and because of their durability, they are useful for such functions as marking territories. Visual signals provide rapid, directional communication over considerable distances. Drawbacks to visual signals include the need for light and the ease with which they can be intercepted by other species. Sound communicates directional information and can travel in complex environments and over long distances, but it does not travel as rapidly as visual signals do. Animals use tactile communication when in close contact with one another. Honey bees dance to communicate the location of a food source in the environment (see Figure 53.18). The specificity of communication signals is enhanced by the use of multiple sensory modalities.

Question 9. A pigeon has been trained to feed at food bins at the eastern end of a circular cage from which it can see the sky. There are food bins at the N, NE, E, SE, S, SW, W, and NW ends of the cage. Based on this information, answer the following questions:

a. If the cage is covered and a fixed light source is presented at the east end of the cage at the time of sunrise, where will the pigeon search for food at noon?

b. If a mirror is used to shift the apparent position of the sun at noon from the south to the northwest, where will the pigeon search for food?

c. The bird has been placed in a light-controlled environment for three weeks and phase-delayed by six hours. If the pigeon is returned to the cage under natural lighting conditions at sunset, where will it search for food?

Question 10. What is the function of a honey bee's waggle dance, and what does it communicate to other bees in a hive?

Question 11. Draw five diagrams showing the orientation of the straight run of the waggle dance that would be performed on the vertical surface of a honeycomb by a foraging honey bee that has discovered a food source at the times and locations described below. Recall that in the northern hemisphere the direction (azimuth) of the sun is due south at noon; assume that the sun rises precisely in the east at 6:00 A.M. and sets precisely in the west at 6:00 P.M. As in Figure 53.18, let the top of the page indicate the "up" direction.

a. Food location: due south of the hive; time: noon
b. Food location: due north of the hive; time: 6:00 A.M.
c. Food location: due north of the hive; time: noon
d. Food location: due west of the hive; time: 9:00 A.M
e. Food location: due east of the hive; time: 6:00 P.M.

53.6 How Does Social Behavior Evolve?

Mating systems maximize the fitness of both partners
Fitness can include more than your own offspring
Eusociality is the extreme result of kin selection
Group living has benefits and costs
Can the concepts of sociobiology be applied to humans?

Sociobiology is the study of the evolution of social behavior. Mating systems evolve to maximize the fitness of both partners. Because males produce vast numbers of sperm that contain next to no resources, and females produce relatively few eggs that are rich in resources, the energetic and opportunity costs of reproduction are greater for females than for males. This disparity is particularly large in mammals because females also bear the costs of gestation. For a female, the best way to maximize her fitness is to invest in her young, so they survive to pass on her genes. Males can maximize their fitness in different ways; some may mate with many females and provide little or no care to their young, whereas others may remain with a single female and help rear their offspring. The best strategy will depend on environmental factors. The mating systems of animals include monogamy (in which one male forms a pair bond with one female and both parents participate in rearing the young), polygyny (one male mates with many females), and polyandry (one female mates with multiple males).

Natural selection sometimes favors altruistic acts—behaviors that reduce the reproductive chances of the individual performing the act but increase the fitness of the helped individual. Typically, such behavior is directed toward a relative of the altruist, with whom it shares alleles. By helping its relatives, an individual can increase the representation of some of its own alleles in the population. An altruistic behavior pattern can evolve if it increases the inclusive fitness of the altruist, which is the fitness derived from an individual's personal reproductive success (individual fitness) plus the fitness derived from the reproductive success of its relatives. The maximization of inclusive fitness underlies kin selection, which is selection for behaviors that increase the reproductive success of relatives even when they have some cost to the performer. According to Hamilton's rule, for an apparent altruistic act to be adaptive, the fitness benefit of that act to the recipient times the degree of relatedness between the performer and the recipient has to be greater than the cost to the performer. Eusocial species are those whose social groups include sterile individuals. In the Hymenoptera (ants, bees, and wasps), kin selection has probably facilitated the evolution of eusociality because of their sex determination system (haplodiploidy). In eusocial species in which both sexes are diploid (termites, naked mole-rats), the difficulty of establishing independent colonies has favored the evolution of eusociality.

Group living has costs and benefits. Costs include increased competition for food and mates, increased conspicuousness to predators, and increased risk of disease transmission. Benefits include enhanced acquisition and defense of food and more rapid detection of predators. The application of sociobiological concepts to human behavior is controversial, given the importance of learning and culture in shaping human behavior.

Question 12. Why has the unusual sex determination system in Hymenoptera predisposed species in this group toward the evolution of eusociality?

Question 13. For each of the two y-axes in the following graph, draw a labeled curve that correctly summarizes observations made on goshawks attacking wood pigeons, as described in the textbook.

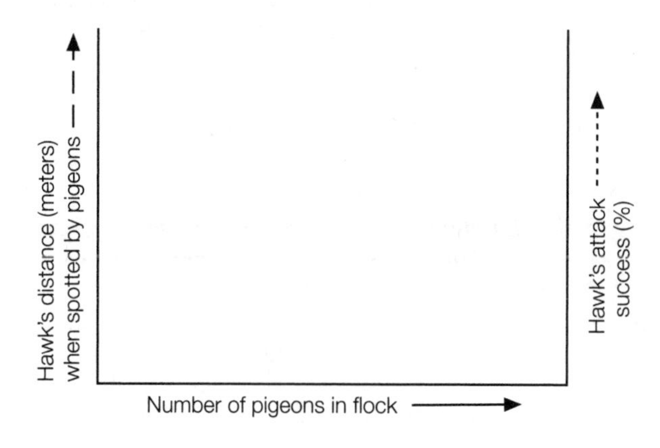

Test Yourself

1. On April mornings, you step outside your front door and notice the singing of a male robin. At that time you also observe that the female of the pair is building a nest in a nearby tree. A month later you observe the pair feeding their nestlings. Which of the following is a question about the ultimate cause of the behavior of these birds?
   a. Did the male bird begin to sing in April because the photoperiod had increased to a critical length?
   b. What were the relative roles of genes and experience in causing the female robin to build her nest out of particular building materials and to place it in a particular location?
   c. What combination of internal physiological factors and external cues stimulates the parents to feed their nestlings?
   d. Do robins that begin nesting in April raise more offspring than they would raise if they delayed nesting until June, and if so, why?
   e. None of the above

   Textbook Reference: *53.1 What Are the Origins of Behavioral Biology?*

2. You train your pet dog to "sit" and "roll over" on command by rewarding it with a dog biscuit whenever it performs the desired behavior. This is an example of
   a. classical conditioning.
   b. stereotypic behavior.
   c. a releaser triggering a fixed action pattern.
   d. a conditioned reflex.
   e. operant conditioning.

   Textbook Reference: *53.1 What Are the Origins of Behavioral Biology?*

3. Which of the following statements about genetic influences on behavior is *false*?
   a. Genes controlling behavior often function in cascades.
   b. A single gene usually codes for complex behavior.
   c. Knockout experiments start with identified genes.
   d. Change in a single gene can alter complex behavior.
   e. Quantitative trait analysis uses genetic markers.

   Textbook Reference: *53.2 How Do Genes Influence Behavior?*

4. Which of the following statements about the development of singing behavior in white-crowned sparrows is *false*?
   a. If a male is deafened after he sings his correct species-specific song, he will continue to sing a normal song.
   b. Female songbirds can be induced to sing by treatment with testosterone.
   c. Females learn their species song, but they do not normally express it under natural conditions.

d. Once a male has learned his song, testosterone is unnecessary for the expression of song.
e. To sing normally as an adult, a male must hear its species-specific song as a nestling.
Textbook Reference: *53.3 How Does Behavior Develop?*

5. A female rat will show female sexual behavior in adulthood if it is
a. given testosterone injections as a newborn and estrogen injections as an adult.
b. spayed as a newborn.
c. spayed as an adult.
d. given testosterone injections as a newborn and as an adult.
e. spayed as a newborn and given estrogen injections as an adult.
Textbook Reference: *53.3 How Does Behavior Develop?*

6. Two species of mice live in the same geographical region, but species 1 prefers open fields whereas species 2 lives in forests. In an experiment, when presented with simulated "fields" and "forests," individuals of each species born in a laboratory prefer the environment in which they naturally live. This experiment illustrates the concept of
a. habitat selection.
b. optimal foraging strategy.
c. territoriality.
d. imprinting.
e. Both a and b
Textbook Reference: *53.4 How Does Behavior Evolve?*

7. Birds spend some of their time scanning the horizon for predators. While scanning, they cannot be foraging for food. This situation illustrates the phenomenon of
a. cooperative hunting.
b. energetic cost.
c. risk cost.
d. optimal foraging strategy.
e. opportunity cost.
Textbook Reference: *53.4 How Does Behavior Evolve?*

8. Which of the following statements about territoriality is *false*?
a. A male defending a territory on a lek is defending a display site for attracting females with which he can mate.
b. Territories do not always include foraging areas.
c. Territorial defense is likely to impose three kinds of costs: energetic, risk, and opportunity.
d. Animals are expected to defend feeding territories when food is widely distributed.
e. Males on central territories in a lek typically mate with the most females.
Textbook Reference: *53.4 How Does Behavior Evolve?*

9. Which of the following statements about circadian rhythms is *false*?
a. In mammals, the master circadian "clock" is located in the suprachiasmatic nuclei of the brain.
b. A circadian clock can be entrained by environmental cues.
c. Free-running circadian clocks are seldom exactly 24 hours long.
d. Genetic mutations can cause changes in the length of the free-running circadian clock.
e. A mammal maintained in complete darkness and at a constant temperature will become arrhythmic.
Textbook Reference: *53.5 What Physiological Mechanisms Underlie Behavior?*

10. Which of the animals listed below would *not* be expected to have an endogenous circannual rhythm?
a. A ground squirrel that hibernates
b. A white-tailed deer living in eastern North America
c. A cave salamander
d. A migratory bird that summers in North America and winters in equatorial South America
e. All of the above would have an endogenous circannual rhythm.
Textbook Reference: *53.5 What Physiological Mechanisms Underlie Behavior?*

11. Which of the following statements about animal orientation is true?
a. Distance-and-direction navigation requires knowledge of latitude and longitude.
b. Bicoordinate navigation requires knowledge of direction and of distance to a destination.
c. The stars offer two sources of information about direction: a fixed point and moving constellations.
d. Piloting involves the use of the sun as a compass.
e. None of the above
Textbook Reference: *53.5 What Physiological Mechanisms Underlie Behavior?*

12. Which of the following statements about pheromones is *false*?
a. They are used for communication between individuals of the same species.
b. They can communicate information about the size, sex, and reproductive status of the signaler.
c. They are unsuitable for rapid exchange of information.
d. They are difficult for predators to intercept.
e. They have low volatility and diffusibility when used to communicate alarm.
Textbook Reference: *53.5 What Physiological Mechanisms Underlie Behavior?*

13. You observe an example of an apparently altruistic act by an animal that seems to reduce its near-term likelihood of reproductive success. Which of the following is the *least* plausible explanation for its behavior?
a. The act aids the reproductive success of individuals sharing a high proportion of genes with the altruistic individual.

b. The act is only apparently altruistic; over the long term, the behavior actually contributes to individual fitness.
c. The act increases the inclusive fitness of the animal performing it.
d. The act is advantageous because it helps the species to survive and reproduce, even if the altruist itself does not.
e. It is impossible to determine the answer from the information provided.

Textbook Reference: *53.6 How Does Social Behavior Evolve?*

14. Some birds give a species-specific vocalization called an "alarm call" when they see a predator, although this call may direct the predator toward them. Other members of this species respond to these calls by taking cover. This would be an example of altruistic behavior that is beneficial to the calling bird if
a. the bird giving the vocalization survives the attack.
b. the inclusive fitness of the bird giving the vocalization is increased.
c. all birds survive the attack.
d. the birds benefiting are offspring of the bird giving the vocalization.
e. All of the above

Textbook Reference: *53.6 How Does Social Behavior Evolve?*

15. Which of the following characteristics is shared by all eusocial species?
a. A sex determination system in which males are haploid and females are diploid
b. Queens that mate with a single male
c. The presence of sterile classes
d. Both a and c
e. All of the above

Textbook Reference: *53.6 How Does Social Behavior Evolve?*

Answers

Key Concept Review

1. The experiment should probably be conducted under laboratory conditions, where variables can be more easily controlled. Design several different artificial models of male sticklebacks, present each to a different male in an aquarium, and determine which model is most effective in releasing aggression. Test several males with each model. The models could vary in presence of blues eyes and presence of the red color on the belly. These characteristics could be placed on realistic models that closely resembled sticklebacks and on crude models that show little resemblance to sticklebacks. Experiments such as these have shown that it is the red color that releases aggression.

2. The question of causation would focus on what stimulates natal dispersal. For example, do peaks in certain hormones stimulate dispersal in young squirrels? The question about development would ask whether dispersal is prompted by changes in an individual's internal environment (e.g., does dispersal occur when a squirrel reaches a certain body mass?) or external environment (e.g., does dispersal occur due to increasing aggression from parents?). Questions about the function of dispersal would focus on how leaving home influences the survival and reproductive success of squirrels. Finally, to understand the evolution of natal dispersal, we might examine patterns of dispersal in other species within the squirrel family.

3. Because genes often function in gene cascades, a change in a single gene can cause major changes in behavior that impact fitness. For example, the product of a given gene may be a transcription factor that controls the expression of many other genes. Changes in any of these genes, or in their expression, can alter behavior. Thus, gene cascades create many opportunities for a change in one gene to cause a change in a behavior.

4. Male mice are typically aggressive toward infants. One way to test the hypothesis that progesterone mediates male aggression toward infants would be to develop progesterone receptor knockout mice. The gene for progesterone receptors would be targeted and inactivated. Thus, progesterone knockout mice would fail to respond to progesterone because they lack the appropriate receptors. If progesterone receptor knockout males were neutral or positive (i.e., not aggressive) to infants, we could tentatively conclude that progesterone mediates the aggression shown by typical male mice toward infants. We could increase our confidence in this conclusion by conducting additional studies.

5. Sexual behavior in male rats is organized by testosterone early in life and activated by testosterone in adulthood. Urinary posture in male dogs is organized by testosterone early in life, but testosterone is not necessary for the expression of urinary behavior in adulthood. This can be illustrated by the fact that neutered male dogs still lift a leg to urinate, even though their testes (the main source of testosterone) have been removed.

6. If a young male songbird is deafened before he has had time to match his song output to the song he has memorized, he will produce an abnormal song. If a male songbird is deafened in adulthood, the deafening will have little, if any, effect on song quality because the song has already crystallized.

7. Migrants might be attracted to residents for a number of reasons. First, residents might indicate high-quality habitat. It might also be efficient to use residents as an indicator of habitat quality because they have had all year to assess the quality of different habitat patches (as compared with migrants, who have a much shorter time period).

8. In studies in which bluegill sunfish were provided with small, medium, and large water fleas at three different prey densities, the fish took equal proportions of the three prey sizes at low densities but mostly large prey at high densities. This result agrees with the prediction from foraging theory that when prey are abundant, a predator should focus on large prey in order to maximize its energy intake.

9. a. The pigeon will eat from the northern bin at noon. Normally at noon the sun is in the south, and the bird would eat from the eastern bin that is 90° to the left of the sun. With the light in the east, the bird will eat in the north, which is 90° to the left of the light.

   b. The bird will go to the bin that is 90° from the sun at noon, and will feed from the southwestern food bin.

   c. At sunset, the eastern bin with food would normally be 180° from the sun in the west. After the bird has been phase-delayed six hours, it will think that the setting sun is the noon sun. At noon the bird usually eats from the eastern bin, which is located 90° to the left of the sun. Therefore, the bird will eat from the southern bin.

10. Honey bees dance to inform other members of the hive about the distance and direction to a food source. The duration of each waggle run indicates distance. The angle of the straight run of the waggle dance indicates direction relative to the sun.

11.

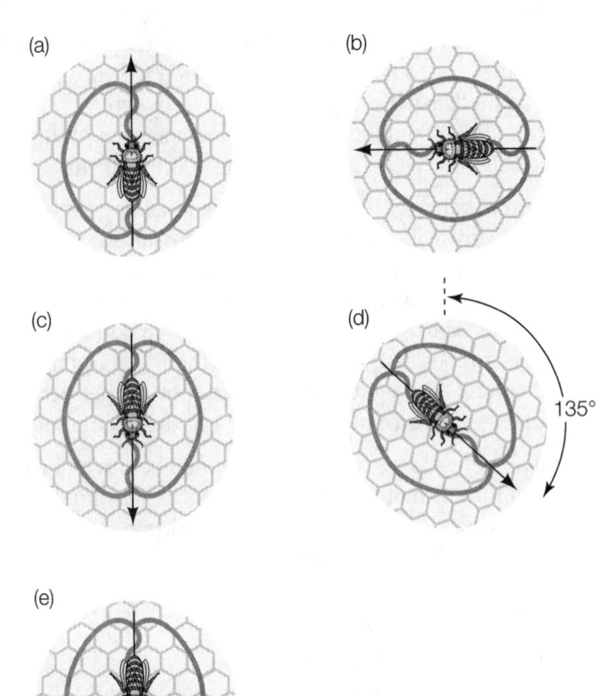

12. All species of the Hymenoptera (ants, bees, and wasps) have a sex determination system in which males are haploid and females are diploid; thus females share 75 percent of their genes with their sisters but only 50 percent of their genes with any offspring they could produce. In species whose sexes are determined in this way, females may increase their inclusive fitness by foregoing reproduction and helping to raise sisters. This explanation does not apply to the evolution of eusociality in species without this mode of sex determination.

13. Your curves should show a positive relationship between the hawk's distance when spotted by the pigeons and pigeon flock size and a negative relationship between the hawk's attack success and pigeon flock size (see below).

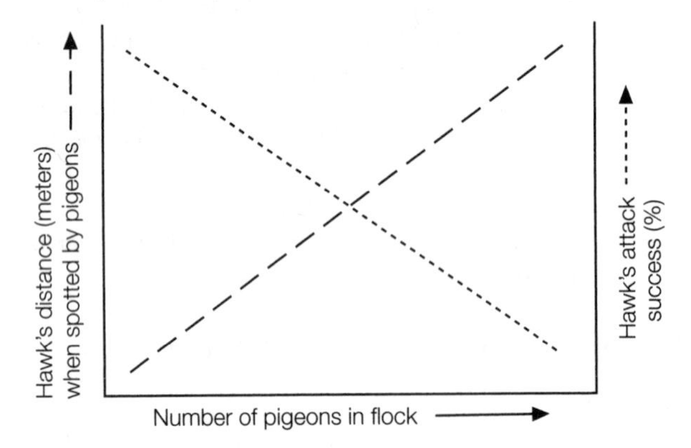

Test Yourself

1. **d.** The first three answer options (**a**, **b**, and **c**) concern the *proximate* mechanisms that underlie a behavior. The fourth question is concerned with the *ultimate* cause of a behavior—the selection pressures that shaped its evolution.
2. **e.** The temporal pairing of a reward with a particular response to a stimulus is a feature of operant conditioning, which is the standard method of training an animal.
3. **b.** Many genes usually control complex behaviors.
4. **d.** Young male songbirds must hear the species' song and then later be able to hear themselves sing. In the adult, the presence of testosterone is needed to increase the size of regions in the brain associated with singing during the breeding season. Absence of testosterone will result in the inability to perform the correct song.
5. **e.** Development of female sexual behavior does not require the presence of estrogen around the time of birth. However, estrogen is required to prompt the expression of female sexual behavior in adulthood. Thus, a female spayed as a newborn and given estrogen injections in adulthood will show female sexual behavior.
6. **a.** The environment in which a species normally lives is its habitat.

7. **e.** The forfeited benefits of behaviors that could not be achieved as a result of performing a different behavior, like scanning, constitute the opportunity cost of the performed behavior.
8. **d.** Food is less likely to be defended when it is widely distributed in space or fluctuates in availability. The costs of defense under these conditions outweigh the benefits.
9. **e.** Mammals maintained in complete darkness and at constant temperature still display daily cycles. Destruction of the suprachiasmatic nuclei will make a mammal arrhythmic.
10. **b.** Hibernators, cave-dwellers, and equatorial migratory species lack the necessary daily cues at some stage during the year, so they most likely have endogenous circannual rhythms. Thus, the white-tailed deer would be the only animal that could use circadian cues.
11. **c.** Many animals use the stars for navigation. The stars present two types of information: a fixed reference point (the point directly above Earth's axis of rotation) and moving constellations, which appear to revolve around the fixed point.
12. **e.** Alarm pheromones are highly volatile and diffusible. Territory- and trail-marking pheromones have low volatility and diffusibility.
13. **d.** For an altruistic behavior to evolve by natural selection, it must increase the inclusive fitness of the individual performing the behavior. Any explanation of an altruistic act based on its purported benefit to the species as a whole violates this principle.
14. **b.** Altruistic behavior is beneficial to the performer when the improvement in the reproductive success of kin (not including offspring) exceeds the reduced reproductive success of the individual performing the act. If this condition is met, then the behavior has improved the inclusive fitness of the performer.
15. **c.** By definition, eusocial species live in social groups with sterile castes. The sex determination mechanism in which males are haploid and females are diploid is found in the Hymenoptera (ants, bees, and wasps) but not in termites and naked mole-rats. The textbook also mentions that naked mole-rat colonies include several reproductive males.

54 Ecology and the Distribution of Life

The Big Picture

- Much evidence supporting the theory of continental drift (described in Chapter 25) is based on the distributions of organisms on Earth. By the same token, the acceptance by geologists and biologists of the validity of continental drift revolutionized the field of biogeography.
- The area phylogeny approach used by biogeographers to understand how the present distribution of groups of organisms came about is a good example of the broad applicability of the modern methods of phylogenetic analysis that were described in Chapters 22 and 24.

Study Strategies

- The relationship between a phylogenic tree and an area phylogeny may be confusing. Study Figure 54.14 to understand how these two types of phylogeny can be used to reconstruct the evolutionary history of a taxon such as the Equidae.
- In attempting to master the information concerning the many biomes described in this chapter, focus on the graphs that summarize the essential points about each one.
- You also should study Figure 54.7 to get a sense of the geographical location and extent of each biome. A useful exercise to help you learn the relative locations of biomes is to imagine a journey (e.g., from equatorial South America to Alaska, or from Maine to California), naming the biomes that you would pass through.
- Go to the Web addresses shown to review the following animated tutorials and activity. You can also find them in BioPortal (yourBioPortal.com), along with many additional learning resources.

  Animated Tutorial 54.1 Biomes (Life10e.com/at54.1)

  Animated Tutorial 54.2 Rain Shadow (Life10e.com/at54.2)

  Activity 54.1 Major Biogeographic Regions (Life10e.com/ac54.1)

Key Concept Review

54.1 What Is Ecology?

Ecology is not the same as environmentalism

Ecologists study biotic and abiotic components of ecosystems

Ecology is the scientific study of the interactions among organisms and between organisms and their physical environment. Named by Ernst Haeckel in the late 1800's, it generates knowledge about interactions in the natural world. The greater our understanding of ecological interactions, the greater the likelihood that we can carry out activities such as growing food and managing pests without causing unintended consequences for ourselves and other organisms.

Whereas ecology is a science, environmentalism uses ecological knowledge and other considerations to inform both personal decisions and public policy relating to the stewardship of natural resources and ecosystems. The environment comprises both abiotic factors (physical and chemical) and biotic factors (living components).

Many new tools—such as mathematical models, molecular techniques, and satellite imaging—are available to ecologists as they strive to understand the biotic and abiotic forces that influence the distribution and abundance of organisms on Earth.

Question 1. Explain the difference between biotic and abiotic factors.

Question 2. What is the difference between ecology and environmentalism?

54.2 Why Do Climates Vary Geographically?

Solar radiation varies over Earth's surface

Solar energy input determines atmospheric circulation patterns

Atmospheric circulation and Earth's rotation result in prevailing winds

Prevailing winds drive ocean currents

Organisms adapt to climatic challenges

The climate of a region is the average of the atmospheric conditions found in that region over the long run, while weather is the short-term state of atmospheric conditions at a particular place and time. Climates may vary geographically, primarily because different latitudes receive different amounts of solar energy. Solar energy input also determines air temperature, and thus climate. Regions near the poles receive less energy per unit of ground area than regions near the equator because of the lower angle of the sun. At high latitudes there is also more variation over the course of a year in both day length and the angle of arriving solar energy than at latitudes closer to the equator (see Figure 54.1). The average air temperature decreases 0.76°C for every degree of latitude (about 110 km). Air temperatures also decrease with elevation, so temperatures at sea level are warmer than temperatures on mountaintops at the same latitude.

Seasonal change is the result of the 23.5-degree tilt of Earth's axis (see Figure 54.2). Unequal heating of the atmosphere at low and high latitudes produces vertical and latitudinal movements of air masses. These movements cause very moist climates to occur at the equator and at 60° north and south latitudes, where air rises, and arid climates to occur at about 30°N and 30°S latitudes and near the poles, where air descends (see Figure 54.3).

The spinning of Earth on its axis causes air mass movement to be deflected to the right in the Northern Hemisphere and to the left in the Southern Hemisphere. Thus, winds blowing toward the equator at low latitudes veer and become the northeast and southeast trade winds, whereas winds blowing away from the equator at mid-latitudes are deflected and become the prevailing Westerlies (see Figure 54.4). Ocean currents are driven primarily by prevailing winds but are deflected by continents. The poleward movement of ocean water warmed in the tropics is a major mechanism of heat transfer to high latitudes. Winds can also influence climate by causing upwellings: areas where colder water from depths below 50 meters rises to mix with and replace warmer surface water.

Organisms must adapt to climatic conditions using metabolic, morphological, or behavioral mechanisms. Metabolic adaptations to an unfavorable environment include resting states characterized by greatly reduced metabolic activity and enhanced physiological resistance to adverse conditions. In mammals, such states may occur in summer (estivation) or winter (hibernation). In invertebrates, diapause is the equivalent of hibernation. To cope with extremely cold temperatures, many organisms produce antifreezes that lower the freezing point of their cell contents or body fluids. Morphological adaptations may include differences in body shape and pigmentation. Particularly among ectotherms, behavioral mechanisms for temperature regulation often complement physiological and morphological adaptations. An example of a behavioral adaptation is moving to a more suitable microclimate—a subset of climatic conditions in a small specific area that differ from those in the environment at large. In response to cyclic environmental changes, organisms may evolve life cycles that include migrations that appear to anticipate those changes.

Question 3. Many organisms move from place to place in an effort to locate and occupy the environment to which they are best adapted. Compare the movement of a bird flying back and forth between its tropical wintering area and its temperate breeding range with the movement of a lizard as it shifts between its burrow and the surface and between shaded and sunny areas on the surface. Is the term "migration" properly applied to both types of movement?

Question 4. Organisms must adapt to climatic conditions. Name three types of adaptation and give an example of each.

54.3 How Is Life Distributed in Terrestrial Environments?

- Tundra is found at high latitudes and high elevations
- Evergreen trees dominate boreal and temperate evergreen forests
- Temperate deciduous forests change with the seasons
- Temperate grasslands are widespread
- Hot deserts form around 30° latitude
- Cold deserts are high and dry
- Chaparral has hot, dry summers and wet, cool winters
- Thorn forests and tropical savannas have similar climates
- Tropical deciduous forests occur in hot lowlands
- Tropical rainforests are rich in species

A biome is a large terrestrial environment defined by its climatic and geographic attributes and characterized by ecologically similar organisms. In biomes that occur in several widely separated areas of the globe, the species occurring in the different locations are unlikely to be identical, but they are likely to share many adaptations to their environment as a result of convergent evolution.

The distribution of biomes on Earth is strongly influenced by annual patterns of temperature and precipitation. Often the boundary between two biomes is somewhat arbitrary because one biome gradually merges into another. We distinguish among the following biomes: tundra, boreal and temperate evergreen forest, temperate deciduous forest, temperate grasslands, hot and cold desert, chaparral, thorn forest and savanna, and tropical deciduous forest and tropical rainforest (see Figure 54.7).

Tundra is found at high latitudes in the Arctic and in high mountains. It is a treeless biome dominated by short perennial plants. In the Arctic, permanently frozen soil called permafrost underlies tundra vegetation.

Boreal forests are located at lower latitudes than Arctic tundra and at lower elevations on temperate-zone mountains. The short summers favor evergreen trees, which are the dominant vegetation. Temperate evergreen forests occur along the coasts of continents at middle to high latitudes, where winters are mild and wet and summers are cool and dry.

Temperate deciduous forests are dominated by deciduous trees, which produce leaves that photosynthesize rapidly during the warm, moist summers and are lost during the cold winters.

Temperate grasslands occur in areas that are relatively dry for much of the year. Grasses dominate this biome because they are well adapted to grazing and fire. Because of their rich topsoil, most temperate grasslands have been turned over to agriculture.

Hot deserts, characterized by very warm and dry conditions year-round, are found in two belts around 30°N and 30°S latitude. Plants and animals of this biome are characterized by adaptations for conserving water.

Cold deserts are found in dry regions at middle to high latitudes. Seasonal changes in temperature are great. They are dominated by a few species of low-growing shrubs.

The chaparral biome, found on the west side of continents at mid-latitudes, is dominated by evergreen shrubs and low trees that are adapted to survive periodic fires. Winters are cool and wet; summers are hot and dry.

Thorn forests and savannas are found in semiarid climates on the equatorial side of hot deserts. Savanna, grasslands punctuated by scattered trees, is maintained by grazing, browsing, and burning. In the absence of these, it will revert to dense thorn forest dominated by small spiny shrubs and trees.

Tropical deciduous forests are dominated by trees that lose their leaves during the long, hot dry season. Because their soils are less leached of nutrients than the soils of wetter areas, most of these forests have been cleared for grazing cattle and growing crops.

Tropical rainforests, found in equatorial regions where annual rainfall exceeds 250 cm annually, have the greatest species richness of all biomes and the highest productivity of all ecological communities. Nevertheless, their soils are poor and usually cannot support long-term agriculture unless massively fertilized. Epiphytes—plants that grow on other plants, deriving their nutrients and water from the atmosphere—thrive in tropical mountain forests. Rainforests are being deforested or converted to agriculture at a high rate.

Biomes may occur in close geographic proximity to one another but differ because geological features alter local temperature and precipitation patterns. The movement of air over mountains often results in a rain shadow—a dry area on the leeward side of the range (see Figure 54.8). The distribution of biomes is determined not only by climate but also by other factors, particularly soil fertility and fire.

Question 5. Tropical rainforests have greater overall productivity and species richness than tropical deciduous forests, yet the latter are more easily converted to productive agricultural land than the former. Why?

Question 6. Identify the two biomes shown in Figures A and B below, and describe the major differences between these two biomes.

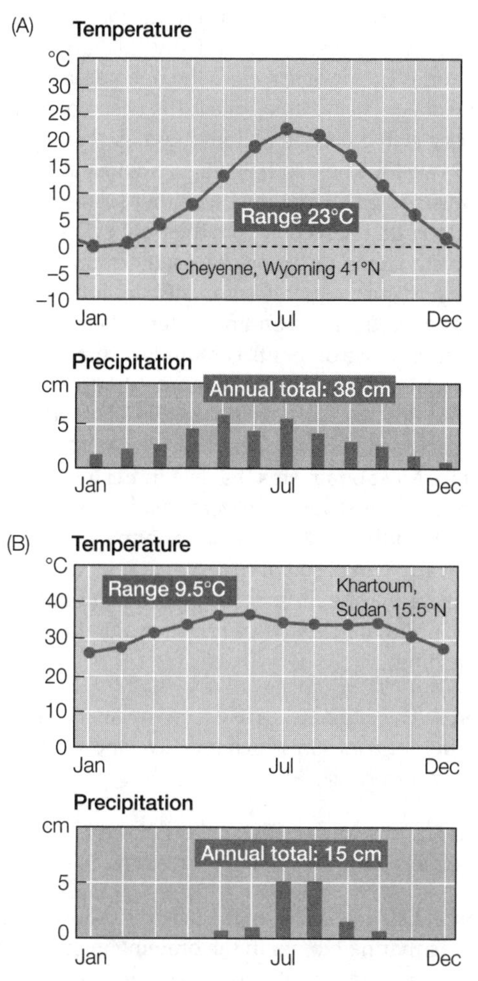

54.4 How Is Life Distributed in Aquatic Environments?

The marine biome can be divided into several life zones

Freshwater biomes may be rich in species

Estuaries have characteristics of both freshwater and marine environments

Aquatic ecosystems can be divided into a number of life zones. Earth's oceans form one interconnected water mass, with only partial barriers to dispersal. Even so, living successfully in a particular region requires that organisms have specific physiological tolerances and morphological characteristics. As a result, most marine organisms have restricted ranges. The marine biome can be divided into several life zones, called the photic zone, coastal zone, intertidal zone, pelagic zone, benthic zone, aphotic zone and abyssal zone (see Figure 54.9).

The photic zone (in both marine and freshwater environments) extends from the surface to the depth at which photosynthesis can no longer occur. Most aquatic life inhabits this zone. The dominant autotrophs are floating microscopic phytoplankton, which are fed upon principally by zooplankton. The coastal zone extends from the shoreline to the edge of the continental shelf. The near-shore region of the coastal zone affected by wave action is the littoral zone, and the portion of this zone between high- and low-tide levels is the intertidal zone. The pelagic zone is the open ocean beyond the coastal zone. The benthic zone is the ocean bottom. The aphotic zone is that portion of both the coastal and pelagic zones that is below the depth at which photosynthesis can occur. Organisms inhabiting this zone either subsist on decaying organic matter that descends from the photic zone or are sustained, directly or indirectly, by chemoautotrophic prokaryotes.

Freshwater environments comprise running water (streams and rivers) and standing water (lakes and ponds). Although these environments contain less than 3 percent of Earth's water, they are the habitat for about 10 percent of all aquatic species. Like oceans, bodies of standing fresh water can be divided into zones based on depth and light penetration. Most organisms that live in fresh water cannot survive in the oceans, and vice-versa, which explains the discontinuity in the ranges of aquatic animals. Estuaries are bodies of water where salt and fresh water mix. Estuarine environments are high in species diversity and important as a resource for humans. Many are now threatened by pollution and overfishing.

Question 7. Trace the life cycle of phytoplankton by describing the different zones in which they live and where they can be found after they die.

Question 8. Compare freshwater and marine biomes, focusing on their size and the organisms residing in each.

Question 9. Draw a "slice" of the ocean to show the life zones of the marine biome. In the drawing include the following: photic zone, aphotic zone, coastal zone, intertidal zone, pelagic zone, abyssal zone, and benthic zone.

54.5 What Factors Determine the Boundaries of Biogeographic Regions?

- Geological history influences the distribution of organisms
- Two scientific advances changed the field of biogeography
- Discontinuous distributions may result from vicariant or dispersal events
- Humans exert a powerful influence on biogeographic patterns

Biogeography is the scientific study of the distributional patterns of populations, species, and ecological communities. Biogeographers divide Earth into a number of biogeographic regions, each containing characteristic assemblages of species. The biotas of the biogeographic regions differ because barriers such as oceans restrict the dispersal of organisms (see Figure 54.11).

Two scientific advances changed the field of biogeography: (1) the acceptance of the theory of continental drift, and (2) the development of phylogenetic taxonomy. Continental drift has influenced the evolution and mixing of species throughout the history of life on Earth. It explains some discontinuous distributions across several biogeographic regions (see Figure 54.12). The development of phylogenetic taxonomy has given biogeographers the ability to convert a taxonomic phylogeny into an area phylogeny, a process that involves replacing the names of the taxa with the names of the places where those taxa live or lived. This method is used to help explain how the current distribution of particular groups of species came about (see Figure 54.14).

Distribution patterns are influenced by vicariance and dispersal. Therefore, a split distribution of a species can be accounted for in two ways: (1) a barrier may arise that splits a species' distribution (a vicariant event), or (2) a species may cross an existing barrier and establish a new population (dispersal). To determine which is more important in a particular case, scientists apply the parsimony principle. That is, they prefer the explanation (as in the case of the distribution of the flightless weevil *Lyperobius huttoni*) that requires the smallest number of unobserved events to account for the pattern (see Figure 54.15). In some cases, vicariance cannot be the explanation for a split distribution because it is clear that no separation of continuous populations could ever have taken place. For example, the occurrence of related species on a continental landmass and on islands never connected to a continent (e.g., the Hawaiian Islands) can only be the result of dispersal.

Humans have deliberately or inadvertently introduced many species to new regions, often with harmful results.

Question 10. More living species of the horse family live in Africa than on any other continent. Yet biogeographers do not believe that the horse family evolved in Africa. What kinds of evidence have biogeographers relied on to explain how the current distributions of horses came about?

Question 11. What is the parsimony principle? Explain how parsimony is helpful in biogeographic studies. Include an example in your answer.

Test Yourself

1. Which of the following is/are *not* usually included within the domain of ecology?
   a. Interactions between conspecifics
   b. Modifications of the environment by organisms
   c. Interactions between humans and domesticated plants and animals
   d. Modifications of the environment by physical processes
   e. Both a and b

   Textbook Reference: *54.1 What Is Ecology?*

2. Which of the following is *not* a difference between an acre of land in Colombia and an acre of land in Michigan?
   a. The angle of the sun reaching the ground during the month of July
   b. The solar energy flux during the month of July
   c. The annual solar energy flux

d. The total hours of daylight per year
e. Both b and c
Textbook Reference: *54.2 Why Do Climates Vary Geographically?*

3. If Earth did not spin on its axis, the northeast trade winds would blow from the
a. northeast.
b. south.
c. north.
d. east.
e. southwest.
Textbook Reference: *54.2 Why Do Climates Vary Geographically?*

4. Which of the following does *not* influence ocean circulation patterns?
a. Circulation of Earth's atmosphere
b. Deflection by land masses
c. Upwelling of deep water
d. Rotation of Earth on its axis
e. All of the above influence ocean circulation patterns.
Textbook Reference: *54.2 Why Do Climates Vary Geographically?*

5. The biome that is maintained by browsers, grazers, or fire is the
a. tundra.
b. temperate grasslands
c. cold desert.
d. tropical deciduous forest.
e. tropical savanna.
Textbook Reference: *54.3 How Is Life Distributed in Terrestrial Environments?*

6. Match the letters of the following biomes with the descriptions that follow.
a. Tundra
b. Boreal forest
c. Hot desert
d. Chaparral
e. Tropical deciduous forest
____ Mostly coniferous, wind-pollinated and wind-dispersed tree species
____ Leaves lost during dry season; agriculturally desirable land
____ Cool winters and hot, dry summers; maritime climate
____ Prominent succulent plants; found at 30°N and 30°S latitude
____ Distribution determined by both the altitude and latitude; permafrost present
Textbook Reference: *54.3 How Is Life Distributed in Terrestrial Environments?*

7. The tropical deciduous forest biome has relatively constant _______, but _______ varies seasonally; the temperate deciduous forest biome has relatively constant _______, but _______ varies seasonally.
a. temperature; rainfall; rainfall; temperature
b. rainfall; temperature; temperature; rainfall
c. rainfall; temperature; rainfall; temperature
d. temperature; rainfall; temperature; rainfall
e. None of the above
Textbook Reference: *54.3 How Is Life Distributed in Terrestrial Environments?*

8. Which of the following biogeographical regions represents the largest area?
a. Nearctic
b. Palearctic
c. Neotropical
d. Oriental
e. Ethiopian
Textbook Reference: *54.5 What Factors Determine the Boundaries of Biogeographic Regions?*

9. The following diagram shows three islands. Islands A and B were connected in the past; Island C was always separate. A species of land snail is found on all three islands.

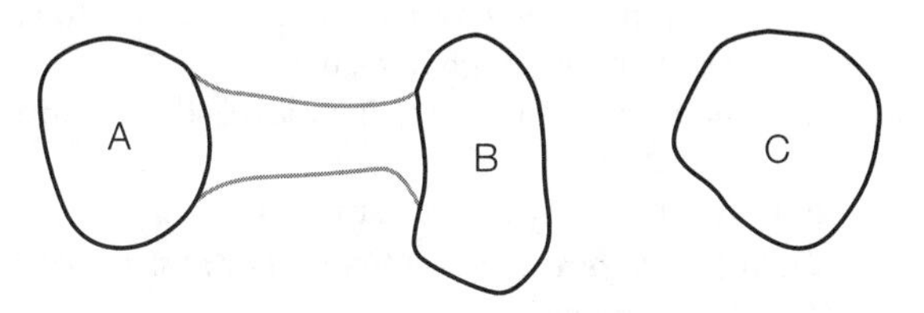

Which of the following does *not* correctly describe the distribution of this snail relative to the three islands?
a. Vicariant distribution relative to A and B
b. Dispersal distribution relative to A and C
c. Dispersal distribution relative to B and C
d. Vicariant distribution relative to A and C
e. All of the above are correct.
Textbook Reference: *54.5 What Factors Determine the Boundaries of Biogeographic Regions?*

10. The region of the ocean that lies close enough to shore to be affected by wave action is the _______ zone.
a. abyssal
b. pelagic
c. benthic
d. aphotic
e. littoral
Textbook Reference: *54.4 How Is Life Distributed in Aquatic Environments?*

11. Which of the following statements comparing freshwater and marine habitats is *false*?
a. About 10 percent of all aquatic species live in freshwater habitats.
b. Unlike the oceans, freshwater lakes and ponds cannot be divided into zones based on depth and light penetration.

c. The global volume of marine habitats is much greater than that of freshwater habitats.
d. Freshwater species richness is less than marine species richness in proportion to the relative extent of the two habitats.
e. Both b and d

Textbook Reference: *54.4 How Is Life Distributed in Aquatic Environments?*

12. An aquatic life zone in which mixing of fresh water and salt water occurs is called a(n)
a. estuary.
b. intertidal zone.
c. littoral zone.
d. pelagic zone.
e. photic zone.

Textbook Reference: *54.4 How Is Life Distributed in Aquatic Environments?*

13. Which of the following statements about estuaries is true?
a. Estuaries have high salt concentrations so they are not beneficial to humans.
b. Estuaries are a type of freshwater biome, and species living in these environments have low salt tolerance.
c. Diversity is very high in estuaries.
d. Estuaries have higher salinity levels than marine biomes.
e. Diversity is very low in estuaries.

Textbook Reference: *54.4 How Is Life Distributed in Aquatic Environments?*

14. Which of the following statements about agriculture in the tropical rainforest biome is true?
a. Only a few hectares of rainforest per year are being cut down or converted to agriculture.
b. These soils are poor in minerals and usually cannot support agriculture without fertilization.
c. Since all seasons in this biome are suitable for plant growth, converting rainforests to agriculture leads to high yield with low input.
d. Rainforests have very low overall productivity among the terrestrial ecological communities; therefore the conversion to agricultural land is encouraged.
e. Turning rainforests into agricultural land does not affect species diversity, since humans can grow a large variety of plants on these lands.

Textbook Reference: *54.3 How Is Life Distributed in Terrestrial Environments?*

15. Which of the following statements about the influence of humans on biogeographic patterns is true?
a. Biographic patterns are too large-scale to be influenced by humans.
b. Humans have very little influence on species distribution, since most species can adapt to new environments.
c. Humans have transported many species to new habitats where those species have become established and compete with native species.
d. All of the above
e. None of the above

Textbook Reference: *54.5 What Factors Determine the Boundaries of Biogeographic Regions?*

Answers

Key Concept Review

1. Biotic factors include the living components of ecosystems, while abiotic factors include the physical and chemical components. The abiotic physical characteristics of Earth's atmosphere, for example, determine surface temperatures and precipitation patterns, which in turn limit where organisms can live.
2. Ecology is a science that generates knowledge about interactions in the natural world; as a field of inquiry, it is not inherently focused on human concerns. Environmentalism is the use of ecological knowledge, along with economics, ethics, and many other considerations, to inform both personal decisions and public policy relating to stewardship of natural resources and ecosystems.
3. Only the bird is engaging in migration, which is the active movement of members of a population, often over long distances, in response to cyclical, predictable environmental events, such as seasonal changes. The lizard, in contrast, travels relatively short distances as it searches for its ideal microclimate, which is a subset of climatic conditions in a small, specific area. For the lizard, temperature is a particularly important aspect of the microclimate. The bird, of course, may also seek out particularly favorable microclimates, not only within its wintering and breeding ranges but also during its migratory journeys.
4. The three adaptation types are metabolic, morphological, and behavioral. An example of metabolic adaptation is the production of antifreezes by temperate-zone insects in order to lower the freezing point of their cell contents or body fluids. An example of morphological adaptation is the proportionally rounder shapes and shorter appendages of endotherms living in cold climates. This adaptation allows them to conserve heat due to reduced surface area. The movement of desert lizards to underground burrows at night is a behavioral adaptation that allows them to maintain their body temperature.
5. The reason for the greater agricultural usefulness of tropical deciduous forests is that their soils are richer in nutrients than the soils of tropical rainforests. Higher precipitation in tropical rainforests causes their soils to lose mineral nutrients very quickly if they are not recycled into the living vegetation, which is in fact where most of the nutrients are located in this type of forest.

6. Figure A shows a cold desert, and Figure B shows a hot desert. Cold deserts are found in dry regions at middle to high latitudes. Seasonal changes in temperature are great. They are dominated by a few species of low-growing shrubs. Hot deserts, characterized by very warm and dry conditions year-round, are found in two belts around the 30°N and 30°S latitudes. Plants and animals of this biome are characterized by adaptations for conserving water.
7. Since phytoplankton are autotroph photosynthetic organisms, while alive, they can be found in the photic zone where they are reached by enough sunlight to support photosynthesis. Within the photic zone, phytoplankton may live in the coastal zone, littoral zone, or pelagic zone. The coastal zone extends from the shoreline to the edge of the continental shelf; it is characterized by relatively shallow, well-oxygenated water and relatively stable temperatures and salinities. The area of the coastal zone at the shoreline that is affected by wave action is called the littoral zone. In the open ocean, or pelagic zone, the principal consumers of phytoplankton are zooplankton. After its death, a phytoplankton will sink to the ocean bottom, or the benthic zone, where it will provide food for other organisms.
8. Salinity is the primary factor that distinguishes the aquatic biomes. The marine biome is characterized by salt water, freshwater biomes are characterized by low salinity, and estuaries are characterized by the mixing of fresh water and salt water. About 70 percent of Earth's surface is covered by saltwater oceans and seas that support abundant life. The small percentage of the aquatic world that consists of fresh water also hosts a significant proportion of Earth's aquatic organisms. In contrast to the vast oceans, bodies of fresh water cover less than 3 percent of Earth's surface, but they are home to about 10 percent of all aquatic species.
9.

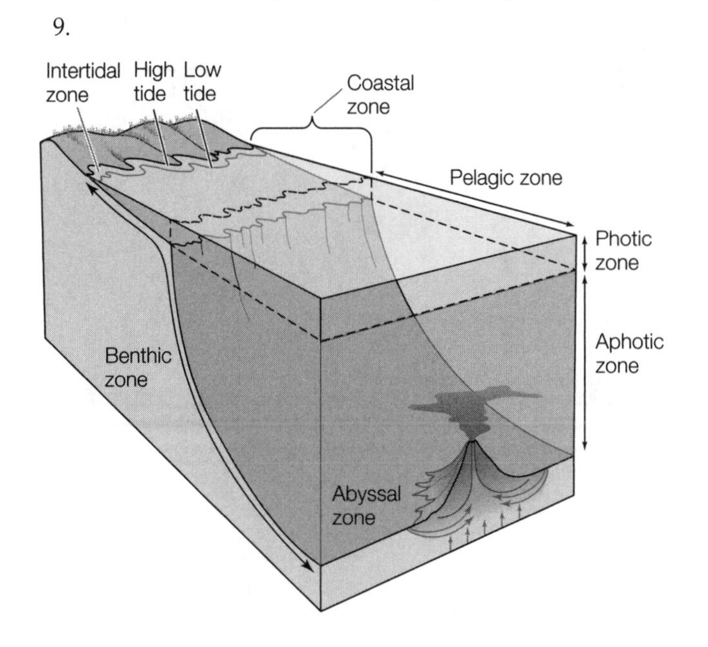

10. The two main sources of evidence used by biogeographers to explain the present distributions of horses are the fossil record and area phylogeny. Fossils clearly indicate that the earliest ancestors of horses evolved in North America. Inasmuch as Przewalski's horse inhabits central Asia (as the area phylogeny in Figure 54.14 shows) and represents the earliest lineage to diverge from the lineages leading to the other living horse species, it is reasonable to assume that the ancestor of all living species of horses first dispersed from North America to Asia. Similar reasoning leads to the conclusion that further speciation events occurred as horses moved from Asia to Africa and that Africa was the site of the speciation of zebras.
11. A split distribution of a species can be accounted for in two ways: (1) a barrier may arise that splits a species' distribution (a vicariant event) or (2) a species may cross an existing barrier and establish a new population (dispersal). To determine which is more important in a particular case, scientists apply the parsimony principle, the explanation that requires the smallest number of unobserved events to account for the pattern. In case of the flightless weevil *Lyperobius huttoni*, for example, its distribution can be explained more simply by a vicariant event than by dispersal. At first glance, its distribution might suggest that, even though this weevil cannot fly, some individuals in the distant past managed to cross Cook Strait, the 25-km-wide body of water that separates the two islands. However, more than 60 other animal and plant species, including other flightless insects, are found on both sides of Cook Strait. Since they cannot fly, wade, or swim, it is unlikely that all 60 of these species made the same ocean crossing independently at different times over the course of their evolutionary history. Thus, a single vicariant event—the separation of the northern tip of South Island from the remainder of the island by the newly formed Cook Strait—could have produced the distribution pattern shared by all 60 species today.

Test Yourself

1. **d.** The modification of the environment by physical factors is not considered part of the domain of ecology; all of the other topics are studied by ecologists.
2. **d.** All areas on Earth receive equal hours of daylight per year. The seasonal distribution of those hours, the solar energy input, and the sun angle do vary latitudinally.
3. **c.** If Earth did not spin on its axis, air flowing south toward the equator would not be deflected to the right. Therefore, the northeast trade winds would blow directly from the north instead of from the northeast.
4. **c.** The upwelling of deep water, especially along the coasts of continents, is an effect of the pattern of ocean current circulation and not a cause of the pattern.
5. **e.** If tropical savanna vegetation is not grazed, browsed, or burned, it typically reverts to dense thorn forest.

6. **b.** Mostly coniferous, wind-pollinated and wind-dispersed tree species

   e. Leaves lost during dry season; agriculturally desirable land

   d. Cool winters and hot dry summers; maritime climate

   c. Prominent succulent plants; found at 30°N and 30°S latitude

   a. Distribution altitudinally or latitudinally determined; permafrost present
7. **a.** The tropical deciduous forest biome has relatively constant temperature, but rainfall varies seasonally; the temperate deciduous forest biome has relatively constant rainfall (or precipitation, because snow may occur in winter), but temperature varies seasonally.
8. **b.** See Figure 54.12 in the textbook.
9. **d.** A vicariant distribution requires that the species once had a continuous distribution in what are now separate areas.
10. **e.** See Figure 54.9 in the textbook.
11. **e.** Life zones occur in bodies of fresh water just as they do in oceans. Even though freshwater habitats occupy a very small global area, about 10 percent of all aquatic species occur in them, which, by proportion, makes them rich in species.
12. **a.** An estuary is the body of water found at the mouth of a river, where salt water mixes with fresh water.
13. **c.** Since they have characteristics of both freshwater and marine systems, estuaries are home to many unique species and play an important role for other species as a conduit between marine and freshwater environments.
14. **b.** Rainforests have the highest overall productivity of all terrestrial ecological communities. However, most mineral nutrients are tied up in the vegetation, so the soils usually cannot support agriculture without massive applications of fertilizers. These forests are home to plants that derive their nutrients and moisture from air and water rather than soil.
15. **c.** Many species have been transported between the continents by humans, either by accident or deliberately. The transporting of these new species often has unintended consequences for the existing species in these regions.

55 Population Ecology

The Big Picture

- An understanding of concepts of population ecology such as exponential growth and carrying capacity is fundamental to the ability to think intelligently about issues such as human population growth, the preservation of biodiversity, and the control of undesirable species.
- Darwin's realization that all populations have the inherent capacity for exponential growth was crucial to the development of his theory of natural selection.

Study Strategies

- Make sure you interpret graphs properly! It is easy to run into problems when interpreting graphs depicting concepts of population ecology. Look carefully at how the axes of a graph are labeled. Ask yourself questions such as: Are the scales arithmetic or logarithmic? Is the fate of a cohort of the population being traced, or is the growth pattern of the entire population being described?
- Redrawing graphs from the textbook is a good way to reinforce your understanding of the concepts being presented.
- Take advantage of laboratory activities involving computer simulations of population growth or other aspects of population ecology.
- Go to the Web addresses shown to review the following animated tutorials and activity. You can also find them in BioPortal (yourBioPortal.com), along with many additional learning resources.

  Animated Tutorial 55.1 Age Structure and Survivorship (Life10e.com/at55.1)

  Animated Tutorial 55.2 Exponential Population Growth (Life10e.com/at55.2)

  Animated Tutorial 55.3 Logistic Population Growth (Life10e.com/at55.3)

  Animated Tutorial 55.4 Habitat Fragmentation (Life10e.com/at55.4)

  Activity 55.1 Logistic Population Growth (Life10e.com/ac55.1)

Key Concept Review

55.1 How Do Ecologists Measure Populations?

Ecologists use a variety of approaches to count and track individuals

Ecologists can estimate population densities from samples

A population's age structure influences its capacity to grow

A population's dispersion pattern reflects how individuals are distributed in space

A population consists of the individuals of a species that interact with one another within a given area at a particular time. Populations have a characteristic age structure (distribution of individuals across age categories) and dispersion pattern (the way those individuals are spread over the environment).

The density of a population is the number of individuals of the population per unit of area. It is a function of the processes that increase the size of a population of individuals by births and immigration (the movement of individuals into the population), and processes that decrease its size, by deaths and emigration (the movement of individuals out of the population). Population dynamics is the patterns and processes of change in populations.

Ecologists use a variety of methods to study populations. For some species, especially those in which individuals are large and population size is small, investigators can perform a full census (complete count) of the population. Ecologists can mark individuals using a variety of methods, from tags and paint to electronic chips. In most species, however, populations are too large for such procedures, so population size must be estimated from representative samples using statistical methods.

Ecologists usually measure the densities of terrestrial organisms as the number of individuals per unit of area. For stationary organisms, ecologists count the number of individuals in sampling plots (quadrats) or along transects and extrapolate the counts to the entire range of the population. For mobile organisms, investigators often use the mark–recapture method.

The age structure of a population describes the proportions of individuals in all age categories. Because reproductive capacity varies with age, the age structure of a population has a profound effect on its potential for growth; it also reveals much about the recent history of births and deaths in the population (see Figure 55.3).

"Dispersion" refers to the spatial distribution of the individuals in a population. Dispersion patterns may be clumped, regular, or random. Abiotic environmental conditions and social interactions such as cooperation and competition are factors that may influence the dispersion pattern of a population.

Question 1. Discuss why you would use different methods to estimate the size of populations made up of large and distinct individuals than you would to estimate the size of populations made up of small, similar-looking individuals.

Question 2. Study the three dispersion types (A–C) below. Name them, briefly explain the conditions under which each occurs, and provide an example of each.

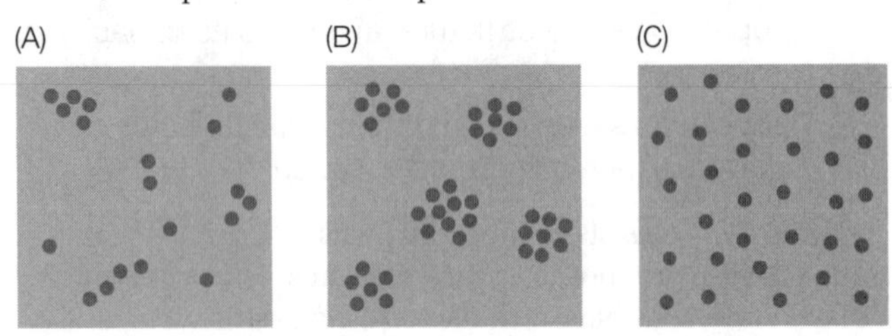

55.2 How Do Ecologists Study Population Dynamics?

Demographic events determine the size of a population

Life tables track demographic events

Survivorship curves reflect life history strategies

The study of population processes is known as demography. Ecologists use multiple estimates of population densities made over time to estimate the rate at which a population is growing or decreasing. Over a given interval of time, the number of individuals in a population increases by the number of individuals added to the population by birth and by immigration, and decreases by the number of individuals lost from the population by death and by emigration.

A life table is a tool used by ecologists to keep track of demographic events (births, deaths, immigration, and emigration) in a population and to determine the rate at which these events occur. A cohort life table is constructed by determining for a cohort (a group of individuals born at the same time) the number still alive at specific times. From these data, investigators can calculate the mortality rate for each age class as well as survivorship (l_x), which is the likelihood of an individual member of the cohort surviving to reach age x (see Table 55.1). The fecundity schedule of a cohort life table tracks the number of offspring produced per female for each age class (see Table 55.1). Data concerning fecundity (m_x) are used to estimate a population's potential for growth. A vertical life table is constructed by sampling a population at a single time.

Survivorship curves show the number of individuals in a cohort still alive at different times over the life span. Graphs of survivorship in real populations often resemble one of three types (see Figure 55.5). In species with a physiological survivorship curve, most individuals survive for most of their potential life span and die at about the same age. Parental care and low fecundity are typical of these species. In species with an ecological survivorship curve, survivorship for individuals is about the same throughout most of the life span. In species with a maturational survivorship curve, survivorship of the young is very low, but it is high for most of the remainder of the life span. Little or no parental care and high fecundity are typical of these species.

Question 3. Draw physiological, ecological, and maturational survivorship curves. For each one, give an example of a species that the curve would apply to.

Question 4. Which type of survivorship curve (physiological, ecological, or maturational) applies to human populations in the United States in the twenty-first century? Would the human survivorship curve 200 years ago in the United States look different? Discuss the reason.

55.3 How Do Environmental Conditions Affect Life Histories?

Survivorship and fecundity determine a population's growth rate

Life history traits vary with environmental conditions

Life history traits are influenced by interspecific interactions

An organism's life history consists of how it allocates its time and energy among growth, reproduction, and other activities. Life histories of different organisms vary dramatically. For example, organisms' life cycles differ in the timing of reproduction and number of offspring produced. Ecologists study life histories because they influence how populations grow and are distributed.

The per capita growth rate of a population is called its intrinsic rate of increase (r). Leaving aside migration, r is the difference between the birth rate (b) and the death rate (d) per individual. It is expressed mathematically by the equation $r = b - d$.

The intrinsic rate of increase of a population is dependent on life history traits that are influenced by environmental conditions. The traits most influenced by these conditions are generation time, number of broods per female, and number of offspring per brood. These factors vary not only between species, but also between populations of the same species.

Iteroparous species are those that reproduce multiple times over the course of their lifetimes. Semelparous species reproduce only once. Predation and other interspecific interactions influence life history strategies.

Question 5. Some species of fish in the salmon family (called salmonids) are anadromous, meaning that they mature in salt water but migrate to bodies of fresh water to breed. These species are usually semelparous. Nonanadromous species in this family live their entire lives in bodies of fresh water and are generally iteroparous. What might be the explanation for the difference in the life history patterns of these two groups of salmonids?

Question 6. Officials of the government of Rongovia, an imaginary country in Eastern Europe, are worried because biologists in the country just published a study having determined that the country's population is declining. The politicians do not understand how this was calculated, or what $r < 0$ means. Briefly explain what $r < 0$ means.

55.4 What Factors Limit Population Densities?

All populations have the potential for exponential growth

Logistic growth occurs as a population approaches its carrying capacity

Population growth can be limited by density-dependent or density-independent factors

Different population regulation factors lead to different life history strategies

Several ecological factors explain species' characteristic population densities

Some newly introduced species reach high population densities

Evolutionary history may explain species abundances

All populations have the potential to grow exponentially. In exponential growth, the per capita rate of increase (r) remains constant per unit of time, whereas the number of individuals added to the population accelerates. This occurs because, as the population grows, more and more individuals are alive to reproduce (see Figure 55.7). Exponential growth is expressed mathematically as $dN/dt = rN$, where dN/dt is the rate of change in the population size (dN = change in number of individuals, dt = change in time, r is the intrinsic rate of growth, and N is the population size). The biotic potential of a population is expressed when it is growing exponentially.

The environmental carrying capacity (K) is the maximum population size that the environment can support indefinitely. Because growth tends to slow as the carrying capacity is approached, an S-shaped growth curve (logistic growth) is characteristic of many populations growing in environments with limited resources. This type of growth can be generated from the exponential growth equation by adding the term $(K - N)/K$, which represents environmental resistance—the reduction in population growth caused by preemption of available resources. Hence the equation for logistic growth is $dN/dt = rN\,((K - N)/K)$. Population growth stops when $N = K$ because then $(K - N)/K = 0$, and thus $dN/dt = 0$ (see Figure 55.8).

Density-dependent regulation factors—such as food supply, predators, and pathogens—cause per capita birth or death rates of a population to change in response to changes in population density. As a result, population growth slows down as density increases. Density-independent factors (such as adverse weather) change per capita birth or death rates of a population irrespective of its density. Abiotic factors tend to be density-independent in their effect on populations, whereas biotic factors are typically density-dependent.

Different population regulation factors and different habitats are associated with different life history strategies (see Figure 55.9). Species termed *r*-strategists have life histories geared to achieve the maximum rate of population increase. They tend to live in a broad range of unpredictable environments and are characterized by a short life span, density-independent mortality, semelparity, and a maturational survivorship curve. Species termed *K*-strategists have life histories that allow them to persist at or near the carrying capacity of their environment. They tend to be specialized in their resource use and live in predictable environments. They are characterized by a long life span, density-dependent mortality, iteroparity, and a physiological or ecological survivorship curve. The life histories of most species combine elements of both strategies.

Some species are more common than others for several reasons. Factors favoring high population density are utilization of abundant resources, small body size, complex social organization, and recent introduction into a new region (see Figure 55.10). In some cases, species have a small population because they have arisen only recently by polyploidy or a founder event, or they are declining toward extinction.

Question 7. The zebra mussel was not found in North America before 1985, but today it is an abundant pest species in the Great Lakes region and Mississippi River drainage. What principles of species abundance would apply to the explosive increase of the zebra mussel?

Question 8. Draw an exponential population growth curve, with time on the x-axis and number of individuals on the y-axis. Then assume that, due to limiting food supply, the environment cannot support more than 5,000 individuals, and edit the graph, leveling it off at 5,000 individuals. What is the term for this maximum number of individuals (in this case that the environment can support)? What is the term for the revised growth curve that levels off at 5,000 individuals?

55.5 How Does Habitat Variation Affect Population Dynamics?

Many populations live in separated habitat patches

Corridors may allow subpopulations to persist

Most populations are divided into geographically separated subpopulations that live in habitat patches—areas of a particular kind of environment that are surrounded by areas of less suitable habitat. The larger population to which the subpopulations belong is called the metapopulation (see Figure 55.10). Because the individual subpopulations are much smaller than the metapopulation, they are prone to fluctuations leading to extinction. In the rescue effect, immigrants from other subpopulations prevent declining subpopulations from becoming extinct.

In any metapopulation, habitat between patches through which organisms can move, known as corridors, plays a critical role in facilitating dispersal to maintain subpopulations, and hence enhance the rescue effect.

Question 9. How does the division of some populations into discrete subpopulations relate to the process known as the rescue effect? How do barriers to immigration influence the rescue effect?

Question 10. One year you see beautiful butterflies in your garden. The next year there is an extreme drought; the flowers that had fed the butterflies do not bloom and the butterflies do not appear. Then, in year 3, there is plenty of rain, the flowers bloom, and the butterflies reappear. How would you explain the reappearance of the butterflies?

55.6 How Can We Use Ecological Principles to Manage Populations?

Management plans must take life history strategies into account

Management plans must be guided by the principles of population dynamics

Human population growth has been exponential

Understanding the life history traits of a commercially valuable species is important for the successful management of its populations. To maximize the number of individuals that can be harvested from a population, the population should be held far enough below carrying capacity to have high birth and growth rates.

Demographic traits, particularly reproductive capacity, determine how heavily a population can be exploited. For commercial reasons, many fish and whale populations have been overharvested in violation of basic principles of population management (see Figure 55.14).

The populations of undesirable species are most effectively controlled by lowering the carrying capacity of the environment for those species. For introduced pest species, biological control—the use of natural enemies to reduce the population density—is sometimes effective, but can be disastrous (see Figure 55.15).

The size of the human population contributes to most of the environmental problems we are facing today, from pollution to extinctions of other species. Earth's carrying capacity for humans has drastically increased because of the development of social systems and communication, the domestication of plants and animals, and ongoing technological advances in food production and disease control. Today the planet supports approximately 7 billion human beings (see Figure 55.16).

Human populations are growing at different rates in different parts of the world. Industrialized countries typically have an age structure that reflects their low birth rates and stable (or even declining) population size. In the developing world, many countries have a high rate of growth and populations skewed toward younger age categories, portending high rates of growth in the future (see Figure 55.17). Earth's carrying capacity for humans depends on our use of resources and the effects of our activities on other species.

Question 11. As an alternative to chemical pesticides, ecologists have often recommended biological control: the introduction of predators or parasites of pests (including non-native species) to bring an undesirable population under control. What are the risks of this strategy?

Question 12. How does the age structure of a country's population affect its potential for future population growth? In your answer, contrast an industrialized country, such as Germany, with a developing country, such as Uganda (see Figure 55.17).

Test Yourself

1. Which of the following is an aspect of population structure studied by ecologists?
   a. Distribution of genotypes
   b. Age distribution
   c. Spacing of population members
   d. Both b and c
   e. All of the above

   Textbook Reference: *55.1 How Do Ecologists Measure Populations?*

2. In an effort to measure the size of a bluegill population in a pond, you capture 40 bluegills, mark each with a tag on the tail fin, and then release them. A week later you return to the pond and catch 40 more, of which 8 have a tag on the tail fin. What is the best estimate of the bluegill population in the pond?
   a. 100
   b. 160
   c. 200
   d. 320
   e. 400

   Textbook Reference: *55.1 How Do Ecologists Measure Populations?*

3. In the life table below, survivorship is greatest during which time interval?

Age (years)	Number Alive
0	800
1	770
2	550
3	125
4	75
5	0

   a. 0–1 years
   b. 1–2 years
   c. 2–3 years
   d. 3–4 years
   e. 4–5 years

   Textbook Reference: *55.2 How Do Ecologists Study Population Dynamics?*

4. Refer to the graph of age distributions below.

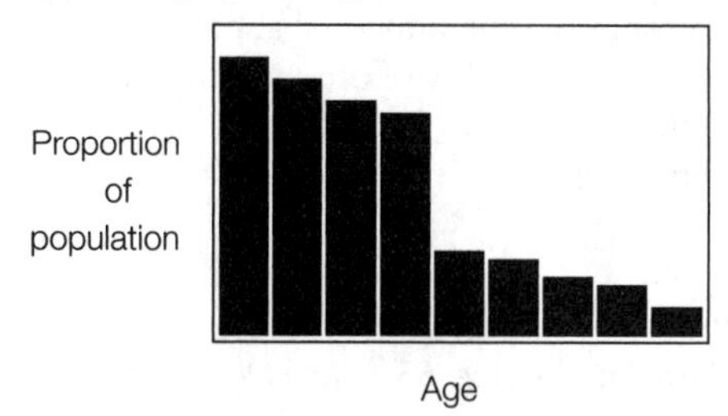

For a population that is stable in size, the age distribution of the graph would indicate
 a. high birth and death rates.
 b. low birth and death rates.
 c. a low birth rate but a high death rate.
 d. a high birth rate but a low death rate.
 e. Either a or d
 Textbook Reference: *55.6 How Can We Use Ecological Principles to Manage Populations?*

5.–6. Refer to the graph below.

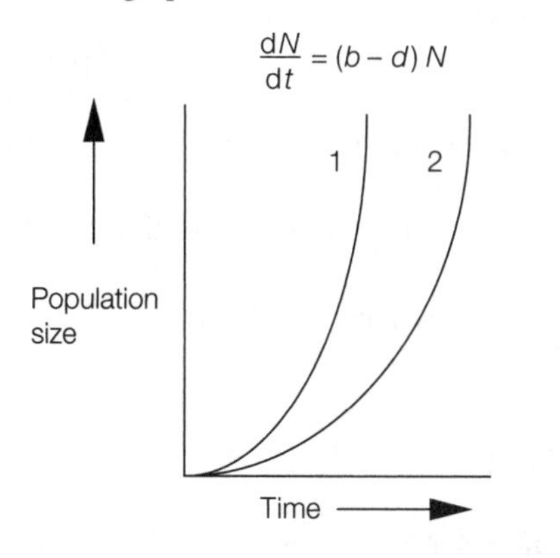

5. In the graph, which of the expressions of the exponential growth equation would need to be increased for curve 1 to become more like curve 2?
 a. *dN*/d*t*
 b. *d*
 c. *b*
 d. (*b* – *d*)
 e. Either b or d
 Textbook Reference: *55.4 What Factors Limit Population Densities?*

6. The intrinsic rates of increase (*r*) of species A = 0.25 and of species B = 0.50. According to the graph, the population growth curve for A should be more like curve _______.
 Textbook Reference: *55.3 How Do Environmental Conditions Affect Life Histories?*

7.–8. Refer to the graph below.

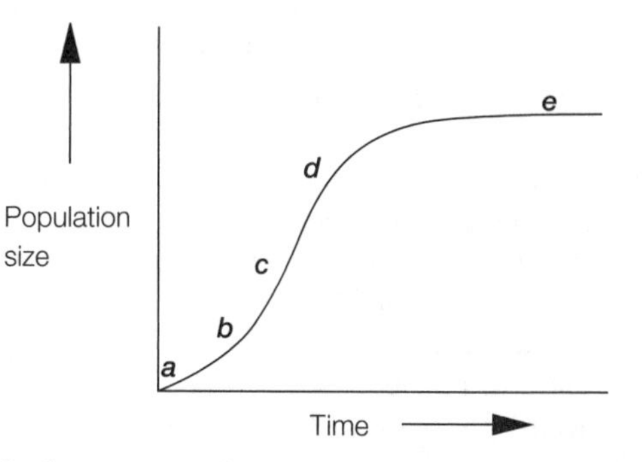

7. In the logistic population growth curve shown in the graph, the rate of growth is greatest at which point?
 Textbook Reference: *55.4 What Factors Limit Population Densities?*

8. At which point in the graph would there be zero population growth ($dN/dt = 0$)?
 Textbook Reference: *55.4 What Factors Limit Population Densities?*

9.–10. Refer to the graph below.

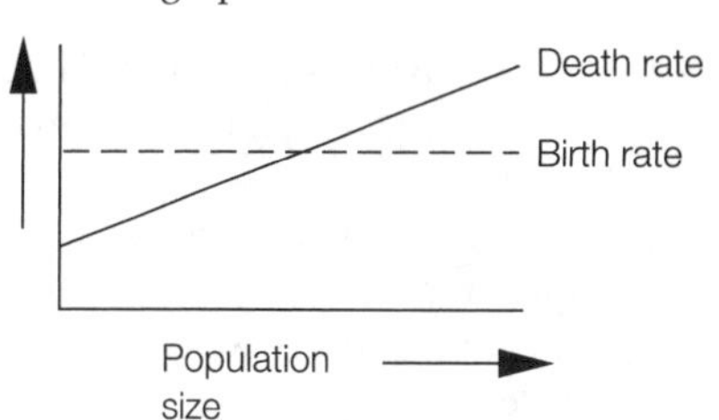

9. Which of the following events is *least* likely to be true of this population?
 a. Pathogens spread between population members.
 b. Increased competition for food causes some individuals to delay reproduction.
 c. The number of predators in the area varies with population density.
 d. Territorial disputes lead to injury and death of some males.
 e. Periods of extreme cold periodically reduce the population density.
 Textbook Reference: *55.4 What Factors Limit Population Densities?*

10. In the graph, the population size at which the curves for the birth and death rates intersect is
 a. an estimate of the environmental carrying capacity.
 b. an estimate of the intrinsic rate of increase.
 c. the point at which density-dependent regulation begins.
 d. the point at which density-independent regulation begins.
 e. an estimate of the biotic potential of the population.
 Textbook Reference: *55.4 What Factors Limit Population Densities?*

11. Which of the following would probably *not* be true of a population whose dynamics are primarily influenced by density-independent factors?
    a. A growth pattern that is similar to the logistic growth curve
    b. A birth rate that is dependent on the nutritional status of its adult females
    c. The highest percentage of deaths caused by unfavorable weather conditions
    d. Both a and b
    e. All of the above
    Textbook Reference: *55.4 What Factors Limit Population Densities?*

12. Which of the following would probably *not* be true of a *K*-strategist?
    a. A long life span
    b. A maturational survivorship curve
    c. Density-dependent mortality
    d. A specialized diet
    e. Iteroparity
    Textbook Reference: *55.4 What Factors Limit Population Densities?*

13. Which of the following tends to be associated with the ability of a species to attain high population densities?
    a. Small body size
    b. Utilization of abundant resources
    c. Complex social organization
    d. Recent introduction into a new region
    e. All of the above
    Textbook Reference: *55.4 What Factors Limit Population Densities?*

14. The best way to decrease the population size of a pest species is to
    a. poison it.
    b. introduce additional predators.
    c. decrease the carrying capacity of its habitat.
    d. introduce competitors.
    e. introduce a pathogen.
    Textbook Reference: *55.6 How Can We Use Ecological Principles to Manage Populations?*

15. Which of the following factors is significant in setting the present carrying capacity of Earth for the human population?
    a. Earth's ability to absorb by-products of human consumption of fossil fuel energy
    b. The availability of water that is suitable for drinking, irrigation, and other human uses
    c. Human willingness to cause extinctions of other species in order to expand the human share of Earth's resources
    d. Both a and c
    e. All of the above
    Textbook Reference: *55.6 How Can We Use Ecological Principles to Manage Populations?*

16. In the dispersion pattern called _______ dispersion, the presence of one individual at any point in space increases the probability of others being near that point.
    a. natural
    b. interspecies
    c. clumped
    d. regular
    e. random
    Textbook Reference: *55.1 How Do Ecologists Measure Populations?*

17. The intrinsic rate of increase is calculated by
    a. $b = r - d$.
    b. $d = r/b$.
    c. $r = d + b$.
    d. $r = b - d$.
    e. $d = r^2$.
    Textbook Reference: *55.3 How Do Environmental Conditions Affect Life Histories?*

18. Which of the following statements about ecological corridors is true?
    a. They limit dispersal and separate subpopulations.
    b. They facilitate dispersal to maintain subpopulations.
    c. They are habitats between patches through which organisms can move.
    d. They are only useful for species that can fly.
    e. Both b and c
    Textbook Reference: *55.5 How Does Habitat Variation Affect Population Dynamics?*

Answers

Key Concept Review

1. In some species, individuals are large and distinct enough, and populations small enough, that investigators can identify all the individuals and count them and a full census can be performed. For example, in Samburu and Buffalo Springs National Reserves in Kenya, a census of the African elephant population was performed by monitoring the elephants for 21 months. The biologists learned to recognize each of the 760 individuals in the population primarily by their unique and distinctive ear markings. For most species, however, recognizing individuals is impossible or impractical. For biologists to identify individuals of such species, the individuals must be marked in some way (e.g., colored leg bands to mark birds).

2. (A) represents random dispersion, (B) represents clumped dispersion, and (C) represents regular dispersion. A dispersion pattern in which the presence of one individual at any point in space does not affect the probability of other individuals being near that point is called random dispersion. For example, plant seeds can be distributed by wind, and will grow in a random distribution. A dispersion pattern in which the presence

of one individual at any point in space increases the probability of others being near that point is called a clumped dispersion. For example, predators such as orcas or lions that hunt together show a clumped dispersion. A dispersion pattern in which the presence of one individual at any point in space reduces the probability of others being near that point is called regular dispersion. For example, seabirds nest at approximately the same distance from each other in a regular dispersion pattern.

3. See Figure 55.5.
4. In the twenty-first century the human survivorship curve in first-world countries looks more like the physiological curve, because most individuals survive to a relatively old age, and the number of survivors at an early age is high. Two hundred years ago, nutrition, medical care, and the social infrastructure (e.g., sanitation) were not as developed. Without medications, vitamins, and sanitization, the childhood death rate was higher, and the number of survivors at a higher age was lower. Therefore a survivorship curve 200 years ago would approach the ecological or even the maturational survivorship curves.
5. Semelparity is typical of organisms that experience very high rates of mortality as adults, whereas iteroparity is typical of organisms whose survival chances increase once they reach adulthood. Anadromous fish die in large numbers in the course of their breeding migration, so the chance that a salmon could make a second successful migration is extremely low. By comparison, nonanadromous salmonids live as adults in a comparatively stable environment in which they may have multiple opportunities to breed.
6. To see how a population is likely to grow, ecologists use life table data to calculate the population's per capita growth rate, which is also known as the intrinsic rate of increase (r). A population's intrinsic rate of increase is the difference between the per capita birth rate (b) and the per capita death rate (d). In other words, it is the average rate of change in population size per individual per unit of time, expressed by the equation $r = b - d$. If the per capita death rate is greater than the per capita birth rate, then $r < 0$ and the population is declining. This is what is happening in Rangovia.
7. The high population density of zebra mussels on this continent illustrates that species introduced into a new region often reach levels of abundance above those found in their native ranges because their normal diseases and predators are absent (see Figure 55.10).
8. An exponential population growth curve should look like Figure 55.7. Any given environment has only enough resources to support a finite number of individuals of a species indefinitely, which is referred to as the environment's carrying capacity (K). In this case $K = 5{,}000$. If the carrying capacity is applied to a population with exponential growth, the result is a logistic growth curve similar to Figure 55.8.
9. The rescue effect occurs if a subpopulation that has undergone a large decline is saved from extinction by the immigration of individuals from other subpopulations. As one would predict, barriers to immigration reduce the rescue effect.
10. The population of butterflies in your garden patch is a subpopulation. It is linked to other populations by the regular movement of individuals between patches. The larger population that includes all such subpopulations is known as a metapopulation. Because the subpopulations are much smaller than the metapopulation, local disturbances (such as the drought) are more likely to cause the extinction of a subpopulation than of the entire metapopulation. Therefore, the butterflies went locally extinct (in your garden), but immigrated again from other patches within the metapopulation.
11. One risk of the biological control strategy is simply that the introduced parasite or predator will fail to control the pest. A more serious hazard is that the introduced species will attack or outcompete other species that are considered valuable. An example of this outcome—the case of the Central American cane toad introduced into Australia—is described in this chapter (see Figure 55.15).
12. Because a high proportion of the German population is relatively old, the death rate in that country is likely to equal or exceed the birth rate in coming years and the population should remain stable or possibly decline. In contrast, the high proportion of prereproductive individuals in the age structure of Uganda portends a high rate of population growth in the future because the birth rate is likely to exceed the death rate for many years to come.

Test Yourself

1. **d.** The aspects of population structure studied by ecologists include the spacing and age distribution of the members of the population but not the distribution of genotypes.
2. **c.** If we assume that the population of bluegills randomly mixed in the interval between the capture of the first sample and the capture of the second sample, and if marked and unmarked individuals were equally likely to survive, then the proportion of tagged bluegills in the second sample can be used to estimate the total bluegill population. Because 8 of the 40 bluegills in the second sample were tagged (i.e., one-fifth of the sample), the best estimate of total population of bluegills is 5×40 (i.e., the size of the first sample) = 200.
3. **a.** The decrease in consecutive age classes is the smallest between 0 and 1 year, so the survivorship is greatest during that interval.

4. **a.** When a population's birth rate and death rate are both high, the age distribution is dominated by young individuals. A similar age distribution would occur in a population with a high birth rate and a low death rate, but such a population would not be stable in size.
5. **b.** Both curves show exponential growth, but the rate of exponential growth for curve 2 is less than the rate for curve 1. Because the rate of growth is determined by the expression $(b - d)$, increasing d, the death rate, would cause curve 1 to be more like curve 2.
6. The population growth curve for A should be more like curve 2. Because $r = (b - d)$, you would expect the growth curve for species A to correspond to the curve with the lower rate of exponential growth.
7. **c.** The curve showing the relationship between population size and time is steepest at point *c*, so the growth rate would be greatest at that point.
8. **e.** The growth rate at *e* would be zero. This is the point, called the environmental carrying capacity, at which the birth and death rates are equal.
9. **b.** The graph indicates that the birth rate is independent of population size, so event *b* would not be true.
10. **a.** The environmental carrying capacity is the equilibrium population size when the birth rate equals the death rate. The birth rate equals the death rate at the intersection of the two curves.
11. **d.** The logistic growth curve describes density-dependent growth because the rate of increase in the size of the population decreases steadily as the carrying capacity is approached. The nutritional status of females in the population would depend on the availability of food and hence would be density-dependent. By contrast, unfavorable weather generally occurs without any predictable relationship to the size of a population and thus is a density-independent factor.
12. **b.** Species with maturational survivorship curves produce many offspring, the vast majority of which die before reaching adulthood. *K*-strategists, in contrast, produce relatively few offspring, many of which are likely to survive to adulthood because of the large parental investment in each one.
13. **e.** All four factors favor the ability to achieve high population densities.
14. **c.** The best way to control a pest species is to reduce the carrying capacity of its habitat by removing the resources it depends on.
15. **e.** All three of these factors contribute to setting the present carrying capacity of Earth for humanity, though water availability is not a factor everywhere.
16. **c.** A clumped dispersion pattern is one in which the presence of one individual at any point in space increases the probability of others being near that point.
17. **d.** A population's intrinsic rate of increase is the difference between the per capita birth rate (b) and the per capita death rate (d).
18. **e.** In any metapopulation, habitat between patches through which organisms can move, known as corridors, play a critical role in facilitating dispersal to maintain subpopulations.

Species Interactions and Coevolution

The Big Picture

- The central role that the concept of competition played in Darwin's thinking as he formulated the theory of natural selection is indicated by his inclusion of the phrase "the preservation of favored races in the struggle for life" in the full title of *The Origin of Species*.
- As you learned in this and previous chapters, some mutualistic relationships have extraordinary evolutionary or ecological significance. Examples described in previous chapters are the role of endosymbiosis in the evolution of eukaryotic cells and the association of nitrogen-fixing bacteria with their host plants.

Study Strategies

- The niche of a species is often confused with its habitat. The latter term refers to the preferred physical environment of a species (e.g., pond versus stream), whereas the former term refers to the entire set of physical and biological conditions a species needs in order to survive and reproduce. Thus the habitat of a species is just one of many attributes of its niche. In addition, the distinction between the fundamental niche of a species and its realized niche can be confusing. Reviewing Connell's classic experiment on barnacles will help clarify the distinction (see Figure 56.15).
- This chapter describes many different categories of interspecific interactions and introduces some terms relating to these categories that may be new to you. To learn the basic types of interactions, study the table presented in Figure 56.1A. Then try creating your own charts or concept maps to organize the material in more detail and to master the terminology.
- Go to the Web addresses shown to review the following animated tutorials and activity. You can also find them in BioPortal (yourBioPortal.com), along with many additional learning resources.

  Animated Tutorial 56.1 Coevolution: Strategies for Survival (Life10e.com/at56.1)

  Animated Tutorial 56.2 Mutualism (Life10e.com/at56.2)

  Activity 56.1 Ecological Interactions (Life10e.com/ac56.1)

Key Concept Review

56.1 What Types of Interactions Do Ecologists Study?

Interactions among species can be grouped into several categories

Interaction types are not always clear-cut

Some types of interactions result in coevolution

Ecologists study several types of interactions between species. Interactions between species in a community fall into seven general categories that reflect whether the outcome of the interactions is positive (+), negative (–), or neutral (0) for each of the species involved (see Figure 56.1).

Antagonistic interactions are those three interactions in which one species benefits and the other is harmed. Such interactions include (1) predation, in which an individual of one species kills and consumes multiple individuals of another species; (2) herbivory, in which an individual of a species consumes part or all of a plant; and (3) parasitism, in which an individual of one species consumes only certain tissues of one or a few individuals of another species (the host). Pathogens are parasites that cause symptoms of disease in their hosts.

Mutualism is a type of interaction between species that benefits both parties. Competition is an interaction in which two or more species use the same resource, and is harmful to all species involved. Commensalism is an interaction that is beneficial to one participant while leaving the other unaffected. Amensalism is harmful to one participant while leaving the other unaffected.

Although it is useful to group interactions among species into categories, the boundaries between categories are not always clear. These different types of interactions are in reality part of a continuum, and their outcomes depend on ecological and evolutionary circumstances.

Some of these interactions result in coevolution, which means that an adaptation in one species leads to the evolution of an adaptation in a species with which it interacts. In a coevolutionary arms race, the evolution of traits that improve the fitness of a predator, herbivore, or parasite exerts selection pressure on its prey or host to counter the consumer's adaptation. The prey or host adaptation, in turn, exerts selection pressure on the consumer to improve its fitness, resulting in an escalating arms race.

Question 1. Two of the main characters in the animated film *Finding Nemo* are anemonefish (also known as clown fish), which live inside of sea anemones and are unaffected by their stings. How is this relationship an example of the ways in which interactions between species are not always easily classifiable?

Question 2. Draw a table with three columns and seven rows. In column 1, list the seven different species interactions discussed in the chapter. In the next two columns, indicate the effects of the interaction, first on species 1 and then on species 2. Indicate whether the effect is positive, negative, or neutral using the symbols "+", "–," or "0." Which of these interactions are antagonistic?

56.2 How Do Antagonistic Interaction Evolve?

Predator–prey interactions result in a range of adaptations

Herbivory is a widespread interaction

Parasite–host interactions may be pathogenic

Antagonistic interactions result in a wide range of adaptations. Predators invariably kill their prey, and over its lifetime one individual predator kills and eats many prey individuals. Predators tend to be less specialized than other consumers, and most are larger than their prey, though a few are smaller. Predators in the former category generally use their strength or swiftness to capture their prey, whereas those in the latter category employ strategies such as venom production.

Prey species have evolved a variey of types of defenses against predators, including swiftness and protective exteriors such as tough skins, shells, and spines. Animals that do not have features such as swiftness or morphological adaptations to protect themselves often rely on chemical defenses to escape or repel predators; these may, in turn, evolve to overcome their prey's defenses. In some cases, a predator may even ingest and sequester its prey's defensive chemicals and use them to protect itself against its own predators. Prey species protected by chemical defenses often advertise their toxicity by warning coloration, or aposematism. Visually hunting predators, especially among the vertebrates, can learn to avoid these species.

Prey species have evolved several methods of avoiding detection. These include crypsis (camouflage), homotypy (resemblance to an inedible object), ceasing to move, and even "playing dead." In Batesian mimicry, a nontoxic prey species (the mimic) evolves a resemblance to an aposematic toxic species (the model). In Müllerian mimicry, two or more aposematic species converge to resemble one another.

Herbivory is the most widespread interaction between species on Earth. Most of the world's herbivores are insects, and they are generally oligophagous—meaning that they feed on one or a few species of plants. In contrast, the great majority of vertebrate herbivores are polyphagous, feeding on many plant species. Though herbivores seldom kill their food plants, they can reduce plant fitness if the plants they attack produce fewer offspring. Although most plants defend themselves against their consumers by producing defensive chemicals (secondary metabolites), many plants have additional defenses. These include thorns, spines, hooked hairs on leaf surfaces, cuticles, and silica.

The great biochemical diversity of the flowering plants and the comparable diversity of herbivorous insects is the consequence of an evolutionary arms race in which plants that have evolved a novel secondary metabolite undergo an adaptive radiation that leads to an adaptive radiation of insects that have evolved resistance to the chemical. With sufficient selection pressure, a resistant herbivore can evolve to use the chemical against its own predators. Large polyphagous herbivores such as deer and horses minimize their exposure to any particular defensive chemical by feeding on a wide variety of plant species and by learning to avoid plants with an unpleasant taste.

Parasitism is an interaction in which one species consumes only certain tissues in one or a few host individuals of another species without necessarily killing them. Microparasites are many orders of magnitude smaller than their hosts and generally live and reproduce inside their hosts. Pathogenic microparasites may cause the death of the host, but because pathogens must continually infect new host individuals, natural selection may favor less deadly strains that can infect a larger number of new hosts. Thus, pathogen and host may reach a state of coexistence that involves increased host resistance and decreased pathogen virulence (ability to cause disease). The hosts of pathogens fall into three classes: (1) susceptible (capable of being infected), (2) infected, and (3) recovered (and thus, in many cases, immune to another infection). When a pathogen invades a population, rates of infection typically rise at first and then fall as the number of susceptible individuals decreases. The rates do not rise again until a sufficiently dense population of susceptible hosts has reappeared. Larger parasites, called macroparasites, rarely cause the kinds of disease symptoms that pathogenic microparasites cause, but they can affect host fitness nevertheless. Ectoparasites (external parasites) are often associated with their hosts only casually. They typically have a number of morphological adaptations that enable them to remain attached to their hosts (see Figure 56.8). Some biologists believe that hairlessness, bipedality, and the opposable thumb evolved in humans as a response to ectoparasites.

Question 3. Some mimicry systems (e.g., those involving tropical butterflies) include several toxic species and several palatable species, all of which are aposematic and resemble one another closely. How would you classify these mimicry systems—Batesian, Müllerian, or both? Why might natural selection have favored the evolution of such complex mimicry systems?

Question 4. After the European rabbit was introduced into Australia in 1859, it quickly became a devastating invasive species, causing the extinction or decline of many native animals and plants, major damage to crops, and soil erosion. To control the exponential growth of the population (which had grown to an estimated 600 million individuals from an original 24), biologists infected rabbits with the myxoma virus in 1950. It is estimated that in the first years after introduction of the virus, as many as 99 percent of infected rabbits died. Over a period of years, however, biologists observed an increase both in the average life span of infected rabbits and in the percentage of rabbits surviving infection. How would you explain these changes?

56.3 How Do Mutualistic Interactions Evolve?

Some mutualistic partners exchange food for care or transport

Some mutualistic partners exchange food or housing for defense

Plants and pollinators exchange food for pollen transport

Plants and frugivores exchange food for seed transport

Like antagonistic interactions, mutually beneficial interactions between species can result in coevolution. Many mutualistic interactions are tightly coevolved. Mutualisms can form virtually without regard to the taxonomic groups to which the participants belong. Many mutualisms involve an exchange of food for housing or defense and arise in environments in which resources are in short supply. Reciprocal adaptations, which ensure that both partners benefit from the exchange, are most likely to arise when an increase in dependency on a partner provides an increase in the benefits realized from the interaction.

About three-quarters of all flowering plant species are animal-pollinated. A mutualistic pollination system requires that (1) the plant provide an attraction or reward to entice the pollinator; (2) the pollinator be able to transport the plant's pollen; and (3) the pollinator visit more than one individual of a plant species. Both plants and pollinators may take advantage of their partners. For example, some plant species have evolved flowers that deceive insects into attempting to copulate with them, and "thieves" that do not transport pollen may nevertheless consume nectar.

Floral characteristics and the timing of flowering can restrict the number of potential pollinators and encourage pollinator fidelity. Nevertheless, most flowers can be pollinated by a number of animal species. Broad suites of floral characteristics that attract certain groups of pollinators exemplify diffuse coevolution: the evolution of similar traits in groups of species experiencing similar selection pressures (see Table 56.1). A few plant–pollinator relationships are much more exclusive and lead to highly specific coevolution.

Many animals that eat fruits (frugivores) disperse the seeds of the plants that produce them. Plants that depend on frugivores for plant dispersal must achieve a balance among three factors: (1) discouraging frugivores from eating fruits before the seeds are mature; (2) attracting frugivores when the seeds are mature; and (3) protecting the seeds from destruction in an animal's digestive tract. Because the mutualism between plants and frugivores is often asymmetric (i.e., one partner benefits more than the other), relatively few highly specialized frugivores exist.

Question 5. Explain diffuse coevolution in pollination.

Question 6. Are most frugivores specialized? Explain your answer.

56.4 What Are The Outcomes of Competition?

Competition is widespread because all species share resources

Interference competition may restrict habitat use

Exploitation competition may lead to coexistence

Species may compete indirectly for a resource

Competition may determine a species' niche

Competition determines which species can persist and which may go extinct, or how many different species can be supported by a particular resource. Competition occurs whenever any resource is not sufficiently abundant to meet the needs of all the organisms with an interest in that resource. Intraspecific competition occurs among individuals of the same species and is a major density-dependent mechanism of population regulation. Interspecific competition occurs among individuals of different species; in addition to limiting population growth, it can influence the persistence and evolution of species. The principle of competitive exclusion states that no two species can coexist for long if they share the same limiting resource. In some cases, the inferior competitor may go extinct locally (competitive exclusion). In other cases, selection pressures resulting from competition may change the ways in which different species use a limiting resource (resource partitioning).

Both intraspecific and interspecific competition occur by means of two major mechanisms. In interference competition, an individual interferes with a competitor's access to a limiting resource. Interference competition may involve restriction of habitat use by a competitor or the production of chemicals that interfere with life processes of a competitor. In exploitation competition, a limiting resource is available to all competitors and the outcome depends on the relative efficiency with which competitors use up the resource. One possible evolutionary outcome of exploitation competition is resource partitioning. Another possible outcome is character displacement, in which morphological attributes of a species vary geographically depending on whether a competitor is present or absent.

Indirect competition may occur if: (1) one species so alters the quality of a resource that it is rendered less usable by other species that encounter it afterward; or (2) the outcome of competition hinges on how the competitors interact with a shared predator.

The niche of a species is the set of physical and biological conditions necessary for its persistence. Every species has a fundamental niche, defined by its physiological capabilities, and a realized niche, defined by its interactions with other species.

Question 7. Review the description of Connell's classic experiment on competition between *Semibalanus* and *Chthamalus* barnacle populations (see Figure 56.15). Is the interaction between these species an example of interference competition or exploitation competition? Why?

Question 8. Describe an experiment that would enable you to assess the degree of interspecific competition that exists between two rodent species living in the same habitat. How would you interpret the results?

Test Yourself

1. Which of the following does *not* describe an antagonistic interaction between two living organisms?
   a. Sheep grazing on grass in a pasture
   b. A robin catching and eating a worm
   c. A flea biting a dog to obtain a blood meal
   d. Lions skirmishing with hyenas to prevent them from feeding on a zebra carcass
   e. All of the above describe antagonistic interactions.
   Textbook Reference: *56.1 What Types of Interactions Do Ecologists Study?*

2. Certain birds follow swarms of foraging army ants and prey upon the insects that the ants flush out. The relationship between these birds and the ants is an example of
   a. competition.
   b. commensalism.
   c. mutualism.
   d. amensalism.
   e. Both a and b
   Textbook Reference: *56.1 What Types of Interactions Do Ecologists Study?*

3. Which of the following type of interaction would be *least* likely to lead to coevolution of the interacting species?
   a. Predator–prey
   b. Commensal
   c. Plant–herbivore
   d. Parasite–host
   e. Mutualistic
   Textbook Reference: *56.1 What Types of Interactions Do Ecologists Study?*

4. If a species of plant that is trampled by an animal eventually evolves sharp spines that prevent trampling, we can say that its association with the animal has changed from _______ to _______.
   a. amensalism; competition
   b. amensalism; commensalism
   c. commensalism; competition
   d. commensalism; mutualism
   e. commensalism; amensalism
   Textbook Reference: *56.1 What Types of Interactions Do Ecologists Study?*

5. As their name suggests, stick insects ("walking sticks") strongly resemble sticks or twigs. The most specific term for this method of avoiding detection by predators is
   a. mimicry.
   b. camouflage.
   c. aposematism.
   d. crypsis.
   e. homotypy.
   Textbook Reference: *56.2 How Do Antagonistic Interactions Evolve?*

6. Skunks, which are noted for their ability to spray would-be predators with a foul-smelling chemical, typically have a coat that is mostly black with contrasting white stripes or spots. This color pattern is an example of
   a. mimicry.
   b. camouflage.
   c. aposematism.
   d. crypsis.
   e. homotypy.
   Textbook Reference: *56.2 How Do Antagonistic Interactions Evolve?*

7. The most common defense of plants against herbivores is
   a. the production of secondary metabolites.
   b. a waxy cuticle covering the surfaces of the leaves.
   c. thorns or spines.
   d. hairy leaves.
   e. leaves containing silica, which wears down the teeth of herbivores.
   Textbook Reference: *56.2 How Do Antagonistic Interactions Evolve?*

8. Which of the following statements about microparasites is true?
   a. Microparasite infection rates typically rise only when a sufficiently dense population of susceptible host individuals is present.
   b. Microparasites include viruses, bacteria, protists, and parasitic worms.
   c. Microparasites are always pathogens.
   d. Both a and c
   e. All of the above
   Textbook Reference: *56.2 How Do Antagonistic Interactions Evolve?*

9. Which of the following statements about ectoparasites is *false*?
   a. Many ectoparasites have evolved adaptations to keep them attached to their hosts.
   b. Some ectoparasites have a relatively brief or casual association with their hosts.
   c. Most ectoparasitic insects are highly specialized, sometimes feeding on only a single host species.
   d. Many ectoparasitic insects have lost their wings during the course of their evolution.
   e. All of the above are true; none is false.
   Textbook Reference: *56.2 How Do Antagonistic Interactions Evolve?*

10. Match the following pollinators with the descriptions of the flowers that follow.
    Bees, Flesh flies
    Beetles, Butterflies
    Moths, Hummingbirds, Bats
    a. Pendant white flowers with a heavy odor
    b. Irregularly shaped purplish flowers with the odor of carrion

c. Tubular red flowers with no perceptible odor
d. Irregularly shaped flowers with a sweet odor
e. Bowl-shaped white flowers with a faint odor
f. Cuplike white flowers with a musty odor
g. Flowers with a tubular platform and a faint odor
Textbook Reference: *56.3 How Do Mutualistic Interactions Evolve?*

11. The suites of floral characteristics listed in question 10 have evolved to attract various groups of pollinators through the process of
a. specific coevolution.
b. interspecific competition.
c. resource partitioning.
d. character displacement.
e. diffuse coevolution.
Textbook Reference: *56.3 How Do Mutualistic Interactions Evolve?*

12. A bird eats the fruit of a plant species. The seeds are not digested and are voided at some distance from the parent plant, where they then germinate. This is an example of
a. predation.
b. competition.
c. commensalism.
d. mutualism.
e. amensalism.
Textbook Reference: *56.3 How Do Mutualistic Interactions Evolve?*

13. In Eurasia, two closely related species of rock nuthatches (small birds that forage for insects and spiders on bare, rocky ground) occur either separately, or in a common habitat. Individuals of both species that exist in the former situation, with no interspecies competition, have bills that are similar in length. But in their common habitat, one species has a longer bill than it has elsewhere and the other species has a shorter bill. This latter situation is an example of _______ competition that has resulted in _______.
a. exploitation; resource partitioning
b. exploitation; character displacement
c. interference; resource partitioning
d. interference; character displacement
e. None of the above
Textbook Reference: *56.4 What Are the Outcomes of Competition?*

14. Which of the following is a possible outcome of competition involving two species?
a. Some individuals of both species experience reduced growth and reproductive rates.
b. One species excludes the other from a habitat it would occupy in the absence of competition.
c. One species restricts the geographic range of the other species.
d. The species evolve to divide up the limiting resource and continue to coexist in the same locality.
e. All of the above
Textbook Reference: *56.4 What Are the Outcomes of Competition?*

15. Which of the following statements about the niche concept is *false*?
a. Compared to species with dissimilar niches, species with similar niches are likely to be stronger competitors with one another.
b. Two species can occupy the same habitat but have different niches.
c. The fundamental niche is defined by the interaction between species.
d. The realized niche of a species is never larger than its fundamental niche, though they may be identical in size.
e. All of the above are true; none is false.
Textbook Reference: *56.4 What Are the Outcomes of Competition?*

Answers

Key Concept Review

1. The anemone fish clearly benefit from the interaction with their host by gaining protection from their enemies. It is less clear that the sea anemones are unaffected (in which case the relationship would be a commensalism). The anemones may gain nutrients from the feces of the fish, or the fish may occasionally steal prey from the anemones, or both. Thus, the interaction could be mutualistic, competitive, or both simultaneously.

2.

Categories of Species Interactions

	Type of interaction	Effect on species 1	Effect on species 2
Antagonistic interactions	Predation (predator-prey)	+	–
Antagonistic interactions	Herbivory (plant-herbivore)	–	+
Antagonistic interactions	Parasitism (parasite/pathogen host)	+	–
	Mutualism	+	+
	Competition	–	–
	Commensalism (commensal-host)	+	0
	Amensalism	0	–

3. This type of mimicry system is both Batesian and Müllerian. The palatable species are Batesian mimics of the toxic species (the models), and the toxic species are Müllerian mimics of one another (each species in a sense being both model and mimic). Natural selection may have favored the convergence of these aposematic species on a common color pattern because it is

probably easier for predators to learn to avoid a single pattern than to learn to avoid multiple patterns.

4. Natural selection favored the survival of those rabbits that by chance had relatively high genetic resistance to the virus. Thus, one would predict that over time, the rabbit population as a whole would evolve to become more resistant. One would also predict selection on the myxoma virus favoring slower or lesser lethality: a mutant strain of the virus that allowed its host to survive for a longer period of time, or to not be killed at all, would be more likely to be transmitted to new hosts than the original strain that killed its host more quickly. Biologists studying this classic example of host–parasite coevolution have documented these evolutionary changes in both the rabbit and virus populations.
5. Most flowers can be successfully pollinated by a number of animal species. The evolution of broad suites of floral characteristics that attract certain groups of pollinators is an example of diffuse coevolution: the evolution of similar suites of traits in species experiencing similar selection pressures
6. Many animals that eat fruits (called frugivores) provide a valuable service to the plants that produce those fruits by dispersing seeds. Because of the often asymmetrical nature of the mutualism between frugivores and plants, relatively few highly specialized frugivores exist.
7. Ecologists regard the interaction between the barnacle populations as an instance of interference competition. *Semibalanus* individuals smother, crush, or dislodge *Chthamalus* individuals but do not directly reduce the food resources available to them.
8. A removal experiment could be done in which one or the other species is removed in two similar areas. You would then look for changes in the abundance of the remaining species. The species with the largest increase in population density would be the one most affected by competition with the removed species.

Test Yourself

1. **d.** Antagonistic interactions are those in which one species benefits and the other species is harmed. They include herbivory, predation, and parasitism. The competition between the lions and hyenas for food is harmful to both species and hence is not classified as an antagonistic interaction.
2. **e.** Both ants and birds are competing for the same food supply; the birds also benefit from the ants' activity in flushing out the insects, but the ants might not be affected.
3. **b.** The types of interactions most likely to lead to coevolution are those that have a strong effect on all of the interacting species. In commensalism, one species benefits from the interaction but the other is unaffected by it.
4. **a.** Initially the plant is harmed and the animal is unaffected (amensalism). Following the evolution of sharp spines, the two species are competitors for space.
5. **e.** Homotypy is the resemblance of a living organism to an object considered inedible by its predators.
6. **c.** Aposematism is warning coloration, frequently exhibited by toxic or otherwise noxious species.
7. **a.** The production of toxic secondary metabolites is by far the most widespread means by which plants defend themselves against herbivores.
8. **a.** Microparasites include viruses, bacteria, and protists—all of which are far smaller than their hosts—but not multicellular parasites such as worms. Only those microparasites that cause disease symptoms are considered pathogens.
9. **e.** The mosquito is an example of an ectoparasite that remains with the host only long enough to feed. Lice and fleas are examples of ectoparasites that lack wings and that are highly specific with regard to their hosts (lice more so than fleas). Figure 56.8 illustrates the winglessness of crab lice as well as their adaptations for remaining attached to their host.
10.
    a. Moths
    b. Flesh flies
    c. Hummingbirds
    d. Bees
    e. Beetles
    f. Bats
    g. Butterflies
11. **e.** Diffuse coevolution is the evolution of similar traits in species experiencing similar selection pressures.
12. **d.** Both species have benefited; the bird dispersed and provided fertilizer for the plant's seed, and the plant provided food for the bird.
13. **b.** Because bill length in these species is clearly related to the type of food resources and foraging sites that the birds utilize, this case must be considered an example of exploitation competition. It is also an example of character displacement, because in both species bill length is dependent on the presence or absence of the competing species. The Iranian populations of the two species have evolved so as to minimize competition between them.
14. **e.** It is generally true that competition, whether intraspecific or interspecific, causes reduced growth and reproductive rates of some individuals participating in the competitive interaction. Interspecific competition in particular may lead to exclusion from a habitat of one of the competing species (i.e., competitive exclusion), or to range restriction, or to resource partitioning.
15. **c.** The fundamental niche is defined by the physiological capabilities of a species, while the realized niche is defined by the interactions among species.

Community Ecology

The Big Picture

- In urging a cautious approach to human manipulation of the natural environment, ecologists like to say that "you can never do just one thing." In this chapter, the descriptions of the effects of keystone species and of other complex community interactions illustrate the meaning of this saying.
- The island biogeographic model, which relates the area of an island to the equilibrium number of species inhabiting the island, has applications in conservation biology. For example, habitat destruction often results in habitat islands which, if small, inevitably lose some of the species that existed in the previously more extensive habitat.
- As discussed in Chapter 59, if we wish to preserve a rare species, in many cases the most important step is to protect its habitat. This often requires an understanding of (and manipulation of) ecological succession to prevent changes in habitat that are unfavorable to the rare species.

Study Strategies

- This chapter contains descriptions of a number of studies of community structure, including several that relate species richness to other community characteristics. As you read about these studies, focus on the particular question that the study was designed to answer and how the results support the conclusions drawn from the study.
- You may have difficulty understanding the difference between energy pyramids and biomass pyramids. Energy pyramids can never be inverted, because there must always be a lower rate of energy flow through any given trophic level than through the trophic level below it. Biomass pyramids can be inverted, because they do not represent energy flow rates, but rather the amount of living matter present in the various trophic levels of the community at a given moment in time.
- The graphical representations of the concepts of island biogeographic theory are sometimes troublesome to learn. Try redrawing the graphs for yourself, paying particular attention to how the axes are labeled.
- Go to the Web addresses shown to review the following animated tutorials and activities. You can also find them in BioPortal (yourBioPortal.com), along with many additional learning resources.

  Animated Tutorial 57.1 Biogeography Simulation (Life10e.com/at57.1)

  Animated Tutorial 57.2 Primary Succession on a Glacial Moraine (Life10e.com/at57.2)

  Activity 57.1 Energy Flow through an Ecological Community (Life10e.com/ac57.1)

  Activity 57.2 The Major Trophic Levels (Life10e.com/ac57.2)

Key Concept Review

57.1 What Are Ecological Communities?

Energy enters communities through primary producers

Consumers use diverse sources of energy

Fewer individuals and less biomass can be supported at higher trophic levels

Productivity and species diversity are linked

An ecological community consists of all the species that live and interact in a particular area. Communities vary greatly in species richness (i.e., the number of species that live in them). Gross primary productivity is defined as the rate at which plants assimilate energy through photosynthesis; the accumulated energy is called gross primary production. Net primary production is the portion of assimilated energy left over after the energy used by plants to power their own metabolism is subtracted. All of the other organisms in an ecosystem derive their energy either directly or indirectly from net primary production, which takes the form of plant growth and reproduction.

The organisms in a community use diverse sources of energy, and organisms are grouped into trophic levels according to the number of steps through which energy passes to reach them. Energy passes, in sequence, from autotrophic primary producers (photosynthesizers) to heterotrophs, beginning with primary consumers (herbivores), followed by secondary consumers (which feed on herbivores), and so on. Detritivores (decomposers) feed on the dead bodies and waste products of other organisms. Omnivores obtain their food from more than one trophic level (see Table 57.1).

A food chain is a sequence of feeding linkages beginning with a primary producer. Food chains in a community are usually interconnected in a complex food web (see Figure 57.3). Ecological efficiency is the ratio of production of a trophic level to production of the trophic level immediately below it. On average, ecological efficiency is about 10 percent. Three factors account for such low ecological efficiency, and are also the reasons that most communities have only three to five trophic levels: (1) organisms use most of the energy they accumulate for respiration and other metabolic processes, and this energy is ultimately dissipated as heat; (2) consumers do not ingest all of the biomass available to them; and (3) consumers cannot assimilate (digest and absorb) all of the biomass they ingest.

An energy pyramid shows how energy decreases in moving from lower to higher trophic levels. A biomass pyramid shows the mass of organisms existing at different trophic levels at a given moment of time. For the same community, these pyramids usually have similar shapes. Major exceptions are most aquatic communities, in which a small mass of rapidly dividing unicellular primary producers can support a much larger biomass of herbivores via photosynthesis (see Figure 57.4). Primary productivity and species richness depend not only on the supply of energy from the sun, but also on the availability of nutrients and water. Annual evapotranspiration—the amount of water released by evaporation from land and water surfaces and by transpiration from plants—is a measure of the availability of water to organisms. Species richness increases with primary productivity up to a point, after which it declines (see Figure 57.5). One hypothesis to explain this decline postulates that interspecific competition becomes more intense when productivity is higher, resulting in competitive exclusion of some species.

Question 1. In nature, where do biomass pyramids like the one shown below occur, and what are the conditions that create them?

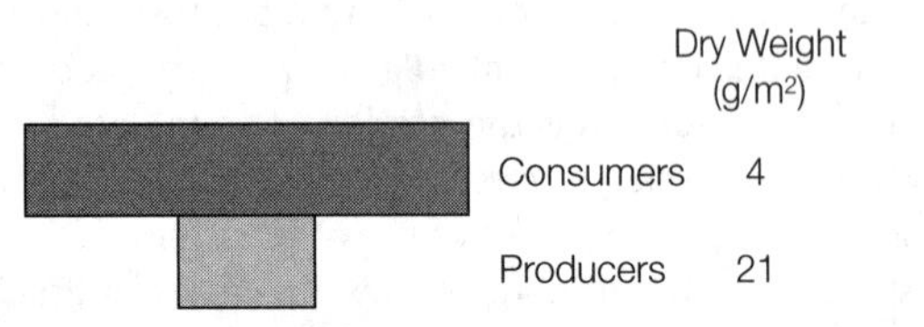

Question 2. Draw a line graph with ecosystem productivity on the *x*-axis and number of species present on the *y*-axis. Show on your graph when the number of species present is at its peak. Provide one hypothesis that may explain this phenomenon.

57.2 How Do Interactions among Species Influence Communities?

Species interactions can cause trophic cascades

Keystone species have disproportionate effects on their communities

Species interactions can influence community structure. In a trophic cascade, predators or herbivores affect many different species, often at lower trophic levels. Ecosystem engineers are organisms that build structures that create environments for other species. Keystone species exert an influence on ecological communities that is out of proportion to their abundance. Examples are predatory species that increase species richness by feeding on dominant competitors (see Figure 57.7) or plant species, such as fig trees in tropical forests, that serve as a food source for many different animals when food is otherwise scarce.

Question 3. Was the reintroduction of wolves to Yellowstone National Park a good ecological decision? Explain your answer.

Question 4. Explain the difference between an ecosystem engineer and a keystone species.

57.3 What Patterns of Species Diversity Have Ecologists Observed?

Diversity comprises both the number and the relative abundance of species

Ecologists have observed latitudinal gradients in diversity

The theory of island biogeography suggests that immigration and extinction rates determine diversity on islands

Ecologists attempt to measure and explain the species diversity of communities. The simplest way is to count the number of species present in a sample; this number is the species richness of the community. The distribution of abundances of individuals across species is called species evenness.

Species diversity in many taxa decreases as distance from the equator increases (see Figure 57.9). Ecologists have proposed at least four hypotheses, each supported by some evidence, to account for latitudinal gradients in diversity. The time hypothesis argues that organisms in tropical regions have had more time to diversify than those in less climatically stable temperate regions have. The spatial heterogeneity hypothesis suggests that tropical regions have more habitat diversity and hence many more species than temperate regions do. The specialization hypothesis proposes that greater competition among tropical species has led to the evolution of narrower niches and correspondingly more species. The predation hypothesis, in contradiction to the specialization hypothesis, proposes that higher predation intensity in the tropics keeps prey populations low, allowing for the persistence of rare species that otherwise would be eliminated by interspecific competition.

The species–area relationship states that within any latitudinal band, larger areas of habitat contain more species than smaller areas. Moreover, continental areas contain more species than oceanic islands of comparable size. According to the theory of island biogeography, equilibrium species richness (the number of species living in an area) on islands is determined by the rates of arrival of new species and extinction of species already present. The island biogeography theory predicts that the equilibrium number of species should increase with island size and decrease with distance from the mainland (see Figure 57.10). Scientists who study island biogeography have confirmed the predictions of the theory both by observation and experiment. The theory of island biogeography can also be applied to habitat islands—isolated patches of suitable habitat surrounded by areas of unsuitable habitat.

Question 5. Although ecologists agree that latitudinal gradients in diversity exist, there are various explanations as to why they exist. Explain the four hypotheses that have been developed to account for latitudinal gradients in diversity.

Question 6. Suppose that a large, natural, unmanaged prairie is surrounded by housing subdivisions. Behind each house is a garden, and this garden can be large or small depending on the size of the family that maintains it. Does MacArthur and Wilson's theory of island biogeography apply to this community? If so, which two factors are likely to affect the species diversity in these gardens?

57.4 How Do Disturbances Affect Ecological Communities?

- Succession is the predictable pattern of change in a community after a disturbance
- Both facilitation and inhibition influence succession
- Cyclical succession requires adaptation to periodic disturbances
- Heterotrophic succession generates distinctive communities

All ecological communities are subjected to a variety of disturbances. A disturbance is disruption to the community that changes the survival rate of one or more species. Small disturbances are much more common than large ones, but a few large events may cause most of the changes in a community. A community's history of disturbance may explain patterns of species diversity that otherwise would be puzzling.

Ecological succession is a change in community composition following a disturbance. In the most common type of succession—directional succession—species come and go in a predictable pattern until a climax community capable of perpetuating itself under local environmental conditions persists for an extended period of time. Primary succession begins on sites that lack living organisms. Disturbances such as glaciers and volcanic activity can initiate this type of succession (see Figure 57.13). Secondary succession begins when some organisms survive a disturbance and reestablish themselves on a disturbed site (see Figure 57.14).

Directional succession is characterized by several trends. In early stages, productivity is high and food webs are simple. As succession proceeds, food webs become more complex and nutrients become more concentrated in biomass than in biotic sources. Colonizing species in early succession tend to be r-strategists with good dispersal ability, whereas late successional species tend to be K-strategists. As succession progresses, species that have already become established may facilitate or inhibit colonization by other species.

In cyclical succession, the climax community is maintained by periodic disturbance, such as forest fires. Heterotrophic succession occurs in detritus-based communities, which are found in dung, carrion, or dead plants. Because these communities cannot generate more energy through photosynthesis, energy resources are depleted in the course of succession, with the result that biomass and biodiversity also decline.

Question 7. Suppose that you are responsible for managing community succession to provide suitable habitat for an endangered animal species that is long-lived and has a low reproductive rate. Would you expect this species to be adapted to early or late successional stages?

Question 8. Generations of Americans have been instructed by Smokey the Bear that fires are always harmful to the environment and should be prevented under all circumstances. In what way are some communities adapted to periodic fires? Explain why fires are beneficial to these communities.

57.5 How Does Species Richness Influence Community Stability?

- Species richness is associated with productivity and stability
- Diversity, productivity, and stability differ between natural and managed communities

Highly productive and stable communities tend to be species-rich. Communities with more species use resources more efficiently and hence tend to be both more productive and less variable in their productivity than communities with fewer species (see Figure 57.16). Population densities of individual species, however, may not be stable, regardless of species richness.

Monocultures—plantings of only a single crop species in a particular area—are notoriously unstable because of their susceptibility to pest outbreaks. Agricultural systems in which two or more crops are grown on the same plot tend to be more stable.

Question 9. Growing crops as monocultures is standard practice in modern agriculture. Why are monocultures unstable? What agricultural practices might result in more stable ecological communities?

Question 10. Draw a graph with the number of species on the x-axis and the relative access to nitrogen in the soil (mg N/kg biomass) on the y-axis. On the graph, draw the line that shows nitrogen use efficiency in relation to number of species. Which plant communities use soil nitrogen more efficiently—less species-rich communities or more species-rich communities?

Test Yourself

1. The ecological community of a pond comprises
   a. all the species living in the pond and the abiotic environment (water, sunlight, etc.) that supports them.
   b. all the species living in the pond.
   c. the invertebrates and vertebrate animals living in the pond.
   d. all of the producers and consumers living in the pond, but not the decomposers.
   e. None of the above
   Textbook Reference: *57.1 What Are Ecological Communities?*

2.–3. Refer to the food web below to answer the questions that follow.

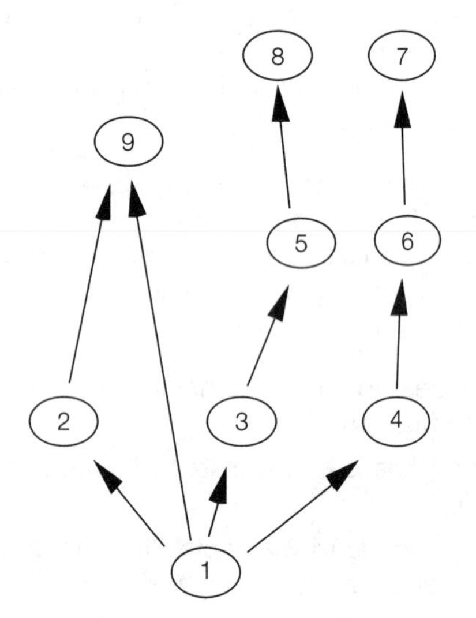

2. Organism 9 is a(n)
   a. herbivore.
   b. primary carnivore.
   c. secondary carnivore.
   d. primary producer.
   e. omnivore.
   Textbook Reference: *57.1 What Are Ecological Communities?*

3. The food web has _______ trophic levels.
   a. two
   b. three
   c. four
   d. five
   e. six
   Textbook Reference: *57.1 What Are Ecological Communities?*

4. In which of the following ecosystems would you expect to observe the lowest biomass of primary consumers relative to the biomass of the primary producers?
   a. Tropical rainforest
   b. Temperate grassland
   c. Freshwater lake
   d. Open ocean
   e. Both c and d
   Textbook Reference: *57.1 What Are Ecological Communities?*

5. A plant in the dark uses 0.02 ml of O_2 per minute. The same plant in sunlight releases 0.14 ml of O_2 per minute. Its rate of gross primary production would be approximately _______ ml of O_2 per minute.
   a. 0.02
   b. 0.12
   c. 0.14
   d. 0.16
   e. 0.18
   Textbook Reference: *57.1 What Are Ecological Communities?*

6. If the net primary productivity of a community is 2,000 kcal/m^2/yr, what would be the best estimate of the productivity (in kcal/m^2/yr) of the secondary consumers in that community?
   a. 4,000
   b. 2,000
   c. 200
   d. 20
   e. 2
   Textbook Reference: *57.1 What Are Ecological Communities?*

7. Which of the following statements about the relationship between species richness and productivity in ecological communities is true?
   a. Species richness is usually highest in the most productive communities.
   b. Species richness is usually highest in the least productive communities.
   c. Species richness is frequently highest at an intermediate level of productivity.
   d. Species richness is frequently lowest at an intermediate level of productivity.
   e. Species richness is not correlated with productivity.
   Textbook Reference: *57.1 What Are Ecological Communities?*

8. Which of the following statements about keystone species is true?
   a. They are very abundant.
   b. Their removal has a great effect on community structure.
   c. They practice herbivory.
   d. They have regular dispersion patterns.
   e. They have clumped dispersion patterns.
   Textbook Reference: *57.2 How Do Interactions among Species Influence Communities?*

9. In comparison with temperate regions at higher latitudes, tropical communities show greater species richness. Which of the following is a hypothesis that has been proposed to explain this difference?

a. Tropical regions have had more time to diversify under stable climatic conditions.
b. Tropical regions have more habitat types.
c. Higher levels of interspecific competition in the tropics have led to narrower niches and hence a greater number of more specialized species.
d. Higher predation intensity in the tropics prevents interspecific competition among prey species and thus allows rare species to persist.
e. All of the above

Textbook Reference: *57.3 What Patterns of Species Diversity Have Ecologists Observed?*

10. Refer to the graph below showing the effect of species number on arrival and extinction rates.

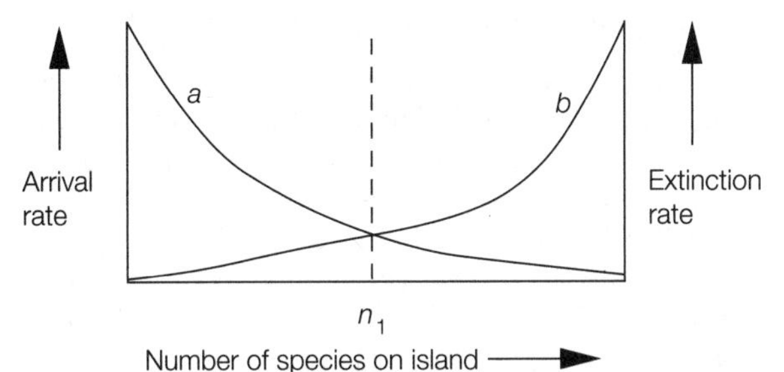

According to the theory of island biogeography, which of the following statements about the graph is true?
a. Curve *a* is the arrival rate curve.
b. Curve *b* is the extinction rate curve.
c. The arrival rate equals the extinction rate at n_1.
d. The extinction rate is zero at n_1.
e. Both a and b

Textbook Reference: *57.3 What Patterns of Species Diversity Have Ecologists Observed?*

11. In the figure below, the species numbers of many islands of different sizes were plotted against their distance from the mainland, and data points for islands of similar size were connected to form the four curves. Circle the letter of the curve corresponding to the group of islands with the smallest size.

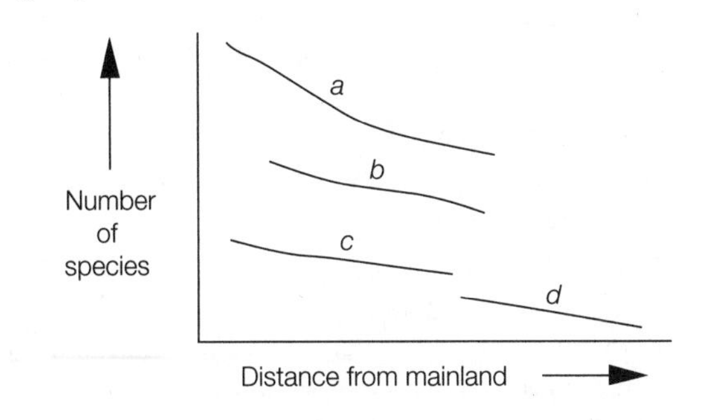

Textbook Reference: *57.3 What Patterns of Species Diversity Have Ecologists Observed?*

12. The theory of island biogeography makes predictions about the effects of species number on the rate of extinction. Refer to the figure below, in which the solid curve shows this relationship for a large island.

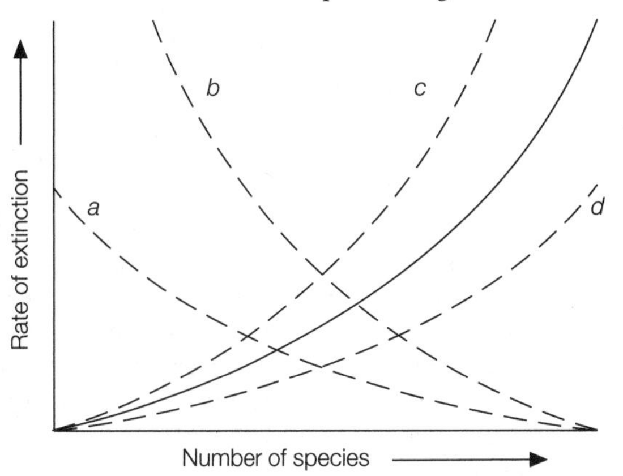

Which of the curves shows the expected relationship for a small island?
a. Curve *a*
b. Curve *b*
c. Curve *c*
d. Curve *d*

Textbook Reference: *57.3 What Patterns of Species Diversity Have Ecologists Observed?*

13. Which of the following places stages of ecological succession occurring on glacial moraines in Alaska in correct chronological sequence, from earliest to latest?
a. Increase in soil nitrogen; arrival of lichens; arrival of willows and alders; arrival of conifers; arrival of bacteria, fungi, and photosynthetic microorganisms
b. Increase in soil nitrogen; arrival of bacteria, fungi, and photosynthetic microorganisms; arrival of lichens; arrival of willows and alders; arrival of conifers
c. Arrival of bacteria, fungi, and photosynthetic microorganisms; increase in soil nitrogen; arrival of lichens; arrival of willows and alders; arrival of conifers
d. Arrival of bacteria, fungi, and photosynthetic microorganisms; arrival of lichens; arrival of willows and alders; increase in soil nitrogen; arrival of conifers
e. Arrival of bacteria, fungi, and photosynthetic microorganisms; arrival of conifers; arrival of willows and alders; arrival of lichens; increase in soil nitrogen

Textbook Reference: *57.4 How Do Disturbances Affect Ecological Communities?*

14. Which of the following represents primary succession?
a. The development of an aquatic community in a newly excavated farm pond, eventually followed by the filling in of the pond with sediment and its conversion to a forest
b. The development of a terrestrial community on lava produced by a volcanic eruption
c. The gradual establishment of a forest following a fire that destroyed the previous community

d. The invasion of weeds followed by shrubs and trees on an abandoned farm field
e. Both a and b

Textbook Reference: *57.4 How Do Disturbances Affect Ecological Communities?*

15. Which of the following statements about succession is *false*?
    a. Primary succession proceeds more rapidly than secondary succession.
    b. Species that are early colonists may either facilitate or inhibit colonization by later-arriving species.
    c. As succession proceeds, a lower proportion of the nutrients in a community are found in abiotic forms.
    d. As succession proceeds, food webs become more complex.
    e. The endpoint of directional succession is the establishment of a self-perpetuating climax community.

    Textbook Reference: *57.4 How Do Disturbances Affect Ecological Communities?*

Answers

Key Concept Review

1. In most aquatic ecosystems, the dominant primary producers are unicellular algae. Their populations grow and multiply so rapidly that they frequently support a herbivore mass that is larger than their own. Note that the diagram depicts an inverted pyramid of biomass. A pyramid of energy can never be inverted.
2. See Figure 57.5. The number of species in a community increases with productivity only up to a point, and after that may decline even if the ecosystem productivity keeps increasing. One hypothesis postulates that interspecific competition becomes more intense when productivity is very high, resulting in competitive exclusion of some species.
3. It was a good decision because the presence of this single predator species influenced not only populations of its prey, but also populations of its prey's food resource and of other species that depended on that resource. In the Lamar Valley of the park, the reintroduction of wolves resulted in a reduction of the population of elk, the wolves' principle prey. In a trophic cascade, the diminution in the number of elk led to increased reproduction of aspen and willow trees, both of which had been heavily browsed by elk. In turn, the increase in the population of streamside willows resulted in an increase in the number of beaver colonies in the valley.
4. Organisms that build structures that alter existing habitats or create new habitats are called ecosystem engineers. Similarly, keystone species are very influential, but they can influence both the number of species and the number of trophic levels in a community.
5. The time hypothesis argues that over evolutionary time, organisms in tropical regions have had more time to diversify under relatively stable climate conditions than have those in more temperate regions. The spatial heterogeneity hypothesis suggests that tropical regions have higher spatial heterogeneity—more different types of microclimates, vegetation, soils, and so forth—and thus contain more different habitats and many more species. The specialization hypothesis attributes latitudinal gradients in diversity to greater interspecific competition in the tropics, which leads to narrower realized niches. The predation hypothesis proposes that predation intensity is greater in the tropics. Where predation is high, it argues, prey populations are held to levels so low that interspecific competition never comes into play, and rare species can persist.
6. Yes, the theory of island biogeography can be applied to habitat islands—isolated patches of suitable habitat surrounded by extensive areas of unsuitable habitat. The size of the gardens may affect species diversity; the smaller the garden and the fewer resources it provides, the greater the potential for competition and the higher the extinction rate will be (see Figure 57.11B). Larger gardens provide greater habitat diversity and can sustain larger populations (which tend to have lower extinction rates than small populations). Distance of the island from the species pool will also affect species diversity. The farther the garden is from the prairie, the lower the immigration rate (the rate at which new species arrive) will be (see Figure 57.11C).
7. Long life span and low reproductive rate are typical characteristics of *K*-strategists. Such species are typical of late successional stages. Pioneer species, in contrast, are likely to be replaced quickly and therefore tend to be *r*-strategists with a high reproductive rate, short life span, and good dispersal ability.
8. The lodgepole pine forests discussed in this chapter are an example of a community that is maintained by periodic fires, which return nutrients to the soil and are required for seed germination. Other examples of fire-dependent or fire-adapted communities include tropical savanna and chaparral. Fire is used as a management tool to protect biodiversity. Fire can remove mature trees weakened by pest infestation, revitalize the soil, and provide favorable conditions for seed germination and new growth.
9. Monocultures are unstable because they are subject to attack by insect pests and pathogens that destroy or damage crops. Experiments have shown that more diverse agroecosystems, in which two or more crops are grown on the same plot, are less subject to pest outbreaks. In corn–sweet potato dicultures, for example, sweet potato pests are reduced because the corn provides a structural barrier, food for protective parasitoid wasps, and chemicals that interfere with the ability of pests to find their host plants.

10.

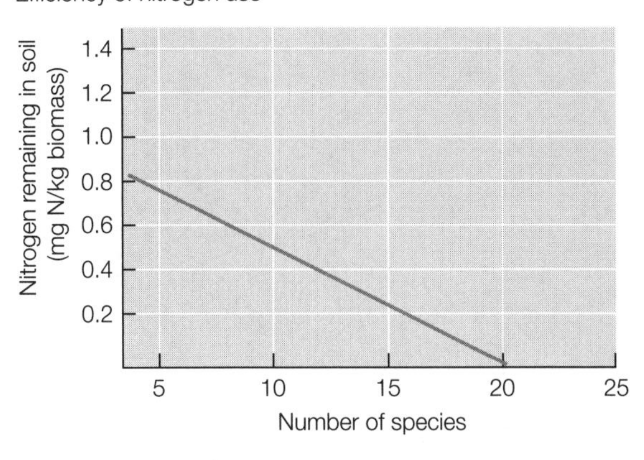

Soil nitrogen is used more efficiently in plots with greater species richness.

Test Yourself

1. **b.** An ecological community comprises all of the populations living in a defined area but excludes its nonliving components.
2. **e.** Because organism 9 eats from both the primary producer level (1) and the herbivore level (2), it is an omnivore.
3. **c.** Four trophic levels are depicted. The levels are primary producer (1), herbivore (2, 3, 4), primary carnivore (5, 6), and secondary carnivore (7, 8). The omnivore (9) occupies either the herbivore or primary carnivore level, depending on whether it is feeding on a primary producer or on an herbivore.
4. **a.** Compared to grasslands or aquatic ecosystems, forests typically have a lower ratio of primary consumer biomass to primary producer biomass because trees (the dominant primary producers) store energy for long periods of time in difficult-to-digest forms.
5. **d.** The amount of production used in maintenance and biosynthesis is added to net primary production to determine gross primary production. Photosynthetic release of O_2 in the light is an estimate of net production; O_2 use in the dark is an estimate of maintenance and biosynthesis costs. The addition of 0.02 ml to 0.14 ml yields an estimate of gross primary production of 0.16 ml per minute.
6. **d.** Net primary productivity is the rate at which energy is incorporated into the bodies of primary producers through growth and reproduction. As a rule, only about 10 percent of the energy at one trophic level is transferred to the next. Because the secondary consumers are two trophic levels above the primary producers, their productivity would be expected to be only 1 percent of net primary productivity.
7. **c.** Both observational and experimental evidence indicate that species richness often peaks in communities of intermediate productivity.
8. **b.** By definition, removal of a keystone species has a larger effect on the community than would be expected based on its abundance.
9. **e.** Community ecologists have not been able to agree on the reasons for the greater species diversity characteristic of tropical communities. All four hypotheses have been proposed, and each has some evidence to support it.
10. **d.** The extinction rate is greater than zero and equals the arrival rate at n_1, which is therefore the equilibrium species number on the island.
11. **d.** Each curve corresponds to a group of similar-sized islands whose species number is plotted against their distance from the mainland. For each curve, species number decreases with distance from the mainland, and the curve that is lowest relative to the vertical axis would be the group of smallest islands.
12. **c.** With less space available, populations of the different species would be smaller. This would subject them to higher extinction rates than would be expected on larger islands.
13. **d.** Of the stages listed, the arrival of bacteria, fungi, and photosynthetic microorganisms would be first. The arrival of lichens would be second, followed by the arrival of willows and alders, an increase in soil nitrogen content, and finally the arrival of conifers.
14. **e.** Primary succession begins on a newly available site where all preexisting living organisms have been stripped away, such as a freshly excavated pond or the surface of lava. Succession that occurs after disturbances such as fire or the conversion of a natural community to farmland is considered secondary.
15. **a.** Because primary succession begins on a site that is devoid of preexising organisms, it is a slower process than secondary succession.

Ecosystems and Global Ecology

The Big Picture

- Keep in mind the different ways in which energy and matter move through ecosystems. Energy flows unidirectionally from producers to higher trophic levels and is ultimately dissipated as heat, whereas elements and water continually cycle through ecosystems.
- The productivity of ecosystems, both terrestrial and aquatic, is relevant to the problem of providing food for the exploding human population (see Chapter 55). The increasing share of global primary production that is appropriated by the expanding human population is also relevant to the problem of maintaining biodiversity (see Chapter 59).
- Knowledge of how materials move through biogeochemical cycles is crucial for understanding and predicting the effects of human alterations of these cycles. Global warming and eutrophication of lakes are two important examples.
- The important roles of nitrogen fixers and other microorganisms in the nitrogen cycle illustrate the sometimes overlooked ways in which rather obscure members of the biotic community may be essential for the healthy functioning of ecosystems.

Study Strategies

- In studying the material on biogeochemical cycles, focus on the following basic questions: Where are the abiotic reserves of an element located? How does the element leave the reserve and enter living organisms? Why do living organisms need the element? How does the element return to its abiotic reserve?
- The reason that eutrophication—the "enrichment" of a body of water with nutrients—often results in a "dead zone" may be puzzling. It is because the additional growth of photosynthetic algae and bacteria caused by added nutrients such as phosphorus and nitrogen results in an increased rate of aerobic respiration by microorganisms as they decompose primary producers. This process may lead to oxygen depletion in part or all of a body of water, such as the deeper waters of a lake.
- Go to the Web addresses shown to review the following animated tutorials and activity. You can also find them in BioPortal (yourBioPortal.com), along with many additional learning resources.

  Animated Tutorial 58.1 Earth's Radiation Balance (Life10e.com/at58.1)

  Animated Tutorial 58.2 The Global Hydrologic Cycle (Life10e.com/at58.2)

  Animated Tutorial 58.3 The Global Carbon Cycle (Life10e.com/at58.3)

  Animated Tutorial 58.4 The Global Nitrogen Cycle (Life10e.com/at58.4)

  Activity 58.1 Concept Matching (Life10e.com/ac58.1)

Key Concept Review

58.1 How Does Energy Flow through the Global Ecosystem?

Energy flows and chemicals cycle through ecosystems

The geographic distribution of energy flow is uneven

Human activities modify the flow of energy

An ecosystem is any ecological system within defined boundaries. It includes all of the organisms within those boundaries as well as the chemical and physical factors that interact with those organisms. Earth is a closed system to chemical elements, but it is an open system to energy. Energy flows through the global ecosystem unidirectionally.

Energy from the sun, combined with the energy of radioactive decay in Earth's interior, drives the processes that move materials around the planet. With very few exceptions, energy flow in ecosystems originates with photosynthesis. Much of the energy that enters each trophic level is used to power metabolism and this energy is eventually dissipated as heat and is lost from the ecosystem. Net primary production is the portion of assimilated energy left over after the energy used by primary producers for their own metabolism is subtracted (see Chapter 57). All of the other organisms in an ecosystem derive their energy directly or indirectly from net primary production, which takes the form of growth and reproduction of the producer organisms. Chemical elements are not altered when they are transferred between organisms and are not lost from the global ecosystem, although they may become unavailable to organisms for long periods.

Production in aquatic ecosystems is limited by light and nutrient availability. Net primary production in the oceans tends to be highest in coastal zones, where runoff from land and upwellings from deeper waters bring nutrients into shallow waters.

Human activities either decrease or increase net primary production; deforestation, for example, decreases net primary production, while the conversion of prairies to agricultural fields increases it. Nevertheless, humans consume about 25 percent of Earth's average annual net primary production.

Question 1. Compare the average net primary production of algal beds with that of oceans. Explain the major differences between them, and the reasons why one contributes more to Earth's net primary production than the other does.

Question 2. Do human activities decrease net primary production? Explain.

58.2 How Do Materials Move through the Global Ecosystem?

- Elements move between biotic and abiotic compartments of ecosystems
- The atmosphere contains large pools of the gases required by living organisms
- The terrestrial surface is influenced by slow geological processes
- Water transports elements among compartments
- Fire is a major mover of elements

In contrast to the energy that powers biological processes, which comes from the sun, the chemical elements that make up the bodies of organisms come from within the Earth system itself. The elements on which life depends cycle among the atmosphere, the oceans, fresh waters, and land.

The rate at which elements move through a system is called its flux. An accumulation of an element in some component of an ecosystem is a pool. In the atmosphere, the chemical composition is influenced by volcanic eruptions, while the activities of living organisms have an enormous impact on the chemical composition of the atmosphere, the water, and the land. The gases of Earth's atmosphere help moderate planetary temperatures and thus keep water in a liquid state.

The lowest layer of the atmosphere is the troposphere, where almost all water vapor is located and where most global air circulation takes place. Above the troposphere is the stratosphere, where ultraviolet radiation is absorbed by the ozone layer (O_3) (see Figure 58.6). Human activities can affect the atmosphere; for example, the depletion of stratospheric ozone caused by the release of chlorinated fluorocarbons (CFCs), such as the refrigerant freon, is a source of concern because increased ultraviolet radiation at Earth's surface is associated with increased rates of skin cancer, cataract formation, and crop damage. The most abundant gases in the atmosphere are N_2 (about 78 percent) and O_2 (about 21 percent). Although CO_2 constitutes only 0.03 percent of the atmosphere, it is the source of the carbon used by terrestrial photosynthetic organisms and of the dissolved carbonate used by marine producers. The greenhouse gases (such as CO_2, water vapor, methane, and nitrous oxide) trap outgoing infrared (heat) radiation emitted by Earth and thus raise Earth's surface temperature (see Figure 58.7). Fires consume the energy of, and release chemical elements from, the vegetation they burn. Biomass burning also contributes a significant quantity of CO_2 and other greenhouse gases to the atmosphere.

Concentrations of mineral nutrients are very low in most ocean waters because most elements that enter the oceans gradually sink to the seafloor. In lakes, surface waters tend to become depleted of nutrients as organisms die and sink to the bottom, whereas deeper waters become depleted of O_2 as organic matter decomposes. Vertical movements called turnover bring nutrients and dissolved CO_2 to the surface and O_2 to deeper water. In deep lakes found in temperate climates, turnover occurs in the spring and fall when the water temperature is uniformly 4°C (see Figure 58.8). As spring and summer progress, the surface water becomes warmer still, and the depth of the warm water layer gradually increases. The depth at which the temperature changes abruptly is called the thermocline.

Question 3. Why do humans deliberately set fire to natural vegetation? In what ways are human-set fires contributing to global warming and to movement of elements through ecosystems?

Question 4. In the figure below of a lake during each of the four seasons, indicate where turnover is occurring and the location of the thermocline.

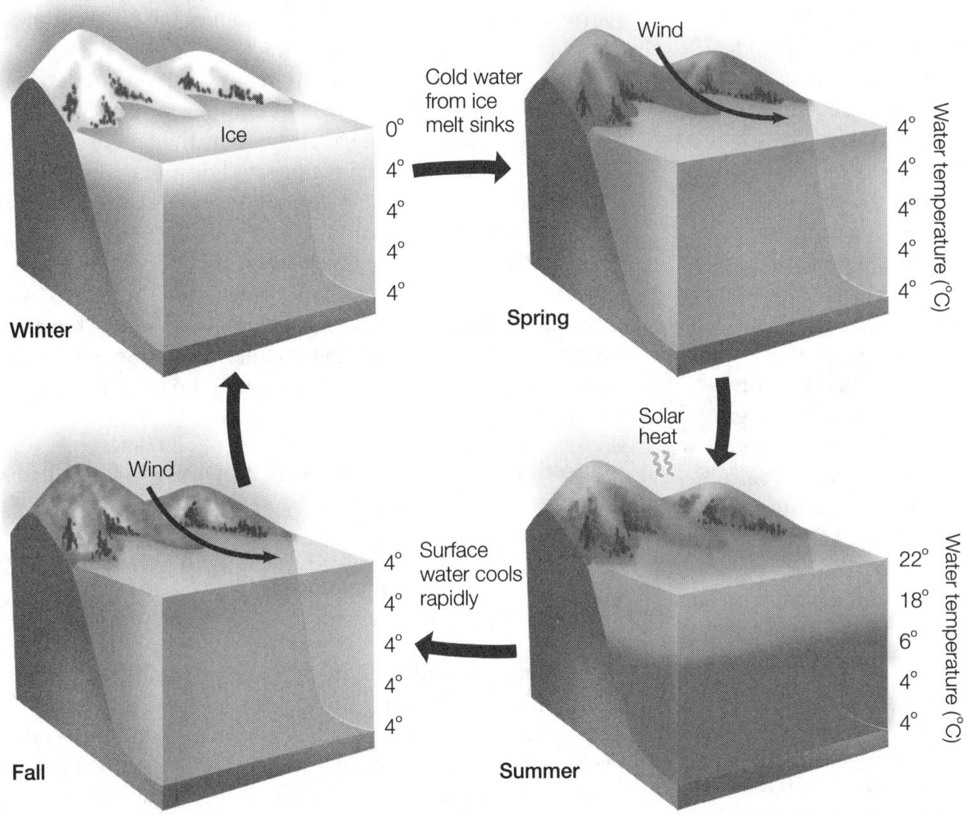

58.3 How Do Specific Nutrients Cycle through the Global Ecosystem?

- Water cycles rapidly through the ecosystem
- The carbon cycle has been altered by human activities
- The nitrogen cycle depends on both biotic and abiotic processes
- The burning of fossil fuels affects the sulfur cycle
- The global phosphorus cycle lacks a significant atmospheric component
- Other biogeochemical cycles are also important
- Biogeochemical cycles interact

Geological, chemical, and biological processes are all important in moving materials. The pattern of movement of an element is called its biogeochemical cycle. Water cycles through the ecosystem in the global hydrologic cycle (see Figure 58.9). The sun powers the hydrologic cycle by causing evaporation, most of it from ocean surfaces. The average residence time of a water molecule in a particular compartment varies greatly, from under a week in organisms to about 3,000 years in the ocean. Water in underground aquifers (sedimentary rocks) has such a long residence time that it plays a small role in the hydrologic cycle. Pumping of this groundwater for irrigation has altered the flux of water from the land to the oceans. If current water consumption trends continue, about half of the world's population will have an inadequate supply of water by 2025.

All of the important macromolecules that make up living organisms contain carbon. Nearly all the carbon in organisms comes from CO_2 in the atmosphere or dissolved carbonate (CO_3^{-2}) or bicarbonate (HCO_3^{-}) in water. Carbon is incorporated into organic molecules by photosynthesis; respiration and combustion break down these organic molecules and return carbon to the atmosphere and water (see Figure 58.10).

Water near the ocean surface is becoming more acidic because of the absorption of quantities of CO_2 that are larger than those absorbed at any time during the past 20 million years. The combination of decreasing pH and increasing water temperature is endangering coral organisms and hence entire reef communities.

Fossil fuels exist because in the remote past, large quantities of organic carbon were removed from the carbon cycle by burial of organisms in sediments lacking oxygen. The ever-increasing rate of burning of fossil fuels over the past 150 years has elevated the CO_2 concentration of the atmosphere (see Figure 58.10). Less than half of the CO_2 released into the atmosphere by human activities remains there; most of the rest dissolves in the oceans, which contain 50 times more dissolved inorganic carbon than the atmosphere contains. Currently the photosynthetic consumption of CO_2 exceeds the metabolic consumption of CO_2, so Earth's terrestrial vegetation is storing carbon that would otherwise be increasing atmospheric CO_2 concentrations. Nevertheless, it is unlikely that terrestrial vegetation can store the extra CO_2 that is being produced by human activities. Historical records and computer models indicate that Earth is warmer when atmospheric CO_2 levels are higher and cooler when they are lower (see Figure 58.12). The global warming caused by a doubling of the atmospheric CO_2 concentration would increase mean annual temperatures worldwide and disrupt current precipitation patterns. It will also melt the polar ice caps, thereby causing the flooding of coastal regions. Global warming has already caused an increase in insect infestations in some temperate forests, and in the future it may cause an increase in the incidence of some human diseases.

Nitrogen gas is the most abundant gas in Earth's atmosphere, but most organisms cannot use nitrogen in its gaseous form. Nitrogen gas (N_2) makes up 78 percent of Earth's atmosphere, but nitrogen can be converted into biologically useful forms by only a few species of microorganisms that can carry out nitrogen fixation. Thus, nitrogen is often in limited supply in ecosystems. Microorganisms also perform denitrification, which removes nitrogen from the biosphere and returns it to the atmosphere (see Figure 58.14). Total nitrogen fixation by humans as a result of the use of fertilizers and burning of fossil fuels equals global natural nitrogen fixation (see Figure 58.15). Eutrophication is an increase in biomass production in a body of water due to inputs of nutrients. When nitrogen applied as fertilizer to cropland enters bodies of water, it can lead to eutrophication, depleting the oxygen in the water and creating a deoxygenated "dead zone." Human-caused perturbations of the nitrogen cycle have a variety of other adverse effects, including increased air pollution, increased atmospheric concentrations of greenhouse gases, and reductions in species richness.

Most mineral nutrients that enter fresh waters are released by the weathering of rocks and are carried to lakes and rivers via groundwater or by surface flow. Most of Earth's sulfur supply is located in rocks on land and as sulfate salts in deep-sea sediments. Sulfur in soil is taken up by plants and incorporated into proteins. Atmospheric sulfur (in the form of dimethyl sulfide released into the atmosphere when phytoplankton decay) plays an important role in the global climate because it is a major component of particles that promote cloud formation. The burning of fossil fuels produces emissions that form sulfuric and nitric acid in the atmosphere. The resulting acid precipitation affects all industrialized countries and has been shown to have harmful effects on lake ecosystems and on terrestrial plants (see Figure 58.16).

The phosphorus cycle differs from the cycles of carbon, nitrogen, and sulfur in that it lacks a gaseous phase. On land, most phosphorus becomes available through the weathering of phosphorus-containing rocks (see Figure 58.17). Because phosphorus is often a limiting nutrient for plant growth, it is a typical component of fertilizer. Because it is also frequently a limiting factor for algal growth in lakes, the addition of phosphorus (from fertilizer runoff, soil erosion, detergents, human wastes, animal manure, and industrial wastes) to freshwater bodies is a major cause of eutrophication. Recovery and recycling of phosphorus from sewage and animal wastes could reduce the amount of phosphorus entering lakes and rivers while supplying much of the needs of the fertilizer and detergent industries.

Organisms may be deficient in any of several minerals that they require in very small amounts. Over much of the ocean, a scarcity of dissolved iron limits the rate of photosynthesis. In some terrestrial regions, iodine and molybdenum are not available in the concentrations necessary to meet the needs of endothermic vertebrates.

Because biochemical cycles interact in significant ways, alterations in any one cycle affect the others. Scientists are constantly discovering previously unknown interactions. An example is the recent finding that elevated atmospheric CO_2 concentrations can cause a decrease in the rate of nitrogen fixation.

Question 5. How does the use of groundwater for irrigation alter the hydrological cycle?

Question 6. In what ways are farming methods that require the large-scale use of manufactured fertilizers contributing to eutrophication and other adverse environmental effects?

Question 7. The phosphorus cycle is very important on Earth. Draw a diagram showing how phosphorus cycles among Earth's crust, soil, lakes (and rivers), and the seafloor. Where appropriate, indicate the presence of living organisms such as algae, animals, and plants at a particular location, and show with arrows how the phosphorus cycles to and from the organisms and those locations, and how long the cycles take to occur.

58.4 What Goods and Services Do Ecosystems Provide?

Among the goods and services provided by ecosystems are food, clean water, clean air, fiber, building materials, fuel, flood control, soil stabilization, pollination, climate regulation, spiritual fulfillment, and aesthetic enjoyment. It would be either impossible or prohibitively expensive to replace these benefits. Modifications of Earth's ecosystems have contributed to human welfare, but the benefits have not been distributed equally. Moreover, short-term increases in some ecosystem goods and services have resulted in the long-term degradation of others.

The most important cause of alterations in ecosystems has been changes in land use as natural ecosystems have been converted to more intensive uses. Though such altered ecosystems provide many benefits (such as food production), they also have resulted in the degradation of other services (such as the ability to provide clean water, habitat for wildlife, and protection from floods). The ecological importance of coastal wetlands was illustrated by Hurricane Katrina, which would not have caused as much flooding in New Orleans if the wetlands surrounding the city had been intact.

Question 8. Wetlands have frequently been regarded as "waste lands" that would have greater economic value if they were drained and converted to agricultural or commercial use. Describe what we learned from Hurricane Katrina, which struck the Gulf Coast in 2005, about the value of intact wetlands for flood protection. Describe two other ways in which the wetlands of this region are valuable.

Question 9. Give at least three examples of goods and three examples of services that ecosystems provide.

58.5 How Can Ecosystems Be Sustainably Managed?

The economic value of a sustainably managed ecosystem is often higher than that of a converted or intensively exploited ecosystem (see Figure 58.19). Because ecosystem services are considered "public goods" that have no market value, government action and incentives may be required to encourage sustainable ecosystem management. In addition, education programs are needed to increase public awareness of how human activities affect ecosystem sustainability.

Question 10. Give several examples of policies and/or incentives that could encourage sustainable ecosystem management.

Question 11. Many ecosystems are converted to human use. Do these ecosystems provide more goods and services after the conversion?

Test Yourself

1. Which of the following statements about the average net primary production of these ecosystems is true?
   a. The average net primary production of open oceans is greater than that of tropical rainforests.
   b. The average net primary production of algal beds and coral reefs is greater than that of any other ecosystem.
   c. The average net primary production of extreme deserts is greater than that of tropical rainforests.
   d. The average net primary production tropical rainforests is greater than that of algal beds and coral reefs.
   e. Upwelling zones have the smallest average net primary production.
   Textbook Reference: *58.1 How Does Energy Flow through the Global Ecosystem?*

2. Which of the following ecosystems makes the greatest contribution to Earth's net primary production?
   a. Tropical rainforest
   b. Algal beds and coral reefs
   c. Extreme deserts
   d. Open oceans
   e. Upwelling zones
   Textbook Reference: *58.1 How Does Energy Flow through the Global Ecosystem?*

3. Which of the following statements is true of oceans but *not* true of freshwater ecosystems?
   a. They receive material from land mostly via groundwater.
   b. Elements are buried in bottom sediments for long periods of time.
   c. There is a seasonal mixing of materials.
   d. Bottom waters frequently lack oxygen.
   e. All of the above are characteristic of both oceans and freshwater ecosystems.
   Textbook Reference: *58.2 How Do Materials Move through the Global Ecosystem?*

4.–5. In the following graph, the temperature of a freshwater lake is plotted against its depth.

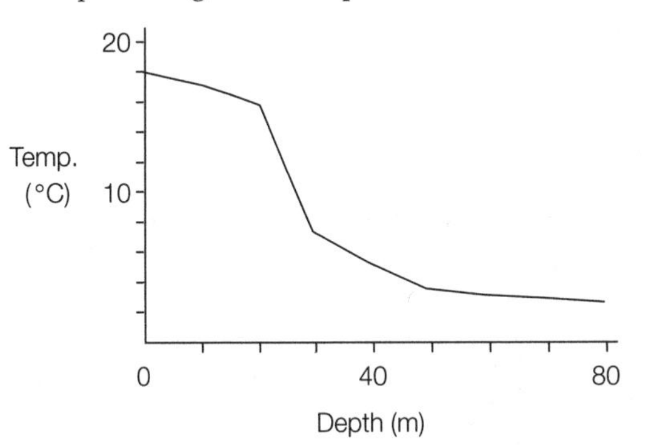

4. The temperature profile of the graph would be most characteristic of the lake during which season?
   a. Winter
   b. Spring
   c. Summer
   d. Fall
   e. Both b and c
   Textbook Reference: *58.2 How Do Materials Move through the Global Ecosystem?*

5. Based on data in the graph, the thermocline for the lake is located between _______ and _______ meters.
   a. 0; 10
   b. 10; 20
   c. 20; 30
   d. 30; 50
   e. 50; 80
   Textbook Reference: *58.2 How Do Materials Move through the Global Ecosystem?*

6. Zones of upwelling near the shores of land masses have high primary production because such zones
   a. bring nutrients from the seafloor up to the surface.
   b. are characterized by clear water that permits light to penetrate to an unusually great depth.
   c. bring warm water to the surface.
   d. trap nutrients washed into the ocean from nearby land masses.
   e. have water of lower salinity than other oceanic regions.
   Textbook Reference: *58.1 How Does Energy Flow through the Global Ecosystem?*

7. Which of the following statements is true of the stratosphere but *not* of the troposphere?
   a. Most greenhouse gases are concentrated here.
   b. Most water vapor resides here.
   c. Most of the mass of the atmosphere lies here.
   d. Circulation of this layer influences ocean currents.
   e. Most ultraviolet radiation is absorbed here.
   Textbook Reference: *58.2 How Do Materials Move through the Global Ecosystem?*

8. Which of the following compartments of the global ecosystem is characterized by very slow movement of materials within it?
   a. Oceans
   b. Fresh waters
   c. Atmosphere
   d. Land
   e. All of the above
   Textbook Reference: *58.2 How Do Materials Move through the Global Ecosystem?*

9. Which of the following environmental factors frequently limits primary production in *both* terrestrial and aquatic ecosystems?
   a. Temperature
   b. Moisture
   c. Light
   d. Nutrient supply
   e. None of the above
   Textbook Reference: *58.1 How Does Energy Flow through the Global Ecosystem?*

10. Which of the following statements about biogeochemical cycles is *false*?
    a. Most elements remain longest in the living portion of their cycle.
    b. Gaseous elements cycle more quickly than elements without a gaseous phase.
    c. Some atoms in the human body may once have been part of a dinosaur.
    d. Biogeochemical cycles all include both organismal and nonliving components.
    e. Perturbations in one biogeochemical cycle can have major effects on other cycles.
    Textbook Reference: *58.3 How Do Specific Nutrients Cycle through the Global Ecosystem?*

11. Next to each of the following descriptions, place a ***c***, ***n***, ***p***, or ***s*** if it is characteristic of the biogeochemical cycle of **c**arbon, ***n***itrogen, ***p***hosphorus, or **s**ulfur. (Note: Some features may apply to more than one cycle.)
    _____ Major reservoir is atmospheric
    _____ Major reservoir is in sedimentary rock
    _____ Often in short supply in ecosystems
    _____ Fossil fuel reserve is part of this cycle
    _____ Involved in cloud formation
    _____ Lacks a gaseous phase
    _____ Includes a form that is a greenhouse gas
    _____ Most fluxes involve organisms
    Textbook Reference: *58.3 How Do Specific Nutrients Cycle through the Global Ecosystem?*

12. Which of the following is *not* one of the major threats to the climate caused by our alterations of the carbon cycle?
    a. The polar ice caps may melt if global warming continues.
    b. The burning of fossil fuels adds compounds of sulfur to the atmosphere.
    c. The increase in atmospheric CO_2 exceeds the ability of the oceans to absorb the increase.
    d. CO_2 is a gas that traps infrared radiation.
    e. Tropical storms are likely to increase in intensity if global warming continues.
    Textbook Reference: *58.3 How Do Specific Nutrients Cycle through the Global Ecosystem?*

13. Which of the following statements about acid precipitation is *false*?
    a. Acids that enter the atmosphere primarily affect ecosystems located less than 100 kilometers from the source of the pollution.
    b. Though lakes are very sensitive to acidification, their pH can return rapidly to normal values.
    c. Regulation of emission sources has raised the pH of precipitation in many parts of the eastern United States in the last two decades.
    d. Sulfuric acid and nitric acid from the burning of fossil fuels are the major causes of acid precipitation.
    e. Organisms lost from freshwater ecosystems because of acidification may not return even decades after pH values have returned to normal.
    Textbook Reference: *58.3 How Do Specific Nutrients Cycle through the Global Ecosystem?*

14. The process by which a lake ecosystem is altered by eutrophication involves several stages. Which of the following represents the correct ordering of stages in the chain of causation?
    a. Algal populations die off; algal blooms occur; oxygen levels drop in deeper water; respiratory demand from decomposers increases; phosphorus input from sewage and agricultural runoff increases
    b. Oxygen levels drop in deeper water; phosphorus input from sewage and agricultural runoff increases; respiratory demand from decomposers increases; algal blooms occur; algal populations die off
    c. Phosphorus input from sewage and agricultural runoff increases; algal blooms occur; algal populations die off; respiratory demand from decomposers increases; oxygen levels drop in deeper water
    d. Phosphorus input from sewage and agricultural runoff increases; algal blooms occur; oxygen levels drop in deeper water; respiratory demand from decomposers increases; algal populations die off

e. Phosphorus input from sewage and agricultural runoff increases; respiratory demand from decomposers increases; oxygen levels drop in deeper water; algal blooms occur; algal populations die off
Textbook Reference: *58.3 How Do Specific Nutrients Cycle through the Global Ecosystem?*

15. Which of the following ecosystem types is being converted to cropland most rapidly at the present time?
a. Deserts
b. Temperate forests
c. Tropical and subtropical biomes
d. Chaparral
e. Temperate grasslands
Textbook Reference: *58.4 What Goods and Services Do Ecosystems Provide?*

16. Which of the following is a major obstacle to the sustainable management of ecosystems?
a. Many ecosystem services are considered "public goods" that have no market value.
b. The total economic value of a sustainably managed ecosystem is almost always lower than that of an intensively exploited ecosystem.
c. Most people are not aware of the extent to which human activities affect the functioning of ecosystems.
d. Both a and c
e. All of the above
Textbook Reference: *58.5 How Can Ecosystems Be Sustainably Managed?*

Answers

Key Concept Review

1. Open oceans have a very low average net primary production, but since they cover 65 percent of Earth's surface, their contribution to Earth's net primary production is large. Algal beds occupy only 0.1 percent of Earth's surface, but their average net primary production is several hundred-fold higher than that of the ocean (see Figure 58.2). However, because of the small surface area, their contribution to Earth's net primary production is minimal.

2. Human activities decrease net primary production when, for example, forests are cut down and replaced by cities. However, human activities can also increase it, as when prairies are converted to extensive agricultural fields.

3. Most fires that occur in forests, savannas, and other biomes are deliberately set by humans to clear land for agriculture. Fires rapidly consume the energy stored in, and release the chemical elements from, the vegetation they burn. Some nutrients, such as nitrogen, are readily vaporized by fire. Nitrogen enters the atmosphere in smoke or is carried into groundwater by rain falling on burned ground. These fires—together with other types of biomass burning, such as combustion of wood and alcohol—account for about 40 percent of the annual influx of CO_2 into the atmosphere and contribute to the production of other greenhouse gases.

4. See Figure 58.8. Turnover occurs during spring and fall. The thermocline (zone of abrupt temperature change several meters below the surface) is shown in the figure during the summer.

5. Groundwater normally has a long residence time and plays a small part in the hydrologic cycle. Thus the extensive use of groundwater for irrigation may result in the depletion of this resource as the water is removed from aquifers more quickly than it can be replaced. Use of this groundwater has also increased flows of water to the ocean and has contributed to the rising sea level of the past century.

6. If farmers apply fertilizers to croplands in quantities that exceed the plants' abilities to absorb them, some of the excess nitrates and phosphates leave the soil and enter bodies of water. This process contributes to eutrophication of lakes and the creation of "dead zones" in the sea, such as the one in the Gulf of Mexico around the mouth of the Mississippi River. Excessive use of nitrogen-containing fertilizers has also caused contamination of groundwater and has been linked to outbreaks of toxic dinoflagellates in estuaries on the Atlantic coast.

7.

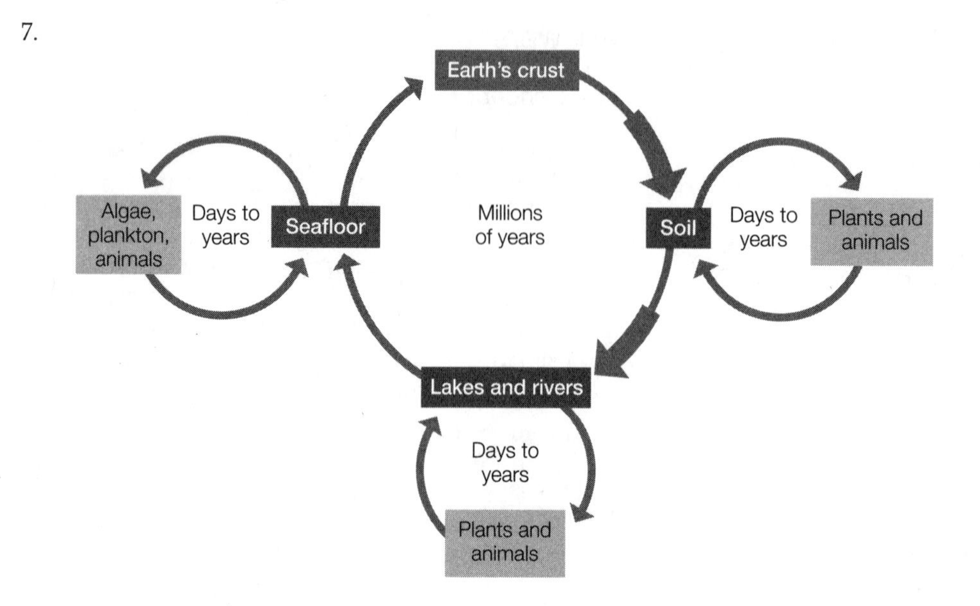

8. If coastal wetlands surrounding New Orleans had been intact, they would have absorbed the storm surge produced by the hurricane and thus protected the city from flooding. Furthermore, even in their reduced capacity, these wetlands remain important both as spawning grounds for many marine organisms, including some of commercial value, and as wintering habitat for migratory birds.
9. The goods include food, clean water, clean air, fiber, building materials, and fuel; the services include flood control, soil stabilization, pollination, and climate regulation.
10. One useful policy would be the elimination of subsidies that promote damaging exploitation of ecosystems, such as subsidies in developing nations for the excessive use of fertilizers. More sustainable use of fresh water from rivers and aquifers could be achieved by charging users the full cost of providing water, developing methods to use water more efficiently, and altering the allocation of water rights. More sustainable use of marine fisheries could be achieved by establishing protected marine reserves and "no-take" zones where fish can grow to reproductive age.
11. After evaluating the net value (dollars per hectare) of the converted land, often the total economic value of a sustainably managed ecosystem is higher than that of a converted or intensively exploited ecosystem (see Figure 58.19).

Test Yourself

1. **b.** The average net primary production of algal beds and coral reefs is greater than that of any other ecosystem. See Figure 58.2.
2. **d.** Open oceans have the highest overall net primary production on Earth because the large surface area compensates for the low average net primary productivity.
3. **b.** Most elements remain in bottom sediments of the ocean for millions of years, until they are elevated above sea level by movements of Earth's crust.
4. **c.** The steep thermocline evident in this temperature profile would typically be established only by mid-to late-summer.
5. **c.** The range of depths over which the temperature changes most abruptly is 20 to 30 meters.
6. **a.** Low concentrations of mineral nutrients limit the primary production of most ocean waters. In zones of upwelling, nutrient-rich water is brought to the surface from the ocean bottom, thereby enhancing the growth of photosynthesizing plankton.
7. **e.** The ozone that shields the surface of Earth from most incoming ultraviolet radiation is located in the stratosphere.
8. **d.** Unlike their behavior in air and water, elements on land move slowly and usually only short distances.
9. **a.** Temperature, nutrient supply, and light can limit production in aquatic ecosystems; temperature and moisture most often limit production on land.
10. **a.** Most elements cycle through organisms more quickly than they cycle through the nonliving world.
11. ***n*** Major reservoir is atmospheric

    c, p Major reservoir is in sedimentary rock

    n, p Often in short supply in ecosystems

    c Fossil fuel reserve is part of this cycle

    s Involved in cloud formation

    p Lacks a gaseous phase

    n, c Includes a form that is a greenhouse gas

    n Most fluxes involve organisms

12. **b.** The addition of compounds of sulfur to the atmosphere has more to do with acid precipitation than it does with global warming.
13. **a.** Acid precipitation is a regional, not a local, environmental problem. Acids may travel hundreds of kilometers before they settle to Earth in precipitation or as dry particles.
14. **c.** The correct sequence is: phosphorus input from sewage and agricultural runoff increases, algal blooms occur, algal populations die off, respiratory demand from decomposers increases, and finally, oxygen levels drop in deeper water.
15. **c.** Conversion to cropland is causing rapid reductions of a number of tropical and subtropical biomes.
16. **d.** Two significant barriers to sustainable ecosystem management are the lack of market value of many ecosystem services and the lack of public understanding of how human activities may harm ecosystems.

59 Biodiversity and Conservation Biology

The Big Picture

- Conservation biology provides an excellent example of the importance of the synthetic approach to science. In this case, knowledge and concepts derived from different areas of biology and the social sciences are being applied to the solution of the problem of species and ecosystem preservation.
- Preserving biological diversity is considered by many informed individuals to be one of the most important challenges facing humanity in the twenty-first century.

Study Strategies

- This chapter refers to and relies on a number of concepts introduced in earlier chapters on ecology, such as: invasive species, habitat fragmentation, and the management of populations (Chapter 55); species richness, disturbance, species–area relationships, and island biogeography (Chapter 57). You may find it useful to review those concepts as you study the material in this chapter.
- It may not be apparent to you why a fragmented habitat is unlikely to support as many species as similar unbroken habitat of equal extent. One reason is that some species may require large expanses of a particular habitat in order to survive. A second reason is that smaller habitat patches are more susceptible to edge effects (see Figure 59.5).
- Go to the Web addresses shown to review the following animated tutorial and activity. You can also find them in BioPortal (yourBioPortal.com), along with many additional learning resources.

  Animated Tutorial 59.1 Edge Effects (Life10e.com/at59.1)

  Activity 59.1 Concept Matching (Life10e.com/ac59.1)

Key Concept Review

59.1 What Is Conservation Biology?

Conservation biology aims to protect and manage biodiversity

Biodiversity has great value to human society

Conservation biology is devoted to protecting and managing Earth's biodiversity. The science of conservation biology draws on concepts and knowledge from ecology, ethology, and evolutionary biology. Conservation biology is guided by three basic principles: (1) Evolution is the process that unites all forms of life; (2) The ecological world is dynamic; (3) Humans are a part of ecosystems.

Extinctions have occurred throughout Earth's history as the environment has changed to favor some species and negatively affect others. The past mass extinction episodes were the result of cataclysmic natural disturbances, whereas today the human population is largely responsible for species extinctions that rival those of the five great mass extinctions of life's history (see Figure 59.1).

There are many reasons that people should value biodiversity. Species are necessary for the functioning of ecosystems and the many benefits and services those ecosystems provide. Many individual species are important because they supply useful products such as food, fiber, and medicinal drugs. Extinctions lessen the aesthetic pleasure many people derive from interacting with other organisms. Extinctions deprive us of opportunities to study the structure and functioning of ecological communities and ecosystems. Extinctions caused by human activities raise ethical issues because other species are judged to have intrinsic value.

Question 1. Aside from the aesthetic benefits of biodiversity, biodiversity provides humans with economic benefits. List three economic reasons that people should be concerned about the escalating pace of species extinctions.

Question 2. What are the three basic principles that guide conservation biologists?

59.2 How Do Conservation Biologists Predict Changes in Biodiversity?

Our knowledge of biodiversity is incomplete

We can predict the effects of human activities on biodiversity

Conservation biologists attempt to predict changes in biodiversity. Accurate estimates of the number of extinctions that will occur in the next century are impossible for four reasons: (1) we do not know exactly how many species exist on Earth today; (2) we do not know where most species live; (3) it is difficult to determine when a species becomes extinct (see Figure 59.2); and (4) we rarely know how the extinction of one species may affect other species.

To estimate the risk of extinction of a particular population, biologists develop models that incorporate information about its size, its genetic variation, and the morphology, physiology, and behavior of its members. Species in imminent danger of extinction are classified as "endangered" or "critically endangered"; those believed to be susceptible to extinction are classified as "vulnerable," and species in any one of these three categories is considered "threatened" (see Figure 59.3). Estimates of current rates of extinction worldwide are based primarily on species–area relationships, the theory of island biogeography, and rates of the disappearing rainforest (see Figure 59.4).

Question 3. According to the International Union for the Conservation of Nature (IUCN), some species fall under the category of endangered, whereas others are deemed vulnerable. What is the distinction between the two categories?

Question 4. As a conservation biologist, you are interviewed by a local television station and asked whether you can predict how many species on Earth are going extinct. Give four reasons why it is very difficult to answer this question.

59.3 What Human Activities Threaten Species Persistence?

Habitat losses endanger species

Overexploitation has driven many species to extinction

Invasive predators, competitors, and pathogens threaten many species

Rapid climate change can cause species extinctions

Human activities that threaten the survival of species include habitat destruction, overexploitation, introduction of exotic species, and climate change.

In the United States, the most important cause of species extinction today is habitat loss, degradation, and fragmentation. As habitats are fragmented into small patches, populations of species that require large areas cannot be maintained. The proportion of a habitat patch subject to detrimental edge effects increases as patch size decreases (see Figure 59.5). For example, species from surrounding habitats may invade the edges of the patch to compete with or prey upon the patch inhabitants. The persistence of species in small patches may be improved if the patches are connected by corridors of suitable habitat through which individuals can disperse.

Overexploitation for commercial reasons (such as use in traditional medicine and trade in pets, ornamental plants, and tropical hardwoods) continues to threaten many species (see Figure 59.7).

Some exotic (non-native) species become invasive, reproducing rapidly and spreading widely to the detriment of native species. Exotic pests, predators, and competitors introduced by humans have driven many species to extinction and threaten many others (see Figure 59.8).

As rapid global warming occurs, species with poor dispersal abilities may become extinct because of their inability to keep pace with climate change by shifting their ranges. Other effects of global warming may include the entire disappearance of some habitats (such as alpine tundra), and the development of new climates (especially at low elevations in the tropics).

Question 5. What are wildlife corridors, and how have experiments demonstrated that they help species persist in patchy environments?

Question 6. Why are exotic species frequently a threat to the biological diversity of the region in which they are introduced?

Question 7. Examine the figure of four forests and their edges below. Evaluate which habitat (A–D) has the largest percentage influenced by edge effects. As a conservation biologist, which one of the four patches would you recommend to be established to support species persistence? Explain your answer.

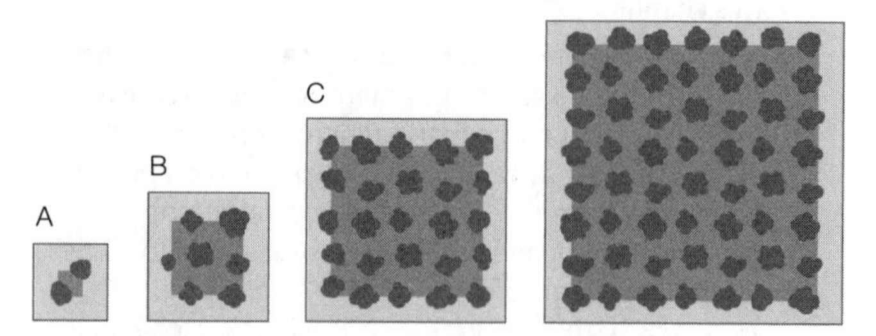

59.4 What Strategies Are Used to Protect Biodiversity?

Protected areas preserve habitat and prevent overexploitation

Degraded ecosystems can be restored

Disturbance patterns sometimes need to be restored

Ending trade is crucial to saving some species

Species invasions must be controlled or prevented

Biodiversity has economic value

Changes in human-dominated landscapes can help protect biodiversity

Captive breeding programs can maintain a few species

Earth is not a ship, a spaceship, or an airplane

Biologists use many strategies to maintain biological diversity. Protected areas preserve habitat, prevent overexploitation, and may serve as nurseries for the replenishment of populations in unprotected areas. Using the criteria of species richness and endemism, biologists have identified a number of "hotspots" for biodiversity protection (see Figure 59.10). To further pinpoint sites with threatened species found nowhere else, conservation biologists have identified 595 "centers of imminent extinction." These are concentrated in tropical forests, on islands, and in mountainous regions (see Figure 59.11).

Practitioners of restoration ecology attempt to return degraded habitats to their natural state. An example is the current project to restore an extensive prairie ecosystem in Montana (see Figure 59.12). Creating new wetlands to substitute for the ones being destroyed by development requires detailed ecological knowledge. Experiments have shown that restoration is most successful when wetlands are planted with a mixture of species instead of just one or two (see Figure 59.14).

Humans often reduce the frequency and intensity of disturbances such as fires and thereby endanger species dependent on these disturbances. Reestablishment of historical disturbance patterns (for example, by controlled burning in forests), may help preserve such species (see Figure 59.15).

Because most endangered species cannot survive further reductions in their breeding populations, it is important to prevent their exploitation. The Convention on International Trade in Endangered Species (CITES) determines which species are banned in international trade. For example, CITES is regulating the international trade in elephant ivory.

Controlling invasions of exotic species is often important in preserving biodiversity. Using the traits found in most invasive species, conservation biologists have developed a "decision tree," which is used to determine whether an exotic plant species should be introduced into North America (see Figure 59.16).

Studies conducted in the field of ecological economics have demonstrated that conserving biodiversity is often economically valuable. Ecotourism, for example, is a major source of income for many developing nations (see Figure 59.17). Intact ecosystems may also be a reliable and relatively inexpensive source of water (see Figure 59.18). Reconciliation ecology is the practice of using land in which people live and extract resources in ways that sustain biodiversity. It is based on the principle that most ecosystem services are provided locally, and that people are more motivated to protect their local interests than they are to work on national or global issues. For a small fraction of endangered species, captive propagation can help prevent extinction by maintaining the species during critical periods and by providing individuals for reintroduction into the wild (see Figure 59.19).

Science can supply information about the accelerating rate of species extinctions, but it is up to society as a whole to determine what rate of species loss due to human activities is acceptable.

Question 8. In recent years, severe forest fires in many parts of the western United States have been used as a forest management tool. At the same time, the U.S. Forest Service tries to prevent forest fires using the iconic mascot Smokey Bear to deliver this message to the public. Evaluate whether fires or the lack of fires is better for the ecosystem.

Question 9. Conservation biologists have developed a "decision tree" based on the traits that characterize plant species that have become invasive. Complete the decision tree below by matching the numbers 1–6 with one of the following: (a) "Deny admission," (b) "Admit," or (c) "Further analysis and monitoring needed."

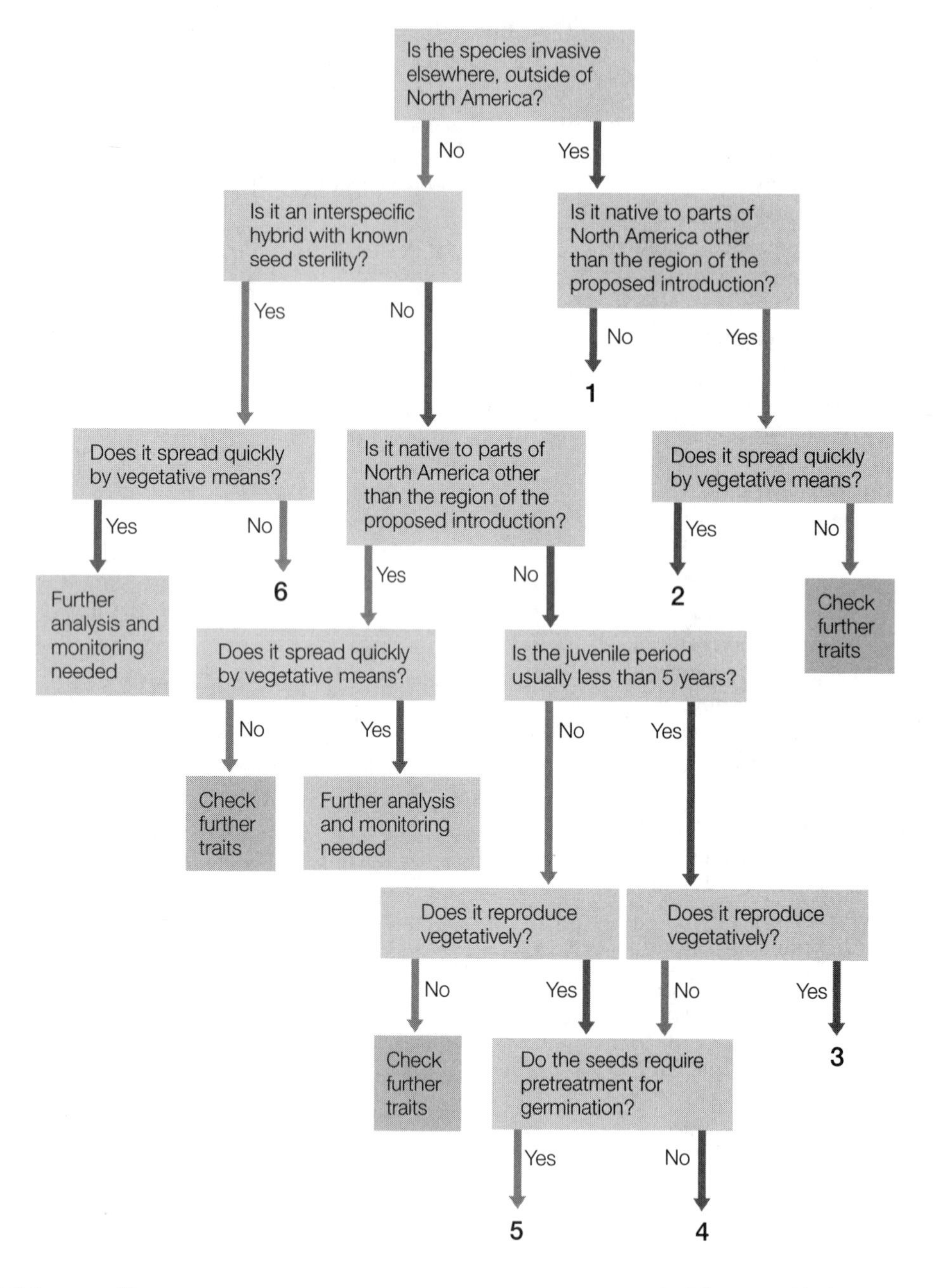

Test Yourself

1. Which of the following fields is *not* one of the sources for the concepts of conservation biology?
   a. Ecology
   b. Evolutionary biology
   c. Population genetics
   d. Immunology
   e. Ethology
   Textbook Reference: *59.1 What Is Conservation Biology?*

2. The most likely cause of the extinctions of many large mammals in North America in the last 14,000 years is
   a. rapid climate change associated with glaciation.
   b. overhunting by humans.
   c. the formation of land bridges to neighboring continents that allowed many new species of competitors and predators to invade these regions.
   d. massive volcanism that caused destruction of the food supply for these mammals.
   e. pathogens introduced by humans.
   Textbook Reference: *59.1 What Is Conservation Biology?*

3. Based on species–area relationships, ecologists predict that
   a. about one million tropical rainforest species may become extinct in the next hundred years.
   b. a 90 percent loss of tropical rainforest habitat will result in loss of about 9 percent of the species living there.
   c. the area required by most tropical rainforest species will decrease as they adapt to the smaller forest size.
   d. the extinction of all tropical rainforest species is inevitable.
   e. None of the above

   Textbook Reference: *59.2 How Do Conservation Biologists Predict Changes in Biodiversity?*

4. Species are considered "threatened" if they fall into the category of
   a. vulnerable.
   b. endangered.
   c. critically endangered.
   d. All of the above
   e. None of the above

   Textbook Reference: *59.2 How Do Conservation Biologists Predict Changes in Biodiversity?*

5. In the graph below, select the curve (*a, b,* or *c*) that correctly shows the expected relationship between habitat patch area and the proportion influenced by edge effects.

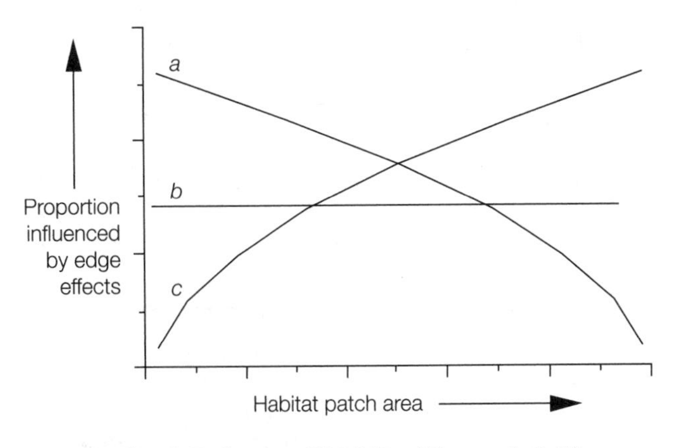

   Textbook Reference: *59.3 What Human Activities Threaten Species Persistence?*

6. Which of the following is *not* currently a major cause of the global reduction in biodiversity?
   a. Overexploitation
   b. Global warming
   c. Habitat destruction
   d. Introduction of foreign predators and disease
   e. All of the above are currently major causes of extinction.

   Textbook Reference: *59.3 What Human Activities Threaten Species Persistence?*

7. Why do conservation biologists believe that global warming may lead to extensive decimation of species?
   a. Since little change in plant community composition has occurred in the past, we cannot expect present communities to adapt to climate change.
   b. The magnitude of climate change will be much greater than any climate change that occurred in the past.
   c. Many sedentary species may not be able to shift their ranges to match the pace at which temperature zones are moving northward.
   d. Both b and c
   e. All of the above

   Textbook Reference: *59.3 What Human Activities Threaten Species Persistence?*

8. A defect of the "hotspot" approach to conserving biodiversity is that it does not direct attention to
   a. regions of low species richness that may nevertheless contain unique species or ecological communities.
   b. marine regions.
   c. regions with a high number of endemic species.
   d. Both a and b
   e. All of the above

   Textbook Reference: *59.4 What Strategies Are Used to Protect Biodiversity?*

9. Which of the following efforts to preserve biodiversity best exemplifies restoration ecology?
   a. A campaign to discourage pesticide use on lawns
   b. A project to convert ranch land into a natural prairie
   c. Designation of the habitat of an endangered species as a protected area
   d. Elimination of commercial trade in products derived from endangered and threatened species
   e. A captive breeding program to maintain an endangered species

   Textbook Reference: *59.4 What Strategies Are Used to Protect Biodiversity?*

10. Which of the following efforts to preserve biodiversity best exemplifies reconciliation ecology?
    a. A campaign to discourage pesticide use on lawns
    b. A project to convert ranch land into a natural prairie
    c. Designation of the habitat of an endangered species as a protected area
    d. Elimination of commercial trade in products derived from endangered and threatened species
    e. A captive breeding program to maintain an endangered species

    Textbook Reference: *59.4 What Strategies Are Used to Protect Biodiversity?*

11. Experiments on wetland restoration have demonstrated that planting a richer mixture of species is associated with
    a. faster accumulation of belowground nitrogen.

b. more complex vegetation structure.
c. more rapid development of vegetation cover.
d. Both b and c
e. All of the above
Textbook Reference: *59.4 What Strategies Are Used to Protect Biodiversity?*

12. To help identify species of exotic plants that have the potential to become invasive if they were to be introduced into North America, conservation biologists make use of a "decision tree." Which of the following plant characteristics might cause biologists to deny approval for a proposed introduction of an exotic plant species?
a. Seeds of the plant do not require pretreatment for germination.
b. The plant spreads quickly by vegetative propagation.
c. The plant is invasive outside of North America.
d. The juvenile period of the plant is less than 5 years.
e. All of the above
Textbook Reference: *59.4 What Strategies Are Used to Protect Biodiversity?*

13. The fynbos shrub community in South Africa
a. is threatened by introduced species of taller, faster-growing plants.
b. is a fire-adapted community.
c. helps maintain a regional supply of high-quality water.
d. includes thousands of plant species found nowhere else in the world.
e. All of the above
Textbook Reference: *59.4 What Strategies Are Used to Protect Biodiversity?*

14. Which of the following statements about elephants and the ivory trade is *false*?
a. African elephants are endangered throughout their range.
b. The geographical source of ivory can be determined by the use of DNA markers.
c. There is a strong demand for ivory in Japan and China because of its use in folk medicines.
d. Online sales of ivory on eBay have been banned.
e. The legal mechanism for prohibiting the ivory trade is the international agreement called the Convention on International Trade in Endangered Species.
Textbook Reference: *59.4 What Strategies Are Used to Protect Biodiversity?*

15. The California condor
a. is being introduced into regions that were not part of its historic geographical range as part of the effort to save the species.
b. became endangered partly because of high mortality that resulted from its eating carcasses containing lead shot.
c. has increased in numbers because of captive propagation, though released captive-bred condors have not yet bred in the wild.
d. has been reintroduced to the wild over the objections of cattle ranchers, who believe correctly that condors kill livestock.
e. All of the above
Textbook Reference: *59.4 What Strategies Are Used to Protect Biodiversity?*

Answers

Key Concept Review

1. The many economic benefits of biodiversity include the provision of natural products (food, fiber, and medicine), services such as fermentation, ecosystems services, and ecotourism.
2. The three principles of conservation biology are (1) The processes of evolution unite all forms of life, so to protect and manage biodiversity, we must understand the evolutionary processes that generate and maintain it; (2) The ecological world is dynamic, so there is no static "balance of nature" that can serve as a goal of conservation activities; and (3) Humans are a part of ecosystems, so our interests and activities must be incorporated into conservation goals and practices.
3. Endangered species are those in imminent danger of extinction over all or a significant portion of their range, whereas vulnerable species are those that are susceptible to extinction in the near future.
4. Accurate estimates of the number of extinctions that will occur in the next century are impossible for four reasons: (1) we do not know exactly how many species exist on Earth today; (2) we do not know where most species live; (3) it is difficult to determine when a species becomes extinct; and (4) we rarely know how the extinction of one species may affect other species.
5. Habitat corridors are relatively thin strips of habitat of a particular type that connect larger patches of the same type of habitat. Their importance in permitting individuals to disperse from one patch to another was demonstrated in experiments. For example, biologists studied plots of tropical rainforest near Manaus, Brazil, before and after they were isolated by forest clearing. Those that were completely isolated lost species more rapidly than did those that were connected to unfragmented forest by corridors.
6. Native species that live in a particular community have evolved to cope successfully with the specific predators, competitors, and diseases that are part of its environment. Introduced species represent a change in the environment for which native species may not have been prepared by natural selection.

7. The smaller a patch of habitat, the greater the proportion of that patch that is influenced by conditions in the surrounding environment. Therefore, habitat patch A will be the most influenced by edge effects. As a conservation biologist you should establish habitat patch D, because small patches cannot maintain populations of species that require large areas.
8. In ecosystems that naturally experience periodic fires, controlled burning prevents the excessive accumulation of fuel (dead branches, leaf litter, etc.) and thereby lessens the danger that catastrophic, tree-consuming canopy fires will occur. In addition, because many species in such ecosystems require periodic fires for successful establishment and reproduction, controlled burning may be necessary to maintain species richness.
9. 1: a; 2: a; 3: a; 4: a ; 5: c; 6: b

Test Yourself

1. **d.** Although immunology is sometimes a useful tool in assessing the amount of genetic variation that exists in a population, it is less important than the other fields listed.
2. **b.** Because these extinctions followed human colonization of North America, most biologists favor the hypothesis that overhunting by humans was the most likely cause.
3. **a.** Based on estimates of current and future disappearance of tropical rainforests, about one million species that live in these communities could become extinct in the next hundred years. This loss is not inevitable if we reduce the rate at which these forests are converted to pasture and cropland.
4. **d.** Vulnerable, endangered, and critically endangered species are all considered threatened by biologists.
5. **a.** Because the edge of a habitat patch equals the perimeter of the patch, it is proportionally greater for smaller habitat patches. Therefore, the relationship between edge effects and habitat patch areas is an inverse relationship, as shown by curve *a*.
6. **b.** Global warming is predicted to be a major cause of extinction in the future, but it is not yet having the kind of impact on populations that overexploitation, habitat destruction, and species introductions are.
7. **c.** Although the expected magnitude of the climate change due to global warming may be similar to past climate changes, the rate of warming will be greater. This may make it impossible for many species to extend their ranges at a rate that keeps up with the northward movement of the temperature zones.
8. **d.** The "hotspot" concept emphasizes the preservation of regions of high species richness and endemism. It overlooks marine regions and terrestrial regions of relatively low species richness.
9. **b.** Restoration ecology focuses on the reestablishment of entire ecosystems to their natural state.
10. **a.** Reconciliation ecology focuses on ways that biodiversity can be sustained in landscapes in which people live.
11. **e.** When a species-rich mixture is introduced, wetland restoration is more successful for all of the reasons listed.
12. **e.** See Figure 59.16 in the textbook.
13. **e.** See Figure 59.18 in the textbook.
14. **a.** A number of countries in southern Africa have so many elephants that some must be killed to prevent them from dispersing to populated regions and damaging crops.
15. **b.** By 1978, the wild population of California condors was heading toward extinction. The cause was mortality due to ingestion of lead shot in animal carcasses eaten by the condors.

DNA

Deoxyribonucleic acid

Double-helix

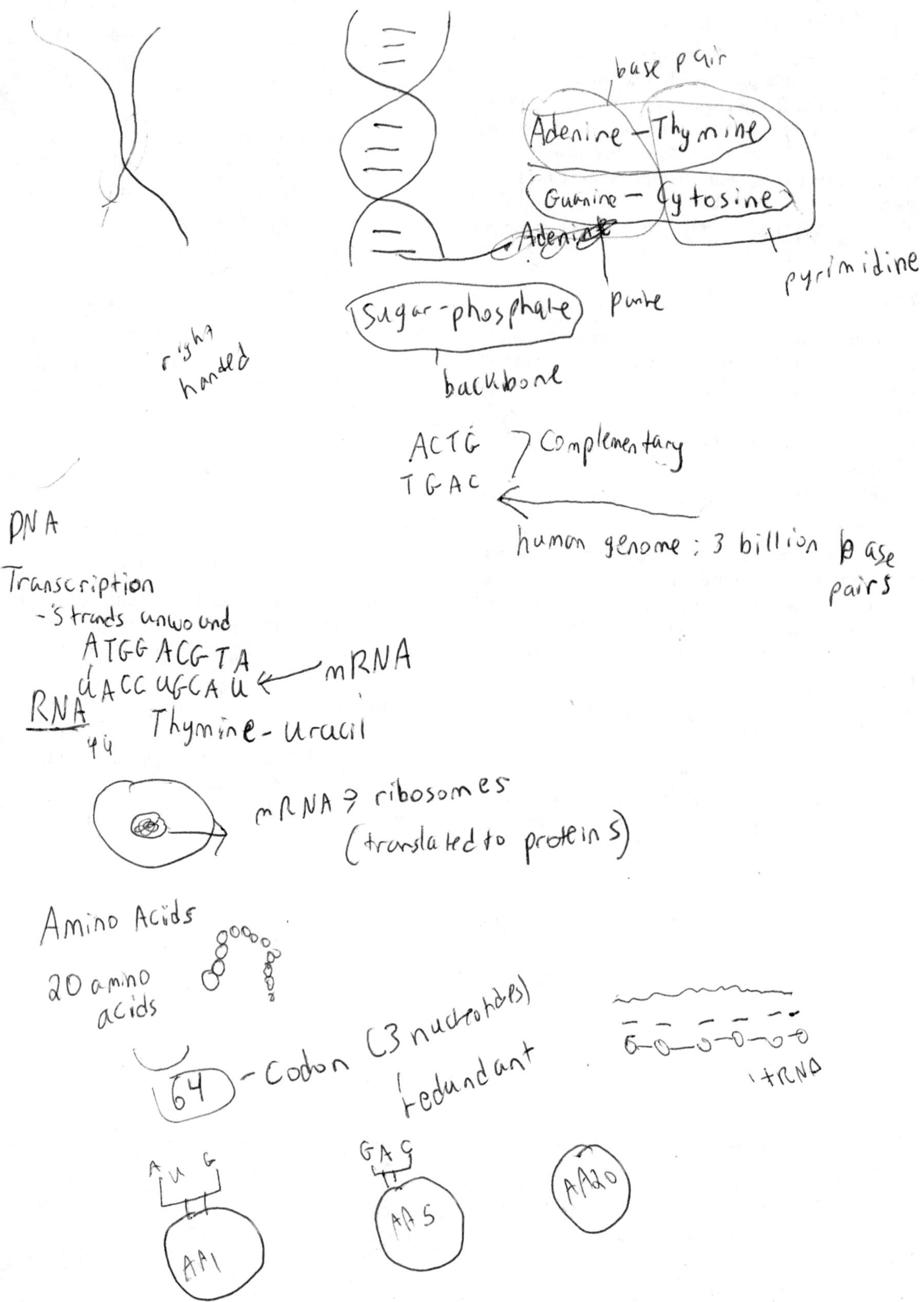